Oribatid Mites

Oribatid Mites
Biodiversity, Taxonomy and Ecology

Valerie Behan-Pelletier
Zoë Lindo

CRC Press is an imprint of the
Taylor & Francis Group, an **informa** business

First edition published 2023
by CRC Press
6000 Broken Sound Parkway NW, Suite 300, Boca Raton, FL 33487-2742

and by CRC Press
4 Park Square, Milton Park, Abingdon, Oxon, OX14 4RN

CRC Press is an imprint of Taylor & Francis Group, LLC

© 2023 Valerie Behan-Pelletier and Zoë Lindo

Reasonable efforts have been made to publish reliable data and information, but the author and publisher cannot assume responsibility for the validity of all materials or the consequences of their use. The authors and publishers have attempted to trace the copyright holders of all material reproduced in this publication and apologize to copyright holders if permission to publish in this form has not been obtained. If any copyright material has not been acknowledged please write and let us know so we may rectify in any future reprint.

Except as permitted under U.S. Copyright Law, no part of this book may be reprinted, reproduced, transmitted, or utilized in any form by any electronic, mechanical, or other means, now known or hereafter invented, including photocopying, microfilming, and recording, or in any information storage or retrieval system, without written permission from the publishers.

For permission to photocopy or use material electronically from this work, access www.copyright.com or contact the Copyright Clearance Center, Inc. (CCC), 222 Rosewood Drive, Danvers, MA 01923, 978-750-8400. For works that are not available on CCC please contact mpkbookspermissions@tandf.co.uk

Trademark notice: Product or corporate names may be trademarks or registered trademarks and are used only for identification and explanation without intent to infringe.

ISBN: 9781032102931 (hbk)
ISBN: 9781032102948 (pbk)
ISBN: 9781003214649 (ebk)

DOI: 10.1201/9781003214649

Typeset in Times
by Evolution Design & Digital Ltd (Kent)

Contents

Preface and Acknowledgments xi
Authors xiii

Chapter 1 Introduction 1

    Summary 1
    Introduction 1
    General Characteristics 1
    Fossil History 2
    Phylogeny and Classification 4
    Higher Classification 4
    Why Are They Important? 4
    Catalogs and Checklists 5

Chapter 2 Form and Function 7

    Summary 7
    Introduction 7
    Morphology of Adult Oribatid Mites 7
        Integument and Body Form 9
        Prodorsum 11
        Notogaster 13
        Ventral Structures 17
        Gnathosoma 20
        Legs 22
        Sensory Structures 24
            Mechanoreceptors 24
            Proprioceptors 26
            Chemoreceptors 26
            Photoreceptors 26
    Function 27
        Respiration 27
        Secretion 27
        Osmoregulation 28
        Hemolymph and Muscles 28
        Nervous System 28
        Digestive System 29
        Reproduction 29

Chapter 3 Taxonomic Keys 33

    Summary 33
    Key to Families 34
    Key to Genera and Species 49
    Infraorder Palaeosomata 49
    Archeonothroidea 50
        Acaronychidae 50
        Archeonothridae 50

Palaeacaroidea	50
Palaeacaridae	50
Ctenacaroidea	51
Aphelacaridae	51
Ctenacaridae	51
Infraorder Enarthronota	52
Brachychthonioidea	53
Brachychthoniidae	53
Atopochthonioidea	62
Atopochthoniidae	62
Pterochthoniidae	62
Hypochthonioidea	62
Eniochthoniidae	62
Hypochthoniidae	64
Mesoplophoridae	65
Protoplophoroidea	66
Cosmochthoniidae	66
Haplochthoniidae	66
Sphaerochthoniidae	67
Heterochthonioidea	67
Arborichthoniidae	67
Trichthoniidae	67
Infraorder Parhyposomata	68
Parhypochthonioidea	68
Gehypochthoniidae	68
Parhypochthoniidae	68
Infraorder Mixonomata	69
Eulohmannioidea	69
Eulohmanniidae	69
Perlohmannioidea	70
Perlohmanniidae	70
Epilohmannioidea	71
Epilohmanniidae	71
Euphthiracaroidea	71
Euphthiracaridae	72
Oribotritiidae	77
Synichotritiidae	80
Phthiracaroidea	81
Phthiracaridae	81
Infraorder Desmonomata	88
Hyporder Nothrina	88
Crotonioidea	88
Crotoniidae	88
Hermanniidae	96
Malaconothridae	97
Nanhermanniidae	99
Nothridae	100
Trhypochthoniidae	103
Hyporder Brachypylina	106
Hermannielloidea	108
Hermanniellidae	108

Plasmobatidae	109
Neoliodoidea	109
Neoliodidae	109
Plateremaeoidea	110
Gymnodamaeidae	110
Licnodamaeidae	114
Plateremaeidae	114
Damaeoidea	114
Damaeidae	114
Cepheoidea	122
Cepheidae	122
Polypterozetoidea	125
Podopterotegaeidae	125
Polypterozetidae	125
Microzetoidea	126
Microzetidae	126
Caleremaeoidea	127
Caleremaeidae	127
Ameroidea	127
Ameridae	128
Damaeolidae	128
Eremobelbidae	129
Eremulidae	129
Hungarobelbidae	130
Zetorchestoidea	130
Eremaeidae	130
Megeremaeidae	138
Gustavioidea	139
Astegistidae	139
Gustaviidae	141
Kodiakellidae	141
Liacaridae	141
Peloppiidae	144
Tenuialidae	149
Carabodoidea	150
Carabodidae	150
Oppioidea	155
Autognetidae	155
Machuellidae	157
Oppiidae	157
Quadroppiidae	163
Thyrisomidae	164
Trizetoidea	166
Suctobelbidae	166
Tectocepheoidea	169
Tectocepheidae	169
Limnozetoidea	171
Hydrozetidae	171
Limnozetidae	172
Ameronothroidea	175
Ameronothridae	175

Podacaridae	176
Selenoribatidae	177
Tegeocranellidae	177
Cymbaeremaeoidea	178
Cymbaeremaeidae	178
Licneremaeoidea	180
Dendroeremaeidae	180
Licneremaeidae	181
Passalozetidae	181
Scutoverticidae	182
Phenopelopoidea	183
Phenopelopidae	183
Unduloribatidae	186
Achipterioidea	186
Achipteriidae	186
Tegoribatidae	190
Oribatelloidea	192
Oribatellidae	192
Oripodoidea	198
Haplozetidae	198
Mochlozetidae	201
Oribatulidae	203
Oripodidae	207
Parakalummidae	207
Scheloribatidae	208
Ceratozetoidea	213
Ceratokalummidae	213
Ceratozetidae	213
Chamobatidae	229
Humerobatidae	230
Euzetidae	230
Punctoribatidae	231
Zetomimidae	240
Galumnoidea	242
Galumnidae	242

Chapter 4 Ecology of Oribatid Mites ... 401

Summary	401
Introduction	401
Feeding Biology	401
Life History Traits	403
Reproduction and Developmental Rates	403
Defence	404
Sclerotization and Mineralization	404
Defensive Setae	406
Ptychoidy	406
Cerotegument and Debris	406
Opisthosomal (Opisthonotal) Glands	406
Pathogens	407
Movement and Dispersal	407

Contents ix

        The Soil Environment 409
            Microhabitats and Non-soil Environments.. 410
            Abiotic Factors and Environmental Gradients 412
        Disturbance 412
        Ecological Roles in Soil 413
        Oribatid Mites of Human Interest 413

Chapter 5 Oribatid Diversity across the Northern North American Landscape 415

        Summary... 415
        Introduction.. 415
        Northern Ecosystems: Taiga, and Arctic Ecozones 417
        Hudson Plains Ecozone 419
        Boreal Ecozones: Boreal Shield, Boreal Plains, Boreal Cordillera and
        Newfoundland Boreal Ecozones 420
        Atlantic Maritime Ecozone 422
        Mixedwood Plains Ecozone 422
        Grassland (Prairie) Ecozones 423
        Montane Cordillera and Western Interior Basin Ecozones...425
        Pacific Maritime Ecozone 426
        Aquatic Habitats..427
        Trends in Oribatid Diversity across the Canadian and Alaskan Landscape............. 428

Abbreviations 429

References 431

Taxonomic index 479

Subject index.. 489

Preface and Acknowledgments

Oribatida, called 'moss mites' or 'beetles mites', are ubiquitous components of the North American landscape. They are rich in diversity and numbers in all soil-type habitats, from east coast maritime forest to western temperate rainforest, from temperate grasslands through boreal forest, subarctic and arctic tundra to the polar desert of the high arctic islands. We consider oribatid mites part of the charismatic microfauna; both adults and immatures are beautiful, and we are privileged to work with them in our research.

Oribatid mites have a short, but rich history of research across northern North America (reviewed in Behan-Pelletier and Lindo 2019). Our goal in this book is to introduce you to the oribatid fauna of northern North American habitats and to give you a framework for the oribatid fauna you can expect in your own biodiversity and ecological studies. Our goal is to aid, but the data are not predictive, as we estimate that only 40% of the oribatid fauna is described on the basis of morphology. Our approach is biogeographic and ecological, because this allows some predictions as to the nature of the oribatid fauna of an ecosystem.

Oribatid mites are small in size, 150 µm–1.5 mm, but adults of most members of the suborder are easily recognizable as such. Our hope is that with our keys you will be able to easily recognize families and identify specimens to genus and, for Canada and Alaska, to species. We add information on biology, where available. Most autecological data are for species that are Holarctic, and data are often from European populations of these species. Whether species are truly Holarctic, or represented by different genetic lineages in the Nearctic, is a subject for future research.

The format of our book is modelled on that of Hornmilben (Oribatida), published in 2006, Gerd Weigmann's masterful treatise on the oribatid mites of Central Europe. We hope that our book does the same for the oribatid fauna of northern North America, with identification keys to species, and with keys to identification of oribatid families and genera that have relevance for the northern Holarctic. Keys to family and genus are based on those presented at the Oribatid Mite Course of the Acarology Summer Program of The Ohio State University, Columbus, Ohio. Roy A. Norton developed this course and presented it for almost 40 years. We were both students and co-teachers of this course, and we are forever grateful to Roy who shared his broad and deep knowledge of these mites with us. Keys to species are based on primary literature and those of David E. Walter, whose contribution to knowledge of the oribatid fauna of Canada through the web-based Almanacs of Alberta Oribatida is immeasurable.

This book would have been impossible without the following:

Professor Roy A. Norton, Professor Emeritus, State University of New York, Syracuse, New York, and Dr. David E. Walter of University of the Sunshine Coast, Queensland, Australia, for their detailed comments on earlier drafts of this book. In addition, Roy Norton shared teaching of the Oribatid Week of the Acarology Summer Program with one or both of us between 2004 and 2018 which helped form the outline for this book. Reviewers of our original proposal to CRC Press who provided excellent and constructive support for this book. Students of the Oribatid Week of the Acarology Summer Program in the years 2004–2018 who contributed to testing the keys to family and genera, and Trevor Pettit, University of Western Ontario, London, who provided us with technical support.

We thank the following colleagues and institutions who gave permission for use of images and those who helped in getting permissions:

Professors: Jun-Ichi Aoki, Japan; Badamdorj Bayartogtokh, National University of Mongolia, Mongolia; L. Borowiec, Biologica Silesiae Publishing House; Shigeo Chinone, Ibaraki Nature Museum, Japan; Sergey Ermilov, Tyumen State University, Tyumen, Russia; Lizel Hugo-Coetzee, National Museum, Bloemfontein, South Africa; Masamichi Ito, Surugadai University, Japan; Juan Carlos Iturronobeitia, Universidad País Vasco, Leioa, Vizcaya, Spain; Wayne Knee and

Monica Young, Canadian National Collection, Agriculture and Agri-Food Canada, Ottawa; Serge Kreiter, Montpellier SupAgro University, Montpellier, France, Editor-in-chief Acarologia; Günther Krisper, University of Graz, Austria; Lisa Lumley, Alberta Biodiversity Monitoring Institute, University of Alberta, Edmonton; Wojciech Ł. Magowski, Adam Mickiewicz University, Poznań, Poland; Olya Makarova, Severtsov Institute of Ecology and Evolution, Russian Academy of Sciences, Moscow, Russia; Ichiro Maruyama, Niigata, Japan; Ladislav Miko, Czech University of Life Sciences, Prague, Czech Republic; Andreas Naegele (www.schweizerbart.de/series/zoologica); Wojciech Niedbała, Adam Mickiewicz University, Poznań, Poland; Roy A. Norton, Professor Emeritus, State University of New York, Syracuse, New York; László Peregovits, Hungarian Natural History Museum, Budapest, Hungary; Tobias Pfingstl, University of Graz, Austria; Vikram Prasad, Indira Publishing House, West Bloomfield, MI; Ina Schaefer, University of Göttingen, Germany; Anna Seniczak, University Museum of Bergen, Bergen, Norway; Stanisław Seniczak, Kazimierz Wielki University, Poland; Satoshi Shimano, Hosei University, Tokyo, Japan; Ingrid Solhøy, University of Bergen, Norway; Luis S. Subías, Complutense University of Madrid, Madrid, Spain; Emma Wahlberg, Editor-in-chief, Entomologisk Tidskrift; David E. Walter, University of the Sunshine Coast, Queensland, Australia; Gerd Weigmann, Berlin, Germany; Xenia Wörle, Zoologica; Zhi-Qiang Zhang, University of Auckland & Researcher, Landcare Research, Auckland, New Zealand.

Authors

Dr. Valerie Behan-Pelletier is an Honorary Research Associate (an Emeritus position) with the Research Branch of Agriculture and Agri-Food Canada, the federal department responsible for maintaining the Canadian National Collection of insects, arachnids, nematodes, fungi and vascular plants. As a global specialist in the taxonomy and ecology of Oribatida for over 45 years, she has published a number of large monographs on oribatid mites for North America and has contributed to multi-chapter works on biodiversity, soil ecology, and ecological assessments. She is a Section Editor of the Oxford University Press Publication 'Soil Ecology and Ecosystem Services' (2012), is a coauthor of the JRC and the Global Soil Biodiversity Initiative (GSBI) 'Global Soil Biodiversity Atlas' (2016) and is coauthor of a chapter on Oribatida in Krantz and Walter 'A Manual of Acarology' Texas Tech. University Press (2009), and of Aquatic Oribatida in Thorp and Covich's (2016) 'Freshwater Invertebrates' (4th ed.), Vol. II: Keys to Nearctic Fauna, Elsevier. She has published over 150 research papers. She is a past Scientific Editor of Memoirs of the Entomological Society of Canada, former Secretary of the International Congress of Acarology, and a Fellow of the Entomological Society of Canada.

Dr. Zoë Lindo is a Full Professor in the Department of Biology at the University of Western Ontario (London, Ontario, Canada). As a specialist in the ecology of soil biodiversity and the taxonomy and ecology of Oribatida for 25 years, her research focuses on ecosystems that are currently undergoing dramatic changes in biodiversity due to habitat loss and fragmentation, pollution, overexploitation, and climate change. Dr. Lindo has worked extensively in Canadian forests including the mixed-wood boreal of Alberta, the subarctic taiga of Quebec, the coastal temperate rainforest of British Columbia, and the black spruce peatlands of Ontario. She has published over 80 research papers and is co-Editor-in-Chief of *Pedobiologia—Journal of Soil Ecology*. Dr. Lindo was a lead author for the UN-FAO Report on the State of Knowledge of Soil Biodiversity: Status, Challenges and Potentialities (2020), and is currently Vice-Chair for the UN-FAO International Network on Soil Biodiversity leading a document on Best Practices to Conserve Soil Biodiversity and Prevent Soil Biodiversity Loss. She is also an active member of the Global Soil Biodiversity Initiative, the Entomological Society of Canada, and the Canadian Society for Ecology and Evolution.

1 Introduction

SUMMARY

Overview of oribatid mites: their relationship to other mites and arachnids; their development, their long life spans, their morphology, their reproduction and genetic systems; their fossil history, based on body fossils, amber inclusions and traces; their modes of defence; their phylogeny and recent classification. What makes oribatid mites special, and why we study them.

INTRODUCTION

Oribatid mites are arachnids, chelicerate arthropods that arose over 500 million years ago in the Cambrian (Schweger 1997). Chelicerata is the sister group of the remaining extant arthropods, the Mandibulata, which includes the Insecta (Giribet and Edgecombe 2019). Most arachnids are fluid-feeding, primarily predatory arthropods. By contrast, most oribatid mites are particle-feeding saprophages and mycophages in soil and litter and produce solid faeces (Norton 2007). In their feeding and roles in decomposition, nutrient cycling, soil formation and soil aggregation, they are more similar to mandibulate arthropods, such as collembola, isopods and millipedes, than to their arachnid relatives.

Their diversity both locally and globally can be overwhelming, and up to 150 species with densities exceeding $100,000\,m^{-2}$ can be found in many forest soil and litter (Norton and Behan-Pelletier 2009). This vast diversity and density are attributed to the diversity of niches present in soil and litter (Walter and Proctor 2013), and to oribatid feeding diversity (Norton 2007). Globally, there are more than 11,000 named species (Subías 2022) representing 172 families. These numbers do not include members of the large oribatid Hyporder Astigmata (=Astigmatina), which because of their radically different biology, many morphological and life cycle innovations, and traditional distinction, are treated separately in taxonomic studies (e.g., Krantz and Walter 2009). Oribatid mites are found in all terrestrial and many aquatic habitats from arctic polar deserts to tropics to Antarctic coastal areas. Among oribatid mites in Canada and Alaska, there are representatives of 580 named species, 249 genera and 99 families, a species richness which we estimate to be at most 40% of the total oribatid fauna of this region (Behan-Pelletier and Lindo 2019).

GENERAL CHARACTERISTICS

As adults, most oribatid mites are medium to dark brown from sclerotization and associated melanization, but they may be colourless or have yellow to red pigment. Their body length is typically 300–700 µm as adults, but collectively they span an order of magnitude in size, from about 150 to 2,000 µm. Oribatid mites have a plesiotypic acariform life cycle, including a calyptostasic (regressed and immobile) prelarva contained within the egg, a mobile hexapod larva, three nymphs (protonymph, deutonymph, tritonymph) and adult. Sexual dimorphism usually is inconspicuous, restricted to slightly smaller size and proportionally smaller genital plates in males, but in some species males can have elaborate modifications (Behan-Pelletier 2015). Fertilization is usually indirect by stalked spermatophore without direct association of male and female. But there are several exceptions including a species of *Pilogalumna* where the spermatophore is deposited directly in the region of the female genitalia, and members of *Collohmannia* that have a complex premating ritual between males and females, although the mode of sperm transfer is as yet unresolved (Norton and Sidorchuk 2014). Oviposition is the rule, but sometimes embryogenesis is completed internally and may even progress to the larval instar in some species, particularly aquatic or semiaquatic taxa, such that active larvae are produced by the female (Norton 1994; Søvik et al. 2003).

DOI: 10.1201/9781003214649-1

Most oribatid mites whose genetic system has been studied are diplodiploid, with haplodiploidy reported only in three brachypyline taxa (Norton et al. 1993). Oribatid mites have the greatest frequency of female parthenogenesis (thelytoky) in arthropods, and much of it exists in taxonomic clusters (Norton and Palmer 1991; Norton et al. 1993; Heethoff et al. 2009; Maraun et al. 2019). There is repeated convergent evolution of parthenogenesis and species-rich parthenogenetic taxa radiated independently at least four times (in Enarthronota, Mixonomata, Nothrina, and Brachypylina) (Pachl et al. 2021). There are isolated parthenogenetic species in otherwise sexual genera, and there are wholly parthenogenetic genera, families and even superfamilies. Oribatid mites are models for studies on the evolutionary implications of asexuality (e.g., Brandt et al. 2017, 2021). This striking aspect of oribatid mites is dealt with more fully in Chapter 4.

Life spans are typically 1–2 years in temperate to boreal regions but can reach 4 or 5 years in polar regions (Block and Convey 1995; Søvik 2004). Although long life spans are found in other arachnids, for example, mygalomorph spiders (Mason et al. 2018), those of oribatid mites are among the longest relative to adult size, and much of the life of an oribatid mite is spent as a juvenile (Mitchell 1977; Luxton 1981a). Long life spans probably represent constraints on secondary production (Norton 1994). The result is that, after taking a long time to reach maturity, adults must live a relatively long time to accrue enough resources for reproduction. This ensures strong selection pressure for defensive mechanisms, which are rich in oribatid mites and particularly in adults. Among these are various forms of protective setae, camouflage, waxy exudates, defensive glands, cuticular hardening and an array of body forms and protective structures (Sanders and Norton 2004; Norton 2007; Raspotnig 2010; Schmelzle et al. 2010, 2012, 2015; Pachl et al. 2012; Brückner et al. 2016; Brückner and Heethoff 2018; Schmelzle and Blüthgen 2019). While these defences seem effective against small predators (Peschel et al. 2006), certain beetles and ants can overcome them (Masuko 1994; Molleman and Walter 2001). These interconnected traits of long life-span and diversity of defence mechanisms, which are unique to oribatid mites among soil invertebrates, are dealt with more fully in Chapter 4.

FOSSIL HISTORY

The fossil record for oribatid mites includes body fossils as the hardened exoskeleton of adults, body fossils as amber inclusions and the characteristic trace fossils these mites leave when boring in plant tissue. Oribatida are an ancient group with their earliest body fossil representatives occurring in modern-looking, multi-trophic ecosystems from the Early (Rhynie Chert) and Middle (Gilboa) Devonian, 410 and 378 mya, respectively (Shear et al. 1984; Norton et al. 1988a; Subías and Arillo 2002). Amber inclusions are more evident in the late Cretaceous and Tertiary, with Baltic and Dominican amber inclusions representing most of the oribatid families found in Canada and Alaska (Labandeira et al. 1997). Trace fossils are particularly rich in Carboniferous plant fossils (Labandeira et al. 1997). The strong resemblance of some Paleozoic fossils to extant species and the presence of all major taxa by the Early Jurassic indicate that oribatid mites became established in soil habitats early in their evolutionary history and survived all mass-extinction events (Krivolutsky and Druk 1986). More recent body fossils (sometimes called subfossils) from the Tertiary and Quaternary to the end of Pleistocene glaciations help in understanding the paleoecology of present-day Canada and Alaska (Klinger et al. 1990; Matthews et al. 2019; Markkula and Kuhry 2020; Barendregt et al. 2021).

There is a lack of fossil evidence for Oribatida for approximately 90 million years after the Cambrian explosion, but molecular data using evidence from 18S rRNA genes suggest that their origin is in the Precambrian (571 ± 37 mya) when the colonization of land started, ca. 150 million years earlier than the oldest fossils of terrestrial ecosystems (Schaefer et al. 2010). These data further imply that these omnivorous and detritivorous arthropods formed a component in early terrestrial food webs, perhaps much as they do today (Schaefer and Caruso 2019) (Figure 1.1A).

Introduction

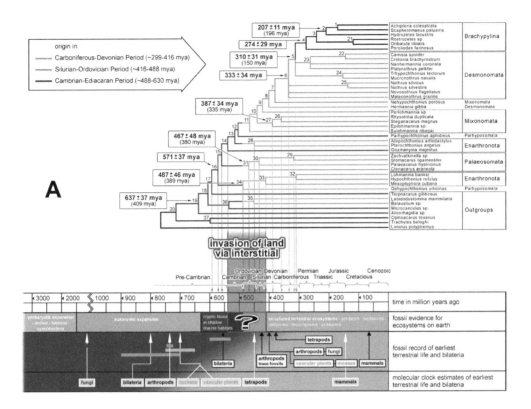

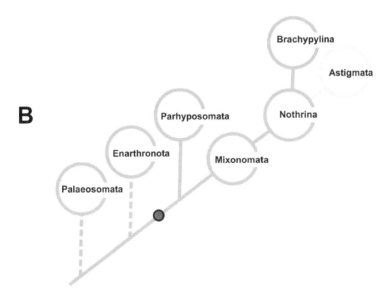

FIGURE 1.1 (A) The Bayesian phylogenetic tree based on 18S rDNA sequences for molecular divergence times of soil-living oribatid mites in relation to major evolutionary events from the fossil record and molecular clock studies (from Schaefer et al. 2010). (B) Hypotheses of the Astigmata–Oribatida phylogenetic relationships based on morphological reconstruction (Norton 1994, 1998); the basal position in oribatid mites is occupied by Palaeosomata and Enarthronota without clearly defined reciprocal relationships; the glandulate Parhyposomata are situated more distally, and Mixonomata and the Hyporder Nothrina are paraphyletic, with Nothrina the sister group for both the monophyletic Brachypylina and Astigmata; the red dot indicates the presence of opisthosomal glands (from Norton 1994).

PHYLOGENY AND CLASSIFICATION

Oribatid mites are members of the Acariformes, one of the two main lineages of the arachnid subclass Acari. We treat Acari as a single group herein, following Lindquist et al. (2009). However, 'Acari' may represent two independent lineages of arachnids, a hypothesis supported by recent molecular analyses (e.g., Pepato and Klimov 2015). Acariformes comprises about two-thirds of known species of Acari and includes two orders: Sarcoptiformes, consisting of Endeostigmata and Oribatida, and Trombidiformes (Dabert et al. 2010).

In our treatment herein, we do not include the Hyporder Astigmata with its about 5,000 described species of parasites, commensals and free-living mites (OConnor 2009). Astigmata is a monophyletic group derived well within the Oribatida (Norton 1998). Based on morphology, Norton (op. cit.) noted 14 apomorphies to support the origin of Astigmata within Oribatida and a further 13 apomorphies supporting the origin of Astigmata at some level within the Desmonomata, probably the Nothrina family Malaconothridae (Figure 1.1B). This is supported by opisthosomal gland chemistry (Sakata and Norton 2001), by mitochondrial DNA (Schäffer et al. 2018) and to some extent by molecular evidence using 18S rRNA and CO1 (Dabert et al. 2010) and nuclear RNA (Pepato and Klimov 2015; Klimov et al. 2017). However, there is conflicting molecular evidence (Domes et al. 2007a), and thus, Schäffer et al. (2018) stressed that future studies need integration of more sarcoptiform genomes, including species from each of the five infraorders of Oribatida.

HIGHER CLASSIFICATION

The higher classification we use herein follows Schatz et al. (2011) with modifications from more recent literature. Much of the classification and terminology we use throughout, and many of the figures, are those of the extraordinary scientist François Grandjean (1882–1975) (see Travé and Vachon 1975 for review). His lucid publications, elegant, detailed observations and pivotal publications, especially his many studies of oribatid families and their higher classification (e.g., Grandjean 1947–1949, 1952a,b, 1954a,b, 1969a), based on studies of all instars, are the foundation for the classification in use today.

Oribatida includes five infraorders: Infraorder **Palaeosomata** Grandjean, 1969, with 3 superfamilies and 6 families; Infraorder **Enarthronota** Grandjean, 1969, with 5 superfamilies and 16 families; Infraorder **Parhyposomata** Grandjean, 1969, with 1 superfamily and 3 families; Infraorder **Mixonomata** Grandjean, 1969, with 7 superfamilies and 9 families; Infraorder **Desmonomata** Woolley, 1973, with 3 Hyporders: Hyporder **Nothrina** van der Hammen, 1982, with 1 superfamily and 6 families; Hyporder **Brachypylina** Hull, 1918 (= Circumdehiscentiae Grandjean, 1954) with 25 superfamilies and 124 families; Hyporder **Astigmata** Canestrini, 1891 (= Astigmatina Krantz and Walter, 2009), with 11 superfamilies.

Diagnoses for superfamilies (other than Astigmata) can be found in Norton and Behan-Pelletier (2009), although caution is warranted as some have been revised subsequently. A short diagnosis is given for adults of each family represented in Canada and Alaska in Chapter 3.

WHY ARE THEY IMPORTANT?

Oribatida are a unique monophyletic lineage with possible answers to questions on (a) the evolution of morphological and physiological constraints in saprophages and fungivores that have a predatory ancestry; (b) origins of the trade-off between sexuality and thelytoky, as Oribatida include thelytokous species, genera and families, and at a frequency of 10–1,000 times that of most animal groups; and (c) what drove the evolution of the diverse and exploitative group Astigmata from within the Oribatida. Unlike the nonavian dinosaurs, whose extinction ca. 66 mya triggered the diversification of the avian dinosaurs (birds), the three most-derived oribatid lineages—the paraphyletic Nothrina, and the monophyletic Brachypylina and Astigmata—have thrived since their putative split, and each expresses very different life-history traits. These questions raised and addressed by Norton (2007) continue to challenge. We know much more about these questions, but the fundamental 'why/how?' remains elusive. For example, questions such as the physiological basis for digestive efficiencies in oribatid mites that are lower than other mandibulate arthropods using similar food; how asexual oribatid clades survived for tens of millions of years; and how soil Astigmata with a diet comparable to other oribatid mites have a production/consumption that is 2–20 times higher?

To summarize, Oribatida provide model organisms for a range of questions in biology and ecology:

- They are a phylogenetically diverse, ancient group, with extant infraorders known from ca. 350–400 mya;
- They inhabit all terrestrial, freshwater and littoral habitats;
- Adults fossilize well, are frequently captured in amber and contribute to defining past environments;
- They have the highest incidence of all female parthenogenesis (thelytoky) in Arthropoda, and among the highest in the animal kingdom;
- They are unusually long-lived for Arthropoda, and most adults have developed an unrivalled diversity of defence mechanisms, including morphological modifications and a diversity of gland secretions;
- Chapter 4 on Ecology highlights the contribution of oribatid mites to biodiversity and ecosystem services such as decomposition, nutrient cycling, soil formation and soil aggregation. This is particularly relevant in habitats where these mites numerically dominate the soil fauna, for example, boreal regions, taiga, arctic and peatlands;
- Many species are important or key bioindicator species, including the ISO (International Organization for Standardization) certified oppiid *Oppia nitens* and the crotoniid *Platynothrus peltifer* (Heethoff et al. 2007);
- In parallel to its importance in tropical ecosystems, the trhypochthoniid species, *Archegozetes longisetosus*, is a model organism for studies on arthropod evolutionary development (Thomas 2002; Barnett and Thomas 2012, 2013; Hartmann et al. 2016), genetics (Bergmann et al. 2018), embryology (Laumann et al. 2010a,b), internal morphology (Heethoff and Cloetens 2008; Heethoff et al. 2008), chemical ecology (Seniczak et al. 2005, 2009) and mechanics (Heethoff and Koerner 2007; Heethoff and Norton 2009; Heethoff et al. 2013).

CATALOGS AND CHECKLISTS

Catalogs and checklists relevant to the Canadian and Alaskan fauna include:

Global: Subías (2004, 2022).
Nearctic region: USA and Canada (Marshall et al. 1987), Canada and Alaska (Behan-Pelletier and Lindo 2019).
Northern Palearctic: Russian Far East (Pan'kov et al. 1997; Ryabinin et al. 2018), Svalbard (Coulson 2008), Greenland (Makarova and Behan-Pelletier 2015), Iceland (Gjelstrup and Solhøy 1994), Finland (Niemi et al. 1997).
Central Palearctic: British Isles (Luxton 1996), Ireland (Arroyo et al. 2017), Austria (Krisper et al. 2017; Schatz 2020), Germany (Weigmann 2006).
Eastern Palearctic: China (Wang et al. 2002a,b; Chen et al. 2010), Korea (Choi 1996), Japan (Fujikawa et al. 1993).

2 Form and Function

SUMMARY

This chapter covers the external morphology of adult oribatid mites and introduces the reader to terms of the prodorsum, notogaster, venter, gnathosoma and legs that are essential for following the identification keys in Chapter 3. The various sensory structures of oribatid mites are explained. Essential functions are reviewed.

INTRODUCTION

Like most acariform mites, oribatid mites have strongly modified the ancestral arachnid body divisions (Figure 2.1A). The anterior **gnathosoma** is derived from the first two primitive somatic segments and carries the primary organs of food acquisition. The **idiosoma** comprises the entire body posterior to the gnathosoma and thus assumes functions parallel to those of the abdomen, thorax and parts of the head of other arthropods. It includes an anterior **propodosoma** and a posterior **hysterosoma** (Figure 2.1A), which may or may not be separated by a sejugal furrow. The anterior two pairs of legs are inserted ventrally or ventrolaterally on the propodosoma, while legs III and legs IV (in postlarval stages) are inserted in the adjacent portion of the hysterosoma. The **podosoma** constitutes the lateral and ventral region of leg-bearing segments. A postpedal furrow may separate the podosoma from the **opisthosoma**, which is that portion of the hysterosoma behind legs IV.

Oribatid mites are conservative in that most maintain the basic attributes of acariform mite development (Walter and Krantz 2009). However, some developmental traits vary among groups and are relevant to taxonomic diagnoses:

Anamorphic Development

Ontogenetic development involves curvature of the idiosoma (**caudal bend**) as a result of anamorphosis, i.e., the addition of terminal body segments during ontogeny, with the protonymph (PN) adding the adanal (AD) segment to the six segments recognized in the larva (C, D, E, F, H and PS (P)), the deutonymph (DN) adding the anal (AN) segment and the tritonymph (TN) adding the peranal (PA) segment (Grandjean 1939b) (Figures 2.1C–F). Other than some members of the infraorders Enarthronota and Parhyposomata (Parhypochthoniidae), all oribatid mites cease to add segments at the deutonymph, such that the anal segment is paraproctal (around the anal aperture) in adults.

Metamorphosis

Many oribatid mites exhibit a metamorphosis between tritonymph and adult that is striking enough to make association of adults and immatures difficult. This is the rule in Brachypylina and in those Enarthronota and Mixonomata having a ptychoid body form, but in most other groups, adults differ from immatures mostly in the degree of sclerotization and the presence of genitalia.

MORPHOLOGY OF ADULT ORIBATID MITES

The following overview of adult morphology is focused on features used in the keys and diagnoses, and relies almost entirely on Norton and Behan-Pelletier (2009), with some updates. Most oribatid terminology was coined by F. Grandjean in the 1930s–1960s (see Travé and Vachon 1975; Hammen 1980 for references).

DOI: 10.1201/9781003214649-2

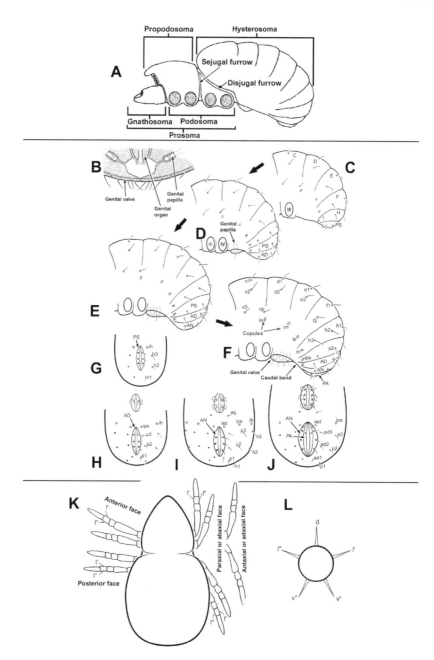

FIGURE 2.1 (A) Diagrammatic lateral aspect of an acariform mite showing major body divisions (based on Krantz and Walter 2009); (B) Schematic section of adult progenital chamber, showing ancestral arrangement of valves, genital papillae and genital organ; (C) schematic hysterosoma of larval acariform mite (left lateral view) showing ancestral segmentation, setae and cupules; (D) same, protonymph; (E) same, deutonymph; (F) same, tritonymph; (G) schematic hysterosoma of larva in ventral view; (H) same, protonymph; (I) same, deutonymph; (J) same, tritonymph. (B after Grandjean 1969b; C–F based on Grandjean 1939; genital setal counts not meaningful), (courtesy of R. A. Norton). Leg chaetotaxy in the Acariformes: (K) schematic acariform mite, dorsal aspect, with legs arranged hypothetically on left and as naturally positioned on right, to show different positions relative to central body axis (paraxial, antiaxial) of metamerically homologous setae according to leg direction (prime indicates anterior face; double prime indicates posterior face); (L) schematic verticil of five setae showing unpaired dorsal, paired lateral and paired ventral setae (courtesy of R. A. Norton).

INTEGUMENT AND BODY FORM

The cuticle consists of thick, chitinous **procuticle** and a thin, overlying **epicuticle**. The procuticle is composed of an underlying **endocuticle** and an outer **exocuticle**, the latter of which may become sclerotized to varying degrees in various regions of the body through orthoquinone tanning. **Pore canals** first appear in the endocuticular layers and move in a helical fashion toward the surface through the exocuticle, where they assume their typical linear, often branched, appearance (Norton et al. 1997). The canals terminate in a profusion of micropores in the cuticulin layer just beneath the epicuticle. When these micropores are extensive, as is common on the **coxisternum** and **prodorsum** (see below), these regions usually represent respiratory surfaces. By contrast, small, localized **porose organs** may be either secretory, with large epidermal cells that are probably involved in cuticular maintenance, or respiratory, with very thin epidermis to maximize gas exchange (Alberti et al. 1997). The cuticular part is called a **porose area** (= area porosa) if it is not invaginated (Figure 2.4F). If invaginated, porose areas have a variety of names: small pouches are **saccules** (Figure 2.4G); large, flat, lamelliform pouches are **platytracheae**; thick, relatively short tubes are **brachytracheae** (Figure 2.3E); and extremely long, filamentous tubes are **tubules** (Figure 3.112C). The epicuticle is a multilayered envelope consisting of an inner epicuticle and an overlying secretion layer, or **cerotegument**, that appears shortly after ecdysis. Cerotegument, which may or may not be conspicuous, consists of an underlying outer epicuticle covered by what appears to be a wax layer. A dense, often highly sculptured **cement layer** completes the epicuticular complex. The wax and cement layers on the body surface may provide some protection against excessive water loss or absorption (Norton et al. 1997). For oribatid mites living in dry microhabitats, Walter (2009) suggested that the ultrastructure of the wax layer may have a water-shedding effect.

Hard cuticular plates may derive from mineralization rather than sclerotization, which means that some rather hard species can be light in colour. Known minerals include calcium carbonate, hydrated calcium oxalate (whewellite) and calcium phosphate, with at least the latter two being deposited in epicuticular chambers (Norton and Behan-Pelletier 1991; Alberti et al. 2001). Articulations between certain plates may be covered by a roof-like extension from the edge of one plate, called a **tectum** (Figures 2.3B; 2.4C; 2.5G); such structures allow mobility while protecting the vulnerable soft cuticle of the articulation from predators.

The cuticle of oribatid mites may be shiny and smooth, but various materials can make it appear grey and irregular. The cerotegument appears as a white or grey coating in live mites and has consistent form within species. In addition to a generally distributed thin coating, there are often excrescences in the form of small tubercles, cylindrical or conical projections, long cottony filaments or reticulations. In some cases, the cerotegument is solid and birefringent in polarized light, either plate-like (Malaconothridae), blocky (e.g., Phenopelopidae, Unduloribatidae, some Gymnodamaeidae) or in an amorphous mass. Various groups of oribatid mites incorporate organic or mineral debris with cerotegument or pack such debris in solid dorsal masses (e.g., Crotoniidae, Damaeidae). Partial exuviae of immature instars are also carried by adults of some species. Such exuviae may represent only the tritonymph (Hermanniellidae, some Crotoniidae) or all previous instars (many Brachypylina), with successive exuviae (**scalps**) stacked like a pagoda.

Immature stages in some higher and all stages in some primitive oribatid mites, particularly Archeonothroidea, have a body form with numerous small plates at the bases of many body setae (Figure 2.2A). However, in the large majority of adult oribatid mites, most visible cuticle is hardened, forming two expansive dorsal plates and various arrangements of ventral plates. The dorsal plate anterior to the sejugal furrow is called a **prodorsum** (or **aspis**, when it is isolated from ventral plates as in ptychoid mites) (Figure 2.2B). The dorsal and lateral cuticle of the hysterosoma (i.e., that portion bearing setal rows c through p) is called the **notogaster** if it is sclerotized, and the **opisthonotum** (=gastronotic region) if it is mostly unsclerotized, as in adults of some early derivative taxa and in most immatures.

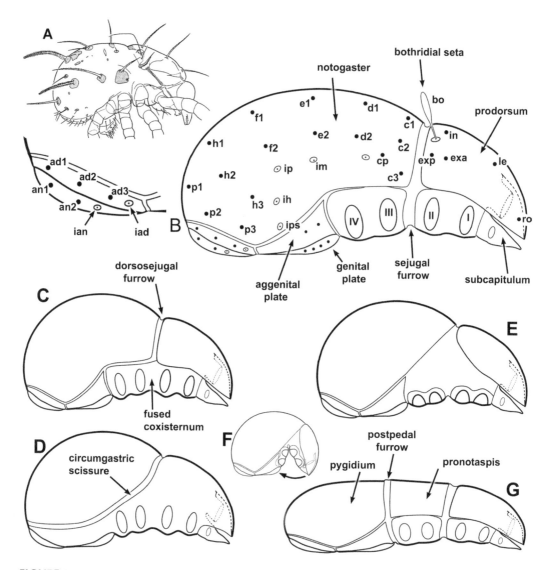

FIGURE 2.2 General body forms of oribatid mites (in lateral aspect; ellipses indicate the base of appendages): (A) *Acaronychus traegardhi*, illustrating primitive acariform mite body without major sclerites (after Grandjean 1954b); (B) schematic dichoid mite showing general cuticular features (enlargement of anal region to left); (C) schematic holoid mite, macropyline type; (D) same, but brachypyline type; (E) schematic ptychoid mite in active posture; (F) same, partially closed; (G) schematic trichoid mite (C–E, G after Norton 2001; F after Sanders and Norton 2004).

Oribatid mites are **dichoid** if the sejugal furrow remains soft, such that the propodosoma and hysterosoma articulate freely (Figure 2.2B). In some dichoid mites, e.g., most Enarthronota, the notogaster comprises several component plates separated by transverse articulations, or **scissures** (see below); in others (many Mixonomata), the notogaster is entire. If a soft postpedal furrow also occurs so that the body has two primary articulations (Figure 2.2G), the mite is **trichoid** (e.g., most Parhyposomata). Several groups (Enarthronota: Mesoplophoridae, Protoplophoridae; and Mixonomata: Euphthiracaroidea, Phthiracaroidea) have a **ptychoid** body form (Figures 2.2E,F; 2.13F), a defensive adaptation allowing the mite to close like a seed when disturbed, giving the common name 'box mite'. Ptychoidy involves use of a set of large muscles that pulls the coxisternum and legs into the opisthosoma (Sanders and Norton 2004; Schmelzle et al. 2009, 2015), which is possible because all podosomal cuticle other than the coxisternum is soft. Another set of muscles deflects the aspis and pulls it toward the opisthosoma, and the legs are captured inside a secondary chamber. In the **holoid** body form, which characterizes most Nothrina (Figure 2.2C) and all Brachypylina (Figure 2.2D), epimeres II and III have fused, eliminating the sejugal furrow ventrally. In holoid mites, the principal body articulation is the more or less horizontal **circumgastric scissure** (Figures 2.2D; 2.4D,E), which separates the notogaster from ventral plates. In Nanhermanniidae, the circumgastric scissure does not exist, and the notogaster and ventral regions merge (Figure 2.13D).

Prodorsum

In most adult oribatid mites, the prodorsum extends far anteriorly as a rostral tectum, or **rostrum**; such a prodorsum is called **stegasime** (Figure 2.3B). The rostrum usually curves ventrally, creating a protected secondary vestibule (**camerostome**), within which the chelicerae operate. Upon disturbance, the chelicerae are fully retracted and the **subcapitulum** is levated to close the vestibule. The edge of a rostrum may be smooth and simple, or variously modified with a reflexed edge, projecting teeth or emarginations of various types. An emargination at the lateral corner of the rostrum, typical of several groups within Brachypylina, is the **genal notch** (Figure 2.3J), which delimits the **genal tooth** (Figure 2.11B). The ancestral condition of the prodorsum in oribatid mites is an inconspicuous aspis that ends abruptly at the articulation with chelicerae, which are fully exposed even when retracted (Figure 2.3A). This **astegasime** condition is characteristic of the Palaeosomata, as well as some Enarthronota, Parhyposomata and Mixonomata. A medial lobe-like **naso** is retained in some Palaeosomata (Archaeonothridae) and Enarthronota (Brachychthoniidae). True eyes are restricted to a few Palaeosomata and Enarthronota.

Ancestrally, six pairs of mechanoreceptive setae insert on the prodorsum (Figure 2.2B). **Rostral setae** (*ro*) are anteriormost; when the rostral tectum is present, they insert on it. Posterior to the rostrals are the **lamellar setae** (*le*), which insert at or near the anterior aspect of the lamellae if the latter are present. The most proximal of the three pairs are the **interlamellar setae** (*in*), which lie in the space between lamellae. A trichobothrium (pseudostigmatic organ) usually inserts in each posterolateral corner of the prodorsum. It consists of a **bothridial seta** (*bo* or *bs*), also called a sensillus (*ss*), and a deep, cup-like **bothridium** in which the seta inserts. The bothridium is a simple inverted cone in primitive oribatid mites (Palaeosomata and some Enarthronota (Figure 2.3C)), but it has a strong, usually S-shaped, curve at its base in most oribatid mites (Figure 2.3D); the shape is mimicked by the inserted basal part of the bothridial seta. Inside the bothridium, the cuticle may invaginate further to form porose saccules, brachytracheae (Figure 2.3E) or short tracheae (Grandjean 1934b; Norton et al. 1997); most such structures are probably respiratory surfaces (Alberti et al. 1997). Bothridial setae vary greatly in shape among oribatid mites, but generally they differ from other prodorsal setae. Ventrolateral to each bothridium are **exobothridial setae**, which ancestrally appear in two pairs; these have been given several designations (*exa/exp*, ex_1/ex_2, *xs/xi* and *xa/xp*). Among prodorsal setae, the exobothridials are most susceptible to loss. Members of Nothrina and Brachypylina always lack at least one pair (the remaining pair is labelled *ex*), and in some cases both are lost. In some groups, lost exobothridial setae are represented by a pore-like alveolar vestige (*exv*) (widespread in Nothrina, and in Eremaeidae, many Damaeidae and Caleremaeidae among Brachypylina).

The surface of the prodorsum is mainly smooth in Palaeosomata, Enarthronota and Parhyposomata. In Brachypylina, the prodorsum usually has tubercles, carinae, ridges or other projecting structures. Many have a pair of longitudinal structures, medially or laterally, called **lamellae**. These can be developed as simple low ridges (ridge-like lamella); or they are blade-like, with at least one free edge (Figures 2.3H–J; 2.11B). **Costulae** are simple secondary supporting ridges in taxa that plesiomorphically lack lamella, e.g., Oppioidea (Figure 2.3G). When a lamella is of sufficient size, it is a protective structure under which leg I is retracted when the mite is disturbed

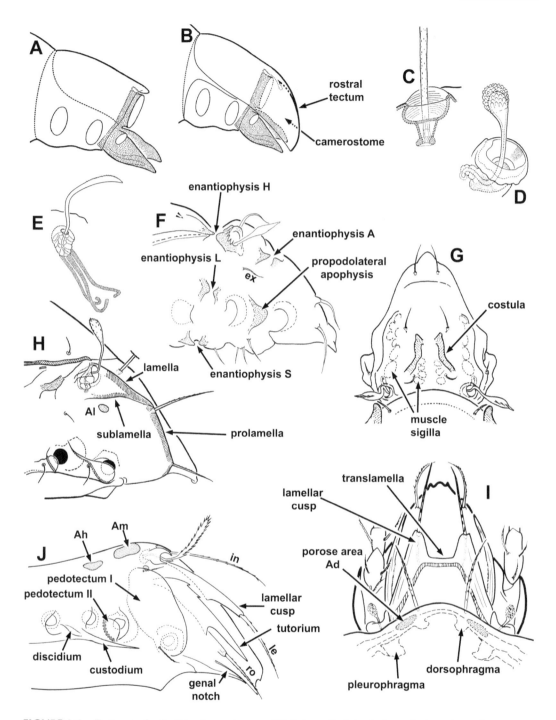

FIGURE 2.3 Features of oribatid mite prodorsum. (A) Schematic lateral view of astegasime mite proterosoma (after Grandjean 1970); (B) same, stegasime mite; (C) primitive trichobothrium, with straight bothridium and base of seta *bo* (after Grandjean 1954b); (D) derived trichobothrium, with bothridium and seta *bo* having basal curve (after Grandjean 1936a); (E) trichobothrium of Phthiracaridae, with three brachytracheae (after Grandjean 1934b); (F–I) anterior half of brachypyline mites in lateral (F, H, J) and dorsal (G, I) views; (F) enantiophyses and other cuticular projections (Norton unpublished); (G) Oppiidae (after Aoki 1983); (H) Scheloribatidae (after Grandjean 1953b); (I) Ceratozetidae (after Behan-Pelletier 1984); (J) Ceratozetidae, prodorsum (after Behan-Pelletier 1986).

(Figures 2.12C; 2.13B). A lamella commonly extends anteriorly as a projecting tooth- or knife-like **lamellar cusp** of various sizes and forms, on which the lamellar seta typically inserts. The lamellae may be independent, may appear connected by a transverse ridge- or blade-like **translamella** (Figure 2.3I) or may fuse for part or all of their length (Figure 2.12C). A narrow carina, the **prolamella**, may run distally from each lamella toward the rostral margin (Humerobatidae and some Scheloribatidae) (Figure 2.3H). Scheloribatidae may also have lamellae that fork posteriorly, with the more ventral branch—the **sublamella**—running below the bothridium. Some Brachypylina have another ridge- or blade-like longitudinal structure, the **tutorium** (Figures 2.3J; 2.11B), on each lateral face of the prodorsum. Like lamellae, tutoria have diverse size and shape and may have a distal cusp. If they are blade-like, the free edge is dorsal, and the distal part of leg I lies in the resulting valley when retracted (Figure 2.13A,B). In Brachypylina, the insertions of legs I and II are often protected by **pedotecta**, scale- or ear-like (auriculiform) structures that project from the body wall immediately posterior to the respective acetabulum (Figures 2.3J; 2.6E). A tooth-, horn- or scale-like lateral projection between legs I and II that is not closely adjacent to an acetabulum is called a **propodolateral apophysis** (P) (Figure 2.6F); such projections do not coexist with pedotecta. A **patronium** is a tectiform projection between legs I and II which may coexist with pedotecta (Figure 3.63E).

An **enantiophysis** may occur at several places on the prodorsum of brachypyline mites (Figures 2.3F; 2.6F). This is a set of two tubercles that oppose each other across an articulation or furrow; typically one tubercle is anterior (a), and the other posterior (p). When present, the **prodorsal enantiophysis** (*eA, A*) spans a transverse or paired groove at mid-length of the prodorsum. The **postbothridial enantiophysis** (*Ba, Bp*) is positioned posteromedial of the bothridium and can be unbalanced with *Bp* almost absent. The **laterosejugal enantiophysis** (*eL, L*) spans the sejugal furrow laterally (Figure 2.3F). The **humeral enantiophysis** (*eH* or *H*) spans the sejugal furrow dorsally, with tubercles on the bothridial wall and the humeral angle of the notogaster (Figure 2.3F). Enantiophyses seem to function in holding and anchoring an air film (plastron) that has contact with the stigmata of the apodemato-acetabular tracheal system (Chen et al. 2004). It is obvious how such localized plastrons are important for intertidal taxa (e.g., Pfingstl 2017), but even for fully terrestrial species they could be important whenever the immediate environment is inundated with water for extended periods of time.

Insertions of the paired cheliceral retractor muscles are often conspicuous on the prodorsum due to their associated **muscle sigilla** (excavations on the internal face of the cuticle where muscles attach). A large group of sigilla is often present in the region between the interlamellar and lamellar setae, and another on the lateral face (Figure 2.3G). In some publications, these are referred to as 'maculae' or 'spots'. In many Brachypylina, cheliceral muscles insert instead on two paired apodemes that project internally from the posterior margin of the prodorsum (Figure 2.3I). The **dorsophragmata** (= dorsophragmatic apophyses) are in the dorsosejugal region. The **pleurophragmata** are more laterally positioned.

Several porose organs—all of which probably are dermal glands—are found on the prodorsum in poronotic superfamilies of Brachypylina (Norton et al. 1997; Alberti et al. 1997). The paired **sublamellar** porose area (*Al*) may be present below the lamella (Figure 2.3H). A humerosejugal series may include one or more of the following: the **dorsosejugal** (*Ad*) and **humerosejugal** (*Aj*) porose areas on the prodorsum; and one (*Ah*) or two (*Am*) **humeral** porose areas in the subhumeral region (Figures 2.3J). In some Brachypylina, *Ah* and/or *Al* are expressed as saccules.

NOTOGASTER

The notogaster in Brachypylina is a single, cap-like sclerite covering the hysterosoma dorsally, separated by the circumgastric scissure from the prodorsum anteriorly, and from the ventral region posteriorly. The anterior separation is variously called the **dorsosejugal scissure**, furrow or groove (Figure 2.2C). In some Nothrina and Brachypylina, it is effaced by a fusion of the prodorsum and notogaster, either imperceptibly or marked by a thickened suture indicating the ancestral line of contact. Outside the Brachypylina, there are many variations in notogastral structure. A paired, longitudinal **suprapleural scissure** often separates a dorsal **notaspis** from a lateral **pleuraspis** (Figure 2.4K). In Enarthronota, the notaspis is usually divided by 1–3 transverse scissures (Figure 2.4K). These may demarcate an anterior **pronotaspis** and a posterior **pygidium** (Figure 2.2G), both of which vary in segmental composition. Three types of transverse scissures are known (Figure 2.5E–G) (Grandjean 1947b; Norton 2001). **Type-E scissures** are simple articulations between plates; edges of the two adjacent plates are unmodified, and the intervening band of soft cuticle is usually narrow (Figure 2.5E). A **type-S scissure** is a compound structure, with the space between two major plates occupied by a transverse series of four closely adjacent intercalary sclerites, each bearing a seta; these small sclerites can be variously combined, depending on the taxon (Figure 2.5F). Ancestrally, setae of type-S scissures are large, erectile and serve a defensive role. A **type-L scissure** is a defensive specialization that permits significant telescoping of two plates, while protecting a broad articulating cuticle. The posterior edge of the more anterior plate is hypertrophied as a tectum (Figure 2.5G).

Notogastral cuticle may be smooth or have various forms of small-scale relief. Longitudinal ridges may be present, either freestanding or connected to anterior tubercles; such a tubercle-ridge complex in Brachypylina is sometimes called a **crista**. The humeral region may be simple or it may have tubercles, or anteriorly directed knife-like processes or laterally directed tecta. A humeral tectum that is well defined and large enough to conceal all or part of the retracted legs is called a **pteromorph** (Figures 2.4A,B,I) and is known only in Brachypylina. In some taxa, the base of a pteromorph is completely or partially desclerotized to form a linear hinge, and some of the dorsoventral musculature is modified to pull the resulting movable pteromorph against the body when the mite is disturbed (Figure 2.4I). The posterior margin of the notogaster may also project as a tectum, overhanging the circumgastric scissure (Figure 2.4B,C). This tectum is incomplete medially in some oribatids where lobe-like projections are separate or overlap (e.g., Punctoribatidae). In some groups, a raised central region is delineated by a circular or U-shaped submarginal depression, the **circum-marginal furrow**.

The most general complement of mechanoreceptive setae on the oribatid notogaster is 16 pairs, a condition known as **holotrichy**. Grandjean (1934a) gave these setae notations according to their presumed segmental origin (Figure 2.5A,B). Except for Hermanniellidae, members of Brachypylina have fewer than 16 setal pairs. The basic number is 15 pairs, but uncertainty about which seta in the middle of the notogaster was lost caused Grandjean (1934a) to create the **unideficiency nomenclature** (Figure 2.5C,D), with the six pairs of setae in the middle of the notogaster given the following notations: three pairs of **dorsocentral setae** (da, dm, dp) flanked by three pairs of **dorsolateral setae** (la, lm, lp). In most cases, adult Brachypylina have either 15 pairs, or a lesser number that can be easily identified with homologous setae of the unideficiency nomenclature. Rarely, the notogaster is neotrichous. Particular notogastral setae may be represented only by vestiges, in which only the pore-like alveolar canal remains or there is a minute setal remnant; for instance, one or both f-row setae often are vestigial in Mixonomata and Nothrina. Setal insertions on the notogaster (and occasionally some leg setae) in Ptyctima and Galumnidae are **apobasic**, that is, the cuticle forms an abrupt depression at the bottom of which is a normal setal insertion. Internal to this 'sunken' insertion is the usual simple cuticular canal; external to this insertion the seta passes through the enveloping tube/cup before emerging at the surface. This differs from setae that insert on a solid, sclerotized structure such as a lamella or tectal limb, where the alveolus is at the surface, and the alveolar canal is greatly elongated to reach the living epidermis (Figure 3.114F).

There are five pairs of notogastral **lyrifissures** (see Sensory Structures: proprioceptors below) in most oribatid mites: ia, im, ip, ih and ips (Figure 2.2B). In the Parhypochthoniidae, the adanal segment becomes incorporated into the notogaster, along with its setae and lyrifissure iad. Except for the Palaeosomata and Enarthronota, most oribatid mites possess a pair of **opisthonotal glands** (also called opisthosomal or oil glands) that open in the mid-lateral region of the notogaster. Usually the opening is inconspicuous, but it may be on a tubercular or funnel-like projection

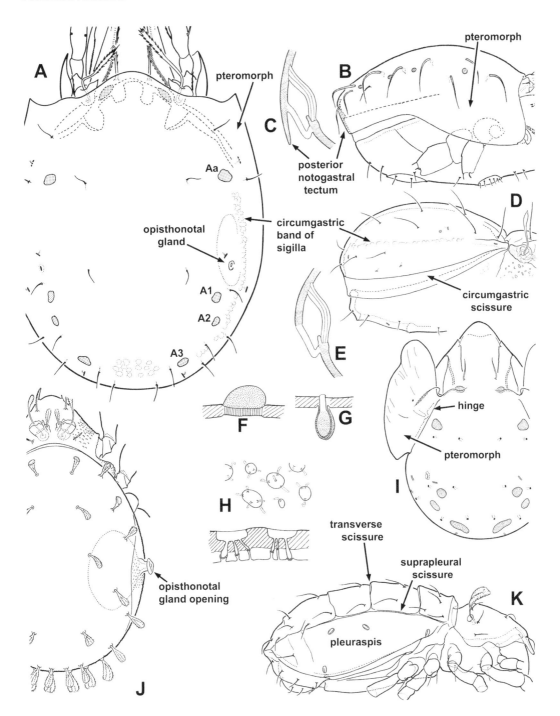

FIGURE 2.4 Features of oribatid mite notogaster. (A) Ceratozetidae, dorsal view (after Behan-Pelletier 1984); (B) Oribatellidae, lateral view (after Grandjean 1956a); (C) schematic cross section of posterior margin with tectum (after Grandjean 1959c); (D) Autognetidae, lateral view (after Grandjean 1960c); (E) schematic cross section of posterior margin without tectum (after Grandjean 1959c); (F) schematic section of cuticular part of porose area (after Norton et al. 1997); (G) schematic section of cuticular part of saccule (after Norton et al. 1997); (H) macropores in Neoliodidae, top, dorsal view (after Grandjean 1934b) and Hermanniellidae, bottom, cross section (after Grandjean 1962a); (I) Galumnidae, dorsal view (after Grandjean 1956f); (J) Hermanniellidae, dorsal view (after Grandjean 1962a); (K) Haplochthoniidae, lateral view (after Grandjean 1947b).

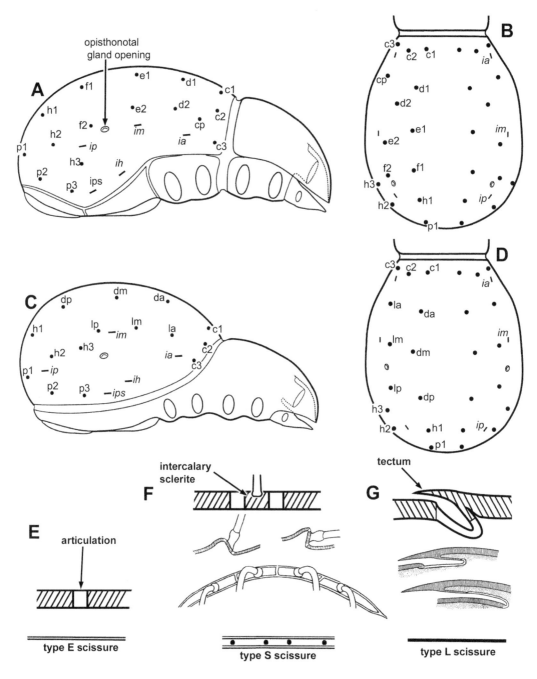

FIGURE 2.5 Comparison of chaetotaxies used for notogastral setae of oribatid mites, shown in lateral (A,C) and dorsal (B,D) views: (A,B) holotrichy; (C,D) unideficiency; (E–G) Structure and distribution of transverse notogastral scissures in Enarthronota; (E) schematic sagittal section of type-E scissure; (F) schematic sagittal sections of type-S scissure showing erect (top) and reclined (middle) positions of erectile setae and dorsal view of same (bottom); (G) schematic sagittal section of type-L scissure (top), and the same showing the range of motion (middle is extended; bottom is contracted) (after Grandjean 1931c, 1934c; Norton 2001).

(Parhypochthoniidae, Hermanniellidae, Plasmobatidae) (Figures 2.4J). In Hermanniellidae and Neoliodidae, numerous small respiratory saccules (**macropores**) may be present (Alberti et al. 1997; Norton et al. 1997) (Figure 2.4H). The notogaster of some Enarthronota and many Brachypylina have well circumscribed secretory porose organs. Usually the cuticular component of such dermal glands is a porose area, but sometimes the porose cuticle is invaginated as a saccule or trachea-like **tubule** (see above under Integument).

When dermal glands occur on the notogaster of Brachypylina, they are typically arranged in four pairs, forming the **octotaxic system** of porose organs (Figure 2.4A). The anteriormost porose area is the **adalar** (*Aa*), with the others being **mesonotic** (*A1, A2, A3*). Saccules and tubules have similar designations, using 'S' or 'T' instead of 'A' as prefix. In rare cases, the octotaxic system may include a mixture of porose areas and saccules. While the typical complement is four pairs of octotaxic organs, higher or lower numbers are not uncommon (Norton and Alberti 1997; Behan-Pelletier 2015).

The **lenticulus** is an unpaired light receptor organ situated near the anterior margin of the notogaster in some Brachypylina (Figures 3.98A; 3.103A). Muscle sigilla may be very conspicuous on the notogaster, especially if the cuticle is well sclerotized; the thinner cuticle of a sigillum shows a lighter colour in transmitted light (thus being sometimes mistaken for a porose area). In Brachypylina, most notogastral muscle sigilla are arranged in a U-shaped curve just inside the margin, indicating the origins of the circumgastric muscle band that controls hemolymph pressure (Figure 2.4A,D).

Ventral Structures

The venter of an oribatid mite exhibits three regions that are usually easily distinguished by sclerotization patterns. Anteriormost is the subcapitulum, a part of the gnathosoma that is discussed below. Behind this, and usually separated from it by a narrow articulation, is the **coxisternum** (or coxisternal region), which forms the floor of the podosoma and serves to support the legs. Behind the coxisternum is the **anogenital region**, in which the anal and genital openings are situated. The four epimera (I–IV) that comprise the coxisternum (Figure 2.6B) always exhibit some level of sclerotization. Minimally, each exhibits a pair of **epimeral plates**, medially separated by a soft longitudinal band and independent of other epimera (Figure 2.6A). Maximally, all are fully sclerotized and collectively fused as a single unit (Figure 2.6E), which characterizes the holoid body form. In dichoid body forms, there are usually two connected groups, I/II and III/IV, separated by the sejugal articulation, but various levels of fusion exist. The anterior edge of epimere I may form a tectum (**mentotectum**) that overhangs the base of the subcapitulum (Figure 2.6C). Typically in Brachypylina, four pairs of blade-like coxisternal apodemes project internally from the ventral plate to serve as attachment locations for muscles: *ap1*, *ap2*, *apsj* and *ap3* (Figure 2.6C). They are often confused with epimeral borders, which are more conspicuous and partly confluent with apodemes when viewed ventrally in transmitted light. The plesiomorphic state for apodemes is a flat, solid cuticular plate, as found in Eremaeidae, Damaeidae and Damaeolidae. A derived trait is for a particular apodeme to be perforated; perforation of *ap2* can be so extensive that the apodeme has the form of an arch, as in Oripodoidea. Epimeral borders (*bo1, bo2, bo3, bo4*) are made conspicuous when the middle area of epimera is internally excavated by many muscle sigilla (Figure 2.6H). The coxisternal setation is usually represented as a formula, with the number of pairs on epimera I–IV in succession. Setae are numbered according to epimere and given letters according to distance from the midline (e.g., *1a*, *3b*; Figure 2.6E); coxisternal setation is neotrichous in a few groups (e.g., Nothridae, Euzetidae). Legs attach at the lateral extent of the epimera, either by a simple articulation with the trochanter, or (in Brachypylina) by a ball and socket joint in which the epimere is invaginated laterally to form the socket, or **acetabulum** (Figure 2.6C,D). This difference in leg attachment is also expressed in the direction in which legs leave the body (Figure 2.6B,C).

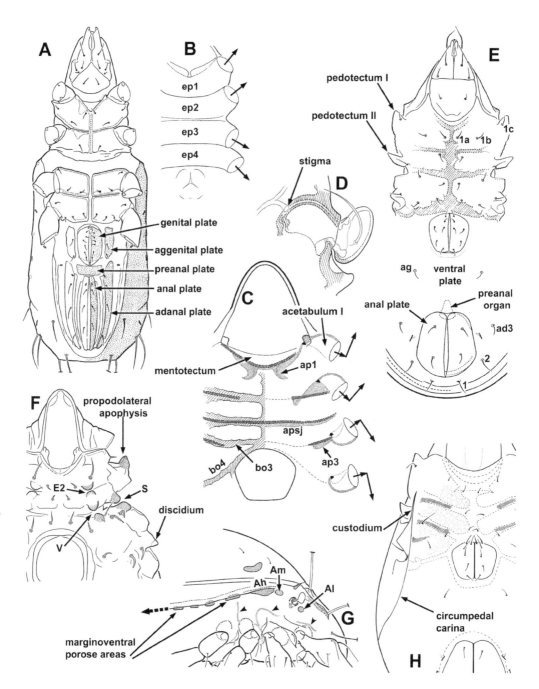

FIGURE 2.6 Ventral and lateral structures of oribatid mites. (A) Generalized macropyline type of ventral structure (after Grandjean 1958a); (B) schematic structure of primitive coxisternum, with arrows indicating the direction of legs as they leave respective epimera (I–IV) (after Grandjean 1952a); (C) schematic structure of brachypyline coxisternum, showing acetabula, apodemes (*ap*) and epimeral borders (*bo*), with arrows indicating the direction of legs as they leave respective acetabula (after Grandjean 1952a); (D) acetabulum I of brachypyline mite, showing opening (stigma) of two-branched trachea (after Grandjean 1959b); (E) brachypyline type of ventral structure (after Grandjean 1960d); (F) Damaeidae, generalized partial venter showing enantiophyses (*S, V, E2*) and other cuticular projections (after Grandjean 1960b); (G) lateral features of poronotic brachypyline mite, showing porose areas and tracheae (arrowheads) (after Grandjean 1971); (H) Ceratozetidae, partial venter (after Behan-Pelletier 1984).

Form and Function 19

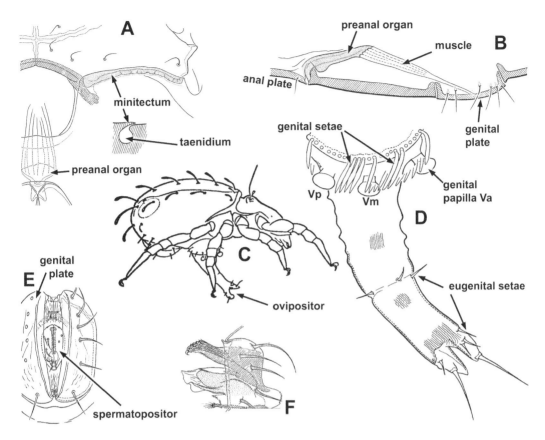

FIGURE 2.7 Further ventral features of oribatid mites. (A) Partial venter of brachypyline mite showing taenidium and minitectum on epimeral border IV, along with preanal organ (after Grandjean 1968); (B) sagittal section of brachypyline venter, showing preanal organ and muscles inserting on genital plate (after Grandjean 1969b); (C) Nothridae, female, lateral aspect, showing extended ovipositor; (D) same, ovipositor in lateral view (after Grandjean 1956a); (E) Damaeidae, genital region of male, showing spermatopositor retracted within genital vestibule; (F) same, lateral view (after Grandjean 1956d).

The anogenital region ancestrally comprises four pairs of plates: the **genital plates**, flanked by the **aggenital plates**, and the **anal plates**, flanked by the **adanal plates** (Figure 2.6A). Each plate bears various numbers of setae having the same respective names: aggenital (*ag*) and genital (*g*) setae are numbered from anterior to posterior; anal (*an*) and adanal (*ad*) setae are numbered in the opposite direction (Figure 2.2B) due to the paraproctal origin of their plates and the caudal bend as a result of the addition of segments during ontogeny (Figure 2.1C–F). Adanal plates usually bear a pair of lyrifissures (*iad*), and anal plates may bear lyrifissure pair *ian*; neither pair exists in any member of Palaeosomata or Enarthronota. The genital plates (which are sometimes transversely divided) close to form the genital vestibule, into which the paired genital papillae and unpaired genital organ are retracted (Figure 2.1B). With rare exception, females have a tubular **ovipositor** (Figure 2.7C,D), and males have a homologous but much smaller **spermatopositor** (also called male organ, male genital sclerite or penis) (Figure 2.7E,F). Setae that insert on these structures are called **eugenital setae**. Usually three pairs of similar **genital papillae** (*Va, Vm, Vp* from anterior to posterior) are present (Figure 2.7D); in some groups one is larger or smaller than others, rarely they are all minute, and rarely one or all are absent. The anal plates close to form the anal vestibule, or rectum. An unpaired **preanal plate** often lies between these (Figure 2.6A), or may be incorporated in the anal vestibule as a **preanal organ** (Brachypylina), which projects internally as an apodeme

(Figure 2.7A,B); both forms serve as origin for muscles that insert on the genital plates. Rarely, the **peranal segment** (PA) is added in the tritonymph (some Enarthronota, some Parhyposomata); if so, a peranal plate usually forms a narrow band at the medial edge of the anal plate.

From this basal anogenital arrangement, a variety of plate fusions have evolved. Various combined terms are also applied to these fusions such as anogenital plates and aggenito-adanal plates. In several groups, particularly Euphthiracaroidea, a pair of large, intercalary **plicature plates**, bearing neither setae nor lyrifissures, lies between the venter and the notogaster proper (Figure 3.21G). In some Nothrina (Hermanniidae) and all Brachypylina, the adanal and aggenital plates are hypertrophied and imperceptibly fused; they also fuse with the coxisternum and usually fuse in front of and behind the anal plates. The result is a strongly constructed **ventral plate** in which the relatively small genital and anal plates open like bomb-bay doors (Figure 2.6E). This is called the **brachypyline** venter, in contrast to the less fused **macropyline** venter of other oribatid mites. The normal setation of a brachypyline venter includes one pair of aggenital and three pairs of adanal setae, but neotrichy occurs in some groups.

Several ventral or lateroventral structures are restricted to (but not universal in) Brachypylina. A spine- or ridge-like **discidium** often occurs between acetabula III and IV, and a paired **circumpedal carina** may curve anteromedially from the edge of the ventral plate, running behind leg IV and forward across epimera IV and III (Figures 2.6H; 2.11A). The circumpedal carina may merge with the discidium and in some cases the anterior end projects as a knife-like **custodium**. Enantiophyses may span any furrow in the coxisternal region (Figure 2.6F). Most common are the **parastigmatic enantiophysis** (*Sa, Sp*), which spans the sejugal furrow ventrolaterally between legs II and III; the **ventrosejugal enantiophysis** (*Va, Vp*), which spans it ventrally; and **aggenital** or **epimeral enantiophysis IV** (*E4a, E4p*). In some groups, a small canal (taenidium) runs laterally from each anteriolateral corner of the genital vestibule to acetabulum IV; a narrow tectum (**minitectum**) overhangs the taenidium from the posterior side (Figure 2.7A).

Various circumscribed porose organs occur on the ventral plate of some Brachypylina. The more obvious ones probably represent dermal glands. A common one is the **postanal porose area**, which in some species develops as a saccule. A **marginoventral series** of porose areas may be present near the circumgastric scissure (Figure 2.6G), resembling an extension of the more common humerosejugal series (see above).

GNATHOSOMA

This is a highly specialized body region with adaptations for sensory reception, food gathering and preoral digestion (Alberti and Coons 1999; Alberti et al. 2011). The gnathosomatic floor and walls (i.e., the subcapitulum and suboral hypostome) provide a channel for the paired **chelicerae** that lie above it and an entrance to the preoral cavity that opens just below the chelicerae. Cheliceral retraction is mediated by well-developed retractor muscles, while cheliceral extrusion is largely driven by hydrostatic pressure generated by contraction of the powerful dorsoventral idiosomatic muscles. Muscle insertions can be identified by muscle scars (**sigilla**) located either in the soft integumental surface or on dorsophragmata and pleurophragmata (Figure 2.3G,I).

Oribatid mites show considerable variation in the structure of all three gnathosomal components; i.e., the unpaired subcapitulum (= infracapitulum) and the paired palps and chelicerae. In ventral aspect (Figure 2.8C), the subcapitulum usually exhibits a large proximal base, the **mentum** (with 1 or 2 pairs of setae); paired anterior lobes, or **genae** (with 1–5 pairs of setae; usually 1 or 2 pairs), the dorsal **labrum**, and, most distally, a pair of **lateral lips** (with up to 3 pairs of **adoral setae**, or_{1-3}); the labrum and paired lateral lips border the mouth. Rarely, an unpaired **ventral lip** (= inferior lip) is present (Archaeonothridae) (Figure 2.8A). There are three subcapitular forms based on the nature of the **labiogenal articulation**, which separates mentum from genae and allows deformation during feeding (Grandjean 1957). An **anarthric** subcapitulum (Figure 2.8B) lacks the labiogenal articulation and is typical of many Enarthronota and members of other groups with highly modified mouthparts. In a **stenarthric** subcapitulum, the labiogenal articulation is V-shaped and runs

Form and Function

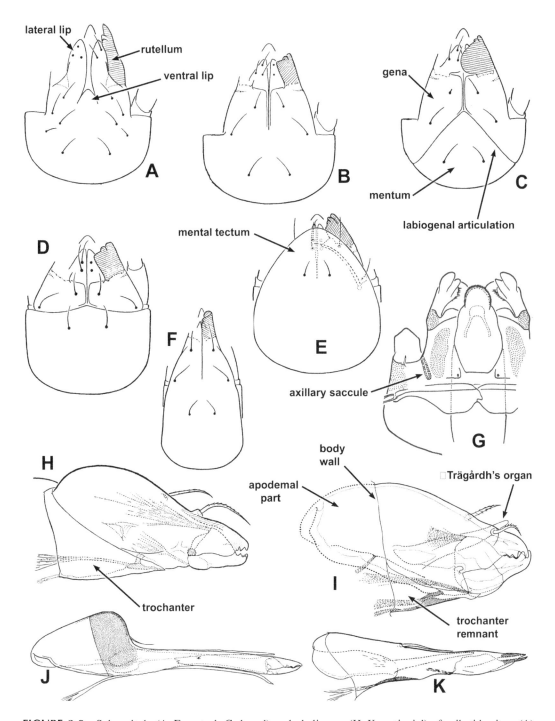

FIGURE 2.8 Subcapitula (A–F ventral, G dorsal) and chelicerae (H–K, antiaxial) of oribatid mites. (A) anarthric, with ventral lip; (B) anarthric, without ventral lip; (C) stenarthric; (D) diarthric; (E) diarthric, with mental tectum; (F) secondarily anarthric, pelopsiform; (G) composite, showing axillary saccule at base of palp; (H) primitive form, with complete trochanter; (I) typical chelate-dentate form; (J) pelopsiform type; (K) attenuate-edentate type (A–F after Grandjean 1957; G after van der Hammen 1968; H–K after Grandjean 1954b, 1947a, 1936a, 1951c, respectively).

obliquely from a central point in the midline to each posterolateral corner (Figures 2.8C; 2.12A, B). This produces a triangular mentum and is typical of most early- to middle-derivative groups. On a **diarthric** subcapitulum (Figures 2.8D; 2.12D,E), the labiogenal articulation runs transversely at a level just posterior to the palps and demarcates a quadrangular mentum (= hypostome) typical of most Brachypylina. Diarthric subcapitula may have a **mental tectum** (= hypostomal tectum) that projects forward to cover the articulation (e.g., Galumnidae; *Tegoribates*) (Figures 2.8E; 2.12C). Distally on each gena of the subcapitulum is a modified seta called the **rutellum** (Figures 2.8A; 2.12A,B). The paired rutella may be setiform in some Enarthronota, but in most species are enlarged, with distal teeth or cutting blades, in which case they are scraping or cutting organs used in conjunction with the chelicerae to remove small particles of food (Figure 2.11C). If chelicerae are elongated (see below), the rutellum may be modified to form half a tube (Figure 2.8F), which when paired serves to guide the chelicerae (Figure 2.11G). Some Brachypylina have a small **axillary saccule** that opens at the base of the palp (Figure 2.8G); Alberti et al. (1997) showed this to be a secretory organ in Phenopelopidae.

The **palps** typically serve as platforms for an array of terminal chemosensory and thigmotactic sensory receptors, much like the antennae of Mandibulata (Figure 2.9A,B). Ancestrally they have five segments (Figure 2.9A), but there are various fusions to form 4, 3 or only 2 functional segments. Palps may be proportionally large in Palaeosomata, but in other groups they are small, inconspicuous, and probably are used as sense organs and for food particle manipulation. The tarsus has a proximodorsal lyrifissure and one solenidion, various normal mechanoreceptive setae and several distal eupathidial setae (eupathidia) (discussed below). More proximal segments have few setae; a typical formula for the basal four segments is 0-2-1-3, with the setal nomenclature indicated in Figure 2.9A. Most setiform organs on the palp are independent, but in some cases distal eupathids are fused at the base to form forked or trifurcate structures and many Brachypylina have a eupathidium (*acm*) fused to the solenidion to form what is often called a '**double horn**' (Figure 2.9B).

Chelicerae are three-segmented, with the movable digit considered a segment, but this is obvious only in primitive groups such as Palaeosomata and Enarthronota (Figure 2.8H). The trochanter is usually greatly regressed, little more than a proximoventral vestige (Figure 2.8I), associated with a change in operating direction from near vertical to near horizontal. In Nothrina and Brachypylina, as well as in Hypochthonioidea (Enarthronota), the body wall attaches rather distally on the chelicerae, so that about one-third of the appendage is internalized as an apodeme. In Nothrina and Brachypylina a movable, fingerlike **Trägårdh's organ** (function unknown) extends distally from near this point on the paraxial face (Figure 2.8I). Most chelicerae have two setae, one dorsal or slightly paraxial (*cha*) and the other usually lower on the antiaxial face (*chb*); one or both may be absent. Chelicerae are typically chelate-dentate (i.e., robust, with strong, conspicuously dentate chelae), but many modifications exist. The greatest diversity of cheliceral form is found in Enarthronota and Brachypylina, but there has been no serious attempt to learn their function or even to classify and name them. An exception is **pelopsiform** (= peloptoid) chelicerae, which are greatly elongated, with small but well-formed dentate chelae (Figure 2.8J; 2.11G), which evolved independently in various brachypyline families. Another exception is **attenuate-edentate** chelicerae best known in Suctobelbidae (Figures 2.8K; 2.11E), but found in other groups as well. Chelicerae of Gustaviidae lack the fixed digit and have a styliform movable digit, with distal serration (Figures 2.11F). The structure of some other forms suggests that they scrape or filter soft substrates.

LEGS

Leg segments are simple tubular structures in most basal groups, with the genu and tibia having similar form and size. The legs of adult Palaeosomata have six free segments, not including the pretarsus (Grandjean 1941) (Figure 2.9C). All other oribatid mites have five; the reduction is due to a fusion of basifemur and telofemur (Figure 2.9E), which even in Palaeosomata do not fully articulate. As in other acariform mites, there is no free coxa. The terminal **pretarsus** typically

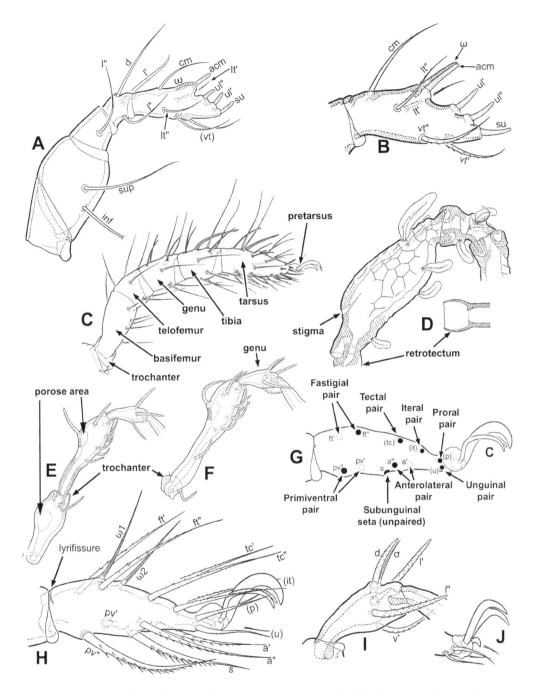

FIGURE 2.9 Palps (A,B) and legs (C–J) of oribatid mites. (A) palp of brachypyline mite showing typical setation; (B) palptarsus of poronotic brachypyline mite showing fused solenidion ω and eupathidium *acm* ('double horn'); (C) palaeosomatid-type leg, with subdivided femur; (D) brachypyline leg with articulations in sockets formed by retrotecta (insert is schematic); (E) proximal half of leg IV of brachypyline mite showing large trochanter; (F) same, but leg I, showing small trochanter; (G) schematic depiction of unique nomenclature and notations of tarsal setae (setal pairs indicated by parentheses); (H) tarsus and tridactylous pretarsus III of brachypyline mite in antiaxial aspect, showing typical setation and normal, large empodial claw; (I) genu of brachypyline mite showing coupled solenidion σ and seta *d*; (J) tridactylous pretarsus with reduced, hooklike empodial claw (A,D after Grandjean 1964; B after Grandjean 1960a; C,J after Grandjean 1954b; E,F,I after Grandjean 1960b; G after Grandjean 1953c; H after Grandjean 1940).

comprises a **basilar sclerite** and a set of distal structures that may include a median **empodium**, a padlike **pulvillus** and paired **claws**. The pretarsus is referred to as an **ambulacrum** when median elements are present. The basilar sclerite serves as a fulcrum that articulates with the tarsus usually via paired, ribbonlike **condylophores** contained in a terminal tarsal stalk. The sclerite is activated by extensor and depressor muscles that arise in the tibia or tarsus. The empodium may be clawlike or padlike or it may be absent, and one or both of the paired claws may be absent. The pretarsus is monodactyl (with only an empodial claw), tridactyl (with empodial and paired lateral claws) or bidactyl (either with only lateral claws or with empodial and one lateral claw). In some basal groups (Palaeosomata, Enarthronota, Parhyposomata), a tridactyl pretarsus may have only a minute, hooklike empodial claw (Figure 2.9J). All claws on a pretarsus may be similar (homotridactylous) or dissimilar (heterotridactylous).

Legs of Brachypylina have much more diverse structure. The genu is usually much shorter (and usually thinner) than the tibia and lacks its own musculature as it serves primarily as a point of flexion (a knee) and as a conduit for tendons (Figure 2.9F). Because legs of Brachypylina insert in acetabula, each has a right-angle bend near its base (Figure 2.6C). The trochanters of legs I and II are very small, mostly hidden within their respective acetabula, and the bend is at the base of the femur (Figure 2.9F). By contrast, the trochanters of legs III and IV are large, and they possess the bend. In some cases, femora have broad ventral keels associated with protecting retracted legs. The tibia and tarsus are sometimes functionally fused by having a very narrow articulation (e.g., Carabodidae). If muscles are concentrated in swollen parts of segments, legs have a **moniliform** appearance; if segments are extremely thin and elongated, the legs appear **filiform**. In some Brachypylina, the proximal part of some segments can have a **retrotectum**, a sclerotized, gauntlet-like covering of the articulation that results in a socket joint (Figure 2.9D). Tarsi I–IV each have a dorsal lyrifissure just distal to the articulation with the tibiae (Figure 2.9H). Leg flexion is by trochanteral intrinsic and extrinsic intersegmental muscles, while leg extension—like that of the chelicerae—relies heavily on hydrostatic pressure (Alberti and Coons 1999). The muscle systems of Phthiracaroidea and Euphthiracaroidea have been studied in some detail using methods including synchrotron X-ray tomography (Sanders and Norton 2004; Schmelzle et al. 2015).

Respiratory porose areas are commonly found on the femora and on trochanters III and IV (Figure 2.9E), and small areas are occasionally found ventrally on the tibia and tarsus. Leg porose areas are internalized as saccules or tracheae in some Brachypylina families (Norton and Alberti 1997). Legs have a number of tactile and sensory setae that generally follow fixed patterns of ontogenetic appearance in a given taxon and are usually small enough in number to establish their identity (Norton 1977; Norton and Sidorchuk 2014) (Figures 2.1K,L; 2.9H).

Sensory Structures

The vast majority of setae are simple mechanoreceptors that respond to tactile stimuli, but some have been identified as chemoreceptors. Most contain an optically active iodophilic material that exhibits birefringence in polarized light. This material, **actinopilin** (or actinochitin), may occur as a solid core or as a layer of material surrounding a protoplasmic extension of the basal setal nerve cell.

Mechanoreceptors

Most mechanoreceptors comprise a shaft, a membranous socket that allows a modicum of shaft movement (Albert and Coons 1999), and a complex of basal receptor cells that transmit signals from the seta to the central nervous system. Mechanoreceptors have a diversity of shapes and sometimes bewildering and conflicting descriptors (some examples are given in Figure 2.10). Tactile setae usually are simple and spinose, but they may be expanded or ornamented. The prodorsal bothridial setae are highly variable in appearance, and they resemble typical acariform tactile setae in having a solid core of actinopilin. They are considered to be vibration and air current receptors, and the wide diversity in shape and complexity are considered to be responses to ecophysiological

Form and Function

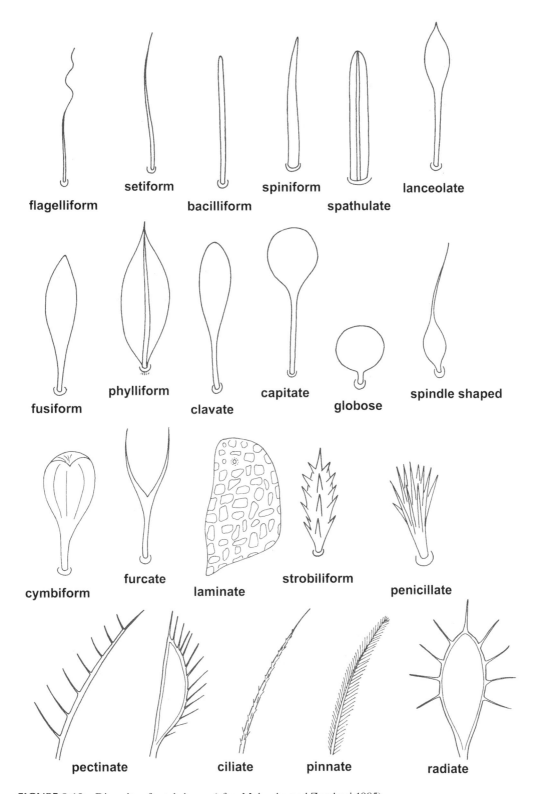

FIGURE 2.10 Diversity of setal shapes (after Mahunka and Zombori 1985).

pressures. For example, clavate bothridial organs of arboreal oribatid mites are thought to function in geotaxis (Aoki 1973; Schäffer et al. 2020), or they may serve to reduce sensitivity in an environment where continuously moving air could overstimulate these receptors (Norton and Palacios-Vargas 1982).

Proprioceptors

These are cuticular and intracuticular organs that receive information on changes in physical tension within the body and then mediate the appropriate response of adjacent subsurface sensory neurons; they may be found on legs, palps and the hysterosoma. These generally take the form of small, membrane-covered clefts, referred to as lyrifissures. **Cupules**, rounded, cup-shaped structures that lie in cuticular depressions, are homologous with lyrifissures (Alberti 1998; Alberti and Coons 1999).

Chemoreceptors

Chemosensory setae generally are found on the terminal segments of legs and on palps. They share with tactile setae the internal characteristic of a nerve-innervated protoplasmic core surrounded, or not, by a sheath of optically active actinopilin. They include:

- **Eupathidia**: (identified by the symbol ζ (zeta)) are smooth, apically rounded or divided structures found on the extremities of tarsi I–II, on genitals and on the tip of the palptarsus. They are homologs of mechanoreceptor setae. They carry a single terminal pore and may function as contact or gustatory chemoreceptors (Alberti 1998; Alberti and Coons 1999).
- **Famulus**: (identified by the symbol e) is a hollow sensory seta, but it tends to be smaller than surrounding setae and generally is found on tarsus I. Its function is unknown.
- **Solenidia**: are porose, thin-walled, terminally rounded or pointed setiform or peglike structures that lack the core or sheath of actinopilin characteristic of eupathidia and famulus, and thus are not birefringent. Solenidia are found on the palptarsus and also occur on tarsus and tibia, less frequently on the genu, of legs I–IV. They are usually inserted in broad and immovable bases and often appear striated because of the cuticular pores (Alberti and Coons 1999). They are considered to be olfactory in function (Evans 1992). In some brachypyline species, one or more tibial solenidia can have a **microcephalic** head (see Figure 3.127A)). Grandjean (1935a, 1939a, 1958b) developed a system of Greek notation for the solenidia of genu (sigma σ), tibia (phi φ) and tarsus (omega ω) (Figure 2.9H, I). If there is more than one solenidion per segment, numbering is in order of appearance during ontogeny.

Photoreceptors

These may be present on the anteromedian or anterolateral aspect of the prodorsum (Alberti 1998). An unpaired anteromedian ocellus or pigmented light-sensitive zone may be found in Acaronychidae, but the median eye or pair of eyes is more commonly carried on the underside of a **naso** (Grandjean 1958c; Alberti 1998). Many brachypyline mites have one or more light-sensitive clear spots or an elevated, lens-like **lenticulus** located just behind the prodorsum on the notogaster. The cuticular cornea of the lenticulus is convex-concave in *Hydrozetes lemnae* and *Scutovertex sculptus* but is biconvex in a species of *Scapheremaeus* (Alberti and Fernandez 1990). Light-sensitive clear spots in Oribatellidae, Chamobatidae and Euzetidae differ from the lenticulus in having their lamellated receptor poles directly attached to the underlying synganglion rather than being subcuticular and well removed from the synganglionic mass (Alberti 1998). Woodring (1966) showed that in adult *Scheloribates parabilis*, which lacks a lenticulus, there is a phototactic response to various wavelengths of light.

FUNCTION

Alberti and Coons (1999) is a general reference source for ultrastructure and general internal morphology and is relied on heavily in the following paragraphs.

Respiration

Exchange of carbon dioxide and oxygen is cuticular, and in adults of Brachypylina, respiration is mediated by a branched tracheal system that opens externally through **stigmata** (Figure 2.6G). These are difficult to see, and the name Cryptostigmata, a synonym of oribatid mites, derives from this trait. There are three pairs of stigmata in most brachypyline families, one within acetabulum I, another in the sejugal furrow and a third within acetabulum III (Figure 2.6G). The first two typically bifurcate close to the stigmata (Figure 2.6D) so that five long pairs of tracheae usually aerate the body (Figure 3.112C). Variations of this **apodemato-acetabular tracheal system** are characteristic of particular families or superfamilies. Respiratory surfaces can also be located in the prodorsal bothridia, where they take the form of short **brachytracheae** (Norton et al. 1997) (Figure 2.3E).

Fingerlike hydrophobic hairs or small cuticular projections may form a plastron on some oribatid mites conserving an air layer for respiration in situations where submersion in ground or seawater may limit normal respiration (Pugh et al. 1987; Pfingstl 2017). In Hydrozetidae, for example, a layer of air is trapped on narrow, rugose bands of cuticle along margins of the notogaster and ventral plate, and these bands connect anterolaterally and are contiguous with the internal respiratory system (Krantz and Baker 1982). Adults of Fortuyniidae have a unique complex system of lateral cuticular taenidia, called **van der Hammen's organ**, which connects the dorsal and ventral plastron areas and is thought to facilitate rapid equilibrium of pressure changes between the respiratory system and the surrounding environment caused by rough wave movements (Pugh et al. 1990; Pfingstl 2017).

Secretion

Some cuticular openings mark the exits of ducts that carry the products of subcuticular gland cells. An aggregation pheromone with probable cuticular origins may be responsible for a substance in *Ceratozetoides cisalpinus* and other oribatid mites that prevents debris from adhering to the cuticle (Woodring and Cook 1962). Alberti et al. (1997) showed that the sac-like organs associated with bothridia of *Eulohmannia ribagai* may have a secretory function, and that the lateral cuticular porose areas of the galumnid *Acrogalumna* and the lohmanniid *Mixacarus* secrete a substance presumed to be a lipid. The humerosejugal porose areas common to many oribatid mites may secrete substances used in the formation of waterproofing cerotegument, while some notogastral porose areas produce sex pheromones (Norton and Alberti 1997).

Oribatid oil glands, the single-celled lateral **opisthonotal glands** (Figure 2.5A), which may appear as large black, brown, yellow or red spots where the integument is transparent, have been intensively studied (Raspotnig 2010; Heethoff et al. 2011a; Heethoff 2012; Merkel et al. 2020). The gland duct exits via a single pore and is protected by a cuticular flap that may serve to control the delivery of gland products to the surface. These secretions contain a variety of volatile compounds (monoterpenes, hydrocarbons, esters, aromatics and hydrogen cyanide) that may vary between species or supraspecific groups and that may serve as alarm pheromones, aggregation and/or sex pheromones, or as defence substances against fungal or predatory attacks (e.g., Raspotnig et al. 2005; Brückner et al. 2017). Species of the brachypyline *Scheloribates* produce alkaloids (pumiliotoxin and related compounds) that may be sequestered in the skin of the poison dart frogs (Dendrobatidae) that eat them (Takada et al. 2005; Saporito et al. 2007; Raspotnig et al. 2011). Presence or absence of certain chemical constituents in opisthonotal gland products may provide a chemical basis for establishing or verifying phylogenetic relationships within higher oribatid mites (Sakata and Norton 2001; Sakata et al. 2003; Raspotnig et al. 2004a,b).

Coxal glands (= nephridia, tubular glands and supracoxal glands of authors) are involved in basic physiological processes, such as ion/water balance, osmoregulation and possibly in excretion (Woodring 1973; Alberti et al. 1996). They are represented by a pair of excretory structures derived from primitive, segmentally arranged nephridia (Woodring 1973). These consist of a coelomic saccule and a coiled duct, or labyrinth (Alberti et al. 1996). Coxal glands of representatives of Lohmanniidae, Eulohmanniidae, Oppiidae, Scutoverticidae, Hydrozetidae, Achipteriidae, Phenopelopidae and Galumnidae were described by Alberti et al. (1996).

The coxal glands are the most posterior elements in a complex of propodosomatic glands that empty into paired lateral secretory ducts or **podocephalic canals**, which gather gland products and discharge them into the gnathosoma (Grandjean 1970, 1971; Alberti and Coons 1999). The paired podocephalic canal starts from an area dorsal or posterodorsal to the insertion of leg I and runs anteriorly, passing above the proximal parts of the pedipalps and terminating dorsomedially at the base of the cervix under the chelicerae. The infracapitular glands, which are independent of the podocephalic gland complex, discharge their secretions through minute openings located close to the base of the labrum (Alberti et al. 2011).

Osmoregulation

Larvae possess a pair of Claparède organs (**urstigmata**), ventrally on epimere I which are considered to be involved in osmoregulation and ion regulation (Walter and Proctor 2013). These are considered metameric homologs to the genital papillae of later ontogenetic instars (Grandjean 1946; Alberti 1979). Genital papillae in most acariform taxa display a sequential development during postlarval ontogeny with one pair in the protonymph, two in the deutonymph and three in the tritonymph and adult (Figure 2.1D–J).

Hemolymph and Muscles

Oribatid mites, like other Acari, lack most antagonistic musculature; thus, the extension of appendages is facilitated through hemolymph pressure that mostly is generated by dorso-ventral compression of the opisthosoma. The clear hemolymph contains a variety of hemocytes, the functions of which may include clotting, macrophagy and tissue dissolution during ecdysis (Alberti and Coons 1999). The hemolymph circulates freely throughout the hemocoel primarily as a result of body movement. Circulation to the legs and other extremities is facilitated by contraction of the dorso-ventral idiosomatic muscles.

General musculature of oribatid mites was studied in *Nothrus palustris* (Akimov and Yastrebtsov 1989). Musculature associated with the defensive process of ptychoidy was described in Euphthiracaridae (Sanders and Norton 2004; Schmelzle et al. 2009) and in Phthiracaridae (Schmelzle et al. 2010, 2012). Schmelzle et al. (2015) compared the modifications in these ptychoid taxa; euphthiracaroids employ a lateral compression of the notogaster using all muscles of the opisthosomal compressor system, whereas phthiracaroids employ a dorsoventral compression generated by the notogaster lateral compressor muscles and the postanal muscle which retract the temporarily unified ventral plates into the hysterosoma.

Nervous System

The compact central nervous system or **synganglion** lies around the oesophagus. It consists of a protocerebral **supraoesophageal ganglion** and a tritocerebral **suboesophageal ganglion** and comprises an outer cortex and an inner neuropile (Hartmann et al. 2016). The acarine brain is confined entirely to the prosoma.

DIGESTIVE SYSTEM

The digestive system is comparatively well known in oribatid mites with the pioneering work of Michael (1883) and the detailed anatomy and histology of *Hypochthonius rufulus* (Tarman 1968). It is best known in *Archegozetes longisetosus* (Alberti et al. 2003), with the gnathosoma thoroughly illustrated in Alberti et al. (2011). It comprises the preoral cavity in front of the mouth and the digestive tract, which is composed of: (a) a cuticle-lined foregut starting with the mouth and continuing with the pharynx and oesophagus; (b) a midgut with ventriculus, one pair of preventricular glands, one pair of caeca, a colon, an intercolon and a postcolon; and (c) a cuticle-lined hindgut composed of the anal atrium and terminating with the anus. The pharynx is crescent shape in cross section and is activated by dilator and depressor muscles. There are also ventral muscles. The cuticle of the pharynx is specialized in a way that probably has functional implications. The oesophagus is a slender tube that passes through the synganglion. It is capable of expansion by means of longitudinal folds in its wall. The oesophagus is predominantly provided with ring muscles. The oesophagus projects into the ventriculus with a pronounced oesophageal valve. The preventricular glands are represented by two pouches formed by cells similar to those of the ventriculus but are filled with dense spherites. The ventriculus frequently contains a food bolus that is surrounded by an indistinct layer of secretion forming an incipient peritrophic membrane. The peritrophic membrane remains in place around the bolus as digestion proceeds, with the resulting faecal pellet finally passing into the hindgut. The muscles involved in defecation of these large faecal pellets were described by Heethoff and Norton (2009).

REPRODUCTION

In males, the basic components of the reproductive system were described in detail by Woodring (1970). They are paired or fused testes and vasa deferentia, one or more accessory glands and an ejaculatory duct. The morphology of spermatozoa is family or infraorder specific (Alberti 2006). Spermatophores and spermatozoa of representatives of Phthiracaridae, Hermanniidae, Hermanniellidae, Liacaridae, Scutoverticidae, Phenopelopidae, Achipteriidae, Chamobatidae and Euzetidae have been described by Fernandez et al. (1991) and Alberti et al. (1991).

Most oribatid males transfer sperm to the female by producing freestanding spermatophores (Woodring 1970). Spermatophore deposition usually is dissociative and proceeds in the absence of females (Woodring 1970; Alberti et al. 1991). However, a trail of stalks on which a sex attractant may be secreted is laid down by males of the galumnid *Pergalumna* sp. as a means of guiding females to their spermatophore deposition sites (Oppedisano et al. 1995), a form of **incomplete dissociation** (Walter and Proctor 2013). Deposition of a spermatophore typically begins with the male lowering its hysterosoma to the substrate and producing a droplet of material from the **spermatopositor** that is produced in the testes. When the male raises its hysterosoma, the droplet is drawn out into a stalk that hardens upon exposure to air or water, and a sperm packet is deposited at its tip (Pauly 1952; Woodring 1970; Alberti et al. 1991; Fernandez et al. 1991). The packet generally is picked up in the genital vestibule of the female or, rarely, placed there through manipulation by the male, and the bare stalk is abandoned.

Archegozetes longisetosus is again the model oribatid mite for understanding the development and function of female genital organs (Bergmann et al. 2008). In females, the ovary is unpaired with the adjoining oviducts typically paired. Secretions from the oviducts produce the eggshell that protects the developing embryo as it passes through the uterus and into the ectodermal vagina (Bergmann and Heethoff 2012). The vagina gives rise to an **ovipositor**, a complex extrusible muscular structure equipped with terminal lobes and sensory setae (Michael 1884; Woodring and Cook 1962) that apparently is capable of placing individual eggs in precise niches (Figure 2.7C,D). The male and female reproductive systems are described and illustrated for *Hermannia gibba* by Liana and Witaliński (2012).

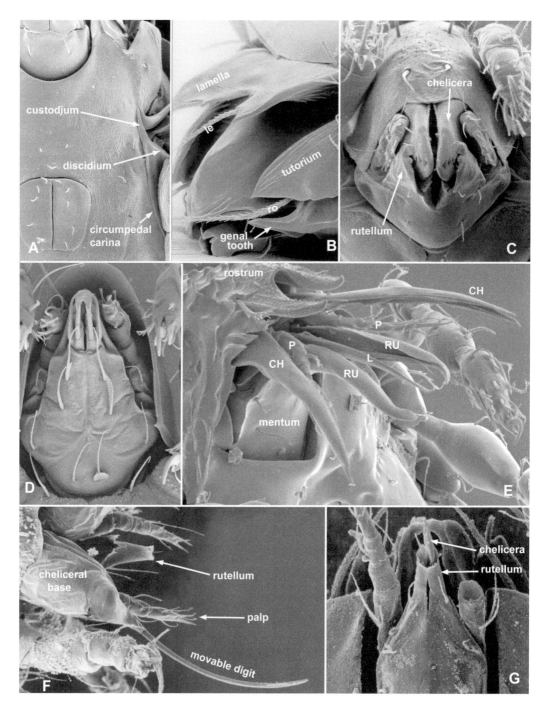

FIGURE 2.11 (A) *Trichoribates incisellus* (Ceratozetidae), coxisternal and genital regions; (B) same, prodorsum, anterolateral; (C) *Melanozetes crossleyi* (Ceratozetidae), anterior view of tritonymph showing gnathosoma; (D) anarthric subcapitulum of *Eniochthonius* sp. (Eniochthoniidae); (E) *Suctobelbella falcata* (Suctobelbidae), anteroventral view showing extended gnathosoma (*CH*, chelicera; *L*, labrum; *P*, palp; *RU*, rutellum); (F) *Gustavia* sp. (Gustaviidae), extended gnathosoma of nymph, with chelicerae pulled away from rutellar tube; (G) *Peloptulus* sp. (Phenopelopidae), ventral view of gnathosoma, with one pelopsiform chelicera partly extending from rutellar tube (A–C from Behan-Pelletier 2000; D courtesy of Sue Lindsay; E courtesy of Maria Minor; F courtesy of Marilyn Clayton; G from Norton and Behan-Pelletier 1986).

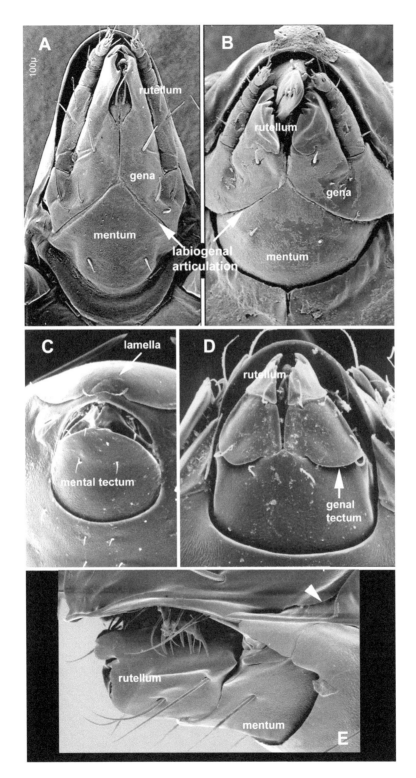

FIGURE 2.12 (A) *Perlohmannia* sp. stenarthric subcapitulum (Perlohmanniidae); (B) *Platynothrus peltifer*, stenarthric subcapitulum (Crotoniidae); (C) *Tegoribates walteri*, subcapitulum with mental tectum (Tegoribatidae); (D) *Mycobates corticeus*, diarthric subcapitulum, with small genal tectum (Punctoribatidae); (E) *Heterozetes minnesotensis*, diarthric gnathosoma, lateral (Zetomimidae) (all Behan-Pelletier unpublished).

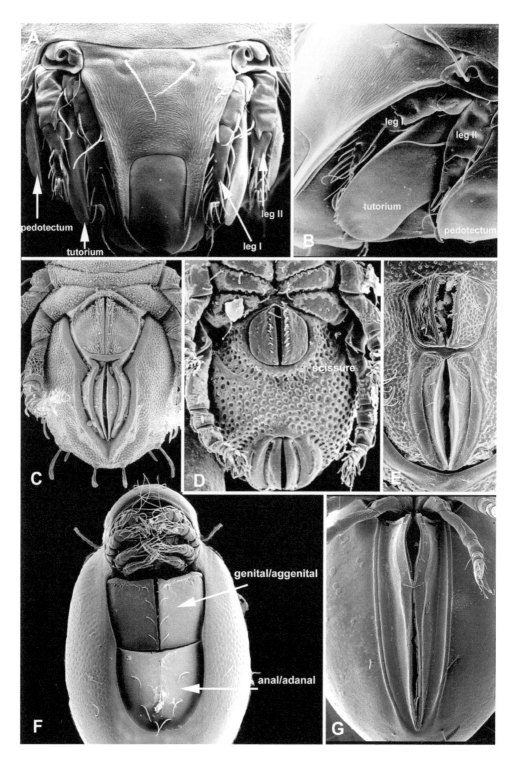

FIGURE 2.13 (A) *Mycobates punctatus*, frontal aspect, with legs I and II at rest (Punctoribatidae); (B) *M. altus*, lateral of prodorsum with legs I and II at rest; (C) *Nothrus anauniensis*, posterior of venter (Nothridae); (D) *Nanhermannia* sp., posterior of venter (Nanhermanniidae); (E) *Hermannia subglabra*, posterior of venter (Hermanniidae); (F) *Hoplophthiracarus* sp. venter (Phthiracaridae); (G) *Oribotritia* sp. venter of anogenital region (Oribotritiidae) (all Behan-Pelletier unpublished).

3 Taxonomic Keys

SUMMARY

Keys are given to the 96 families, 250 genera and 580 described species in published records from Canada and Alaska. Diagnosis is given for adults of each family. Habitat, distribution data, biology, literature and a figure for most species is included. Genera with undescribed or unidentified species from regions of Canada and Alaska are indicated.

INTRODUCTION

The following identification keys for oribatid mites relate to published records for Canada and Alaska (updated from Behan-Pelletier and Lindo 2019, as of January 2022). They consist of a key to families, followed by a key to genera within each family, followed by a key to species for each genus. Genera and species are treated alphabetically. When a family is represented by only one genus in Canada and Alaska, the genus is indicated in the key to family. Taxa are listed according to Infraorder.

For each family the following information is given:

- **Brief diagnosis** for the adult. References concerning juveniles, if available, are given in Norton and Ermilov (2014).
- **Information** on general biology and ecology, where available.
- **Notes or Remarks**: where needed.

For each genus the following information is given:

- **Information** on general biology, including ecology, where available.
- **Literature** on recent genus-level taxonomic studies.
- **Notes or Remarks**: where needed.

For each identified species the following information is given:

- **Species name** and author, and reference to an illustration, where available.
- **Combination and Synonymy**: the original name is given if this varies from the present name. Most commonly used combinations and synonymies are given; further combinations and synonymies are in Behan-Pelletier and Lindo (2019). Common names for many species are given in Walter (2013) and Walter et al. (2014).
- **Biology**: information on general biology, including ecology, where available.
- **Distribution**: includes records for Alaska and for the Provinces and Territories of Canada, listed primarily from West to East. Distribution in other states of the USA and general distribution, following Subías (2022) and Schatz (2020), is given in [ ], with states of the USA, other than Alaska, listed in alphabetical order.
- **Literature**: Marshall et al. (1987) and Behan-Pelletier and Lindo (2019) are important references for most species herein, but are not repeated in the text. The former provides references on Description, Taxonomy, Combinations and Synonymy, Key, and Biology that were published prior to 1986. The latter is a checklist for Canada and Alaska up to date as

of March 2019. Additional, relevant literature, other than the original description, is given under each taxon, where available.
- **Figures**: these are taken from various sources, as indicated in the captions. No scale bar is associated with these figures, as the known range in size of species is given in the key couplet.
- **Notes or Remarks**: where needed.

ABBREVIATIONS: CANADIAN PROVINCES, TERRITORIES AND STATES OF USA

Canada: YT, Yukon; NT, Northwest Territories; NU, Nunavut; BC, British Columbia; AB, Alberta; SK, Saskatchewan; MB, Manitoba; ON, Ontario; QC, Quebec; NL, Newfoundland and Labrador; PE, Prince Edward Island; NB, New Brunswick; NS, Nova Scotia; USA: AK, Alaska; AL, Alabama; AR, Arkansas; AZ, Arizona; CA, California; CO, Colorado; CT, Connecticut; DC, District of Columbia; DE, Delaware; FL, Florida; GA, Georgia; ID, Idaho; OK, Oklahoma; IA, Iowa; IL, Illinois; IN, Indiana; KS, Kansas; KY, Kentucky; LA, Louisiana; MA, Massachusetts; MD, Maryland; ME, Maine; MI, Michigan; MN, Minnesota; MO, Missouri; MS, Mississippi; MT, Montana; NC, North Carolina; NE, Nebraska; NH, New Hampshire; NJ, New Jersey; NM, New Mexico; NY, New York; OH, Ohio; OK, Oklahoma; OR, Oregon; PA, Pennsylvania; SC, South Carolina; SD, South Dakota; TN, Tennessee; TX, Texas; VA, Virginia; VT, Vermont; WA, Washington; WI, Wisconsin; WV, West Virginia; WY, Wyoming.

Abbreviations in text:

CNC, Canadian National Collection of Insects, Arachnids and Nematodes; VBP, V. Behan-Pelletier; ZL, Zoë Lindo; *auct*, auctorum, of authors; *ca*, approximately; *cf*, comparable with; *[!]*, misspelling of author; *mya*, million years ago; *sic*., thus; *s.l.* sensu lato; SEM, scanning electron micrograph; DIC, differential interference contrast; OSU, The Ohio State University.

KEY TO FAMILIES OF ORIBATID MITES OF CANADA AND ALASKA
(MODIFIED FROM NORTON AND BEHAN-PELLETIER 2009)

1	Genua of all legs noticeably shorter than tibiae, differently shaped, functioning as knee. Legs articulate with body in deep pockets (acetabula), as "ball and socket" joint; trochanters I and II almost totally contained within acetabula. Venter of "brachypyline" type: comprising unified rigid plate with only subcapitulum and paired genital valves and anal valves distinct and moveable (i.e., epimeral, aggenital and adanal regions fused into a single unit that carries their respective setae). (The "higher"oribatid mites) BRACHYPYLINA 33
–	Genua of all legs similar to tibiae in size and shape; not reduced to knee-like segment. Leg articulation not in deep acetabula, at most in shallow depressions; trochanters I and II small, but clearly external. Venter of various form, but rarely as above; paired aggenital and adanal plates often distinguishable. Epimeral region often transversely divided by sejugal articulation and usually not fused to aggenital region. (The "lower" or "macropyline" oribatid mites) 2
2(1)	Body form ptychoid: legs can be fully withdrawn as a group and aspis (prodorsum) rotated ventrally such that it touches the notogaster and ventral plate (thus fully covering retracted legs) 3
–	Body form not ptychoid: legs longer or, if short, clearly not capable of being withdrawn and covered by prodorsum 7
3(2)	Hysterosoma covered dorsally by an apparent notogaster with only 6–8 pairs of setae (rows *c*, *d*, and *e*); body region bearing setae of rows *f*, *h*, *p* (and *ad* in *Mesoplophora*), shifted

	ventrally, incorporated into a "ventral plate" in which genital and anal (and sometimes adanal) plates insert. Subcapitulum anarthric. Genu I with 1 solenidion (Figure 3.16A–E) .... **MESOPLOPHORIDAE** (sensu lato; includes Archoplophoridae) (Pg. 65)
–	Hysterosoma covered with normally formed notogaster, bearing 14 or 15 (rarely more) pairs of setae (rows *c*, *d*, *e*, *h*, *p* present; row *f* represented only by inconspicuous, vestigial alveoli). No separate, seta-bearing ventral plate in which anal or genital plates insert. Subcapitulum stenarthric. Genu I with 2 solenidia (PTYCTIMA) 4
4(3)	Combined anogenital region broad, length approximately twice width; composed of two fully independent pairs of plates (fused genital/aggenital plates, and fused anal/adanal plates). Without plicature plates. With 3 tubular internal tracheal organs descending into proterosoma from base of bothridium (Figures 2.13F; 3.16F,G) .... **PHTHIRACARIDAE**, sensu lato (Pg. 81) **Note**: Some authors distinguish two families: Phthiracaridae, containing only *Phthiracarus*, and Steganacaridae, containing all other genera of Phthiracaroidea.
–	Combined anogenital region narrow, length usually greater than three times maximum width. Paired, narrow plicature plates (without setae) present between notogaster and aggenital/adanal plates; rarely absent. Tracheal organs descending from bothridium present or absent, but never as described above (Euphthiracaroidea) 5
5(4)	Anal plates (bearing anal setae) discrete although can be very narrow posteriorly; not fused to adanal plate, or to adanal portion of aggenito-adanal plate. Genital plate of various form, fused to aggenital region or not (Figures 2.13G; 3.18A,B) .... **ORIBOTRITIIDAE** (Pg. 77)
–	Neither anal nor genital plates discernable; i.e., genital plates fused to aggenitals, and anal plates fused to adanals 6
6(5)	Paired ventral plates connected across midline in preanal position by short region of interdigitating ridges (in transmitted light appearing as a dark, sclerotized "interlocking triangle"). Genital plates open normally, with genital organ extruding between them. Genital setae spread the length of genital region, all located on medial edge (only aggenital setae not on medial edge). All genital papillae of approximately equal size. Palp 3-segmented (trochanter/femur/genu fused). Bothridial tracheoles (tracheal organs) very fine, numerous. Opisthonotal gland present (Figure 3.18C,D) .... **EUPHTHIRACARIDAE** (Pg. 72)
–	Preanal region without interlocking triangle. Paired plates attached to each other medially in genital region, such that genital organ emerges anteriorly, behind coxisternum IV. Genital setae concentrated anteriorly, not restricted to medial region. Posterior genital papilla greatly reduced, half size of anterior and middle papilla. Palp 4 (femur/genu fusion) or 5 (no fusions) segmented. No bothridial tracheoles. Opisthonotal gland absent (Figures 3.18E,F; 3.29) .... **SYNICHOTRITIIDAE**: *Synichotritia* (Pg. 80)
7(2)	Legs with 6 free segments past epimere: all femora divided. Prodorsum astegasime, with no rostral tectum; with or without naso. Bothridium and base of bothridial seta simple, without obvious bend; bothridial seta always smooth. Body white, at least in anterior half; usually with inconspicuous, thin sclerites on hysterosoma .... **PALAEOSOMATA** 8
–	Legs with 5 free segments past epimere: femora undivided. Prodorsum usually stegasime, with well-developed rostral tectum (astegasime in some Parhyposomata, some Enarthronota); naso not evident (except Enarthronota: *Paralycus*). Bothridium and base of bothridial seta with conspicuous S-shaped bend (absent in some Enarthronota); bothridial seta of diverse form, smooth or barbed 12
8(7)	Famulus large, erect, ciliated, the most proximal setiform organ on tarsus I, adjacent to lyrifissure. Exobothrial seta *xa* (= *xi*) large, barbed, similar to interlamellar seta. Main part

of chelicera oriented almost vertically, with narrow, elongated digits ("ornithocephalic" form). Palp tarsus short, little longer than palp tibia. Hysterosoma without aesthenic zone; dorsum with series of small, thin, mostly paired sclerites, each bearing 4 or fewer setae. Tibia II with 2 solenidia (Acaronychoidea, Archeonothroidea) 9

– Famulus large or small, but smooth and inserted more distally (1 or more solenidia more basal than famulus, and closer to lyrifissure). Exobothrial seta xa (= xi) clearly smaller than interlamellar seta (usually about half its length). Chelicera oriented anteriorly or anteroventrally and digits not narrow, elongated. Palp tarsus more elongated, at least twice palp tibial length. Hysterosoma with aesthenic zone (wide zone of soft, telescoping cuticle immediately posterior to prodorsum). Posterior region of hysterosoma with weak, but distinct pygidial shield bearing all setae posterior to row e, or with larger, indistinct shield which includes all setae posterior to row c. Tibia II with 1 solenidion...10

9(8) Famulus broadened distally, somewhat spatulate. Inferior lip minute, difficult to distinguish. Prodorsum without "false lamellae" (Figure 3.1A–D) ....
.... **ACARONYCHIDAE**: *Acaronychus* (Pg. 50)

– Famulus not broadened distally. Inferior lip of mouth large, easily distinguished. Prodorsum often with so-called "false lamellae" (vague internal sclerotized thickenings, as pair, or connected to form a U or square shape) (Figure 3.1E,F) ....
....**ARCHEONOTHRIDAE**: *Zachvatkinella* (Pg. 50)

10(8) Tarsi II–IV bidactylous. Famulus large, approximately equal in length to solenidia or only slightly shorter; inserted on conspicuous tubercle in proximal quarter or third of tarsus. All genital papillae of normal size. Anterior adoral seta (or_1) simple, smaller than others. Distal eupathidia of palp clearly barbed. Bothridial seta filiform (Figure 3.1G–I) ....
....**PALAEACARIDAE**: *Palaeacarus* (Pg. 50)

– Tarsi II–IV tridactylous, with empodial claw well developed or much reduced. Famulus much smaller than solenidia; inserted at midpoint or in anterior half of tarsus. Anterior pair of genital papillae greatly reduced or absent. Adoral seta or_1 strongly barbed or pectinate, larger than others. Distal eupathidia of palp smooth. Bothridial seta slightly to greatly expanded (rarely filiform)... 11

11(10) Two pairs of hysterosomal setae (d_2 and e_1) enlarged, longest of dorsal setae, darkly pigmented and erectile. Body form normal; hysterosoma about 1.5 times longer than broad. With broad region of striated integument between setal rows c and d (folded in contracted specimens) (Figure 3.2A–C) ....**CTENACARIDAE**: *Beklemishevia* (Pg. 51)

– Hysterosomal setae of various lengths, but none stand out from others as unusually large, dark, erectile. Body elongated; hysterosoma about twice as long as broad. With only a groove (no striations) between setal rows c and d (Figure 3.2D–F)....................................
....**APHELACARIDAE**: *Aphelacarus* (Pg. 51)

12(7) Notogaster subdivided transversely by 1–3 complete scissures or grooves 13
– Notogaster not subdivided transversely; usually well sclerotized......................................24

13(12) Opisthonotal gland present. Notogaster with a single transverse scissure or groove, lying between setal rows d and e; row e inserted well posterior to scissure. Rather weakly sclerotized mites, colour white to pale yellow. Lyrifissures iad and ian present. Rutellum strongly developed, distally broad and toothed. Legs with well developed lateral claws; empodial claw reduced, hooklike, or absent (PARHYPOSOMATA) 14

– Opisthonotal gland absent. Notogaster with 1–3 transverse scissures or grooves [if only 1, then: mites more strongly pigmented, not soft-bodied; scissure runs either anterior to row d or posterior to e or contains row e on narrow intercalary sclerite; legs monodactylous].

Taxonomic Keys 37

Lyrifissures *ian* and *iad* absent. Rutellum various, but rarely strongly developed and distally broadened (ENARTHRONOTA in part) 15

14(13) Opisthonotal gland opening on flared, funnel-shaped protuberance bearing seta f_2. With 4 pairs of anal, 4 pairs of adanal setae. Peranal segment (PA) present (paraproctal), bearing 1 pair of setae. One pair of epimere II setae. Genu II and tarsus III with 2 and 1 solenidia, respectively (Figure 3.15A,B) .... **PARHYPOCHTHONIIDAE**: *Parhypochthonius* (Pg. 68)
– Opisthonotal gland opening without conspicuous protuberance; seta f_2 represented only by inconspicuous alveolus near gland opening. With 3 pairs of adanal, 2 pairs of anal setae. Peranal segment absent; anal segment paraproctal. Two pairs of epimere II setae. Genu II and tarsus III with 1 and 0 solenidia, respectively (Figure 3.15C–G) .... **GEHYPOCHTHONIIDAE**: *Gehypochthonius* (Pg. 68)

15(13) Notogaster with 1 transverse scissure lying posterior to setal row *c*, *d* or *e*. [Note: telescoping, type-L scissures can appear complex by transparency, due to overlapping structures; some Sphaerochthoniidae have vestiges of 1 or 2 more posterior scissures, but they appear simply as thin ridges, fused to surrounding cuticle; Eniochthoniidae have a transverse sulcus (groove) near setal row *d*, that is easily confused with a scissure] 16
– Notogaster with 2–3 transverse scissures or grooves, separating 3–4 dorsal plates or regions 19

16(15) Single transverse scissure of type-S: with intercalary sclerite containing large, erectile setae of row *e*. Setae f_1, f_2 also erectile, stout, with long branches, inserted on intercalary sclerites in pair of circular unsclerotized depressions; seta p_1, p_2 phylliform. Prodorsum astegasime. Legs bidactylous; empodial claw absent (Figure 3.11A) .... **ARBORICHTHONIIDAE**: *Arborichthonius* (Pg. 67)
– Single transverse scissure of other type; either telescoping type-L or, if type-S (with intercalary sclerite) then setae of row *e* (those on sclerite) smaller than nearby notogastral setae, often vestigial. No hysterosomal setae conspicuously enlarged 17

17(16) Transverse scissure, telescoping, type-L, positioned posterior to setal row *c*, such that only 4 pairs of setae insert on anterior notogastral plate (pronotaspis). At least some dorsal setae (of notogaster and/or prodorsum) heavily ciliate, T-shaped. Integument distinctly patterned (foveolate/reticulate) (Figure 3.11D) .... **SPHAEROCHTHONIIDAE**: *Sphaerochthonius* (Pg. 67)
– Transverse scissure of type-L or type-S, but located posterior to setal row *d*, such that 6–8 pairs of setae borne on pronotaspis. Dorsal setae simple, setiform, or flattened, lanceolate. Integument smooth or sculpted, but without foveae or reticulation 18

18(17) Single transverse scissure of type-L, located behind setal row *e*; pronotaspis with 8 pairs of setae (setae *e* similar to others). Pronotaspis with transverse, medially incomplete sulcus at level of setal row *d*. Setae h_2, h_3 on separate, longitudinally elongated lateral plate. Both genital and aggenital plates transversely divided; a single aggenital seta on anterior aggenital plate (Figure 3.13) .... **ENIOCHTHONIIDAE**: *Eniochthonius* (Pg. 62)
– Single transverse scissure of modified type-S, with undivided intercalary plate bearing small or vestigial setae e_1, e_2. Pronotaspis with 6 pairs of setae; without sulcus. No isolated lateral plate bearing setae. Genital plate transversely divided or not, but aggenital plate and aggenital seta absent (Figure 3.14) ...... **HYPOCHTHONIIDAE**: *Hypochthonius* (Pg. 64)

19(15) With 2 or 3 simple, type-E scissures or grooves. Notogastral setae generally homogeneous, similar in shape and size (among species this varies from setiform to narrowly phylliform, smooth to ciliate); none extremely broad (plate-like) or unusually large or erectile. Bothridial seta clavate 20

– Either 1 or 2 scissures of type-S, with intercalary sclerite and large, erectile setae; or notogastral setae extremely broad, plate-like, covering body like shields. Bothridial seta of various form... 21

20(19) Hysterosoma with 3 transverse type-E scissures. Four pairs of setae on pronotaspis (c row only). With cupule immediately lateral to each mid-dorsal seta (c_1, d_1, e_1, f_1), in addition to normal complement. Genital plates twice size of anal plates. Aggenital seta absent; anal plates with 4 pairs of setae; peranal plate and seta absent. (Figure 3.11B,C)....................... .... **HAPLOCHTHONIIDAE**: *Haplochthonius* (Pg. 66)

– Hysterosoma with 2 transverse type-E scissures. Six pairs of setae on anteriormost plate (c and d rows). Without extra cupules immediately lateral to mid-dorsal setae. Genital plates and combined anal/peranal plates subequal in size. One pair of aggenital setae; anal plates with 2 pairs of setae; peranal plates present, with 1 pair of setae. (Figures 3.3–3.10)......... .... **BRACHYCHTHONIIDAE** (Pg. 53)

21(19) Some or all setae of prodorsum and notogaster broad, shield-like, sometimes covering all or most of dorsal surface; setal surface with distinct pattern of foveae or reticulation; without barbs or cilia. Colourless mites, without noticeable sclerotization. With 6–8 pairs of genital setae (Atopochthonioidea)... 22

– Dorsal setae of various size and ornamentation—smooth, ciliate or pectinate—but not foveate or reticulate. Four pairs of setae (e_1, e_2, f_1, f_2) erectile, longer than others, inserted on intercalary sclerites in 2 type-S scissures. Cuticle usually with noticeable sclerotization, light brown colour. Ten pairs of genital setae 23

22(21) Legs monodactylous. All dorsal setae shield-like, similar in shape and mostly of similar size, none erectile or inserted on large tubercles. Chelicerae unusual; with large, functional trochanter and distinct ventral bend; fixed digit broad, poorly defined, with teeth directed anteriorly, rake-like, and movable digit simple, narrow, without teeth. Palp tarsus terminating in long, ribbon-shaped eupathidia, about equal to length of distal 3 palpal segments combined (Figure 3.12A,B) .... **PTEROCHTHONIIDAE**: *Pterochthonius* (Pg. 62)

– Legs bidactylous, with strong lateral claws but without empodium. One or two pairs of notogastral setae different from others in shape and orientation. Chelicerae narrow, straight, with both digits similar: simple, tapering, and with small, indistinct teeth. Palp tarsus with forked distal eupathid, similar in length to solenidion (Figure 3.12C,D)........... .... **ATOPOCHTHONIIDAE**: *Atopochthonius* (Pg. 62)

23(21) Hysterosomal dorsum with 3 transverse scissures; two type-S scissures bearing erectile setal rows e and f, and another just anterior to small setae of row d. Integument often ornamented (foveolate, reticulate or punctate); rostrum with distinct fenestrations. Prodorsal setae, including bothridial seta, densely ciliate. Adanal plates broadly joined posteriorly to form U-shape. With 4 pairs of anal setae. Genital papillae greatly reduced, inconspicuous. Epimere II with 2 or 3 pairs of setae. Legs with 2 or 3 claws (Figure 3.12E,F) .... **COSMOCHTHONIIDAE**: *Cosmochthonius* (Pg. 66)

– Hysterosomal dorsum with only 2 transverse type-S scissures; none at or near setal row d (although inconspicuous groove maybe present). Integument without distinct ornamentation; rostrum without fenestrations. Prodorsal setae simple or barbed, but not densely ciliate. Adanal plates not joined posteriorly. With 2 pairs of anal setae. Genital papillae of normal size. Epimere II with 1 pair of setae. Legs monodactylous (Figure 3.11E,F)........... .... **TRICHTHONIIDAE**: *Gozmanyina* (Pg. 67)

24(12) Dichoid mites: with full sejugal articulation, such that proterosoma is capable of horizontal and lateral movement relative to hysterosoma, often slightly retractable into it; epimeres II and III not fused, separated by band of unsclerotized cuticle. Exobothridial setae 2 pairs 25

– Holoid mites: epimere II and III usually entirely fused, such that sejugal articulation present only in dorsal and sometimes lateral region; proterosoma not retractable into hysterosoma (exception: narrow band of unsclerotized cuticle connects epimeres II and III in Malaconothridae and some Trhypochthoniidae, allowing slight dorso-ventral movement of proterosoma). Exobothridial setae 0 or 1 pair 28

25(24) With characteristic body shape: ventral surface flat, with lateroventral depressions to accommodate retracted legs; dorsal surface strongly arched; prodorsum posteriorly equally as broad as notogaster. Anterior margin of notogaster with large, thin tectum overhanging posterior region of prodorsum. Ten pairs of genital setae, 6 in medial line (Figure 3.17A,B) ....**LOHMANNIIDAE**
Note: Family unknown for Canada or Alaska, but known from North and South Carolina (Norton et al. 1978).

– Body differently shaped: venter usually not noticeably flattened, and without depressions to accommodate retracted legs (if flat, then notogaster also rather flat). Prodorsum usually narrower than notogaster. Notogaster without anterior tectum. Fewer than 10 pairs of genital setae 26

26(25) Total length over 800μm. Epimeres III and IV with paired plates, separated medially by unsclerotized cuticle. Genital plates immediately posterior to epimere IV, located in anterior half of hysterosoma. Tarsus I with more than 35 setae. Palp with 5 free segments (Figure 3.19) ....**PERLOHMANNIIDAE** (Pg. 70)

– Total length usually less than 800μm. Epimeral plates III and IV fused medially, with or without demarcation. Genital plates well posterior to epimere IV, in posterior half of hysterosoma. Tarsus I with fewer than 25 setae. Palp with 2 or 4 free segments 27

27(26) Body elongated, cylindrical; proterosoma noticeably constricted behind level of bothridium. Colour light yellow to orange. With much fusion of plates in hysterosoma; except for paired adanal, anal and genital plates; U-shaped scissure passes between latter two plates. Epimeres III, IV and aggenital region fused without demarcation. Legs usually heterotridactylous with strong lateral claws and minute empodium; rarely monodactylous (Figure 3.17C–E) .... **EULOHMANNIIDAE**: *Eulohmannia* (Pg. 69)

– Body only moderately elongated; proterosoma without noticeable posterior constriction. Colour light to dark brown. Epimeres III, IV outlined by borders and apodemes. Adanal plates broadly fused anterior to anal plates, separating anal plates from genital plates; aggenital and adanal region separated by transverse scissure or indistinguishably fused. Legs monodactylous (Figure 3.17F,G).... **EPILOHMANNIIDAE**: *Epilohmannia* (Pg. 71)

28(24) Body cylindrical, somewhat elongated. Notogaster completely fused to ventral region in posterior half, such that adanal plates are distinguishable only by differences in cuticular pattern; cuticle delimiting notogastral margin in anterior half curving ventrally toward genital plates, such that in ventral aspect pair of crescent-shaped scissures or depressions directed toward wide space between genital and anal plates. Prodorsum posteriorly with pair of tooth-like or ridge-like tubercles projecting over dorsosejugal groove (Figure 3.42) ....**NANHERMANNIIDAE**: *Nanhermannia* (Pg. 99)

– Body not cylindrical, often globose or dorsally flattened. Notogaster distinct from ventral region throughout its length. Adanal plates distinct; genital and anal plates adjacent or only narrowly separated. Prodorsum without posteriorly directed tubercles 29

29(28) Ventral region brachypyline, similar to that of "higher" oribatid mites; adanal and aggenital regions broad, completely fused, and aggenital region fused with epimere IV. With 9 pairs of genital setae. Lyrifissure *ian* absent (Figure 3.40) ....
.... **HERMANNIIDAE**: *Hermannia* (Pg. 96)
– Ventral region not brachypyline; adanal and aggenital plates (when latter distinct) relatively narrow; if sclerotized, aggenital region never fully fused to epimere IV. With 4–24 pairs of genital setae. Lyrifissure *ian* present.. 30

30(29) Epimere II with 3 or more pairs of setae. With 9 pairs of genital setae, 1 posterior seta far lateral to others, away from medial edge of plate. Aggenital setae absent. Preanal plate distinct. Rostrum with short medial incision. Bothridium with group of internal bulbous tubules attached to its base; bothridial seta setiform, longer than interlamellar seta (Figure 2.13C; 3.43) .... **NOTHRIDAE**: *Nothrus* (Pg. 100)
– Epimere II with 1 (usually) or 0 pair of setae. With 4–24 pairs of genital setae, but all near medial edge. Aggenital setae present or absent. Preanal plate present or absent. Rostrum without medial incision. Bothridium without internal tubules (saccule may be present); bothridial seta often clavate, length usually equal to or less than that of interlamellar seta 31

31(30) Medial margin of genital plates, bearing genital setae, delimited from rest of plate by distinct carina. Anal plate wider or only slightly narrower than adanal plate, clearly not greatly reduced. Aggenital setae present (Figures 3.36–3.39) ....
.... **CROTONIIDAE** (incl. Camisiidae auct.) (Pg. 88)
– Genital plate without carina that delimits marginal seta-bearing region. Anal plates usually reduced, much narrower than adanal plate, sometimes inconspicuous. Aggenital setae absent... 32

32(31) Bothridial seta and bothridium lost without trace. Subcapitulum anarthric, without labiogenal articulation. Tibia IV without solenidion; tarsus II with 1 solenidion; genu III without solenidion. Genua III and IV with 1 seta. Cerotegument present in form of large, thin plates that appear waxy, strongly birefringent in polarized light (Figure 3.41) ....
.... **MALACONOTHRIDAE** (Pg. 97)
– Bothridial seta present, usually with well-developed bothridium (in some populations of *Trhypochthoniellus* these are vestigial or absent). Subcapitulum with labiogenal articulation. Tibia IV with 1 solenidion; tarsus II with 2 solenidia; genu III with 1 solenidion. Genua III and IV with 2 or 3 setae. Cerotegument, if present, of different form, not birefringent (Figure 3.44) .... **TRHYPOCHTHONIIDAE** (Pg. 103)
(sensu lato; includes Trhypochthoniellidae, Allonothridae and Mucronothridae of authors)

33(1) Notogaster without octotaxic system of porose areas or saccules; pteromorphs absent. Epimeres II, III and IV usually rectangular in shape; apodemes 2 and sejugal apodeme transverse. Discidium and circumpedal carina usually absent. Humerosejugal porose organ *Am* absent, *Ah* usually absent 34
Note: Peloppiidae is keyed more than once because genera in this family have either diarthric or anarthric subcapitulum. Thyrisomidae is keyed more than once because the presence/absence of apodeme 4 is genus-specific in this family. Cymbaeremaeidae and Hydrozetidae are keyed more than once because their respective coxisternal region can be misinterpreted.
– Notogaster usually with octotaxic system of porose areas or saccules; pteromorphs present or absent. Epimera II, III and IV never rectangular in shape; apodemes 2 and sejugal apodeme angled towards genital plate. Discidium and circumpedal carina usually present (absent in Passalozetidae). Humerosejugal porose organs *Am* and *Ah* usually present (absent in Microzetidae) 78

34(33)	Paired opisthonotal glands on notogaster opening on distinct, funnel-shaped or spout-shaped tubes or on large apophyses (tubercles)	35
–	Paired opisthonotal glands opening directly on notogaster, funnel-shaped tubes or apophyses absent	36
35(34)	Gnathosoma modified, chelicerae pelopsiform. Paired opisthonotal glands opening on large anteriorly directed apophyses. Rostrum with deep medial incision. Lamellar setae small, hardly visible. Trochanters and femora I and II with retrotecta. Adults with or without scalps, scalps when present concentrically arranged. Length: 330–440µm (Figure 3.47A–E) ......**PLASMOBATIDAE**: *Plasmobates* (Pg. 109)	
	Gnathosoma normal, chelicerae chelate-dentate. Paired opisthonotal glands opening on distinct, funnel-shaped tubes. Rostrum without incision. Lamellar setae well-developed. Trochanters and femora I and II without retrotecta. Adults with only inconspicuous tritonymphal scalp adhering closely to notogaster. Length: 550–800µm (Figure 3.46)......**HERMANNIELLIDAE**: *Hermanniella* (Pg. 108)	
36(34)	Genital plates with transverse scissure; 7 or 8 (exceptionally up to 18) pairs of genital setae. Adults with strongly sclerotized scalps. Femora I–IV and trochanters III, IV with brachytrachea. Length: 800–1,100µm (Figure 3.48)......**NEOLIODIDAE** (Pg. 109)	
–	Genital plates without transverse scissure; usually with 3–7 pairs of genital setae. Adults with or without scalps, scalps if present weakly sclerotized. Femora I–IV and trochanters III, IV generally with porose areas; rarely with filamentous tracheae	37
37(36)	Tarsus I with 3 solenidia, pretarsus with very small claw, tarsal pulvillus present. Legs IV modified for jumping, or not. Cerotegument with large globules. Length: 330–550µm (Figure 3.47F–I) ......**ZETORCHESTIDAE**: *Zetorchestes*	
	Note: Family known from southern QC (CNC record).	
–	Tarsus I with 2 solenidia, pretarsus with medium to large claw, tarsal pulvillus present or absent. Legs IV not modified for jumping. Cerotegument with large or small globules or microtuberculate	38
38(37)	Notogaster with combination of tubercles on anterior margin; with strong topography consisting of relatively flat lateral region and two strong bulges (transverse anterior bulge and longitudinal posterior bulge) separated by foveate transverse sulcus. Pedotectum II absent. Tarsus II with 1 solenidion. Length: 300–475µm (Figure 3.65A–D)......**CALEREMAEIDAE**: *Caleremaeus* (Pg. 127)	
–	Notogaster with or without tubercles on anterior margin; without strong topography, as given above and without foveate transverse sulcus. Pedotectum II present or absent. Tarsus II usually with 2 solenidia	39
39(38)	Prodorsum without blade-like lamellae; with or without ridge-like lamella	40
–	Prodorsum with blade-like lamellae	64
40(39)	Notogaster with 2–9 pairs of setae, positioned posteromarginally; dorsocentral notogastral setae absent or only 1 pair present; notogaster usually flat-topped and elliptical. Thick, ornamented cerotegument usually present	41
–	Notogaster with 9–15 pairs of setae or setal alveoli, only 3 pairs of setae positioned posteromarginally; notogaster usually convex. Ornamented cerotegument usually absent	43
41(40)	Legs thin, filiform; femora with porose areas. Pedotectum I present. Propodolateral apophysis absent. Length: 480–700µm (Figures 3.49–3.51F) ......**GYMNODAMAEIDAE** (Pg. 110)	
–	Legs never filiform; femora with porose areas or with tracheae. Pedotectum I present or absent. Propodolateral apophysis present or absent	42

42(41)		Femur I with minimum of 8 setae; femur IV with minimum of 5 setae; trochanter IV with 3 setae. Bothridial seta filiform. Epimeral region neotrichous. Length: 600–800µm (Figure 3.51G,H).. **PLATEREMAEIDAE**: *Allodamaeus* (Pg. 114)
–		Femur I with 5 setae; femur IV with 3 setae; trochanter IV with 1 seta. Head of bothridial seta flat leaf-shaped, fusiform, or club-shaped. Epimeral setation 3-1-3-3 (Figure 3.51I,J) . ... **LICNODAMAEIDAE**: *Licnodamaeus* (Pg. 114)
43(40)		Anal plates with 2–9 pairs of setae 44
–		Anal plates with 2 pairs of setae or their alveoli 46
44(43)		Anal plates with 3 pairs of setae. Epimeral setation 4-1-3-3. Length: 750–900µm (Figure 3.76E,F) .... **KODIAKELLIDAE**: *Kodiakella* (Pg. 141)
–		Anal plates with 2–9 pairs of setae. Epimeral setation 3-1-3-3 45
45(44)		Tutorium present lateral to each ridge-like lamella on prodorsum. Notogaster with 2 pairs of tubercles anteriorly. 2–3 pairs of anal setae. Length: 550–1,100µm (Figure 3.74) .... **MEGEREMAEIDAE**: *Megeremaeus* (Pg. 138)
–		Tutorium absent. Notogaster without tubercles anteriorly. 2–9 pairs of anal setae. Length: 400–800µm (Figures 3.68–3.73) .... **EREMAEIDAE** (Pg. 130)
46(43)		Gnathosoma modified: either rutellum very large, and/or thin and leaf-like, or paired rutella forming tube; chelicera with unusually large teeth, or pelopsiform, or digits tapered, elongate. Chelicera with 0, 1 or 2 setae 47
–		Gnathosoma normal, chelicera broad with dentate chelae; rutella normally developed. Chelicera with 2 setae 49
47(46)		Subcapitulum diarthric. Rutellum very large, thin and leaf-like. Chelicera with unusually large teeth. Palpgenu without seta, fused or not to femur; palptarsus with 3–5 setae. Length: 215–250µm (Figure 3.66B,E–I) .... **DAMAEOLIDAE** (Pg. 128)
–		Subcapitulum anarthric. Rutellum either cupped, with pair forming tube, or narrow, terminally rounded, with thin, spatulate ventral expansion 48
48(47)		Chelicera attenuate-edentate. Paired rutella forming tube (Figures 3.92D; 3.94–3.96) .... **SUCTOBELBIDAE** (Pg. 166)
–		Chelicera weakly sclerotized, elongate with narrow digits. Anterior and lateral portion of gena expanded forming tectum covering palpal trochanter and base of rutellum. Rutellum narrow, terminally rounded; with thin, spatulate, ventral expansion (Figure 3.65E–I)........ .... *Veloppia* Hammer (Pg. 127)
		Note: Family placement of this genus is under study.
49(46)		Posterior of prodorsum and anterior of notogaster strongly flattened to concave; usually fused. Longer setae of notogaster often arched, with lp and h_3 curved toward midline, and h_1 and h_2 curved posteriorly. Rostrum usually with pair of incisions. Length: 600–1,050 µm (Figure 3.66A) .... **AMERIDAE**: *Gymnodampia* (Pg. 128)
–		Posterior of prodorsum and anterior of notogaster convex, fused or not. Notogastral setae of similar size and shape, usually none arching medially. Rostrum with single incision or incisions absent 50
50(49)		Epimeral and ventral plate setae branched. Length: 275–500µm (Figure 3.66C,D) .... **EREMULIDAE**: *Eremulus* (Pg. 129)
–		Epimeral setae branched or not; ventral plate setae not branched 51
51(50)		Adanal neotrichy present (more than 3 pairs of setae). Tibiae and tarsi I–IV with retrotecta. Femora I–IV and trochanters III, IV with saccules. Length: 300–550µm (Figure 3.67C–F) .... **EREMOBELBIDAE**: *Eremobelba* (Pg. 129)

–	Adanal setation 3 pairs. Tibiae and tarsi I–IV without retrotecta. Femora I–IV and trochanters III, IV with porose areas	52
52(51)	Epimeral neotrichy present (2–3 pairs on epimere II). Aggenital neotrichy present (2–3 pairs of setae). With distinct medial prodorsal protuberance, appearing as round or quadrangular in dorsal view. Anterior margin of notogaster truncate. Propodolateral apophysis present. Length: 335–415μm (Figure 3.67A,B) **HUNGAROBELBIDAE**: *Hungarobelba* (Pg. 130)	
–	Epimeral and aggenital neotrichy absent. Prodorsal protuberance absent. Propodolateral apophysis present or absent. Anterior margin of notogaster convex	53
53(52)	Notogastral setae (other than *p* series) arranged in 2 conspicuous longitudinal rows. Well developed parastigmatic enantiophysis present. Legs moniliform (tarsal segments swollen proximally, tapered distally; femora and tibiae narrow proximally, swollen distally). Length: 450–1,300μm (Figures 3.52–3.60) **DAMAEIDAE** (Pg. 114)	
–	Notogastral setae not arranged in 2 longitudinal rows. Parastigmatic enantiophysis absent. Legs moniliform (Oppioidea) or not	54
54(53)	Epimera III and IV distinctly delineated by intervening border	55
–	Epimera III and IV not distinctly delineated: border of epimere III not visible	59
55(54)	Lenticulus present, circular, distinctly convex. Seta *d* inserted on proximal fifth of femora I–III, proximal to other femoral setae. 13, or 15 to 17 pairs of notogastral setae (that is, may have extra *h* setae). Bothridial seta reduced or absent (frequently broken). Length: 400–700μm (Figure 3.98) **HYDROZETIDAE** (pars): *Hydrozetes* (Pg. 171)	
–	Lenticulus, if present, not distinctly convex. Seta *d* not most proximal seta on femora I–III. 10 to 15 pairs of notogastral setae, (3 pairs of *h* setae). Bothridial seta normal, reduced or absent	56
56(55)	Epimeral setation 1-0-1-1. Tarsus II with 1 solenidion. Genital plates with 3 pairs of setae. Length: 370–420μm (Figure 3.103D–F) **SELENORIBATIDAE**: *Thalassozetes* (Pg. 177)	
–	Epimeral setation 3-1-3-3, 3-1-3-4, 3-1-2-2, 3-1-2-3 or 3-1-3-2. Tarsus II with 2 solenidia. Genital plates with 3–6 pairs of setae	57
57(56)	Integument dark, but weakly sclerotized. Bothridial seta very small or absent. Border of epimere IV without minitectum	58
–	Integument well sclerotized. Bothridial seta normally developed. Border of epimere IV with minitectum. Length: 300–500μm (Figure 3.93) **THYRISOMIDAE** (pars) (Pg. 164)	
58(57)	Integument dark, with wrinkled pattern. Dorsosejugal scissure medially incomplete. Bothridial seta present or absent; if rudimentary or absent, exobothridial seta present; if bothridial seta present, exobothridial seta absent. Pedotecta I, II absent or represented by vestiges. Mono- or tridactylous. Coxisternal setation 3-1-2-2. Length: 550–800μm (Figure 3.101) **AMERONOTHRIDAE**: *Ameronothrus* (Pg. 175)	
–	Integument without wrinkled pattern. Dorsosejugal scissure complete. Bothridial seta present; exobothridial setae present. Pedotectum I present. Tridactylous. Coxisternal setae 3-1-2-3 or neotrichy on epimere I (Figure 3.102) **PODACARIDAE**: *Alaskozetes* (Pg. 176)	
59(54)	Lenticulus present. Notogaster with distinct marginal zone and separate central area. Eupathidium *acm* of palptarsus on tubercle (Figure 3.97G–J) **CYMBAEREMAEIDAE** (pars) (Pg. 178)	
–	Lenticulus absent. Notogaster without distinct marginal zone and separate central area. Eupathidium *acm* of palptarsus not on distinct tubercle	60

60(59) Border of epimere IV with minitectum. Genital and anal plates very large, almost touching, separated by distance of less than half length of genital plate. Length: 290–500µm (Figure 3.93) ....**THYRISOMIDAE** (pars) (Pg. 164)
– Border of epimere IV generally without minitectum. Genital and anal plates separated by distance greater than half length of genital plate 61

61(60) Tibia I with large dorsodistal tubercle overhanging tarsus I and bearing solenidion φ_1. Rostrum usually with straight, deep, medial incision. Humeral enantiophysis usually present. Length: 250–350µm (Figure 3.88).................................**AUTOGNETIDAE** (Pg. 155)
– Tibia I without large dorsodistal tubercle overhanging tarsus I. Rostrum usually without medial incision. Humeral enantiophysis present or absent.. 62

62(61) Epimeral setation 2-2-4-5, or up to 17 pairs of epimeral setae directed medially towards ventrosejugal region, forming "basket" in thick layer of epimeral cerotegument. Anterior margin of epimere I with medial tooth. Length: 150–250µm (Figure 3.92E,F)..................
.... **MACHUELLIDAE**: *Machuella* (Pg. 157)
– Epimeral setation usually 3-1-3-3 or setation reduced; epimeral setae not directed medially towards ventrosejugal region, not forming "basket" in thick layer of cerotegument. Anterior margin of epimere I without tooth.. 63

63(62) Costulae, if present, shorter than half length of prodorsum. Crista if present, not longer than third length of notogaster. Length: 120–550µm (Figures 3.89–3.91)............................
....**OPPIIDAE** (Pg. 157)
– Costulae longer than half length of prodorsum. Crista third length of notogaster, or longer. Length: 150–230µm (Figure 3.92A–C) .... **QUADROPPIIDAE**: *Quadroppia* (Pg. 163)

64(39) Scalps present. Cerotegument thin, without adherent dirt. Lamellae very large and fused proximomedially, not thickened. Patronium absent. Subcapitulum diarthric; mentum and gena not extending anteriorly over rutella. Chelicerae chelate-dentate; cheliceral seta *cha* curved and coiled many times back over digit. Length: 350–400µm (Figure 3.63I–K) ......
....**PODOPTEROTEGAEIDAE**: *Podopterotegaeus* (Pg. 125)
– Scalps absent. Patronium present or absent. Subcapitulum anarthric or diarthric; mentum and gena extending or not anteriorly over rutella. Chelicerae chelate-dentate or highly modified; cheliceral seta *cha* setiform... 65

65(64) Cerotegument very thick and bearing dirt mass. Lamellae very large and touching medially, medially thickened. Patronium present. Subcapitulum anarthric, mentum and gena extending anteriorly over rutella like tectum. Rutellum in form of large bifurcate seta. Chelicerae highly modified with strainer-like teeth. Length: 400–560µm (Figure 3.63D–H)...............
.... **POLYPTEROZETIDAE**: *Polypterozetes* (Pg. 125)
– Cerotegument thin and not bearing dirt mass. Lamellae not touching medially, or medially thickened. Patronium absent. Subcapitulum anarthric or diarthric; mentum and gena not extending anteriorly over rutella. Rutellum not in form of large bifurcate seta. Chelicerae highly modified or not; but without strainer-like teeth 66

66(65) Subcapitulum anarthric. Cheliceral digits highly modified, or chelicerae pelopsiform or styliform (moveable digit serrate distally).. 67
– Subcapitulum diarthric (can appear almost stenarthric). Chelicerae chelate-dentate 68

67(66) Chelicerae styliform, moveable digit serrate distally, fixed digit absent. Border of epimere IV with taenidium and minitectum extending from genital plate to circumpedal carina. Length: ca. 600µm (Figure 3.76A–D) ....**GUSTAVIIDAE**: *Gustavia* (Pg. 141)

Taxonomic Keys 45

– Chelicerae chelate-dentate or pelopsiform. Border of epimere IV with taenidium or minitectum extending from close to genital plate to circumpedal carina, or taenidium absent. Length: 400–800 µm (Figures 3.80–3.83) ....**PELOPPIIDAE** (pars) (Pg. 144)

68(66) Notogaster with long, triangular or subrectangular humeral process, projecting anteriorly to bothridium, or projecting dorsally over bothridium 69
– Notogaster without long, triangular humeral process, if humeral tubercles present, not reaching level of bothridium 70

69(68) Subcapitular mentum with tectum. Palptarsal eupathidium *acm* inserted on large tubercle (Figure 3.104B). Notogaster strongly overhanging ventral plate. Length: 350–800µm (Figure 3.97G–J).. **CYMBAEREMAEIDAE** (pars) (Pg. 178)
– Subcapitular mentum without tectum. Palptarsal eupathidium not on tubercle. Notogaster not strongly overhanging ventral plate (notogaster may have reflexed rim). Length: 800–1200µm (Figure 3.84).. **TENUIALIDAE** (Pg. 149)

70(68) Integument of genital area usually darker than surrounding cuticle. Epimeral setae *4a*, *4b* closely adjacent. Pedotectum I in two parts (narrow dorsal part and wider ventral part). Length: 250–500µm (Figure 3.103A–C)...
....**TEGEOCRANELLIDAE**: *Tegeocranellus* (Pg. 177)
– Integument of genital area same colour as anal region. Epimeral setae *4a*, *4b* not closely adjacent. Pedotectum I undivided.. 71

71(70) Subcapitular mentum with tectum. Eupathidium *acm* of palptarsus on tubercle. Lamellar cusps touching medially, fused or not, covering rostrum medially. Lamellar seta arising anteroventrally on cusp. Length: 350–800µm (Figure 3.104)..
.... **CYMBAEREMAEIDAE** (pars) (Pg. 178)
– Subcapitular mentum without tectum. Eupathidium *acm* of palptarsus not on tubercle. Lamellar cusps if touching medially, not covering rostrum. Lamellar seta arising anteriorly or dorsally on lamella or lamellar cusp 72

72(71) Humeral region with 2 pairs of small, closely adjacent setae *c* (c_1 and c_2). Length: 500–1,160µm (Figures 3.77–3.79).. **LIACARIDAE** (Pg. 141)
– Humeral region with 0 or 1 pair of seta *c*... 73

73(72) Notogastral setae clearly visible and in marginal and posteromarginal rows; with 0 or 1 pair of setae positioned centrodorsally. Aggenital enantiophysis present lateral to genital aperture, or absent 74
– Notogastral setae, other than *p* series, absent from marginal and posteromarginal rows (setal alveoli present or not); with 0 or 2 pairs of setae positioned centrodorsally. Aggenital enantiophysis absent 75

74(73) Rostrum with strongly dentate margin (Figure 3.63A–C) ....
....*Nemacepheus dentatus* Aoki (Pg. 170)
Note: Included in Tectocepheidae by Aoki (1968), and Balogh and Balogh (1992), but in Nodocepheidae by Fujikawa (2001) and Subías (2004). It lacks key character states of both families.
– Rostrum smooth or with single pair of teeth (Figures 3.61; 3.62)...... **CEPHEIDAE** (Pg. 122)

75(73) Integument smooth and shiny. Lamellae converging or parallel, if parallel, not positioned close to lateral margin of prodorsum.. 76
– Integument distinctly sculptured with coarse ridges, tubercles, or areoles. Lamellae usually broad, almost parallel, close to lateral margin of prodorsum 77

76(75)	Genital plates large; distance between genital and anal plates usually less than length of anal plates. Notogaster glabrous or with 10 pairs of setae. Length: 250–600µm (Figure 3.75)... **ASTEGISTIDAE** (Pg. 139)
–	Genital plates medium in size; distance between genital and anal plates usually greater than, or equal to, length of anal plates. Notogaster with 8 pairs of setae or setae vestigial. Length: 400–800µm (Figures 3.80–3.83) .... **PELOPPIIDAE** (pars) (Pg. 144)
77(75)	Genal incision present. Circumpedal carina partially developed. Tarsi and tibiae articulate normally. Anal plate triangular in shape. Interlamellar seta positioned in interbothridial region. Length: 225–300µm (Figure 3.97A–F) .... **TECTOCEPHEIDAE**: *Tectocepheus* (Pg. 169)
–	Genal incision absent. Circumpedal carina absent. Leg tarsi functionally fused to tibiae. Anal plate rectangular in shape. Interlamellar seta usually positioned well anterior of bothridium or on lamella. Length: 350–600µm (Figures 3.85–3.87) .... **CARABODIDAE** (Pg. 150)
78(33)	Pteromorphs present, moveable, auriculate (ear-like), i.e., extending anteriorly and posteriorly past articulation with notogaster such that all legs covered when pteromorphs closed. Seta *c* positioned on pteromorph or on body of notogaster 79
–	Pteromorphs present or absent; if present, fixed or moveable, but not auriculate. Seta *c* positioned on body of notogaster 81
79(78)	Lamellae absent, at most narrow carina present (line L). Pteromorph with alary furrow, appearing bilobed. Subcapitular mentum with tectum. Tutorium reduced to line (S) or absent. Notogaster with posterior tectum. Length: 300–1,100µm (Figures 3.153, 3.154)... **GALUMNIDAE** (Pg. 242)
–	Lamella present. Pteromorph without alary furrow. Subcapitular mentum without tectum. Tutorium expressed as carina, or absent. Notogaster with or without posterior tectum. 80
80(79)	Lamella usually narrow, directed anteriorly from bothridium as simple carina. Seta *c* positioned on body of notogaster. Notogaster without posterior tectum. Tibiae and tarsi I–IV with porose areas ventrally. Adalar porose organ (*Aa*) expressed as porose area or saccule, other 3 (mesonotic porose organs) always expressed as saccules. Length: 300–820µm (Figure 3.125E–G)..**PARAKALUMMIDAE** (Pg. 207)
–	Lamellae broad, with distinct cusps. Seta *c* positioned on pteromorph. Notogaster with posterior tectum. Tibiae and tarsi I–IV without porose areas ventrally. Octotaxic system expressed as porose areas. Length: 250–350µm (Figure 3.129A–C)... **CERATOKALUMMIDAE**: *Cultrobates* (Pg. 213)
81(78)	Epimeral border IV clearly evident, extending transversely anterior to genital plates; border of epimere IV with or without tectum. Prodorsum subequal in length to notogaster; notogaster as wide as long, or wider than long. Tarsi II–IV with short, thick setae (*p*). Length: 150–550µm (Figure 3.64) .... **MICROZETIDAE** (Pg. 126)
–	Epimeral border IV not clearly evident, or if present, not transverse, but angled anteriorly to meet genital plates. Border of epimere IV without tectum. Prodorsum shorter than notogaster; notogaster longer than wide. Tarsi II–IV with setiform setae (*p*)......................... 82
82(81)	Epimera II–IV clearly delineated, distinctly angled towards genital plates. Bothridial seta often weakly developed, reduced or absent... 83
–	Epimera II–IV weakly delineated. Bothridial seta well developed 84
83(82)	Lenticulus absent. Notogaster with small or large pteromorphs; 10 pairs of notogastral setae. Lamella and tutorium well developed. Length: 260–380µm (Figures 3.99, 3.100) .... **LIMNOZETIDAE**: *Limnozetes* (Pg. 172)

| – | Lenticulus present, circular, strongly convex. Notogaster without pteromorphs; 13, or 15 to 17 pairs of notogastral setae (i.e., extra h setae). Lamella and tutorium weakly expressed. Length: 400–700µm (Figure 3.98).......... **HYDROZETIDAE** (pars): *Hydrozetes* (Pg. 171) |

84(82) Cerotegument well developed, very thick, with or without blocky structure, birefringent in polarized light. Pteromorphs present 85
– Cerotegument indistinct, not blocky in structure, never birefringent. Pteromorphs present or absent.. 86

85(84) Octotaxic system absent. Notogastral setal pair h_1 divergent. Pedotectum I without transverse carina. Anterior notogastral margin undulate. Posteromarginal region of notogaster undulate medially. Mentum with reflexed anterior rim, with or without medial carina. Interlamellar seta setose. Chelicera normal. Length: 550–670µm (Figure 3.107A–D) .... **UNDULORIBATIDAE**: *Unduloribates* (Pg. 186)
– Octotaxic system of porose areas; when present, all closely associated with setae. Notogastral setal pair h_1 convergent or directed posteriorly. Pedotectum I with transverse carina. Posteromarginal region of notogaster rounded medially. Mentum without reflexed anterior rim, without medial carina. Interlamellar seta large, spatulate or short, setose. Chelicera pelopsiform or normal. Length: 400–1,000µm (Figure 3.107E–J) .... **PHENOPELOPIDAE** (Pg. 183)

86(84) Pteromorphs present, with or without long, knife-like anterior projection. Bothridium with internal ring-like thickenings. Genu IV longer than genu III, concave dorsally, usually longer than tibia IV. Length: 390–850µm (Figures 3.109–3.111).. .... **ACHIPTERIIDAE** (Pg. 186)
– Pteromorphs present or absent; if present, without knife-like projection. Bothridium without internal ring-like thickenings. Genu IV subequal in length to genu III, straight dorsally, shorter than tibia IV 87

87(86) Lenticulus sharply defined, oval. Lamella usually absent; ridge-like if present. Lamellar setae positioned closely adjacent to rostral setae, not associated with ridge-like lamellae (when present). Tibiae with retrotecta. Pteromorphs absent. Length: 230–460µm (Figure 3.106A–C)...................................... **PASSALOZETIDAE**: *Passalozetes* (Pg. 181)
– Lenticulus, if present, not sharply defined, rectangular to subtriangular in shape. Lamella present or absent. Lamellar setae borne on lamellae or lamellar cusps or on adjacent prodorsum; lamellar setae not closely adjacent to rostral setae. Tibiae without retrotecta. Pteromorphs present or absent 88

88(87) Pedotectum I at most weakly developed as small lamina. Prodorsum with or without genal incision.. 89
– Pedotectum I well-developed as large lamina. Prodorsum generally with genal incision... 96

89(88) Prodorsum with genal incision. Axillary saccule of subcapitulum present. Octotaxic system as 4 pairs of saccules. Pteromorphs absent. Length: 375–435µm (Figure 3.105A–C) .. .. **DENDROEREMAEIDAE**: *Dendroeremaeus* (Pg. 180)
– Prodorsum without genal incision. Axillary saccule of subcapitulum absent. Octotaxic system as porose areas or saccules Pteromorphs present or absent 90

90(89) Genital and anal plates closely adjacent, separated by less than half length of genital plates. Pedotectum II usually absent. Octotaxic system as 2 pairs of small porose areas ($Aa, A2$). Bothridial seta large, distinctly widened distally, leaf-shaped. Notogaster extending medially between bothridia and interlamellar seta, forming acute angle. Pteromorphs absent. Length: 200–250µm (Figure 3.105D–F).. .... **LICNEREMAEIDAE**: *Licneremaeus* (Pg. 181)

– Genital and anal plates separated by more than half length of genital plate. Pedotectum II present. Notogaster with 0 to many pairs of porose areas or saccules. Notogaster not extending medially between bothridia and interlamellar seta, not forming acute angle. Pteromorphs present or absent 91

91(90) Octotaxic system reduced to 1 to 3 pairs of minute saccules (*Sa* and either *S2* or *S3*). Tibiae I–IV with brachytracheae or saccules. Notogastral tectum present posteriorly. Length: 350–500µm (Figure 3.106D–I) .... **SCUTOVERTICIDAE** (Pg. 182)
– Octotaxic system present, reduced or absent, but never expressed as 1 to 3 pairs of minute saccules. Tibiae I–IV with porose areas or respiratory surface absent. Posterior notogastral tectum absent 92

92(91) Pteromorphs usually well developed, with hinge and insertions of adductor muscles for pteromorphs evident on notogaster; or without hinge. Dorsophragmata narrow and well separated. Discidium and custodium present, maybe small. Length: 250–650µm (Figures 3.118–3.120) .... **HAPLOZETIDAE** (Pg. 198)
– Pteromorphs if present, without hinge; pteromorph adductor muscles absent. Dorsophragmata adjacent medially. Discidium and custodium present or absent 93

93(92) Genital setation 1–4 pairs. Tarsus I strongly truncate. Rostral tectum not completely covering chelicerae dorsally. Bothridium inserted lateroposteriorly, posterior to level of anterior notogastral margin; anterolateral border of notogaster partially or completely covering bothridium. Anal, adanal setae often long, flagelliform. Length: 250–550µm (Figure 3.125A–D) .... **ORIPODIDAE** (Pg. 207)
– Genital setation 4 (or more) pairs. Tarsi not truncate. Rostral tectum covering chelicerae dorsally. Bothridium inserted anterior to notogastral margin. Anal, adanal setae never long and flagelliform 94

94(93) Notogaster with 4 or more pairs of porose areas, their limits often well defined by thickened border; porose areas usually sexually dimorphic in male (extra porose areas). Notogaster without conspicuous setae; 3 pairs of short setae in *p* row and 7 pairs of alveoli. Prodorsum fused with notogaster medially. Length: 500–800µm (Figure 3.121) ....
.... **MOCHLOZETIDAE** (Pg. 201)
– Notogaster usually with 4 or less pairs of porose areas or saccules, without thickened border. Notogaster with 10 to 14 pairs of well-developed setae. Prodorsum fused, or not, with notogaster 95

95(94) Pteromorphs or humeral projections usually present. Notogaster with 2–4 pairs of porose areas or saccules. Genital setation 1 or 4 pairs. Prodorsum with prolamella often present; sublamella present. Sternal furrow often present on coxisternum, extending from genital plates anteriorly. Distal eupathia on palptarsus in same plane (comb-like) (Figure 3.128E). Length: 200–650µm (Figures 3.126–3.128) .... **SCHELORIBATIDAE** (Pg. 208)
– Pteromorphs or humeral projections generally absent (small projections present in *Oribatula*). Notogaster with 4 or 5 pairs of porose areas. Genital setation 4 or 5 pairs. Prodorsum without prolamella and sublamella. Sternal furrow absent. Distal eupathidia on palptarsus not in same plane (trowel-like) (Figure 3.124D). Length: 250–900µm (Figures 3.122–3.124) .... **ORIBATULIDAE** (Pg. 203)

96(88) Lamellae well-developed, fused at base of cusps; lamellar cusps almost covering entire prodorsum, usually deeply incised, forming large medial and lateral teeth, or large lateral tooth; lamellar setae inserted at base of incision. Pedotectum I with distinct depression ventrally. Pteromorph without hinge. Length: 240–750µm (Figures 3.114–3.117) ....
.... **ORIBATELLIDAE** (Pg. 192)

Taxonomic Keys

- Lamellae well- or weakly-developed, usually not fused at level of base of cusps; lamellar cusps usually not covering entire prodorsum (exceptions in Ceratozetidae); lamellar cusps usually not deeply incised; medial and lateral teeth if present, small. Pedotectum I without depression ventrally. Pteromorph with or without hinge...97

97(96) Legs I and IV usually with different number of claws. Genital opening often displaced anteriorly so that epimeral setae *2a* and *3a* in almost transverse alignment. Aquatic or semiaquatic. Length: 250–750µm (Figure 3.152).....................**ZETOMIMIDAE** (Pg. 240)
- Legs I and IV with same number of claws. Genital opening not displaced anteriorly; epimeral setae *2a* and *3a* not in transverse alignment. Usually terrestrial 98

98(97) Notogaster with posterior tectum. 10 pairs of notogastral setae 99
- Notogaster without posterior tectum. 10 to 15 pairs of notogastral setae 102

99(98) Lamellae completely fused medially or touching medially, or separate. Anterior of notogaster with or without hexagonal pattern. Length: 280–580µm (Figures 3.112–3.113) ....
.... **TEGORIBATIDAE** (Pg. 190)
- Lamellae present or absent, if present, never completely fused. Anterior of notogaster without hexagonal pattern 100

100(99) Lamellar seta inserted on prodorsal surface. Lamellar cusp absent; lamellar tooth present. Length: 290–700µm (Figure 3.142A–C) .... **CHAMOBATIDAE**: *Chamobates* (Pg. 229)
- Lamellar seta inserted on lamellae or on lamellar cusp. Lamellar cusps usually present, rarely absent... 101

101(100) Prolamella extending from tip of lamellae to rostral margin, positioned medial to rostral seta. Tibiae and tarsi I–IV with distoventral and proximoventral porose areas. Subcapitular mentum without tectum. Length: 600–936µm (Figure 3.142D–F) ....
.... **HUMEROBATIDAE**: *Humerobates* (Pg. 230)
- Prolamella absent. Tibiae and tarsi I–IV without porose areas. Subcapitular mentum with or without tectum. Length: 400–650µm (Figures 3.143–3.151)..
.... **PUNCTORIBATIDAE (=MYCOBATIDAE)** (Pg. 231)

102(98) Subcapitular mentum with tectum. Epimere IV neotrichous (4 or more pairs of setae). 10 pairs of minute notogastral setae or setae reduced to alveoli. Length: 1,000–1,200µm (Figure 3.142G,H).. **EUZETIDAE**: *Euzetes* (Pg. 230)
- Subcapitular mentum without tectum. Epimere IV with 2 or 3 pairs of setae. 10 to 15 pairs of minute to long notogastral setae. Length: 270–990µm (Figures 3.129D,E–3.141) ....
.... **CERATOZETIDAE** (Pg. 213)

FAMILIES, GENERA AND SPECIES OF ORIBATID MITES OF CANADA AND ALASKA

Infraorder Palaeosomata Grandjean, 1969 (from Norton and Behan-Pelletier 2009)

This group (also called Palaeosomatides, Bifemorata, and Bifemoratina) is assumed to comprise the most primitive Oribatida. They are small to relatively large (150–700µm), pale, and lack the strong cuticular sclerotization that characterizes most adult oribatid mites. However, all have inconspicuous, weak sclerites that are small to extensive, according to group, and these create several recognizable body forms.

Diagnosis: **Aesthenic** zone (wide dorsal region of soft, malleable cuticle between prodorsum and pronotaspis) present or absent. Setae birefringent only at their base. Prodorsum astegasime; narrow rostral tectum present; naso present or absent. Bothridial seta simple, straight at its base, as is bothridium. Lyrifissures as cupules; *ian* and *iad* always absent. If notogaster seems present (rarely), it is transversely divided. With rare exception, at least 17 pairs of opisthonotal setae present (p_4 present, pygidial neotrichy may occur), often with some setae much larger than others and conspicuously darkened. Setae d_2 and e_1 erectile in some groups and individually movable. Opisthonotal glands absent. Ventral plates poorly defined or absent; preanal plate absent. Legs with six free segments; all femora divided. Subcapitulum stenarthric or anarthric. Chelicerae chelate-dentate; sometimes with elongated digits. Chelicerae oriented vertically or obliquely; their bases not inserted into body as apodemes. Trägårdh's organ absent.

Biology: Males are unknown in the Palaeacaridae (Norton and Palmer 1991), but all other families appear to be bisexual.

Superfamily Archeonothroidea Grandjean, 1932

Family Archeonothridae Grandjean, 1932

Diagnosis: *Adult:* Naso present, with ventral protuberance. Prodorsum in midregion often with vague, internal sclerotized thickenings forming U or square shape. Larger exobothridial seta (*xa*) barbed, positioned below other (*xp*) and similar in length to interlamellar seta. Aesthenic zone usually absent. Opisthonotum with 17 pairs of setae, some large, darkly pigmented, none erectile. Notogaster absent, series of small, mostly paired sclerites present, each bearing 4 or fewer setae. Setae c_1 and c_2 on single unpaired sclerite. Subcapitulum with large unpaired ventral lip. Main part of chelicerae oriented almost vertically, with narrow, elongated digits. Legs tridactylous. Famulus large, erect, not broadened distally; basally positioned on tarsus I, adjacent to lyrifissure. Tibia II with 2 solenidia.

Biology: Walter and Proctor (1998) found fungal hyphae, spores and remains of small animals in the guts of some Australian archeonothrids.

Zachvatkinella Lange, 1954 (Figure 3.1E,F)
Type-species: *Zachvatkinella belbiformes* Lange, 1954
Note: Unidentified *Zachvatkinella* species known from BC.

Family Acaronychidae Grandjean, 1932

Diagnosis: *Adult:* Naso present, with ventral protuberance. Aesthenic zone usually absent. Prodorsum without thickenings. Larger exobothridial seta (*xa*) barbed, positioned below other (*xp*) and similar in length to interlamellar seta. Hysterosoma relatively short. Opisthonotum with series of small, weak, mostly paired sclerites, each bearing 4 or fewer setae; with 17 pairs of setae, some large and darkly pigmented, but none erectile. Setae of pair c_2 on individual small sclerites. Subcapitulum with small unpaired ventral lip. Cheliceral trochanter complete, main part of chelicera oriented almost vertically, with narrow, elongated digits. Palptarsus short, little longer than tibia. Legs tridactylous. Famulus large, erect, broadened distally; most basally positioned setiform organ on tarsus I, adjacent to lyrifissure. Tibia II with 2 solenidia.

Acaronychus Grandjean, 1932
Type-species: *Acaronychus trägårdhi* Grandjean, 1932

Acaronychus traegardhi Grandjean, 1932 (Figure 3.1A–D)
Distribution: BC [CO, NC; Holarctic, Oriental, Neotropical]

Superfamily Palaeacaroidea Grandjean, 1932

Family Palaeacaridae Grandjean, 1932

Diagnosis: *Adult:* Body dichoid. Naso absent. Aesthenic zone present; large. Bothridial seta filiform. Larger exobothridial seta (*xa*) positioned anterior to other (*xp*) and usually about half length of interlamellar seta. Opisthonotum with distinct pygidial shield bearing all dorsal setae posterior to row *e*. Several pairs of opisthonotal setae large, darkly pigmented, with d_2 and e_1 largest, erectile; h_1 and p_1 short, swollen, or spinelike. Most anterior pair of genital setae strongly modified. Subcapitulum without unpaired ventral lip. Chelicerae chelate-dentate, oriented obliquely, anteroventrally; digits not narrowed. Palptarsus at least twice tibial length; distal eupathidia barbed. Legs II–IV bidactylous. Famulus of tarsus I large, approximately equal in length to solenidia; inserted on conspicuous tubercle in proximal quarter or third of tarsus. Tibia II with 1 solenidion.

Palaeacarus Trägårdh, 1932
   Type-species: *Palaeacarus hystricinus* Trägårdh, 1932

Palaeacarus hystricinus Trägårdh, 1932 (Figure 3.1G–I)
   Biology: From forests, shrub tundra, mixed-wood boreal forest. Considered an edaphic species. This species has been recorded from bird feathers (Krivolutsky and Lebedeva 2004). The ribosomal 18S region (18S) and the nuclear elongation factor 1 alpha (*ef1*) genes of this species is discussed in Domes et al. (2007).
   Distribution: AK, YT, BC, ON, QC, NL [MD, NY; Holarctic, Oriental]
   Literature: Schatz (2004); Maraun et al. (2019).
   Note: Unidentified *Palaeacarus* species from AK, YT, QC, NS.

Superfamily Ctenacaroidea Grandjean, 1954

Family Aphelacaridae Grandjean, 1954
   Diagnosis: *Adult:* Body trichoid. Naso absent. Aesthenic zone present. Bothridial seta abruptly broadened, flattened distally, straight to strongly curved, shorter than interlamellar seta. Larger exobothridial seta (*xa*) positioned anterior to other (*xp*) and usually about half length of interlamellar seta or less. Opisthonotum with weakly defined notogaster; posterior shield bears all setae posterior to row *c*, with latter on separate anterior sclerite; separation between sclerites narrow or broad. Groove between setal rows *c* and *d*; c_1 longer than c_2. Opisthonotum usually neotrichous in pygidial region; rarely with 16 pairs. Notogastral setae large, d_2 and e_1 not erectile. Anterior pair of genital papillae absent. Adoral seta or_1 strongly barbed or pectinate, usually larger than other 2 setae. Chelicerae chelate-dentate, oriented obliquely, anteroventrally; digits not narrowed. Palptarsus elongated, at least twice tibial length. Palpfemur with 1 seta. Famulus of tarsus I much smaller than solenidia; inserted at midpoint or in distal half of tarsus. Tibia II with 1 solenidion. Tarsus III with 1 solenidion.

Aphelacarus Grandjean, 1932
   Type-species: *Parhypochthonius acarinus* Berlese, 1910

Aphelacarus acarinus (Berlese, 1910) (Figure 3.2D–F)
   Comb./Syn.: *Parhypochthonius acarinus* Berlese, 1910
   Biology: Common and often abundant in soils of deserts and dry coniferous forests, and in prairie soils. The elongated, trichoid body seems adapted to moving among soil pores. It also occurs in buildings in some more northerly regions.
   Distribution: BC, AB [CA, OR; Semicosmopolitan]
   Literature: Pachl et al. (2012); Maraun et al. (2019); Walter and Lumley (2021).

Family Ctenacaridae Grandjean, 1954
   Diagnosis: *Adult:* Body dichoid. Naso absent. Aesthenic zone present. Larger exobothridial seta (*xa*) positioned anterior to other (*xp*) and about half length of interlamellar seta or less. Opisthonotum with weakly defined notogaster; posterior shield bearing all setae posterior to

row c; with latter on separate anterior sclerite; separation between sclerites narrow or broad. Setae of various form, large and darkly pigmented or not, d_2 and e_1 erectile; c_1 shorter than c_2; setae h_1 and p_1 not unusually short. Anterior pair of genital papillae absent. Adoral seta or_1 strongly barbed or pectinate, usually larger than other 2 setae. Chelicera chelate-dentate, oriented obliquely, anteroventrally; digits not narrowed. Palptarsus elongated, at least twice tibial length; eupathidia forked (fused). Palpfemur with 1 seta. Famulus of tarsus I much smaller than solenidia; inserted at midpoint or in distal half of tarsus. Tibia II with 1 solenidion. Tarsus III with 1 solenidion. Legs II–IV tridactylous, with empodial claw well developed or much reduced.

Beklemishevia Zachvatkin, 1945 (Figure 3.2A–C)
Type-species: *Beklemishevia galeodula* Zachvatkin, 1945
Note: Unidentified *Beklemishevia* species known from ON.

Infraorder Enarthronota Grandjean, 1969 (modified from Norton and Behan-Pelletier 2009)
This group of early-derivative oribatid mites (= Enarthronotides, Arthronota, Arthronotina) shows extensive plasticity in body form (Norton 2001). They range widely in size (100–1,000µm), but most are small. The infraorder includes oribatid mites with strong yellow, orange, or red pigmentation. Notogaster subdivided by transverse scissures to form plates, which are variously regressed, hypertrophied, or fused in different ways, and articulations between plates have been functionally modified (Norton 2001). For example, all transverse articulations are fused in members of Lohmanniidae, a ptychoid body form has evolved independently in Mesoplophoridae and Protoplophoridae, and paedomorphic loss of most sclerotization, including all notogastral plates, characterizes the Pediculochelidae.
Diagnosis: *Adult*: Body form typically dichoid; two families ptychoid. Prodorsum usually with well-developed rostral tectum but may be astegasime. Eyes weakly developed or absent. Bothridium simple and straight, or with strong proximal curve. Notogaster with longitudinal suprapleural scissure usually partly or fully isolating paired pleuraspis from dorsal notaspis. Notaspis comprising 2 to 4 dorsal plates separated by 1–3 transverse scissures of 3 possible types in various combinations (Norton 2001). In Brachychthoniidae and Haplochthoniidae all scissures are type-E: simple, narrow articulations (Figure 2.4E). Type-S scissures (Figure 2.4F) ancestrally carrying hypertrophied erectile setae, but setae reduced or vestigial in Hypochthoniidae. Erectile setae present in one (Atopochthoniidae) or, usually, 2 rows (e, f); in 3 rows in Nanohystricidae (e, f, h). These directed posteriorly when at rest, but quickly raised in concert when mite is disturbed. Telescoping, type-L scissures (Figure 2.4G) present in two nonptychoid families, Eniochthoniidae and Sphaerochthoniidae, each of which is sister group of a ptychoid family (Mesoplophoridae and Protoplophoridae, respectively). Most families with 16 pairs of notogastral setae, although some may be vestigial; neotrichy only in Lohmanniidae. Opisthonotal glands absent. Notogastral lyrifissures well formed. Epimera fused in various ways, but coxisternum never fused across sejugal furrow into united sclerite. Preanal plate present in some groups. Peranal segment, along with its paired paraproctal plate and seta, present or absent. Lyrifissures *ian* and *iad* always absent. Legs with 5 free segments, with flagellate solenidia and solenidion-seta coupling occurring in some groups. Subcapitulum anarthric, with few exceptions. Trägårdh's organ absent.
Biology: Fungi and minute organic fragments appear to be common food items, but necrophagy is also reported. Most enarthronotan families are parthenogenetic, with sexuality known only in most members of Mesoplophoridae and Protoplophoroidea (Norton et al. 1993), and in Nanohystricidae (Norton and Fuangarworn 2015).

The monophyly of Enarthronota has only weak morphological support (Norton 1984); molecular evidence suggests paraphyly with respect to Palaeosomata (Pachl et al. 2012).

Superfamily Brachychthonioidea Thor, 1934

Family Brachychthoniidae Thor, 1934

Diagnosis: *Adult:* Body dichoid, pale or pigmented, often yellow, orange, or red; usually measuring less than 270µm in length. Prodorsum stegasime. Naso not evident, but remnant may exist in form of eyes underneath rostrum; paired lateral eye present or absent in exobothridial region. Bothridium and base of bothridial seta sharply bent. One pair of exobothridial setae present. Notogaster with 16 pairs of setae, none erectile. With 2 transverse type-E scissures between setal rows *d–e* and *e–f*, respectively, such that pronotaspis bears 6 pairs of setae (*c* and *d* setae); pleural region comprised of multiple plates. Epimere II with 1 pair of setae; epimere III fused to IV but medially divided. Genital plates and combined anal/peranal plates subequal in size. Narrow peranal plates with 1 pair of setae. With 3 pairs of genital papillae. With 1 pair of aggenital and 2 pairs of anal setae. Preanal plate absent. Chelicerae with teeth absent or very small. Legs monodactylous. Trochanters I–II without setae. Tarsal solenidia ω_2 and ω_3 coupled, fully attached; tibia I solenidion flagellate; genu I with 2 solenidia; seta *d* coupled to solenidion on most genua and tibiae.

Biology: Brachychthoniid mites are diverse and abundant in moist forest soil and litter, and are common in disclimax communities, newly created habitat and grassland. Luxton (1972) considered members of this family microphytophagous. Walter (1987) cultured a species of *Brachychthonius* through several generations on algae and fungi.

KEY TO GENERA OF BRACHYCHTHONIIDAE OF CANADA AND ALASKA
(MODIFIED FROM NORTON OSU MITE KEYS)

1   Rostrum anteriorly with 3–5 large anteromedial teeth in addition to small marginal serration and large compound tooth. Dorsal hystersomal shields and prodorsum distinctly ornamented with concavities, irregular borders of which accentuated by small pits appearing clear in transmitted light. Genital setation usually 6 pairs, 3 in medial row. Suprapleural plate SpE present and divided by more or less vertical suture. Pleural plate connected to pygidium, with suture clearly marking attachment (Figure 3.3) ....
.... ***Synchthonius*** van der Hammen

–   Rostrum without large anterior teeth, with or without marginal serration and lateral compound tooth. Dorsal ornamentation different, or absent. Genital setation 7 pairs, 4 in medial row. Suprapleural plate SpE absent, or if present, not divided by suture. Pleural plate separated from pygidium by narrow or broad unsclerotized band 2

2(1)   Seta d_2 inserted near margin of shield NA. Aggenital plate present, usually bearing seta *ag*, occasionally indistinct. Adanal plates independent posteriorly. With well sclerotized prodorsum and hysterosomal plates, usually ornamented (not in *Eobrachychthonius*). Rostrum usually with marginal serration 3

–   Seta d_2 inserted more mediad, approximately half-way between d_1 and margin of shield NA. Aggenital plate absent; seta *ag* inserted on unsclerotized integument. Adanal plates usually fused posteriad of anal plates (independent in *Neoliochthonius*). Prodorsum and hysterosomal plates only weakly sclerotized, smooth; only rarely with poorly developed concavities (*Neobrachychthonius*). Rostrum without marginal serration 6

3(2)   Without distinct dorsal ornamentation (muscle sigilla may appear by transparency as pale spots on sclerotized plates). With lateral border of shields NA, NM and pygidium thickened, produced as distinct sutures separating region bearing setae c_2, *cp*, d_2, e_2 and f_2 from remainder of respective shield. Four pairs of suprapleural plates on hysterosoma (SpD present); with small prosomal supraplural plate and 2 podopleural plates. Aggenital plate very large, triangular; seta *ag* inserted at its posterior tip, or on small, adjacent but separate sclerite (Figure 3.4).. ***Eobrachychthonius*** Jacot

– Usually with distinct pattern of dorsal concavities distributed in medial and paired lateral regions; lateral pattern of shield NA often in form of rosette, with central pit ("cuticular ring"). Without distinct, sharp ridge near lateral margin of dorsal shields, but shield edges (bearing the setae) often marginally thickened and somewhat raised. Three (SpD absent) or 2 (SpD and SpF absent) pairs of suprapleural plates; without prosomal or podosomal plates in pleural region. Aggenital plate medium to small, or if large, *ag* inserted nearer middle 4

4(3) Central region of shield NA with concavities united longitudinally into 1 principal long pattern. Seta p_2 laterad or ventrolaterad of p_1; pygidium with only 3 pairs of setae in medial row. Hysterosoma without distinct "ridge and valley" structure; posterior "tubercles" absent (Figure 3.5A–C) .... ***Poecilochthonius*** Balogh

– Central region of shield NA with concavities in distinct groups, usually separated by 1 or more narrow regions of undepressed integument. Seta p_2 shifted posteromediad, below p_1; pygidium in dorsal aspect with 4 pairs of setae in medial row. Hysterosomal dorsum with medial paired setal rows (c_1, d_1, e_1, etc.) on low ridges separating "valleys" of central and lateral concavity regions; posteriorly, these ridges usually appearing as pygidial "tubercles" 5

5(4) Hysterosoma with 2 suprapelural plates (SpF absent). Adanal setae ad_2 and ad_3 thin or only slightly thickened, relative to ad_1; ad_1 as long as, or only slightly shorter than, ad_2 (Figure 3.6) .... ***Brachychthonius*** (sensu stricto) Berlese

– Hysterosoma with 3 suprapleural plates (SpF present). Adanal setae ad_2 and ad_3 enlarged, saberlike, with thin unilateral vane; ad_1 thin, simple, much shorter than ad_2 (Figure 3.7) .. .. ***Sellnickochthonius*** Krivolutsky

6(2) Genua III and IV each with 3 setae; *d* (coupled with solenidion σ), *l'* and *v'*. Epimere IV with 4 pairs of setae; subcapitulum venter with 4 pairs (not including adoral setae), m_2 present. Without separate suprapleural plate SpC; seta c_3 inserted on shield NA, usually on anterolateral lobe (representing SpC) fully connected to NA 7

– Genua III and IV each with 2 setae (*v'* absent). Epimere IV with 3 pairs of setae; subcapitulum venter with 3 pairs (not including adorals), m_2 absent. With separate suprapleural plate SpC, bearing seta c_3 9

7(6) Adanal seta ad_2 similar to ad_1 and ad_3, inserted approximately midway between them. Genital plate longer than anal-peranal group.. 8

– Adanal seta heterogeneous: ad_2 enlarged, saber-like, with unilateral vane; ad_1 and ad_3 simple, setiform; ad_3–ad_2 distance half or less that of ad_2–ad_1. Genital plates shorter than anal-peranal group (Figures 3.8, 3.9A–D)......................... ***Liochthonius*** van der Hammen

8(7) All dorsal setae ciliate, feather-like. Seta p_2 inserted laterad of p_1. Pygidium with 3 pairs of setae in medial row. Shield NA without anterolateral lobe-like projection (Figure 3.9E–H)... ***Mixochthonius*** Niedbała

– All dorsal setae simple; piliform or somewhat lanceolate. Seta p_2 inserted ventrad of p_1. Pygidium with 4 pairs of setae in medial row (including p_2). Shield NA with anterolateral lobe, bearing seta c_3 (Figure 3.5D–F)... ***Verachthonius*** Moritz

9(6) Prodorsum distinctly flattened dorsoventrally; rostral setae inserted on dome-like convexity. Pygidium dorsally flattened and laterally angular, with resulting ridge-like edge reaching posteriad to near seta p_2; lateral angle continued anteriad as projecting margins of shields NM and NA. Suprapleural plate SpE present. Adanal plates fused behind anal plates. Dorsal hysterosomal setae usually lanceolate; interlamellar seta distinctly shorter than rostral seta; bothridial seta broadly spindle-shaped. Prodorsum and hysterosomal shields with shallow, inconspicuous ornamentation, distributed as in *Brachychthonius*. Subcapitulum with 2 pairs of adoral setae, or_1 large, thick distally, or_2 minute and pointed (Figure 3.5G–I)... ***Neobrachychthonius*** Moritz

– Prodorsum not flattened; without rostral "dome". Pygidium rounded laterally; margins of NA and NM not projecting. Suprapleural plate SpE absent. Adanal plates separate, independent posteriorly. Dorsal hysterosomal setae attenuate or somewhat lanceolate, but longer; interlamellar seta approximately equal in length to rostral seta; bothridial seta subcapitate. Prodorsum and hysterosomal dorsum with essentially smooth, even surface. Subcapitulum with single tapered adoral seta (Figure 3.10A–E)....... ***Neoliochthonius*** Lee

Brachychthonius Berlese, 1910
    Type-species: *Brachychthonius berlesei* Willmann, 1928
    Comb./Syn.: *Brachychochthonius* Jacot, 1938; *Poecilochthonius* Balogh, 1943, in part

KEY TO ADULT *BRACHYCHTHONIUS* OF CANADA AND ALASKA

1   Prodorsal setae (other than bothridial seta) and notogastral setae ciliate. Bothridial seta strongly clavate with rows of cilia. NA with 2 central fields. Length: 157–173µm (Figure 3.6A,B) .... ***B. bimaculatus*** Willmann
–   Prodorsal setae (other than bothridial seta) and notogastral setae smooth. Bothridial seta strongly clavate or fusiform, but with at most short barbs. NA with 3 central fields 2

2(1)   Notogastral shield PY without 4 distinct medial cells in transverse row. Length: 175–186µm (Figure 3.6C). .... ***B. pius*** Moritz
–   Notogastral shield PY with 4 distinct medial cells in transverse row 3

3(2)   Median paired fields of notogastral shield NA partially separated (with middle line) 4
–   Median paired fields of notogastral shield NA fused, without middle line 5

4(3)   Notogastral setae acuminate. Notogastral shield NM with 3 central fields. Bothridial seta strongly clavate. Length: 175–192µm (Figure 3.6D) .... ***B. impressus*** Moritz
–   Notogastral setae tapered to clavate. NM with 2 pairs of central fields. Bothridial seta setiform. Length: ca. 160µm (Figure 3.6E) .... ***B. jugatus*** (Jacot)

5(3)   Bothridial seta clavate. Length: 196–212µm (Figure 3.6F) .... ***B. berlesei*** Willmann
–   Bothridial seta setiform. Length: ca. 174µm (Figure 3.6G) .... ***B. berlesei erosus*** (Jacot)
     Note: Unidentified and possibly undescribed *Brachychthonius* species from BC, AB, ON, QC, NS.

Brachychthonius berlesei Willmann, 1928 (Figure 3.6F)
    Comb./Syn.: *Brachychthonius brevis* Berlese, 1910; non Michael, 1888; *Brachychochthonius jugatus* Jacot, 1938
    Biology: Found in tundra, swamps, forests, fescue prairie. Microphytophagous (Borcard 1991). Specimens have been found in the feathers of birds in Russia (Krivolutsky and Lebedeva 2004).
    Distribution: NU, AB [NC, NY; semicosmopolitan]
    Literature: Weigmann (2006); Maraun et al. (2019).

Brachychthonius berlesei erosus (Jacot, 1938) (Figure 3.6G)
    Comb./Syn.: *Brachychochthonius berlesei erosus* Jacot, 1938
    Biology: From mixed deciduous and coniferous litter.
    Distribution: QC [NC; Nearctic]

Brachychthonius bimaculatus Willmann, 1936 (Figure 3.6A,B)
    Comb./Syn.: *Brachychthonius helveticus* Schweizer, 1956
    Biology: From coniferous litter; fescue prairie.
    Distribution: BC, AB, ON, QC [CA; Holarctic]
    Literature: Pachl (2010).

Brachychthonius impressus Moritz, 1976 (Figure 3.6D)
    Biology: Boreal species.
    Distribution: BC [Holarctic, Ethiopia]
    Literature: Weigmann (2006); Pachl (2010); Maraun et al. (2019).

Brachychthonius jugatus (Jacot, 1938) (Figure 3.6E)
    Comb./Syn.: *Brachychochthonius jugatus* Jacot, 1938; *Sellnickochthonius jugatus* (Jacot, 1938)
    Biology: From tundra; mixed deciduous and coniferous forest.
    Distribution: NU, AB, QC [MI, NC, NY; Nearctic]

Brachychthonius pius Moritz, 1976 (Figure 3.6C)
    Comb./Syn.: *Brachychochthonius berlesei erosus* sensu Hammer, 1952; non Jacot, 1938
    Biology: From arctic and subarctic; from pasture.
    Distribution: NT, NU, AB [NY; Holarctic]
    Literature: Maraun et al. (2019).

Eobrachychthonius Jacot, 1936
    Type-species: *Brachychthonius latior* Berlese, 1910 (=*Eobrachychthonius sexnotatus* Jacot, 1936)
    Literature: Moritz (1976a,b).

KEY TO ADULT *EOBRACHYCHTHONIUS* OF CANADA AND ALASKA

1   Aggenital plate whole, not subdivided. Length: 270–370µm (Figure 3.4A,B) ....
    .... *E. latior* (Berlese)
–   Aggenital plate divided, aggenital setae on smaller posterior part 2

2(1) Notogaster with few depressions. Interlamellar setae shorter than their mutual distance.
      Length: 199–215µm (Figure 3.4C,D) .... *E. borealis* Forsslund
–   Notogaster with many depressions. Interlamellar setae longer than their mutual distance.
      Length: 250–335µm (Figure 3.4E,F)............................... *E. oudemansi* van der Hammen
      Note: Unidentified and possibly undescribed *Eobrachychthonius* species from BC.

Eobrachychthonius borealis Forsslund, 1942 (Figure 3.4C,D)
    Biology: From arctic and subarctic.
    Distribution: AK, YT, NT [Holarctic]
    Literature: Seniczak and Seniczak (2017).

Eobrachychthonius latior (Berlese, 1910) (Figure 3.4A,B)
    Comb./Syn.: *Brachychthonius latior* Berlese, 1910; *Brachychthonius brevis glabra* Thor, 1930; *Eobrachychthonius sexnotatus* Jacot, 1936; *Brachychthonius grandis* Sellnick, 1944
    Biology: Arctic and subarctic species; also found in oligotrophic swamps and areas with ericaceous shrubs (Borcard 1991). Gravid females carry 1 egg.
    Distribution: AK, YT, NT, AB, MB, ON, QC [CA, FL, NC, NY, VA; Holarctic, India]
    Literature: Seniczak and Seniczak (2017); Maraun et al. (2019); ABMI (2019).

Eobrachychthonius oudemansi van der Hammen, 1952 (Figure 3.4E,F)
Comb./Syn.: *Brachychthonius oudemansi* (Hammen, 1952); *Eobrachychthonius argentinensis* Hammer, 1958; *Brachychthonius laetepictus* sensu Willmann, 1931; non Berlese, 1910
Biology: From moist, rich tundra. Found close to springs (Borcard 1991). West (1982) described a two-year life cycle in the Antarctic. Gravid females carry 1 egg.
Distribution: AK, NT [Semicosmopolitan]
Literature: Seniczak and Seniczak (2017).

Liochthonius van der Hammen, 1959
  Type-species: *Hypochthonius brevis* Michael, 1888
  Comb./Syn.: *Brachychthonius*, in part

KEY TO ADULT *LIOCHTHONIUS* OF CANADA AND ALASKA

1	Notogastral setae bilaterally strongly widened, usually phylliform. Length: 156–159µm (Figure 3.8A) ....***L. forsslundi*** (Hammer)	
–	Notogastral setae not widened, or with narrow marginal membrane unilaterally.............. 2	
2(1)	Bothridial seta fusiform, slightly rounded distally or terminating in a single point 3	
–	Bothridial seta clavate, truncate distally or excised, forming 2 terminal points 7	
3(2)	Notogastral setae long, such that seta c_1 extending well beyond insertion of d_1. Length: 175–218µm (Figure 3.8B) .... ***L. hystricinus*** (Forsslund)	
–	Notogastral setae shorter, such that seta c_1 at most approaching insertion of d_1 4	
4(3)	Notogastral setae very thin, acicular; c_1 approaching insertion of d_1. Length: 192–205µm (Figure 3.8C) .... ***L. tuxeni*** (Forsslund)	
–	Notogastral setae thicker, more robust; c_1 short of insertion of d_1 5	
5(4)	Notogastral seta c_1 longer than half distance between c_1 and d_1 6	
–	Notogastral setae c_1 shorter than half distance c_1 to d_1. Length: 166–177µm (Figure 3.8D) ....***L. simplex*** (Forsslund)	
6(5)	Interlamellar setae distinctly longer than their mutual distance. Length: 160–223µm (Figure 3.8E)..***L. brevis*** (Michael)	
–	Interlamellar setae at most equal to their mutual distance. Length: 160–175µm (Figure 3.8F)..***L. leptaleus*** Moritz	
7(2)	Prodorsum with ridge between lamellar setae. Transverse ridge present posterior to setae f_1. Body brownish. Length: 200–237µm (Figure 3.9A,B) .... ***L. sellnicki*** (Thor)	
–	Prodorsum without ridge between lamellar setae. Transverse ridge absent posterior to setae f_1.. 8	
8(7)	Head of bothridial seta very long, longer than stalk and 3× longer than wide. Distance between notogastral setal pair c_1 subequal to that of notogastral setal pair e_1. Length: 200–228µm (Figure 3.9C) ....***L. muscorum*** Forsslund	
–	Head of bothridial seta subequal in length to stalk and 2.5× longer than wide. Distance between notogastral setal pair c_1 greater than that of notogastral setal pair e_1. Length: 175–200µm (Figure 3.9D) ....***L. lapponicus*** (Trägårdh)	

Note: Unidentified and undescribed *Liochthonius* species from BC, AB, MB, ON, QC, NS.

Liochthonius brevis (Michael, 1888) (Figure 3.8E)
    Comb./Syn.: *Hypochthonius brevis* Michael, 1888; *Brachychthonius perpusillus* Berlese, 1910
    Biology: From tundra and forest litter.
    Distribution: AK, NT, NU, BC, ON, QC [CA, FL, NC, NY; Holarctic, northern Neotropical]
    Literature: Seniczak and Seniczak (2017); Maraun et al. (2019); Pachl et al. (2021).

Liochthonius forsslundi (Hammer, 1952) (Figure 3.8A)
    Comb./Syn.: *Brachychthonius forsslundi* Hammer, 1952
    Biology: From tundra.
    Distribution: AK, YT, NT [Nearctic]

Liochthonius hystricinus (Forsslund, 1942) (Figure 3.8B)
    Comb./Syn.: *Brachychthonius hystricinus* Forsslund, 1942; *Brachychthonius ocellatus* Hammer, 1952
    Biology: From tundra; hemlock forest.
    Distribution: NT, QC [NY; Holarctic, Chile]
    Literature: Weigmann et al. (2015).

Liochthonius lapponicus (Trägårdh, 1910) (Figure 3.9D)
    Comb./Syn.: *Hypochthonius brevis lapponica* Trägårdh, 1910
    Biology: From tundra; mixed deciduous and coniferous forest; fescue prairie. This species has been found on a gull, *Larus hyperboreus*, at Spitsbergen (Lebedeva et al. 2012).
    Distribution: AK, YT, NT, NU, BC, AB, MB, ON, QC, NL [MI, NY; Holarctic]
    Literature: Weigmann et al. (2015).

Liochthonius leptaleus Moritz, 1976 (Figure 3.8F)
    Biology: From fescue prairie.
    Distribution: AB [Holarctic, Chile]
    Literature: Weigmann et al. (2015).

Liochthonius muscorum Forsslund, 1964 (Figure 3.9C)
    Biology: From coniferous forest litter; arctic tundra.
    Distribution: BC [Holarctic]
    Literature: Bayartogtokh et al. (2011).

Liochthonius sellnicki (Thor, 1930) (Figure 3.9A,B)
    Comb./Syn.: *Brachychthonius sellnicki* Thor, 1930; *Brachychthonius scalaris* Forsslund, 1942; *Brachychthonius nodosus* Willmann, 1952; *Brachychthonius brevis* sensu Hammer, 1944; non Berlese, 1910
    Biology: From tundra; coniferous litter. Specimens found in the feathers of birds in Russia (Krivolutsky and Lebedeva 2004).
    Distribution: AK, YT, NT, NU, BC, AB, MB, ON, QC [NY; Holarctic, China]
    Literature: Maraun et al. (2019).

Liochthonius simplex (Forsslund, 1942) (Figure 3.8D)
    Comb./Syn.: *Brachychthonius simplex* Forsslund, 1942
    Biology: From tundra; coniferous forest litter.
    Distribution: AK, YT, NT, NU, BC, AB, QC [Semicosmopolitan]
    Literature: Weigmann et al. (2015); Maraun et al. (2019).

Liochthonius tuxeni (Forsslund, 1957) (Figure 3.8C)
    Comb./Syn.: *Brachychthonius hystricinus tuxeni* Forsslund, 1957
    Biology: From coniferous forest litter.
    Distribution: BC [Holarctic]
    Literature: Weigmann et al. (2015); Maraun et al. (2019).

Mixochthonius Niedbała, 1972
   Type-species: *Brachychthonius pilososetosus* Forsslund, 1942

KEY TO ADULT *MIXOCHTHONIUS* OF CANADA AND ALASKA

1 | Interlamellar setae reaching only half way to insertion of lamellar setae. NM with pair of large depressions; pleural region with deep indentation. Length: 173–180µm (Figure 3.9E) ....*M. concavus* (Chinone)
– | Interlamellar setae almost reaching insertion of lamellar setae. NM without pair of depressions; pleural region without indentation. Length: 145–161µm (Figure 3.9F–H) ....*M. pilososetosus* (Forsslund)
   Note: *Mixochthonius* nr. *concavus* found in AB (Lindo and Visser 2004).

Mixochthonius pilososetosus (Forsslund, 1942) (Figure 3.9F–H)
   Comb./Syn.: *Brachychthonius pilososetosus* Forsslund, 1942; *Brachychthonius perpusillus* sensu Willmann, 1928; non Berlese, 1910
   Biology: From oligotrophic swamps; in mixed deciduous and coniferous litter. Microphytophagous (Borcard 1991).
   Distribution: QC [Holarctic]
   Literature: Corral-Hernández et al. (2016).

Neobrachychthonius Moritz, 1976
   Type-species: *Brachychthonius marginatus* Forsslund, 1942 (Figure 3.5G–I)
   Note: Unidentified and possibly undescribed *Neobrachychthonius* species from BC; one found in large numbers in canopy litter (Lindo and Stevenson 2007).

Neoliochthonius Lee, 1982 (=*Paraliochthonius* Moritz, 1976, preoccupied name)
   Type-species: *Brachychthonius piluliferus* Forsslund, 1942

KEY TO ADULT *NEOLIOCHTHONIUS* OF CANADA AND ALASKA

1 | Interlamellar seta shorter than head of bothridial seta. Length: 123–166µm (Figure 3.10A–D)..*N. piluliferus* (Forsslund)
– | Interlamellar seta longer than head of bothridial seta. Length: 182–210µm (Figure 3.10E) ....*N. occultus* (Niedbała)

Neoliochthonius occultus (Niedbała, 1971) (Figure 3.10E)
   Comb./Syn.: *Liochthonius occultus* Niedbała, 1971
   Biology: From coniferous and deciduous forest.
   Distribution: BC [Holarctic]
   Literature: Seniczak et al. (2019).

Neoliochthonius piluliferus (Forsslund, 1942) (Figure 3.10A–D)
   Comb./Syn.: *Brachychthonius piluliferus* Forsslund, 1942; *Paraliochthonius piluliferus* (Forsslund, 1942)
   Biology: From fescue prairie, and deciduous forest.
   Distribution: AB [NY; Holarctic]
   Literature: Seniczak et al. (2019).

Poecilochthonius Balogh, 1943
   Type-species: *Brachychthonius brevis italicus* Berlese, 1910
   Comb./Syn.: *Brachychochthonius*, in part

Poecilochthonius spiciger (Berlese, 1910) (Figure 3.5A–C)
  Comb./Syn.: *Brachychthonius brevis spiciger* Berlese, 1910; *Brachychochthonius italicus spiciger* (Berlese, 1910)
  Biology: From boreal forest.
  Distribution: ON, QC [FL, NY; Holarctic, Neotropical, Oriental]
  Literature: Weigmann et al. (2015); Maraun et al. (2019).

Sellnickochthonius Krivolutsky, 1964
  Type-species: *Brachychthonius zelawaiensis* Sellnick, 1928
  Comb./Syn.: *Brachychochthonius*, in part; *Brachychthonius*, in part; *Poecilochthonius*, in part

KEY TO ADULT *SELLNICKOCHTHONIUS* OF CANADA AND ALASKA

1 Some notogastral setae widened like blade, or phylliform. Length: 148–171µm (Figure 3.7A) .. .. *S. zelawaiensis* (Sellnick)
– Notogastral setae thin, at most slightly fusiform, sparsely ciliate 2

2(1) Notogastral shield NA with median field pairs coalesced forming 2 larger fields 3
– Notogastral shield NA with at least 4 of 6 median field pairs separated in midline 4

3(2) Median notogastral fields with sinuous and sharply defined borders. Length: 136–153µm (Figure 3.7B,C) .... *S. lydiae* Jacot
– Median notogastral fields with borders not sharply defined. Length: 170–195µm (Figure 3.7D) .... *S. immaculatus* (Forsslund)

4(2) Prodorsum laterally with deep incision behind rostrum. Length: 175–205µm (Figure 3.7E).. .... *S. rostratus* (Jacot)
– Prodorsum laterally without incision behind rostrum 5

5(4) Notogastral fields with smooth integument. Length: 145–155µm (Figure 3.7F) .... *S. furcatus* Weis-Fogh
– Notogastral fields with dotted integument. Length: 157–162µm (Figure 3.7G) .... *S. suecicus* Forsslund

  Note: Unidentified and possibly undescribed *Sellnickochthonius* species from BC, AB, NS.

Sellnickochthonius furcatus (Weis-Fogh, 1948) (Figure 3.7F)
  Comb./Syn.: *Brachychthonius* (*Brachychochthonius*) *furcatus* Weis-Fogh, 1948
  Biology: From fescue prairie.
  Distribution: AB [Holarctic]
  Literature: Weigmann et al. (2015); Corral-Hernández et al. (2016).

Sellnickochthonius immaculatus (Forsslund, 1942) (Figure 3.7D)
  Comb./Syn.: *Brachychochthonius immaculatus* Forsslund 1942; *Brachychthonius semiornatus* Evans, 1952; *Brachychochthonius arcticus* Hammer, 1952
  Biology: From tussock tundra; coniferous and deciduous forest litter.
  Distribution: AK, YT, NT, NU, BC, AB, QC [MI, NY; Holarctic, Neotropical]
  Literature: Pachl (2010); Corral-Hernández et al. (2016); Maraun et al. (2019).

Sellnickochthonius lydiae (Jacot, 1938) (Figure 3.7B,C)
  Comb./Syn.: *Brachychochthonius lydiae* Jacot, 1938
  Biology: From coniferous and deciduous forest soils (Reeves and Marshall 1971).
  Distribution: QC [NH, NC, NY; Holarctic]
  Literature: Maraun et al. (2019).

Taxonomic Keys

Sellnickochthonius rostratus (Jacot, 1936) (Figure 3.7E)
    Comb./Syn.: *Brachychthonius rostratus* Jacot, 1936
    Biology: From boreal forest; mixed deciduous and coniferous forest.
    Distribution: YT, NT, AB, QC [CA, NC; Holarctic, China]
    Literature: Weigmann et al. (2015); Corral-Hernández et al. (2016).

Sellnickochthonius suecicus (Forsslund, 1942) (Figure 3.7G)
    Comb./Syn.: *Brachychochthonius jugatus suecica* Forsslund, 1942; *Brachychthonius jugatus* sensu Niedbała, 1972; non Jacot, 1938
    Biology: From tundra; deciduous and coniferous forest.
    Distribution: YT, NT, NU, BC, AB, ON, QC [NY; Semicosmopolitan]
    Literature: Maraun et al. (2019).

Sellnickochthonius zelawaiensis (Sellnick, 1928) (Figure 3.7A)
    Comb./Syn.: *Brachychthonius zelawaiensis* Sellnick, 1928
    Biology: From mixed deciduous and coniferous forest. In oak litter (Lamoncha and Crossley 1998).
    Distribution: ON, QC [NY, NC; Holarctic, Neotropical]
    Literature: Maraun et al. (2019).

Synchthonius van der Hammen, 1952
Type-species: *Brachychochthonius crenulatus* Jacot, 1938

KEY TO ADULT *SYNCHTHONIUS* OF CANADA AND ALASKA

1   Notogastral seta c_1 at most reaching half distance to insertion of d_1 and shorter than distance c_1–c_1. Lateral rostral teeth serriform. Length: 195–206µm (Figure 3.3A–C)............
    ....***S. crenulatus*** (Jacot)
–   Notogastral seta c_1 reaching insertion of d_1 and equal in length to c_1–c_1. Lateral rostral teeth ressembling fleur-de-lis. Length: 205–230µm (Figure 3.3D–F)................................
    .... ***S. elegans*** Forsslund
    Note: Unidentified *Synchthonius* species from BC.

Synchthonius crenulatus (Jacot, 1938) (Figure 3.3A–C)
    Comb./Syn.: *Brachychochthonius crenulatus* Jacot, 1938; *Synchthonius boschmai* Hammen, 1952
    Biology: From mixed deciduous and coniferous forest.
    Distribution: BC, AB, ON, QC, NL [MI, NC, NY; Holarctic]
    Literature: Maraun et al. (2019); Schatz (2021).

Synchthonius elegans Forsslund, 1957 (Figure 3.3D–F)
    Biology: From forest litter.
    Distribution: AK, BC, AB [Holarctic]

Verachthonius Moritz, 1976
    Type-species: *Brachychthonius laticeps* Strenzke, 1951

Verachthonius montanus (Hammer, 1952) (Figure 3.5D–F)
    Comb./Syn.: *Eobrachychthonius montanus* Hammer, 1952
    Biology: From mixed deciduous and coniferous forest; fescue prairie.
    Distribution: AB, ON [Nearctic, Neotropical]

Superfamily Atopochthonioidea Grandjean, 1949

Family Atopochthoniidae Grandjean, 1949
Diagnosis: *Adult*: Dichoid; body pale; sclerotization weak to absent. Prodorsum astegasime or weakly stegasime; naso not evident. Interlamellar setae shield-like, as broad as long, foveate. Bothridium and bothridial seta with sharp bend at base. Exobothridial setae 2 pairs. Notogaster with 2 transverse depressions; most notogastral setae short, broad; with 2 pairs of erectile setae inserted on large, independent tubercles. Epimere II with 1 pair of setae. Genital setation 6 pairs. Anal setation 3 pairs; adanal setation 4 pairs; peranal plates absent, preanal plate absent. Chelicerae narrow, straight, with both digits simple, tapering, and with small, indistinct teeth. Palp tarsus with forked distal eupathid, similar in length to solenidion. Rutellum elongated, with narrow base, broad and toothed distally. Tarsus III with or without solenidion. Legs bidactylous.

Atopochthonius Grandjean, 1949
Type-species: *Atopochthonius artiodactylus* Grandjean, 1949

Atopochthonius artiodactylus Grandjean, 1949 (Figure 3.12C,D)
Biology: From forest floor litter.
Distribution: BC, ON, NS [NY, OR; Holarctic, Oriental]

Family Pterochthoniidae Grandjean, 1950
Diagnosis: *Adult*: Dichoid; body pale; sclerotization weak. Prodorsum astegasime or weakly stegasime; naso not evident. Bothridium and bothridial seta with sharp bend at base. Exobothridial setae 2 pairs. Notogaster with 3 transverse scissures. All dorsal setae (other than p_1, p_2) shield-like, similar in shape and mostly of similar size, none erectile or inserted on large tubercles. Chelicerae with large, functional trochanter and distinct ventral bend; fixed digit broad, poorly defined, with teeth directed anteriorly, rake-like, and movable digit simple, narrow, without teeth. Palptarsus terminating in long, ribbon-shaped eupathidia. Rutellum minute, conical, smooth. Epimere II with 1 pair of setae. Genital setae 8 pairs. Anal setae 3 pairs; adanal setae 4 pairs. Peranal plates present; preanal plate absent. Leg genua apparently without solenidia (vestige may be present on genu I). Legs monodactylous.

Pterochthonius Berlese, 1913
Type-species: *Cosmochthonius angelus* Berlese, 1910

Pterochthonius angelus (Berlese, 1910) (Figure 3.12A,B)
Comb./Syn.: *Cosmochthonius angelus* Berlese, 1910
Biology: This minute, ornate mite is common in aspen forest litter.
Distribution: BC, AB, QC [MO, NY; Semicosmopolitan]
Literature: Maraun et al. (2019).

Superfamily Hypochthonioidea Berlese, 1910
Biology: All known species considered thelytokous (Fuangarworn and Norton 2013).

Family Eniochthoniidae Grandjean, 1947
Diagnosis: *Adult:* Dichoid. Integument with light sclerotization and epicuticular chambers mineralized with birefringent whewellite (monohydrous calcium oxalate: $CaC_2O_4 \cdot H_2O$). Prodorsum fully articulated with proterosomal coxisternum. Notogaster with transverse sulcus at level of setal row *d* and tectiform, telescoping, type–L transverse scissure posterior to row *e*; pronotaspis bordered laterally by elongated suprapleural plate, bearing setae h_2 and h_3; pleural region continuous with pygidium. Pleural carina present dorsal to legs III, IV. Normal 5 pairs of notogastral lyrifissures present. Genital plate large, pentagonal as pair, subdivided transversely by narrow band; with 10 pairs of setae in two longitudinal rows. Aggenital plate

divided at same level as genital plate: posterior part narrow, elongated; anterior part large, triangular, with or without small middle plate; single aggenital seta on anterior, posterior or middle plate (if subdivided). Anal and adanal plates distinct, with 2 and 3 pairs of setae, respectively; anal plates anteriorly with adjacent pair of horn-like apodemes. Paired plicature plate present, bearing lyrifissure *ips*. Subcapitulum anarthric, with groove in labiogenal position. Palp setation 0-2-1-3-11, plus solenidion. Legs monodactylous.

Biology: The mineralization pattern of *E. minutissimus* was described by Norton and Behan-Pelletier (1991) and Alberti et al. (2001). Whewellite is deposited in small epicuticular chambers. Since the same mineral is found in the related ptychoid species *Archoplophora rostralis* (Willmann), and in the more distant *Prototritia major* (Jacot), it probably also fills the chambers of all *Eniochthonius* species. Probably the hardening caused by mineralization is a form of predator defence, one of many defences known in oribatid mites (Norton 2007; Pachl et al. 2012). Norton and Behan-Pelletier (1991) suggested that the mineral may be obtained from fungal food, since calcium oxalate often accumulates on the outside of hyphae as a waste product of incomplete carbohydrate metabolism.

All species of *Eniochthonius* are probably parthenogenetic. This remains unproven for any species, but Norton and Behan-Pelletier (2007) found no males in any sample of the three species that they studied.

Eniochthonius Grandjean, 1933
    Comb./Syn.: *Hypochthoniella* Berlese, 1910; *Arthrochthonius* Ewing, 1917
    Type-species: *Hypochthonius minutissimus* Berlese, 1903

KEY TO ADULT *ENIOCHTHONIUS* OF CANADA AND ALASKA
(MODIFIED FROM NORTON AND BEHAN-PELLETIER 2007)

1   Notogastral outline only moderately arched in lateral aspect, obovate in dorsal aspect; setae relatively long, f_1 reaching insertion of h_1 in dorsal aspect. Bothridial seta with 2–4 long tines. Aggenital region with 3 plates; aggenital seta on small middle plate. Length: 309–343µm (Figure 3.13A-C,F) .... *E. mahunkai* Norton and Behan-Pelletier

–   Notogastral outline strongly arched in lateral aspect, roughly diamond-shaped in dorsal aspect; setae shorter, f_1 not reaching insertion of h_1. Bothridial seta either with more than 9 long tines or with none. Aggenital region with 2 plates; aggenital seta on larger anterior plate. Length: 360–400µm 2

2(1)   Bothridial seta pectinate, with 9 or more long tines in addition to small barbs. Thicker branch of forked famulus distally with slight knob, shorter than other (narrower, usually attenuate) branch. Length: 360–380µm (Figure 3.13D,Eb) .... *E. minutissimus* (Berlese)

–   Bothridial seta with small barbs, without long tines. Thicker branch of forked famulus distally with minute knob, slightly longer than other (narrower, distally blunt) branch. Length: 380–400µm (Figure 3.13E,Ea,G) .... *E. crosbyi* (Ewing)

Eniochthonius crosbyi (Ewing, 1909) (Figure 3.13E,Ea,G)
    Comb./Syn.: *Hypochthonius crosbyi* Ewing, 1909; *Eniochthonius borealis* Jacot, 1939
    Biology: From deciduous and coniferous forest litter and decaying wood, moss, bracket fungi and muskeg grass clumps.
    Distribution: AB, ON, QC, NB, NS, NL [MO, MI, NC, NH, NY; Nearctic]
    Literature: Norton and Behan-Pelletier (2007); ABMI (2019).

Eniochthonius mahunkai Norton and Behan-Pelletier, 2007 (Figure 3.13A-C,F)
    Biology: Collected from peatlands, with sphagnum moss being a consistent component of the microhabitat. Collected with *E. minutissimus*, although the microhabitat preferences of these

species may differ. Ventricular and fecal boluses of both adult and immature *E. mahunkai* often contain unidentifiable organic fragments, and both clear and pigmented fungal hyphae.
Distribution: AB, NB, NS [NY, WI; Nearctic]
Literature: ABMI (2019).

Eniochthonius minutissimus (Berlese, 1903) (Figure 3.13D,Eb)
Comb./Syn.: *Hypochthonius minutissimus* Berlese, 1903; *Hypochthonius pallidulus* sensu Michael, 1888; non C.L. Koch, 1836; *Arthrochthonius pallidulus* sensu Ewing, 1917; *Hypochthoniella pallidulus* sensu Sellnick, 1928; *Eniochthonius pallidulus* sensu Grandjean, 1933; *Eniochthonius grandjeani* van der Hammen, 1952
Biology: The food of *E. minutissimus* is fungal hyphae (Pande and Berthet 1973; Anderson 1975; Schneider et al. 2004). In a cafeteria experiment testing preference for ubiquitous vs. specialized saprotrophic fungi from pine litter this species significantly preferred the ubiquitous ascomycete *Cladosporium herbarum* which colonizes dead organic matter (Koukol et al. 2009). This species has been recorded from bird feathers (Krivolutsky and Lebedeva 2004).
Distribution: NT, BC, AB, MB, QC, NB, NS, NL [GA, NH, NY, VA, VT, WI; Cosmopolitan]
Literature: Norton and Behan-Pelletier (2007); Seniczak et al. (2009); Maraun et al. (2019); ABMI (2019).

Family Hypochthoniidae Berlese, 1910
Diagnosis: *Adult:* Dichoid. Body moderately sclerotized or mineralized; colour ranging from pale yellow to reddish brown. Mineralized epicuticular chambers either widespread or localized in patches on body and legs. Rostral tectum with simple or serrate margin. Prodorsum fused to coxisternal region or not; lateral region relatively uniform, with or without formation for coaptation of retracted legs. Bothridium and base of bothridial seta with sharp bend. Base of bothridium porose, with or without internalized saccule. Exobothridial setae, 2 pairs. Setae of row *e* usually vestigial, alveoli on narrow, ribbon-like intercalary sclerite lying between pronotaspis and pygidium; or setae of row *e* large or small, but conspicuous; their narrow sclerite either conspicuous and isolated from pronotaspis and pygidium by scissures, or fused to latter sclerites. Genital plates with or without projecting, posterolateral tectum. Anal plates usually clearly separated from adanal plates by intervening scissure. Anal setae present or absent. Legs monodactylous.

Hypochthonius C.L. Koch, 1835
Type-species: *Hypochthonius rufulus* C.L. Koch, 1835

KEY TO ADULT *HYPOCHTHONIUS* OF CANADA AND ALASKA

1 Dorsal setae relatively short. Notogastral seta c_1 not reaching insertion of d_1, d_1 just reaching transverse scissure. Length: 447–650µm (Figure 3.14A).............. *H. luteus* Oudemans
– Most dorsal setae relatively long. Notogastral seta c_1 approaching or passing insertion of d_1, d_1 extending well past scissure. Length: 550–800µm (Figure 3.14B–D).......................
....*H. rufulus* C.L. Koch

Hypochthonius luteus Oudemans, 1917 (Figure 3.14A)
Comb./Syn.: *Hypochthonius rufulus* sensu Oudemans, 1914; non C.L. Koch, 1836
Biology: Gut contents of adults from mixed deciduous forest in Europe included: animal setae, animal claws (probably Collembola), fungal hyphae and spores, and small amounts of plant material (Schuster 1956).
Distribution: AB [NC, OH, UT; Holarctic, New Zealand]
Literature: Maraun et al. (2019); ABMI (2019).

Hypochthonius rufulus C.L. Koch, 1835 (Figure 3.14B–D)
> **Comb./Syn.**: *Hypochthonius pallidulus* C.L. Koch, 1835; non Michael, 1888
> **Biology**: This species is considered an omnivore. Riha (1951) observed it scavenging dead animals, particularly collembolans, and feeding on macerated fallen leaves when starving. Behan-Pelletier and Hill (1983) found plant, fungal, moss fragments, pine pollen, collembolan and mite fragments and enchytraeid setae in gut contents of adult and juveniles, with collembolan remains comprising up to 90% of the gut contents in some specimens. Adults will feed on the entomopathogenic nematode *Steinernema feltiae* in culture (Heidemann et al. 2011). High $^{15}N/^{14}N$ ratios of *H. rufulus* indicate that this species predominantly lives on an animal diet (Schneider et al. 2004), and this was associated with chelicera with low leverage index and little space for levator muscles (Perdomo et al. 2012). In a cafeteria experiment with a choice of eight common soil fungi, *H. rufulus* fed on all fungi presented, but significantly preferred the dematiacous *Phialophora verrucosa* (Schneider et al. 2005). This species sequesters calcium phosphate in its integument, as a possible defence mechanism (Pachl et al. 2012).
> **Distribution**: AK, YT, NT, AB, MB, ON, QC, NS, NL [IL, IO, MI, NC, NH, NY, OH, TN, VA; Semicosmopolitan]
> **Literature**: Seniczak et al. (2009); Maraun et al. (2019); Konecka and Olszanowski (2021).

Family Mesoplophoridae Ewing, 1917
> **Diagnosis**: *Adult*: Ptychoid; small mites; body moderately hardened or mineralized; stegasime; naso not evident. Bothridium and base of bothridial seta with sharp bend. Exobothridial setae, 2 pairs. Type-L scissure terminal, such that hysterosoma covered dorsally by an apparent notogaster with only 6–8 pairs of setae (rows *c*, *d*, and *e*); body region bearing setae of rows *f*, *h*, *p* (and *ad* in *Mesoplophora*), shifted ventrally, incorporated into ventral plate in which genital and anal (and sometimes adanal) plates insert. Subcapitulum anarthric; rutella narrow. Legs monodactylous. Trochanters I and II without setae. Genu I with 1 solenidion. Solenidia and setae *d* of tibiae and genua either coupled, or *d* absent.

KEY TO GENERA OF MESOPLOPHORIDAE OF CANADA AND ALASKA

1   Adanal plates (each with 3 setae) distinct, although medially very narrow. Genital plates contiguous with anal and aggenital plates. Six pairs of setae on dorsal hysterosomal sclerite (apparent "notogaster"). Proral setae present on all tarsi (Figure 3.16A,B) ....*Archoplophora* van der Hammen

–   Adanal plates indistinguishable, incorporated (along with adanal setae) into "ventral plate". Genital and anal plates widely separated. Eight pairs of setae on dorsal hysterosomal sclerite (apparent "notogaster"). Proral setae absent from all tarsi (Figure 3.16C–E)................. ....*Mesoplophora* Berlese

Archoplophora van der Hammen, 1959
> **Type-species**: *Phthiracarulus laevis* Jacot, 1938

Archoplophora rostralis (Willmann, 1930) (Figure 3.16A,B)
> **Comb./Syn.**: *Phthiracarulus rostralis* Willmann, 1930; *Phthiracarulus laevis* Jacot, 1938; *Archoplophora villosa* Aoki, 1980
> **Biology**: The integument is mineralized with calcium oxalate (Norton and Behan-Pelletier 1991).
> **Distribution**: BC, MB, ON, QC, NB, NS [CN, MI, NH, NY; Semicosmopolitan]
> **Literature**: Alberti et al. (2001); Pachl et al. (2012).

Mesoplophora Berlese, 1904
Type-species: *Mesoplophora michaeliana* Berlese, 1904
Comb./Syn.: *Phthiracarulus* Berlese, 1920

Mesoplophora japonica Aoki, 1970 (Figure 3.16C–E)
Distribution: AK, ON [Holarctic].
Literature: Liu (2018).
Note: Unidentified *Mesoplophora* species from NS.

Superfamily Protoplophoroidea Ewing, 1917
Note: Subcapitulum with unique pharyngeal complex; large dorsal muscles originate on cupola-like capitular apodeme and insert on small sclerotized dorsal region of pharynx. The feeding habits of protoplophoroid mites are virtually unstudied. Nothing is known about how their unique pharyngeal complex is employed in feeding, although its structure suggests that very powerful aspiration can be generated.

Family Cosmochthoniidae Grandjean, 1947
Diagnosis: *Adult*: Dichoid. Pale to light brown. Integument often ornamented (foveolate, reticulate or punctate); rostrum with distinct fenestrations. Prodorsum stegasime. Bothridium and base of bothridial seta with sharp bend, with seta usually expanded distally. Exobothridial setae, 2 pairs. Hysterosomal dorsum with 3 transverse scissures; two type-S scissures bearing erectile setal rows e and f, and type-E scissure just anterior to small setae of row d. Setae of rows c–f occupy anterior 1/2 of notogaster; no large area without setae; setal row d not enlarged, similar to row c. Coxisternum with 2 pairs of epimere II setae; without sternal apodeme. Two pairs of genital tracheae present, with bulbous terminations. Genital plate not divided; aggenital plates small, not fused to coxisternum. Adanal plates fused posteriorly to form a U shape. Legs bi- or tridactylous. Genual solenidia absent on all legs. Solenidia ω_3 and ω_2 of tarsus I absent; seta m'' present. Solenidion present on tibia IV.

Cosmochthonius Berlese, 1910 (Figure 3.12E,F)
Type-species: *Hypochthonius lanatus* Michael, 1885
Literature: Norton et al. (1983); Seniczak et al. (2011); Maraun et al. (2019).
Note: *Cosmochthonius* nr. *lanatus* (Michael, 1885), and unidentified *Cosmochthonius* sp. recorded from BC.

Family Haplochthoniidae van der Hammen, 1959
Diagnosis: *Adults*: Dichoid. Integument pale to light brown. Prodorsum stegasime or astegasime. Bothridium and base of bothridial seta with sharp bend, seta usually expanded distally. Exobothridial setae, 2 pairs. Notogaster with 3 transverse scissures of type-E. Lyrifissure *im* inserted on pleuraspis or in soft lateral cuticle. Additional cupules on notogaster associated with setal pairs c_1, d_1, e_1, f_1, h_1. Coxisternum with 2 pairs of epimere II setae; without sternal apodeme. Aggenital plates, if present, not fused to epimere IV. Genital plates large and long. Anterior and middle pairs of genital papillae, often small and inconspicuous, or genital papillae absent. Two distinct pairs of tracheae, with bulbous terminations, opening into large genital vestibule at its anterior and posterior limits, respectively, or only the anterior pair present. Preanal plate present or absent, peranal segment not formed. Legs mono- or tridactylous. Solenidia ω_3 and ω_2 of tarsus I absent; seta m'' present or absent. Solenidion absent from tibia IV. Genual solenidia absent on all legs.

Haplochthonius Willmann, 1930
Type-species: *Cosmochthonius* (*Haplochthonius*) *simplex* Willmann, 1930 (Figure 3.11B,C)
Comb./Syn.: *Tetrochthonius* Hammer, 1958
Note: Unidentified *Haplochthonius* species from AB (Osler et al. 2008).

Family Sphaerochthoniidae Grandjean, 1947
Diagnosis: *Adult*. Dichoid. Integument pale to light brown. Prodorsum stegasime. Rostrum normal. Bothridium and base of bothridial seta with sharp bend, with seta usually expanded distally. Exobothridial setae, 2 pairs. Transverse telescoping, type-L scissure, positioned posterior to setal row *c*; only 4 pairs of setae insert on anterior notogastral plate. Some dorsal setae (of notogaster and/or prodorsum) heavily ciliate, T-shaped. Integument distinctly patterned (foveolate/reticulate). Coxisternum with 2 or 3 pairs of epimere II setae; without sternal apodeme. Genital plate of normal porportions. Aggenital plates fused with genital plates. Genital papillae *Va*, *Vm* normal; posterior pair (*Vp*) small and inconspicuous. Preanal plate absent, peranal segment not formed. Legs heterotridactylous. Genual solenidia absent on all legs. Solenidion ω_3 of tarsus I retained. Solenidion present on tibia IV.

Sphaerochthonius Berlese, 1910 (Figure 3.11D)
Type-species: *Hypochthonius splendidus* Berlese, 1904
Literature: Maraun et al. (2019).
Note: *Sphaerochthonius* cf. *splendidus* (Berlese, 1904) and unidentified *Sphaerochthonius* reported from AB (Osler et al. 2008; Walter and Lumley 2021).

Superfamily Heterochthonioidea Grandjean, 1954
Literature: Norton and Fuangaworn (2015).

Family Arborichthoniidae Balogh and Balogh, 1992
Diagnosis: *Adult:* Dichoid. Integument pale to light brown; astegasime. Medial and lateral eyes absent. Bothridium and base of bothridial seta straight. Exobothridial setae, 2 pairs. Single transverse scissure of type-S, with intercalary sclerite with large, erectile setae of row *e*. Setae f_1, f_2 erectile, stout, with long branches, inserted on intercalary sclerites in pair of circular unsclerotized depressions; setae p_1, p_2 phylliform. Coxisternum with 1 pair of epimere II setae; without sternal apodeme. Subcapitulum anarthric. With 4 pairs of subcapitular setae. Rutellum narrow, tridentate. Sclerotized pharyngeal cupola present, dorsad of pharynx; without sagittal carina. Chelicera narrow, each digit with 3 teeth. Palp setation 0-2-1-3-9(1); some eupathidia bifid. Genital tracheae absent. Legs bidactylous; empodial claw absent. Solenidion ω medium sized, baculiform, inserted on proximal half of tarsus, not reaching distally to tip of segment. Famulus with bract.

Arborichthonius Norton, 1982
Type-species: *Arborichthonius styosetosus* Norton, 1982

Arborichthonius styosetosus Norton, 1982 (Figure 3.11A)
Biology: From deciduous forest litter.
Distribution: ON [NH, Japan, China]
Literature: Norton (2010); Lotfollahi et al. (2016).

Family Trichthoniidae Lee, 1982
Diagnosis: *Adult*. Dichoid. Integument pale to light brown. Prodorsum stegasime; naso not evident; medial and lateral eyes absent. Bothridium and base of bothridial seta straight. Exobothridial setae, 2 pairs. Notogaster with 2 type-S scissures with large erectile setae of rows *e* and *f*. Erectile setae of various forms, simple and smooth to broadly phylliform. Intercalary sclerite of each seta separate. Coxisternum with 1 pair of epimeria II setae; without sternal apodeme. Aggenital plates not fused to epimere IV; genital plates not transversely divided. Preanal plate absent. Peranal segment not formed. Subcapitulum anarthric. Chelicerae chelate-dentate; base never inserted into body as apodeme. Legs monodactylous. Tarsal seta *m"* present.
Note: Family is not considered monophyletic (Norton and Fuangarworn 2015).

Gozmanyina Balogh and Mahunka, 1983 (=*Marshallia* Gordeeva, 1980, preoccupied name)
   Type-species: *Trichthonius majestus* Marshall and Reeves, 1971

Gozmanyina majestus (Marshall and Reeves, 1971) (Figure 3.11E,F)
   Comb./Syn.: *Trichthonius majestus* Marshall and Reeves, 1971
   Biology: From mixed deciduous and coniferous forest litter; from peatland.
   Distribution: MB, ON, QC, NS, NL [NH, NY; Nearctic]
   Literature: Pachl et al. (2012, 2017); Maraun et al. (2019).

Infraorder Parhyposomata Grandjean, 1969

This group (also called Parhypochthonata, or Monofissurae) comprises the superfamily Parhypochthonioidea, with 2 families. Due to the absence of males, all members are presumed parthenogenetic (Norton and Palmer 1991; Norton et al. 1993).

Superfamily Parhypochthonioidea Grandjean, 1932

Family Gehypochthoniidae Strenzke, 1963
   Diagnosis: *Adults*: Trichoid, or nearly so, with postpedal furrow weakly or strongly developed. Prodorsum astegasime. Bothridium and bothridial seta with sharp proximal bend, seta pectinate or clavate with barbs. Exobothridial setae, 2 pairs. Notogaster with 1 simple transverse scissure or groove, lying between setal rows *d* and *e*; row *e* inserted well posterior to scissure. Lyrifissures *iad* and *ian* present. Opisthonotal gland opening without conspicuous protuberance; seta f_2 absent, represented by inconspicuous alveolus near gland opening. Adanal setation 3 pairs; anal setation 2 pairs; peranal segment absent. Subcapitulum stenarthric; rutellum strongly developed, distally broad and toothed; adoral setation 3 pairs. Chelicera normal, chelate-dentate, with base not inserted into body as apodeme. Legs with well-developed lateral claws; empodial claw reduced, hooklike.

Gehypochthonius Jacot, 1936
   Type-species: *Gehypochthonius rhadamanthus* Jacot, 1936

KEY TO ADULT *GEHYPOCHTHONIUS* OF CANADA AND ALASKA

1   Genital setation 5 or fewer pairs (3 medial + 2 lateral); 1 pair of aggenital setae. Length: ca. 237µm (Figure 3.15F,G) .... ***G. gracilis*** Pan'kov
–   Genital setation 9 pairs (6 medial +3 lateral); 2 pairs of aggenital setae. Length: 260–275µm (Figure 3.15C–E) ....*G. rhadamanthus* Jacot
       Note: Unidentified *Gehypochthonius* species from BC, AB.

Gehypochthonius gracilis Pan'kov, 2002 (Figure 3.15F,G)
   Biology: From boreal forest.
   Distribution: AB [Russia Far East]
   Literature: Walter et al. (2014).

Gehypochthonius rhadamanthus Jacot, 1936 (Figure 3.15C–E)
   Biology: From subarctic; boreal forest; from sphagnum.
   Distribution: YT; ON, QC [NC, NH; Semicosmopolitan]
   Literature: Weigmann et al. (2015); Maraun et al. (2019).

Family Parhypochthoniidae Grandjean, 1932
   Diagnosis: *Adult*. Trichoid, or nearly so, with postpedal furrow weakly or strongly developed. Integument weakly sclerotized, white to pale yellow; cuticle smooth. Prodorsum astegasime. Bothridium and bothridial seta with sharp proximal bend, seta pectinate or clavate with barbs.

Exobothridial setae, 2 pairs. Notogaster with 1 simple transverse scissure or groove, lying between setal rows *d* and *e;* row *e* inserted well posterior to scissure. Epimere III fused to IV or not; pairs fused medially or not. Opisthonotal gland opening on flared, funnel-shaped protuberance bearing seta f_2. Adanal setation 4 pairs; anal setation 4 pairs. With 1 pair of epimere II setae. Peranal segment developed, bearing 1 pair of setae; adanal segment incorporated into notogaster. Subcapitulum stenarthric, rutella strongly developed, distally broad and toothed; 3 pairs of adoral setae. Chelicerae normal, chelate-dentate, with base not inserted into body as apodeme. Legs with well-developed lateral claws; empodial claw reduced, hooklike or absent.

Parhypochthonius Berlese, 1904
   Type-species: *Parhypochthonius aphidinus* Berlese, 1904

KEY TO ADULT *PARHYPOCHTHONIUS* OF CANADA AND ALASKA

1	Bothridial seta with 8 tines. Posterior notogastral setae *p* longer than their mutual distance. Length: ca. 470µm (including chelicera)........................***P. aphidinus octofilamentis*** Jacot
–	Bothridial seta with 9–10 tines. Posterior notogastral setae *p* shorter than their mutual distance. Length: ca. 520µm (including chelicera) (Figure 3.15A,B)..........***P. aphidinus*** Berlese

Parhypochthonius aphidinus Berlese, 1904 (Figure 3.15A,B)
   Biology: Alberta collections are from farmland/pasture. Feeding of this species is not well studied, but large food boluses are easily observed through the translucent cuticle. Opisthonotal gland chemistry was studied by Sakata and Norton (2001).
   Distribution: AB, ON, QC [Cosmopolitan]
   Literature: Raspotnig (2010); Pachl et al. (2012); Schäffer et al. (2020).

Parhypochthonius aphidinus octofilamentis Jacot, 1938
   Biology: From deciduous logs.
   Distribution: QC [IL, NC; Nearctic]
   Note: Species nr. *P. aphidinus octofilamentis* from BC (Lindo and Winchester 2006); unidentified *Parhypochthonius* species from NS.

Infraorder Mixonomata Grandjean, 1969 (modified from Norton and Behan-Pelletier 2009)
   Mixonomata includes dichoid and ptychoid mites in seven superfamilies, with all, other than Nehypochthonioidea and Collohmannioidea, present in Canada and Alaska. Most are stegasime, all have bothridial setae and bothridia with sharp proximal bend and undivided notogaster, and many have opisthonotal glands. Lyrifissure *iad* is generally present (*ian* present or absent). Peranal segment never forms. Cheliceral base not inserted noticeably into body as apodeme.
   Note: In its present configuration, Mixonomata is paraphyletic in that the closest relatives of the Infraorder Desmonomata almost certainly lie within this group (Haumann 1991; Pachl et al. 2012). However, precise relationships remain uncertain (Norton and Sidorchuk 2014).

Superfamily Eulohmannioidea Grandjean, 1931

Family Eulohmanniidae Grandjean, 1931
   Diagnosis: *Adult:* Dichoid, light yellow, elongated (570–850µm); proterosoma noticeably constricted behind level of bothridium. Prodorsum weakly stegasime; exobothridial setae, 2 pairs. Bothridial seta pectinate. Bothridium with saccules. Proterosomal-hysterosomal articulation present. Notogaster without suprapleural scissure; with 15 pairs of setae, f_1 absent. Opisthonotal gland absent. Epimeres III, IV and aggenital region fully fused, unbroken by

epimeral borders or noticeable apodemes. Epimere IV and aggenital region neotrichous, together with ~15 or more pairs of setae. Genital aperture small, oval, displaced far posteriorly; with 7-10 (usually 9) pairs of genital setae. Anal setation 4 pairs (if anal segment present); adanal setation 4 pairs. Subcapitulum stenarthric; rutella strongly dentate. Adoral setae, 3 pairs. Supracoxal spine present. Chelicerae chelate-dentate; Trägårdh's organ and seta *cha* absent. Palp 4-segmented with femur and genu fused. Legs usually heterotridactylous with strong lateral claws and minute empodium (appearing bidactylous); rarely monodactylous. Solenidia long, finely tapered, none coupled to seta d on any segment.

Eulohmannia Berlese, 1910
Type-species: *Lohmannia (Eulohmannia) ribagai* Berlese, 1910
Comb./Syn.: *Arthronothrus* Trägårdh, 1910

Eulohmannia ribagai (Berlese, 1910) (Figure 3.17C–E)
Comb./Syn.: *Lohmannia (Eulohmannia) ribagai* Berlese, 1910; *Arthronothrus biunguiculatus* Trägårdh, 1910
Distribution: AK, YT, NT, BC, AB (CNC record), MB, ON, QC, NS [NY; Holarctic, Oriental]
Literature: Pachl et al. (2012, 2017); Maraun et al. (2019).

Superfamily Perlohmannioidea Grandjean, 1954

Family Perlohmanniidae Grandjean, 1954
Diagnosis: *Adult*: Dichoid, moderately sclerotized; medium to large (620–1,400µm). Prodorsum stegasime; exobothridial setae, 2 pairs; bothridial seta pectinate, longer than lamellar seta. Hysterosoma subrectangular in dorsal aspect and dorsoventrally flattened. Notogaster with 15 pairs of similar, short setae (f_1 represented by alveolar vestige): 8 setae of rows *c*, *d*, *e* usually in 2 longitudinal rows, with lateral row on carina; suprapleural scissure developed under carina or not. Opisthonotal gland present. Epimera III-IV fused but well delineated, each separated medially by unsclerotized cuticle; not fused to notogaster or to aggenital plates; 1 pair of epimere II setae. Genital plate divided transversely or not. Paired aggenital plates distinct, with 2 pairs of setae. Anal and adanal plates long, narrow, each with 3 pairs of setae. Lyrifissures *iad* and *ian* present. Preanal plate absent, unpaired postanal plate present; paired plicature plates weakly developed between adanal plates and notogaster. Subcapitulum stenarthric, with strong, toothed rutella. Palps with 5 free segments. Chelicerae without Trägårdh's organ. Adoral setae, 3 pairs. Legs monodactylous (usually) or tridactylous; tarsus I rich in ventral setae. Seta *d* of various genua and tibiae fully developed and independent from solenidion, or reduced, closely adjacent to solenidion.

KEY TO GENERA OF PERLOHMANNIIDAE OF CANADA AND ALASKA

1   Genital plates entire, each bearing 8–9 setae (Figure 3.19C) ....
    ....*Hololohmannia* Kubota and Aoki
–   Genital plates each divided transversely by narrow soft cuticle, such that 6 pairs of genital
    setae on anterior part, and 2 pairs on posterior part (Figure 3.19A,B).................................
    ....*Perlohmannia* Berlese

Hololohmannia Kubota and Aoki, 1998
Type-species: *Hololohmannia alaskensis* Kubota and Aoki, 1998

Hololohmannia alaskensis Kubota and Aoki, 1998 (Figure 3.19C)
Biology: From alpine tundra.
Distribution: AK [Alaskan endemic]

Perlohmannia Berlese, 1916
   Type-species: *Lohmannia insignis* Berlese, 1904
   Comb./Syn.: *Apolohmannia* Aoki, 1960; *Neolohmannia* Bulanova-Zachvatkina, 1960

Perlohmannia sp nr. *coiffaiti* Grandjean, 1961
   Distribution: AK, YT, NT, NS
   Note: Undescribed *Perlohmannia* sp. from BC (Lindo 2020).

Superfamily Epilohmannioidea Oudemans, 1923

Family Epilohmanniidae Oudemans, 1923
   Diagnosis: *Adult*: Dichoid, with unusually broad sejugal articulation; somewhat cylindrical, moderately elongated. Proterosoma without noticeable posterior constriction. Small to large mites (320–800μm). Colour yellowish to dark reddish brown, moderately sclerotized. Prodorsum stegasime, but rostrum sometimes strongly emarginated laterally. Exobothridial setae, 2 pairs. Bothridial seta filiform to clavate; bothridium with or without brachytracheae or porose pockets. Notogastral setae f_1, f_2 absent, represented only by inconspicuous alveoli. Opisthonotal gland present. Epimeres III and IV fused medially and to each other, outlined by borders and apodemes; epimere IV fused to hypertrophied aggenital plates, which themselves broadly fused anterior to genital plates. Adanal plates fused anterior to anal plates. Aggenital and adanal region separated by transverse scissure (sometimes called **schizogastry**) or indistinguishably fused; often with 3 pairs of aggenital setae, or neotrichous; with 3 pairs each of anal and adanal setae. Lyrifissure *ian* present. Subcapitulum stenarthric; mentum fused with coxisternum or not. Palps with 2–3 free segments, with at least palptrochanter, femur, and genu fused. Chelicerae chelate-dentate, without Trägårdh's organ. Rutella strongly developed, distally toothed. Adoral seta 3 pairs. Legs monodactylous. Seta *d* of genua and tibiae not independent; absent, or minute, inconspicuous, and coupled to solenidion.

Epilohmannia Berlese, 1910
   Type-species: *Lohomannia* [!] *cylindrica* Berlese, 1904
   Comb./Syn.: *Lesseria* Oudemans, 1917

Epilohmannia cylindrica (Berlese, 1904) (Figure 3.17F,G)
   Comb./Syn.: *Lohomannia* [!] *cylindrica* Berlese, 1904; *Lesseria szanisloi* Oudemans, 1917; *Epilohmannia verrucosa* Jacot, 1934
   Biology: From mixed deciduous and coniferous litter.
   Distribution: QC [TN; Cosmopolitan]
   Literature: Corral-Hernandez et al. (2016); Maraun et al. (2019).
   Note: Unidentified or undescribed *Epilohmannia* species from YT, BC, AB, SK, ON, QC. At least two undescribed North American species of *Epilohmannia* feed on decaying wood or bark as adults (Seastedt et al. 1989). The genetics of one of these species was studied by Pachl et al. (2012).

Superfamily Euphthiracaroidea Jacot, 1930
   Euphthiracaroidea comprises the three ptychoid families Oribotritiidae, Euphthiracaridae and Synichotritiidae. Euphthiracaroid mites show trends toward fusion of ventral plates that relate to biomechanical factors (Sanders and Norton 2004; Heethoff et al. 2018; Schmelzle et al. 2015). During normal activity muscles effect a lateral compression that creates the hydrostatic pressure needed to support the legs. When the mite is disturbed, the muscles relax (collapsing the high pressure), the legs are pulled into the opisthosoma, and the prodorsum deflects and retracts to cover the secondary space. Ventral plates that are fused are probably more efficient in containing hydrostatic pressure. Consequently, in most Euphthiracaridae, the original 4 pairs of plates are fused into a single large pair of holoventral plates. The same

seems to have occurred in some Synichotritiidae, which evolved ventral plate fusion independently from euphthiracarid mites. These fuse the paired plates into a single unit, anterior to the anal aperture (Norton and Lions 1992). Earlier stages in plate fusions are seen in the Oribotritiidae, a paraphyletic family containing species that retain some degree of ventral plate separation (Haumann 1991). Both hydraulic efficiency and resistance are improved, at least in some groups, by mineralization of the cuticle (Norton and Behan-Pelletier 1991; Sanders and Norton 2004); unlike the ptychoid enarthronotids, the mineral involved is calcium carbonate.

Biology: Euphthiracaroid mites are considered primarily macrophytophagous, feeding on the decaying parts of higher plants. All studied juveniles are endophagic, creating individual burrows in dead wood or woody structures of plants that they leave only after maturing (Michael 1888).

Family Euphthiracaridae Jacot, 1930

Diagnosis: *Adult*: Ptychoid, pale to dark brown; medium to large mites (ca. 400–2,000µm when fully extended). Prodorsum stegasime; usually with paired **manubrium** (narrow, projecting posterolateral apodeme); often with sagittal apodeme; exobothridial setae, 0–2 pairs. Bothridial seta various, never pectinate; bothridium with respiratory saccules or fine trachea; never with 3 brachytracheae. Notogaster laterally compressed; anterior margin extended as collarlike tectum. Usually with 14 pairs of notogastral setae; f_1 and f_2 represented only by alveolar vestiges. Opisthonotal gland present. Combined anogenital region narrow, length usually greater than 3 times maximum width; component plates separate or variously fused. Paired ventral plates connected across midline in preanal position by short region of interdigitating ridges (in transmitted light appearing as dark interlocking triangle). Genital plates not attached medially, genital setae spread the length of genital region, located on medial edge. All genital papillae of approximately equal size. Subcapitulum stenarthric; rutella broad, massive; palp 3-segmented (trochanter/femur/genu fused). Legs mono- or tridactylous.

KEY TO GENERA OF EUPHTHIRACARIDAE OF CANADA AND ALASKA
(MODIFIED FROM NIEDBAŁA 2002)

1   Bothridial scale positioned ventrad of bothridium. With 2 interlocking triangles, at both the anterior and posterior ends of anal region. Plicature and anogenital plates usually sculptured (ridged or foveolate). Genu IV with solenidion (Figures 3.18C,D; 3.21–3.25E).. ....*Euphthiracarus* Ewing

–   Bothridial scale dorsad, or somewhat posteriad of bothridium. With 1 interlocking triangle at anterior end of anal region. Plicature and anogenital plates with little or no sculpturing. Genu IV without solenidion. 2

2(1)   Genital setation 8 or 9 pairs, usually 3 of which are concentrated near anteromedial corner of plate. Interlamellar seta barbed, usually longer than bothridial seta. Exobothridial seta small, but usually present. Trochanters III and IV each with 2 setae. Legs tridactylous (usually) or monodactylous (Figure 3.20) ....*Acrotritia* Jacot

–   Genital setation 4 or 5 pairs, none concentrated in group near anteromedial corner. Interlamellar seta thin, smooth, much shorter than bothridial seta. Exobothridial seta usually represented only by alveolus. Trochanters III and IV each with 1 seta. Legs monodactylous (Figure 3.25F–I)..*Microtritia* Märkel

Acrotritia Jacot, 1923
  Type-species: *Phthiracarus americanus* Ewing, 1909
  Comb./Syn.: *Rhysotritia* Märkel and Meyer, 1959

KEY TO ADULT *ACROTRITIA* OF CANADA AND ALASKA
(MODIFIED FROM NIEDBAŁA 2002)

Note: Lengths usually are not given for these species because, being ptychoid, they are hard to measure.

1	Lateral carina of prodorsum not forked distally	2
	Lateral carina of prodorsum forked distally (Figure 3.20A,B)	*A. vestita* (Berlese)
2(1)	Surface of notogaster reticulate (Figure 3.20C,D)	*A. parareticulata* (Niedbała)
–	Surface of notogaster punctate	3
3(2)	Leg I bidactylous; legs II–IV heterotridactylous (Figure 3.20E–G)	*A. ardua* (C.L. Koch)
–	At least one leg monodactylous	4
4(3)	Only leg IV monodactylous; leg I bidactylous, legs II–III heterotridactylous (Figure 3.20H)	*A. diaphoros* (Niedbała)
–	All legs monodactylous	5
5(4)	Interlamellar, lamellar and rostral setae similar in length; bothridial seta without distinct head (Figure 3.20I)	*A. scotti* Walker
–	Interlamellar setae considerably longer than lamellar and rostral setae; bothridial seta with distinct head (Figure 3.20J)	*A. curticephala* (Jacot)

Note: Unidentified *Acrotritia* species from BC, QC, NS.

Acrotritia ardua (C.L. Koch, 1841) (Figure 3.20E–G)

Comb./Syn.: *Hoplophora ardua* C.L. Koch, 1841; *Rhysotritia ardua* (C.L. Koch, 1841); *Hoplophora arctata* Riley, 1874; *Phthiracarus canestrinii* Michael, 1898; *Phthiracarus americanus* Ewing, 1909; *Phthiracarus pectinatus* Ewing, 1917; *Oribotritia loricata* sensu Willmann, 1931; non Hammer, 1955

Biology: Found in many habitats including peatlands, rotton wood and twigs. Parthenogenetic; gravid females carry up to 3 eggs. Riha (1951) observed this species (as *Oribotritia loricata*) feeding on fallen leaves and moist dead wood. Hayes (1965) found preference for stage of decomposition, rather than tree species In a mixed coniferous forest juvenile *A. ardua* excavated cavities and fed inside decomposing needles of spruce (*Picea abies*) (Hågvar 1998). Microsporidian parasites were more abundant than gregarines in this species (Purrini and Bukva 1984).

Distribution: YT, NT, BC, AB, SK, MB, ON, QC, NB, NS, NL [AL, AR, AZ, CO, CT, FL, GA, IA, IL, IN, KY, LA, MA, MD, ME, MI, MN, MO, MS, MT, NC, NH, NM, NE, NJ, NY, OH, OK, PN, SC, SD, TX, UT, VA, VT, WI, WV, WY; cosmopolitan]

Literature: Shimano (2005); Schmelzle et al. (2008); Niedbała and Liu (2018); ABMI (2019).

Acrotritia curticephala (Jacot, 1938) (Figure 3.20J)

Comb./Syn.: *Pseudotritia ardua curticephala* Jacot, 1938; *Rhysotritia curticephala* (Jacot, 1938)

Biology: From deciduous and coniferous litter; tall grass prairie litter.

Distribution: ON, NB, NS [AL, AR, CO, CT, GA, IN, KY, MA, MS, MT, NH, OK, PN, TX, VT, WI; semicosmopolitan]

Literature: Niedbała (2002); Niedbała and Liu (2018).

Acrotritia diaphoros (Niedbała, 2002) (Figure 3.20H)

Comb./Syn.: *Rhysotritia diaphoros* Niedbała, 2002

Biology: From deciduous and coniferous litter.

Distribution: AB, ON [AR, IA, IL, IN, MN, MO, OH, PA, TN, WI; Nearctic, Neotropical]

Literature: Niedbała and Liu (2018); ABMI (2019).

Acrotritia parareticulata (Niedbała, 2002) (Figure 3.20C,D)
 Comb./Syn.: *Rhysotritia parareticulata* Niedbała, 2002
 Biology: From southern boreal forest.
 Distribution: ON, NS [Canada, Tanzania, Argentina].
 Literature: Niedbała and Liu (2018).

Acrotritia scotti (Walker, 1965) (Figure 3.20I)
 Comb./Syn.: *Rhysotritia scotti* Walker, 1965
 Biology: From coniferous forest litter.
 Distribution: BC [CA; Nearctic].
 Literature: Niedbała (2002); Niedbała and Liu (2018).

Acrotritia vestita (Berlese, 1913) (Figure 3.20A,B)
 Comb./Syn.: *Hoploderma vestitum* Berlese, 1913; *Rhysotritia comteae* Mahunka, 1983; *Rhysotritia anchistea* Niedbała, 1998
 Biology: According to Niedbała and Liu (2018) this species is Pantropical, and was probably introduced to Ontario, where it is found in oak, red pine and spruce woodland.
 Distribution: ON [Pantropical].

Euphthiracarus Ewing, 1917
 Type-species: *Phthiracarus flavus* Ewing, 1908.
 Comb./Syn.: *Pseudotritia* Willmann, 1919

KEY TO ADULT *EUPHTHIRACARUS* OF CANADA AND ALASKA
(MODIFIED FROM NIEDBAŁA 2002)

1	Bothridial seta with spindle-shaped or fusiform head	2
–	Bothridial seta without distinct head, but may be slightly swollen or spiculate distally	7
2(1)	Legs monodactylous	3
–	Legs heterotridactylous	4
3(2)	Bothridial seta strongly fusiform, covered with small spines. Holoventral plate without lateral pockets. Plicature plate without noticeable sculpturing other than narrow lateral band of fine transverse striations restricted to region near articulation with notogaster (Figure 3.21A–D)	*E. monodactylus* (Willmann)
–	Bothridial seta slightly enlarged, covered with distinct barbs. Holoventral plate with pair of conspicuous lateral pockets at level of setae g_8, g_9. Plicature plate with sculpturing in form of strong transverse ridges running across its width (Figure 3.21E–G)	*E. monyx* Walker
4(2)	Prodorsum with 2 pairs of lateral carinae	5
–	Prodorsum with 1 pair of lateral carinae	6
5(4)	Integument with distinct foveae	7
–	Integument punctate or porose (Figure 3.21H–J)	*E. fulvus* (Ewing)
6(4)	Surface of body covered by mosaic pattern (Figure 3.22A–D)	*E. jacoti* Niedbała and Liu
–	Surface of body punctate or porose (Figure 3.22E–G)	*E. fusulus* Niedbała
	Note: Known from Northeast USA (ME); possibly in Canada.	
7(5)	Bothridial seta with head almost smooth (Figure 3.22H–L)	*E. depressculus* Jacot
–	Bothridial seta with small, strongly barbed head (Figure 3.23A–E)	*E. crassisetae* Jacot
8(1)	Bothridial seta with tuft of spines distally	9
–	Bothridial seta rod-like, covered with small spines in distal half	10

Taxonomic Keys

9(8) Dorsal setae of prodorsum almost equal in length; rostral setae longer than interlamellar setae (Figure 3.23F–H).. ***E. longirostralis*** Walker
– Dorsal setae of prodorsum unequal in length; interlamellar setae considerably longer than lamellar and rostral setae (Figure 3.23I–K) .... ***E. cernuus*** Walker

10(8) Notogastral integument foveolate or with mosaic pattern.. 11
Notogastral integument punctate.. 13

11(10) Notogastral integument foveolate 12
– Notogastral integument with mosaic pattern (Figure 3.24A,B) .... ***E. pulchrus*** Jacot

12(11) Notogastral seta c_3 much shorter than c_2 and more or less in line with $c_{1–2}$. Aggenital seta ag_2 (medial seta) about 2 × length ag_1 (lateral). Genu I with 5 setae (Figure 3.24C–F).......
....***E. cribrarius*** (Berlese)
– Notogastral seta c_3 inserted more anteriorly than $c_{1–2}$ and about 80% as long as c_2. Aggenital seta ag_2 (medial seta) similar in length to ag_1 (lateral). Genu I with 4 setae (Figure 3.24G,H) ....***E. flavus*** (Ewing)

13(10) Genital setae, all equal in length (Figure 3.25A–C) ....***E. vicinus*** Niedbała
Note: Specimens similar to *E. vicinus* known from QC (Déchêne and Buddle 2010).
– Genital setae g_1 and g_2 longest genital setae (Figure 3.25D,E)............. ***E. tanythrix*** Walker
Note: Unidentified *Euphthiracarus* species from AK, BC, MB, NS, NL.

Euphthiracarus cernuus Walker, 1965 (Figure 3.23I–K)
Biology: From mixed temperate deciduous and coniferous forest; this species was found in the canopy of Western redcedar (Lindo 2010).
Distribution: BC, NS [CA, OR, UT, WA; Nearctic]
Literature: Niedbała (2002); Niedbała and Liu (2018).

Euphthiracarus crassisetae Jacot, 1938 (Figure 3.23A–E)
Distribution: BC [AL, CA, IA, IL, LA, NC,NY, MO, OH, OR, PN, WV; Nearctic]
Literature: Niedbała (2002); Niedbała and Liu (2018).

Euphthiracarus cribrarius (Berlese, 1904) (Figure 3.24C–F)
Comb./Syn.: *Phtiracarus* [!] *cribrarius* Berlese, 1904; *Phthiracarus punctulatus* Berlese, 1913
Biology: From mixed deciduous and coniferous forest litter.
Distribution: BC, AB, MB, NS [CO, IA, ID, ME, MI, MN, OR, PN, TN, VA, VT, WA, WI, WV, WY; Holarctic, Oriental]
Literature: Niedbała (2002); Niedbała and Liu (2018); Maraun et al. (2019).

Euphthiracarus depressculus Jacot, 1924 (Figure 3.22H–L)
Biology: From mixed deciduous and coniferous litter.
Distribution: BC, NB, NS [CT, NY, TX, IA, KY, ME, VT, NH, PN, WV, AR, AL, SC, GA, FL, Nearctic]
Literature: Niedbała (2002); Niedbała and Liu (2018).

Euphthiracarus flavus (Ewing, 1908) (Figure 3.24G,H)
Comb./Syn.: *Phthiracarus flavus* Ewing, 1908
Biology: Gravid females carry up to 3 eggs. Gut boli have mixtures of coarse organic matter, dark and light hyphae, or all hyphae (Walter et al. 2014). Rohde (1955) noted that eggs were deposited in the feeding passages of the adult.
Distribution: BC, AB, MB, ON, QC, NS [IL, MS, NC, NY, OR, PA, TN, VA, WV; Nearctic]
Literature: Niedbała (2002); Niedbała and Liu (2018); ABMI (2019).

Euphthiracarus fulvus (Ewing, 1909) (Figure 3.21H–J)
    Comb./Syn.: *Phthiracarus fulvus* Ewing, 1909
    Biology: From boreal forest litter.
    Distribution: AB, QC [CO, WI, IA, IL, OH, KY, ME, MN, VT, NH, NY, FL; Nearctic]
    Literature: Niedbała (2002); Niedbała and Liu (2018); ABMI (2019).

Euphthiracarus jacoti Niedbała and Liu, 2018 (Figure 3.22A–D)
    Comb./Syn.: *Euphthiracarus punctulatus* Jacot, 1930 (name changed because of homonymy with *E. punctulatus* (Berlese, 1913)
    Biology: From beaver lodge.
    Distribution: NS [CT, ME, MS, NY, PA, WV; Nearctic]

Euphthiracarus longirostralis Walker, 1965 (Figure 3.23F–H)
    Biology: From temperate forest litter.
    Distribution: BC [CA, OR; Nearctic]
    Literature: Niedbała (2002); Niedbała and Liu (2018).

Euphthiracarus monodactylus (Willmann, 1919) (Figure 3.21A–D)
    Comb./Syn.: *Tritia (Pseudotritia) monodactyla* Willmann, 1919
    Biology: From deciduous and coniferous litter.
    Distribution: ON [MI, OH; Holarctic, New Guinea, India]
    Literature: Niedbała (2002); Weigmann et al. (2015); Niedbała and Liu (2018).

Euphthiracarus monyx Walker, 1965 (Figure 3.21E–G)
    Biology: From temperate coniferous forest; this species was found in the canopy of Western redcedar (Lindo 2010).
    Distribution: BC [CA, OR; Nearctic]
    Literature: Niedbała and Liu (2018).

Euphthiracarus pulchrus Jacot, 1930 (Figure 3.24A,B)
    Comb./Syn.: *Euphthiracarus flavum pulchrum* Jacot, 1930
    Biology: From deciduous litter.
    Distribution: BC, ON, QC [CT, FL, GA, IN, KY, MO, MN, MS, NC, NY, OH, TN, VA, WV; Nearctic]
    Literature: Niedbała (2002); Niedbała and Liu (2018).

Euphthiracarus tanythrix Walker, 1965 (Figure 3.25D,E)
    Biology: From spruce litter and soil.
    Distribution: ON [CA, TN; Nearctic]
    Literature: Niedbała and Stary (2015); Niedbała and Liu (2018).

Microtritia Märkel, 1964
    Type-species: *Phthiracarus minimus* Berlese, 1904

KEY TO ADULT *MICROTRITIA* OF CANADA AND ALASKA

1   Aggenital setation 2 pairs (Figure 3.25F,G) .... ***M. minima*** (Berlese)
–   Aggenital setation 1 pair (Figure 3.25 H–J) .... ***M. simplex*** (Jacot)
    Note: Unidentified *Microtritia* species from BC.

Microtritia minima (Berlese, 1904) (Figure 3.25F,G)
    Comb./Syn.: *Phthiracarus minimus* Berlese, 1904; *Tritia (Pseudotritia) minuta* Willmann 1919

Biology: From peatland. Pande and Berthet (1973) found adults and nymphs feeding inside decomposing needles and wood. This species fed 100% on higher plant material (Kaneko 1988).
Distribution: BC, ON, QC, NS [NY, NC, OR, VA; Semicosmopolitan]
Literature: Niedbała and Liu (2018); Maraun et al. (2019).

Microtritia simplex (Jacot, 1930) (Figure 3.25 H–J)
Comb./Syn.: *Pseudotritia simplex* Jacot, 1930; *Rhysotritia simplex* (Jacot, 1930); *Rhysotritia paeneminima* Walker, 1965
Biology: From boreal mixedwood forest. From sphagnum.
Distribution: NT, BC, MB, ON, QC, NS [AL, AZ, CA, CT, FL, GA, IA, IL, KY, LA, MA, MD, ME, MI, MO, MS, NH, NJ, NY, OK, OR, PA, SD, TN, TX, VT, UT, WY; Nearctic, Neotropical]
Literature: Niedbała and Stary (2015); Niedbała and Liu (2018).

Family Oribotritiidae Grandjean, 1954
Diagnosis: *Adult*: Ptychoid, pale to dark brown; medium to large mites (about 400–2,000μm when fully extended). Notogastral integument usually smooth, uniformly shiny in reflected light. Prodorsum stegasime; with paired manubrium (narrow, projecting posterolateral apodeme); often with sagittal apodeme; exobothridial setae 0–2 pairs. Bothridial seta various, never pectinate; bothridium with or without respiratory saccules or fine trachea. Notogaster laterally compressed; anterior margin extended as collarlike tectum. Usually with 14 pairs of notogastral setae; f_1 and f_2 represented by alveolar vestiges. Opisthonotal gland present. Combined anogenital region narrow, length usually greater than 3 times maximum width. Anal plates (bearing anal setae) discrete, although may be very narrow posteriorly, not fused to adanal plate or to adanal portion of aggenito-adanal plate. Genital plate of various form, fused to aggenital region or not. Anterior genital papillae never reduced; posterior pair rarely reduced. Subcapitulum stenarthric; rutella broad, massive; palp 3–5 segmented. Legs mono- or tridactylous. Seta *d* of genua and tibiae absent or apparently so (coupled to solenidion but minute, inconspicuous).

KEY TO GENERA OF ORIBOTRITIIDAE OF CANADA AND ALASKA
(MODIFIED FROM NORTON OSU MITE KEYS)

1 Bothridial scale positioned dorsal to bothridium. Fused aggenital-adanal plate with oblique, unsclerotized cleft (*trv*) originating from junction of anal and genital plates. Palp 5-segmented (Figures 3.18A,B; 3.28) .... ***Oribotritia*** Jacot
– Bothridial scale positioned ventral to bothridium. Fused aggenital-adanal plate without cleft, or at most with slight transverse notch at anogenital junction. Palp 3-segmented (trochanter, femur and genu fused) or rarely 4-segmented (trochanter and femur fused) 2

2(1) Aspis with 1 pair of lateral carinae. Lamellar seta displaced laterally, inserted near margin of aspis; mutual distance of lamellar setae subequal to that between interlamellar seta, or greater. Genu IV without solenidion (Figure 3.27A–D) .... ***Mesotritia*** Forsslund
– Aspis without lateral carina. Lamellar seta not laterally displaced; mutual distance of interlamellar setae distinctly greater than that of lamellar setae. Genu IV with 1 solenidion..... 3

3(2) Smaller species; notogaster usually less than 400μm, usually light in colour. Legs monodactylous. Trochanters III and IV each usually with 2 setae (Figure 3.27E–H) ....
.... ***Protoribotritia*** Jacot

– Larger species; notogaster usually more than 600µm long (up to 1,500µm), medium to dark brown colour. Legs tridactylous. Trochanters III and IV each with 3 setae (Figure 3.26) ... . *Maerkelotritia* Hammer

Maerkelotritia Hammer, 1967
Type-species: *Markelotritia alaskensis* Hammer, 1967

KEY TO ADULT *MAERKELOTRITIA* OF CANADA AND ALASKA
(MODIFIED FROM NORTON OSU MITE KEYS)

1 Anal setation 3 pairs; an_3 short, positioned anteriorly on anal plate. Length prodorsum ca. 429µm; length notogaster ca. 737µm (Figure 3.26D,E) .... *M. cryptopa* (Banks)
– Anal setation 1 pair, positioned medially on anal plate. Length prodorsum 500–590µm; length notogaster ca. 1,040–1,130µm (Figure 3.26A–C) .... *M. kishidai* (Aoki)
   Note: Unidentified *Maerkelotritia* species from BC, QC.

Maerkelotritia cryptopa (Banks, 1904) (Figure 3.26D,E)
   Comb./Syn.: *Phthiracarus cryptopus* Banks, 1904; *Oribotritia gibbera* Walker, 1965; *Phthiracarus maximus* Ewing, 1913
   Biology: From coastal coniferous forests.
   Distribution: BC [CA, OR, UT, WA; Holarctic]
   Literature: Niedbała (2002); Niedbała and Liu (2018).

Maerkelotritia kishidai (Aoki, 1958) (Figure 3.26A–C)
   Comb./Syn.: *Oribotritia kishidai* Aoki, 1958; *Maerkelotritia kishidai* Aoki 1980; *Maerkelotritia alaskensis* Hammer, 1967: *Oribotritia sellnicki* Walker, 1965: *Oribotritia* ? *loricata* sensu Hammer, 1955; non Rathke, 1799
   Biology: From coastal coniferous forests.
   Distribution: AK, BC [CA, NC, OR, UT, WA; Holarctic]
   Literature: Liu et al. (2009); Niedbała and Liu (2018).

Mesotritia Forsslund, 1963
Type-species: *Mesotritia testacea* Forsslund, 1963

KEY TO ADULT *MESOTRITIA* OF CANADA AND ALASKA
(MODIFIED FROM NIEDBAŁA 2002)

1 Anal setation 1 pair (Figure 3.27A) ....*M. flagelliformis* (Ewing)
– Anal setation 2 pairs (Figure 3.27B–D) .... *M. nuda* (Berlese)
   Note: Unidentified *Mesotritia* species from BC, NS, NL.

Mesotritia flagelliformis (Ewing, 1909) (Figure 3.27A)
   Comb./Syn.: *Phthiracarus flagelliformis* Ewing, 1909; *Mesotritia testacea* Forsslund, 1963
   Biology: From deciduous and coniferous litter.
   Distribution: ON, QC, NS [CA, FL, GA, IL, IN, MS, OR, ME, MO, NC, NH, NM, NY, SC, TN, VA, VT, WV, WY; Holarctic]
   Literature: Niedbała (2002); Liu and Chen (2010); Niedbała and Liu (2018).

Mesotritia nuda (Berlese, 1887) (Figure 3.27B–D)
   Comb./Syn.: *Tritia nuda* Berlese, 1887; *Oribotritia brachythrix* Walker, 1965; *Mesotritia elastica* Sergienko, 1987; *Oribotritia grandjeani* Feider and Suciu, 1957; *Mesotritia* (*Entomotritia*) *piffli* Märkel, 1964
   Biology: From deciduous and coniferous litter.

Distribution: BC, AB, ON, QC, NB [AR, AZ, CA, FL, GA, IA, IN, LA, MN, MS, NM, NV, SC, SD, TN, TX, UT, VT, WI, WY; Semicosmopolitan]
Literature: Niedbała and Liu (2018); ABMI (2019); Huang et al. (2021).

Oribotritia Jacot, 1924
Type-species: *Hoplophora decumana* auct. non C.L. Koch, 1836 (=*Phthiracarus berlesei* Michael, 1898)
Comb./Syn.: *Plesiotritia* Walker, 1965

KEY TO ADULT *ORIBOTRITIA* OF CANADA AND ALASKA
(MODIFIED FROM NIEDBAŁA 2007b)

1	Anal plates without setae. Length prodorsum ca. 590μm; length notogaster ca. 1,116μm (Figure 3.28A–C) .... ***O. banksi*** (Oudemans)	
–	Anal plates with 2–3 pairs of setae	2
2(1)	Anal setation 2 pairs	3
–	Anal setation 3 pairs	4
3(2)	Interlamellar seta erect; genital setation 8 pairs; lyrifissure *iad* posterior to seta ad_3. Length prodorsum ca. 634μm; length notogaster ca. 1,302μm (Figure 3.28D–F) .... ***O. carolinae*** Jacot	
–	Interlamellar seta procumbent; genital setation 7 pairs; lyrifissure *iad* anterior to seta ad_3. Length prodorsum ca. 545μm; length notogaster ca. 1,040μm .... ***O. paracarolinae*** Niedbała	
4(2)	One pair of lateral carinae present on prodorsum. Length prodorsum ca. 699μm; length notogaster ca. 1,316μm (Figure 3.28G–I) .... ***O. henicos*** Niedbała	
–	More than 1 pair of lateral carinae present on prodorsum. Length prodorsum ca. 642μm; length notogaster ca. 1,190μm (Figure 3.28J,K) .... ***O. megale*** Walker	

Note: Unidentified *Oribotritia* species from BC.

Oribotritia banksi (Oudemans, 1916) (Figure 3.28A–C)
Comb./Syn.: *Tritia banksi* Oudemans, 1916
Biology: The ptychoid defence mechanism of Euphthiracaroidea was studied in this species among others by Schmelzle et al. (2008, 2009).
Distribution: NS [CA, CN, FL, GA, MD, ME, NC, NJ, NY, PA, VA, WV; Nearctic]
Literature: Niedbała (2002); Niedbała and Liu (2018).

Oribotritia carolinae Jacot, 1930 (Figure 3.28D–F)
Biology: From deciduous and coniferous litter.
Distribution: QC [NC, WV; Nearctic]
Literature: Niedbała (2002); Niedbała and Liu (2018).

Oribotritia henicos Niedbała, 2002 (Figure 3.28G–I)
Biology: From boreal mixedwood forest.
Distribution: QC [Canada]
Literature: Niedbała and Liu (2018).

Oribotritia megale (Walker, 1965) (Figure 3.28J,K)
Comb./Syn.: *Plesiotritia megale* Walker, 1965
Biology: From coastal temperate rainforest. Walker (1965) studied the life history of specimens from *Sequoia sempervirens* forest.
Distribution: BC [CA, OR; Nearctic]
Literature: Niedbała (2002); Niedbała and Liu (2018).

Oribotritia paracarolinae Niedbała, 2007
   Biology: From mixed deciduous and coniferous forest.
   Distribution: BC, NS [WA; Nearctic]
   Literature: Niedbała and Stary (2015); Niedbała and Liu (2018).

Protoribotritia Jacot, 1938
   Type-species: *Protoribotritia canadaris* Jacot, 1938

KEY TO ADULT *PROTORIBOTRITIA* OF CANADA AND ALASKA
(MODIFIED FROM NIEDBAŁA 2002)

1   Adanal setation 3 pairs; an_1 and an_2 inserted closely adjacent, between setae ad_{1-2} (Figure 3.27E,F) ....***P. oligotricha*** Märkel
–   Adanal setation 4 pairs; an_1 and an_2 inserted distant from each other (Figure 3.27G) ....***P. canadaris*** Jacot
   Note: Unidentified species from YT, BC, AB, QC.

Protoribotritia canadaris Jacot, 1938 (Figure 3.27G)
   Biology: Gut boli with hyphae or indistinct or coarse organic matter. Gravid females carry a single large egg (Walter et al. 2014).
   Distribution: YT, AB, ON, QC, NS, NL [CT, NM, NY, OR; Nearctic]
   Literature: Niedbała (2002); Niedbała and Liu (2018).

Protoribotritia oligotricha Märkel, 1963 (Figure 3.27E,F)
   Biology: From tundra; boreal mixedwood forest.
   Distribution: YT, BC, AB, ON, QC [ME, NH, SD, UT, WY; Holarctic, Oriental]
   Literature: Niedbała (2002); Niedbała and Liu (2018).

Family Synichotritiidae Walker, 1965
   Diagnosis: *Adult*: Ptychoid, pale to dark brown; medium to large mites (ca. 400–1,250μm). Body cuticle mineralized. Prodorsum stegasime; paired manubrium absent; sagittal apodeme well developed. Aspis with single pair of lateral carinae; bothridial scale posterodorsal to bothridium; no elongate brachytracheae or other tubular structures descending from bothridium. Notogaster laterally compressed; anterior margin extended as collarlike tectum. Notogaster with 14 or 15 pairs of notogastral setae; p_4 present or absent; f_1 and f_2 represented by alveolar vestiges. Opisthonotal gland absent. With genital plates connected in sagittal plane, either fused without trace or connected by narrow longitudinal band of articulating cuticle. Genital setae not arranged in medial row; all or most on anterior genital region. Progenital chamber opening in soft cuticle between genital plates and coxisternum IV. Posterior genital papillae highly regressed. Complete fusion between anal and adanal plates. Subcapitulum stenarthric; rutella broad, massive. Palp genu free or fused to femur (palp with 5 or 4 free segments). Legs mono-or tridactylous.
Literature: Norton and Lions (1992).

Synichotritia Walker, 1965
Type-species: *Synichotritia caroli* Walker, 1965

KEY TO ADULT *SYNICHOTRITIA* OF CANADA AND ALASKA

1   Notogastral setation 14 pairs. Length prodorsum ca. 369μm; length ventral plate ca. 505μm (Figures 3.18E,F; 3.29A,B) ....***S. caroli*** Walker
–   Notogastral setation 15 pairs. Length prodorsum ca. 273μm; length ventral plate ca. 202μm (Figure 3.29C–E) ....***S. spinulosa*** Walker

Synichotritia caroli Walker, 1965 (Figures 3.18E,F; 3.29A,B)
Biology: From temperate rainforest.
Distribution: BC [CA, MS, NC, OR; Nearctic]
Literature: Lions and Norton (1998); Niedbała (2002).

Synichotritia spinulosa Walker, 1965 (Figure 3.29C–E)
Biology: From temperate rainforest.
Distribution: BC [CA, MS, OR; Nearctic, Russian Far East]
Literature: Lions and Norton (1998); Niedbała (2002); Niedbała and Liu (2018).

Superfamily Phthiracaroidea Perty, 1841

Family Phthiracaridae Perty, 1841
  Diagnosis: *Adult*: Ptychoid, pale to nearly black, sometimes opalescent; medium to large mites (about 400–1,500µm when fully extended). Prodorsum stegasime; without manubrium; often with sagittal apodeme; exobothridial setae, 0–1. Bothridial seta various, never pectinate; 3 brachytracheae open into each bothridium. Notogaster convex, never laterally compressed; margins extended as deep, nearly continuous tectum. Usually with 15 pairs of notogastral setae (seta p_4 present; f_1 and f_2 represented by alveolar vestiges); sometimes with weak to strong neotrichy. Opisthonotal gland absent. Combined anogenital region broad, length approximately twice width; with imperceptibly fused genital/aggenital plates, and fused anal/adanal plates), without plicature plates. Anterior genital papillae (*Va*) reduced. Subcapitulum stenarthric; rutella broad, massive. Chelicerae robust, chelate-dentate. Palps with 3 segments (trochanter, femur, tibia fused). Legs monodactylous.
  Biology: Phthiracarid cuticle is hardened by mineralization to a greater or lesser degree. Where identified, the mineral is calcium carbonate (Pachl et al. 2012). Many species are mineralized to the point of having brittle cuticle, sometimes with a pale colour and an opalescent sheen in reflected light. In all known instances, eggs are retained by females to the completion of embryogenesis; i.e., prelarvae are deposited rather than eggs.
  Immature phthiracarid mites are endophages in partially decayed woody tissues of logs, twigs, seed-bearing structures, or in coniferous needles (Edsberg and Hågvar 1999). In wood, they create characteristic linear, frass-filled burrows that usually align with the wood grain. Juveniles are rarely found outside such materials, but adults are common in leaf litter and consume leaf litter and fungi (Anderson 1975). Phthiracarid mites are considered important decomposers, especially in coniferous forests, where they can contribute substantially to respiratory metabolism of the oribatid community (Luxton 1981a). Feeding biology is comparatively well-studied, e.g., in a detailed study of mouthparts using SEM, Dinsdale (1974) showed *Phthiracarus* sp. feeding by using its chelicera to tear material from the "spongy mesophyll" of decaying spruce needles. *Steganacarus magnus* (Nicolet), the best-studied member of the family, is tolerant of drought and heat extremes (Siepel 1996). It has a demonstrated supercooling point between −7°C and −38°C, made possible by the accumulation of cryoprotectants, low molecular weight solutes such as sugars and polyols (Webb and Block 1993). This species is sexual and its mitochondrial genome was studied by Domes et al. (2008). Life history traits of *S. magnus* were summarized by Lebrun et al. (1991). Adult females chew shallow pits in woody substrates such as pine cones and the cupules of beech (*Fagus* sp.) to deposit prelarvae, from which larvae burrow into the substrate (Harding and Easton 1984; Webb 1991). This large species takes more than a year to become mature and lives for more than three years (Webb 1989). Other Phthiracaridae may take up to five years for a generation (Soma 1990).

KEY TO GENERA OF PHTHIRACARIDAE OF CANADA AND ALASKA
(MODIFIED FROM NORTON OSU MITE KEYS)

1   Two (an_1, an_2) of the 5 pairs of anal/adanal plate setae (or their vestiges, well marked by alveoli) inserted on medial margin and 3 (ad_1, ad_2, ad_3) in oblique row; if 3 appear to be on margin, the posterior 2 (ad_1, an_1) are widely spaced, more than 4 alveolar diameters apart..2

–   Anal/adanal plate setae differently distributed, with 3 (an_1, an_2, ad_1) or 4 (plus ad_2) pairs closely adjacent on medial margin (at least posterior 2 setae less than 3 alveolar diameters apart)..4

2(1) Interlamellar seta erect, barbed, much larger than lamellar seta, 3× as long or more, and usually distinctly thicker. Notogaster porose and foveolate, at least anteriorly. Notogastral setae barbed. (Figure 3.30A–F)...***Hoplophthiracarus*** Jacot

–   Interlamellar seta similar to lamellar seta, at most 2× its length but still of equal thickness; both setae procumbent, thin, simple and may be inconspicuous. Notogaster rarely foveolate, usually without discernable surface sculpture, although densely porose....................3

3(2) Notogastral setae usually smooth, fine (Figures 3.16G; 3.31–3.35A–C)..................................
....***Phthiracarus*** Perty (sensu stricto)

-   Notogastral setae 15 pairs, rigid, covered with small spines; often neotrichous ....
....***Austrophthiracarus*** Balogh and Mahunka

4(1) Anal/adanal plate with 3 closely spaced setae on medial margin; seta ad_2 well mediad. Notogastral setae usually short, smooth, often spoon shaped (Figure 3.30G–I)..................
....***Hoplophorella*** Berlese

–   Anal/adanal plate with 4 closely spaced setae on the medial margin, including seta ad_2. Notogastral setae of various forms, but not spoon-shaped...5

5(4) Smaller species, notogaster usually less than 500μm long. Tibia IV with 1 distinct seta (v') and 1 solenidion (φ); seta d minute and coupled to φ, as on tibiae I–III (Figure 3.16F).......
....***Atropacarus*** Ewing

–   Larger species, notogaster usually over 500μm long. Tibia IV with 2 distinct setae (d, v') and 1 solenidion (φ); d coupled to φ on tibiae I–III, but not on tibia IV (Figure 3.35D–K).
....***Steganacarus*** Ewing

Atropacarus Ewing, 1917
   Type-species: *Hoplophora stricula* C.L. Koch, 1835
   Comb./Syn.: *Helvetacarus* Mahunka, 1993

Atropacarus striculus (C.L. Koch, 1835) (Figure 3.16F)
   Comb./Syn.: *Hoplophora stricula* C.L. Koch, 1835; non Nicolet, 1855; *Steganacarus diaphanus* Jacot, 1930; *Steganacarus striculus diaphanus* Jacot, 1938; *Steganacarus senex* Aoki, 1958
   Biology: From Sitka spruce forest floor on Vancouver Island; from oak litter. It was one of the common species in sphagnum mires in Western Norway (Seniczak et al. 2019).
   In a series of comparative experiments with different foods provided independently Hartenstein (1962a) found adults of this species (as *Steganacarus diaphanum*) fed primarily on decaying leaves, needles and wood, but also the fungus *Cladosporium*. In his study of the relationship between cheliceral shape and feeding biology, Kaneko (1988) determined that *Atropacarus striculus* adults fed 100% on higher plant material. Ponge (1991) used thin sections of soil to determine that this species (as *S. diaphanum*) always preferred the xylem of coniferous needles and of the petiole of decaying leaves, leaving vascular bundles untouched.

Juveniles of *A.* cf. *striculus* excavated cavities and fed inside decomposing needles of spruce (*Picea abies*) (Hågvar 1998). Using a no-choice feeding experiment and corroboration with CO1 (*cox1* gene) of two nematode species, the parasitic bacterivore *Phasmarhabditis hermaphrodita* and the entomopathogenic *Steinernema feltiae*, Heidemann et al. (2011) found this species fed only on *P. hermaphrodita* in laboratory experiments.

Gravid females can carry at least 4 eggs. Specimens [as *Steganacarus*] were found in the feathers of birds in Russia (Krivolutsky and Lebedeva 2004). Genetics of this species was studied by Kreipe et al. (2015) using mitochondrial and nuclear markers.

Distribution: YT, NT, BC, AB, MB, ON, QC, NB, NS, NL [AZ, CA, CN, CO, GA, IA, IL, IN, MA, MT, ME, MI, MN, NC, NH, NJ, NM, NY, OH, OR, PA, SC, SD, TN, UT, VA, VT, WA, WI, WV, WY; Semicosmopolitan]
Literature: Corral-Hernandez et al. (2016); ABMI (2019); Maraun et al. (2019).

Austrophthiracarus Balogh and Mahunka, 1978
    Type-species: *Austrophthiracarus radiatus* Balogh and Mahunka, 1978
    Comb./Syn.: *Hoplophthiracarus* Jacot, 1923 (in part)

Austrophthiracarus olivaceus (Jacot, 1929)
    Comb./Syn.: *Phthiracarus olivaceum* Jacot, 1929; *Notophthiracarus olivaceus* (Jacot, 1929); *Phthiracarus erinaceus* Jacot, 1930
    Biology: From mixed deciduous and coniferous litter.
    Distribution: ON, QC [CN, IA, IL, IN, KY, MI, MN, MO, NC, NH, NY, OH, OR, PA, TX, WI, WV; Nearctic]
    Literature: Niedbała (2002); Niedbała and Liu (2018).

Hoplophorella Berlese, 1923
    Type-species: *Hoploderma cucullatum* Ewing, 1909

Hoplophorella cucullatus (Ewing, 1909) (Figure 3.30G–I)
    Comb./Syn.: *Hoploderma cucullatum* Ewing, 1909; *Hoploderma licnophorum* Berlese, 1913; *Hoplophorella cucullata cuculloides* Jacot, 1933
    Biology: From mixed deciduous and coniferous forest litter.
    Distribution: ON [AL, AR, AZ, FL, GA, KY, LA, MA, MO, MS, OK, SC, TX; Semicosmopolitan]
    Literature: Niedbała and Liu (2018); Maraun et al. (2019).
    Note: Unidentified *Hoplophorella* species from MB, QC, NS.

Hoplophthiracarus Jacot, 1933
    Type-species: *Hoploderma histricinum* Berlese, 1908

KEY TO ADULT *HOPLOPHTHIRACARUS* OF CANADA AND ALASKA

1   Surface of body punctate (Figure 3.30A–C) ....*H. illinoisensis* (Ewing)
–   Surface of body covered with small concavities (Figure 3.30D–F)......................................
        .... *H. histricinus* (Berlese)
    Note: Unidentified *Hoplophthiracarus* species from ON, QC.

Hoplophthiracarus histricinus (Berlese, 1908) (Figure 3.30D–F)
    Comb./Syn.: *Hoploderma histricinus* Berlese, 1908; *Hoplophthiracarus robustior* Jacot, 1933; *Hoplophthiracarus pavidus*: sensu Aoki 1980
    Note: Niedbała (2002) doubted records of this species (as *H. pavidus*) from the Nearctic.
    Biology: From wide variety of forest and prairie habitats.

Distribution: ON [AL, AZ, FL GA, IL, IN, KY, LA, MO, MN, MS, OK, TX, WI, WV; Holarctic, Neotropical]
Literature: Niedbała and Liu (2018).

Hoplophthiracarus illinoisensis (Ewing, 1909) (Figure 3.30A–C)
Comb./Syn.: *Hoploderma illinoisensis* Ewing, 1909; *Hoplophthiracarus paludis* Jacot, 1938
Biology: From variety of habitats, including spruce and sphagnum (Niedbała 2002). Gut boli contain mostly coarse dark material with some fungal hyphae. Gravid females can carry at least 4 eggs (Walter et al. 2014).
Distribution: YT, AB, MB, ON, QC, NB, NS [AL, IL, CN, NY, OK, TX, WI, MO, MI, IN, KY, MA, LA, GA, FL, Semicosmopolitan]
Literature: Niedbała and Liu (2018); Maraun et al. (2019); ABMI (2019).

Phthiracarus Perty, 1841
Type-species: *Phthiracarus contractilis* Perty, 1841
Comb./Syn.: *Hoploderma* Michael, 1898; *Ginglymacarus* Ewing, 1917; *Peridromotritia* Jacot, 1923
Note: We use *Phthiracarus* in the broad sense (cf. Parry 1979) and not in the restricted sense of Balogh and Mahunka (1979).

KEY TO ADULT *PHTHIRACARUS* OF CANADA AND ALASKA
(MODIFIED FROM NIEDBAŁA 2002)

1	Bothridial seta short and wide, length not more than 10 times width	2
–	Bothridial seta long and narrow, length more than 10 times width	13
2(1)	Adanal setae ad_1 and ad_2 vestigial or minuscule	3
–	All setae of ano-adanal plates well developed	9
3(2)	Setae *d* on femur I located in middle of segment	4
–	Setae *d* on femur I located in distal part of segment	5
4(3)	Femora with 3 setae; seta *l'* on tibia IV absent (Figure 3.31A–D) .... ***P. longulus*** (C.L. Koch)	
–	Femora with 4 setae; seta *l'* on tibia IV present (Figure 3.31E–H) .... ***P. validus*** Niedbała	
5(3)	Prodorsum with dorsal carina (Figure 3.31I–M) .... ***P. globosus*** (C.L. Koch)	
–	Prodorsum without dorsal carina	6
6(5)	Seta f_1 of notogaster located anterior to seta h_1 (Figure 3.31N–P) .... ***P. nitidus*** Niedbała	
–	Seta f_1 posterior to h_1	7
7(6)	Setae *h* of mentum longer than mutual distance; bothridial seta sharply tapering distally (Figure 3.32A–C) .... ***P. compressus*** Jacot	
–	Setae *h* of mentum shorter than mutual distance; bothridial seta not sharply tapering distally	8
8(7)	Exobothridial seta considerably longer than bothridial seta (Figure 3.32D–F) .... ***P. irreprehensus*** Niedbała	
–	Exobothridial seta and bothridial seta equally long (Figure 3.32G–I) ... ***P. japonicus*** Aoki	
9(2)	Adanal setae ad_1 and ad_3 displaced proximally and almost forming a row with anal setae (Figure 3.35A–C) .... ***P. anonymus*** Grandjean **Note**: Known from Northeast USA (NY); undoubtedly in Canada.	
–	Adanal setae remote from proximal margin and not forming row with anal setae	10

Taxonomic Keys

10(9)	Four pairs of lyrifissures present: *ia*, *im*, *ip*, *ips*	11
–	Two pairs of lyrifissures present: *ia*, *im*	12
11(10)	Notogastral setation 16 pairs; bothridial seta lanceolate, curved anteriorly (Figure 3.33A–C)	***P. brevisetae*** Jacot
–	Notogastral setation 15 pairs; bothridial seta fusiform, pointed distally	***P. modestus*** Niedbała
12(10)	Notogastral setae long, $c_1 > c_1-d_1$. Seta *d* of femur I located in middle of segment (Figure 3.33D–F)	***P. bryobius*** Jacot
–	Notogastral setae short, $c_1 < c_1-d_1$. Seta *d* of femur I located distally on segment (Figure 3.33G,H)	***P. luridus*** (Ewing)
13(1)	Four pairs of lyrifissures, *ia*, *im*, *ip*, *ips* present. Genital setation 7 pairs (Figure 3.33I–K)	***P. boresetosus*** Jacot
–	Two pairs of lyrifissures *ia* and *im* present. Genital setation 9 pairs	14
14(13)	Bothridial seta smooth, swollen in proximal end, tapering distally	15
–	Bothridial seta not swollen, covered with small spines distally (Figure 3.34A–D)	***P. setosus*** (Banks)
15(14)	Adanal setae ad_1 and ad_2 vestigial (Figure 3.34E–G)	***P. lentulus*** (C.L. Koch)
–	Adanal setae ad_1 and ad_2 well developed	16
16(15)	Vestigial seta f_1 positioned anteriorly of h_1 seta. Genital setation 7 pairs (Figure 3.34H–J)	***P. aliquantus*** Niedbała
–	Vestigial seta f_1 positioned posteriorly of h_1 seta. Genital setation 6 pairs (Figure 3.34K,L)	***P. cognatus*** Niedbała

Note: Unidentified *Phthiracarus* species from BC, ON, QC, NS, NL.

Phthiracarus aliquantus Niedbała, 1988 (Figure 3.34H–J)
Biology: From moist litter in west coast forests.
Distribution: YT, BC [AZ, CA, KY, OR, PA, TN, WA; Nearctic]
Literature: Niedbała and Liu (2018).

Phthiracarus borealis (Trägårdh, 1910)
Comb./Syn.: *Hoploderma boreale* Trägårdh, 1910
Note: The type of this species is unknown. It was considered a 'species inquirendum' by Niedbała and Liu (2018), who considered it a Palaearctic species. Walter et al. (2014) considered specimens from Alberta as being this species and to be consistent with previous records from Alaska and Ontario. However, all these records may prove to be misidentifications of *P. boresetosus* Jacot, and thus, we exclude *P. borealis* from the key to species.
Biology: In a cafeteria experiment among twigs from four different deciduous trees, this species had a preference for hornbeam and oak over birch and beech (Berthet 1964). This species, and three other oribatid species consumed 0.22% of the annual primary production of an alpine meadow in Europe (Reutimann 1987). Gravid females carry at least 5 eggs. Specimens have been found in the feathers of birds in Russia (Krivolutsky and Lebedeva 2004).
Distribution: AK, AB, ON [MI; Holarctic]
Literature: Beck et al. (2014); Niedbała and Liu (2018); ABMI (2019).

Phthiracarus boresetosus Jacot, 1930 (Figure 3.33I–K)
Biology: Gut boli have coarse organic matter, clear material, and some hyphae. Gravid females carry at least 6 eggs.
Distribution: YT, BC, AB, MB, ON, QC, NB, NS, NL [AL, AZ, CA, CN, CO, CT, IA, ME, MI, MN, NC, NH, NM, NY, ON, OR, PA, SD, UT, VT, WA, WI, WV, WY; Semicosmopolitan]
Literature: Walter et al. (2014); Beck et al. (2014); Niedbała and Liu (2018); ABMI (2019).

Phthiracarus brevisetae Jacot, 1930 (Figure 3.33A–C)
Biology: From mixed temperate forest litter.
Distribution: ON, QC [AL, AZ, CA, CN, IA, IL, KY, MO, MS, NC, NH, NY, OH, OK, VA, VT, WI, WV; Nearctic, Mesoamerica]
Literature: Niedbała (2002); Niedbała and Liu (2018).

Phthiracarus bryobius Jacot, 1930 (Figure 3.33D–F)
Comb./Syn.: *Phthiracarus setosellum bryobium* Jacot, 1930
Biology: From mixed temperate forest litter.
Distribution: BC [CA, CN, IA, NY, SD, WI, WY; Holarctic]
Literature: Niedbała (2002); Niedbała and Liu (2018); Maraun et al. (2019).

Phthiracarus cognatus Niedbała, 1988 (Figure 3.34K,L)
Biology: From forest litter.
Distribution: BC, SK, ON, QC [CA, OR, TN, VT, WA; Nearctic]
Literature: Niedbała (2002); Niedbała and Liu (2018).

Phthiracarus compressus Jacot, 1930 (Figure 3.32A–C)
Biology: From forest litter (Niedbała 2002).
Distribution: NL [CA, CT, IA, ME, MS, NC, NH, NY, TX, VT, WI; Holarctic, Oriental]
Note: Niedbała and Liu (2018) rejected the synonymy of this species with *P. commutabilis* Niedbała, 1983.

Phthiracarus globosus (C.L. Koch, 1841) (Figure 3.31I–M)
Comb./Syn.: *Hoplophora globosa* C.L. Koch, 1841; *Phthiracarus globus* Parry, 1979
Biology: Common species in sphagnum mires (Seniczak et al. 2019). Its exoskeletal and muscular adaptations to ptychoidy were studied using synchrotron X-ray microtomography by Schmelzle et al. (2012).
Distribution: BC, MB, ON, QC, NS [AL, AZ, FL, GA, IA, IL, IN, KY, MA, ME, MI, MO, NC, NJ, NY, OH, OR, PA, VA, VT, WI, WV; Holarctic, Neotropical, Borneo]
Literature: Beck et al. (2014); Niedbała and Liu (2018); Maraun et al. (2019).

Phthiracarus irreprehensus Niedbała, 1988 (Figure 3.32D–F)
Biology: Collected from early decomposing conifer branch shavings.
Distribution: BC [CA, WA; Nearctic]
Literature: Niedbała (2002); Niedbała and Liu (2018).

Phthiracarus japonicus Aoki, 1958 (Figure 3.32G–I)
Comb./Syn.: *Phthiracarus miyamaensis* Fujikawa, 2004
Biology: From mixed temperate forest litter.
Distribution: BC, ON, NS [CA, KY, MA, ME, NC, NY, OR, PA, VA, VT, WA, WV; Holarctic, Oriental]
Literature: Niedbała (2002); Niedbała and Liu (2018).

Phthiracarus lentulus (C.L. Koch, 1841) (Figure 3.34E–G)
Comb./Syn.: *Hoplophora lentula* C.L. Koch, 1841; *Phthiracarus conformis* Sergienko, 1987; *Phthiracarus incertus* Niedbała, 1983; *Phthiracarus rotundus* Berlese, 1923
Biology: From mixed forest litter.
Distribution: ON, QC, NS [IL, IN, KY, NY, MI, OR, PA, TX, WV; Holarctic, Ethiopia]
Literature: Beck et al. (2014); Niedbała and Liu (2018); Maraun et al. (2019).

Phthiracarus ligneus Willmann, 1931
Comb./Syn.: *Archiphthiracarus ligneus* (Willmann, 1931); *Phthiracarus sellnicki* Feider and Suciu, 1957

Note: Niedbała (2002) considered this name a junior synonym of *Phthiracarus ferrugineus* (C.L. Koch, 1841), and considered its record from Alaska doubtful, thus, we exclude *P. ligneus* from the key to species.
Distribution: AK [Holarctic]
Literature: Niedbała (2002); Beck et al. (2014); Niedbała and Liu (2018).

Phthiracarus longulus (C.L. Koch, 1841) (Figure 3.31A–D)
Comb./Syn.: *Hoplophora longula* C.L. Koch, 1841; *Phthiracarus apiculatus* Jacot, 1939; *Phthiracarus flexisetosus* Parry, 1979; *Phthiracarus tardus* Forsslund, 1956.
Biology: From boreal mixedwood forest litter; peatlands. Its exoskeletal and muscular adaptations to ptychoidy were studied using synchrotron X-ray microtomography by Schmelzle et al. (2010, 2012).
Distribution: BC, ON, QC [AL, CA, FL, KY, MI, MS, NC, NM, NY, PA, TN, UT; Holarctic, Uruguay].
Literature: Niedbała (2002); Beck et al. (2014); Niedbała and Liu (2018); Maraun et al. (2019).

Phthiracarus luridus (Ewing, 1909) (Figure 3.33G,H)
Comb./Syn.: *Hoploderma lurida* Ewing, 1909; *Ginglymacarus lurida* (Ewing, 1909)
Biology: From temperate rainforest; boreal forest.
Distribution: BC, QC [CA, IL, OR; Nearctic]
Literature: Niedbała and Liu (2018).

Phthiracarus modestus Niedbała, 1988
Biology: From moist deciduous and coniferous litter.
Distribution: BC, ON [CA, NY, OR, SD; Nearctic]
Literature: Niedbała (2002); Niedbała and Liu (2018).

Phthiracarus nitidus Niedbała, 1986 (Figure 3.31N–P)
Biology: From deciduous and coniferous litter.
Distribution: BC, ON, QC, NS [UT; Nearctic]
Literature: Niedbała (2002); Niedbała and Liu (2018).

Phthiracarus piger (Scopoli, 1763)
Comb./Syn.: *Acarus piger* Scopoli, 1763; *Phthiracarus contractilis* Perty, 1841
Note: Niedbała (2002) considered this species a *nomen dubium*, because the diagnosis is insufficient for correct identification. We exclude it from the key.
Distribution: AK [VA; Holarctic]
Literature: Berg et al. (1990); Neidbała (2002); Beck et al. (2014); Niedbała and Liu (2018).

Phthiracarus setosus (Banks, 1895) (Figure 3.34A–D)
Comb./Syn.: *Hoplophora setosa* Banks,1895; *Metaphthiracarus bacillatus* Aoki, 1980
Biology: Kaneko (1988) determined that this species (as *Phthiracarus bacillatus*) fed 100% on higher plant material.
Distribution: NT, ON, QC, NS [CN, IA, IL, KY, MA, ME, MI, NH, NJ, NY, OH, PA, TN, TX, VA, VT, WI, WV; Holarctic]
Literature: Niedbała (2002); Beck et al. (2014); Niedbała and Liu (2018).

Phthiracarus validus Niedbała, 1986 (Figure 3.31E–H)
Biology: Reported from moss, lichen, pine litter.
Distribution: AK, BC, AB, ON, QC, NB, NS [AL, AZ, CA, CO, GA, IA, MA, ME, MI, NC, NH, NM, NY, SD, VT, WI, VA, UT, WV; Nearctic]
Literature: Niedbała (2002); Niedbała and Liu (2018); ABMI (2019).

Steganacarus Ewing, 1917
Type-species: *Hoploderma anomala* Berlese, 1883

KEY TO ADULT *STEGANACARUS* OF CANADA AND ALASKA

1   Bothridial seta enlarged distally and covered with small spines. Notogastral setae covered with small spines in distal half. Seta *l"* on femur I located at level of seta *v"* (Figure 3.35D–H) .... *S. thoreaui* Jacot
–   Bothridial seta tapered and pointed distally, smooth. Notogastral setae covered with small spines along length. Seta *l"* on femur I located proximally of seta *v'* (Figure 3.35I–K) .... *S. granulatus* (Banks)

Steganacarus granulatus (Banks, 1902) (Figure 3.35I–K)
    Comb./Syn.: *Hoploderma granulata* Banks, 1902
    Biology: Genetics of this species were discussed by Pachl et al. (2012, 2017).
    Distribution: ON [ME; Nearctic]
    Literature: Niedbała (2002); Niedbała and Liu (2018); Schäffer et al. (2020).

Steganacarus thoreaui Jacot, 1930 (Figure 3.35D–H)
    Comb./Syn.: *Hoplophorella thoreaui* (Jacot, 1930); *Rhacaplacarus* (*R.*) *thoreaui* (Jacot, 1930)
    Biology: From arctic to temperate habitats.
    Distribution: NT, ON, QC, NB, NS, NL [CT, FL, ME, MI, MS, NC, NH, NY, TN, TX, VT, WV; Nearctic]
    Literature: Niedbała (2002); Niedbała and Liu (2018).

Infraorder Desmonomata Woolley, 1973 (modified from Norton and Behan-Pelletier 2009)
    Desmonomata encompass most of the diversity in Oribatida (Norton 1998), focused in 2 of its 3 Hyporders, one of which, the Astigmata, is not dealt with herein, because of its different evolutionary trajectory and historical and traditional separation. The Hyporder Brachypylina comprises the bulk of species and family-group taxa of traditional oribatid mites. The Hyporder Nothrina includes an assemblage of seven families, with uncertain interrelationships. Because the monophyletic taxa Astigmata and Brachypylina probably both originated among these families, the Nothrina can be considered doubly paraphyletic.
    Diagnosis. *Adult*: Holoid; with macropyline or brachypyline venter. With few exceptions, opisthonotal glands present. Well-sclerotized representatives with stegasime prodorsum; bothridial seta and bothridium with strong proximal bend. Subcapitulum stenarthric, diarthric, or secondarily anarthric. Chelicerae with only vestige of trochanter, and basal portion (often the proximal third) inserted into body as apodeme, paraxially emarginated to increase angles of muscle insertion. Trägårdh's organ usually present (absent in Astigmata).

Hyporder Nothrina van der Hammen, 1982
    Nothrina is a paraphyletic assemblage also known as Nothronata (Johnston 1982), Desmonomata (Woolley 1973), Holosomata (Balogh and Mahunka 1979; Fujikawa 1991), and Nothroidea *senso lato* (Grandjean 1954b; Travé et al. 1996; Woas 2002).

Superfamily Crotonioidea Thorell, 1876

Family Crotoniidae Thorell, 1876 (incl. Camisiidae *auct.*)
    Diagnosis: *Adult*: Holoid. Prodorsum stegasime, not fused to coxisternum. Bothridial seta various but never pectinate. Ventral region macropyline, with large plates and little or no distance between genital and anal apertures. Notogaster with opisthonotal gland; setation 15 pairs (f_1 represented by vestige). Bothridial respiratory saccules or tubules present or absent. Body with cerotegument and adherent debris in many species; tritonymphal

exuvium fragment retained or not on caudal region of adult. Epimeral setation 3-1-3-3, or sparingly neotrichous (typically 1 or 2 extra pairs of setae). Epimeres contiguous; subequal in size and shape (elongated, trapezoid or subtriangular, tapering medially). Genital setation 4–24 pairs, all positioned on medial edges of genital plates. Aggenital setae present or absent. Preanal plate distinct, inconspicuous, or absent. Anal plate subequal in breadth to adanal plate. Subcapitulum usually stenarthric; rutella well developed, with distal teeth or carina. Chelicerae chelate-dentate; usually with Trägårdh's organ. Palptibia fused with tarsus dorsally. Palp tarsus with 7 or 8 setae. Adoral setation 3 pairs. Legs monodactylous or homotridactylous.

Biology. Members of this family are found worldwide and many species are camouflaged by adherent organic debris.

KEY TO GENERA OF CROTONIIDAE OF CANADA AND ALASKA
(MODIFIED FROM NORTON OSU MITE KEYS)

1 Anal setation 2 pairs; genital setation 9–24 pairs; aggenital setation 2 pairs, inserted on medial edge of aggenital plates (latter completely fused to adanal plates). Epimeral plates medially fused, with or without medial notch in posterior margin of epimere IV. Legs usually monodactylous...2

– Anal setation 3 pairs; genital setation 9 pairs; aggenital setation 2 pairs, inserted in unsclerotized cuticle laterad of genital plates (aggenital plates often not well defined). Epimeral plates not fused medially. Legs usually tridactylous (Figures 3.36, 3.37) ....
....***Camisia*** von Heyden

2(1) Posterior notogastral setae inserted on distinct apophyses; apophyses longer than wide; dorsocentral setal pairs d_1, d_2, e_1 short, not as long as mutual separation. Notogaster without medial pair of carinae. Whole integument usually with coating of organic debris (Figure 3.38A–D) ....***Heminothrus*** Berlese

– Posterior notogastral setae at most on small tubercles; tubercles as wide as long. Either notogaster with pair of medial longitudinal carinae or, if carinae absent, then dorso-central setal pairs d_1, d_2, e_1 long, about equal to or clearly longer than their mutual separation. Integument without coating of organic debris, or debris restricted to posterior region 3

3(2) Notogaster almost parallel-sided, with pair of medial longitudinal carinae; setae short, slightly lanceolate; setae d_1, e_1 inserted mediad of carinae. Bothridium without internal saccule. Genital plates uniformly sclerotized (Figure 3.38E,F)........***Neonothrus*** Forsslund

– Notogaster usually ovate, widened posteriorly, with or without paired longitudinal carinae; setae narrow, not noticeably lanceolate; if carinae present, setae d_1, e_1 inserted laterad of them. Genital plates often with broad transverse band of unpigmented cuticle (vague in cleared or weakly sclerotized specimens). Bothridium internally with saccule attached to its base (Figures 3.38G; 3.39) ....***Platynothrus*** Berlese

Camisia von Heyden, 1826
   Type-species: *Notaspis segnis* Hermann, 1804
   Comb./Syn.: *Nothrus* C.L. Koch, 1836, sensu Berlese, 1885; *Uronothrus* Berlese, 1913
   Note: Subías (2004) recognized the subgenera *Camisia* and *Ensicamisia;* we do not recognize these subgenera herein.
   Biology: Species of *Camisia* are primarily arboreal, living on the trunks of trees or in the canopy, or among mosses and lichens on rock surfaces (Travé 1963).

KEY TO ADULT *CAMISIA* OF CANADA AND ALASKA
(MODIFIED FROM COLLOFF (1993); Additional key to Nearctic species in Bromberek and Olszanowski (2012))

1	Bothridial seta completely contained within bothridium. Length: 840–1,005µm (Figure 3.36A,B)	*C. abdosensilla* Olszanowski and Clayton
–	Bothridial seta extending beyond edge of bothridium	2
2(1)	Notogastral setae narrowly phylliform, denticulate. Epimeral setae denticulate. Length: 971–985µm (Figure 3.36C)	*C. oregonae* Colloff
–	Notogastral setae setiform, spinose. Epimeral setae smooth	3
3(2)	Genital setation 18–20 pairs. Length: 708–826µm (Figure 3.36D)	*C. foveolata* Hammer
–	Genital setation 16 pairs or fewer	4
4(3)	Notogastral microsculpture strongly reticulate. Length: 751–854µm (Figure 3.36E)	*C. dictyna* Colloff
–	Notogastral microsculpture tuberculate	5
5(4)	Marginal notogastral setae mostly short, sessile; only h_2 may be on strong apophysis	6
–	All marginal notogastral setae whip-like, inserted on strong apophyses. Length: 940–1,140µm (Figure 3.36F)	*C. spinifer* (C.L. Koch)
6(5)	Seta h_2 inserted on or near corner of notogaster, not associated with lateral lobe-like process or exuvial remnants	7
–	Seta h_2 inserted on apophysis medial to small or large lobe-like lateral process. Lateral process covered or not by remnant of tritonymphal exuviae	8
7(6)	Lateral notogastral tubercle lobe-like. Tritonymphal exuvial fragment absent. Notogastral seta c_3 subequal to c_1; d_1–d_2 longer than e_1. Length: 1,190–1,220µm (Figure 3.36G)	*C. orthogonia* Olszanowski, Szywilewska and Norton
–	Lateral notogastral tubercle longer than wide. Tritonymphal exuvial fragment usually retained. Notogastral seta c_3 shorter than c_1; d_1–d_2 slightly shorter than e_1. Length: 950–1,240µm (Figure 3.37A)	*C. biurus* (C.L. Koch)
8(6)	Interlamellar setae minute	9
–	Interlamellar setae well developed, reaching half or more distance to insertion of lamellar setae	10
9(8)	Posterior notogastral seta h_2 (105–110µm) inserted on short apophysis distant from posterior corner of notogaster. Setae p_1 each inserted in funnel-like recess and positioned relatively close to each other. Transverse median ridge absent between setae e_1. Length: 1,040–1,150µm (Figure 3.37B)	*C. biverrucata* (C.L. Koch)
–	Seta h_2 on short apophysis near posterior corner of notogaster. Setae p_1 inserted on caudal ledge and relatively widely separated. Transverse carina present between setae e_1. Length: 825–960µm (Figure 3.37C)	*C. horrida* (Hermann)
10(8)	Monodactylous. Posterior notogastral margin with caudal ledge bearing setae p_1. Notogastral setae at least somewhat expanded to leaf-shaped. Interlamellar seta sessile. Length: 740–780µm (Figure 3.37D)	*C. lapponica* (Trägårdh)
–	Tridactylous. Posterior notogastral margin concave between setae h_2, caudal ledge absent. Notogastral setae setiform. Interlamellar setae on short apophyses, elongate and reaching insertion of lamellar setae. Length: 830–900µm (Figure 3.37E,F)	*C. segnis* (Hermann)

Note: Unidentified *Camisia* species from BC, AB, MB, ON, QC, NB.

Camisia abdosensilla Olszanowski, Clayton and Humble, 2002 (Figure 3.36A,B)
: **Biology**: From arboreal habitats in coniferous forests of the Pacific Northwest.
Distribution: BC [Canada]

Camisia biurus (C.L. Koch, 1839) (Figure 3.37A)
: **Comb./Syn.**: *Nothrus biurus* C.L. Koch, 1839; *Nothrus segnis* C.L. Koch, 1839: non Hermann, 1804; *Nothrus furcatus* C.L. Koch, 1839; *Nothrus pigerrimus* C.L. Koch, 1844; *Camisia exuvialis* Grandjean, 1939; *Uronothrus kochi* Willmann, 1943
Biology: A xerophilous mite found in peatlands; moss on rotten wood; grass moss turf; mixed pine-poplar litter. In the laboratory they preferred *Pleurococcus* algae as food (Ermilov and Chistyakov 2008). Evolutionary trends in opisthonotal gland secretions were studied in this species among others (Raspotnig 2010). It is included in molecular studies by Domes et al. (2007) and Schaffer et al. (2020). Only females are known.
Distribution: AK, YT, NT, NU, AB, QC, NS, NL [MI, MT, NY, WV; Holarctic, China]
Literature: Colloff (1993); Ermilov (2007); Maraun et al. (2019); ABMI (2019).

Camisia biverrucata (C.L. Koch, 1839) (Figure 3.37B)
: **Comb./Syn.**: *Nothrus biverrucatus* C.L. Koch, 1839; *Nothrus horridus:* sensu Nicolet, 1855 and Berlese, 1885; *Camisia fischeri* Oudemans, 1900; *Camisia nicoletii* Oudemans, 1900; *Camisia berlesei* Oudemans, 1900
Biology: Known from prairie grass and sod; moss on rotten wood; juniper litter; vertical moss mats; and spruce duff. Only females are known. Gravid females carry at least 6 eggs and the eggs may contain fully developed prelarvae. In the laboratory they preferred *Pleurococcus* algae as food (Ermilov and Chistyakov 2008). Gut contents include coarse dark organic matter with some fragments of hyphae and spores (Walter et al. 2014).
Distribution: AK, AB [IL, TN, UT; Holarctic]
Literature: Colloff (1993); ABMI (2019); Schäffer et al. (2020).

Camisia dictyna Colloff, 1993 (Figure 3.36E)
: **Biology**: Mainly arctic species.
Distribution: YT, NU [CA; Eastern Russia]
Literature: Bayartogtokh et al. (2011); Makarova et al. (2015).

Camisia foveolata Hammer, 1955 (Figure 3.36D)
: **Biology**: From arctic coastal tundra.
Distribution: AK, YT [northern Holarctic, Chile]
Literature: Seniczak (1991); Colloff (1993); Hågvar et al. (2009).

Camisia horrida (Hermann, 1804) (Figure 3.37C)
: **Comb./Syn.**: *Notaspis horridus* Hermann, 1804; *Nothrus mutilis* C.L. Koch, 1839; *Nothrus bistriatus* C.L. Koch, 1839; *Nothrus sinuatus* C.L. Koch, 1839; *Nothrus angulatus* C.L. Koch: sensu Berlese, 1885
Biology: Gravid females carry up to 7 eggs. Known from moss and rotten wood; moss on rocks; fallen squirrel nests; liverwort mat; mixed pine-poplar litter; and bracket fungi. Known from branch and lichen samples from Amabilis fir and Western hemlock on Vancouver Island (Winchester et al. 2008). Recorded from epiphytic lichens in the Adirondack Mountains (Root et al. 2007). Specimens have been found in the feathers of birds in Russia (Krivolutsky and Lebedeva 2004). Evolutionary trends in opisthonotal gland secretions were studied in this species among others (Raspotnig et al. 2008; Raspotnig 2010).
Distribution: AK, YT, NT, NU, BC, AB, MB, NB, NS [IL, NY, OR, VA; Holarctic, Oriental, Neotropical, Ethiopia]
Literature: Colloff (1993); Maraun et al. (2019); ABMI (2019); Schäffer et al. (2020).

Camisia lapponica (Trägårdh, 1910) (Figure 3.37D)
   Comb./Syn.: *Nothrus lapponicus* Trägårdh, 1910; *Camisia labradorica* Behan, 1978
   Biology: From mixed forests and peatlands.
   Distribution: NT, BC, QC, NL [UT; Holarctic]
   Literature: Seniczak (1991); Colloff (1993); Seniczak et al. (2019); Maraun et al. (2019).

Camisia oregonae Colloff, 1993 (Figure 3.36C)
   Biology: From coastal temperate rainforest.
   Distribution: BC [OR; Nearctic]

Camisia orthogonia Olszanowski, Szywilewska and Norton, 2001 (Figure 3.36G)
   Biology: From primarily old-growth coniferous litter. Parthenogenetic; gravid females carry 5–7 eggs (some with prelarvae).
   Distribution: BC, AB [AZ, CA, CO, NM, OR, WY; Nearctic]

Camisia segnis (Hermann, 1804) (Figure 3.37E,F)
   Comb./Syn.: *Notaspis segnis* Hermann, 1804; *Nothrus bicarinatus* C.L. Koch, 1839; *Nothrus ventricosus* C.L. Koch; *Nothrus rostratus* C.L. Koch, 1839; *Nothrus excisus* Banks, 1895.
   Biology: Parthenogenetic; gravid females carry up to 6 eggs. Known from branch and lichen samples from Amabilis fir and Western hemlock and from forest litter on Vancouver Island (Winchester et al. 2008). This mite was reported as occurring in the hair of small mammals in Slovakia (Miko and Stanko 1991). Grandjean (1950) observed this species feeding on grey lichen in nature. This species produced eggs and larvae when fed on a mixture of green algae scraped from tree trunks (Littlewood 1969). Higher plant remains and conifer pollen, fungal and arthropod fragments including collembolan scales were found in gut contents of adult and juveniles (Behan-Pelletier and Hill 1983). Genetics of this species have contributed to studies of oribatid phylogeny (Palmer and Norton 1992; Schäffer et al. 2020).
   Distribution: AK, NT, BC, AB, ON, QC, NB, NS [WA; Cosmopolitan]
   Literature: Colloff (1993); Maraun et al. (2019).

Camisia spinifer (C.L. Koch, 1835) (Figure 3.36F)
   Comb./Syn.: *Nothrus spinifer* C.L. Koch, 1835; *Nothrus echinatus* C.L. Koch, 1835; *Nothrus sordidus* C.L. Koch, 1839; *Nothrus taurinus* Banks, 1906.
   Biology: This species produced eggs and larvae when fed on a mixture of green algae scraped from tree trunks (Littlewood 1969). Hartenstein (1962) found *C. spinifer* to feed almost equally on decaying plant material and fungi. In the laboratory they preferred *Pleurococcus* algae as food (Ermilov and Chistyakov 2008). This mite is known from lodgepole pine litter and mixed pine litter. It is found in sphagnum mires in Western Norway (Seniczak et al. 2019). Specimens have been found in the feathers of birds in Russia (Krivolutsky and Lebedeva 2004) ,and in the hair of small mammals in Slovakia (Miko and Stanko 1991). Evolutionary trends in opisthonotal glands was studied in this species among others (Raspotnig et al. 2008).
   Distribution: AK, YT, BC, AB, ON [NY, NC, OR, TN, VA; Semicosmopolitan]
   Literature: Colloff (1993); Maraun et al. (2019); ABMI (2019); Schäffer et al (2020).

Heminothrus Berlese, 1913
   Type-species: *Nothrus targionii* Berlese, 1885
   Comb./Syn.: *Heminothrus* (*Capillonothrus*) Kunst, 1971; *Ovonothrus* Kunst, 1971

KEY TO ADULT *HEMINOTHRUS* OF CANADA AND ALASKA

1	Bothridial seta linear to slightly expanded distally. Prodorsum foveolate	2
–	Bothridial seta with rhombic head. Prodorsum finely punctate, without foveae. Length: 530–550μm (Figure 3.38A,B)	***H. minor*** Aoki

2(1) Posterior tubercles longer than wide. Median notogastral setae relatively short, not passing next most posterior seta in series 3
– Posterior tubercles about as long as wide. Median notogastral setae very long, passing next most posterior seta in series. Length: ca. 805µm (Figure 3.38G) ....
.... *H. (Capillonothrus)/Platynothrus thori* (Berlese)
Note: Keyed here, but considered a *Platynothrus*.

3(2) Setae on margins of notogaster long, extending well beyond next seta in series. Epimeral plates IV medially separated by soft cuticle. Length: ca. 675µm (Figure 3.38C)................
....*H. longisetosus* Willmann
– Setae on margins of notogaster shorter, just passing next seta in series. Epimeral plates IV fused medially. Length: 936–972µm (Figure 3.38D) ....*H. targionii* (Berlese)
Note: Unidentified *Heminothrus* species from BC, QC, NS.

Heminothrus longisetosus Willmann, 1925 (Figure 3.38C)
  Biology: From boreal forest and subarctic. In the laboratory they preferred *Pleurococcus* algae as food (Ermilov and Chistyakov 2008).
  Distribution: AK, YT, BC, AB, ON, QC, NS, NL [MT; Holarctic, Taiwan]
  Literature: Domes et al. (2007); Maraun et al. (2019); ABMI (2019); Schäffer et al (2020).

Heminothrus minor Aoki, 1969 (Figure 3.38A,B)
  Biology: Fed primarily on fungi and undetermined fine particulates (Kaneko 1988).
  Distribution: AB [Holarctic, Oriental]

Heminothrus targionii (Berlese, 1885) (Figure 3.38D)
  Comb./Syn.: *Nothrus targionii* Berlese, 1885; *Nothrus princeps* Berlese, 1916
  Biology: Evolutionary trends in opisthonotal glands was studied in this species among others (Raspotnig et al. 2008; Raspotnig 2010).
  Distribution: AB, QC [Holarctic, Oriental]
  Literature: Maraun et al. (2019); ABMI (2019).

Neonothrus Forsslund, 1955
  Type-species: *Neonothrus humicola* Forsslund, 1955

Neonothrus humicola Forsslund, 1955 (in Sellnick and Forsslund 1955) (Figure 3.38E,F)
  Biology: Parthenogenetic; gravid females carry 1–2 eggs. Gut boli have coarse organic matter, including some fungal hyphae.
  Distribution: AK, BC, AB [MT; Holarctic]
  Literature: ABMI (2019).

Platynothrus Berlese, 1913
  Type-species: *Nothrus peltifer* C.L. Koch, 1839
  Note: Some authors, e.g., Subías (2004), subsume *Platynothrus* into *Heminothrus* and vary in the placement of many species between the two genera.

KEY TO ADULT *PLATYNOTHRUS* OF CANADA AND ALASKA

1 Notogastral setae relatively long, most reaching or passing insertion of the next more posterior seta 2
– Notogastral setae short, at most reaching only half distance to next more posterior seta. Length: ca. 864µm (Figure 3.39A) .... *P. punctatus* (L. Koch)

2(1) Interlamellar setae relatively short, not passing insertion of lamellar setae. Notogastral setae of moderate length, e_1 not reaching level of h_2. Body often clean 3

	Interlamellar setae long, passing insertion of lamellar setae. Notogastral setae very long, extending well past next more posterior seta, e_1 reaching level of h_2. Organic debris often adherent to posterior notogaster and legs 4
3(2)	Median notogastral carinae well developed and separate anteriorly. Length: 958–1,067µm (Figure 3.39B) .... ***P. peltifer*** (C.L. Koch)
–	Median carinae fused into single ridge anteriorly. Length: 790–920µm (Figure 3.39C) .... ***P. banksi*** (Michael)
4(2)	Lateral notogastral carinae absent 5
–	Lateral notogastral carinae well developed 6
5(4)	Bothridial seta long, longer than length of interlamellar seta. Notogaster irregularly punctuate medially. Genital setation 14–16 pairs. Length: 954–1,010µm (Figure 3.38G) .... ***P. thori*** (Berlese)
–	Bothridial seta short, less than a third length of interlamellar seta, Notogaster with reticulate pattern. Genital setation 19–21 pairs. Length: 960–1,060µm (Figure 3.39D) .... ***P. capillatus*** (Berlese)
6(4)	Median carinae present. Trochanters III with 4 setae on outer margin. Length: ca. 767µm (Figure 3.39E) .... ***P. sibiricus*** Sitnikova
–	Median carinae absent. Trochanters III with 3 setae on outer margin. Length: 650–690µm (Figure 3.39F) .... ***P. yamasakii*** (Aoki)

Note: Unidentified *Platynothrus* species from BC, AB, QC, NB, NS.

Platynothrus banksi (Michael, 1898) (Figure 3.39C)
Comb./Syn.: *Nothrus banksi* Michael, 1898; *Nothrus furcatus* Banks, 1895
Biology: From coastal temperate rainforest. This species was reared experimentally by Palmer and Norton (1990).
Distribution: BC [OR, WA; Nearctic, China]
Literature: Norton (1979).

Platynothrus capillatus (Berlese, 1914) (Figure 3.39D)
Comb./Syn.: *Angelia capillata* Berlese, 1914; *Heminothrus septentrionalis* (Sellnick, 1944)
Biology: From forest litter; peatlands.
Distribution: BC [CA; Holarctic]
Literature: Weigmann (2006); Seniczak et al. (1990); Seniczak et al. (2019).

Platynothrus peltifer (C.L. Koch, 1839) (Figure 3.39B)
Comb./Syn.: *Nothrus peltifer* C.L. Koch, 1839; *Nothrus palliatus* C.L. Koch, 1839; *Angelia palliata* (C.L. Koch, 1839); *Nothrus cirrosus* Canestrini and Fanzago, 1876
Biology: This is the best-studied crotoniid mite; it is widespread both geographically and ecologically. It inhabits forest soil and litter, mosses, peatlands, various freshwater habitats, and has been found in benthic habitats (Schatz and Gerecke 1996). Adults are relatively tolerant of drought and heat extremes (Siepel 1996). Median lethal time submerged under water was 151 days; animals were active, moving around and feeding on the filter paper. Larviparous reproduction of *P. peltifer* under water was observed as two juveniles appeared after 13 days and survived 16 and 108 days (Bardel and Pfingstl 2018). Generation time in temperate European forests is probably at least 1 year (Weigmann 1975). Palmer and Norton (1992) studied the genetic diversity of this parthenogentic species. Gravid females have up to 14 eggs. Opisthonotal gland secretions of this species are diverse (Raspotnig 2010).

Heethoff et al. (2007) found 7 major clades of *P. peltifer* based on the *cox1* gene from two sites in North America (NY, WA), two in Japan, and 12 in Europe. The clades were geographically separate and had relatively low divergence internally (<2%); however, genetic distances between clades averaged 56%. These authors considered that *P. peltifer* has been around for perhaps 100 million years and its present distribution may be largely a result of continental drift. Molecular studies of the ribosomal internal transcribed spacer region 1 (ITS1) indicated that *P. peltifer* has a general-purpose genotype adapted to a wide range of habitats (Heethoff 2000).

Platynothrus peltifer has been widely used in ecotoxicological studies; e.g., Dennemann and Van Straalen (1991), Crommentuijn et al. (1995), Lebrun and Van Straalen (1995), Wei et al. (2017), and has been found to have good tolerance to metal contamination from smelters (Zaitsev and Van Straalen 2001).

According to many authors *P. peltifer* is a non-specialized feeder, for example, Berthet (1964) reported it as mainly fungivorous, while Wallwork (1967) found it mainly macrophytophagous. In a series of comparative experiments with different foods provided independently Hartenstein (1962) found this species to feed almost equally on decaying plant material and fungi. Behan-Pelletier and Hill (1983) found moss fragments, pine pollen, fungal and arthropod fragments, with conifer pollen comprising 30–40% of gut contents, seasonally. Gut microbiota of this species was studied by Gong et al. (2018). Heidemann et al. (2011) found *P. peltifer* fed on live and dead specimens of the parasitic nematodes, *Phasmarhabditis hermaphrodita and Steinernema feltiae*. In the laboratory they had a wide food spectrum, but preferred *Pleurococcus* algae as food (Ermilov and Chistyakov 2008). Schneider (2005) placed *P. peltifer* in the primary decomposer feeding guild based on stable isotope analysis. Associated with this it has chelicerae with a high leverage index (Perdomo et al. 2012).

Specimens have been found in the feathers of birds in Russia (Krivolutsky and Lebedeva 2004). This mite also occurs in the hair of small mammals in Slovakia (Miko and Stanko 1991). Specimens are known from tertiary fossil sites on Meigen Island (NU) (Matthews et al. 2019).

Distribution: AK, YT, NT, NU, BC, AB, ON, QC, NS, NL [IL, KY, MI, MO, NC, NY, OH, TN, VA, WV; Semicosmopolitan]
Literature: Domes et al. (2007); Maraun et al. (2019); ABMI (2019).

Platynothrus punctatus (L. Koch, 1879) (Figure 3.39A)
Comb./Syn.: *Nothrus punctatus* L. Koch, 1879; *Hermannia carinata* Kramer, 1897; *Heminothrus valentianus* Hull, 1916
Biology: Common in arctic soils. It has been found in the feathers of birds (Lebedeva et al. 2006).
Distribution: AK, YT, NT, NU, MB, ON, NS [Holarctic]
Literature: Makarova et al. (2015); Seniczak et al. (2019); Maraun et al. (2019).

Platynothrus sibiricus Sitnikova, 1975 (Figure 3.39E)
Biology: Females can carry 4–5 eggs (Walter et al. 2014).
Distribution: AK, BC, AB [Holarctic]
Literature: Wang and Norton (1988); ABMI (2019).

Platynothrus thori (Berlese, 1904) (Figure 3.38G)
Comb./Syn.: *Angelia thori* Berlese, 1904; *Heminothrus* (*Capillonothrus*) *thori* (Berlese, 1904); *Heminothrus thori* (Berlese, 1904)
Note: This species is keyed under its synonyms in recent publications, e.g., as *Heminothrus thori* in Walter et al. (2014).
Biology: Found in sphagnum mires in Western Norway (Seniczak et al. 2019). Specimens have been found in the feathers of birds in Russia (Krivolutsky and Lebedeva 2004). In the laboratory they preferred *Pleurococcus* algae as food (Ermilov and Chistyakov 2008). Genetic variability was studied by Heethoff et al. (2000).
Distribution: AB, MB, QC, NL [Holarctic, Oriental]
Literature: Seniczak (1990).

Platynothrus yamasakii (Aoki, 1958) (Figure 3.39F)
Comb./Syn.: *Heminothrus yamasakii* Aoki, 1958
Biology: From forest soil.
Distribution: AB, ON, NB [Holarctic, Oriental]
Literature: Wang and Norton (1988); ABMI (2019).

Family Hermanniidae Sellnick, 1928
Diagnosis: *Adults*: Holoid. Ventral region brachypyline. Bothridial seta various but never pectinate; bothridium without respiratory saccules or tubules. Notogaster with opisthonotal gland; holotrichous, with 16 pairs of notogastral setae, f_1 present. Epimere II with 1 pair of setae. Aggenital setae present; genital plate with 6 pairs of setae; preanal plate absent. Subcapitulum diarthric; rutella well developed, with distal teeth or carina. Chelicerae chelate-dentate; with Trägårdh's organ. Adoral setation 3 pairs. Palptibia fused with tarsus dorsally. Legs mono- or tridactylous.

Hermannia Nicolet, 1855
Type-species: *Nothrus gibbus* C.L. Koch, 1839 (= *Hermannia crassipes* Nicolet, 1855)

KEY TO ADULT *HERMANNIA* OF CANADA AND ALASKA

1	Notogaster with conspicuous nodules throughout	2
–	Notogaster with reticulate sculpturing throughout, or present anterior to setae *e*	3
2(1)	Notogastral nodules large, distinct. Bothridial seta about 100μm. Interlamellar seta broadened distally, spatulate. Notogastral seta f_1 overlapping base of seta h_1. Epimeral setation 3-1-3-4. Length: 780–940μm (Figure 3.40A,B) .... ***H. gibba*** (C.L. Koch)	
–	Notogastral nodules granular, very small, dispersed, may collasce to form longitudinal lines. Bothridial seta clavate, 60–70μm. Interlamellar seta short, truncate, setiform. Notogastral seta f_1 not overlapping base of seta h_1. Epimeral setation 3-1-4/5-6/8. Length: 1,000–1,150μm (Figure 3.40C) .... ***H. scabra*** (L. Koch)	
3(1)	Notogastral integument reticulate; sculpturing with large cells. Epimeral setation 2/3-1/2-2–5-3–6. Length: 734–940μm (Figure 3.40D)............***H. reticulata*** Thorell	
–	Notogastral integument punctate, or sculpturing with small cells. Epimeral setation either 3-1-5-5 or 3-1-5-6 4	
4(3)	Sculpturing weakly reticulate, cells 15–20μm. Notogastral seta f_1 not overlapping base of seta h_1. Epimeral setation 3-1-5-5. Length: 1,010–1,220μm (Figure 3.40E,F)................. ***H. subglabra*** Berlese	
–	Sculpturing reticulate, cells 7–10μm, anterior to seta e_1; sculpturing forming longitudinal low ridges posteriorly. Epimeral setation 3-1-5-6. Length: 790–900μm (Figure 3.40G) ....***H. hokkaidensis*** Aoki and Ohnishi	

Note: Unidentified *Hermannia* sp. from AK: Lava Camp, Seward Peninsula (Late Miocene: 5.7 mya) (Matthews et al. 2019). Unidentified extant *Hermannia* species from QC, NS.

Hermannia gibba (C.L. Koch, 1839) (Figure 3.40A,B)
Comb./Syn.: *Nothrus gibbus* C.L. Koch, 1839; *Hermannia crassipes* Nicolet, 1855
Biology: Unspecialized feeder on decaying litter and fungi (Nannelli 1990). Morphology, and biology of this species was studied by Bäumler (1970a,b); musculature by Yastrebstov (1987) and their reproductive system and microorganisms were studied by Liana and Witaliński (2010a,b; 2012). Palmer and Norton (1992) studied the genetic diversity of this parthenogenetic species.
Distribution: AK, BC [MI, WA; Holarctic]

Hermannia hokkaidensis Aoki and Ohnishi, 1974 (Figure 3.40G)
   Biology: From coastal tundra and beach debris.
   Distribution: YT, BC [Japan]

Hermannia reticulata Thorell, 1871 (Figure 3.40D)
   Comb./Syn.: *Hermannia quadriseriata* Banks, 1899
   Biology: From coastal tundra; beach wrack. Morphology and population biology of this species was described for Svalbard by Seniczak et al. (2017a,b).
   Distribution: AK, YT, NT, NU, BC, QC [Holarctic]
   Literature: Ermilov et al. (2012); Maraun et al. (2019).

Hermannia scabra (L. Koch, 1879) (Figure 3.40C)
   Comb./Syn.: *Nothrus scabra* L. Koch, 1879; *Lohmannia scabra* (L. Koch, 1879); *Hermannia gigantea* Sitnikova, 1975
   Biology: From tussock tundra. Population ecology of this species was described for Svalbard by Seniczak et al. (2017a,b) and Ermilov et al. (2019).
   Distribution: AK, NU [VA; Holarctic]

Hermannia subglabra Berlese, 1910 (Figure 3.40E,F)
   Comb./Syn.: *Hermannia pulchella* Willmann, 1952
   Biology: From shrub tundra.
   Distribution: AK, YT, NT, NU, NS [Holarctic]
   Literature: Seniczak et al. (2017).

Family Malaconothridae Berlese, 1916
   Diagnosis: *Adult*: Holoid, but with narrow ventrosejugal articulation; with macropyline venter. Cerotegument birefringent. Prodorsum stegasime. Bothridial seta absent. Notogaster with centrodorsal ridges and M-shaped posterior ridges, or ridges secondarily lost. Body setae typically smooth, occasionally barbed. Notogaster with opisthonotal gland; with 15 pairs of setae. Anterolateral margin of epimeral plates I rounded or with pronounced projecting spur. Ventral region with little or no distance between genital and anal apertures. Caudal region of hysterosoma typically U-shaped. Genital setation 4–9 pairs. Lyrifissure *ian* present. Subcapitulum anarthric. Trägårdh's organ absent. Palptarsus subequal in length to palptibia; palpfemur without seta. Adoral setation 3 pairs. Legs monodactylous or tridactylous. Tarsi, particularly tarsus I, short, broad, typically no more than 1.5–2× longer than maximum width. Tibia I with 1 solenidion.
   Note: A phylogenetic analysis of this family, based on morphological analysis of all species, identified two major clades (Colloff and Cameron 2013). Malaconothridae are considered a sister group to Astigmata in Norton (1998) which is a key reference for this family.
   Biology: Species are considered freshwater inhabitants that need saturated air to reproduce; they are found in wet moss, wet meadows, sodden organic debris, and freshwater (Schatz and Behan-Pelletier 2008).

KEY TO GENERA OF MALACONOTHRIDAE OF CANADA AND ALASKA
(MODIFIED FROM NORTON OSU MITE KEYS)

1   Prodorsal carina strongly incurved anteriorly, S-shaped, broad, enclosing lamellar seta. Interlamellar seta very long, flagelliform, at least 3–5× the length of exobothridial seta. Caudal region of hysterosoma usually broader than anterior region. Genital setation 5–12 pairs. Leg tarsi gradually narrowed distally. Seta *ft"* of tarsi I–III setiform, attenuate, usually curved distad (Figure 3.41C–E)...***Tyrphonothrus*** Knülle

– Prodorsal carina narrow, straight or curved, extending as far as rostral seta; never enclosing lamellar seta. Interlamellar seta typically 2–3× longer than exobothridial seta, or occasionally both setae short, sub-equal. Caudal region of hysterosoma usually U-shaped, as broad or narrower than anterior region; notogastral margins parallel. Genital setation 4–9 pairs. Leg tarsi short, broad, not noticeably tapering. Seta *ft"* of tarsi I–III not noticeably curved, short, blunt, spiniform (Figure 3.41A,B) .... ***Malaconothrus*** Berlese

Malaconothrus Berlese, 1904
Type-species: *Nothrus monodactylus* Michael, 1888

Malaconothrus mollisetosus Hammer, 1952 (Figure 3.41A)
Note: This is considered a synonym of *M. monodactylus* (Michael, 1888) in many treatments, e.g., Subías (2004), however, Walter et al. (2014) considered it distinct, pending further study. We retain this species, herein.
Biology: This species was described as common in moss and litter in wet meadows at Yellowknife and Churchill (Hammer 1952a). Gut contents of Alberta specimens include extensive brown fungal hyphae and bits of organic matter (Walter et al. 2014). Parthenogenetic: gravid females carry a single large egg.
Distribution: AK, YT, NT, AB, MB, ON, QC [MT; Nearctic]
Literature: ABMI (2019).
Note: Unidentified or undescribed *Malaconothrus* species from AK, BC, MB, ON, QC, NB, NS, NL.

Tyrphonothrus Knülle, 1957
Type-species: *Malaconothrus novus* Sellnick, 1921 (= *Malaconothrus maior* Berlese, 1910).
Comb./Syn.: *Trimalaconothrus* (*Tyrphonothrus*) Knülle, 1957; *Fossonothrus* Hammer, 1962

KEY TO ADULT *TYRPHONOTHRUS* OF CANADA AND ALASKA

1 Notogastral seta d_1 inserted closer to c_1 than to e_1. Notogastral seta h_2 longer than seta h_1. Length: 520–670µm (Figure 3.41D) ....***Ty. maior*** (Berlese)
 Notogastral seta d_1 inserted at equal distance between c_1 and e_1. Notogastral seta h_2 shorter than seta h_1. Length: 380–440µm (Figure 3.41E) ....***Ty. foveolatus*** Willmann
 Note: Unidentified or undescribed *Tyrphonothrus* species from AK, BC, AB, ON, NB, NS.

Tyrphonothrus foveolatus (Willmann, 1931) (Figure 3.41E)
Comb./Syn.: *Trimalaconothrus foveolatus* Willmann, 1931
Biology: An aquatic mite known from bogs and peatlands. In the laboratory it had a wide food spectrum, but preferred *Pleurococcus* algae as food (Ermilov and Chistyakov 2008). Parthenogenetic.
Distribution: AK, AB, ON, NB [Holarctic]
Literature: ABMI (2019); Schäffer et al. (2020).

Tyrphonothrus maior (Berlese, 1910) (Figure 3.41D)
Comb./Syn.: *Malaconothrus maior* Berlese, 1910; *Malaconothrus novus* Sellnick, 1921; *Trimalaconothrus intermedius* Cooreman, 1941; *Malaconothrus sphagnicola* Trägårdh, 1910
Biology: An aquatic mite found in peatlands. In the laboratory it had a wide food spectrum, but preferred *Pleurococcus* algae as food (Ermilov and Chistyakov 2008). Parthenogenetic; apparently ovoviviparous, females carry up to 4 larvae internally.
Distribution: AK, NT, NU, AB, ON, QC, NB, NS, NL [FL, MT, NH, NY, VA; Holarctic; Australian; Neotropics]
Literature: ABMI (2019); Schäffer et al. (2020).

Family Nanhermanniidae Sellnick, 1928
Diagnosis: *Adult*: Holoid; Body cylindrical, somewhat elongated. Prodorsum stegasime; posteriorly with tubercles or ridges directed over dorsosejugal scissure. Bothridium without respiratory saccules or tubules. Notogaster indistinguishably fused to ventral region in posterior 1/2, giving appearance of pair of crescent-shaped scissures directed between widely separated genital and anal plates (**digastry**). Notogaster with opisthonotal gland; with 15 pairs of setae. Epimere II with 1 pair of setae. Aggenital setae present or absent. Genital setation 4–24 pairs. Preanal plate absent. Lyrifissures *iad* and *ian* present. Subcapitulum stenarthric; rutella well developed. Chelicerae usually robust, chelate-dentate; Trägårdh's organ present.

Nanhermannia Berlese, 1913
Type-species: *Nothrus nanus* Nicolet, 1885.
Remarks: *Nanhermannia* of North America are in need of revision, especially to resolve the possible synonymy of *N. coronata* with *N. dorsalis* by Jacot (1937a) (see: Norton and Kethley (1990)). *Nanhermannia dorsalis* is a common Eastern U.S. species, found in forests and variable in the form of prodorsal ridges. Following Norton (2021, pers. comm.), we consider *Nanhermannia dorsalis* distinct from *N. coronata*, and that both *N. dorsalis* and *N. nana*, common bog species in northeastern North America, may coexist. Data from Marshall et al. (1987) are outdated, and it is unclear which records of *N. nana* (Nicolet) from North America are valid.

KEY TO ADULT *NANHERMANNIA* OF CANADA AND ALASKA

1 Curved ventral scissures nearly meeting medially, separated by less than width of genital aperture. Notogastral sculpturing of large round to angular foveae; largest more posteriorly on notogaster. Central area of prodorsum equally foveate medial and lateral to lamellar setae. Body about 2.3–2.4× longer than wide. Length: 510–600µm (Figure 3.42A,B) .... ***N. elegantula*** Berlese
– Curved ventral scissures separated at least by width of genital aperture. Notogastral sculpturing of small, round foveae; subequal in size. Central area of prodorsum foveate medially, almost smooth laterally. Body about 1.9–2.2× longer than wide 2

2(1) Prodorsal tubercles strongly triangular, terminating with one strong point. Length: ca. 490µm (Figure 3.42C,D) .... ***N. nana*** (Nicolet)
– Prodorsal tubercles broad, irregular or subtriangular, terminating without strong point .. 3

3(2) Central region of prodorsum broad, weakly foveate. Genua I and II with distally broadened, shallowly bifurcating setae. Length: 610–660µm (Figure 3.42E,F)........................... ....***N. comitalis*** Berlese
Note: Species near *N. comitalis* from BC (Lindo and Winchester 2006).
– Central region of prodorsum narrow, distinctly foveate. Genua I and II without large bifurcating setae distally.. 4

4(3) Prodorsal tubercles in form of transverse ridge, with 6-7 blunt knobs, barely overhanging dorsosejugal scissure. Notogastral setae thin, attenuate, flexible. Length: 480–520µm (Figure 3.42G,H) .... ***N. coronata*** Berlese
Prodorsal tubercles broadly subtriangular, well overhanging dorsosejugal scissure. Notogastral setae thick. Length: ca. 580µm (Figure 3.42I–K).............. ***N. dorsalis*** (Banks)
Note: Undescribed *Nanhermannia* species from YT, BC, AB, MB, QC, NS, NL.

Nanhermannia coronata Berlese, 1913 (Figure 3.42G,H)
   Biology: From bogs and peatlands.
   Distribution: No distribution records, but species possibly is widely present in eastern North America (see Remark under Genus). [Holarctic]
   Literature: Norton and Kethley (1990).

Nanhermannia dorsalis (Banks, 1896) (Figure 3.42I–K)
   Comb./Syn.: *Carabodes dorsalis* Banks, 1896
   Biology: In oak litter (Lamoncha and Crossley 1998). This species was reared experimentally by Palmer and Norton (1990) and Palmer and Norton (1992) studied the genetic diversity of this parthenogenetic species.
   Distribution: AK, YT, NT, ON, QC, NS [NC, NY, WV; Nearctic]
   Literature: Norton and Kethley (1990); Maraun et al. (2019).

Nanhermannia elegantula Berlese, 1913 (Figure 3.42A,B)
   Comb./Syn.: *Nanhermannia areolata* Strenzke, 1953
   Biology: Macrophytophage (Schuster, 1956; Hartenstein, 1962). Over a 20 month period feeding preference was primarily for leaf material (Anderson 1975). Using soil thin sections Ponge (1991) observed that adults show a marked preference for needles which are very decomposed and usually contain growing fungi. This species was reared experimentally by Palmer and Norton (1990).
   Distribution: BC, QC, NS [NC, OR; Holarctic]
   Literature: Weigmann (2006); Maraun et al. (2019).

Nanhermannia nana (Nicolet, 1855) (Figure 3.42C,D)
   Comb./Syn.: *Nothrus nanus* Nicolet, 1855; *Hermannia nana* (Nicolet, 1855); *Nanhermannia dorsalis elegantula* Jacot, 1937; *Nanhermannia elegantula* Willmann, 1931, sensu Strenzke, 1953
   Note: It is unclear which records of *N. nana* (Nicolet) from North America are valid.
   Biology: *Nanhermannia nana* was successfully reared on lichen (Grandjean 1950) and *Protococcus* algae and moss (Sengbusch 1954). Behan-Pelletier and Hill (1983) found grass pollen, fungal and arthropod fragments including collembolan scales in guts of adults and juveniles.
   Distribution: MB, QC [Semicosmopolitan]
   Literature: Weigmann (2006); Maraun et al. (2019).

Family Nothridae Berlese, 1896
   Diagnosis: *Adult*: Holoid. Prodorsum stegasime. Bothridial seta rod-like or whip-like; bothridium with respiratory saccules or tubules. Ventral region macropyline, with large plates and little or no distance between genital and anal apertures. Notogaster with opisthonotal gland; with 16 pairs of setae (f_1 present). Epimeres I with 3–9 pairs of setae; II with 3–6; III with 3–6; IV with 3–7. Aggenital setae absent. Genital setation 7 pairs. Preanal plate present. Lyrifissure *iad* and *ian* present. Subcapitulum stenarthric; rutella well developed, with distal teeth or carina. Chelicerae chelate-dentate; with Trägårdh's organ. Palp setation: 0-1-1-3-9(1). Adoral setation 2 pairs. Palptibia fused with palptarsus dorsally.

Nothrus C.L. Koch, 1835
   Type-species: *Nothrus palustris* C.L. Koch, 1839
   Comb./Syn.: *Angelia* Berlese, 1885; *Gymnonothrus* Ewing, 1917

KEY TO ADULT *NOTHRUS* OF CANADA AND ALASKA

1	Legs monodactylous or bidactylous	2
–	Legs tridactylous	6

2(1)		Median notogastral setae (c_1–f_1) and h_1 relatively short acicular to foliate or club-shaped. Lamellar setae shorter than distance between insertions. Monodactylous..........................3
–		Median notogastral setae and h_1 relatively long, slightly swollen distally; setae h_2 very long. Lamellar setae relatively long, as long as or longer than distance between insertions. Monodactylous or bidactylous. Length: 710–810μm (Figure 3.43A,B)..*N. silvestris* Nicolet
3(2)		Setae on posterior margin of notogaster leaf-like to club-shaped; not much longer than other setae..........................4
–		Posterior setae of notogaster narrow, not leaf-like; h_2 much longer than other setae (up to 150μm). Length: 760–930μm (Figure 3.43C) ....*N. pratensis* Sellnick
4(3)		Lamellar seta club-shaped..........................5
–		Lamellar seta thick, not tapering. Length: 735–785μm .... *N. monodactylus* (Berlese)
5(4)		Notogaster convex posteriorly. Length: 675–710μm (Figure 3.43D,E) ....*N. parvus* Sitnikova
–		Notogaster strongly truncate posteriorly. Length: 920–935μm.............*N. truncatus* Banks
6(1)		Notogastral setae h_2 very long, whip-like, about half length of body; setae c_2 inserted near c_3. Length: 990–1,200μm (Figure 3.43F)...*N. palustris* C.L. Koch
–		Notogastral setae h_2 much shorter, expanded; setae c_2 inserted distant from c_3 7
7(6)		Setae on posterior margin of notogaster leaf-like to club-shaped. Interlamellar seta club-like. Lamellar seta thick, not tapering. Bothridial seta slightly swollen distally. Often with adherent debris; grey brown. Length: 700–1,050μm (Figure 3.43G).. *N. anauniensis* Canestrini and Fanzago
–		Setae on posterior margin of notogaster longer, fine, bacilliform. Bothridial seta setiform, not swollen distally. Red brown. Length: 890–1,020μm (Figure 3.43H)..*N. borussicus* Sellnick

Note: Unidentified *Nothrus* species from BC, AB, ON, QC, NB, PE.

Nothrus anauniensis Canestrini and Fanzago, 1876 (Figure 3.43G)
Comb./Syn.: *Nothrus biciliatus* sensu Sellnick and Forsslund, 1955; non C.L. Koch, 1841
Biology: Parthenogenetic. This species was reared experimentally by Palmer and Norton (1990). A widely distributed and eurytopic mite from mesophilous habitats, bogs, wet forests, compost. Hartenstein (1962) found this species to feed primarily on decaying leaves, needles and wood. Kaneko (1988) determined that this species fed 100% on higher plant material. In the laboratory they had a wide food spectrum, but preferred *Pleurococcus* algae as food (Ermilov and Chistyakov 2008). Specimens have been found in the feathers of birds in Russia (as *Nothrus biciliatus*) (Krivolutsky and Lebedeva 2004).
Distribution: AK, BC, AB, ON, QC, NL [NY; Semicosmopolitan]
Literature: Norton and Kethley (1990); Olszanowski et al. (2007); Ermilov and Liao (2018); Maraun et al. (2019); ABMI (2019).

Nothrus borussicus Sellnick, 1928 (Figure 3.43H)
Comb./Syn.: *Nothrus biciliatus* sensu Trägårdh, 1904; non C.L. Koch, 1841; *Nothrus silvestris* sensu Jørgensen, 1934; non Nicolet, 1855
Biology: Parthenogenetic; gravid females with up to 4 eggs. Gut contents consist of coarse fragments of organic material and fungal hyphae (Walter et al. 2014). In a cafeteria experiment, specimens from an alpine meadow preferred the oldest leaves of *Sesleria*, whole shoots of *Carex* and older leaf litter of *Dryas* (Reutimann 1991). Adults also fed on protococcal algae, and on microorganisms on the leaf surfaces. Females laid their eggs randomly on plant material, and juveniles preferred the distal parts of the leaves as food.

Distribution: AK, YT, NT, BC, AB, MB, ON, QC [Holarctic]
Literature: Maraun et al. (2019); ABMI (2019); Schäffer et al. (2020).

Nothrus monodactylus (Berlese, 1910)
Comb./Syn.: *Angelia anauniensis monodactylus* Berlese, 1910; *Nothrus terminalis* Banks, (1910)
Biology: From leaf mould in the riparian (Norton and Kethley 1990).
Distribution: ON, QC [FL, TX; Nearctic]
Literature: Norton and Kethley (1990).

Nothrus palustris C.L. Koch, 1839 (Figure 3.43F)
Comb./Syn.: *Angelia palustris* (C.L. Koch, 1839)
Biology: Parthenogenetic. Subject of numerous studies on its population dynamics, feeding habits, and influence of temperature on development (e.g., Lebrun 1968, 1970, 1984; Lebrun et al. 1991; Grishina 1993). Akimov and Yastrebstov (1989) described its musculature. The opisthonotal gland secretions of adult *N. palustris* are composed mainly of a monoterpene, dehydrocineole, and a hydrocarbon. In contrast, nymphs produce neral and geranial, the latter of which functions as an alarm pheromone (Shimano et al. 2002; Raspotnig 2010). *N. palustris* is quite tolerant of metal contamination from smelters (Zaitsev and Van Straalen 2001). Schneider (2005) placed this mite in the primary decomposer feeding guild (i.e., feeds mostly on litter). Associated with this it has chelicerae with a high leverage index (Perdomo et al. 2012).
Distribution: AK, AB, QC, NL [MI, MT; Holarctic]
Literature: Maraun et al. (2019).

Nothrus parvus Sitnikova, 1975 (Figure 3.43D,E)
Comb./Syn.: *Nothrus pulchellus* (Berlese, 1910)
Distribution: AB [Holarctic]
Literature: ABMI (2019).

Nothrus pratensis Sellnick, 1928 (Figure 3.43C)
Biology: From peatlands. Parthenogenetic; gravid females with up to 2 eggs. Gut bolus with coarse organic material or hyphae (Walter et al. 2014).
Distribution: AK, YT, NT, NU, AB, QC [MI, NY, VA; Holarctic]
Literature: Maraun et al. (2019); ABMI (2019).

Nothrus silvestris Nicolet, 1855 (Figure 3.43A,B)
Comb./Syn.: *Angelia sylvestris* (Nicolet, 1855); *Nothrus anauniensis* sensu Berlese, 1885; *Angelia anauniensis* sensu Lombardini, 1936
Biology: Schuster (1956) considered this species an unspecialized feeder. Grandjean (1950) and Sengbusch (1954) successfully reared this species on lichen, and *Protococcus* algae and moss, respectively. Specimens from a chestnut and beech forest showed feeding preference primarily for leaf material with lesser representation of fungal hyphae, fungal spores and amorphous material (Anderson 1975). Gut contents of adults and juveniles included plant material, moss fragments, pine pollen, fungal material, arthropod fragments, and protists (Behan-Pelletier and Hill 1983). This species significantly preferred the ascomycete *Ceuthospora pinastri* which colonizes conifer needles, and the ubiquitous *Cladosporium herbarum* which colonizes dead organic matter, in a cafeteria experiment (Koukol et al. 2009). Schneider et al. (2005) showed that ectomycorrhizal fungi are preferentially consumed. Priming a beech forest soil with two species of parasitic nematodes, *Phasmarhabditis hermaphrodita* and *Steinernema feltiae*, Heidemann et al. (2011) found adults fed preferentially on live and dead *P. hermaphrodita*, indicating that this species can be a facultative carnivore and scavenger. Adults were occasionally eaten by the mesostigmatan *Pergamasus septentrionalis* in a no-choice feeding experiment (Peschel et al. 2006).

The metabolism of this species was studied by Webb (1969, 1970). Purrini (1984) described some coccidian parasites. Gut microbiota was studied by Gong et al. (2018). Genetics of this species have contributed to studies of oribatid phylogeny (Palmer and Norton 1992; Domes 2007; Schäffer et al. (2020).
Distribution: BC, MB, ON, QC [CN, MI, NC, NY, TN, VA; Holarctic]
Literature: Norton and Kethley (1990); Siepel (1990); Petrzik et al. (2016); Maraun et al. (2019).

Nothrus truncatus Banks, 1895
Biology: From sphagnum (Donaldson 1996); Palmer and Norton (1992) established that this species is genetically distinct.
Distribution: QC [NH, NY; Nearctic].
Literature: Norton (1979); Norton and Kethley (1990).

Family Trhypochthoniidae Willmann, 1931
Diagnosis: *Adults*: Holoid. Prodorsum stegasime. Ventral region macropyline. Rostrum rounded or laterally excised to form a projecting "mucro" bearing rostral setae. Body with narrow ventrosejugal articulation. Cerotegument not birefringent. Bothridial seta various but never pectinate, rarely highly regressed or absent. Bothridium without respiratory saccules or tubules. Exobothridial seta absent in nymphs and adults. Ventral region with large plates and little or no distance between genital and anal apertures. Notogaster with opisthonotal gland; with 15 pairs of setae (f_1 represented by vestige). Epimere II with 1 pair of setae. Aggenital setae present or absent. Genital setation 4–24 pairs. Preanal plate distinct, inconspicuous, or absent. Lyrifissure *iad*, *ian* present. Subcapitulum usually stenarthric, with 1 pair of vestigial *m* setae; rarely anarthric or diarthric; rutella well developed. Adoral setation 3 pairs. Chelicerae usually robust, chelate-dentate; Trägårdh's organ present. Palptibia fused with tarsus dorsally. Legs monodactylous or tridactylous.

KEY TO GENERA OF TRHYPOCHTHONIIDAE OF CANADA AND ALASKA
(MODIFIED FROM NORTON OSU MITE KEYS)

1	Rostral setae closely adjacent, inserted on anterior mucro formed by laterally excised rostrum. Bothridial seta small, setiform; its alveolus narrow, simple, bothridium located on small, disk-like elevation (Figure 3.44A–C) ....***Mucronothrus*** Trägårdh
–	Rostral setae well separated, mutual distance usually equal to or greater than that of lamellar setae; rostrum not excised laterally to form anterior mucro. Bothridial seta and bothridium well formed, typical trichobothrium, or (rarely) highly regressed, represented by vestiges or absent (some *Trhypochthoniellus*) 2
2(1)	Subcapitulum diarthric. Anal/adanal setation 1/2 or 0/2. Body strongly flattened. Anal plates unusually narrow (Figure 3.44D–G).. ....***Trhypochthoniellus*** (= *Hydronothrus*) Willmann
–	Subcapitulum stenarthric or anarthric. Anal/adanal setation 1/3 or 2/3. Body deep or only slightly flattened. Anal plates variously developed 3
3(2)	Anal setation 1 pair. Genital setation 6–18 pairs. Epimere IV setation 3 pairs. Subcapitular setae *m* vestigial. Labiogenal scissure posteriorly complete, fully dividing gena and mentum. Genua setation (I–IV, solenidia not included): 5-5-3-3 (Figures 3.44J; 3.45) ....***Trhypochthonius*** Berlese
–	Anal setation 2 pairs. Genital setation 6 pairs. Epimere IV setation 2 pairs. Two pairs of subcapitular setae *m*, closely adjacent. Labiogenal articulation incomplete posteriorly, such that gena and mentum partially fuse. Genua setation (I–IV, solenidia not included): 4-4-2-2 (Figure 3.44H,I)...***Mainothrus*** Choi

Mainothrus Choi, 1996
Type-species: *Mainothrus aquaticus* Choi, 1996
Comb./Syn.: *Altrhypochthonius* Weigmann, 1997

Mainothrus badius (Berlese, 1905) (Figure 3.44H,I)
Comb./Syn.: *Trhypochthonius badius* Berlese, 1905
Biology: Found in wet moss, sodden debris, and freshwater (Schatz and Behan-Pelletier 2008). No males are known: gravid females carry up to 6 eggs. Gut contents are comprised primarily of dark fungal hyphae, spores, and coarse granular material. In the laboratory ate *Pleurococcus* algae and sphagnum as food (Ermilov and Chistyakov 2008). This species was reared experimentally by Palmer and Norton (1990). It was included in the study of phylogenetic relationships by Domes et al. (2007), and Schäffer et al. (2020).
Distribution: AB, MB, ON, QC [NY; Holarctic]
Literature: Seniczak et al. (1998); Kuriki et al. (2000); Seniczak and Seniczak (2009a); ABMI (2019).

Mucronothrus Trägårdh, 1931
Type-species: *Mucronothrus nasalis* (Willmann, 1929) (= *Mucronothrus rostratus* (Tragårdh, 1931)
Note: Genus is considered truly aquatic by Norton et al. (1996).

Mucronothrus nasalis (Willmann, 1929) (Figure 3.44A–C)
Comb./Syn.: *Malaconothrus* (?) *nasalis* Willmann, 1929; *Mucronothrus rostratus* Trägårdh, 1931
Biology: This species inhabits spring-fed lotic environments and spring seepages, and is globally widespread, known from every continent except Africa and Antarctica (Norton et al. 1996). Hammer (1965) suggested that the extensive distribution of this species is relictual, and that the species predates the breakup of Pangaea ca. 200 million years ago. Knowledge of the biology of this species is based on a population from a small springbrook near Toronto (Norton et al. 1988). The species is parthenogenetic; gravid females carry up to 4 eggs. It is an unspecialized feeder (Norton and Palmer 1991). This species was reared experimentally by Palmer and Norton (1990) and Palmer and Norton (1992) studied its genetic diversity.
Distribution: BC, AB, SK, ON, QC, NB, NL [CO, OR; Cosmopolitan]
Literature: Domes et al. (2007); Maraun et al. (2019); ABMI (2019); Schäffer et al. (2020).

Trhypochthoniellus Willmann, 1928
Type-species: *Trhypochthonius* (*Trhypochthoniellus*) *setosus* Willmann, 1928
Comb./Syn.: *Hydronothrus* Aoki, 1964
Note: Species are not keyed because of questions about interpretation of species in this genus (Weigmann 1997a,b; 1999).

Trhypochthoniellus longisetus (Berlese, 1904) (Figure 3.44D–G)
Comb./Syn.: *Trhypochthonius longisetus* Berlese, 1904; *Camisia excavata* Willmann, 1919, *Trhypochthoniellus excavatus* (Willmann, 1919); *Nothrus crassus* Warburton and Pearce, 1905, *Hydronothrus crispus* Aoki, 1964.
Remarks: (from Weigmann 1999) Two well-known morphs of *Trhypochthoniellus* in Europe: one with developed bothridia and bothridial seta in adults (=*T. trichosus* (Schweizer, 1922)), and one without (=*T. setosus* Willmann, 1928) are regarded as junior synonyms of the polymorphic and thelytokous species *Trhypochthoniellus longisetus* (Berlese, 1904). Three populations have been studied morphometrically. In one population all degrees of trichobothridial regression have been found, many of the specimens being asymmetrical. Thus this species is considered to include morphologically different clones.
Biology: From fen habitats. Opisthonotal gland secretions studied by Raspotnig (2010). Found on skin, fins and gills of fish (Olmeda et al. 2011). It is common in aquatic habitats that are rich

in organic material, such as mires, high-mounted drinking-water tanks, and swimming pools (Tagami et al. 1992; Kuriki 2005).
Distribution: AB [Cosmopolitan].
Literature: Weigmann (2006); Maraun et al. (2019).

Trhypochthoniellus setosus canadensis Hammer, 1952
Comb./Syn.: *Trhypochthonius setosus canadensis* (Hammer, 1952)
Note: The European *Trhypochthoniellus longisetus* forma *setosus* Willmann, 1928, while not reported from North America (Marshall et al. 1987), may be the same as Hammer's subspecies.
Biology: This aquatic mite is found in peatlands and moss along lake margins. No males are known.
Distribution: AK, YT, NT, AB, NB, NS, NL [Canada]
Literature: Weigmann (1997a,b); ABMI (2019).

Trhypochthonius Berlese, 1904
Type-species: *Hypochthonius tectorum* Berlese, 1896
Comb./Syn.: *Tumidalvus* Ewing, 1908a; *Trilohmannia* Willmann, 1923
Literature: Weigmann and Raspotnig (2009).

KEY TO ADULT *TRHYPOCHTHONIUS* OF CANADA AND ALASKA (MODIFIED FROM WALTER ET AL. 2014)

1 Notogastral setae (c_1, $d_{1\text{-}2}$, e_1) smooth, acuminate; lateral and posterior setae smooth acuminate or barbed distally 2
– All dorsal and lateral notogastral setae densely barbed, most expanded distally and with bushy appearance, some may be reduced and rod-like 3

2(1) Lateral and posterior notogastral setae smooth acuminate. Length: 480–565µm (Figure 3.45A).. *T. cladonicola* (Willmann)
Lateral and posterior notogastral setae barbed distally. Length: 524–584µm (Figure 3.45C,D)..*T. silvestris* Jacot
Known from NC, possibly occurs in southern Canada.

3(1) Dorsal setae of various lengths, especially those in rows *c*, *d*, and *e*, marginal setae relatively long; notogastral seta c_2 about twice as long as c_1, d_1 subequal to or about twice as long as d_2. Interlamellar seta longer than either the bothridial seta or rostral seta 4
– Most dorsal setae short and bushy, setae within rows only moderately different in lengths; notogastral seta c_{1-2} and d_{1-2} very short (<20µm), subequal. Interlamellar seta much shorter than both bothridial seta and rostral seta. Length: 525–590µm (Figures 3.44J, 3.45B).. *T. nigricans* Willmann (=*T. sphagnicola* Weigmann)

4(3) Posterior notogastral seta p_1 ca. 75µm; f_2 and p_2 ca. 50µm. Genital setation 7–11 (usually 9–11) pairs. Length: 599–700µm (Figure 3.45E–G) .... *T. tectorum tectorum* (Berlese)
– Posterior notogastral seta p_1 very long (ca. 90µm); f_2 and p_2 >60µm. Genital setation 6–8 (usually 7) pairs. Length: 570–680µm (Figure 3.45H) .... *T. americanus* (Ewing)
Note: Unidentified *Trhypochthonius* species from BC, ON, NB.

Trhypochthonius americanus (Ewing, 1908) (Figure 3.45H)
Comb./Syn.: *Tumidalvus americanus* Ewing, 1908
Biology: From moss; often in pine litter. Reared experimentally by Palmer and Norton (1990).
Distribution: ON, QC [CA, CO, IL, MO, NC, NY, TN, UT; Nearctic]
Literature: Seniczak and Norton (1994); Domes et al. (2007); Weigmann and Raspotnig (2009); Maraun et al. (2019); Schäffer et al. (2020).

Trhypochthonius cladonicola (Willmann, 1919) (Figure 3.45A)
  Comb./Syn.: *Camisia cladonicola* Willmann, 1919; *Trilohmannia cladonicola* (Willmann, 1923); *Trhypochthonius cladonicolus* (Willmann, 1919)
  Biology: Associated with lichens, conifer litter, and muskeg. Parthenogenetic; gravid females carry 1–3 eggs. Gut contents include dark hyphae and pale to brown granular material.
  Distribution: AB, ON [Holarctic]
  Literature: Szywilewska-Szczykutowicz and Olszanowski (2007); Weigmann and Raspotnig (2009); ABMI (2019).

Trhypochthonius nigricans Willmann, 1928 (Figures 3.44J, 3.45B)
  Comb./Syn.: *Trhypochthonius sphagnicola* Weigmann, 1997
  Biology: Associated with moss.
  Distribution: AB [Holarctic]
  Literature: ABMI (2019).

Trhypochthonius silvestris Jacot, 1937 (Figure 3.45C,D)
  Distribution: [NC; Holarctic]
  Literature: Seniczak and Norton (1994); Weigmann and Raspotnig (2009).

Trhypochthonius tectorum (Berlese, 1896) *s.l.* (Figure 3.45E–G)
  Comb./Syn.: *Hypochthonius tectorum* Berlese, 1896; *Nothrus tectorum* (Berlese, 1896); *Trhypochthonius nigricans* sensu Jørgensen, 1934
  Note: In Alberta *T. tectorum* is a morphologically and genetically (*cox1* gene) variable parthenogenetic form with some populations showing similarity to *T. americanus* and possibly to *T. silvestris* or one of the species described from Japan (e.g., *T. septentrionalis* Fujikawa, 1995). Thus Walter et al. (2014) treats these as a species complex (i.e., *tectorum s.l.*), as we do herein.
  Biology: A mesophilous mite found in forest soils, peatlands; wind swept alpine; juniper, *Cabresia* turf, aspen litter. Gravid females carry up to 7 eggs. Gut contents include pigmented fungal hyphae, conidia, arthrospores, and bits of organic matter (Ermilov et al. 2004). They reported that *T. tectorum* takes 97–116 days to develop from egg to adult at 17°C and 49–60 days at 20°C. Meier et al. (2002) found that *T. tectorum* did better on the lichen *Xanthoria parietina* than on non-lichenized fungi, and that viable spores of both the lichen and its alga *Trebouxia arboricola* were passed in fecal pellets. They also report ovovivipary in *T. tectorum*, or at least the eggs overwinter in dead females, hatch within the female, and the newly hatched babies "first feed on the viscera of the dead female prior to switching, as deutonymphs and adults, to a diet consisting of [putrefying] plant material and lichens." Opisthonotal gland secretions of this species were studied by Raspotnig (2010).
  Distribution: AK, YT, NT, BC, AB, MB, ON, QC, NS [MT, NC, OR; Semicosmopolitan]
  Literature: Weigmann (1997a,b); Fujikawa (2000); Raspotnig et al. (2004); Weigmann and Raspotnig (2009); Heethoff et al. (2011); Maraun et al. (2019); ABMI (2019); Schäffer et al. (2020).

Hyporder Brachypylina Hull, 1918 (= Circumdehiscentiae Grandjean, 1954)
  Note: Brachypylina are the largest, most taxonomically rich group of traditional oribatid mites. This monophyletic lineage is defined by the combination of a holoid body, brachypyline venter, distinct acetabula with trochanters I–II almost totally contained within them, and genua of legs I–III (and usually IV) shorter than tibiae and lacking intrinsic musculature. Most adults have a typical apodemato-acetabular system of tracheae, but all have some type of internalized respiratory surface (Norton et al. 1997). Lyrifissure *ian* is usually absent or reduced in size. The basic chelate-dentate chelicera has evolved into elongated or filtering forms in families of Polypterozetoidea, Ameroidea, Gustavioidea, Oppioidea, Trizetoidea, Phenopelopoidea, Oripodoidea and Galumnoidea and the subcapitulum is often modified accordingly.

Juvenile Brachypylina generally have a more conservative morphology than do adults. There usually is a striking metamorphosis between tritonymph and adult that is otherwise found only in ptychoid taxa. This prevents easy association of adult with immature specimens based on morphology alone. However, Norton and Ermilov (2014) present excellent suggestions on how to make adult-juvenile associations with confidence, including rearing methods and visual association. Most importantly, they provide recent data on references for known immatures. Brachypyline mites are circumdehiscent when molting (with minor variations); posteriorly, the line of dehiscense (θ) passes just above the anal region and is directed toward the dorsosejugal region anteriorly. Depending on the group, θ may be circular—that is, complete in the dorsosejugal region—or it may stop short on each side, assuming a U shape. In the latter case the exuviae remains intact. If θ is circular, a dorsal cap, or **scalp**, breaks away as the mite ecloses. In most such cases, the emerging nymphs (and adults in some genera) retain the scalps from all previous instars, which accumulate in a pagoda-like fashion.

Grandjean (1954) recognized the potential value of juvenile Brachypylina as indicators of natural taxonomic groups. Characters that have been given priority are:

Nature of dehiscence and scalp retention. These are:

apheredermous, which means that nymphs (and adults) do not retain scalps regardless of whether or not line θ is complete.

eupheredermous, meaning that θ is complete and that nymphs retain scalps in a pagoda-like fashion, tightly pressed to their bodies, but adults may or may not retain them.

apopheredermous; that is, θ is complete and nymphs retain scalps, but the scalps are held away from the body by setae (found in Oribatellidae).

opsiopheredermous; nymphs of Hermanniellidae do not retain scalps; however, the adult retains the tritonymphal scalp, although it is pressed tightly to the notogaster and is difficult to discern.

Attachment of scalps. Some oribatid mites whose nymphs carry exuvial scalps have evolved structural specialization for attaching the scalps. The best-known mechanism is in Damaeidae, in which a distinct projecting cornicle on the opisthonotum inserts into the cornicle of the previous exuvium, forming a nested attachment structure. In Caleremaeidae the cornicle is a simple tubercle or papilla. In Oribatellidae opisthonotal seta *dp* is either structurally modified or it inserts in a sheath-like callosity under the previous exuvium. In some Tegoribatidae (*Tegoribates*) a small bulge around seta *dp* forms a convexity in exuvia that nest together.

Opisthonotal setation (i.e., setae in the region of the presumptive notogaster). The number of setae that are missing from the ancestral oribatid mite complement of 16 pairs (13 in larva, in which segment PS does not yet contribute to the opisthonotal region) is counted as a deficiency. *Uni–*, *bi–*, *quadri–*, and *quinquedeficient* nymphs would therefore have 15, 14, 12, and 11 pairs of setae, respectively. Greater losses are usually indicated as *multideficiency*. Unideficiency is most common. Adults of species that have unideficient juveniles may retain the 15 pairs, but often there are further losses (setal regressions). Eupheredermous nymphs are generally quadrideficient, lacking the dorsocentral pairs *da*, *dm*, and *dp*; scalps are located directly over the central region normally occupied by these mechanoreceptive setae. Adults in eupheredermous taxa never reform these setae. Rarely, apheredermous taxa are similarly quadrideficient; gustavioid families like Tenuialidae, Liacaridae and Anderemaeidae are examples. Hermanniellidae are uniquely holotrichous in Brachypylina.

Hysterosoma cuticle (particularly in the opisthonotal region). One widespread nymphal condition is for the cuticle to be somewhat tanned and leathery and provided with many folds; referred to as *plicate nymphs*. Large sclerites are present on the opisthonotal cuticle in some assemblages, including some with plicate nymphs (e.g., some Ameronothroidea). When these sclerites occupy considerable parts of the surface, they are called *macrosclerites*, and are typical of the poronotic groups Ceratozetoidea and Galumnoidea. Some or all of the opisthonotal setae of immature Oripodoidea have a dermal gland at their base, and both the setal

insertion and the porose area associated with the gland are on small *excentrosclerites* (= microsclerites). Rarely, porose areas are similarly adjacent to setae on a macrosclerite, as in Humerobatidae.
Literature: Grandjean (1954b, 1965a); Norton et al. (1993, 1997); Norton (1998); Woas (2002); Weigmann (2006); Pachl et al. (2012, 2017).

Superfamily Hermannielloidea Grandjean, 1934

Family Hermanniellidae Grandjean, 1934
Diagnosis: *Adults*: Non-poronotic Brachypylina. **Opsiopheredermous**, setae of notogaster covered by tritonymphal setal homologues. Rostrum with or without medial incision. Apodemato-acetabular system abnormal. Respiratory bladder (air bag) attached to cotyloid wall IV or simple pocket present distorting this wall. Dorsophragmata and pleurophragmata absent. Notogaster with pores which may be saccules; edge of notogastral shield with row of saccules. Anal saccules present in some genera at lateral edge of anal plates. Notogastral setation 15 or 16 pairs; seta c_3 present or absent. Opisthonotal glands opening on large apophyses laterally on notogaster. Chelicera chelate-dentate. Rutellum **pantelobasic** (base of rutella reaches infrabuccal fissure).

Hermanniella Berlese, 1908
Type-species: *Hermanniella granulata* Nicolet, 1855

KEY TO ADULT *HERMANNIELLA* OF CANADA AND ALASKA

1 Genital setation 7 pairs, with g_3 offset laterally (Figure 3.46A–E)...***H. punctulata*** Berlese
– Genital setation 6 pairs, with g_4 offset laterally, or all genital setae in line.........................2

2(1) Genital setae g_4 offset laterally from other genital setae. Prodorsum with 2 humps anterior of bothridia, giving appearance of ridges when specimen flattened. Notogaster foveate with foveae not merging. Length: 800µm (Figure 3.46F).......................***H. robusta*** Ewing
– Genital setae g_4 in line with other genital setae. Prodorsum without humps anterior of bothridia. Notogaster foveate or not, if foveate, cells merged, or not. Length: 550–670µm 3

3(2) Smaller species, about 560µm..***H. subnigra*** (Ewing)
– Larger species, about 660µm (Figure 3.46G,H).............................. ***H. occidentalis*** Ewing
Note: Unidentified species from AK, BC, QC, NS, NL.

Hermanniella occidentalis Ewing, 1918 (Figure 3.46G,H)
Comb./Syn.: *Hermanniella punctulata occidentalis* Ewing, 1918
Biology: Recorded from old-growth temperate rainforest (Winchester et al. 1999) and from beech litter (Sylvain and Buddle 2010).
Distribution: BC, QC [OR; Nearctic]

Hermanniella punctulata Berlese, 1908 (Figure 3.46A–E)
Comb./Syn.: *Nothrus piceus* C.L. Koch, 1839?
Distribution: BC [VA; Holarctic]
Literature: Toluk et al. (2009); Maraun et al. (2019).
Note: We follow Weigmann (2006) and Krisper and Lazarus (2014) in questioning the synonymy of *Nothrus piceus* C.L. Koch, 1839 with *Hermanniella punctulata* Berlese.

Hermanniella robusta Ewing, 1918 (Figure 3.46F)
Comb./Syn.: *Hermanniella punctulata robusta* Ewing, 1918
Biology: Associated with mesophilous habitats and peatlands. Gut contents have brown hyphae, coarse organic material (Walter et al. 2014).

Distribution: BC, AB [OR, TN; Nearctic]
Literature: ABMI (2019).

Hermanniella subnigra (Ewing, 1909)
Comb./Syn.: *Hermannia subnigra* Ewing, 1909; *Hermanniella punctulata columbiana* Berlese, 1910
Biology: From forest litter.
Distribution: BC [IL, MO; Nearctic, Australian]
Literature: Norton and Kethley (1990).

Family Plasmobatidae Grandjean, 1961
Diagnosis: *Adults*: Non-poronotic Brachypylina. Bearing concentric nymphal exuviae (scalps). Cerotegument well developed; prodorsal band of cerotegument remains attached to each scalp. Rostrum with deep incision. Tracheal system abnormal with trachea I, trachea *sj* with saccule, trachea IV present, trachea III absent. Lamella, tutorium, pedotecta and discidium absent. Dorsophragmata and pleurophragmata absent. Opisthonotal glands opening on large lateral apophyses. Macropores present on notogaster. Lyrifissure *ian* present or vestige present. Chelicera weakly pelopsiform; palpal eupathidium *acm* separate from solenidion. Legs monodactylous. Retrotecta on trochanters and femora I and II; tendancy to tibial-tarsal fusion.

Plasmobates Grandjean, 1929
Type-species: *Plasmobates pagoda* Grandjean, 1929 (Figure 3.47A–E)
Note: Unidentified *Plasmobates* spp. from BC, QC.

Superfamily Neoliodoidea Sellnick, 1928

Family Neoliodidae Sellnick, 1928
Diagnosis: *Adults*: Non-poronotic Brachypylina. Bearing concentric nymphal exuviae (scalps). Large mites (1,000–2,000μm), heavily sclerotized. Apodemato-acetabular system abnormal and variable at the genus level. Lamella and tutorium absent. Lamellar seta inserted on prodorsum or absent. Dorsophragmata and pleurophragmata absent. Genital plates with transverse scissure, setation 7–8 pairs. Preanal organ with brachytrachea or platytrachea. Lyrifissure *ian* present. Chelicera chelate-dentate. Rutellum pantelobasic. Palpal eupathidium *acm* separate from solenidion, inserted on tubercle or mound. Leg segments with saccules or brachytracheae. Porose areas present on tibiae and tarsi. Seta *d* present on tibiae I–IV and genua I–III. Tibia of legs with (*Platyliodes*) or without retrotectum.

KEY TO GENERA OF NEOLIODIDAE OF CANADA AND ALASKA

1 Notogaster flat; ventral plate not fused posterior to anal plates. Tibia of legs with retrotectum (Figure 3.48A–D) ....***Platyliodes*** Berlese
– Notogaster convex; ventral plate fused posterior to anal plates. Tibia of legs without retrotectum (Figure 3.48E) ....***Teleioliodes*** Grandjean
Note: *Poroliodes* Grandjean, 1934 and *Neoliodes* Berlese, 1888 probably occur in Canada, although there are no published records.

Platyliodes Berlese, 1916
Type-species: *Nothrus doderleini* Berlese, 1883

KEY TO ADULT *PLATYLIODES* OF CANADA AND ALASKA

1. Rostral setae palmate. Medial setae on posterior apophysis about 7× length brush-like lateral setae. Length: ca. 846μm (Figure 3.48A,B) ***P. macroprionus*** Woolley and Higgins
– Rostral setae brush-like. Medial setae on posterior apophysis about 3× length palmate lateral setae. Length: 968–1,259μm (Figure 3.48C,D).................... ***P. scaliger*** (C.L. Koch)
    Note: Unnamed species from QC, NB.

Platyliodes macroprionus Woolley and Higgins, 1969 (Figure 3.48A,B)
    Biology: From moss in forests. This species is a resident of the canopy of old growth temperate rainforests of the Pacific Northwest.
    Distribution: BC [WA; Nearctic, Japan]
    Literature: Seniczak et al. (2018).

Platyliodes scaliger (C.L. Koch, 1839) (Figure 3.48C,D)
    Comb./Syn.: *Nothrus scaliger* C.L. Koch, 1839
    Biology: Arboreal mite inhabiting moss and bark in mesophilous habitats and peatlands. Collected from the fur of small mammals, a possible mode of short distance dispersal (Miko 1990). This species was included in Lienhard et al. (2013) molecular study on reverse evolution and cryptic diversity.
    Distribution: AB [MI, NY; Holarctic]
    Literature: Weigmann (2006); Seniczak et al. (2018).

Teleioliodes Grandjean, 1934
    Type-species: *Teleioliodes madininensis* Grandjean, 1934
    Note: Unidentified species from BC (Lindo and Winchester 2006).

Superfamily Plateremaeoidea Trägårdh, 1931

Family Gymnodamaeidae Grandjean, 1954
    Diagnosis: *Adults*: Non-poronotic Brachypylina. Cuticle smooth to lightly rugose and with thick coating of cerotegument with emergent, crystalloid pustules on body, legs, and setae. With subnormal apodemato-acetabular system, lacking trachea I; with or without porose sac opening into acetabulum. Lamellae absent. Parastigmatic enantiophysis, dorsophragmata, and pleurophragmata absent. Notogaster typically flattened to concave and with raised rim (somewhat convex in *Adrodamaeus*). Maximum of 6 pairs of notogastral setae (h_{1-3}, p_{1-3}) (h_3 present in only one known species). Pedotectum I auriculate; pedotectum II auriculate, rounded or drawn-out and pointed distally. Propodolateral apophysis (P) absent. Postbothridial tubercle (*Bp*) variously developed. Epimeral setation 3-1-3-3. Anal and genital openings separated by band of ventral plate (ano-genital bridge) covering preanal organ or confluent with narrow band of soft cuticle between plates. Anal setation 2 pairs; genital setation 6–7 pairs. Subcapitulum diarthric, base of gena usually covered by well developed tectum. Legs tridactylous; pretarsi inserted on stalk-like pedicel. Tibia I with 2 solenidia borne on strong distal apophysis. Femora III–IV with retrotecta; tibiae with or without retrotecta. Tarsus II with 1–2 solenidia. Leg segments with or without internalized porose organs.
    Note: Gymnodamaeidae is one of up to eight currently recognized families in the Plateremaeoidea (Schatz et al. 2011) and the only diverse plateremaeoid family in Canada and Alaska (Marshall et al. 1987, Walter et al. 2014). Gymnodamaeid mites are characteristic of dry sites including grasslands, alpine and tundra soils, and dry patches of soil, litter, and moss in more mesic environments.

KEY TO GENERA OF GYMNODAMAEIDAE OF CANADA AND ALASKA
(AFTER WALTER 2009)

1   Genital and anal plates contiguous, with openings confluent, rather than separated by sclerotized ano-genital bridge 5
–   Genital and anal plates separated by distinct band of sclerotized cuticle (the ano-genital bridge) 2

2(1)   Distal segments of legs with cup-like retrotectum forming ball-and-socket joint with previous segment. Lamellar seta inserted laterally near rostral seta. Notogaster gently convex dorsally (Figure 3.49A–D) .... ***Adrodamaeus*** Paschoal
–   Distal segments of legs without cup-like retrotectum. Lamellar seta inserted dorso-medially, distant from rostral seta. Notogaster flattened to concave 3

3(2)   Interlamellar setae inserted between bothridia, on relatively small apophyses. Notogastral setation 4 pairs (h_1, p_{1-3}). Medial margin of genital shields smooth, without interlocking teeth. Small mites with ventral lengths <400µm 4
–   Interlamellar setae inserted anterior to bothridia, on strong apophyses. Notogastral setation 5 pairs (h_{1-2}, p_{1-3}). Medial margin of genital shields each with few to many teeth interlocking in zipper-like fashion. Large mites with ventral length: 730–900µm (Figure 3.49E–G) .... ***Odontodamaeus*** Paschoal

4(2)   Apophyses of posterior notogastral setae h_1, p_1 strongly developed and clustered medially. Genital setation 7 pairs, g_2 posterior and lateral to g_1. Notogastral cerotegument with raised cell in mature adults (Figure 3.49H–J) .... ***Jacotella*** Banks
–   Apophyses of posterior notogastral setae variously developed, but not clustered medially. Genital setation 6 pairs, g_2 inserted posterior to g_1 and in line with other genital setae. Notogastral cerotegument forming series of longitudinal folds in cerotegument (Figure 3.50A–D) .... ***Joshuella*** Paschoal

5(1)   Notogastral setation maximum of 5 pairs, seta h_3 absent. Bothridial rim without posterior tooth 6
–   Notogastral setation 6 pairs including h_3 inserted about midway between lyrifissures *im* and *ip*. Bothridial rim with posterior tooth (Figure 3.50E,F) ....
     .... ***Donjohnstonella*** Walter (= *Johnstonella* Paschoal)
     Note: Known from Washington; not as yet from Canada or Alaska.

6(5)   Notogastral cerotegumental pustules spherical. Prodorsum with interlamellar setae opposed to distinct median tubercle 7
–   Notogastral cerotegumental pustules longer than their diameter, bullet-shaped to digitiform. Interlamellar setae not opposed to median tubercle, (which poorly developed or absent) (Figure 3.50G–L) .... ***Roynortonella*** Walter (= *Nortonella* Paschoal)

7(6)   Cerotegument on notogaster without raised pattern of ridges or cells, but with anterior median patch of strong tubercles. Seta h_1 inserted posteriorly on notogaster. Pedotectum II acuminate distally (Figure 3.51A–C) .... ***Pleodamaeus*** Paschoal
–   Cerotegument on notogaster with lateral folds and with median pattern of raised ridges (sometimes forming closed cells in mature adults). Notogastral seta h_1 inserted on dorsal surface of notogaster. Pedotectum II with rounded margin (Figure 3.51D–F) ....
     .... ***Gymnodamaeus*** Kulczynski

Adrodamaeus Paschoal, 1982
     Type-species: *Damaeus magnisetosus* Ewing, 1909

Adrodamaeus magnisetosus (Ewing, 1909) (Figure 3.49A–D)
   Comb./Syn.: *Damaeus magnisetosus* Ewing, 1909; *Heterodamaeus magnisetosus* (Ewing, 1909)
   Biology: From maple forest litter.
   Distribution: ON, QC [IL; Nearctic]
   Literature: Paschoal (1984); Walter (2009).

Gymnodamaeus Kulczynski, 1902
   Comb./Syn.: *Heterodamaeus* Ewing, 1917.
   Type-species: *Damaeus bicostatus* Koch, 1935
   Note: Subías (2004) placed *Johnstonella* Paschoal (=*Donjohnstonella* Walter), *Odontodamaeus* Paschoal and *Pleodamaeus* Paschoal in synonymy with *Gymnodamaeus*, without justification. Following Walter (2009) these synonymies are rejected herein.

Gymnodamaeus of Canada and Alaska
   Three species of *Gymnodamaeus* are known for Canada and Alaska, but only *G. ornatus* Hammer is recognizable (Walter 2009). Its diagnosis is: distinct pattern on notogaster with oval and crescent-shaped areas bordered by smooth ridges. Bothridial seta with long stalk and flattened head. Interlamellar setae minute. Lamellar setae long, extending past tip of rostrum. Ventral plate with pair of large lateral pits; anogenital bridge incomplete (Figure 3.51D,E)
   Note: Unidentified species from Alaska: Lava Camp, Seward Peninsula (Late Miocene: 5.7 mya) (Matthews et al. 2019). Unidentified species known from QC, NB.

Gymnodamaeus ornatus Hammer, 1952 (Figure 3.51D,E)
   Comb./Syn.: *Allodamaeus ornatus* (Hammer, 1952)
   Biology: The type series of 7 adults was collected from dead leaves on cushion plants (*Pulsatilla*, *Saxifraga tricuspidata*, *Artemisia frigida*) on a south slope at Reindeer Station (62° 42'N, 134° 08'W), YT. This species was included in Newton and Proctor's (2013) analysis of weight-estimation models for soil mites.
   Distribution: YT, NT, BC, AB, ON, QC [Canada]
   Literature: Walter (2009); Bayartogtokh and Schatz (2009); ABMI (2019).

Gymnodamaeus saltuensis Paschoal, 1982; *species inquirenda*
   Note: This species is considered a *species inquirenda* by Walter (2009).
   Distribution: MB
   Literature: Paschoal (1982a); Walter (2009).

Gymnodamaeus taedaceus Paschoal, 1982; *species inquirenda*
   Note: This mite was described from a single female from Alberta "in duff layer around *Pinus contorta*" and not illustrated. The type specimen is crushed and most of the characters cannot be discerned and Paschoal's description has numerous misinterpretations. Both Bayartogtokh and Schatz (2009) and Walter (2009) considered *G. taedaceus* to be unrecognizable, and that it should be considered a *species inquirenda*.
   Gymnodamaeus bicostatus (C.L. Koch, 1835) does not occur in North America according to Paschoal (1982) and the North American record in Powell and Skaley (1975) is a specimen Paschoal subsequently used as the holotype of *Gymnodamaeus taedaceus* Paschoal, 1982.
   Distribution: AB
   Literature: Paschoal (1982a); Walter (2009).

Jacotella Banks, 1947
   Type-species: *Gymnodamaeus quadricaudicula* Jacot, 1937

Jacotella quadricaudicula (Jacot, 1937) (Figure 3.49H–J)
   Comb./Syn.: *Gymnodamaeus quadricaudicula* Jacot, 1937

Biology: From oak litter (Lamoncha and Crossley 1998). Gravid females with 3 eggs (Walter 2009).
Distribution: AB, ON, NB [NC; Nearctic]
Literature: Paschoal (1983b); Walter (2009); ABMI (2019).

Joshuella Wallwork, 1972
Type-species: *Joshuella striata* Wallwork, 1972

Joshuella agrosticula Paschoal, 1983 (Figure 3.50A–D)
Biology: Grassland inhabitant; it was described from the shortgrass prairie in Colorado. Gut contents include brown fungal hyphae and arthroconidia. Gravid females with up to 2 relatively large eggs (Walter et al. 2014).
Distribution: YT, AB [CO; Nearctic]
Literature: Walter (2009); ABMI (2019).
Note: Unidentified species from BC.

Odontodamaeus Paschoal, 1982
Type-species: *Gymnodamaeus veriornatus* Higgins, 1961.

Odontodamaeus veriornatus (Higgins, 1961) (Figure 3.49E–G)
Comb./Syn.: *Gymnodamaeus veriornatus* Higgins, 1961
Biology: From deciduous litter.
Distribution: BC [CA, ID, OR, UT, WY; Nearctic]
Literature: Walter (2009).

Pleodamaeus Paschoal, 1982
Type-species: *Gymnodamaeus plokosus* Woolley and Higgins, 1973

Pleodamaeus plokosus (Woolley and Higgins, 1973) (Figure 3.51A–C)
Comb./Syn.: *Gymnodamaeus plokosus* Woolley and Higgins, 1973
Biology: From dry deciduous and coniferous litter; grass sod among pines.
Distribution: AB, SK [CO, SD, ID, UT, WY; Nearctic]
Literature: Walter (2009); ABMI (2019).
Note: Unidentified species from YT, AB, ON.

Roynortonella Walter, 2009
Comb./Syn.: *Gymnodamaeus, Paschoalia, Nortella, Nortonella*
Type-species: *Gymnodamaeus gildersleeveae* Hammer, 1952

KEY TO ADULT *ROYNORTONELLA* OF CANADA AND ALASKA (AFTER WALTER 2009)

1 Cerotegument on notogaster forming 2 large distinct depressions medially, 1 anterior to the other. Length: 390–410µm (Figure 3.50J–L) ....***R. gildersleeveae*** (Hammer)
– Cerotegument on notogaster forming pair of large distinct depressions anteriorly, 2 large distinct depressions medially, 1 anterior to the other, and 4 pairs of depressions lateral to these. Length: >700µm (Figure 3.50G–I) ....***R. victoriae*** (Paschoal)

Roynortonella gildersleeveae (Hammer, 1952) (Figure 3.50J–L)
Comb./Syn.: *Gymnodamaeus gildersleeveae* Hammer, 1952; *Nortonella gildersleeveae* (Hammer, 1952)
Biology: Type specimens were collected from moss, liverworts, and lichens under spruce with a *Vaccinium vitisidae* understory near Yellowknife, NT. Paschoal's (1982) collections were from a variety of coniferous litters from Colorado to Washington. Walter et al. (2014) found this mite in bare spots in open areas on south slopes in aspen parkland.

Distribution: AK, YT, NT, AB, MB, ON [CA, CO, ON, UT, WA; Nearctic]
Literature: ABMI (2019).

Roynortonella victoriae (Paschoal, 1982) (Figure 3.50G–I)
Comb./Syn.: *Gymnodamaeus victoriae* Paschoal, 1982
Biology: Collected from Garry oak litter; from foliose lichen in upper canopy of *Abies amabilis*.
Distribution: BC [Canada]
Literature: Walter (2009).

Family Plateremaeidae Trägårdh, 1931
Diagnosis: *Adults*: Non-poronotic Brachypylina. Cuticle smooth to foveolate with thick coating of cerotegument. Prodorsum with lamellar setae positioned far laterally and close to rostral setae; interlamellar setae very short. Bothridium opening dorsally and positioned close to notogastral edge, cup- or funnel-shaped. Pedotecta absent; propodolateral apophysis P present. Parastigmatic enantiophysis, dorsophragmata, and pleurophragmata absent. Notogaster typically flattened to concave; with centro-dorsal setae absent and with 6 pairs of postero-lateral notogastral setae; dorsal lyrifissures small. Epimere II with more than 2 pairs of setae. Genital and anal apertures almost circular. Genital setation 7 pairs. Anal setation 3–8 pairs. Adanal setae lateral to anal plates. Rutellum pantelobasic. Legs tridactylous with small claws, median claw strongest. Leg segments with retrotecta. Tarsi I and II with distal apophyses; tarsal pedicels long and narrow. Insertion area of tarsal famulus on leg I lying inside an incavation. Tibia I with large, dorsal, distad pointing, hornlike projection.

Allodamaeus Banks, 1947
Type-species: *Allodamaeus ewingi* Banks, 1947 (Figure 3.51G,H)
Note: Unidentified *Allodamaeus* species known from AK, MB, ON, QC.

Family Licnodamaeidae Grandjean, 1954
Diagnosis: *Adults*: Non-poronotic Brachypylina. With subnormal apodemato-acetabular system, lacking trachea I; trachea III and *sj* present. Prodorsum with lamellar setae positioned very laterally, at same level as, and close to rostral seta; interlamellar setae very short, or absent. Exobothridial seta present or absent. Bothridium opening dorsally and positioned close to notogastral edge, cup- or funnel-shaped; bothridial seta long, flattened. Pedotecta absent; propodolateral apophysis P present. Parastigmatic enantiophysis, dorsophragmata, and pleurophragmata absent. Notogaster flattened or slightly concave, highly sculptured; with centro-dorsal setae absent and with 4–6 pairs of postero-lateral notogastral setae. Dorsal lyrifissure *im* large. Epimeral setation 3-1-3-3. Genital setation 5 or 6 pairs. Anal setation 2 pairs. Adanal setation usually 2, exceptionally 3 pairs. Leg tarsus, tibia, genu with retrotectum, retrotectum of femur complete or incomplete; tarsi I and II with distal apophyses; tarsal pedicels long and narrow; insertion area of tarsal famulus on leg I not lying inside an incavation; tibia I with large, dorsal, distad pointing, hornlike projection. Legs tridactylous with small claws, median claw strongest.

Licnodamaeus Grandjean, 1931 (Figure 3.51I,J)
Type-species: *Licneremaeus undulatus* Paoli, 1908
Note: Unidentified species known from AB, NS.

Superfamily Damaeoidea Berlese, 1896

Family Damaeidae Berlese, 1896
Diagnosis: *Adults*: Non-poronotic Brachypylina. Bearing concentric nymphal exuviae, or not; sometimes with mass of debris on scalps or notogaster. Cerotegument often dense and

conspicuous, covering body, and setae. Mostly long-legged, middle-sized to large species. Prodorsum roughly triangular. Prodorsal lamellae absent. Tutorium absent. Propodolateral apophysis and spina adnata present or absent. Well-developed parastigmatic enantiophysis present. Dorsophragmata and pleurophragmata absent. Bothridium funnel-like, expanded distally. Notogastral setation 11 pairs, setae of rows *c*, *l* and *h* arranged in two, more or less parallel, longitudinal rows. Epimeral setation 3-1-3-4 or neotrichous. Genital setation 6 pairs. Subcapitulum diarthric; palpal eupathidium *acm* separate from solenidion. Rutellum atelobasic, with distal globular hyaline expansions. Leg segments clavate or moniliform.

Biology: Most data support fungivory in this family. Walter (1987) cultured a species of *Epidamaeus* through several generations on fungi. There is also evidence for feeding on *Protococcus* algae (Littlewood 1969).

KEY TO GENERA OF DAMAEIDAE OF CANADA AND ALASKA

1	Seta *d* absent from tibiae II–IV	2
–	Seta *d* present and coupled to solenidion on at least one of tibiae II–IV	5
2(1)	Propodolateral apophysis well developed. Large (usually over 600µm) nearly black species (Figure 3.52A)	***Damaeus*** C.L. Koch
–	Propodolateral apophysis usually absent, if present, not well-developed. Smaller, usually light to medium brown in colour	3
3(2)	Spina adnata absent (Figure 3.52B,C)	***Parabelbella*** Bulanova–Zachvatkina
–	Spina adnata present	4
4(3)	Prodorsal tubercles (*Ba*, *La*), present, *Da* absent. Spina adnata distally sharply deflected laterally. Propodolateral apophysis P developed usually as tip or tubercle perpendicular to body, or absent (Figure 3.54F,G)	***Kunstidamaeus*** Miko
–	Prodorsal tubercle *La* absent, *Ba* and *Da* present or absent. Spina adnata distally usually straight, not sharply deflected laterally. Propodolateral apophysis P usually absent (Figure 3.56–3.60)	***Epidamaeus*** Bulanova–Zachvatkina
5(1)	Tibiae II–IV with seta *d* present	6
–	Seta *d* absent from at least one of tibiae II–IV	10
6(5)	Spina adnata absent	***Belba*** von Heyden
–	Spina adnata present (can be small)	7
7(6)	Prodorsum laterally with propodolateral apophysis (P). Adult with exuvial scalps (Figure 3.54A–C)	***Weigmannia*** Miko and Norton
–	Prodorsum without propodolateral apophysis. Adult without exuvial scalps	8
8(7)	Setation of trochanters 1-1-2-1, femora 7-6-4-4, genua 4-4-3-3 (Figure 3.53A)	***Belbodamaeus*** Bulanova-Zachvatkina
–	Setation other than above	9
9(8)	Adults carrying organic debris. Trochanteral setation 1-1-2-2 (*l'* present on IV); femoral setation 7-7-4-4; genual setation 4-4-3-3 (Figure 3.52D–F)	***Belba*** (***Protobelba***)
–	Organic debris absent. Trochanteral setation 1-1-3-1 (seta *d* present on III); genual setation 4-4-4-4 (seta *v"* present on III–IV) (Figure 3.54D,E)	***Quatrobelba*** Norton
10(5)	Seta *d* present on tibia III and IV. Spina adnata large. Cerotegument highly filamentous (Figure 3.53B–E)	***Lanibelba*** Norton
–	Seta *d* present on tibia II and III. Spina adnata small or absent. Cerotegument various	11

11(10) Notogaster without organic debris. Spina adnata small or absent. Prodorsum often with several pairs of posterior tubercles and/or small propodolateral apophysis (Figure 3.55A–E).. ....***Dyobelba*** Norton

Note: *Belba longitarsalis* Hammer should key out here; but this species is an *Epidamaeus* (Norton and Ryabinin 1994) and is keyed below. Bayartogtokh and Norton (2007) do not include it in their key to world species of *Dyobelba*.

– Notogaster with organic debris. Spina adnata absent. Prodorsum with single pair of posterior tubercles; without propodolateral apophysis (Figure 3.55F–J)............................ ....***Caenobelba*** Norton

Belba von Heyden, 1826
 Type-species: *Notaspis corynopus* Hermann, 1804
 Note: Unidentified species known from late Pliocene/early Pleistocene fossil deposits in AK (Matthews et al. 2019). Unidentified species known from AK, BC, AB, QC.

Belba (Protobelba) californica (Banks, 1904) (Figure 3.52D–F)
 Comb./Syn.: *Oribata californica* Banks, 1904; *Damaeus californica* (Banks, 1904)
 Biology: Known habitats are moist, cool forests dominated by Red alder, or a variety of conifers, such as Western hemlock, Sitka spruce, Western redcedar, Douglas-fir, or coast redwood (Norton 1979a).
 Distribution: BC [CA, OR, WA; Nearctic].
 Literature: Norton (1980).

Belbodamaeus Bulanova-Zachvatkina, 1960
 Type-species: *Belbodamaeus tuberculatus* Bulanova-Zachvatkina, 1960
 Note: Unidentified species known from YT, BC.

Caenobelba Norton, 1980
 Type-species: *Caenobelba alleganiensis* Norton, 1980

Caenobelba alleganiensis Norton, 1980 (Figure 3.55F–J)
 Biology: From mixed forest, dominated by Red pine, Red oak, and Eastern hemlock (Norton 1980).
 Distribution: AK, YT, BC, AB, ON [NY; Nearctic]
 Literature: Ermilov (2012); ABMI (2019).

Damaeus C.L. Koch, 1835
 Type-species: *Damaeus auritus* C.L. Koch, 1835
 Note: *Damaeus* sp. nr. *gracilipes* Kulczynski, 1902 known from QC (Smith 1978).

Dyobelba Norton, 1978
 Type-species: *Oribata carolinensis* Banks, 1947 (Figure 3.55A-E)
 Note: Unnamed species known from BC, AB.
 Literature: Norton and Ryabinin (1994); Bayartogtokh and Norton (2007); Ermilov (2012).

Epidamaeus Bulanova-Zachvatkina, 1957
 Type-species: *Oribata bituberculatus* Kulczynski, 1902
 Literature: Miko (2006); Mourek et al. (2011); Norton and Ermilov (2021).
 Note: Subías (2004) considered *Epidamaeus* (sensu stricto) as a subgenus of *Damaeus*, and *Akrodamaeus* Norton, 1979 as a subgenus of *Metabelbella*, rather than a subgenus of *Epidamaeus* as proposed by Norton (*op. cit.*). We follow Miko et al. (2011) who rejected Subías' proposals, and considered *Akrodamaeus* a junior synonym of *Parabelbella* Bulanova-Zachvatkina, 1967.

KEY TO ADULT *EPIDAMAEUS* OF CANADA AND ALASKA
(MODIFIED FROM NORTON AND ERMILOV 2021)

1	Leg IV twice or more body length	2
–	Leg IV less than 2 times body length	3
2(1)	Prodorsal tubercle *Ba* absent; parallel ridges perpendicular to sejugal scissure present. Tibia IV (excluding claw) subequal to or longer than tarsus IV. Seta *d* absent from tibiae II and III. Scalps absent. Length: 797–842µm (Figure 3.56A)	***E. tenuissimus*** Hammer
–	Prodorsal tubercle *Ba* present; parallel ridges absent. Tarsus IV (excluding claw) 25–30% longer than tibia IV. Seta *d* retained on tibiae II and III. Scalps present. Length: ca. 700µm	***E. longitarsalis*** (Hammer)
3(1)	Trochanters III, IV unusually broad and flattened, with distinct keel; legs somewhat stocky, petioles short. Length: 359–430µm (Figure 3.56B,C)	***E. coxalis*** (Hammer)
–	Trochanters III, IV not flattened or with keel-like margin; leg segments slender, strongly petiolate	4
4(3)	Prodorsum with distinct projection over rostrum. Length: 382–426µm (Figure 3.56D,E)	***E. nasutus*** Behan-Pelletier and Norton
–	Prodorsum without projection over rostrum	5
5(4)	Notogastral setae in *c*, *l* and *h* rows longitudinally oriented; tubercle *Va* present	6
-	Notogastral setae in *c*, *l* and *h* rows radially oriented; *Va* present or absent	11
6(5)	Propodolateral apophysis P present	7
–	Propodolateral apophysis P absent	8
7(6)	Tectum of podocephalic fossa strongly developed, produced laterally as strong tooth (t) overreaching trochanter I. Spina adnata small (8µm), more widely separated than tubercle *Ba*. Notogastral setae with 1 to several barbs on outer curve; setae c_1 directed laterad. Bothridial seta with small barbs. Length: 338–395µm (Figure 3.56F,G)	***E. mackenziensis*** (Hammer)
–	Tectum of podocephalic fossa not produced laterally as strong tooth. Spina adnata longer (18µm), less separated than tubercle *Ba*. Notogastral setae smooth; setae c_1 directed anteriad. Bothridial seta smooth. Length: 385–453µm (Figure 3.57A)	***E. bakeri*** (Hammer)
8(6)	Spina adnata conspicuous, medium-sized (18–20µm)	9
–	Spina adnata minute to small, easily overlooked	10
9(8)	Cerotegument bottle-shaped, conical, or columnar erinose. Bothridial seta almost straight, gradually tapering, with small barbs. Coxisternal pit (cp) circular. Length: 430–494µm. (Figure 3.58C)	***E. kodiakensis*** Hammer
–	Cerotegument vermiform filaments. Bothridial seta subflagellate, smooth. Coxisternal pit (cp) as inverted Y. Length: 365–426µm (Figure 3.59A,B)	***E. craigheadi*** (Jacot)
10(8)	Prodorsal tubercle *Ba* present. Bothridial seta sparsely barbed, whip-like. Spina adnata ca. 7µm. Leg IV ca. 1.5× body length. Coxisternal pit circular. Length: 510–580µm (Figure 3.58B)	***E. arcticola*** (Hammer)
–	Prodorsal tubercle *Ba* absent. Bothridial seta elongate, smooth, acuminate. Spina adnata ca. 11µm. Leg IV equal body length. Coxisternal pit absent (Figure 3.60C)	***E. hammerae*** Behan-Pelletier and Norton
11(5)	Notogastral setae large and leaf-like to knife-like, or dark brown. Ventral tubercle *M* present or absent	12

–	Notogastral setae usually setiform, curved or straight, but without strong blade-like margins; never dark brown. Ventral tubercle *M* absent 15
12(11)	Posteriorly directed prodorsal tubercles *Da* and *Ba* both present. Ventral tubercle *M* absent. Length: 528–627µm (Figure 3.57B) .... ***E. tritylos*** Behan-Pelletier and Norton
–	Prodorsal tubercles (*Ba*) present or absent; *Da* absent. Ventral tubercle *M* present 13
13(12)	Notogastral setae in *c*, *l*, and *h*-series leaf-like, with well-developed vanes on either side of rachis. Propodolateral apophysis weakly developed. Length: 630–730µm (Figure 3.57C).. ....***E. puritanicus*** Banks
–	Notogastral setae in *c*, *l*, and *h*-series with vane at most on one side of rachis, or vane absent. Propodolateral apophysis absent 14
14(13)	Prodorsal tubercles *Ba* and *Bp* present. Bothridial seta straight, barbed, acuminate. Notogastral setae in *c*, *l*, and *h*-series with vane unilaterally. Length: 716–795µm (Figure 3.57D) .... ***E. fortispinosus*** Hammer
–	Prodorsal tubercles *Ba* and *Bp* absent. Bothridial seta straight, barbed, lanceolate distally. Notogastral setae in *c*, *l*, and *h*-series dark brown. Length: 640–650µm (Figure 3.60A,B). ....***E. hastatus*** Hammer
15(11)	Prodorsal tubercle *Ba* present. Ventral tubercles *V* present, *E2* present or absent 16
–	Prodorsal tubercle *Ba* absent. Ventral tubercles *V* and *E2* absent 18
16(15)	Bothridial setae bacilliform, with or without weakly lanceolate head. Prodorsal tubercle *Ba* prominent, conical. Parastigmatic enantiophysis with *Sa* and *Sp* similar, horn-like, converging. Length: 448–563µm (Figure 3.58D).. ***E. michaeli*** Ewing
–	Bothridial seta flagellate, without barbs. Prodorsal tubercle *Ba* small. Parastigmatic enantiophysis with *Sa* horn-like, *Sp* short tubercle 17
17(16)	Ventral enantiophyses *E2* absent. Discidium narrow. Length: 390–434µm (Figure 3.58A) ....***E. floccosus*** Behan-Pelletier and Norton
–	Ventral enantiophyses *E2* present. Discidium absent. Length: 398–448µm (Figure 3.59C) .... ***E. olitor*** (Jacot)
18(15)	Bothridial seta smooth, strongly acuminate. Length: 560–630µm (Figure 3.60D)............. ....***E. koyukon*** Behan-Pelletier and Norton
–	Bothridial seta tapered, barbed along length 19
19(18)	Coxisternal pit absent. Interlamellar setae subequal to mutual distance. Parastigmatic tubercle *Sp* pointed. Discidium pointed tubercle. Male length ca. 457µm (Figure 3.60E,F) ***E. gibbofemoratus*** (Hammer)
–	Coxisternal pit present. Interlamellar setae longer than mutual distance. Parastigmatic tubercle *Sp* rounded. Discidium claw-like. Male length 752–879µm (Figure 3.59D)............ ....***E. globifer*** (Ewing)

Note: Unidentified species from AK: Lava Camp, Seward Peninsula (Late Miocene: 5.7 mya) (Matthews et al. 2019). Undescribed species known from BC, AB.

Epidamaeus arcticola (Hammer, 1952) (Figure 3.58B)
Comb./Syn.: *Belba arcticola* Hammer, 1952
Biology: From alpine tundra, moss, bogs, riverine litter, wet meadows, mixed litters, lichens. Known from Tertiary fossil sites on Meigen Island (NU) (Matthews et al. 2019).
Distribution: AK, YT, NT, NU, BC, AB [OR, NH; Eastern Palaearctic]
Literature: Behan-Pelletier and Norton (1983); ABMI (2019); Norton and Ermilov (2021).

Epidamaeus bakeri (Hammer, 1952) (Figure 3.57A)
  Comb./Syn.: *Belba bakeri* Hammer, 1952
  Biology: From alpine tundra, moss, bogs, riverine litter, wet meadows, mixed litters, lichens.
  Distribution: AK, YT, NT [VA; Holarctic]; the Virginia record is questionable (Behan-Pelletier and Norton 1983).
  Literature: Behan-Pelletier and Norton (1983); Bayartogtokh (2000a).

Epidamaeus coxalis (Hammer, 1952) (Figure 3.56B,C)
  Comb./Syn.: *Belba coxalis* Hammer, 1952
  Biology: Gravid females carry at least 6 large eggs. Gut contents include pigmented and unpigmented fungal hyphae. Originally described from moist birch litter.
  Distribution: AK, YT, NT, AB [OR; Holarctic]
  Literature: Behan-Pelletier and Norton (1983); Bayartogtokh (2000); ABMI (2019).

Epidamaeus craigheadi (Jacot, 1939) (Figure 3.59A,B)
  Comb./Syn.: *Metabelba craigheadi* Jacot, 1939
  Biology: From wide range of habitats in deciduous and coniferous forests.
  Distribution: AB, ON [CO, ME, MN, MT, NH, NY, SD, VT, WA; Holarctic]
  Literature: Norton and Ermilov (2021)

Epidamaeus floccosus Behan-Pelletier and Norton, 1985 (Figure 3.58A)
  Biology: From spruce litter.
  Distribution: AK, YT, BC, AB [Nearctic]
  Literature: Bayartogtokh (2000a); ABMI (2019).

Epidamaeus fortispinosus Hammer, 1967 (Figure 3.57D)
  Biology: Gravid females carry at least 6 eggs and may have eggs attached to scalps. From spruce litter, mosses, birch litter, grasses, and heath. Boreal to subarctic species.
  Distribution: AK, YT, ON, QC [Holarctic]; known from Tertiary fossil sites on Meigen Island (NU) (Matthews et al. 2019).
  Literature: Behan-Pelletier and Norton (1983); ABMI (2019).

Epidamaeus gibbofemoratus (Hammer, 1955) (Figure 3.60E,F)
  Comb./Syn.: *Belba gibbofemorata* Hammer, 1955
  Biology: From sphagnum and moss and spruce needles.
  Distribution: AK, YT, NT [Nearctic].
  Literature: Bayartogtokh (2000a).

Epidamaeus globifer (Ewing, 1913) (Figure 3.59D)
  Comb./Syn.: *Damaeus globifer* Ewing, 1913; *Belba globifer* (Ewing, 1913)
  Biology: From coniferous and deciduous forests, and from litter and decaying woody substrates.
  Distribution: ON [IA, MI, MN, ND, NH, NY, SD, VA, WA: Nearctic].
  Literature: Norton and Ermilov (2021)

Epidamaeus hammerae Behan-Pelletier and Norton, 1983 (Figure 3.60C)
  Biology: From moss and alder litter on north facing slope with *Betula nana*, *Ledum*, and *Vaccinium*; from wet moss and *Rhododendron* litter and from alder and *Pinus* litter.
  Distribution: BC [OR; Russian Far East]
  Literature: Bayartogtokh (2000a).

Epidamaeus hastatus Hammer, 1967 (Figure 3.60A,B)
  Biology: From gravel along river in subarctic.
  Distribution: AK [Alaska]
  Literature: Behan-Pelletier and Norton (1985).

Epidamaeus kodiakensis Hammer, 1967 (Figure 3.58C)
Biology: From broad bog with patches of *Sphagnum*, *Empetrum*, *Betula nana*, *Oxycoccus*, *Rubus*; from upland tundra.
Distribution: AK [Holarctic]
Literature: Behan-Pelletier and Norton (1985); Bayartogtokh (2000a).

Epidamaeus koyukon Behan-Pelletier and Norton, 1985 (Figure 3.60D)
Biology: From south-facing bluff, with grass clumps, aspen litter, litter with lichens.
Distribution: AK, YT, NT, BC, AB, QC [OR; Nearctic; Greenland]
Literature: Hammer (1967); Bayartogtokh (2000a); ABMI (2019).

Epidamaeus longitarsalis (Hammer, 1952)
Comb./Syn.: *Belba longitarsalis* Hammer, 1952
Biology: From tussock tundra; Balsam fir forests.
Distribution: AK, YT, NT, NU, NL [Holarctic]
Literature: Behan-Pelletier and Norton (1985); Bayartogtokh (2000a).

Epidamaeus mackenziensis (Hammer, 1952) (Figure 3.56F,G)
Comb./Syn.: *Belba mackenziensis* Hammer, 1952
Biology: From tundra; from *Vaccinium*, lichens, *Carex*, *Dryas* litter; from grass meadow; from dry willow litter.
Distribution: AK, YT, NT [Nearctic]
Literature: Behan-Pelletier and Norton (1983); Bayartogtokh (2000a).

Epidamaeus michaeli (Ewing, 1909) (Figure 3.58D)
Comb./Syn.: *Damaeus michaeli* Ewing, 1909; *Oribata canadensis* Banks, 1909; *Epidamaeus canadensis* (Banks, 1909)
Biology: From under bark of ironwood. In oak litter. In decaying wood in hardwood forests.
Distribution: ON, QC, NB [AR, IN, MD, MA, MI, NH, NC, NY, OH, VA, VT, WV; Nearctic]
Literature: Norton and Ermilov (2021).

Epidamaeus nasutus Behan-Pelletier and Norton, 1985 (Figure 3.56D,E)
Biology: From Black spruce, White spuce and birch forest. Boreal to subarctic species.
Distribution: AK, YT [Nearctic]
Literature: Bayartogtokh (2000a).

Epidamaeus olitor (Jacot, 1937) (Figure 3.59C)
Comb./Syn.: *Belba olitor* Jacot, 1937
Biology: From deciduous and coniferous litter, and successional habitats.
Distribution: ON [NC, NH, NY; Nearctic]
Literature: Norton and Ermilov (2021).

Epidamaeus puritanicus (Banks, 1906) (Figure 3.57C)
Comb./Syn.: *Oribata puritanica* Banks, 1906; *Damaeus puritanicus* (Banks, 1906)
Note: Norton and Ermilov (2021) rejected the synonymy of *E. grandjeani* Bulanova-Zachvatkina, 1957 with *E. puritanicus*.
Biology: From coniferous and deciduous forests, bogs and prairies.
Distribution: MB, ON [IL, MA, MN, NH, NY, VT; Nearctic]
Literature: Norton and Ermilov (2021).

Epidamaeus tenuissimus Hammer, 1967 (Figure 3.56A)
Biology: Alpine species; collected from pitfall trap and from litter at treeline.
Distribution: AK, BC [Nearctic]
Literature: Behan-Pelletier and Norton (1985); Bayartogtokh (2000a).

Epidamaeus tritylos Behan-Pelletier and Norton, 1983 (Figure 3.57B)
    Distribution: AK, YT, NT, AB [Russian Far East]
    Biology: From willow litter, moss, alder litter.
    Literature: Bayartogtokh (2000a); ABMI (2019).

Kunstidamaeus Miko, 2006
    Type-species: *Belba lengersdorfi* Willmann, 1932
    Comb./Syn.: *Kunstidamaeus lengersdorfi* (Willmann, 1932)

Kunstidamaeus arthurjacoti Norton, Miko and Ermilov, 2022 (Figure 3.54F,G)
    Distribution: ON [CT, IL, IN, IO, MI, MN, MO, NY, ND, OH: Nearctic]
    Biology: From deciduous and coniferous forest litter.

Lanibelba Norton, 1980
    Comb./Syn.: *Belbodamaeus* Bulanova-Zachvatkina, 1960, *Belba* (*Lanibelba*) Norton, 1980
    Type-species: *Lanibelba pini* (Norton, 1980)

Lanibelba pini Norton, 1980 (Figure 3.53B–E)
    Comb./Syn.: *Belbodamaeus* (*Lanibelba*) *pini* Norton, 1980
    Distribution: ON, NB [GA; Nearctic]
    Biology: From litter of Eastern cedar, Balsam fir, oaks.
    Literature: Ermilov (2012).
    Note: Unnamed species from YT, AB.

Parabelbella Bulanova-Zachvatkina, 1972
    Type-species: *Parabelbella elisabethae* Bulanova-Zachvatkina, 1967
    Comb./Syn.: *Akrodamaeus* Norton, 1979a

KEY TO ADULT *PARABELBELLA* OF CANADA AND ALASKA
(FROM ERMILOV AND RYABININ 2020)

1   Prodorsal tubercle *Ba* developed. Length: 531–611µm (Figure 3.52B,C) ....
    ....***P. longiseta*** (Banks)
–   Prodorsal tubercle *Ba* not developed. Length: 381–415µm...
    ....***P. pseudoinaequipes*** Ermilov and Ryabinin
    Note: Known from Northeast USA (NY); possibly in Canada.

Parabelbella longiseta (Banks, 1906) (Figure 3.52B,C)
    Comb./Syn.: *Oribata longiseta* Banks, 1906; *Damaeus longiseta* (Banks, 1906); *Metabelbella* (*Akrodamaeus*) *longiseta* (Banks, 1906); *Akrodamaeus longiseta* (Banks, 1906)
    Biology: From litter of a wide variety of forest types.
    Distribution: ON [AR, DE, ME, MO, NH, NY, PA, TX, VA, WA; Holarctic]
    Literature: Norton (1979a); Miko et al. (2011); Schäffer et al. (2020).

Quatrobelba Norton, 1980
    Type-species: *Quatrobelba montana* Norton, 1980

Quatrobelba montana Norton, 1980 (Figure 3.54D,E)
    Biology: Gravid females with at least three large eggs. Gut contents include pigmented and hyaline fungal hyphae and spores.
    Distribution: AB [CA, CO, ID, MT, SD, WA, WY; Nearctic]
    Literature: Ermilov (2012); ABMI (2019).

Weigmannia Miko and Norton, 2010
    Type-species: *Porobelba parki* Jacot, 1937

Weigmannia parki (Jacot, 1937) (Figure 3.54A–C)
   Comb./Syn.: *Porobelba parki* Jacot, 1937
   Biology: From litter and soil in decaying wood; tree hole; rotting wood; ex Ringed bill gull nest.
   Distribution: MB, ON, NB [IN, IL, NC, NY, MD, ME, MI, MO, OH, VA; Holarctic]
   Literature: Miko and Norton (2010); Ryabinin and Ermilov (2021).

Superfamily Cepheoidea Berlese, 1896

Family Cepheidae Berlese, 1896
   Diagnosis: *Adults*: Non-poronotic Brachypylina. Prodorsum with or without lamellae and lamellar cusps. Tutorium present or absent. Parastigmatic enantiophysis absent. Dorsophragmata and pleurophragmata absent. Pedotectum I deeply incised or normal; pedotectum II present. Notogaster with or without well-developed humeral apophysis; notogastral setation 9–10 pairs, often in marginal position. Epimeral setation 3-1-3-3. Circumpedal carina, discidium present. Custodium present or absent. Genital setation 6 pairs. Paired aggenital taenidium and minitectum present (*Conoppia*) or absent. Aggenital enantiophysis present or absent. Subcapitulum diarthric, chelicerae chelate-dentate; palpal eupathidium *acm* associated with solenidion, fusion often incomplete. Legs mono- or tridactylous. Sharp spines on trochanter IV, and sometimes trochanter III present or absent.

KEY TO GENERA OF CEPHEIDAE OF CANADA AND ALASKA

1   Bothridial seta globular, often black in colour, with very short stalk, situated within, or hardly emerging from bothridium. Lamellae subparallel, translamella absent. Tridactylous (Figure 3.61A,B) ....***Ommatocepheus*** Berlese
–   Bothridial seta extending far beyond bothridium; never globular, but may have distinct clubbed head. Lamellae distinctly converging; translamella present or absent; Mono- or tridactylous 2

2(1)   Lamellar cusps broad, projecting well beyond rostrum, truncate, medially oblique, or with medial shoe- or knife-like process. Interlamellar setae minute or spine-like (Figure 3.62A–D) ....***Eupterotegaeus*** Berlese
–   Lamellae with lamellar cusps not extending, or just extending beyond rostrum. Interlamellar setae small to long 3

3(2)   Legs tridactylous 4
–   Legs monodactylous 5

4(3)   Integument smooth. Most notogastral setae represented only by alveoli. Lamellae narrow; translamella linear, weakly developed (Figure 3.61C) ....***Conoppia*** Berlese
–   Integument sculptured. Notogastral setae present. Lamellae and translamella ribbon-shaped (Figure. 3.61F) ....***Sphodrocepheus*** Woolley and Higgins

5(3)   Notogastral setae long, curved, barbed, tapering and extending beyond body margin. Interlamellar seta inserted almost at level of bothridium and reaching only about half length of lamellae (Figure 3.61D,E) ....***Oribatodes*** Banks
–   Notogastral setae short, smooth or barbed. Interlamellar setae usually inserted about middle of prodorsum and almost reaching tips of lamellar cusps (Figure 3.62E–G) ....
   ....***Cepheus*** C.L. Koch

Cepheus C.L. Koch, 1835
   Type-species: *Cepheus latus* C.L. Koch, 1835
   Comb./Syn.: *Tegeocranus* Nicolet, 1855

KEY TO ADULT *CEPHEUS* OF CANADA AND ALASKA
(MODIFIED FROM WALTER ET AL. 2014)

1 Notogaster with pattern of connected ridges that enclose cells (40–100µm); with reticulate sculpturing. Interlamellar seta smooth, thick, short. Bothridial seta somewhat recurved, with spiculate, club-shaped head. Length: ca. 680µm (Figure 3.62E,F)........*C. corae* Jacot
– Notogastral sculpturing consisting of dark, irregular ridges separate from each other. Interlamellar seta slender, bristle-like. Bothridial seta clavate, barbed. Length: 670–900µm (Figure 3.62G) .... *C. latus* C.L. Koch
Note: Unidentified species from AK: Lava Camp, Seward Peninsula (Late Miocene: 5.7 mya), from Tertiary fossil sites on NU, Prince Patrick Island and from late Pliocene/early Pleistocene fossil deposits in AK (Matthews et al. 2019). Unnamed species from YT, BC, AB, ON, QC, NB, NS.

Cepheus corae Jacot, 1928 (Figure 3.62E,F)
Biology: This species was reported from Boreal forest (Lindo and Visser 2004) and from forest floor under fir and hemlock, Sitka spruce and Western redcedar on Vancouver Island (Winchester et al. 2008). *Cepheus* sp. nr. *corae* was found under rust-infected Lodgepole pine (Powell and Skaley 1975). This species is known from late Pliocene/early Pleistocene fossil deposits in AK (Matthews et al. 2019).
Distribution: AK, YT, BC, AB, QC, NL [ME, MI, NY; Nearctic]
Literature: Walter et al. (2014); Schäffer et al. (2020).

Cepheus latus C.L. Koch, 1835 (Figure 3.62G)
Comb./Syn.: *Carabodes latus* (C.L. Koch, 1835); *Tegeocranus latus* (C.L. Koch, 1835)
Biology: Reported from arboreal fruiting bodies of the fungus *Daldinia concentrica* in England (Hingley 1971). Reported as macrophytophagous (Mitchell 1978).
Distribution: NU, AB [Holarctic]
Literature: Bernini and Bernini (1990); Walter et al. (2014); Maraun et al. (2019).

Conoppia Berlese, 1908
Type-species: *Oppia microptera* Berlese, 1885 (Figure 3.61C)
Comb./Syn.: *Phyllotegeus* Berlese, 1913
Note: Unidentified *Conoppia* sp. from NT, BC.

Eupterotegaeus Berlese, 1916
Type-species: *Tegeocranus ornatissimus* Berlese, 1908
Comb./Syn.: *Diodontocepheus* Mihelčič, 1958
Note: A few minutes in concentrated (6%) sodium hypochlorite dissolves the cerotegument and produces specimens where underlying cuticular characters can be clearly seen (Walter et al. 2014).

KEY TO ADULT *EUPTEROTEGAEUS* OF CANADA AND ALASKA
(FROM WALTER ET AL. 2014)

1 Tridactylous. Humeral enantiophysis absent. Aggenital enantiophysis absent 2
– Monodactylous. Humeral enantiophysis present. Aggenital enantiophysis present, bracketing the genital opening laterally. Ventral length: 400–600µm (Figure 3.62A) ....
.... *E. rhamphosus* Higgins and Woolley

2(1) Lamellar process broad, knife-shaped. Anterior margin of notogaster without dens 3

–	Lamellar process narrow, needle-like, "shoe-shaped". Anterior margin of notogaster with dens. Length: 659–755μm (Figure 3.62B) .... ***E. ornatissimus*** (Berlese)
3(2)	Translamella absent; median spine absent. Rostrum with small mucro. Notogaster foveolate. Length: 612–700μm (Figure 3.62C) .... ***E. rostratus*** Higgins and Woolley
–	Translamella present and bearing strong median spine. Rostrum truncate. Notogaster reticulate. Length: 594μm (Known from UT; not presently known for Canada or Alaska) (Figure 3.62D) .... ***E. spinatus*** Higgins and Woolley

Note: Unidentified *Eupterotegaeus* sp. from AK, BC, AB, NB, NS.

Eupterotegaeus ornatissimus (Berlese, 1908) (Figure 3.62B)
  Comb./Syn.: *Tegeocranus ornatissimus* Berlese, 1908; *Scutovertex ornatissimus* (Berlese, 1908)
  Biology: From Black spruce litter; *Sphagnum*; boreal mixedwood forest.
  Distribution: ON, QC [Holarctic]
  Literature: Weigmann (2006).

Eupterotegaeus rhamphosus Higgins and Woolley, 1963 (Figure 3.62A)
  Biology: From Lodgepole pine litter. Gravid females carry at least 2 large eggs. Gut contents include dark hyphae.
  Distribution: BC, AB [MT, OR, UT; Nearctic]
  Literature: Walter et al. (2014); ABMI (2019).

Eupterotegaeus rostratus Higgins and Woolley, 1963 (Figure 3.62C)
  Biology: Found in litter layer of aspen woodland (Mitchell 1978). Gut contents of field collected animals consisted primarily of pigmented hyphae; in the laboratory, adults fed on the fungus *Phoma exiga* and unidentified pigmented hyphae (Mitchell and Parkinson 1976).
  Distribution: BC, AB [CA, CO, WA, WY; Nearctic]
  Literature: ABMI (2019).

Oribatodes Banks, 1895
  Type-species: *Oribatodes mirabilis* Banks, 1895

Oribatodes mirabilis Banks, 1895 (Figure 3.61D,E)
  Comb./Syn.: *Oribata mirabilis* (Banks, 1895); *Cepheus lamellatus* Banks, 1906; *Carabodes lamellatus* (Banks, 1906); *Tegeocranus longisetus* Berlese, 1910
  Biology: From boreal forest litter.
  Distribution: AB, ON, QC [CN, DC, IL, MA, NY; Nearctic]
  Literature: ABMI (2019).

Ommatocepheus Berlese, 1913
  Type-species: *Cepheus ocellatus* Michael, 1882

KEY TO ADULT *OMMATOCEPHEUS* OF CANADA AND ALASKA

1	Interlamellar setae clavate. Length: ca. 474μm (Figure 3.61A).. .... ***O. clavatus*** Woolley and Higgins
–	Interlamellar seta setose, tapered. Length: 583–620μm (Figure 3.61B) .... ***O. ocellatus*** (Michael)

Ommatocepheus clavatus Woolley and Higgins, 1964 (Figure 3.61A)
  Biology: Recorded from epiphytic lichens in the Adirondack Mountains (Root et al. 2007).
  Distribution: QC [NC, NY; Nearctic]
  Literature: Erickson et al. (2003).

Ommatocepheus ocellatus (Michael, 1882) (Figure 3.61B)
   Comb./Syn.: *Cepheus ocellatus* Michael, 1882; *Ommatocepheus pulcherrimus* Berlese, 1913
   Biology: Travé (1963) recovered specimens from lichen biotopes and found that nymphs and adults tended to inhabit the surface layers of crustose lichens. The mites remained inactive in dry conditions, feeding and burrowing only when the lichen thallus was moist. Specimens contained food balls and faecal pellets which consisted almost entirely of *Trebouxia* (Colloff 1984).
   Distribution: BC [NC; Holarctic]
   Literature: Wunderle (1991); Corral-Hernandez et al. (2016); Maraun et al. (2019).

Sphodrocepheus Woolley and Higgins, 1963
   Type-species: *Sphodrocepheus tridactylus* Woolley and Higgins, 1963
   Note: Unidentified species from AK: Lava Camp, Seward Peninsula (Late Miocene: 5.7 mya) (Matthews et al. 2019).

Sphodrocepheus anthelionus Woolley and Higgins, 1968 (Figure 3.61F)
   Biology: Recorded from temperate rainforest on Vancouver Island (Winchester et al. 1999).
   Distribution: BC [OR, UT, WA; Nearctic]

Tritegeus Berlese, 1913
   Type-species: *Cepheus bifidatus* Nicolet sensu Berlese, 1913 = *Tritegeus bisulcatus* Grandjean, 1953
   Literature: Maraun et al. (2019).
   Note: Unidentified species from BC (Fagan et al. 2005), are probably misidentified *Sphodrocepheus* sp. Unidentified specimen from late Pliocene/early Pleistocene fossil deposits in AK (Matthews et al. 2019).

Superfamily Polypterozetoidea Grandjean, 1959

Family Polypterozetidae Grandjean, 1959
   Diagnosis: *Adults*: Non-poronotic Brachypylina. Bearing nymphal exuviae and thick debris. Thick, reticulate cerotegument present. Prodorsum with broad lamellae and lamellar cusps, touching medially, but not fused. Translamella absent. Lamellar seta borne anteromedially on cusp. Bothridium expanded distally. Tutorium present. Pedotectum I present; pedotectum II absent. **Patronium** present (scalelike, tectiform projection in similar position to propodolateral apophysis) subequal in length to pedotectum I. Apophysis present ventral to acetabula II and III. Custodium, discidium and circumpedal carina absent. Dorsophragmata and pleurophragmata absent. Notogaster with humeral apophysis; often obscured by adherent dirt attached to cerotegument; setation 11 pairs. Epimeral setation 3–2–3–3. Genital setation 6 pairs. Lyrifissure *ian* present. Seta ad_1 absent. Subcapitulum anarthric; chelicera modified, with harpoon-like tooth distally, surrounded by dentate lamina. Rutellum developed as thick seta, bifid distally. Subcapitular mentum with tectum. Palp setation 0-2-1-3-9(1); palpal eupathidium *acm* separate from recumbent solenidion. Legs short, monodactylous. Seta *d* absent from genua I–III and tibiae I–IV. Tarsus II with 1 solenidion. Trochanters III, IV with large ventral carina.

Polypterozetes Berlese, 1916
   Type-species: *Polypterozetes cherubin* Berlese, 1916 (Figure 3.63D–H)
   Note: Undescribed *Polypterozetes* sp. from BC (Battigelli et al. 2004).

Family Podopterotegaeidae Pifﬂ, 1972
   Diagnosis: *Adults*: Non-poronotic Brachypylina. Bearing concentric nymphal exuviae; scalps easily detached. Prodorsum mostly covered by lamellae. Lamellae and lamellar cusps fused

proximally, cusps separated by narrow to wide notch distally; lamellar seta arising medially on cusps. Bothridium large, expanding (vase-shaped) distally, with humeral enantiophysis. Tutorium and subtutorium present. Pedotectum I large, extending dorsally almost to seta *ex*; pedotectum II present. Dorsophragmata and pleurophragmata absent. Notogaster convex anteriorly; setation 11 pairs; pit at base of humeral apophysis bearing setae c_1, c_2. Epimeral setation 3-1-3-3. Epimeral region with apophyses. Custodium and circumpedal carina absent. Genital setation 5 pairs. Genital papillae *Va* wide; *Vm* and *Vp* slender and tapered. Aggenital or adanal neotrichy absent. Subcapitulum diarthric. Chelicera chelate-dentate with seta *cha* coiled 2 to many times over back of chelicera. Palp setation: 0-2-1-3-9(1); palpal eupathidium *acm* separate from recumbent solenidion. Trochanters and tibiae III, IV with broad ventral carina. Seta *d* of genua II–IV and tibiae I–IV associated with solenidion.

Podopterotegaeus Aoki, 1969
   Type-species: *Podopterotegaeus tectus* Aoki, 1969

Podopterotegaeus tectus Aoki, 1969 (Figure 3.63I–K)
   Biology: From maple forest litter, primarily associated with decaying logs (Déchêne and Buddle 2010).
   Distribution: QC, NL [NY; holarctic]
   Literature: Maraun et al. (2019).

Superfamily Microzetoidea Grandjean, 1936

Family Microzetidae Grandjean, 1936
   Diagnosis: *Adults*: Non-poronotic Brachypylina. Proterosoma subequal to hysterosoma in length, so that notogaster as wide as, or wider than long. Prodorsum with broad lamellae, translamella present or absent. Tutorium present. Dorsophragmata and pleurophragmata absent. Dorsosejugal scissure present or absent. Pedotectum I well developed; pedotectum II present. Posterior notogastral tectum and pteromorphs present. Notogastral setation 9 pairs. Epimeral setation 3-1-3-3. Epimeral border IV clearly evident, extending transversely anterior to genital plates; with tectum. Genital setation 6 pairs. Circumpedal carina present. Subcapitulum diarthric, eupathidium *acm* separate from solenidion; chelicerae dorsally with bacilliform or brush-like tubercle. Tarsus II with 2 solenidia; proral setae of tarsi II–IV short and thick.

KEY TO GENERA OF MICROZETIDAE OF CANADA AND ALASKA

1   Interlamellar setae long, at least half as long as lamellae (Figure 3.64A,B) ....
.... ***Berlesezetes*** Mahunka
–   Interlamellar setae short, hardly discernible (Figure 3.64C,D) .... ***Kalyptrazetes*** Balogh

Berlesezetes Mahunka, 1980
   Type-species: *Microzetes auxiliaris* Grandjean, 1936

Berlesezetes appalachicolus (Jacot, 1938) (Figure 3.64A,B)
   Comb./Syn.: *Microzetes auxiliaris appalachicola* Jacot, 1938; *Microzetes appalachicola* Jacot, 1938
   Biology: From maple forest litter.
   Distribution: QC [KY, LA, MS, NC, NM, TN; Nearctic]

Kalyptrazetes Balogh, 1972 (= *Allozetes* Higgins, 1965, preoccupied name)
   Type-species: *Allozetes harpezus* Higgins, 1965
   Note: Species nr. *Kalyptrazetes* from BC (Lindo and Clayton 2011).

Superfamily Caleremaeoidea Grandjean, 1965

Family Caleremaeidae Grandjean, 1965
   Diagnosis: *Adult*: Non-poronotic Brachypylina. Integument with enveloping cerotegument, usually with dense, dome- to mushroom-shaped excrescences; sclerotized procuticle partly foveate to foveolate. Prodorsum with or without paired, ridge-like lamella, tutorium and prodorsal enantiophysis. Rostrum with strongly developed submarginal crest. Bothridial seta with basal stalk and flattened, expanded head. Dorso- and pleurophragmata absent. Notogaster anterior margin nearly straight, with small dentate tubercles or knots and distinct humeral process; humeral enantiophysis present. Notogaster with strong topography consisting of relatively flat lateral region and two strong bulges (transverse anterior bulge and longitudinal posterior bulge) separated by foveate transverse sulcus; setation 10 pairs of marginal to submarginal setae. Pedotectum I present, II absent; propodolateral apophysis absent; distinct discidium present or absent; circumpedal carina absent; lateral, parastigmatic and aggenital enantiophyses present. Rutellum atelobasic. Tibiae I, II unusually large, with narrow basal stalk and swollen distal bulb, tibia I with dorsodistal process. Legs monodactylous; seta *d* absent from genua I–III and all tibiae.
   Literature: Norton and Behan-Pelletier (2020); Lienhard and Krisper (2021).

Caleremaeus Berlese, 1910
   Type-species: *Notaspis monilipes* Michael, 1882

KEY TO ADULT *CALEREMAEUS* OF CANADA AND ALASKA

1 Lamella and tutorium distinct. Lamellar seta inserted on strong cusp. Enantiophysis *eA* present. Length: 306–340μm (Figure 3.65A,B) .... ***C. retractus*** (Banks)
– Lamella and tutorium weakly developed. Lamellar seta inserted on low transverse ridge. Enantiophysis *eA* absent. Length: 311–353μm (Figure 3.65C,D) ....***C. arboricolus*** Norton and Behan-Pelletier
   Note: Unidentified species from BC.

Caleremaeus arboricolus Norton and Behan-Pelletier, 2020 (Figure 3.65C,D)
   Biology: Consistently associated with arboreal microhabitats, including bark, twigs and lichens; Adults and juveniles ingest diverse fungal material.
   Distribution: ON [AL, ME, NJ, NY, TN, VA; Nearctic]

Caleremaeus retractus (Banks, 1947) (Figure 3.65A,B)
   Comb./Syn.: *Carabodoides retracta* Banks, 1947
   Biology: From soil microhabitats. Adults and juveniles ingest diverse fungal material.
   Distribution: QC [NC; Nearctic]; see Norton and Behan-Pelletier (2020) on possibility of a 'retractus' species group, including the record from QC.

Superfamily Ameroidea Bulanova-Zachvatkina, 1957 (= Amerobelboidea Grandjean, 1954; = Eremuloidea Grandjean, 1965)

Veloppia Hammer, 1955
   Type-species: *Veloppia pulchra* Hammer, 1955
   Note: *Veloppia* presently is an unplaced genus. It was tentatively placed in Caleremaeidae by Norton (1978) and subsequent authors (Subías 2004; Seniczak and Seniczak 2019); it is possibly a member of the Ameroidea.

KEY TO ADULT *VELOPPIA* OF CANADA AND ALASKA

1. Isotropic tip of bothridial seta attenuate. Prodorsal enantiophysis and ridged lamella absent. Anterior notogastral margin with 1 pair of tubercles. Notogastral setae relatively thick with barbs. Length: 290µm (Figure 3.65E) .... *V. pulchra* Hammer
– Isotropic tip of bothridial seta narrowly rounded. Prodorsal enantiophysis and ridged lamella present. Anterior notogastral margin with 2 pair of tubercles. Notogastral setae smooth. Length: 281–306µm (Figure 3.65F–I) .... *V. kananaskis* Norton

Veloppia kananaskis Norton, 1978 (Figure 3.65F–I)
Biology: From organic soil horizons in woodland dominated by aspen and Balsam poplar (Norton 1978).
Distribution: BC, AB [Canada]

Veloppia pulchra Hammer, 1955 (Figure 3.65E)
Biology: Boreal species; in woods, lichens and grasses.
Distribution: AK, YT, QC [Nearctic]
Literature: Norton (1978).

Family Ameridae Grandjean, 1965
Diagnosis: *Adults*: Non-poronotic Brachypylina. Rostrum with deep incisions. Prodorsum and anterior of notogaster strongly depressed. Prodorsum with or without enantiophyses; with or without lamellae. Tutorium absent. Dorsophragmata and pleurophragmata absent. With constriction in dorsosejugal region. Dorsosejugal scissure complete or partially absent. Notogaster with or without humeral depressions; setation 10 pairs. Humeral apophysis present. Circumpedal carina, custodium, and discidium absent. Pedotectum I present; pedotectum II present or absent (*Gymnodampia*). Aggenital neotrichy present or absent. Subcapitulum diarthric, eupathidium *acm* separate from recumbent solenidion. Chelicerae chelate-dentate. Axillary saccule of subcapitulum present or absent. Proral setae of tarsi II–IV short and spiniform.

Gymnodampia Jacot, 1937
Type-species: *Amerobelba setata* Berlese, 1916

Gymnodampia setata (Berlese, 1916) (Figure 3.66A)
Comb./Syn.: *Amerobelba setata* Berlese, 1916
Biology: From hemlock litter; moist peaty areas. In oak litter (Lamoncha and Crossley 1998).
Distribution: QC [AL, AR, MI, MO, NC, OH, TN; Nearctic]
Literature: Chen et al. (2004).

Family Damaeolidae Grandjean, 1965
Diagnosis: *Adults*: Non-poronotic Brachypylina. Trachea I expressed as brachytrachea. Prodorsum without lamellae. Pedotectum I very small or absent; pedotectum II absent. Tutorium absent. Dorsophragmata and pleurophragmata absent. With constriction in sejugal region. Dorsosejugal scissure present. Notogastral setation 11 pairs. Lyrifissures *im* and *ip* absent. Humeral apophysis absent. Circumpedal carina, custodium, and discidium absent. Preanal organ thin, small, flat. Ovipositor absent; only vulva present. Epimeral setation 3-1-3-3. Aggenital neotrichy present. Subcapitulum diarthric; with large, diaphanous, leaf-like rutellum. Chelicera strongly toothed, with 0 or 1 seta. Palpgenu without setae, fused or not with femur; palptibia with 0 or 1 seta; palptarsus with 3–5 setae. Axillary saccule of subcapitulum absent. Seta *d* present on genua I–III and tibiae I–IV, not associated with solenidion except on leg I. Proral setae setiform. Tarsi with very large lyrifissure.

KEY TO GENERA OF DAMAEOLIDAE OF CANADA AND ALASKA

1   Notogaster with 4 semicircular depressions (Figure 3.66E–I) .... ***Fosseremus*** Grandjean
–   Notogaster without semicircular depressions (Figure 3.66B) .... ***Damaeolus*** Paoli

Damaeolus Paoli, 1908
  Type-species: *Dameosoma asperatum* Berlese, 1904
  Note: Unidentified species from BC.
  Literature: Seniczak et al. (2020).

Fosseremus Grandjean, 1954
  Type-species: *Damaeosoma laciniatus* Berlese, 1905

Fosseremus laciniatus (Berlese, 1905)
  Comb./Syn.: *Damaeosoma laciniatus* Berlese, 1905
  Distribution: AB [Holarctic]
  Literature: Baran et al. (2010); Maraun et al. (2019); Walter and Lumley (2021).

Family Eremobelbidae Balogh, 1961
  Diagnosis: *Adults*: Non-poronotic Brachypylina. Prodorsum with large tubercles separated by depression which form curved shapes anterior to bothridium. Small ridges present bearing lamellar seta anteriorly. Tutorium absent. Dorsophragmata and pleurophragmata absent. Constriction present in sejugal region. Pedotecta I and II well-developed. Notogastral setation 11 pairs. Humeral apophysis present. Circumpedal carina, custodium, and discidium absent. Genital papillae *Va* larger than *Vm* and *Vp*. Some epimeral setae branched. Aggenital and adanal neotrichy present. Subcapitulum diarthric. On palptarsus eupathidium *acm* separate from recumbent solenidion. Chelicerae chelate-dentate. Axillary saccule of the subcapitulum absent. Tibiae and tarsi I–IV with retrotecta. Femora I–IV and trochanters III, IV with 3–5 saccules in row. Proral setae of tarsi II–IV short and spiniform.

Eremobelba Berlese, 1908 (Figure 3.67C,D)
  Type-species: *Eremaeus leporosus* Haller, 1884

Eremobelba gracilior Berlese, 1908
  Comb./Syn.: *Eremobelba neosota* (Banks, 1909); *Eremobelba leporosus flagellaris* Jacot, 1938; *Eremobelba nervosa* Hartenstein, 1962
  Biology: In oak litter (Lamoncha and Crossley 1998). This species (as *E. nervosa*) fed preferentially on fungi; it did not feed on fresh or decaying leaves, needles or wood (Hartenstein 1962).
  Distribution: ON, QC [NC, MO, NY, VA; Nearctic]
  Literature: Norton and Kethley (1990); Maraun et al. (2019); Ermilov (2021).

Family Eremulidae Grandjean, 1965
  Diagnosis: *Adults*: Non-poronotic Brachypylina. Ridge-like lamella present, parallel or slightly divergent anteriorly. Lamellar seta inserted anteriorly on lamella. Genal incision absent. Prodorsal, bothridial, lateral, and sejugal apophyses absent. Tutorium absent. Dorsophragmata and pleurophragmata absent. Pedotectum I present, pedotectum II absent. With constriction in sejugal region. Notogastral setation 11 pairs. Humeral apophysis absent. Circumpedal carina, custodium, and discidium absent. Epimeral, genital and aggenital setae branched. Epimeral setation 3-1-3-3. Genital setation 6 pairs. Aggenital neotrichy present. Genital papillae *Va* larger than *Vm*, *Vp*, and distal integument thicker. Lyrifissure *iad* anterior to seta ad_2 or at same level. Subcapitulum diarthric, eupathidium *acm* separate from recumbent solenidion. Chelicera chelate-dentate. Axillary saccule of subcapitulum absent. Seta *d* of tibia IV large, not associated with solenidion. Seta *d* of tibiae I–III associated with solenidion. Proral setae on tarsi II–IV small, thin, setiform.

Eremulus **Berlese, 1908**
  Type-species: *Eremulus flagellifer* Berlese, 1908

Eremulus cingulatus Jacot, 1937 (Figure 3.66C)
  Biology: Found in areas of high organic matter.
  Distribution: QC [NC, OH; Nearctic]

Family Hungarobelbidae Miko and Travé, 1996
  Diagnosis: *Adults*: Non-poronotic Brachypylina. Body, proximal part of bothridial seta and notogastral setae covered by granular, tubercular or amorphous cerotegument. Brachytrachea present in place of trachea I. Bothridium funnel-like, bothridial seta usually setiform, flagelliform distally. Distinct prodorsal groove present, tending to form prodorsal enantiophysis. Lamellae and tutorium absent. Propodolateral apophysis present. Postbothridial apophysis, lateral enantiophysis and parastigmatic enantiophysis present. Pedotecta I and II absent. Discidium present. Notogaster with straight or truncate anterior border. Humeral apophysis usually present. Notogastral setation 10 or 11 pairs. Setae c, l, h, and p_1 in approximately longitudinal rows. Epimere II neotrichous or not. Genital setation 6 pairs. Aggenital neotrichy usually present. Legs monodactylous, moniliform, shorter than body. Solenidion of tibia I with companion seta *d*. All genual solenidia with companion setae. Solenidia of tibiae II–IV, usually with companion setae. Two solenidia on tarsus II. Proral setae setiform.
  Literature: Bayartogtokh and Ermilov (2021).

Hungarobelba Balogh, 1943
  Type-species: *Belba visnyai* Balogh, 1938

Hungarobelba nortonroyi Bayartogtokh and Ermilov, 2021 (Figure 3.67A,B)
  Biology: From coniferous forest litter.
  Distribution: BC [WA; Nearctic]
  Note: Unidentified *Hungarobelba* sp. known from BC, AB.

Superfamily Zetorchestoidea Michael, 1898 (= Eremaeoidea Oudemans, 1900)
  Note: Zetorchestoidea is paraphyletic based on molecular studies by Lienhard et al. (2014). Species in this superfamily primarily inhabit periodically dry habitats. Norton and Alberti (1997) suggested a correlation between this living habit and the presence of the apodematoacetabular tracheal system in juveniles as well as adults in this superfamily.

Family Eremaeidae Oudemans, 1900
  Diagnosis: *Adults*: Non-poronotic Brachypylina. Cerotegument filiform or formed of minute hexagonal platelets, tracing, or not, reticulate pattern of integument. Ridge-like lamella present or absent; with or without translamella; lamellar setae positioned anteriorly on ridge-like lamella, or arising from prodorsum, well anteriad of lamella. Tutorium absent. Alveolus of second exobothridial seta present. Dorsophragma and pleurophragma absent. Notogastral setation 10–18 pairs. Pedotecta I and II present; pedotectum II distally broad, forked or truncate in dorsoventral aspect. Discidium present or absent. Custodium and circumpedal carina absent. Epimere II with 1–3 pairs of setae; neotrichy present or absent on epimera III and IV. Anal and adanal neotrichy present or absent. Lyrifissure *ian* present or absent. Subcapitulum diarthric; palp setation 0-2-1-3-9(1); palptarsal solenidion arising at level of setae (*lt*), lying parallel to surface of palp, extending anteriorly to base of eupathidium *acm*; adoral setae strongly barbed. Axillary saccule of subcapitulum generally absent; occasionally present. Legs tridactylous. Femoral neotrichy present or absent. Seta *d* of genua I–III and tibiae I–IV present or absent. Insertion of setae (*p*) on leg tarsi on soft cuticle. Porose areas on tibiae and tarsi. Sharp spines on trochanter IV, and sometimes trochanter III present or absent.
  Note: The family is represented by 3 extant genera in North America: *Eremaeus*, *Eueremaeus* and *Tricheremaeus;* the latter from only one dead specimen in BC. The eremaeid genus

Proteremaeus which is well represented in Eastern and Far Eastern Russia is known from North America only on the basis of fossil records from Pliocene deposits on Ellesmere Island (Behan-Pelletier and Ryabinin 1991).

Biology: Species of Eremaeidae are found primarily in the litter layer, or in moss and lichens on the soil surface; a few species are arboreal. Occasionally a large percentage of the population may be found in the fermentation layer; but there was no seasonal pattem of vertical migration, and no significant correlation between soil moisture and horizontal distribution (Mitchell 1978). It is clear from data in Behan-Pelletier (1993) that most North American species of Eremaeidae prefer dry habitats. Apparently these habitats need to be well structured and/or stable, because only *Eueremaeus aridulus*, an inhabitant of shortgrass prairie, has been recorded from cultivated soil.

Species of Eremaeidae are known fungivores (Schuster 1956; Mitchell and Parkinson 1976). Mitchell and Parkinson (1976) found that in culture *Eremaeus* spp. would feed only on *Phoma exigua* of seven species of fungi offered.

KEY TO GENERA OF EREMAEIDAE OF CANADA AND ALASKA

1	Notogastral setation 15–20 pairs	***Tricheremaeus*** Berlese
–	Notogastral setation 10–11 pairs	2
2(1)	Ventral plate rounded posteriorly, without projection. Notogaster without posteromarginal sclerite; setation usually 11 pairs (Figures 3.68; 3.69)	***Eremaeus*** C.L. Koch
–	Ventral plate with postanal projection. Notogaster usually with posteromarginal sclerite; setation usually 10 pairs (Figure 3.70–3.73)	***Eueremaeus*** Mihelčič

Eremaeus C.L. Koch, 1835

Type-species: *Eremaeus hepaticus* C.L. Koch, 1835

Biology: No species of *Eremaeus* has a transcontinental distribution. Of the 18 species recognized in North America, 3 are distributed in the eastern deciduous vegetation zone: *E. brevitarsus* is found throughout the eastern deciduous and transitional vegetation zones, *E. grandis* occurs in the Western arctic, and *E. translamellatus* occurs in the Western arctic, subarctic, and boreal zones. The remaining 12 species are distributed in the Cordillera and western coastal vegetation zones.

KEY TO ADULT *EREMAEUS* OF CANADA AND ALASKA

1	Notogastral setation 11 pairs, seta c_2 present	2
–	Notogastral setation 10 pairs, seta c_2 absent	11
2(1)	Ridge-like lamella absent, or expression very faint. Bothridial seta clavate, shorter than mutual distance of bothridia	3
–	Ridge-like lamella well developed. Bothridial seta setiform or slightly expanded distally, longer than mutual distance of bothridia	4
3(2)	Notogastral length to width ratio about 1.2:1. Interlamellar setae less than 50µm long. Tubercles *La* absent. Prodorsum with 2 or 3 rows of small tubercles posteriorly. Length: 616–113µm (Figure 3.68B)	*E. kananaskis* Mitchell
–	Notogastral length to width ratio 1.4:1. Interlamellar setae 70–80µm long. Tubercles *La* present. Prodorsum with pair of ridges posteriorly. Length: 648–687µm (Figure 3.68C)	*E. salish* Behan-Pelletier
4(2)	Ridge-like lamella longer than 65µm, parallel; outline of ridges smooth, or crenulate	5
–	Ridge-like lamella short, less than 65µm, concave to parallel; outline of ridges smooth	8

5(4)		Ridge-like lamella smooth. Adanal setation 6 pairs. Anal setation usually 8 (7–10) pairs. Length: 642–829μm (Figure 3.68D,E) .... ***E. plumosus*** Woolley
–		Ridge-like lamella crenulate. Adanal setation 5 pairs, or fewer. Anal setation fewer than 8 pairs 6
6(5)		Prodorsum with 8–14 pairs of tubercles posteromedially 7
–		Prodorsum with 2 pairs of tubercles posteromedially, or with 2–3 pairs of large U-shaped ridges posteromedially. Length: 661–718μm (Figure 3.68A) .... ***E. appalachicus*** Behan-Pelletier
7(6)		Femora I and II with cusp ventrodistally. Prodorsum with about 14 pairs of tubercles posteromedially. Length: 661–739μm (Figure 3.68F,G) .... ***E. occidentalis*** Behan-Pelletier
–		Femora I and II without cusp ventrodistally. Prodorsum with 8–11 pairs of tubercles posteromedially. Length: 622–680μm (Figure 3.69A) .... ***E. kevani*** Behan-Pelletier
8(4)		Notogastral seta c_1 with porose ring or pores at base. Anal setation 5 pairs. Bothridial seta 80–110μm long. Length: 512–544μm (Figure 3.69B) .... ***E. walteri*** Behan-Pelletier
–		Notogastral seta c_1 without pores or porose ring at base. Bothridial seta 109–186μm long. Anal setation usually 6 pairs (occasionally 5 pairs). 9
9(8)		Notogastral setae of c series longer than 110μm. Interlamellar setae longer than 80μm ... 10
–		Notogastral setae of c series 80–90μm. Interlamellar setae 42μm long. Length: 580–680μm (Figure 3.69C) .... ***E. brevitarsus*** (Ewing)
10(9)		Notogaster foveate, not pitted. Bothridial seta only slightly expanded distally. Lamellar ridge concave medially, translamella present or absent. Length: 518–745μm (Figure 3.69D) .... ***E. translamellatus*** Hammer
–		Notogaster pitted. Bothridial seta almost clavate. Lamellar ridge short, parallel or diverging posteriorly. Length: 674–718μm (Figure 3.69E) .... ***E. grandis*** Hammer
11(1)		Length of males <670μm. Interlamellar seta 60–90μm long. Bothridial seta 110–130μm long. Length: 609–667μm (Figure 3.69F) .... ***E. monticolus*** Behan-Pelletier **Note**: Known from western USA (AZ, CA, NM, NV, UT, TX); possibly in Canada.
–		Length of males >690μm. Interlamellar seta 110–130μm long. Bothridial seta 130–190μm long. Length: 700–778μm (Figure 3.69G) .... ***E. boreomontanus*** Behan-Pelletier

Eremaeus appalachicus Behan-Pelletier, 1993 (Figure 3.68A)
    Biology: From maple forest litter.
    Distribution: ON; [NH, NC; Nearctic]

Eremaeus boreomontanus Behan-Pelletier, 1993 (Figure 3.69G)
    Note: Considered sister species of *E. monticolus*.
    Biology: From dry moss; conifer litter.
    Distribution: BC, AB [Canada]
    Literature: ABMI (2019).

Eremaeus brevitarsus (Ewing, 1917) (Figure 3.69C)
    Comb./Syn.: *Damaeus brevitarsus* Ewing, 1917
    Biology: From dry alder, poplar, Burr oak, Jack pine, White pine, cedar, maple litter.
    Distribution: BC, MB, ON, QC [DC, IA, NH; Nearctic]

Eremaeus grandis Hammer, 1952 (Figure 3.69E)
    Biology: From mesic to very dry arctic tundra. Gravid females with 2 to 3 eggs.
    Distribution: AK, YT, NT [Nearctic]

Note: Collection records for this species indicate that it has a more northern distribution than that of *E. translamellatus*. *Eremaeus grandis* and *E. translamellatus* are sympatric in northern Alaska, Yukon, and Northwest Territories, whereas *E. grandis* is absent from collections from the latitude of Fairbanks south.

Eremaeus kananaskis Behan-Pelletier, 1993 (Figure 3.68B)
  Biology: From moss, lichens on alpine scree. Gravid females with 2 eggs.
  Distribution: AB [Canada]
  Literature: Mitchell (1977a, 1978).

Eremaeus kevani Behan-Pelletier, 1993 (Figure 3.69A)
  Biology: From poplar, maple, pine, hemlock litter.
  Distribution: ON, QC [ME, NH, VT; Nearctic]

Eremaeus occidentalis Behan-Pelletier, 1993 (Figure 3.68F,G)
  Biology: From moss, lichens, rotten wood on trunks of Douglas fir; Douglas fir, hemlock, cedar, Sitka spruce; maple, alder, Ponderosa pine litter.
  Distribution: BC [CA, OR, WA; Nearctic]

Eremaeus plumosus Woolley, 1964 (Figure 3.68D,E)
  Biology: From spruce, Douglas fir, hemlock, Lodgepole pine litter
  Distribution: BC, AB [CO, NM, UT, WA, WY; Nearctic]
  Literature: Behan-Pelletier (1993).

Eremaeus salish Behan-Pelletier, 1993 (Figure 3.68C)
  Biology: From *Silene*, *Antennaria*, *Saxifraga* on quartz.
  Distribution: BC [Canada]

Eremaeus translamellatus Hammer, 1952 (Figure 3.69D)
  Biology: Associated with various forest and open litter habitats, logs, moss, rodent nests, lichen heath. Inhabitant of rich mesic to dry tundra, rich subarctic and arctic subalpine, and Lodgepole pine, Engelmann spruce, Douglas fir, alder, aspen, and Paper birch litter.
  Distribution: AK, YT, NT, BC, AB, SK [Holarctic, Brazil]. Known from Tertiary fossil sites on Meigen Island (NU) (Matthews et al. 2019).
  Literature: Behan-Pelletier (1993); Matthews and Tekla (1997); ABMI (2019).

Eremaeus walteri Behan-Pelletier, 1993 (Figure 3.69B)
  Biology: From dry alpine tundra and forest litter.
  Distribution: AB [CO, NV UT; Nearctic]
  Literature: ABMI (2019).

Eueremaeus Mihelčič, 1963
  Type-species: *Eueremaeus oblongus* (C.L. Koch, 1835)
  Biology: Only 2 of the 25 recognized species in North America (*Eu. marshalli* and *Eu. tetrosus*) are transcontinental in distribution, both species extending south along the Cordillera as far as New Mexico. Three species are found in the eastern deciduous vegetation zone; *Eu. yukonensis* occurs in the western arctic, and *Eu. quadrilamellatus* and *Eu. foveolatus* occur in the western arctic, subarctic, and boreal zones. As with *Eremaeus*, the greatest diversity of *Eueremaeus* (13 species) occurs in dry habitats in the west and southwest of North America.

KEY TO ADULT *EUEREMAEUS* OF CANADA AND ALASKA

1 Anal setation 3 pairs. Adanal setation 3 pairs. Length: 467–510µm (Figure 3.70A,B) ....***Eu. trionus*** (Higgins)
– Anal setation 4 or 5 pairs. Adanal setation more than 3 pairs 2

2(1)	Anal setation 4 pairs. Adanal setation 4 pairs.	16
–	Anal setation 5 pairs. Adanal setation 4 or 5 pairs	3
3(2)	Prodorsum with single, small flattened tubercle posteromedially. Adanal setation 4 pairs. Epimeral setation 3-1-3-3. Posteromarginal sclerite of notogaster strongly developed. Trochanter III with 2 setae	19
–	Prodorsum smooth, or with 2 or more tubercles posteromedially. Adanal setation usually 5 pairs. Epimeral setation 3-2-3-3 or 3-3-3-3. Posteromarginal sclerite of notogaster weakly subtriangular, subrectangular, or crescent-shaped. Trochanter III with 3 setae	4
4(3)	Notogastral setation 10 pairs, seta c_2 absent. Ridge-like lamella present (absent in *Eu. aridulus* and *Eu. acostulatus*). Seta *s* on tarsus I eupathidial	5
–	Notogastral setation 11 pairs, seta c_2 present. Ridge-like lamella absent. Seta *s* on tarsus I setiform. Length: 441–505μm (Figure 3.70C,D) ... ***Eu. nahani*** Behan-Pelletier	
5(4)	Prodorsum with 2 or 3 rows of small tubercles posteromedially. Posteromarginal sclerite weakly developed, subtriangular	6
–	Prodorsum smooth posteromedially or with tubercles, but not in multiple rows. Posteromarginal sclerite well or weakly developed, subrectangular, crescentic, or subtriangular	9
6(5)	Interlamellar setae <65μm. Ridge-like lamella parallel, or slightly concave medially	7
–	Interlamellar setae 80–94μm. Ridge-like lamella converging posteriorly. Length: 512–517μm (Figure 3.70E) ... ***Eu. michaeli*** Behan-Pelletier	
7(6)	Interlamellar seta isodiametric, blunt distally, three-quarters length of bothridial seta. Length: 486–538μm (Figure 3.70F,G) ... ***Eu. masinasin*** Behan-Pelletier	
–	Interlamellar seta tapered, about half length of bothridial seta	8
8(7)	Translamella present. Length: about 480μm ... ***Eu. quadrilamellatus*** (Hammer)	
–	Translamella present or absent. Length: 530–720μm (Figure 3.70H,I) ... ***Eu. marshalli*** Behan-Pelletier	
9(5)	Notogaster, genital and anal plates smooth to foveate	10
–	Notogaster, genital and anal plates punctate. Length: 440–518μm (Figure 3.71A,B) ... ***Eu. woolleyi*** (Higgins)	
	Note: Known from western USA (CO, UT); possibly in Canada.	
10(9)	Ridge-like lamella long, length more than 2× minimum mutual distance	11
–	Ridge-like lamella short, subequal in length or shorter than their minimum mutual distance	13
11(10)	Area between and laterad of ridge-like lamella foveate; ridges converging anteriorly. Interlamellar seta about half length of bothridial seta. Femora I and II foveate antiaxially. Length: 473–558μm (Figure 3.71C–E) ... ***Eu. foveolatus*** (Hammer)	
–	Area between and lateral of ridge-like lamella smooth; ridges crenulate, not converging anteriorly. Interlamellar seta about quarter length of bothridial seta. Femora I and II without foveae	12
12(11)	Tibiae I and II with proximally directed, broad, concave cusp ventrodistally. Length: 460–518μm (Figure 3.71F,G) ... ***Eu. proximus*** (Berlese)	
–	Tibiae I and II without cusp ventrodistally. Length: 473–557μm (Figure 3.71H,I) ... ***Eu. nemoralis*** Behan-Pelletier	
	Note: Known from Eastern USA (NJ, VA, WV); possibly in Canada.	
13(10)	Ridge-like lamella well developed. Posteromarginal sclerite crescentic or subrectangular	14

–	Ridge-like lamella absent or weakly developed. Posteromarginal sclerite weakly subtriangular... 15
14(13)	Posteromarginal sclerite distinctly subrectangular, almost twice as wide as long. Postanal process broadly flattened posteriorly. Length: 428–499μm (Figure 3.71J,K)....................... ....*Eu. columbianus* (Berlese)
–	Posteromarginal sclerite small, oval. Postanal process distinctly V-shaped posteriorly. Length: 577–661μm (Figure 3.72A) .... *Eu. yukonensis* Behan-Pelletier
15(13)	Dorsal and ventral porose areas on femora I and II connected by porose bands. Posteromarginal sclerite subtriangular. Length: 447–505μm. East of Rocky Mountains (Figure 3.72B,C) ....*Eu. aridulus* Behan-Pelletier
–	Dorsal and ventral porose areas on femora I and II not connected by porose bands. Posteromarginal sclerite crescentic. Length: 473–540μm. West of Rocky Mountains (Figure 3.72D,E) ....*Eu. acostulatus* Behan-Pelletier
16(2)	Posteromarginal sclerite present. Tibiae II–IV with 4-3-3 setae, respectively. Notogastral setae distinctly longer than interlamellar seta 17
–	Posteromarginal sclerite absent. Tibiae II–IV with 3-2-2 setae, respectively. Notogastral setae subequal to or slightly longer than interlamellar setae. Length: 499–570μm (Figure 3.72F) .... *Eu. tetrosus* (Higgins)
17(16)	Notogaster and prodorsum foveate. Femora I–IV with antiaxial and paraxial carinae; without dorsal porose area.. 18
–	Notogaster and prodorsum areolate. Femora I–IV without carina, with dorsal and ventral porose areas. Length: 642–687μm (Figure 3.72G,H).............*Eu. higginsi* Behan-Pelletier
18(17)	Postanal process broadly flattened, U-shaped. Mutual distance of notogastral setae *lm* less than that of setae *la*. Length: 460–525μm............................*Eu. carinatus* Behan-Pelletier
–	Postanal process strongly V-shaped. Mutual distance of notogastral setae *lm* greater than that of setae *la*. Length: 680–758 (Figure 3.73A,B) ....*Eu. stiktos* (Higgins)
19(3)	Notogastral setation 10 pairs 20
–	Notogastral setation 11 pairs. Length: 467–525μm (Figure 3.73C) ....*Eu. lindquisti* Behan-Pelletier
	Note: Known from Western USA (AZ, CO, OR); possibly in Canada.
20(19)	Notogaster with humeral projections laterad of seta *c*. Length: 596–706μm (Figure 3.73D)... *Eu. chiatous* (Higgins)
	Notogaster without humeral projections laterad of seta *c* 21
21(20)	Ridge-like lamella well developed; foveae present between ridges. Notogaster with transverse row of depressions anteriad of setae *c*... 22
–	Ridge-like lamella faint to absent; foveae absent between ridges. Notogaster without transverse row of depressions anteriad of setae *c*. Length: 467–517μm (Figure 3.73E) ....*Eu. magniporus* (Wallwork)
	Note: Known from Western USA (CA); possibly in Canada.
22(21)	Length of ridge-like lamella subequal to minimum mutual distance. Bothridial seta shorter than mutual distance of bothridia..23
–	Length of ridge-like lamella twice minimum mutual distance. Bothridial seta 130–150μm long, narrowly clavate, subequal to mutual distance of bothridia. Length: about 713μm (Figure 3.73I) ....*Eu. osoyoosensis* Behan-Pelletier
23(22)	Interlamellar seta plumose. Femora I and II without carina, without crenulate margin. Length: 544–592μm (Figure 3.73F) ....*Eu. alvordensis* Behan-Pelletier

Note: Known from Western USA (CA, CO, NV, OR, UT); possibly in Canada.
- Interlamellar setae fusiform to tapered. Femora I and II with narrow carina, with crenulate margin. Length: 512–622µm (Figure 3.73G,H) ....*Eu. aysineep* Behan-Pelletier

Eueremaeus acostulatus Behan-Pelletier, 1993 (Figure 3.72D,E)
Biology: Found primarily on lichens in *Abies amabilis* and *Tsuga heterophylla* forests of western BC.
Distribution: BC [WA; Nearctic]

Eueremaeus aridulus Behan-Pelletier, 1993 (Figure 3.72B,C)
Biology: From shortgrass prairie with *Opuntia*.
Distribution: AB, SK [Canada]
Literature: ABMI (2019).

Eueremaeus aysineep Behan-Pelletier, 1993 (Figure 3.73G,H)
Biology: In very dry birch and juniper litter and moss. Also in arboreal habitats in Sitka spruce.
Distribution: BC, AB [Canada]

Eueremaeus chiatous (Higgins, 1979) (Figure 3.73D)
Comb./Syn.: *Kartoeremaeus chiatous* Higgins, 1979
Biology: An inhabitant of moss and lichens on bark and litter of maples, hemlock, Western juniper, aspen, Whitebark pine, Jeffrey pine, Douglas fir, and spruce. On Vancouver Island this species has been reported from the forest floor under Amabilis fir, Western hemlock, Sitka spruce and Western redcedar and from branch and lichen from Amabilis fir and Western hemlock (Lindo and Winchester 2006; Winchester et al. 2008). Pollen grains of conifers were noted in the gut of specimens (Behan-Pelletier 1993).
Distribution: BC, AB [CA, UT; Nearctic]
Literature: Behan-Pelletier (1993); ABMI (2019).

Eueremaeus columbianus (Berlese, 1916) (Figure 3.71J,K)
Comb./Syn.: *Eremaeus columbianus* Berlese, 1916; *Kartoeremaeus reevesi* Higgins, 1979; *Eremaeus politus* Banks, 1947
Biology: From mixed forest litter: cedar, pine, Sugar maple; fungi, moss, and lichens on tree trunks. Recorded from epiphytic lichens in the Adirondack Mountains (Root et al. 2007).
Distribution: ON, QC [DE, KY, ME, MO, NC, NH, NY, VA, WV; Nearctic]
Literature: Norton and Kethley (1990).

Eueremaeus foveolatus (Hammer, 1952) (Figure 3.71C–E)
Comb./Syn.: *Eremaeus foveolatus* Hammer, 1952
Biology: From rich mesic tundra, subarctic forest litter, and litter under herbs and aspen.
Distribution: AK, YT, NT, NU, BC, AB, SK, QC [Nearctic]
Literature: Behan-Pelletier (1993); ABMI (2019).

Eueremaeus higginsi Behan-Pelletier, 1993 (Figure 3.72G,H)
Biology: From juniper litter.
Distribution: BC [CA; Nearctic]

Eueremaeus marshalli Behan-Pelletier, 1993 (Figure 3.70H,I)
Biology: Gravid females carry 2 eggs. Gut contents are of pigmented and unpigmented fungal hyphae and some spores. Found in mesic tundra, moss, lichen, fungi, and mixed coniferous and deciduous litters. From Western redcedar canopy (Lindo and Winchester 2006, 2007); in arboreal habitats in Sitka spruce forests of BC (Winchester et al. 1999), and in the

Balsam fir forests of Newfoundland (Dwyer et al. 1997). Recorded from epiphytic lichens in the Adirondack Mountains (Root et al. 2007).
Distribution: AK, YT, NT, BC, AB, SK, MB, ON, QC, NL [ME, NM, NV, NY, PA, VT, WA; Nearctic]
Literature: Walter et al. (2014); ABMI (2019).

Eueremaeus masinasin Behan-Pelletier, 1993 (Figure 3.70F,G)
Biology: A grassland species reported from sod, moss, assorted litter, dry shortgrass prairie, fescue prairie and alpine grasslands. Gravid females can carry 4 eggs. Gut contents usually include fungal hyphae and sometimes spores. This species was included in an analysis of weight-estimation models for soil mites (Newton and Proctor 2013).
Distribution: AB, SK, MB [CO; Nearctic]
Literature: Walter et al. (2014); ABMI (2019).

Eueremaeus michaeli Behan-Pelletier, 1993 (Figure 3.70E)
Biology: From dry larch, Ponderosa pine, hemlock, cedar litter
Distribution: BC [Canada]

Eueremaeus nahani Behan-Pelletier, 1993 (Figure 3.70C,D)
Biology: From very dry Lodgepole pine litter; tundra vegetation.
Distribution: YT [CO; Nearctic]

Eueremaeus oblongus (C.L. Koch, 1836)
Comb./Syn.: *Eremaeus oblongus* C.L. Koch, 1836; *Notaspis oblonga* Michael, 1888
Note: Species is not keyed, because records for North America are questionable (Behan-Pelletier 1993).
Biology: Important inhabitant of epiphytic lichens. While this is a xeric species, minimum survival of this species under flooding was five days, maximum survival 129 days, median lethal time was reached after 52 days (Bardel and Pfingstl 2018).
Distribution: MB [IL, NC; Holarctic]
Literature: Maraun et al. (2019).

Eueremaeus osoyoosensis Behan-Pelletier, 1993 (Figure 3.73I)
Biology: From dry wood litter of Ponderosa pine, Whitebark pine, and spruce.
Distribution: BC, AB [Canada]
Literature: ABMI (2019).

Eueremaeus proximus (Berlese, 1916) (Figure 3.71F,G)
Comb./Syn.: *Eremaeus proximus* Berlese, 1916
Biology: From mixed-woodland litter.
Distribution: ON, QC, NB, NS [GA, KY, ME, MO, NH; Nearctic]
Literature: Norton and Kethley (1990).

Eueremaeus quadrilamellatus (Hammer, 1952)
Comb./Syn.: *Eremaeus quadrilamellatus* Hammer, 1952
Biology: Gravid females carry 2 large eggs. Gut contents are composed primarily of pigmented hyphae.
Distribution: AK, YT, NT, AB [Nearctic]
Literature: Walter et al. (2014).

Eueremaeus stiktos (Higgins, 1962) (Figure 3.73A,B)
Comb./Syn.: *Eremaeus stiktos* Higgins, 1962
Biology: From forest litter: oak, maple, Ponderosa pine. Douglas fir; White pine.
Distribution: BC [CA, ID, OR, UT; Nearctic]
Literature: Behan-Pelletier (1993).

Eueremaeus tetrosus (Higgins, 1979) (Figure 3.72F)
   Comb./Syn.: *Kartoeremaeus tetrosus* Higgins, 1979
   Biology: From meadows, moss, rotten logs, and the litter of aspen, Lodgepole pine, willow, spruce.
   Distribution: AB, SK, MB, ON, QC, NB, NS, NL [AZ, CO, NH, NM, UT; Nearctic]
   Literature: Behan-Pelletier (1993); Walter et al. (2014).

Eueremaeus trionus (Higgins, 1979) (Figure 3.70A,B)
   Comb./Syn.: *Kartoeremaeus trionus* Higgins, 1979
   Biology: From under bark of pine.
   Distribution: AB, NS [CO; Western Siberia]. This species has an unusual distribution in North America. It is only known from Alberta, Young's Gulch in Colorado (type material) and from under the bark of White pine (*Pinus strobus* L.) in Kouchibouguac National Park, NS, a distribution that may be a relict of a much broader range during the Tertiary period.
   Literature: Behan-Pelletier (1993); ABMI (2019).

Eueremaeus yukonensis Behan-Pelletier, 1993 (Figure 3.72A)
   Biology: Tundra species.
   Distribution: YT [Canada]

Tricheremaeus Berlese, 1908
   Type-species: *Notaspis serrata* Michael, 1885
   Note: Unidentified dead specimen collected from canopy habitat in BC (Lindo and Clayton 2011).

Family Megeremaeidae Woolley and Higgins, 1968
   Diagnosis: *Adult*: Non-poronotic Brachypylina. Granular cerotegument covering all body and leg surfaces. Rostrum broadly rounded, with narrow carina following margin of rostrum laterally, absent medially. Lamellae ridge-like. Tutorium ridge-like. Transverse ridge connecting lamella and tutorium distally usually present. Tutorium generally with, occasionally without, small apophysis posteriorly, which opposes anteriorly directed apophysis on prodorsum. Pedotecta I, II and toothlike discidium present. Dorsoventral aspect of pedotectum II distally broad, forked or truncate. Tubercle present posterolaterad of exobothridial seta opposing tubercle dorsad of trochanter III. Notogastral setation 10 pairs; anterior margin with 2 pairs of tubercles. Epimeral setation 3-1-3-3. Six pairs of genital, 1 pair aggenital, 3 pairs adanal, and 2 or 3 pairs of anal setae. Anal plate with lyrifissure *ian*. Infracapitulum diarthric. Axillary saccule of subcapitulum absent. Palp setation: 0-2-1-3-9(1). Palptarsal solenidion arising at level of setae *lt'*, lying parallel to surface of palp, extending anteriorly to base of eupathidium *acm*. Legs heterotridactylous. Companion seta *d* closely associated with solenidion on genua I–III and tibiae I–IV. Porose areas present on femora I–IV and trochanters III, IV; absent from tibiae and tarsi. Sharp spines present on trochanters III and IV.

Megeremaeus Higgins and Woolley, 1965
   Type-species: *Megeremaeus montanus* Higgins and Woolley, 1965

KEY TO ADULTS OF EXTANT *MEGEREMAEUS* OF CANADA AND ALASKA

1   Anal setation 2 pairs..2
–   Anal setation 3 pairs. Length: ca. 1,002µm (Figure 3.74A)..
   ....*M. montanus* Higgins and Woolley

2(1)   Discidium blunt distally. Trochanter III with very short dorsodistal spine. Notogastral setae short, at least c_2, *la*, *lm*, *lp* and/or h_3 not reaching insertion of next more posterior setae. Bothridial seta capitate. Length: 972–1,076µm (Figure 3.74B) ....
   ....*M. kootenai* Behan-Pelletier

| – | Discidium pointed distally. Trochanter III with well-developed dorsodistal spine. At least one of notogastral setae c_2, *la*, *lm*, *lp* reaching insertion of next more posterior seta. Bothridial seta gradually expanded distally .... 3 |

3(2) Bothridial seta 71–91μm long, about equal in length to interlamellar seta. Notogastral seta c_2 extending well posteriad of insertion of *la*. Length: 901–972μm (Figure 3.74C)............. ....***M. keewatin*** Behan-Pelletier
– Bothridial seta 140–190μm long, 1.4–2.5× length of interlamellar seta. Notogastral seta c_2 not extending posteriad of insertion of *la*... 4

4. Translamellar ridge extending between lamellae at basal quarter of their length; bothridial seta about 1.5× length seta of isodiametric interlamellar seta. Length: 973–1,063μm (Figure 3.74D,E) .... ***M. hylaius*** Behan-Pelletier
– Translamellar ridge absent; bothridial seta about 2.5× length of clavate interlamellar seta. Length: 858–930μm .... ***M. ditrichosus*** Woolley and Higgins

Megeremaeus ditrichosus Woolley and Higgins, 1968
Distribution: [OR; Nearctic]; this species and unidentified species known from AK: Lava Camp, Seward Peninsula (Late Miocene: 5.7 mya) (Matthews et al. 2019).
Literature: Behan-Pelletier (1990).

Megeremaeus hylaius Behan-Pelletier, 1990 (Figure 3.74D,E)
Biology: From bark of elm, *Abies* and *Betula* covered with lichen and moss.
Distribution: ON, QC [NY; Nearctic]
Literature: Root et al. (2007); Sidorchuk and Behan-Pelletier (2017); Ryabinin and Wu (2018).

Megeremaeus keewatin Behan-Pelletier, 1990 (Figure 3.74C)
Biology: From dry tundra litter.
Distribution: AK, YT [Nearctic]; known from Tertiary fossil sites on Meigen Island (NU) (Matthews et al. 2019).
Literature: Sidorchuk and Behan-Pelletier (2017); Ryabinin and Wu (2018).

Megeremaeus kootenai Behan-Pelletier, 1990 (Figure 3.74B)
Biology: From sifted moss and bark of standing dead Douglas-fir and spruce.
Distribution: BC, AB [Canada]
Literature: Sidorchuk and Behan-Pelletier (2017).

Megeremaeus montanus Woolley and Higgins, 1968 (Figure 3.74A)
Biology: From western temperate rainforest litter.
Distribution: BC [OR, WA, WY; Nearctic]; subfossil specimens, approximately 10 000 years old, from Haida Gwaii Islands, Cape Ball, BC.
Literature: Behan-Pelletier (1990); Sidorchuk and Behan-Pelletier (2017).

Superfamily Gustavioidea Oudemans, 1900

Family Astegistidae Balogh, 1961
Diagnosis: *Adults*: Non-poronotic Brachypylina. Prodorsum with lamellae, tutorium, with or without translamella. Dorsophragmata, pleurophragmata present or absent. Pedotectum I without deep incision. Notogaster with or without humeral process, if process present, never knife-like; pteromorph absent; setation 9–10 pairs. Custodium absent, circumpedal carina and discidium present. Genital and anal plates large and closely adjacent or well-separated. Genital setation 5–6 pairs. Aggenital taenidium and minitectum present. Subcapitulum diarthric, chelicerae chelate-dentate. Palpal eupathidium *acm* and solenidion separate. Axillary saccule of subcapitulum absent. Seta *d* absent on tibiae and genua when respective solenidion is present.

KEY TO GENERA OF ASTEGISTIDAE OF CANADA AND ALASKA

1 Notogastral setae longer than lamellar setae. Dorsosejugal scissure incomplete medially (Figure 3.75A,B) .... *Astegistes* Hull
– Notogastral setae shorter than lamellar setae. Dorsosejugal scissure complete medially .. 2

2(1) Rostrum rounded to pointed. Interlamellar setae long, extending anterior of translamella. Larger species (>600μm) (Figure 3.75G,H) .... *Furcoribula* Balogh
– Rostrum with medial and lateral teeth or dentate. Interlamellar setae short or long, extending anterior of translamella. Smaller species (<400μm) (Figure 3.75C–F)..........................
.... *Cultroribula* Berlese

Astegistes Hull, 1916
Type-species: *Zetes pilosus* C.L. Koch, 1841 (=*Liacarus bicornis* Warburton and Pearce, 1905) (Figure 3.75A,B)
Note: Unidentified *Astegistes* spp. from AB, ON, NB.

Cultroribula Berlese, 1908
Type-species: *Notaspis juncta* Michael, 1885

KEY TO ADULT *CULTRORIBULA* OF CANADA AND ALASKA

1 Lamellae diverging distally. Rostrum with 2 cusps distally. Length: 230–240μm...............
.... *C. divergens* Jacot
– Lamellae directed anteriorly, not diverging distally. Rostrum with 3 cusps or multiple teeth 2

2(1) Rostrum with 3 attenuate cusps distally. Length: 214–224μm (Figure 3.75C,D) ....
.... *C. bicultrata* (Berlese)
– Rostrum with long cusp distally and ca. 6 pairs of shorter cusps (dens) laterally on rostrum. Length: ca. 320μm (Figure 3.75E,F) .... *C. dentata* Willmann
Note: Unidentified *Cultroribula* sp. from BC, NB, NS.

Cultroribula bicultrata (Berlese, 1905) (Figure 3.75C,D)
Comb./Syn.: *Dameosoma bicultratum* Berlese, 1905; *Cultroribula trifurcata* Jacot, 1939; *Cultroribula falcata* Evans, 1952; *Cultroribula trifurcata rotundata* Krivolutsky, 1962
Biology: From temperate deciduous and coniferous forest litter.
Distribution: BC, ON, QC, NS, NL [CA, NC, NY, OR; Holarctic]
Literature: Weigmann (2008); Maraun et al. (2019).

Cultroribula dentata Willmann, 1950 (Figure 3.75E,F)
Distribution: AK, YT, NT [Holarctic]
Literature: Bayartogtokh (2012).

Cultroribula divergens Jacot, 1939
Biology: From grass fields and shrubs; from *Sphagnum*.
Distribution: ON, QC [KY, NC, NH, NY; Nearctic]
Literature: Maraun et al. (2019).

Furcoribula Balogh, 1943
Type-species: *Notaspis furcillata* Nordenskiöld, 1901

Furcoribula furcillata (Nordenskiöld, 1901) (Figure 3.75G,H)
Comb./Syn.: *Notaspis furcillata* Nordenskiöld, 1901; *Cultroribula furcillata* (Nordenskiöld, 1901); *Cultroribula confinis magna* Ewing, 1917

Distribution: NS [IL; Holarctic]
Literature: Ermilov and Kolesnikov (2012); Maraun et al. (2019).

Family Gustaviidae Oudemans, 1900
Diagnosis: *Adults*: Non-poronotic Brachypylina. Prodorsum with lamellae, prolamella, tutorium; without translamella. Dorsophragmata and pleurophragmata present. Pedotectum I without deep incision, extending dorsally almost to exobothridial seta. Circumpedal carina, discidium present; custodium absent. Notogaster without knife-like humeral process; pteromorph absent; setation 10 pairs; many with just alveoli expressed. Genital setation 6 pairs. Aggenital taenidium and minitectum present. Chelicera styliform; fixed digit of chelicera reduced; moveable digit serrate distally. Palpal eupathidium *acm* and solenidion attached in imperfect double horn. Axillary saccule of subcapitulum absent. Seta *d* absent on tibiae and genua when respective solenidion present.

Gustavia Kramer, 1879
Type-species: *Leiosoma microcephala* Nicolet, 1855 (=*Gustavia sol* Kramer, 1879)
Comb./Syn.: *Serrarius* Michael, 1883; *Neozetes* Berlese, 1885

Gustavia parvula (Banks, 1909)
Comb./Syn.: *Liacarus parvulus* Banks, 1909
Distribution: ON [MI; Nearctic]
Literature: Smith et al. (1998).
Note: Unidentified *Gustavia* sp. from AK: Lava Camp, Seward Peninsula (Late Miocene: 5.7 mya) (Matthews et al. 2019). Unidentified *Gustavia* sp. from BC, AB, MB.

Family Kodiakellidae Hammer, 1967
Diagnosis: *Adults*: Non-poronotic Brachypylina. Prodorsum with lamellae. Tutorium, prolamella and carina absent. Dorsophragmata and pleurophragmata absent. Pedotectum I large without deep incision or medial tooth. Circumpedal carina, custodium and discidium absent. Notogaster without knife-like humeral process; pteromorph absent; setation 10 pairs, setae developed or only alveoli present. Genital setation 6 pairs. Anal setation 3 pairs. Aggenital taenidium and minitectum absent. Subcapitulum diarthric, chelicerae chelate-dentate. Palpal eupathidium *acm* and solenidion attached or separate. Axillary saccule of subcapitulum absent. Seta *d* absent on tibiae and genua when respective solenidion present.

Kodiakella Hammer, 1967
Type-species: *Kodiakella lutea* Hammer, 1967

Kodiakella lutea Hammer, 1967 (Figure 3.76E)
Biology: From forest soil.
Distribution: AK, BC [OR; Nearctic]

Family Liacaridae Sellnick, 1928
Diagnosis: *Adults*: Non-poronotic Brachypylina. Prodorsum with lamellae, tutorium; with or without translamella. Dorsophragmata and pleurophragmata present. Pedotectum I without deep incision, extending to exobothridial seta. Custodium, discidium and circumpedal carina absent. Notogaster without knife-like humeral process; pteromorph absent; setation 8, 10, or 11 pairs. Sejugal apodeme aligned transversely. Genital setation 5–6 pairs. Aggenital taenidium and minitectum present or absent. Subcapitulum diarthric, chelicerae chelate-dentate. Palpal eupathidium *acm* and solenidion separate; solenidion pressed to surface of palptarsus. Axillary saccule of subcapitulum absent. Seta *d* absent from tibiae and genua when respective solenidion is present.
Biology: Liacaridae comprise a large number of species with smooth and shiny body surfaces that display extraordinary anti-wetting properties (Brückner et al. 2015). Liacarid

unwettability is not related to micro-structured surfaces as present in many oribatid mites ("Lotus effect") but to the formation of raincoat-like lipid layers covering the epicuticle (Raspotnig and Leis 2009; Raspotnig and Matischek 2010). Cuticular lipids are of two types comprising two chemically distinguishable systems, with that of *Adoristes* distinct from that of *Liacarus* and *Dorycranosus*.

Species in this family are primarily saprophagous, but feed to a lesser extent on fungi and can be scavengers or carnivorous. There is strong evidence for species to be endophagous (reviewed in Labandeira et al. 1997). Trávníček (1989) described the rearing and biology of species of *Liacarus*, *Dorycranosus* and *Adoristes*, including details of preferred food of both juveniles and adults. Although no species that occur in Canada and Alaska were studied, his results are probably widely applicable to members of these genera.

KEY TO GENERA OF LIACARIDAE OF CANADA AND ALASKA

1 Integument strongly sculptured (Figure 3.77D) .... ***Xenillus*** Robineau-Desvoidy
– Integument mostly smooth 2

2(1) Lamellae well separated, not connected with translamella, not fused apically. Genital setation 5 pairs. Anterior notogastral margin slightly concave medially (Figure 3.77A) ....
.... ***Adoristes*** Hull
– Lamellae either connected by translamella or touching or fused. Genital setation 5 or 6 pairs. Anterior notogastral margin straight medially or slightly convex 3

3(2) Bothridial seta setiform or bacilliform (Figure 3.77B) .... ***Rhaphidosus*** Woolley
– Bothridial seta clavate or spindle-shaped 4

4(3) Bothridial seta spindle-shaped (Figure 3.77C,G) .... ***Liacarus*** Michael
– Bothridial seta clavate (Figures 3.77E,F; 3.78) .... ***Dorycranosus*** Woolley
  Note: *Rhaphidosus* and *Dorycranosus* were considered subgenera of *Liacarus* by Subías (2004).

Adoristes Hull, 1916
  Type-species: *Oribates ovatus* C.L. Koch, 1839

Adoristes ammonoosuci Jacot, 1938 (Figure 3.77A)
  Comb./Syn.: *Adoristes ovatus ammonoosuci* Jacot, 1938
  Biology: From spruce litter.
  Distribution: QC [NH, NY, SC; Nearctic]
  Literature: Gourbière et al. (1985); Lions and Gourbière (1988, 1989); Seniczak et al. (2017).
  Note: Unidentified *Adoristes* spp. from ON, QC, NS, NL.

Dorycranosus Woolley, 1969
  Type-species: *Liacarus abdominalis* Banks, 1906
  Note: We consider *Procorynetes* Woolley, 1969 a junior synonym of *Dorycranosus* as the genera differ only in the shape of bothridial seta. Unidentified species known from AK, NT, BC, QC.

KEY TO ADULT *DORYCRANOSUS* OF CANADA AND ALASKA

1 Translamella very short, without medial dens. Length: ca. 1,000µm (Figure 3.78C)...........
.... ***D. abdominalis*** (Banks)
– Translamella short or long, with medial dens 2

2(1)	Medial dens of translamella half length of lamellar cusps. Length: ca. 500µm (Figure 3.78A,B)	***D. altaicus*** Krivolutsky
–	Medial dens of translamella subequal or shorter in length than lamellar cusps	3
3(2)	Medial dens of translamella equal to width of translamella basally; equal in length to lamellar cusp. Length: 850–1,060µm (Figure 3.78D–F)	***D. acutidens*** (Aoki)
–	Medial dens of translamella much shorter than width of translamella basally; shorter than length of lamellar cusp. Length: 700–800µm (Figure 3.78G,H)	***D. parallelus*** (Hammer)

Note: Unidentified *Dorycranosus* spp. from YT, NT, BC, AB, NB, NL.

Doycranosus abdominalis (Banks, 1906) (Figure 3.78C)
  Comb./Syn.: *Liacarus abdominalis* Banks, 1906
  Distribution: AK, NS [CA; Nearctic]

Dorycranosus acutidens (Aoki, 1965) (Figure 3.78D–F)
  Comb./Syn.: *Liacarus acutidens* Aoki, 1965
  Biology: Feeds on higher plant material and fungi (Kaneko 1988).
  Distribution: AK, AB, NS [Holarctic]
  Literature: ABMI (2019).

Dorycranosus altaicus Krivolutsky, 1974 (Figure 3.78A,B)
  Distribution: AK, NT, NU, QC [Holarctic]

Dorycranosus parallelus (Hammer, 1967) (Figure 3.78G,H)
  Comb./Syn.: *Liacarus parallelus* Hammer, 1967
  Biology: From forest litter.
  Distribution: AK, YT, AB [Holarctic]
  Literature: ABMI (2019).

Liacarus Michael, 1898 (=*Leiosoma* Nicolet, 1855, preoccupied name)
  Type-species: *Oribata nitens* Gervais, 1844

KEY TO ADULT *LIACARUS* OF CANADA AND ALASKA

1	Lamellar cusps shorter than width of lamella	2
–	Lamellar cusps longer than width of lamella	3
2(1)	Lamellar cusp with small medial dens. Distal flagellum of bothridial seta subequal in length to elliptical middle. Length: ca. 552µm (Figure 3.79A,B)	***L. cidarus*** Woolley
–	Lamellar cusp without medial dens. Distal flagellum of bothridial seta almost 2× length elliptical middle. Length: ca. 534µm (Figure 3.79C)	***L. detosus*** Woolley
3(1)	Interlamellar setae longer than bothridial setae. Translamella with medial dens. Length: ca. 1,000µm (Figure 3.79D)	***L. robustus*** Ewing
–	Interlamellar setae shorter than bothridial setae. Translamella without medial dens	4
4(3)	Rostrum smooth without medial extension. Translamella absent. Lamellar cusps meeting proximally. Length: ca. 1,070µm (Figure 3.79E)	***L. latus*** Ewing
–	Rostrum with medial, flattened extension. Short translamella present. Length: ca. 852–910µm (Figure 3.79F–H)	***L. bidentatus*** Ewing

Note: Unidentified species from AK: Lava Camp, Seward Peninsula (Late Miocene: 5.7 mya) (Matthews et al. 2019). Unidentified *Liacarus* spp. from BC, QC, NS.

Liacarus bidentatus Ewing, 1918 (Figure 3.79F–H)
   Biology: From coniferous forest litter.
   Distribution: AK, YT, NT, BC [CA, NC, OR, WA, WY; Nearctic]
   Literature: Walter (1985).

Liacarus cidarus Woolley, 1968 (Figure 3.79A,B)
   Biology: Arlian and Woolley (1969, 1970) cultured this species, described its development, feeding habits, biology and described the juveniles.
   Distribution: BC [CA, CO, MT, UT, WY; Nearctic]

Liacarus detosus Woolley, 1968 (Figure 3.79C)
   Biology: Endophagous species; immatures in oak petioles and woody debris (Hansen 1999).
   Distribution: BC [NC; Nearctic]

Liacarus latus Ewing, 1909 (Figure 3.79E)
   Biology: From coniferous forest soil. In oak litter (Lamoncha and Crossley 1998).
   Distribution: BC [IL, NC, OR; Nearctic]

Liacarus robustus Ewing, 1918 (Figure 3.79D)
   Biology: From forest soil.
   Distribution: BC? [OR, WA; Nearctic]
   Literature: Walter (1985).

Rhaphidosus Woolley, 1969
   Type-species: *Liacarus carolinensis* Banks, 1906
   Note: Unidentified *Rhaphidosus* spp. from BC.

Xenillus Robineau-Desvoidy, 1839
   Type-species: *Xenillus clypeator* Robineau-Desvoidy, 1839
   Comb./Syn.: *Banksia* Oudemans and Voigts, 1905 (= *Kochia* Oudemans, 1900, preoccupied name); *Pseudocepheus* Jacot, 1928
   Note: Unidentified species from AK: Lava Camp, Seward Peninsula (Late Miocene: 5.7 mya) (Matthews et al. 2019). Unidentified *Xenillus* sp. from MB.

Family Peloppiidae Balogh, 1943 (= Metrioppiidae Balogh, 1943; = Ceratoppiidae Grandjean, 1954)
   Diagnosis: *Adults*: Non-poronotic Brachypylina. Prodorsum with lamella, tutorium; with or without translamella; with or without prolamella. Dorsophragmata present or absent; pleurophragmata generally present. Parietal carina present or absent. Circumpedal carina present or absent. Custodium absent. Pedotectum I large; with or without tooth medially. Notogaster without knife-like humeral process; pteromorph absent; setation 9–11 pairs, or their alveoli; 2 or 3 pairs of posterior notogastral setae often prominent. Crista present or absent. Epimeral border IV with transverse or semioval furrow, incomplete adjacent to genital plate. Minitectum present, may be incomplete. Epimeral setation: 3-1-3-3/3-1-3-5. Genital setation 5–6 pairs. Subcapitulum diarthric with chelate-dentate chelicera, or anarthric with pelopsiform chelicera. Palpal eupathidium *acm* and solenidion attached at least distally, or not. Axillary saccule of subcapitulum absent. Seta *d* occasionally present on tibiae and genua when respective solenidion is present. Porose areas on tibiae and tarsi present, or absent.

KEY TO GENERA OF PELOPPIIDAE OF CANADA AND ALASKA

1   Subcapitulum anarthric. Chelicera pelopsiform 2
–   Subcapitulum diarthric. Chelicera chelate-dentate 3

2(1)	Lamellae either connected by translamella or nearly touching; cusps parallel (Figure 3.81A)...*Paenoppia* Woolley and Higgins	
–	Lamellae and cusps converging, but well separated (Figure 3.80B–F) .... *Metrioppia* Grandjean	
3(2)	Legs monodactylous. Lamellar cusps short (Figure 3.81B) ....*Parapyroppia* Pérez-Íñigo and Subías	
–	Legs tridactylous. Lamellar cusps long, or short, or absent 4	
4(3)	Bothridial seta capitate. One (h_1) or 2 pairs (h_1, p_1) of knife-like posterior notogastral setae. Lamellae linear, almost parallel; lamellar cusps absent (Figure 3.80A).................... ....*Dendrozetes* Aoki	
–	Bothridial seta setiform or lanceolate. Posterior notogastral setae setiform. Lamellae ribbon-like, converging; lamellar cusps present 5	
5(4)	Bothridial seta short with slightly lanceolate head. Notogaster obtusely pointed posteriorly (Figure 3.81C–F) .... *Pyroppia* Hammer	
–	Bothridial seta long, setiform. Notogaster rounded posteriorly (Figure 3.82, 3.83) .... *Ceratoppia* Berlese	

Ceratoppia Berlese, 1908
Type-species: *Notaspis bipilis* Hermann, 1804

KEY TO ADULT *CERATOPPIA* OF CANADA AND ALASKA

1	Mentum of subcapitulum with 2 pairs of hypostomal setae (*h*) 2	
–	Mentum of subcapitulum with 1 pair of hypostomal setae 4	
2(1)	Posterior notogastral setae (p_1, p_3) long... 3	
–	Posterior notogastral setae (p_1, p_3) short. Length: 540–670µm (Figure 3.82A).................. .... *C. valerieae* Lindo	
3(2)	Length: 600–1,100µm (Figure 3.82B)... *C. bipilis* (Hermann)	
–	Length: 480–800µm...*C. bipilis spinipes* (Banks)	
4(1)	Two or 3 pairs of conspicuous notogastral setae 5	
–	Notogastral setae reduced, much shorter than length of rostral setae................................. 9	
5(4)	Three pairs of notogastral setae long, conspicuous (h_1, p_2, p_3). Length: 650–750µm (Figure 3.82C) .... *C. sexpilosa* Willmann	
–	Two pairs of conspicuous notogastral setae 6	
6(5)	Rostrum deeply indented with inset medial tooth. Interlamellar setae distinctly shorter than length of lamellae. Length: 560–650µm (Figure 3.82D,E)............*C. indentata* Lindo	
–	Rostrum not deeply indented, rounded, dentate or with large medial tooth. Interlamellar setae as long as, or longer than length of lamellae 7	
7(6)	Rostrum with strong medial tooth and lateral denticles. Length: 500–600µm (Figure 3.83A) .... *C. quadridentata* (Haller)	
–	Rostrum rounded anteriorly, dentate, but without medial tooth.. 8	
8(7)	Large (700–1,000µm); dark (almost black) in colour. Rostral bump absent. Length: 700–1,000µm (Figure 3.83B) .... *C. sphaerica* (L. Koch)	
–	Smaller species (550–650µm), reddish in colour, with prominent rostral bump in lateral view. Length: 550–650µm (Figure 3.83C,D) ....*C. offarostrata* Lindo	

9(4) Three pairs of reduced notogastral setae (p_1, p_2, p_3), in particular setae p_2 minute, not discernable in all specimens. Lamellae long, reaching insertion of rostral setae, with 2/3 free cusps. Lamellar setae reduced, much shorter than rostral setae. Length: 600–770µm (Figure 3.83E).. *C. longicuspis* Lindo

– Two pairs of reduced notogastral setae (p_1, p_3). Lamellae with short cusps, not reaching insertion of rostral setae. Length: 530–580µm (Figure 3.83F)........... *C. tofinoensis* Lindo
Note: Unidentified species from AK: Lava Camp, Seward Peninsula (Late Miocene: 5.7 mya) and *C. rotundirostris* Drouk, 1982 and unidentified species from late Pliocene/early Pleistocene fossil deposits in AK (Matthews et al. 2019).

Ceratoppia bipilis (Hermann, 1804) (Figure 3.82B)
Comb./Syn.: *Notaspis bipilis* Hermann, 1804; *Ceratoppia herculeana* Berlese, 1908
Note: *Ceratoppia bipilis* is a highly variable species as presently conceived; at least two morphological forms are present in North America, both co-occurring in northern, boreal forests of Canada alongside *C. quadridentata arctica* (Lindo 2011). The main qualitative difference among these morphological variants is size and colour (darkness). Larger, darker forms are similar to European species of *C. bipilis*, while smaller, lighter forms are nearly indistinguishable from *C. quadridentata arctica* with the exception of number of hypostomal setae. Grandjean (1936) noted that the variation in many characters of *C. bipilis* may be due to geographical variation and ecology.
Biology: Adults with varied diet. They were considered necrophagous by Littlewood (1969), but produced eggs and larvae when fed on a mixture of green algae scraped from tree trunks. Gut contents of adult and juveniles included protists, grass and conifer pollen, fungal and arthropod fragments including collembolan scales and enchytraeid setae (Behan-Pelletier and Hill 1983). Hartenstein (1962) found this species to feed preferentially on fungi, but not on fresh or decaying leaves, needles or wood. This species was successfully reared on *Pleurococcus* algae and some lichen (Ermilov and Łochyńska 2008).
Distribution: AK, YT, NT, NU, BC, AB, MB, ON, QC, NB, NS, NL [CA, MI, MN, MT, NC, NY, OR, TN, VA; Semicosmopolitan]. Specimens known from late Pliocene/early Pleistocene fossil deposits in AK and from Tertiary fossil sites on Meigen Island (NU) (Matthews et al. 2019).
Literature: Maraun et al. (2019); ABMI (2019).

Ceratoppia bipilis spinipes (Banks, 1906)
Comb./Syn.: *Oppia spinipes* Banks, 1906
Distribution: ON [VA; Nearctic]
Note: Numerous subspecies have been proposed for *C. bipilis* including *C. bipilis spinipes* originally described from Church Falls, Virginia (Banks 1906). There is high variation in size among the *C. bipilis spinipes* type specimens themselves. While the observed specimens are generally smaller (480–800µm) than *C. bipilis* (600–1,100µm), the size range overlaps, and at present, no single diagnostic character, or unique set of character states differentiate *C. bipilis spinipes* from the smaller *C. bipilis* present in northern North America (Lindo 2011).

Ceratoppia indentata Lindo, 2011 (Figure 3.82D,E)
Biology: Dominant *Ceratoppia* on the forest floor throughout most of the Pacific Northwest coastal temperate rainforest. Habitat is mixed or single litter from both conifer and deciduous trees, moss, and lichens; often collected near beaches, small creeks, river mouths or ravines suggesting a moist habitat preference.
Distribution: BC [CA, OR, WA; Nearctic]

Ceratoppia longicuspis Lindo, 2011 (Figure 3.83E)
Biology: Dominant *Ceratoppia* in arboreal habitats, primarily epiphytic bryophytes, throughout the Pacific Northwest coastal temperate rainforest, but also co-occurring in lesser abundance in forest floor habitats with other *Ceratoppia* species.
Distribution: BC [CA, OR, WA; Nearctic]

Ceratoppia offarostrata Lindo, 2011 (Figure 3.83C,D)
Biology: Occurs in low frequency and low abundance in the Walbran Valley on Vancouver Island, BC. Specimens collected from type locality in Walbran Valley are associated exclusively with bark scrapings on Western redcedar. The distribution range of *C. offarostrata* appears limited to coastal locations on Vancouver Island and Haida Gwaii. *C. offarostrata* may be southern variant subspecies of *C. sphaerica* (Lindo 2011).
Distribution: BC [Canada]

Ceratoppia quadridentata (Haller, 1882) (Figure 3.83A)
Comb./Syn.: *Notaspis bipilis quadridentata* Haller, 1882; *Ceratoppia quadridentata arctica* Hammer, 1955
Note: Both *Ceratoppia quadridentata* and the subspecies *C. quadridentata arctica* are listed for northern North America (Marshall et al. 1987), and are primarily differentiated by the length of the lamellar cusps.
Biology: Successfully reared on *Pleurococcus* algae and some lichen (Ermilov and Łochyńska 2008).
Distribution: BC, AB, ON [Holarctic]. Known from late Pliocene/early Pleistocene fossil deposits in AK and from Tertiary fossil sites on Meigen and Ellesmere Islands (NU) (Matthews et al. 2019).
Literature: Lindo (2011, 2018); Maraun et al. (2019).

Ceratoppia quadridentata arctica Hammer, 1955
Note: The validity of *C. quadridentata arctica* as a subspecies has been questioned. Marshall et al. (1987) noted that although they list *C. quadridentata arctica* as a subspecies, *C. quadridentata* contained a number of geographical forms that are difficult to separate. It is likely that records of *C. quadridentata* and *C. quadridentata arctica* in Canada represent the same species (Lindo 2011).
Biology: From arctic and subarctic meadows; peatlands; sub-Boreal and Boreal forest. *C. quadridentata arctica* occurs on the west coast of Vancouver Island, BC within temperate rainforest in alpine and high elevation areas. Hammer (1955) first described the subspecies from Alaska.
Distribution: AK, YT, NT, NU, BC, AB, QC, NL [Holarctic]
Literature: Lindo (2011, 2018); ABMI (2019).

Ceratoppia sexpilosa Willmann, 1938 (Figure 3.82C)
Biology: Common in sphagnum mires in Western Norway (Seniczak et al. 2019).
Distribution: AK, YT [Holarctic]
Literature: Lindo (2011); Maraun et al. (2019).

Ceratoppia sphaerica (L. Koch, 1879) (Figure 3.83B)
Comb./Syn.: *Oppia sphaerica* L. Koch, 1879; *Ceratoppia bipilis sphaerica* (L. Koch, 1879)
Note: Lack of a type specimen and poor original description for *C. sphaerica* are noted in the literature (Grandjean 1936). These issues may indicate that there is variation within the species among the diagnostic character states used, and/or that *C. sphaerica* may represent a more variable species complex than previously thought, which may or may not include *C. offarostrata* (Lindo 2011).
Distribution: AK, YT, NT [Holarctic]
Literature: Lindo (2011); Maraun et al. (2019).

Ceratoppia tofinoensis Lindo, 2011 (Figure 3.83F)
Biology: Occurring with highest densities within Clayoquot Sound UNESCO Biosphere reserve near the town of Tofino. Possible arboreal specialist.
Distribution: BC [OR, WA; Nearctic]

Ceratoppia valerieae Lindo, 2011 (Figure 3.82A)
   Biology: Occurs frequently throughout southern Vancouver Island with low abundance in forest floor samples; dominant microarthropod collected in canopy malaise traps in the Upper Carmanah Valley. Its range follows a north-south coastal temperate rainforest distribution, however, it also occurs in interior zones of British Columbia and Washington. It is the dominant *Ceratoppia* of interior British Columbia occurring as far east as Waterton Lakes National Park, Alberta where it frequently co-occurs with *C. bipilis*.
   Distribution: BC, AB [CA, OR, WA; Nearctic]

Dendrozetes Aoki, 1970
   Type-species: *Dendrozetes caudatus* Aoki, 1970

Dendrozetes jordani Lindo, Clayton and Behan-Pelletier, 2010 (Figure 3.80A)
   Biology: Habitat arboreal, similar to that of *D. caudatus*, i.e., associated with tree species primarily from the family Pinaceae (Aoki 1970b; Fujikawa et al. 1993). *Dendrozetes jordani* was collected in high densities on branch tips, but with lower observed abundance within lichens on branches (Winchester et al. 2008). *Dendrozetes jordani* may be host-tree dependent. The abundances of *D. jordani* adults and immature instars throughout the active growing season indicate that populations studied have continuous reproduction throughout the season, or possibly two hatching events per year.
   Distribution: BC [Canada]

Metrioppia Grandjean, 1931
   Type-species: *Metrioppia helvetica* Grandjean, 1931

KEY TO SPECIES OF *METRIOPPIA* OF CANADA AND ALASKA

1   Interlamellar setae long. Rostral setae longer than lamellar setae. Length: 375–420µm (Figure 3.80B–D) ....*M. walbranensis* Lindo
–   Interlamellar setae minute. Rostral setae shorter than lamellar setae.................................2

2(1)   Three pairs of notogastral setae fully developed. Length: 420–450µm (Figure 3.80E).......
   ....*M. helvetica* Grandjean
–   Three pairs of notogastral setae expressed as microsetae. Length: ca. 360µm (Figure 3.80F)..*M. oregonensis* Woolley and Higgins

Metrioppia helvetica Grandjean, 1931 (Figure 3.80E)
   Biology: From tussock tundra.
   Distribution: AK, YT, NT [Holarctic; Neotropical; Oriental].
   Literature: Lindo (2015, 2018); Matthews et al. (2019).

Metrioppia oregonensis Woolley and Higgins, 1969 (Figure 3.80F)
   Biology: Abundance of this species was found to significantly decline under forest thinning (Peck and Niwa 2005).
   Distribution: BC [CA, OR, WA; Nearctic]
   Literature: Lindo (2015).

Metrioppia walbranensis Lindo, 2015 (Figure 3.80B–D)
   Biology: Collected predominantly from forest litter under Western redcedar in the Walbran Valley on Vancouver Island, BC, and from a suspended soil habitat in the canopy of Western redcedar at the same location (Lindo and Winchester 2008). Also collected from well-developed forest floor soils with distinct litter-fermentation humus horizons that had extensive fungal hyphae mats.
   Distribution: BC [Canada]

Paenoppia Woolley and Higgins, 1965
   Type-species: *Paenoppia forficula* Woolley and Higgins, 1965 (Figure 3.81A)
   Note: Unidentified *Paenoppia* species from BC.

Parapyroppia Pérez-Íñigo and Subías, 1979
   Type-species: *Parapyroppia monodactyla* Pérez-Íñigo and Subías, 1979 (Figure 3.81B)

Parapyroppia transitoria (Berlese, 1908)
   Comb./Syn.: *Protoribates transitorius* Berlese, 1908; *Notaspis lamellata* Ewing, 1909; *Parapyroppia lamellata* (Ewing, 1909)
   Distribution: AK, YT [MO; Nearctic]
   Literature: Norton and Kethley (1990); Lindo (2018).
   Note: Unidentified *Parapyroppia* sp. from AB (Walter et al. 2014).

Pyroppia Hammer, 1955
   Type-species: *Pyroppia lanceolata* Hammer, 1955

KEY TO ADULT *PYROPPIA* OF CANADA AND ALASKA

1   Rostral margin with single medial dens and single dens laterally. Bothridial seta fusiform. Length: 620–670µm (Figure 3.81C) ....*P. lanceolata* Hammer
–   Rostral margin with multiple dentes. Bothridial seta capitate 2

2(1)   Rostral margin with medial dentes larger than lateral dentes. Interlamellar seta extending almost to tip of rostrum. Length: ca. 500µm (Figure 3.81D)............*P. dentata* Krivolutsky
–   Rostral margin with medial dentes subequal to lateral dentes. Interlamellar seta extending to tip of lamellar cusp. Length: ca. 600µm (Figure 3.81E,F)............. *P. serrifrons* (Banks)
   Note: Unidentified *Pyroppia* spp. from BC, AB, MB.

Pyroppia dentata Krivolutsky, 1974 (Figure 3.81D)
   Distribution: AK [Russian Far East]
   Literature: Krivolutsky and Ryabinin (1974).

Pyroppia lanceolata Hammer, 1955 (Figure 3.81C)
   Biology: From tussock tundra.
   Distribution: AK, YT [Holarctic]

Pyroppia serrifrons (Banks, 1923) (Figure 3.81E,F)
   Comb./Syn.: *Notaspis serrifrons* Banks, 1923
   Distribution: AK [Nearctic]

Family Tenuialidae Jacot, 1929
   Diagnosis: *Adults*: Non-poronotic Brachypylina. Medium to large (700–1,100µm) with smooth shiny integument. Lamellae well developed, with distinct cusp. Dorsophragmata and pleurophragmata present. Pedotectum I well developed, incised or not; pedotectum II small or absent. Tutorium, discidium and posterior circumpedal carina present; custodium absent. Notogaster without true pteromorphs; anterolateral corner with anteriorly directed, pointed scapular process; setation 0, 1, 9 pairs, or their alveoli. Genital setation 5–6 pairs. Epimeral border IV with or without taenidium and minitectum. One pair of aggenital, 2 pairs of anal, and 3 pairs of adanal setae present. Lateral podosomal region with concavities for reception of retracted legs. Subcapitulum diarthric, chelicerae chelate-dentate. Palpal eupathidium *acm* and solenidion separate. Axillary saccule of subcapitulum absent. All femora, and trochanters III, IV, with ventral blade-like processes (sometimes poorly developed on femur I). Legs tridactylous.

KEY TO GENERA OF TENUIALIDAE OF CANADA AND ALASKA

1   Body compressed (dorsoventrally flattened). Notogaster with reflexed rim; with 9 pairs of setae (seta p_3 absent). Lamellae widely separated, hiding lateral contour of prodorsum in dorsal aspect. Translamella absent. Exobothridial seta present (Figure 3.84A–C) .... ***Peltenuiala*** Norton
–   Body strongly convex; notogaster without reflexed rim. Notogastral setae absent, or with 1 pair of posteromarginal setae (p_1) present. Lamellae widely separated, or positioned more medially. Translamella present or absent. Exobothridial seta absent 2

2(1)   Epimeral setation 4-1-3-3. Notogastral seta p_1 present. Taenidium and minitectum of epimeral border IV absent (Figure 3.84D) .... ***Tenuialoides*** Woolley and Higgins
–   Epimeral setation 3-1-3-3. Notogastral seta p_1 absent; only alveolus present. Taenidium and minitectum of epimeral border IV present 3

3(2)   Humeral process of notogaster long, extending anteriorly of anteromedial notogastral margin; usually with anteromedial serration (Figure 3.84E) .... ***Tenuiala*** Ewing
–   Humeral process of notogaster short, extending to or slightly beyond anteromedial notogastral margin; without anteromedial serration (Figure 3.84F,G) .... ***Hafenferrefia*** Jacot

Hafenferrefia Jacot, 1939
   Type-species: *Galumna nitidula* Banks, 1906
   Comb./Syn.: *Hafenrefferiella* Sellnick, 1952

Hafenferrefia nitidula (Banks, 1906) (Figure 3.84F,G)
   Comb./Syn.: *Galumna nitidula* Banks, 1906
   Biology: From maple forest litter.
   Distribution: QC, NS, NL [NH; Nearctic]
   Literature: Norton (1983).
   Note: Unidentified *Hafenferrefia* species from AB, QC.

Peltenuiala Norton, 1983.
   Type-species: *Peltenuiala pacifica* Norton, 1983

Peltenuiala pacifica Norton, 1983 (Figure 3.84A–C)
   Comb./Syn.: *Hafenferrefia nitidula* sensu Higgins and Woolley, 1957
   Distribution: BC [OR, WA; Nearctic]
   Literature: Lindo and Winchester (2009).

Tenuiala Ewing, 1913
   Type-species: *Tenuiala nuda* Ewing, 1913 (Figure 3.84E)
   Note: Unidentified *Tenuiala* species from YT, BC, AB, QC, NS.

Tenuialoides Woolley and Higgins, 1965
   Type-species: *Tenuialoides medialis* Woolley and Higgins, 1965 (Figure 3.84D)
   Note: Unidentified *Tenuialoides* species from QC.

Superfamily Carabodoidea C.L. Koch, 1837

Family Carabodidae C.L. Koch, 1837
   Diagnosis: *Adults:* Non-poronotic Brachypylina. Integumental birefringence present or absent. Prodorsum with or without lamellae, positioned laterally; with tutorium. Dorsophragmata and pleurophragmata absent. Prodorsum and notogaster fused or not; dorsosejugal depression present or absent. Notogastral setation 10–15 pairs; deep depressions on notogaster present or absent; circumgastric furrow present or absent. Humeral apophysis present or absent.

Circumpedal carina and custodium absent, discidium present or absent. Pedotecta I, II present. Epimeral setation 3-1-3-3, with occasional loss. Aggenital setae 0–2 pairs. Genital setation 3–6 pairs. Subcapitulum diarthric, eupathidium *acm* separate from recumbent solenidion; chelicerae chelate-dentate.

KEY TO GENERA OF CARABODIDAE OF CANADA AND ALASKA

1 Notogastral setation 8–10 pairs. Body slightly longer than wide. Notogaster anteriorly without anteriorly directed setae (Figure 3.85–3.87) .... ***Carabodes*** C.L. Koch
– Notogastral setation 12–15 pairs. Body elongate, twice as long as wide. Notogaster anteriorly with 2–4 pairs of anteriorly directed setae (Figure 3.86H) .... ***Odontocepheus*** Berlese

Carabodes C.L. Koch, 1835
 Type-species: *Carabodes coriaceus* C.L. Koch, 1835
 Comb./Syn.: *Neocepheus* Willmann, 1936
 Biology: *Carabodes* are primarily endophagous fungivores, though they have been reported as macro-, or micro-, or panphytophages (Schuster 1956; Hartenstein 1962a; Kaneko 1988); there is no evidence for carnivory or necrophagy.
 Literature: Reeves (1998) for all North American species.

KEY TO ADULT *CARABODES* OF CANADA AND ALASKA
(MODIFIED FROM REEVES AND BEHAN-PELLETIER 1998)

1 Central notogastral region with tubercles arranged in groups, coalesced into ridges, or with irregular pattern of longitudinal ridges. Notogastral seta c_2 inserted nearer to shoulder than to midline, anterolateral to *lm*. Interlamellar setae similar to notogastral setae 2
– Central notogastral region with foveae or unconnected tubercles. Notogastral seta c_2 inserted nearer to midline than to shoulder and anterior to *lm*. Interlamellar setae usually different from notogastral setae... 6

2(1) Notogastral setae clavate 3
– Notogastral setae spiniform or bacilliform 4

3(2) Prodorsal protuberances present in interlamellar region and separated by deep sinus. Dorsosejugal depression wide, deep. Interlamellar, notogastral, ad_1 and ad_2 setae clavate. Notogastral tubercles coalesced into irregular ridges. Epimeral, genital, and aggenital setae barbed. Genital setation 4–6 pairs. Length: 700–900µm (Figures 3.85A; 3.87A) ......
 .... ***C. cochleaformis*** Reeves
– Prodorsal protuberances absent in interlamellar region. Dorsosejugal depression narrow, shallow. Interlamellar, notogastral, ad_1 and ad_2 setae narrowly clavate. Notogastral tubercles distinct but arranged in groups like bunches of grapes. Epimeral, genital, and aggenital setae glabrous. Genital setation 4 pairs. Length: 605–805µm (Figures 3.85B; 3.87B) ...
 .... ***C. wonalancetanus*** Reeves

4(2) Notogastral tubercles coalesced into irregular ridges of single tubercle width. Dorsosejugal depression narrow. Interlamellar region without protuberances or medial depression; bothridial seta directed laterally. Genital setation 4 pairs. Length: 430–580µm (Figures 3.85C,D; 3.87C) .... ***C. labyrinthicus*** (Michael)
– Notogaster with irregular, long longitudinal ridges separated by shorter transverse ridges. Dorsosejugal depression wide. Interlamellar region with lateral protuberances and medial depression; bothridial seta directed dorsally. Genital setation 4–7 pairs........................... 5

5(4)	Prodorsal surface mostly granulate, foveae poorly developed. Elevation in interlamellar region well-developed with distinct medial depression, without inverted Y-shaped indentation in center (Figures 3.85E; 3.87D) .... ***C. rugosior*** Berlese
–	Prodorsal foveae well-developed, especially in depression medial to lamellae. Elevation in interlamellar region chevron-shaped, medial depression weakly developed, posterolateral corners overhanging dorsosejugal depression, inverted Y-shaped indentation in center. Length: 515–710µm (Figure 3.85F) .... ***C. hoh*** Reeves and Behan-Pelletier
6(1)	Notogaster tuberculate; lateral surface between acetabula and bothridial/notogastral rims granulate. Notogastral setae spiniform, pencil-shaped, or narrowly clavate and fluted. Small (length usually <400µm), body brown 7
–	Notogaster foveate; small tubercles on lateral surface above acetabula and below bothridial and notogastral rims. Notogastral setae usually clavate, narrowly clavate, penicilliform (shaped like hair pencil), phylliform, bacilliform, or fusiform, rarely spiniform, or fluted. Large (length usually >400µm), body brown, dark brown, or black 9
7(6)	Notogastral tubercles closely packed with little or no space between. Length of interlamellar setae 2/3 or less the distance between their insertions. Bothridial seta short 8
–	Notogastral tubercles scattered, often arranged in circular groups. Interlamellar setae longer than 2/3 distance between their insertions. Bothridial seta long. Length: 285–415µm (Figure 3.85G) .... ***C. dickinsoni*** Reeves and Behan-Pelletier
8(7)	Central notogastral setae pencil-shaped or slightly enlarged distally, posteromarginal setae setiform. Length of interlamellar setae approximately 2/3 distance between their insertions. Length: 340–450µm (Figure 3.85H) .... ***C. willmanni*** Bernini
–	Central and posteromarginal setae all setiform. Length of interlamellar setae approximately 1/2 distance between their insertions. Length: 315–390µm (Figures 3.86A; 3.87E) .... ***C. higginsi*** Reeves
9(6)	Interlamellar and notogastral setae penicilliform. Notogastral foveae rosettiform. Length: ca. 400µm (Figure 3.87F) .... ***C. granulatus*** Banks
–	Interlamellar setae spiniform, notogastral setae variable (spiniform, fusiform, or pencil-shaped). Notogastral foveae usually ovate (rosettiform only in *spiniformis*) 10
10(9)	Interlamellar setae directed anteriorly, usually not grooved. Genital setation 4 or 5 pairs, short, less than 1/4 length of genital plate, directed medially or posteriorly. Central notogastral setae spiniform, penicilliform, fusiform; notogastral setal dimorphism usually absent. Dorsosejugal depression usually a narrow slit. Interlamellar region without medial depression separating lateral protuberances 11
–	Interlamellar setae directed medially, grooved. Genital setation 4 pairs; long, g_2 more than 1/3 length of genital plate, g_1 directed posteriorly, g_2, g_3, and g_4 directed laterally. Central notogastral setae usually clavate and spoon-shaped, rarely phylliform, spiniform, bacilliform, or penicilliform. Notogastral setae usually dimorphic. Dorsosejugal depression usually wide, deep. Interlamellar region often depressed medially, with distinct lateral protuberances 13
11(10)	Prodorsal surface between lamellae evenly convex. Circumgastric depression almost disappearing posteriorly. Notogastral setae pencil-shaped 12
–	Prodorsal surface between lamellae with medial ridge and lateral depression. Circumgastric depression complete. Notogastral setae spiniform, fusiform, or pencil-shaped. Length: 525–680µm (Figure 3.86B) .... ***C. colorado*** Reeves and Behan-Pelletier
12(11)	Aggenital setation 1 pair. Notogastral foveae small, separated by more than pit diameter. Notogastral setae long, minutely barbed distally. Bothridial seta capitate. Length: 390–513µm (Figures 3.86C; 3.87G) .... ***C. radiatus*** Berlese

Taxonomic Keys 153

–   Aggenital setation 2 pairs. Notogastral foveae larger, most foveae separated by less than pit diameter. Notogastral setae shorter, barbed along shaft. Bothridial seta distally similar to opening flower bud. Length: 380–500μm (Figures 3.86D; 3.87H)......*C. chandleri* Reeves

13(10)   Posteromarginal setae either clavate or spoon-shaped and erect, or spiniform and approximately at 45° angle to margin of notogaster in dorsal view 14
–   Posteromarginal setae of various shape, at 30° angle or less to margin of notogaster in dorsal view.. 15

14(13)   Central notogastral setae penicilliform; posteromarginal setae spiniform, at approximately 45° angle to notogastral margin. Bothridial seta capitate. Body light brown. Length: 450–580μm (Figure 3.86E) ....*C. brevis* Banks
–   Central and posteromarginal notogastral setae clavate, spoon-shaped; posteromarginal setae erect. Bothridial seta narrowly clavate. Body black. Length: 445–590μm (Figure 3.86F)...*C. erectus* Reeves

15(13)   Central notogastral setae narrowly clavate, or bacilliform. Anterior margin of dorsosejugal depression straight. Interlamellar region without medial depression (or poorly developed). Tarsus I setae (*u*) short. Length: 475–610μm (Figure 3.86G) ....*C. polyporetes* Reeves
–   Central notogastral setae clavate, or phylliform. Anterior margin of dorsosejugal depression indented medially. Interlamellar region usually with lateral protuberances separated by distinct medial depression. Tarsus I setae (*u*) long, attenuate. Length: ca. 500μm (Figure 3.87I).. *C. niger* Banks

Carabodes brevis Banks, 1896 (Figure 3.86E)
    Biology: From a wide variety of forest habitats including leaf litter, rotten wood, moss, lichens, bark and fungi. Recorded from epiphytic lichens in the Adirondack Mountains (Root et al. 2007). Co-occurs with *Carabodes niger* (Reeves 1988).
    Distribution: ON, QC, NB, NS, NL [DE, GA, KY, ME, NC, NH, NJ, NY, PA, VA, VT, WV; Nearctic]

Carabodes chandleri Reeves, 1992 (Figures 3.86D; 3.87H)
    Biology: From coniferous and hardwood leaf litter, rotten wood, lichens, moss and fungi, but shows preference for arboreal habitat (Reeves 1992). Endophagous species (Hansen 1999).
    Distribution: QC [DE, GA, KY, MD, NC, NH, NJ, PA, VA, WV; Nearctic]

Carabodes cochleaformis Reeves, 1990 (Figures 3.85A; 3.87A)
    Biology: From coniferous leaf litter or rotten logs, hardwood leaf litter, and fungi. Reeves (1990) kept adults alive for several months on the polyporous fungus *Trametes versicolor*.
    Distribution: NB [KY, ME, NC, NH, PA, VT; Nearctic]

Carabodes colorado Reeves and Behan-Pelletier, 1998 (Figure 3.86B)
    Biology: Associated with polypores, lichens, moss, rotten wood, litter.
    Distribution: BC, AB [AZ, ID, NM, WY; Nearctic]

Carabodes dickinsoni Reeves and Behan-Pelletier, 1998 (Figure 3.85F)
    Biology: From leaf litter, rotten wood, moss, and mycorrhizal mats.
    Distribution: BC [CA, OR, WA; Nearctic]

Carabodes erectus Reeves, 1992 (Figure 3.86F)
    Biology: From leaf litter. bark, rotten wood, lichens, river drift, and several kinds of fungi, especially polypore fungi (Reeves 1992). In oak litter (Lamoncha and Crossley 1998). Juveniles were cultured on the mossy maple polypore, *Oxyparus populinus*, multicolour gill

polypore, *Lenzites betulina*, and chicken mushroom, *Laetiporus sulphureus*. The life-cycle required at least 12 weeks to complete at room temperature.
Distribution: ON [GA, KY, LA, MI, MS, NC, NJ, OH, OK, PA, VA, WV, SC, TN; Nearctic]

Carabodes granulatus Banks, 1895 (Figure 3.87F)
Biology: *Carabodes* with the widest distribution in eastern North America. Among the most abundant species in leaf litter and rotten wood samples but may also be found in sphagnum and other mosses, lichens, bark, grass sod and fungi. Gravid females carry up to 2 large eggs. Endophagous species (Hansen 1999).
Distribution: AB, ON, QC, NB, NS, NL [AL, CA, FL, GA, IL, KY, LA, MA, ME, MI, MO, MS, NC, NH, NJ, NY, OK, PA, SC, TN, VA, VT, WV; Nearctic]
Literature: ABMI (2019).

Carabodes higginsi Reeves, 1988 (Figures 3.86A; 3.87E)
Biology: From conifer leaf litter, bark and branches; fir and spruce branches and White pine duff. Recorded from epiphytic lichens in the Adirondack Mountains (Root et al. 2007). Endophagous species (Hansen 1999).
Distribution: NB NS [AL, NC, NH, NY NJ, PA, VA; Nearctic]

Carabodes hoh Reeves and Behan-Pelletier, 1998 (Figure 3.85F)
Biology: From leaf litter, rotten wood, moss, bracket and polyporous fungi, and lichens.
Distribution: BC [CA, OR, WA; Nearctic]

Carabodes labyrinthicus (Michael, 1879) (Figures 3.85C,D; 3.87C)
Comb./Syn.: *Tegeocranus labyrinthicus* Michael, 1879; *Cepheus heimi* Oudemans, 1903; *Carabodes vermiculatus* Berlese, 1916
Biology: This species is considered an indicator of past permafrost conditions (Markkula 2020). Travé (1963) found this species more common in moss and lichens on rock and tree surfaces than in soil. Prinzing and Wirtz (1997) recorded this species living and feeding on and in the lichen *Evernia prunastri*. When specimens from a chestnut and beech forest were analysed over a 20 month period feeding preference was fungal hyphae > amorphous material > fungal spores, with almost no leaf material eaten (Anderson 1975). Schneider (2005) placed this mite in the phycophage/fungivore feeding guild (i.e., feeds on lichens and algae). Wunderle (1992) found this mite commonly on tree bark. Specimens have been found in the feathers of birds in Russia (Krivolutsky and Lebedeva 2004).
Distribution: AK, NT, YT, AB, MB, ON, QC, NB, NS, PE, NL [NJ, NY, PA, VA, WV, ME, NH, VT; Holarctic; Mexico]
Literature: Prinzing et al. (2004); Maraun et al. (2019); ABMI (2019); Seniczak et al. (2019).

Carabodes niger Banks, 1895 (Figure 3.87I)
Biology: From oak litter (Lamoncha and Crossley 1998).
Distribution: MB, ON, QC, NB, NS, NL, PE [AL, AR, DE, GA, IL, KY, MA, MO, NJ, NY, PA, TN, TX, VA, WV; Nearctic]
Literature: Prinzing et al. (2004).

Carabodes polyporetes Reeves, 1991 (Figure 3.86G)
Biology: From diverse forest habitats and most abundant in leaf litter, rotten wood, and polypore fungi but may also be found in moss and lichens. Collected from beaver lodges, mouse nest litter, moose dung, wet sphagnum, and pitcher plant "pitchers". Reeves (1991) cultured this species on the polypores *Oxyporus populinus*, *Lenzites betulina* and *Laetiporus sulphureus*. At room temperature development took 10–12 weeks. Gravid females carry up to 4 eggs.
Distribution: AB, MB, ON, QC, NB, NS, PE, NL [AL, AR, IN, ME, NC, NH, NY, OH, PA, SD, VA, VT, WV; Nearctic]
Literature: ABMI (2019).

Carabodes radiatus Berlese, 1916 (Figures 3.86C; 3.87G)
    Comb./Syn.: *Carabodes dendroetus* Reeves, 1987
    Biology: Arboreal species; on bark and branches of conifers and deciduous trees. Endophagous species (Hansen 1999). Known from *Sphagnum* (Donaldson 1996)
    Distribution: NB, NS [CA, KY, MO, NC, NH, PA, WV; Nearctic]
    Literature: Root et al. (2007).

Carabodes rugosior Berlese, 1916 (Figures 3.85E; 3.87D)
    Biology: From a mixture of moss, leaf litter and rotten wood from the forest floor.
    Distribution: QC, NB, NS, NL [GA, NC, NH, PA, SC, VT; Holarctic]
    Literature: Maraun et al. (2019).

Carabodes willmanni Bernini, 1975 (Figure 3.85H)
    Biology: Bellido (1979) found all stages burrowing into lichens. Gut content analysis of adults showed fungivory with occasional fragments of moss and plant material, grass pollen, and arthropod fragments (Behan-Pelletier and Hill 1983).
    Distribution: MB, NB, NS, NL [MA, NH, NJ; Holarctic]
    Literature: Reeves and Behan-Pelletier (1998); Maraun et al. (2019).

Carabodes wonalancetanus Reeves, 1990 (Figures 3.85B; 3.87B)
    Biology: From conifer and deciduous litter and logs, polypores, moss, fungi (Reeves 1990).
    Distribution BC, AB, ON, QC, NB, NS, NL [NH, VT; Nearctic]
    Literature: ABMI (2019).

Odontocepheus Berlese, 1913 (Figure 3.86H)
    Type-species: *Tegeocranus elongatus* Michael, 1879

Odontocepheus oblongus (Banks, 1895)
    Comb./Syn.: *Carabodes oblonga* Banks, 1895
    Biology: From forest litter.
    Distribution: ON, QC, NB, NS [FL, NC, NY, SC, TX; Holarctic]
    Remarks: This name was considered a synonym of *Odontocepheus elongatus* (Michael, 1879). Subsequent studies of type and topotype specimens show consistent differences between British and New York populations; so the synonymy was rejected by Marshall et al. (1987).

Superfamily Oppioidea Grandjean, 1951

Family Autognetidae Grandjean, 1960
    Diagnosis: *Adults*: Non-poronotic Brachypylina. Rostrum with straight, deep, medial incision. Prodorsum with long, usually parallel ridge-like lamella, bearing seta *le* anteriorly; longitudinal ridge present or not lateral of ridge-like lamella, dorsal to acetabulum I; defined tutorium absent. Prodorsum with enantiophysis *eA* present, well, or weakly developed, or absent. Humeral enantiophysis present, with bothridial apophysis well-developed. Pedotecta I and II developed, small. Notogaster without pteromorphs; with U-shaped furrow extending from seta *la* posteriorly, well or weakly developed; setation 10 pairs. Sternal apodeme well or weakly developed. Genital setation 5 or 6 pairs. Genital papillae *Vm* and *Vp* broadly rounded distally, *Va* positioned deeper in genital chamber and strongly tapered distally. Gnathosoma diarthric; chelicera chelate dentate. Axillary saccule absent, but porose region on mentum laterally, facing palp trochanter. Palp setation, 0-2-1-3-9(1), eupathidium *acm* positioned proximal and separate from solenidion. Legs monodactylous. Tibia I with large apophysis overhanging tarsus I, bearing solenidion φ_1.

KEY TO GENERA OF AUTOGNETIDAE OF CANADA AND ALASKA

1   Ridge-like lamellae with prominent cusps (Figure 3.88F)......................***Eremobodes*** Jacot
–   Ridge-like lamellae without distinct cusps..2

2(1)   Genital papilla *Va* with long, sclerotized stylet distally. Prodorsum lacking enantiophysis *eA* (Figure 3.88G,H)... ***Rhaphigneta*** Grandjean
–   Genital papilla *Va* strongly tapered distally; sclerotized stylet absent. Prodorsum with enantiophysis *eA*..3

3(2)   Bothridial seta setiform, unilaterally denticulate (Figure 3.88E) . ***Conchogneta*** Grandjean
–   Bothridial seta dilated distally (Figure 3.88A–D).. ***Autogneta*** Hull

Autogneta Hull, 1916
    Type-species: *Autogneta longilamellata* (Michael, 1885)
    Comb./Syn.: *Notaspis longilamellata* Michael, 1885

KEY TO ADULT *AUTOGNETA* OF CANADA AND ALASKA

1   In both sexes, apophysis posteriorly on prodorsum expressed as pair of flattened ridges. Circular cluster of tubercles present lateral of lamellar seta. Males with oblong porose area posteromedially positioned between lyrifissures *ips*, bearing unmodified seta h_1. Length: 243–288µm (Figure 3.88A) ....***A. amica*** Jacot
–   In both sexes, apophysis posteriorly on prodorsum expressed as pair of tubercles. Cluster of tubercles lateral of lamellar seta absent. Males with either long porose area posteromedially, or oblong porose area with setae h_1, p_1 modified..2

2(1)   In both sexes, notogastral setae thick, weakly barbed, tapered. Bothridial setae clavate to fusiform. Males with long, narrow porose area, extending between setae h_3–h_3, not bearing either of setae h_1 and p_1; setae h_1 and p_1 not modified. Length: 278–336µm (Figure 3.88C) ....***A. longilamellata*** (Michael)
–   In both sexes, notogastral setae thin, acuminate. Bothridial setae clavate or capitate. Males with setae p_1 and h_1 spatulate, positioned almost in transverse alignment on wide porose area. Length: ca. 320µm (Figure 3.88D) ....***A. flaheyi*** Behan-Pelletier
    Note: Unidentified species known from BC, AB, NS, NB.

Autogneta amica Jacot, 1938 (Figure 3.88A)
    Comb./Syn.: *Autogneta longilamellata amicus* Jacot, 1938
    Biology: Sexually dimorphic species. Gravid females carry up to 2 eggs. Gut contents are primarily darkly pigmented fungi.
    Distribution: ON [AL, NC, NY, VA; Nearctic]
    Literature: Behan-Pelletier (2015).

Autogneta flaheyi Behan-Pelletier, 2015 (Figure 3.88D)
    Biology: Sexually dimorphic species. Gravid females carry up to 2 large eggs. Gut contents are primarily lightly pigmented fungal hyphae and spores.
    Distribution: BC, AB [Canada]

Autogneta longilamellata (Michael, 1885) (Figure 3.88C)
    Comb./Syn.: *Notaspis longilamellata* Michael, 1885; *Dameosoma longilamellatum* (Michael, 1885)
    Biology: Sexually dimorphic species. Main habitat is rotting wood of stumps or fallen trunks in forests (Grandjean 1963). Many specimens from eastern North America have been collected

from similar microhabitats and from bracket fungi. This species is among many considered characteristic of deadwood by Sokołowska et al. (2009). In their study of Oribatida of 11 habitats ranging from forest to grassland to seashore, Penttinen et al. (2008) found this species to be associated only with decaying wood and Skubała and Duras (2008) found this species primarily associated with early stages of decay in fallen spruce logs. Huhta et al. (2012) found this species only associated with tree trunk and stumps among the microhabitats they examined. This species has also been recorded from the sporocarps of *Fomitopsis pinicola* (Maraun et al. 2014). Although primarily associated with decaying wood, this species has also been recorded occasionally from general forest litter, e.g., Déchêne and Buddle (2010). Lebedeva and Lebedev (2008) found this species in bird nests on the Murmansk Coast, Russia.
Distribution: AK, YT, NT, BC, AB, ON, QC, NB, NS, NL [AL, VA; Holarctic]
Literature: Behan-Pelletier (2015); ABMI (2019).

Conchogneta Grandjean, 1963
Type-species: *Autogneta delacarlica* Forsslund, 1947

Conchogneta traegardhi (Forsslund, 1947) (Figure 3.88E)
Comb./Syn.: *Autogneta traegardhi* Forsslund, 1947
Distribution: QC, NL [Holarctic]

Rhaphigneta Grandjean, 1960
Type-species: *Rhaphigneta numidiana* Grandjean, 1960 (Figure 3.88G,H)
Note: Considered subgenus of *Autogneta* by Subías (2004). Unidentified species from BC.

Eremobodes Jacot, 1937
Type-species: *Eremobodes pectinatus* Jacot, 1937 (Figure 3.88F)
Note: Unidentified species from NL.

Family Machuellidae Balogh, 1983
Diagnosis: *Adults*: Non-poronotic Brachypylina. Prodorsum without ridge-like lamella. Tutorium absent. Dorsophragmata and pleurophragmata absent. Body with constriction in sejugal region. Prodorsum and notogaster separate. Notogaster with ridges or crista anteriorly; setation 10 pairs. Humeral apophysis absent. Circumpedal carina, custodium, and discidium absent. Pedotecta I and II absent. Epimera III and IV fused. Epimeral setation 2-2-4-5, or up to 17 pairs of epimeral setae directed medially towards ventrosejugal region, forming "basket" in thick layer of epimeral cerotegument. Anterior margin of epimere I with medial tooth. Genital setation 5–6 pairs; aggenital and adanal neotrichy absent. Subcapitulum diarthric; palptarsus with 8–9 setae; eupathidium *acm* separate from recumbent solenidion; chelicerae chelate-dentate. Legs moniliform; monodactylous.

Machuella Hammer, 1961
Type-species: *Machuella ventrisetosa* Hammer, 1961
Note: Undescribed species from southern ON (CNC record).

Family Oppiidae Grandjean, 1951
Diagnosis: *Adults*: Non-poronotic Brachypylina. Prodorsum with or without ridge-like lamella. Tutorium absent. Dorsophragmata and pleurophragmata absent. Body with constriction in sejugal region. Prodorsum and notogaster separate, or dorsosejugal scissure absent medially. Notogaster often with tubercles or crista anteriorly; setation 9–14 pairs. Humeral apophysis present or absent. Circumpedal carina, custodium, and discidium absent. Pedotectum I usually present, small, scale-like; pedotectum II absent. Epimera III and IV fused. Epimeral setation 3-1-3-3. Genital setation 4–7 pairs. Ventral neotrichy absent. Subcapitulum diarthric; palptarsus with 8–9 setae; eupathidium *acm* separate from recumbent solenidion; chelicerae

chelate-dentate. Legs moniliform; monodactylous. Proral setae of tarsi II–IV short and spine-like, or absent.

Biology: Species feed primarily on fungi (Anderson 1975). Based on stable isotope analysis a species of *Amerioppia* is a predator in Ecuadoran rainforest (Illig et al. 2005).

KEY TO GENERA OF OPPIIDAE OF CANADA AND ALASKA
(MODIFIED FROM MIKO (2006) AND WALTER AND LUMLEY (2021))

1	Rostral setae inserted on medial protuberance on rostrum. Genital setation 5 pairs. (Figure 3.89J)	***Dissorhina*** Hull
–	Rostral setae not inserted on medial rostral protuberance. Genital setation 4–6 pairs.	2
2	Crista present or seta c_2 subequal to other notogastral setae. Genital setation 4–6 pairs. Lyrifissure *iad* parallel to anal shield (paranal)	3
–	Crista absent or weakly developed; c_2 either absent, or shorter than other notogastral setae. Genital setation 4–5 pairs. Lyrifissure *iad* paranal or apoanal (angled)	4
3(2)	Costulae absent (lamellar lines maybe present)	MEDIOPPIINAE
–	Costulae present	OPPIELLINAE
4(2)	Notogaster with 1 pair of small humeral processes or spines anteriorly, or interbothridial region with or without costulae; if without, then faint lamellar and translamellar lines present. Lyrifissure *iad* paranal or apoanal	OXYOPPIINAE
	(One genus in North America ***Subiasella*** Balogh) (Figure 3.90E)	
–	Notogaster without humeral spines or processes anteriorly. Costulae absent. Lamellar and translamellar lines present or absent. Lyrifissure *iad* usually paranal	5
5(4)	Lamellar and translamellar lines absent. Bothridial seta lanceolate or fusiform, never pectinate, radiate or ciliate	OPPIINAE
–	Lamellar and/or translamellar lines present, or if absent, bothridial seta either radiate, pectinate or ciliate	MULTIOPPIINAE

MEDIOPPIINAE

1	Without lines or sclerotized apophyses running from dorsosejugal scissure to basal part of prodorsum (Figure 3.89A)	***Discoppia*** (***Cylindroppia***) Subías and Rodríguez
–	Sclerotized apophyses running from dorsosejugal scissure to basal part of prodorsum present. Genital setation usually 4 pairs (Figure 3.89B)	***Microppia*** Balogh

OPPIELLINAE

1	Dorsosejugal scissure straight or slightly arched. Bothridial seta usually fusiform, ciliate	2
–	Dorsosejugal scissure strongly convex, parabolic or semicircular. Bothridial seta either radiate or globular and aciculate. Rostrum tridentate (Figure 3.90G)	***Berniniella*** Balogh
2(1)	Genital setation 4–5 pairs. Crista with 1–2 tubercles forming enantiophysis with bothridial tubercle. Rostrum without teeth	3
–	Genital setation 6 pairs. Humeral enantiophysis usually not formed. Rostrum rounded or with strong median and weaker lateral teeth (Figure 3.89F,G)	***Oppiella*** (***Rhinoppia***) Balogh
3(2)	Genital setation 5 pairs. Prodorsal ridges usually well developed. Prodorsal pits usually present. Notogastral anterior margin with crista usually clearly developed, straight or with S-shaped side branches (Figure 3.89C,D)	***Oppiella*** (***Oppiella***) Jacot

– Genital setation 4 pairs (*O. translamellata* with 5 pairs). Anterior margin of notogaster without crista. Humeral apophysis sometimes present (Figures 3.90A; 3.91G)..................... ....*Oppiella* (***Moritzoppia***) Subías and Rodríguez

OPPIINAE

1	Bothridial seta setiform, long, bifurcate (Figure 3.90B)... ....***Sphagnoppia*** J. Balogh and P. Balogh	
–	Bothridial seta not bifurcate 2	
2(1)	Bothridial seta globular or clavate (Figure 3.90C,D)............................ ***Aeroppia*** Hammer	
–	Bothridial seta setiform, lanceolate, or elongate fusiform 3	
3(2)	Bothridial seta setiform or lanceolate. Notogastral heterotrichy present, with 5 or 6 pairs of long notogastral setae (Figure 3.91H)..***Lasiobelba*** Aoki	
–	Bothridial seta elongate fusiform or lanceolate. Notogastral heterotrichy (other than *p* series setae) absent (Figure 3.91C,D)...***Oppia*** C.L. Koch	

MULTIOPPIINAE

1 Notogaster with 10–12 pairs of setae; c_2 present.. 2
– Notogaster with 9 pairs of setae; c_2 absent..4

2(1) Notogaster with 10 pairs of setae. Translamellar line absent (Figure 3.90F) ....***Anomaloppia*** Subías
– Notogaster with 12 pairs of setae. Translamellar line weakly developed or absent............3

3(2) Genital setation 4 pairs. Seta ad_3 posterior to aggenital seta (Figure 3.91F)....................... ....***Graptoppia*** (***Stenoppia***) Balogh
– Genital setation 5 pairs Seta ad_3 anterior or posterolateral to aggenital seta (Figure 3.91A,B) ....***Multioppia*** Hammer

4(1) Translamellar line present. Rostral setae separated (Figure 3.90H,I)..................................... ....***Brachioppiella*** Hammer
– Translamellar line present or absent. Rostral setae often closely adjacent, usually slightly arched or geniculate (Figure 3.91E) ....***Ramusella*** (***Ramusella***) Hammer

Aeroppia Balogh, 1965
 Type-species: *Aeroppia peruensis* Hammer, 1961
 Note: Undescribed species from AB (Walter et al. 2014).

Anomaloppia Subías, 1978
 Type-species: *Anomaloppia canariensis* Subías, 1978

Anomaloppia manifera (Hammer, 1955) (Figure 3.90F)
 Comb./Syn.: *Oppia manifera* Hammer, 1955
 Biology: From mixed conifer-hardwood forest; Balsam fir forests.
 Distribution: AK, ON, NL [Holarctic]

Berniniella Balogh, 1983 (Figure 3.90G)
 Type-species: *Oppia aeoliana* Bernini, 1973
 Note: Unidentified species from NS.

Brachioppiella Hammer, 1962
 Type-species: *Brachioppiella periculosa* Hammer, 1962

Brachioppiella periculosa Hammer, 1962 (Figure 3.90H,I)
 Distribution: QC [Pantropical]

Discoppia (*Cylindroppia*) Subías and Rodríguez, 1986
Type-species: *Oppia minus cylindrica* Pérez-Íñigo, 1965 (Figure 3.89A)
Note: Unidentifed species from BC, ON.

Dissorhina Hull, 1916
Type-species: *Dissorhina ornata* (Oudemans, 1900)

Dissorhina ornata (Oudemans, 1900) (Figure 3.89J)
Comb./Syn.: *Eremaeus ornatus* Oudemans, 1900; *Cosmoppia ornata* (Balogh, 1983); *Dameosoma tricarinatum* Paoli, 1908; *Damaeosoma captator* Hull, 1915
Distribution: NL [Holarctic]
Literature: Maraun et al. (2019).

Graptoppia (*Stenoppia*) Balogh, 1983
Type-species: (*Oppia heterotricha* Bernini, 1969) =*Oppia italica* Bernini, 1973

Graptoppia (*Stenoppia*) *italica* (Bernini, 1973)
Comb./Syn.: *Oppia heterotricha* Bernini, 1969
Biology: From grass fields (Cianciolo and Norton 2006).
Distribution: AB [NY; Holarctic]
Note: Unidentified species from QC.

Lasiobelba Aoki, 1959
Type-species: *Lasiobelba remota* Aoki, 1959 (Figure 3.91H)
Note: Unidentified species from ON.

Microppia Balogh, 1983
Type-species: *Dameosoma minus* Paoli, 1908

KEY TO ADULT *MICROPPIA* OF CANADA AND ALASKA

1 Area between bothridia without sculpturing. Head of bothridial seta ovate with 3–4 barbs. Length: ca. 225 μm ....*M. simplissimus* (Jacot)
– Area between bothridia with sculpturing (pair of ridges). Head of bothridial seta ovate with 0–3 barbs. Length: 170–215μm (Figure 3.89B)................................*M. minus* (Paoli)

Microppia minus (Paoli, 1908) (Figure 3.89B)
Comb./Syn.: *Dameosoma minus* Paoli, 1908; *Oppia minus* (Paoli, 1908); *Oppiella minus* (Paoli, 1908)
Biology: From coniferous forest litter. In oak litter (Lamoncha and Crossley 1998).
Distribution: AK, NT, AB, MB, ON, QC [CA, DC, KY, NC; holarctic]
Literature: Maraun et al. (2019).

Microppia simplissimus (Jacot, 1938)
Comb./Syn.: *Oppia minus simplissimus* Jacot, 1938; *Opiella* [sic] *simplissimus* (Jacot, 1938)
Biology: From Douglas-fir litter; maple, beech forest litter.
Distribution: BC, AB, QC [IL; Nearctic]

Multioppia Balogh, 1965
Type-species: *Multioppia radiata* Hammer, 1961

Multioppia carolinae (Jacot, 1938) (Figure 3.91A,B)
   Comb./Syn.: *Oppia carolinae* Jacot, 1938
   Biology: From mixed deciduous coniferous litter.
   Distribution: ON, QC [NC; Nearctic]
   Note: Unidentified *Multioppia* species from BC, AB, NS.

Oppiella (***Moritzoppia***) Subías and Rodríguez, 1986
   Type-species: *Oppia keilbachi* Moritz, 1969
   Note: *Moritzoppia* is considered a subgenus of *Oppiella* by Miko (2006), whom we follow herein.

KEY TO ADULT *OPPIELLA (MORITZOPPIA)* OF CANADA AND ALASKA

1. Anterior margin of notogaster distinctly flattened; bothridial seta strongly curved medially, fusiform. Length: 280–310µm (Figure 3.91G) ....
    .... ***Oppiella (M.) translamellata*** (Willmann)
–   Anterior margin of notogaster gently flattened; bothridial seta straight, clavate. Length: 284–315µm (Figure 3.90A).. ***Oppiella (M.) clavigera*** (Hammer)

Oppiella (*Moritzoppia*) *clavigera* (Hammer, 1952) (Figure 3.90A)
   Comb./Syn.: *Oppia clavigera* Hammer, 1952; *Moritzella clavigera* (Hammer, 1952)
   Note: Subías and Arillo (2001) considered *clavigera* a subspecies of *Moritzoppia unicarinata* (Paoli, 1908); we retain it as a separate species, pending a morphological and molecular revision of *Oppiella*.
   Biology: From tussock tundra; boreal forest litter.
   Distribution: AK, YT, NT, NU, BC, AB, MB, ON [MT, OR; Nearctic]
   Literature: Colloff and Seyd (1991).

Oppiella (*Moritzoppia*) *translamellata* (Willmann, 1923) (Figure 3.91G)
   Comb./Syn.: *Dameosoma translamellata* Willmann, 1923; *Oppia translamellata* (Willmann, 1923)
   Note: We follow Miko (2006) in considering this species distinct from *Oppiella* (*Moritzoppia*) *neerlandica* (Oudemans, 1900).
   Biology: from tussock tundra heath; boreal mixedwood forest.
   Distribution: AK, YT, NT, NU, QC [Holarctic]
   Literature: Maraun et al. (2019).
   Note: Unidentified *Oppiella* (*Moritzoppia*) spp. from BC, AB.

Oppiella (***Oppiella***) Jacot, 1937
   Type-species: *Eremaeus novus* Oudemans, 1902

KEY TO ADULT *OPPIELLA (OPPIELLA)* OF CANADA AND ALASKA

–   Notogaster without humeral apophysis. Crista absent to weakly developed. Bothridial seta extending laterally. Length: 245–270µm (Figure 3.89H,I) ....
    .... ***Oppiella (O.) maritima*** (Willmann)
–   Notogaster with humeral apophysis. Crista present. Bothridial seta, curved strongly medially. Length: 220–320µm (Figure 3.89C,D) .... ***Oppiella (O.) nova*** (Oudemans)
    Note: Unidentified *Oppiella* species from BC, AB.

Oppiella (*Oppiella*) *maritima* (Willmann, 1929) (Figure 3.89H,I)
   Comb./Syn.: *Dameosoma falcatum maritima* Willmann, 1929; *Oppia maritima* (Willmann, 1929); *Oppia fissurata* Hammer, 1952; *Lauroppia maritima* (Willmann, 1929)
   Biology: From boreal forest; tussock tundra.
   Distribution: AK, YT, NT, NU, QC [Holarctic]
   Literature: Miko (2006); Maraun et al. (2019).

Oppiella (*Oppiella*) *nova* (Oudemans, 1902) (Figure 3.89C,D)
   Comb./Syn.: *Eremaeus novus* Oudemans, 1902; *Oppia nova* (Oudemans, 1902); *Dameosoma corrugatum* Berlese, 1904; *Dameosoma uliginosum* Willmann, 1919; *Oppia uliginosum* (Willmann, 1919); *Dameosoma neerlandicum* sensu Sellnick, 1928; non Oudemans, 1900; *Oppia neerlandica* sensu Willmann, 1931; non Oudemans, 1900; *Oppia washburni* Hammer 1952
   Biology: Feeds preferentially on *Aspergillus niger* and to a lesser extent other fungi (*Phialophora*, *Cladosporium*, *Hormodendrum* and *Stemphylium*), but not on fresh or decaying leaves, needles or wood (Hartenstein 1962). Surprisingly, *A. niger* and an unknown species of *Penicillium* repelled all oribatids except *O. nova* and some other Oppiidae, and both the adult and its larvae fed upon this antibiotic-producer. Walter (1987) cultured this species through several generations on algae and also fungi. Woodring and Cook (1962) reared this species on lichen. Specimens from forest and grassland, assessed genetically showed that these habitats were inhabited by distinct genetic lineages with transitional habitats being colonized by both genetic lineages; notably, individuals of grasslands were significantly larger than individuals in forests (von Saltzwedel et al. 2014). Brandt et al. (2021) provided genetic evidence for long-term asexuality in this species.
   Distribution: AK, YT, NT, NU, BC, AB, MB, ON, QC, NB, NL [CA, FL, KY, MI, MO, MT, NH, NY, TX, UT, VA; Cosmopolitan]
   Literature: Maraun et al. (2019); ABMI (2019).

Oppiella (*Rhinoppia*) Balogh, 1983
   Type-species: *Oppia nasuta* Moritz, 1965
   Comb./Syn.: *Lauroppia* Subías and Mínguez, 1986; *Medioppia* Subías and Mínguez, 1985: Miko (2006)
   Note: Unnamed species from BC, AB (as *Lauroppia*).

Oppiella (*Rhinoppia*) *subpectinata* (Oudemans, 1900) (Figure 3.89F,G)
   Comb./Syn.: *Eremaeus subpectinatus* Oudemans, 1900, *Dameosoma subpectinatum* (Oudemans, 1900), *Oppia subpectinata* (Oudemans, 1900), *Dameosoma clavipectinatum* sensu Paoli, 1908; non Michael, 1885; *Medioppia subpectinata* (Oudemans, 1900)
   Biology: From maple, beech, conifer litter.
   Distribution: QC [NY, TN; Holarctic]
   Literature: Miko (2006); Maraun et al. (2019); Brandt et al. (2021).

Oppia C.L. Koch, 1835
   Type-species: *Oppia nitens* C.L. Koch, 1835
   Comb./Syn.: *Dameosoma* Berlese, 1892; *Dissorhina* Hull, 1916; *Zetobelba* Hull, 1916

KEY TO ADULT *OPPIA* OF CANADA AND ALASKA

1 Bothridial seta fusiform, about half length of prodorsum. Length: 470–540µm (Figure 3.91C,D) .... *O. nitens* C.L. Koch
– Bothridial seta slightly pectinate, about equal to length of prodorsum. Length: ca. 460µm .... *O. rigida* (Ewing)
   Note: Unidentified *Oppia* species from BC, AB, MB, ON, QC, NB, NS, PE.

Oppia minuta (Banks, 1895)
>   **Comb./Syn.**: *Belba minuta* Banks, 1895; *Oribata minuta* (Banks, 1895); *Oppia perolata* (Banks, 1909)
>   **Biology**: from mixed deciduous coniferous forest litter, White birch and Trembling aspen litter; corn stubble.
>   **Distribution**: ON [IL, NJ, NY; Nearctic]
>   **Note**: This species is probably a junior synonym of *O. nitens*, as indicated by Ewing (1909), though Marshall et al. (1987) retained it as a distinct species. The illustration by Banks (1904) looks like *O. nitens*. As there are no further images, we do not key this species.

Oppia nitens C.L. Koch, 1835 (Figure 3.91C,D)
>   **Comb./Syn.**: *Damaeus nitens* (C.L. Koch, 1835); *Dameosoma nitens* (C.L. Koch, 1835)
>   **Biology**: From forest litter; freshly decaying leaves. *Aspergillus flavus* is toxic to vertebrates yet Shereef (1970) observed that it was a favorite diet for this species. Farahat (1966) reported fungal spores commonly found packing the guts of *O. nitens*. This species is used as a soil toxicity test species (Princz et al. 2010, 2018; Owojori and Siciliano 2012; Ardestani et al. 2020; Fajana et al. 2020).
>   **Distribution**: ON, NL [IL, NY; Holarctic]

Oppia rigida (Ewing, 1909)
>   **Comb./Syn.**: *Damaeus rigida* Ewing, 1909
>   **Biology**: From forest litter; freshly decaying leaves; *Sphagnum*.
>   **Distribution**: ON, QC? [IL, NH; Nearctic]

Ramusella Hammer, 1962
>   **Type-species**: *Ramusella puertomonttensis* Hammer, 1962 (Figure 3.91E)
>   **Note**: Miko (2006) placed *Ramusella* (*Insculptoppia*) Subías, 1980 and *Ramusella* (*Rectoppia*) Subías, 1980 in synonymy with *Ramusella* and we follow this author herein.
>   Unidentified *Ramusella* spp. from BC, AB, NS.

Ramusella clavipectinata (Michael, 1885)
>   **Comb./Syn.**: *Notaspis clavipectinata* Michael, 1885, *Dameosoma clavipectinatum* (Michael, 1885); *Eremaeus clavipectinata* (Michael, 1885); *Oppia clavipectinata* (Michael, 1885); *Xenillus clavipectinatus* (Michael, 1885)
>   **Distribution**: NS? [CA; Holarctic]
>   **Literature**: Miko (2006).

Subiasella Balogh, 1983
>   **Comb./Syn.**: *Subiasella* (*Lalmoppia*) Subías and Rodríguez
>   **Type-species**: *Oppia exiguus* Hammer, 1971

Subiasella (*Lalmoppia*) *maculata* (Hammer, 1952) (Figure 3.90E)
>   **Comb./Syn.**: *Oppia maculata* Hammer, 1952 (=*Oppia ventronodosa* Hammer, 1962)
>   **Biology**: Abundant in *Sphagnum*, a few in moss in wet biotopes; cushion of reindeer lichen on rocks; heath-like vegetation with *Salix*, *Arctostaphylos*, *Vaccinium*, *Rhododendron*, *Dryas*, *Polygonum*.
>   **Distribution**: AK, YT, NT, MB, QC [Nearctic]
>   **Note**: Unidentified *Subiasella* spp. from AB, QC, NS, NL.

Family Quadroppiidae Balogh, 1983
>   **Diagnosis**: *Adults*: Non-poronotic Brachypylina. Rostral structure (ridge-like, posteriorly directed) present or absent. Costula convergent anteriorly, concave and trapezoid, with strongly sclerotized cusps; forming or not elongate closed ellipsoidal ring, horseshoe-shaped. Small translamella present. Lamellar seta positioned on prodorsum below lamellar cusp. Pleural

ridges present laterally. Lateral carina present above legs I and II, possibly homologous to tutorium, ending in triangular tip. Pedotectum I large; pedotectum II absent. Notogaster arched in dorsal view, anterior border straight; with two pairs of cristae arising from elevated lateral and median knots on dorsosejugal scissure. With groove between median and lateral cristae, continuing as circumpleural cuticular furrow around dorsal of notogaster. Notogastral setation 9 pairs; c_2 on lateral edge of median knots. Anterior sternal groove present on epimere I; posterior sternal groove present on epimera III and IV anterior to genital plates. Epimeral setation 3-1-3-3. Genital setation 5 pairs. Legs monodactylous. Tarsus II with 1 or 2 solenidia.

Quadroppia Jacot, 1939
Type-species: *Notaspis quadricarinata* Michael, 1885

KEY TO ADULT *QUADROPPIA* OF CANADA AND ALASKA

1	Notogastral cristae continuing posteriorly and joining posteromedially to form horse-shoe shape. Length: ca. 165 µm	*Q. ferrumequina* (Jacot)
–	Notogastral cristae continuing posteriorly at most 3/4 length notogaster; not joining posteromedially	2
2(1)	Rostrum sculptured anterior of translamella. Length: 190–230µm (Figure 3.92A,B)	*Q. quadricarinata* (Michael)
–	Rostrum smooth anterior of translamella. Length: ca. 180µm (Figure 3.92C)	*Q. skookumchucki* Jacot

Quadroppia ferrumequina (Jacot, 1938)
  Comb./Syn.: *Oppia quadricarinata ferrumequina* Jacot, 1938
  Biology: From Douglas fir litter.
  Distribution: BC [NC; Nearctic]

Quadroppia quadricarinata (Michael, 1885) (Figure 3.92A,B)
  Comb./Syn.: *Notaspis quadricarinata* Michael, 1885; *Dameosoma quadricarinatum* (Michael, 1885); *Oppia quadricarinata* (Michael, 1885)
  Biology: From forest soils.
  Distribution: AK, YT, NT, NU, BC, AB, MB, ON, QC [CA, MO, MT, NY, OR; Cosmopolitan]
  Literature: Miko (2006); Maraun et al. (2019).

Quadroppia skookumchucki Jacot, 1939 (Figure 3.92C)
  Distribution: QC? (Déchêne and Buddle 2010) [NH; Nearctic]

Family Thyrisomidae Grandjean, 1954 [=Banksinomidae Kunst, 1971; Oribellidae Kunst, 1971; Pantelozetidae Kunst, 1971]
  Diagnosis: *Adults*: Non-poronotic Brachypylina. Surface of body smooth. Prodorsum with narrow ridge-like lamellae without prominent cusp. Pedotectum I present; pedotectum II absent. Tutorium absent. Dorsosejugal scissure present. Dorsophragmata and pleurophragmata absent. Notogaster with 10, 11 or 14 pairs of setae. Circumpedal carina, custodium, and discidium absent. Anal and genital plates large, closely adjacent. Genital setation 5–6 pairs. Paired aggenital taenidium and minitectum present. Ventral plate without neotrichy. Epimeral setation 3-1-3-3. Epimera III and IV delineated, or not, by distinct epimeral border III. Subcapitulum diarthric; eupathidium *acm* separate from recumbent solenidion; chelicerae chelate-dentate. Legs moniliform; monodactylous, occasionally tridactylous. Palptarsus with 8–9 setae. Trochanter III with 1–2 setae; tarsus II with 2 solenidion.

KEY TO GENERA OF THYRISOMIDAE OF CANADA AND ALASKA

1 Epimera III and IV not delineated by distinct epimeral border III. Ridge-like lamellae strongly converging, shorter than half length of prodorsum, almost meeting apically or with spines between apices (Figure 3.93A–E) .... ***Banksinoma*** Oudemans
– Epimera III and IV delineated by distinct epimeral border III. Ridge-like lamellae weakly converging; apically far removed from each other (Figure 3.93F–H)..................................
.... ***Pantelozetes*** Grandjean

Banksinoma Oudemans, 1930
    Comb./Syn.: *Thyrisoma* Grandjean, 1953
    Type-species: *Notaspis castaneus* Hermann, 1804 sensu Oudemans 1930 (=*Notaspis lanceolata* Michael, 1885)

KEY TO *BANKSINOMA* OF CANADA AND ALASKA

1 Interlamellar area subtriangular, smooth between apices 2
– Interlamellar area trapezoidal, ending in field of denticles. Length: ca. 400µm (Figure 3.93A,B) ....***B. spinifera*** (Hammer)

2(1) Interlamellar seta short, ca. 1/2 length of lamellar seta. Bothridial seta acuminate distally. Rostrum with distal mucro 3
– Interlamellar seta long, subequal to lamellar seta. Bothridial seta spindle-shaped, tapered distally. Rostrum rounded, without mucro. Length: 442–466µm (Figure 3.93C) ....
....***B. setosa*** (Ryabinin)

3(2) Rostral mucro rounded. Length: ca. 350µm (Figure 3.93D) ....***B. lanceolata*** (Michael)
– Rostral mucro acuminate. Length: ca. 350µm (Figure 3.93E)...
....***B. lanceolata canadensis*** Fujikawa
Note: Unidentified species from AK, BC, NB, NS.

Banksinoma lanceolata canadensis Fujikawa, 1979 (Figure 3.93E)
    Biology: Gravid females carry 1 or 2 large eggs. Gut contents include hyaline to brown hyphae and fine to coarse organic material (Walter et al. 2014).
    Distribution: YT, NT, AB, ON, QC [Canada]
    Literature: ABMI (2019).

Banksinoma lanceolata (Michael, 1885) (Figure 3.93D)
    Comb./Syn.: *Notaspis lanceolata* Michael, 1885, *Oribella lanceolata* (Michael, 1885), *Oribella castanea* sensu Willmann, 1931; non Hermann, 1804; *Xenillus castaneus* sensu Sellnick, 1928; non Hermann, 1804
    Biology: From tussock tundra; Western redcedar trunk; mixed deciduous coniferous forest; Balsam fir forest.
    Distribution: AK, YT, NT, BC, ON, QC, NS, NL [Holarctic]
    Literature: Weigmann (2006); Ahaniazad et al. (2016); Maraun et al. (2019).

Banksinoma setosa Ryabinin, 1974 (Figure 3.93C)
    Distribution: QC [Holarctic]

Banksinoma spinifera (Hammer, 1952) (Figure 3.93A,B)
    Comb./Syn.: *Oribella spinifera* Hammer, 1952; *Thyrisoma spinifera* (Hammer, 1952)
    Biology: From Boreal forest with White spruce, Black spruce, Jack pine and Trembling aspen; moss, lichens.

Distribution: YT, AB, MB, QC, NS [CA, NM; Nearctic; Neotropical]
Literature: Bayartogtokh (2006); ABMI (2019).

Pantelozetes Grandjean, 1953
Type-species: *Xenillus paolii* Oudemans, 1913
Comb./Syn.: *Xenillus paolii* Oudemans, 1913; *Oribella paolii* (Oudemans, 1913); *Eremaeus pectinatus* sensu Oudemans, 1900

KEY TO ADULT *PANTELOZETES* OF CANADA AND ALASKA

1   Tridactylous. Rostral margin without teeth. Length: 410–455μm (Figure 3.93F,G).............
.... ***P. alpestris*** (Willmann)
–   Monodactylous. Rostral margin toothed. Length: 360–425μm (Figure 3.93H).................
....***P. paolii*** (Oudemans)
     Note: Unidentified *Pantelozetes* species from AK, YT, BC, AB.

Pantelozetes alpestris (Willmann, 1929) (Figure 3.93F,G)
Comb./Syn.: *Xenillus alpestris* Willmann, 1929; *Montizetes alpestris* (Willmann, 1929)
Biology: From peatland; wood, lichen, grass.
Distribution: AK [Holarctic]
Literature: Ermilov et al. (2015).

Pantelozetes paolii (Oudemans, 1913) (Figure 3.93H)
Comb./Syn.: *Xenillus paolii* Oudemans. 1913
Distribution: AK [Holarctic]
Literature: Ermilov et al. (2015); Maraun et al. (2019).

Superfamily Trizetoidea Ewing, 1917

Family Suctobelbidae Jacot, 1938
     Diagnosis: *Adults*: Non-poronotic Brachypylina. Rostrum with or without teeth. Prodorsum with or without costula, or lamellar knobs; large, paired flat regions (**tectopedial fields**) present or absent. Tutorium, custodium, circumpedal carina absent; discidium small or absent. Dorsophragmata and pleurophragmata absent. Body with constriction in sejugal region. Dorsosejugal scissure present. Notogaster often with tubercles or condyles anteriorly; notogastral setation 9–14 pairs. Humeral apophysis absent. Pedotectum I usually present, small, scalelike; pedotectum II absent. Epimera III and IV fused. Epimeral setation 3-1-3-3, or epimera III and IV neotrichous. Genital setation 4–7 pairs. Subcapitulum anarthric; palptarsus with 8–9 setae; eupathidium *acm* separate from recumbent solenidion; chelicerae and rutella modified (suctobelbid type), with chelicerae attenuate and edentate or with fine teeth distally; 2 cheliceral setae; axillary saccule absent. Legs moniliform; monodactylous. Trochanter III with 1–2 setae; tarsus II with 2 solenidia; proral setae of tarsi II–IV short and spinelike.

KEY TO GENERA OF SUCTOBELBIDAE OF CANADA AND ALASKA

1   Anterior margin of notogaster with 1–2 pairs of tubercles or cristae (Figure 3.95, 3.96).....
.... ***Suctobelbella*** Jacot
–   Anterior margin of notogaster without tubercles or crista 2

2(1)   Rostrum almost twice as long as wide; with 1 pair of large lateral teeth, elongately protruding (Figure 3.94A–C) .... ***Rhinosuctobelba*** Woolley and Higgins
–   Rostrum only a little longer than wide; without lateral teeth or with more than 1 pair 3

3(2)	Tectopedial fields absent (Figure 3.94D)...***Rhynchobelba*** Willmann
–	Tectopedial fields (pair of flattened elliptical areas on prodorsum) present 4

4(3)	Rostrum with distinct lateral incision, with 2–3 pairs of teeth laterally. Bothridium with posterior tubercle; body small, 190–290μm (Figure 3.94G)..................... ***Suctobelba*** Paoli
–	Rostrum without lateral incision, with 1 or 3–4 pairs of teeth laterally. Bothridium without posterior tubercle; body larger: >350μm (Figure 3.94E,F)............... ***Allosuctobelba*** Moritz

Allosuctobelba Moritz, 1970
    Type-species: *Suctobelba grandis* Paoli, 1908

Allosuctobelba gigantea (Hammer, 1955) (Figure 3.94E,F)
    Comb./Syn.: *Suctobelba gigantea* Hammer, 1955
    Distribution: AK, AB [Nearctic]
    Literature: Walter et al. (2014); ABMI (2019).
    Note: Unidentified species from BC, AB, ON, QC, NL.

Rhinosuctobelba Woolley and Higgins, 1969
    Type-species: *Rhinosuctobelba dicerosa* Woolley and Higgins, 1969

Rhinosuctobelba dicerosa Woolley and Higgins, 1969 (Figure 3.94A–C)
    Biology: From coastal temperate rainforest.
    Distribution: BC [WA; Nearctic]

Rhynchobelba Willmann, 1953
    Type-species: *Rhynchobelba inexpectata* Willmann, 1953
    Note: Unidentified species from BC.

Suctobelba Paoli, 1908
    Type-species: *Notaspis trigona* Michael, 1888
    Note: Unidentified species from AB, QC, NS.

Suctobelbella Jacot, 1937
    Type-species: *Suctobelbella serratirostrum* Jacot, 1937

KEY TO ADULT *SUCTOBELBELLA* OF CANADA AND ALASKA

1	Notogastral setae *la* and *lp* extending beyond insertion of next more posterior seta..........2
–	Notogastral setae *la* and *lp* not reaching insertion of next more posterior seta 3

2(1)	Notogastral setae long, setae *la* and *lp* extending 3/4 length beyond insertion of next most posterior seta. Rostrum with lateral tooth. Length: ca. 246μm (Figure 3.95A,B) ....***S. frothinghami*** Jacot
–	Notogastral setae *la* and *lp* extending at most 1/2 length beyond insertion of next most posterior seta. Rostrum with 2 lateral teeth. Length: ca. 200μm (Figure 3.95C,D) .... ***S. hurshi*** Jacot

3(1)	Bothridial seta spindle-shaped, with narrowed tip subequal in length to expanded area. Length: 260μm (Figure 3.95E)..***S. setosoclavata*** (Hammer)
–	Bothridial seta fusiform or clavate, not spindle shaped 4

4(3)	Bothridial seta very long, narrowly fusiform. Length: ca. 320μm (Figure 3.95F-H) ....***S. punctata*** (Hammer)
–	Bothridial seta broadly fusiform to clavate. Length: 170–310μm 5

5(4) Tectopedial field and anterior of prodorsum granulate. Length: 195–220µm (Figure 3.95I,J) .... *S. acutidens* (Forsslund)
– Tectopedial field smooth, at most medial and anterior area of prodorsum granulate 6

6(5) Rostrum elongated; without medial incision. Length: 262–308µm (Figure 3.95K,L) ....
.... *S. longirostris* (Forsslund)
– Rostrum not elongated. Rostrum with V or U-shaped medial incision 7

7(6) Rostrum with U-shaped medial incision, with flattened tooth lateral to incision 8
– Rostrum with V-shaped incision, or gently curved 9

8(7) Rostrum with 2 distinct teeth posterior to flattened tooth. Bothridial seta fusiform with barbs laterally. Length: 180–225µm (Figure 3.96A) .... *S. arcana* Moritz
– Rostrum with 3 distinct teeth posterior to flattened tooth. Bothridial seta fusiform, barbed laterally and ventrally. Length: 190–245µm (Figure 3.96B) .... *S. sarekensis* (Forsslund)

9(7) Rostrum with very long, narrow tooth laterally, extending anteriorly of rostrum in dorsal view. Bothridial seta fusiform, obliquely truncate. Length: ca. 200µm (Figure 3.96C,D) ...
.... *S. longicuspis* Jacot
– Rostrum with 3 lateral teeth, none very long, narrow. Bothridial seta fusiform, not obliquely truncate. Length: 170–270µm 10

10(9) Bothridial seta long narrowly fusiform. Rostral teeth very small, separated by shallow incision same width as teeth. Length: 210–230µm (Figure 3.96E–G) ....
.... *S. hammerae* (Krivolutsky)
– Bothridial seta broadly fusiform to clavate. Rostral teeth of variable length 11

11(10) Bothridial seta clavate. Length: 230–270µm (Figure 3.96H,I) .... *S. palustris* (Forsslund)
– Bothridial seta fusiform with row of spines laterally. Length: 170–190µm (Figure 3.96J,K) .... *S. laxtoni* Jacot
Note: *S.* nr. *subcornigera* (Forsslund 1941) from BC. At least 20 unidentified species from BC, AB, QC, NB, NS, NL.

Suctobelbella acutidens (Forsslund, 1941) (Figure 3.95I,J)
    Comb./Syn.: *Suctobelba acutidens* Forsslund, 1941
    Biology: From tussock heath.
    Distribution: AK, YT, NT, NU, BC, QC [Holarctic]
    Literature: Maraun et al. (2019).
    Note: *S.* nr. *acutidens* from BC, AB.

Suctobelbella arcana Moritz, 1970 (Figure 3.96A)
    Biology: Found in peatlands (Borcard 1994).
    Distribution: AB [Japan]

Suctobelbella frothinghami Jacot, 1937 (Figure 3.95A,B)
    Comb./Syn.: *Suctobelba forthinghami* [sic] (Jacot, 1937)
    Biology: From maple litter.
    Distribution: QC [VA, NC; Nearctic]

Suctobelbella hammerae (Krivolutsky, 1965) (Figure 3.96E–G)
    Comb./Syn.: *Suctobelba hammerae* Krivolutsky, 1965; *Suctobelba sarekensis* sensu Hammer 1952a; non Forsslund 1941
    Biology: From lichen, moss; tussock tundra; peatlands.
    Distribution: AK, YT, NT, NU, MB, QC [Holarctic]

Suctobelbella hurshi Jacot, 1937 (Figure 3.95C,D)
   Comb./Syn.: *Suctobelba hurshi* (Jacot, 1937)
   Biology: From grass fields, shrub and forest (Cianciolo and Norton 2006). From *Carex* fen.
   Distribution: ON, QC [NC, NY; Nearctic]

Suctobelbella laxtoni Jacot, 1937 (Figure 3.96J,K)
   Comb./Syn.: *Suctobelba laxtoni* (Jacot, 1937)
   Biology: From *Sphagnum* fen; from forest litter.
   Distribution: BC, ON, QC [NC; Nearctic]

Suctobelbella longicuspis Jacot, 1937 (Figure 3.96C,D)
   Comb./Syn.: *Suctobelba longicuspis* (Jacot, 1937)
   Biology: From forest litter.
   Distribution: QC [VA, NC; Nearctic]

Suctobelbella longirostris (Forsslund, 1941) (Figure 3.95K,L)
   Comb./Syn.: *Suctobelba longirostris* Forsslund, 1941
   Biology: From peatlands (Borcard 1994).
   Distribution: AK, ON [Holarctic]
   Literature: Blackford et al. 2014.

Suctobelbella palustris (Forsslund, 1953) (Figure 3.96H,I)
   Comb./Syn.: *Suctobelba palustris* Forsslund, 1953
   Biology: From grass fields and shrubs (Cianciolo and Norton 2006). From *Sphagnum* and reed beds.
   Distribution: NT, BC, ON [NY, TN, VA; Holarctic]

Suctobelbella punctata (Hammer, 1955) (Figure 3.95F-H)
   Comb./Syn.: *Suctobelba punctata* Hammer, 1955
   Biology: From lichen, grass.
   Distribution: AK, AB [Nearctic]
   Literature: ABMI (2019).

Suctobelbella sarekensis (Forsslund, 1941) (Figure 3.96B)
   Comb./Syn.: *Suctobelba sarekensis* Forsslund, 1941, *nomen novum* for *Suctobelba cornigera* sensu Trägårdh, 1910; non Berlese, 1902
   Biology: From forest litter; *Carex* fen.
   Distribution: AK, NT, MB, ON [VA, Holarctic]
   Literature: Beckmann (1988); Hägvar et al. (2009).

Suctobelbella setosoclavata (Hammer, 1952) (Figure 3.95E)
   Comb./Syn.: *Suctobelba setosoclavata* Hammer, 1952; *Suctobelbella setosclavata* [sic] (Hammer, 1952)
   Biology: From tussock heath, reindeer moss.
   Distribution: AK, YT, NT, NU [Holarctic]
   Literature: Leonov and Rakhleeva (2020).

Superfamily Tectocepheoidea Grandjean, 1954

Family Tectocepheidae Grandjean, 1954
   Diagnosis: *Adults*: Non-poronotic Brachypylina. Granular cerotegument present. Prodorsum with lamellae and lamellar cusps, translamella present or absent. Genal incision present.

Dorsophragmata and pleurophragmata present. Dorsosejugal scissure usually incomplete. Discidium, custodium, and postanal porose area absent; circumpedal carina incomplete. Pedotectum I large, extending to insertion of exobothridial seta, or small, barely covering acetabulum I; pedotectum II present. Notogastral setation 10 pairs (rarely 11 pairs). Pteromorphs present, small; without hinge, or absent; posterior tectum absent. Genital setation 5–6 pairs. Anal plates strongly triangular. Chelicerae chelate-dentate; palp eupathidium *acm* and solenidion incompletely coupled; subcapitular mentum without tectum. Axillary saccule of subcapitulum usually absent.

Nemacepheus Aoki, 1968
Type-species: *Nemacepheus dentatus* Aoki, 1968 (Figure 3.63A–C)
Note: This species, known from canopy habitats was included in Tectocepheidae by Aoki (1968), and Balogh and Balogh (1992), but in Nodocepheidae by Fujikawa (2001) and Subías (2004). The family concept has varied among authors and needs reassessment.

Nemacepheus dentatus Aoki, 1968
Distribution: BC [OR, Japan].

Tectocepheus Berlese, 1896
Type-species: *Tegeocranus velatus* Michael, 1880
Note: Using nucleotide sequences of 18S, 28S, *ef-1a* and *hsp82* Laumann et al. (2007) showed *Tectocepheus minor*, *T. velatus* and *T. sarekensis* to be distinct species.

KEY TO ADULT *TECTOCEPHEUS* OF CANADA AND ALASKA

1   Lamellar cusp weakly tapered. Notogaster with one pair of depressions. Length: 280–317µm (Figure 3.97A–D) ....*T. velatus* (Michael)
–   Lamellar cusp not tapered. Notogaster with with 4 pairs of depressions. Length: 302–349µm (Figure 3.97E,F) ....*T. sarekensis* Trägårdh

Tectocepheus sarekensis Trägårdh, 1910 (Figure 3.97E,F)
Comb./Syn.: *Tectocepheus velatus sarekensis* Trägårdh, 1910; *Tectocepheus velatus angulatus* Mihelčič, 1957; *Tectocepheus velatus ibericus* Mihelčič, 1957; *Tectocepheus velatus inflexus* Mihelčič, 1957
Biology: Parthenogenetic; gravid females carry 3 large eggs (Domes et al. 2007). This species can be used as a bioindicator for Zn pollution in soils (Stamou and Argyropoulou 1995). Median lethal time submerged under water was 80 days; activity was reduced to a certain extent but not completely ceased, animals either sat quietly on the substrate or moved around normally (Bardel and Pfingstl 2018).
Distribution: AK, YT, BC, AB [NC, NH; Holarctic]; known from late Pliocene/early Pleistocene fossil deposits in AK and from Tertiary fossil sites on Meigen Island (NU) (Matthews et al. 2019).
Literature: Maraun et al. (2019); ABMI (2019).

Tectocepheus velatus (Michael, 1880) (Figure 3.97A–D)
Comb./Syn.: *Tegeocranus velatus* Michael, 1880; *Scutovertex velatus* (Michael, 1880)
Biology: Parthenogenetic; gravid females carry 2 large eggs. Specimens have been found in the feathers of birds in Russia (Krivolutsky and Lebedeva 2004). *T. velatus* is among the earliest colonists of new soil exposed by retreating glaciers (Hågvar et al. 2009). Observed to feed on fungi (Riha 1951; Anderson 1975). Stable isotope analysis indicated that this species is a primary decomposer (Schneider et al. 2004). Associated with this, it has chelicerae with a high leverage index (Perdomo et al. 2012). Cysticercoids were found in this species (Fritz 1995).

Distribution: AK, YT, NT, NU, BC, AB, SK, MB, ON, QC, NB, NS, PE, NL [CA, FL, IL, KY, MI, MN, MT, NC, NH, NM, NY, OR, VA; Cosmopolitan]
Literature: Iordansky and Stein-Margolina (1993); Fujikawa (1988, 1999); Maraun et al. (2019); ABMI (2019).

Superfamily Limnozetoidea Thor, 1937

Family Hydrozetidae Grandjean, 1954
Diagnosis: *Adults*: Non-poronotic Brachypylina. Prodorsum with ridge-like lamellae, without lamellar cusps. Tutorium absent. Genal incision absent. Dorsophragmata and pleurophragmata present. Bothridial seta present, reduced or absent (frequently broken). Notogaster with 10–11 or 15–17 pairs of notogastral setae (if more than 15, several are neotrichous *h* setae); pteromorphs absent; lenticulus present. Epimera II–IV clearly delineated; epimeral borders II–III often fused anterolateral of genital plates, and meeting epimeral border IV at, or anterior to, genital aperture. Subcapitulum diarthric. Palpal eupathidium associated with solenidion, forming 'double-horn'. Chelicerae chelate-dentate. Axillary saccule of subcapitulum absent. Seta *d* inserted on proximal fifth of femora I–III, proximal to other femoral setae.

Hydrozetes Berlese, 1902
Type-species: *Notaspis lacustris* Michael, 1882
Biology: Parthenogenetic or sexual; some species show sexual dimorphism (Behan-Pelletier 2015). Members are considered truly aquatic (i.e., reproduce in water and all stages live in fresh water or at its margins) (Schatz and Behan-Pelletier 2008; Seniczak et al. 2009). Immature *Hydrozetes* generally burrow within aquatic vegetation such as duckweed (*Lemna* spp.) and adults feed on the surface of vegetation. Large numbers can be collected from wet moss on lake margins, muskeg, and other peatlands. Members of this genus are often present in palaeoecology studies, and can prove to be important in interpreting past environments (Erickson et al. 2003). An undescribed Palaeocene (~58 mya) fossil species of *Hydrozetes* is known from the Paskapoo Formation in the Red Deer River Valley of Alberta; others are known from Lower Jurassic deposits (~190 mya) in Sweden and Oligocene-Miocene deposits in Mexico (~23 mya) (Baker and Wighton 1984).
Literature: Krause et al. (2016); Lehmitz and Decker (2017).

KEY TO ADULT *HYDROZETES* OF CANADA AND ALASKA

1 Tarsus IV with 2 claws, empodial claw thick, lateral claw very thin. Bothridial seta well developed, clavate. Genital setation 6 pairs 2
– Tarsus IV with only empodial claw, lateral claws absent. Bothridial seta weakly developed, setiform or absent. Genital setation 7 or more pairs (rarely 6 unilaterally) 3

2(1) Notogastral seta *da* not reaching base of seta *dm*. Femur IV with 3 setae. Length: 515–640μm (Figure 3.98A) .... *H. thienemanni* Strenzke
– Notogastral seta *da* reaching base of seta *dm*; seta *dm* reaching insertions of *dp*. Femur IV with 2 setae. Length: 408–512μm .... *H. lemnae* (Coggi)
Note: Known from Northeast USA (NY); undoubtedly in Canada.

3(1) Notogastral seta *lm* inserted posterior to opisthonotal gland opening (*gla*), seta *lm* very close to *lp*; three pairs of *h*-series setae present. Length: 480–515μm (Figure 3.98D).........
.... *H. octosetosus* Willmann
– Notogastral seta *lm* inserted medial or anterior to *gla*; *lm* and *lp* more distant; setae in *h*-series various 4

4(3) Three pairs of *h*-series setae present. Length: 450–510μm (Figure 3.98E) ....
....*H. lacustris* (Michael)
– More than three *h*-series setae present on each side (maybe asymmetrical). Length: 450–500μm (Figure 3.98F) ....*H. parisiensis* Grandjean
Note: Unidentified specimens known from late Pliocene/early Pleistocene fossil deposits in AK, and from Tertiary fossil sites on Prince Patrick, Meigen and Ellesmere Islands (NU) (Matthews et al. 2019). Unnamed and undescribed *Hydrozetes* species known from AK, YT, AB, QC, NB, NS.

Hydrozetes lacustris (Michael, 1882) (Figure 3.98E)
Comb./Syn.: *Notaspis lacustris* Michael, 1882; *Scutovertex lacustris* (Michael, 1882); *Notaspis speciosus* Chinaglia, 1917
Distribution: BC, MB [Holarctic]
Literature: Maraun et al. (2019); Seniczak et al. (2019).

Hydrozetes lemnae (Coggi, 1897)
Comb./Syn.: *Notaspis lemnae* Coggi, 1897
Biology: Population dynamics on duckweed were studied by Athias-Binche and Fernandez (1986). Fine structure of the lenticulus was elucidateded by Alberti and Fernandez (1988). Known to infest farmed eels in New Zealand (Fan and Heath 2019).
Distribution: Cosmopolitan
Literature: Maraun et al. (2019); Seniczak et al. (2019).

Hydrozetes octosetosus Willmann, 1931 (Figure 3.98D)
Comb./Syn.: *Hydrozetes lacustris octosetosus* Willmann, 1931
Distribution: AB [Holarctic]
Literature: Maraun et al. (2019); ABMI (2019); Seniczak et al. (2019).

Hydrozetes parisiensis Grandjean, 1948 (Figure 3.98F)
Comb./Syn.: *Hydrozetes lacustris parisiensis* Grandjean, 1948
Distribution: AB [Holarctic]
Literature: Deichsel (2005); Weigmann (2006); Seniczak et al. (2009); Walter et al. (2014).

Hydrozetes thienemanni Strenzke, 1943 (Figure 3.98A)
Comb./Syn.: *Hydrozetes incisus* Grandjean, 1948
Distribution: AK [Holarctic]
Literature: Seniczak and Seniczak (2008); Walter et al. (2014).

Family Limnozetidae Grandjean, 1954
Diagnosis: *Adults*: Non-poronotic Brachypylina. Prodorsum with lamella generally narrow, occasionally wide; with or without cusp; translamella present or absent. Bothridial seta reduced or absent (frequently broken). Taenidium present on paraxial surface of well-developed genal process. Genal incision present. Tutorium present, with or without tooth distally. Pedotectum I large, with convex dorsal margin, covering acetabulum I and most of exobothridial seta. Pedotectum II present. Discidium present; custodium absent. Circumpedal carina incomplete. Porose area *Ah* present, *Am* and *Al* absent. Dorsophragmata and pleurophragmata present. Notogastral setation 10 pairs; lenticulus absent. Narrow to wide pteromorphs present, strongly curved ventrally, without line of desclerotization; posterior tectum absent. Epimeral setation generally 3-1-2-2, occasionally 3-1-3-3; epimeral apodeme IV clearly evident; epimeral apodeme III fusing with sejugal apodeme. Genital setation 6–7 pairs. Chelicera

chelate-dentate. Palp setation: 0-2-1-3-9(1); eupathidia on palptarsus short, with *acm* fused to solenidion distally, and seta *l'* barbed, setose. Axillary saccule absent. Legs heterotridactylous. Seta *d* inserted on proximal fifth of femora I–III, proximal to other femoral setae. Genu IV with 0 or 2 setae.

Limnozetes Hull, 1916
   Type-species: *Acarus ciliatus* Schrank, 1803 (=*Oribata sphagni* Michael, 1884)
   Comb./Syn.: *Vietobates* Mahunka, 1987
   Biology: Species of *Limnozetes* of eastern Canada were discovered as a result of intensive studies on the oribatid mites of semiaquatic habitats (Behan-Pelletier 1989a; Behan-Pelletier and Bissett 1994). Similar, targeted collecting has yet to be undertaken in these habitats in other parts of North America. The diversity of *Limnozetes* species found in such a defined and seemingly simple habitat as *Sphagnum* bogs suggests that some species in this genus may be useful indicators of subtle changes in this habitat, e.g., climate warming (Barreto et al. 2021). In addition, members of this genus are often present in palaeoecology studies and can prove to be important in interpreting past environments (Erickson et al. 2003).
   Literature: Krause et al. (2016); Lehmitz and Decker (2017).

KEY TO ADULT *LIMNOZETES* OF CANADA AND ALASKA

1	Genu IV with 2 setae (*d*, *l'*). Tarsus IV with 11 setae. Notogastral sculpturing with small or shallow pits or shallow swellings, without ridges	2
–	Genu IV without setae. Tarsus IV with 10 setae. Notogastral sculpturing with ridges, without small pits	8
2(1)	Rostral indentations present. Notogaster longer than wide, ratio 1.2–1.3:1. Pteromorph longer than wide, width to length ratio less than 1:1	3
–	Rostral indentations absent. Notogaster almost as wide as long, length to width ratio 1.1:1. Pteromorph wider than long, ratio about 1.2:1. Length: ca. 365µm (Figure 3.99A) ***L. atmetos*** Behan-Pelletier	
3(2)	Notogastral sculpturing shallow pitting. Ratio tutorium length to that of tutorial cusp 8:1, or cusp absent. Pteromorph width to length ratio 0.6–0.7:1	4
–	Notogastral sculpturing shallow swellings. Ratio tutorium length to that of tutorial cusp 3:1. Pteromorph width to length ratio 0.9:1. Length: 343–365µm (Figure 3.99B,C) ***L. lustrum*** Behan-Pelletier	
4(3)	Notogastral sculpturing small, circular, shallow pits or puncta. Tarsus III with 13–14 setae. Length: >310µm	5
–	Notogastral sculpturing shallow, elongated pits. Tarsus III with 12 setae. Length: 266–292µm (Figure 3.99D) ***L. foveolatus*** Willmann	
5(4)	Tutorial cusp present. Maximum lamellar width about 10µm. Tarsus III with 13 setae	6
–	Tutorial cusp absent. Maximum lamellar width about 6µm. Tarsus III with 14 setae. Length: 343–373µm (Figure 3.99E) ***L. amnicus*** Behan-Pelletier	
6(5)	Genital setation 6 pairs. Bothridial seta clavate (often broken). Length: 311–343µm (Figure 3.99G–I) ***L. onondaga*** Behan-Pelletier	
–	Genital setation 7 pairs	7
7(6)	Translamella complete medially. Length: 270–330µm (Figure 3.99F) ***L. ciliatus*** (Schrank)	

| – | Translamella incomplete medially; only partially developed. Length: about 370μm... ***L. canadensis*** Hammer |

8(1) Lamellar cusps present; lamella narrow, maximum width about 10μm. Pteromorph almost as wide as long to wider than long, ratio 0.9–1.4:1. Femur III with 3 setae (*d, ev', l'*); Tarsi I–III with 15, 14, and 13 setae, respectively 9
– Lamellar cusps absent; lamella very broad, maximum width about 30μm. Pteromorph much wider than long, ratio about 1.8:1. Femur III with 2 setae (*d, ev'*). Tarsi I–III with 14, 11, and 11 setae, respectively. Length: 324–343μm (Figure 3.100A–C).. ***L. latilamellatus*** Behan-Pelletier

9(8) Notogastral sculpturing with ridges and irregular, shallow pits. Pteromorphs almost as wide as long, ratio 0.9:1. Pair of ridges extending from level of lamellar cusps almost to rostral indentations. Length: 330–370μm (Figure 3.100D–F) .... ***L. guyi*** Behan-Pelletier
– Notogastral sculpturing with ridges, without shallow pits. Pteromorphs wider than long, ratio 1.4:1. Medial ridge extending from anteriad of lamellar cusps almost to rostral indentations. Length: 343–376μm (Figure 3.100G–I)........................***L. borealis*** Behan-Pelletier
Note: Possible new species from BC (Lindo 2020).

Limnozetes amnicus Behan-Pelletier, 1989 (Figure 3.99E)
   Distribution: NB, NS [Canada]
   Literature: Seniczak and Seniczak (2009b).

Limnozetes atmetos Behan-Pelletier, 1989 (Figure 3.99A)
   Distribution: ON, NL [Canada]
   Literature: Seniczak and Seniczak (2009b).

Limnozetes borealis Behan-Pelletier, 1989 (Figure 3.100G–I)
   Distribution: AK, QC, NB, NL [Nearctic]
   Literature: Seniczak and Seniczak (2009b).

Limnozetes canadensis Hammer, 1952
   Biology: From wet subarctic meadows. Described on the basis of single specimen. Considered very similar to, but larger than, *L. ciliatus* (Hammer 1952a).
   Distribution: AK, YT, AB, MB [MT; Nearctic]
   Literature: Behan-Pelletier (1989a); ABMI (2019); Markkula and Kuhry (2020).

Limnozetes ciliatus (Schrank, 1803) (Figure 3.99F)
   Comb./Syn.: *Acarus ciliatus* Schrank, 1803; *Oribata sphagni* Michael, 1880
   Biology: Gut content analysis of adults over a 3 month period showed them to be primarily fungivorous, but with fragments of moss and plant material, grass pollen, and arthropod fragments as well as enchytraeid setae; in one month conifer pollen made up between 60–70% of gut content (Behan-Pelletier and Hill 1983).
   Distribution: QC, NL [NH; Holarctic]
   Literature: Behan-Pelletier (1989a); Kuriki (2008); Seniczak and Seniczak (2009b, 2020); Maraun et al. (2019).

Limnozetes foveolatus Willmann, 1939 (Figure 3.99D)
   Comb./Syn.: *Limnozetes palmerae* Behan-Pelletier, 1989
   Distribution: AK, QC, NB, NL [NH, NY; Holarctic]
   Literature: Seniczak and Seniczak (2009b, 2010).

Limnozetes guyi Behan-Pelletier, 1989 (Figure 3.100D–F)
   Biology: From *Sphagnum* peatlands; from *Carex* fen.
   Distribution: ON, QC, NB, NL [Canada]
   Literature: Seniczak and Seniczak (2009b, 2010); Barreto and Lindo (2018).

Limnozetes latilamellatus Behan-Pelletier, 1989 (Figure 3.100A–C)
   Distribution: QC, NB, NL [Canada]
   Literature: Seniczak and Seniczak (2009b, 2010).

Limnozetes lustrum Behan-Pelletier, 1989 (Figure 3.99B,C)
   Distribution: ON, QC, NB [NH; Holarctic]; Known from Tertiary fossil sites on Meigen Island (NU) (Matthews et al. 2019).
   Literature: Seniczak and Seniczak (2009b, 2010).

Limnozetes onondaga Behan-Pelletier, 1989 (Figure 3.99G–I)
   Biology: From *Carex* fens.
   Distribution: ON [NY; Nearctic]
   Literature: Seniczak and Seniczak (2009b, 2010); Barreto and Lindo (2018); Barreto et al. (2021).

Superfamily Ameronothroidea Vitzthum, 1943

Family Ameronothridae Vitzthum, 1943
   Diagnosis: *Adults*: Non-poronotic Brachypylina. Lamella, tutorium, pteromorphs absent. Cuticle relatively soft to moderately sclerotized; cerotegument dense, darkly coloured and often with even darker inclusions or nodules, exposing pale procuticle when removed. Rostrum without proximolateral genal incision or taenidium. Interlamellar setae usually absent. Pedotecta I, II absent; discrete discidium absent, but with cuticular fold between acetabula III, IV. Dorsosejugal articulation complete or medially incomplete. Dorsophragmata and pleurophragmata absent. Notogaster with 10–15 pairs of setae; with or without ascleritic lateral band; pteromorphs absent. Anterior margin of genital plates aligned approximately with apodeme III; genital papillae flattened or not. Ovipositor short, about equal to body depth when fully extended; spermatopositor of normal, short form, or strongly elongated. Subcapitulum with labiogenal articulation complete or incomplete laterally; mental tectum and axillary saccule absent; palp seta *acm* inserted on tubercle or not, *acm* not coupled with solenidion. Legs with porose organs surficial or invaginated as saccules or brachytrachea; tibia I with or without distal tubercle bearing solenidion φ_1; solenidion ω_2 of tarsus II present or absent.
   Biology: Schulte (1976) recorded all species of Ameronothridae in his study feeding on microphyta (algae and lichens), with a variable degree of specialization. Green algae are the food of species living on intertidal rocks, while salt marsh species feed primarily on fungi.
   Literature: Krause et al. (2016).

Ameronothrus Berlese, 1896
   Comb./Syn.: *Hygroribates* Jacot, 1934
   Type-species: *Eremaeus lineatus* Thörell, 1871
   Literature: Schulte et al. (1975); Schulte and Weigmann (1977); Pfingstl (2021).

KEY TO ADULT *AMERONOTHRUS* OF CANADA AND ALASKA

1   Tridactylous 2
–   Monodactylous. Length: 676–765μm (Figure 3.101A)...........*A. nigrofemoratus* (L. Koch)

2(1)　　Labiogenal articulation complete. Seta *d* present on tibiae. Length: 517–660μm (Figure 3.101B,C) .... ***A. maculatus*** (Michael)

–　　Labiogenal articulation incomplete. Seta *d* absent from tibiae. Length: 615–763μm (Figure 3.101D–F) .... ***A. lineatus*** (Thörell)
　　Note: Unidentified species from AB, NL.

Ameronothrus maculatus (Michael, 1882) (Figure 3.101B,C)
　　Comb./Syn.: *Scutovertex maculatus* Michael, 1882; *Scutovertex maculatus groenlandica* Trägårdh, 1904; *Scutovertex maculatus insularis* Hull, 1916; *Scutovertex maculatus pseudomaculatus* Hull, 1916; *Scutovertex pseudomaculatus* Hull, 1914; *Scutovertex pseudomaculatus angularis* Hull, 1914
　　Biology: Laundon (1967) recorded adults burrowing under the cortex of the lichen *Caloplacetum heppianae* and eating the centre; Schulte (1976) recorded them feeding on soridia of crustose lichens, and Gilbert (1976) found *Xanthoria parietina* heavily infested with *A. maculatus*.
　　Distribution: MB [Holarctic]
　　Literature: Pfingstl et al. (2019).

Ameronothrus lineatus (Thörell, 1871) (Figure 3.101D–F)
　　Comb./Syn.: *Eremaeus lineatus* Thörell, 1871; *Scutovertex lineatus* (Thörell, 1871); *Scutovertex corrugatus* Michael, 1888; *Ameronothrus corrugatus* (Michael, 1888)
　　Biology: Søvik et al. (2003) maintained populations on a 1 mm layer of cyanobacteria taken from water-logged mud flats on which the mites were the only arthropod and on which they fed in culture. In temperate estuaries of northern Germany, this species has a 2–year life cycle (Bücking et al. 1998), whereas populations living on cyanobacterial mats or in saltmarshes in Svalbard have a minimum 5–year life cycle (Søvik et al. 2003). Synchronous larviposition and molting has been observed (Søvik et al. 2003). Serial sections and computer-assisted reconstructions were used for establishing the 3-D structure of the gut of this species (Bücking 2002).
　　Distribution: AK, YT, NT, MB [Holarctic]
　　Literature: Søvik (2003, 2004).

Ameronothrus nigrofemoratus (L. Koch, 1879) (Figure 3.101A)
　　Comb./Syn.: *Nothrus nigro-femoratus* L. Koch, 1879; *Scutovertex nigrofemoratus* (L. Koch, 1879); *Ameronothrus lineatus nigrofemorata* (L. Koch, 1879); *Ameronothrus lineatus brevipes* Willmann, 1937
　　Distribution: AK, NT, BC, MB [Holarctic]
　　Literature: Maraun et al. (2019).

Family Podacaridae Grandjean, 1955
　　Diagnosis: *Adults*: Non-poronotic Brachypylina. Prodorsal lamella, tutorium absent; carina present in position of tutorium; transverse carina present dorsal of acetabula I and II. Cuticle moderately sclerotized. Rostrum without genal incision or taenidium. Interlamellar setae present. Pedotecta I, II present; discrete discidium absent, but with cuticular fold between acetabula III, IV. Dorsosejugal articulation complete. Dorsophragmata and pleurophragmata weakly developed. Notogastral setation 14 pairs, occasionally c_3 present; with or without ascleritic lateral band; pteromorphs absent. Genital papillae strongly laterally compressed, or not. Ovipositor relatively short, about equal to body depth when fully extended, with or without coronal setae; spermatopositor short. Subcapitulum with labiogenal articulation complete or incomplete laterally; without mental tectum or axillary saccule; palp seta *acm* inserted on low tubercle, *acm* coupled with solenidion, forming perfect 'double horn'. Legs with porose areas surficial; tibia I without distal tubercle bearing solenidion φ_1; solenidion ω_2 of tarsus II present. Legs tridactylous; lateral claws like ice-axe distally.

Alaskozetes Hammer, 1955
Type-species: *Alaskozetes coriaceus* Hammer, 1955
Biology: The congeneric, *Alaskozetes antarcticus* (Michael) lives in extreme environments of the Antarctic. *Alaskozetes antarcticus* displays low enzyme activation energies and elevated metabolic rates at low temperatures and associated low optimum temperatures for activity, feeding and growth in comparison with temperate species (Block and Convey 1995). It has a life cycle of 5–6 years (Convey 1994). Although considered general herbivores/detritivores, these mites can be selective in the type of microalgae they choose (Worland and Lukešová 2000).

Alaskozetes coriaceus Hammer, 1955 (Figure 3.102D)
Biology: From terrestrial habitats.
Distribution: AK
Literature: Schulte and Weigmann (1977); Pfingstl (2017).

Family Selenoribatidae Schuster, 1963
Diagnosis: *Adult:* Non-poronotic Brachypylina. Apherdermous. Tracheal system normal; only sejugal trachea divided in two; trachea III opens on anterodorsal extremity of cotyloid wall, close to opening of acetabulum. Genal incision absent. Tutorium, custodium and posterior circumpedal carina absent. Pedotectum I very small; pedotectum II absent. Dorsophragma and pleurophragma absent. Notogastral setation 13–15 pairs, with 3 pairs of centrodorsal setae always present. Notogaster with posterior tectum; pteromorphs absent. All epimeral borders evident, except *bo4*. Sternal border present. Epimeral setation 1-0-1-1/2. Genital setation 3 pairs. Aggenital seta absent. Lyrifissure *iad* anterior to *ad₃*. On palptarsus *acm* separated from solenidion. Monodactylous; partial fusion of tibia-tarsus. Seta *d* absent from genua and tibiae; tarsus II with 1 solenidion.
Literature: Pfingstl (2013); Pfingstl and Krisper (2014); Pfingstl and Lienhard (2017).

Thalassozetes Schuster, 1963 (Figure 3.103D–F)
Type-species: *Thalassozetes riparius* Schuster, 1963
Note: Unidentified species from BC.

Family Tegeocranellidae Balogh and Balogh, 1988
Diagnosis: *Adults*: Non-poronotic Brachypylina. Lamella and lamellar cusp present; cusps broad or narrow, medially touching or not; lateral border of lamella curving on itself, appearing thickened. Translamella absent. Prodorsal surface between lamellae flattened. Exobothridial seta reduced to alveolus. Genal incision absent. Tutorium, discidium present; custodium and circumpedal carina absent. Pedotectum I large, bipartite. Humeral porose area *Ah* present. Dorsophragmata and pleurophragmata present. Notogastral setation 10 or 12 pairs. Lenticulus usually present. Notogastral depressions present lateral to lenticulus, or absent. Notogaster without posterior tectum. Epimeral borders clearly evident. Epimeral setation 2-1-2-3. Sternal depression present medially on border of epimera III and IV, or absent. Genital setation 6 pairs. Genital and anal plates closely adjacent; genital plates large, combined plates wider than long, wider than anal plates; integument darker, or not, than that of anal plates. Chelicera chelate-dentate. Axillary saccule absent. Palp setation: 0-2-1-3-9(1). Palptarsal solenidion arising at level of seta *lt"*, lying parallel to tarsal surface, extending anteriorly to base of eupathidium *acm*. Legs monodactylous. Seta *d* absent from genua and tibiae. Tarsus II with 1 solenidion.

Tegeocranellus Berlese, 1913
Type-species: *Tegeocranus laevis* Berlese, 1905

Tegeocranellus muscorum Behan-Pelletier, 1997 (Figure 3.103A–C)
Biology: From moss, herbs, and substrate on rocks at creek edge; moss under cattails in swamp; wet sphagnum at edge of pond; decaying vegetation at edge of slough in oak-maple parkland.
Distribution: ON [NY, NC, MS, FL; Nearctic]

Superfamily Cymbaeremaeoidea Sellnick, 1928

Family Cymbaeremaeidae Sellnick, 1928

Diagnosis: *Adult:* Non-poronotic Brachypylina. Genal incision absent. Lamellae, lamellar cusps, interlamellar setae, tutorium present or absent. Pedotecta I and II present. Dorsophragmata and pleurophragmata generally absent (present in *Scapheremaeus*). Notogaster with or without lenticulus; with or without humeral projection; with or without tectal border laterally; without posterior tectum. Notogastral setae normal, or apobasic (*Scapheremaeus*); setation 7, 9, 10, 13, or 14 pairs; *d* series present or absent. Epimeral setation 3-1-2-2 or 3-1-3-3 or 3-1-2-3 or 3-1-1-2 or 2-1-2-2. Genital setation 4 or 6 pairs. Adanal setation 2–3 pairs. Anal setation 2–3 pairs. Palpal eupathidium *acm* not attached to solenidion, inserted on large tubercle (exception, few species of *Scapheremaeus*). Mentum with or without tectum. Chelicerae chelate-dentate. Trochanters III, IV and tibiae I–IV with or without porose areas or saccules; femora I–IV and tarsi I–IV with either porose areas or saccules; Tarsus II with 2 solenidia (1 in *Scapheremaeus*). Setae *d* generally absent from adult genua and tibiae (expressed in two species of *Ametroproctus*).

KEY TO GENERA OF CYMBAEREMAEIDAE OF CANADA AND ALASKA

1 Lenticulus present. Genital and anal apertures separated by area equal to half length of anal plate. Anal setae not inserted on medial margin of anal plates. Lamella, lamellar cusps, tutorium absent. (Figure 3.97G,H) .... ***Scapheremaeus*** Berlese

– Lenticulus absent. Genital and anal apertures closely adjacent. Anal setae inserted on medial margin of anal plates. Lamellae, lamellar cusps, tutorium present 2

2(1) Notogaster with large humeral process. Epimeral setae *3c* and *4c* present (Figure 3.97I,J) .... ***Scapuleremaeus*** Behan-Pelletier

– Notogaster without humeral process. Epimeral setae *3c* and *4c* absent (Figure 3.104) .... ***Ametroproctus*** Higgins and Woolley

Ametroproctus Higgins and Woolley, 1968

Type-species: *Ametroproctus oresbios* Higgins and Woolley, 1968

Biology: *Ametroproctus* species are found in dry, primarily subalpine or alpine habitats, and in the canopy of conifers (Behan-Pelletier 1987a). Species are probably mycophages, and in some habitats they can be the numerically dominant oribatid mites.

Note: Two subgenera are currently recognized, *Ametroproctus* Higgins and Woolley, 1968 and *Coropoculia* Aoki and Fujikawa, 1972.

KEY TO ADULT *AMETROPROCTUS* OF CANADA AND ALASKA

1 Pedotectum I not extending to seta *ex*. Genital setation 4 pairs. Anterior margin of notogaster convex medially; with or without humeral projection. Femur III with 2 setae (seta *l'* absent). .... Subgenus ***Ametroproctus*** 2

– Pedotectum I extending to seta *ex*. Genital setation 6 pairs. Anterior margin of notogaster flattened; with humeral projection. Femur III with 3 setae (setae *l'* present)....................... .... Subgenus ***Coropoculia*** 3

2(1) Notogastral sculpturing forming short to long irregular ridges. Body cerotegument thick, 5–8μm deep. Ratio of fixed portion of tutorium to cusp about 4:1. Length: 603–667μm (Figure 3.104A,B) .... ***A. (Ametroproctus) oresbios*** Higgins and Woolley

Taxonomic Keys

– Notogastral sculpturing forming large tubercles. Body cerotegument thin, 1–2μm deep. Ratio of fixed portion of tutorium to cusp about 7.5:1. Length: 648–714μm (Figure 3.104C,D) .... *A. (Ametroproctus) tuberculosus* Behan-Pelletier

3(1) Pedotectum I with incision. Femur I with 5 setae (seta *v"* present). Dorsosejugal scissure complete or interrupted medially 4

– Pedotectum I without incision. Femur I with 4 setae (seta *v"* absent). Dorsosejugal scissure complete. Length: 591–616μm (Figure 3.104G) ....
.... *A. (Coropoculia) beringianus* Behan-Pelletier

4(3) Notogaster with subreticulate sculpturing, many walls between foveae incomplete. Dorsosejugal scissure interrupted medially. Rostral and lamellar setae about 20μm and 28μm long, respectively; interlamellar seta present. Femora I with 3–5 saccules, II with 4–5. Length: 518–583μm (Figure 3.104E,F) ....
.... *A. (Coropoculia) reticulatus* (Aoki and Fujikawa)

– Notogaster without sculpturing. Dorsosejugal scissure complete. Rostral and lamellar setae about 50μm and 48μm long, respectively; interlamellar seta absent. Femora I with 1–2 saccules, II with 1 saccule. Length: 583–624μm (Figure 3.104H) ....
.... *A. (Coropoculia) canningsi* Behan-Pelletier

Ametroproctus (A.) oresbios Higgins and Woolley, 1968 (Figure 3.104A,B)
 Biology: Collected from a variety of litter types in Western North America.
 Distribution: AB [AZ, CA, WA, OR, UT; Nearctic]
 Literature: Behan-Pelletier (1987a); ABMI (2019); Behan-Pelletier and Ermilov (2020).

Ametroproctus (A.) tuberculosus Behan-Pelletier, 1987 (Figure 3.104C,D)
 Biology: From moss, lichens, and litter at high elevations in SW Alberta.
 Distribution: AB [Canada]
 Literature: Behan-Pelletier and Ermilov (2020).

Ametroproctus (C.) beringianus Behan-Pelletier, 1987 (Figure 3.104G)
 Biology: From moss and lichens among purple shale rocks; *Senecio*, *Potentilla*, *Minuartia*, lichens, moss among rocks; mixed vegetation on rocky hill.
 Distribution: YT [Canada]
 Literature: Behan-Pelletier and Ermilov (2020).

Ametroproctus (C.) canningsi Behan-Pelletier, 1987 (Figure 3.104G)
 Biology: From moss, cushion plants, and litter at 1,863–2,063m in the Cascade and Rocky Mountains.
 Distribution: BC, AB [Canada]
 Literature: Behan-Pelletier and Ermilov (2020).

Ametroproctus (C.) reticulatus (Aoki and Fujikawa, 1972) (Figure 3.104E,F)
 Comb./Syn.: *Coropoculia reticulatus* Aoki and Fujikawa, 1972
 Biology: From moss and a variety of litter types at high elevations in Western North America and Northern Japan.
 Distribution: BC, AB [WA, Japan]
 Literature: Behan-Pelletier (1987a); ABMI (2019); Behan-Pelletier and Ermilov (2020).

Scapheremaeus Berlese, 1910
 Type-species: *Cymbaeremaeus (Scapheremaeus) patella* Berlese, 1910
 Biology: Species of the genus *Scapheremaeus* prefer epilithic and epiphytic habitats and are among the dominant corticolous mites in temperate and tropical regions of the world (Colloff 2009).

KEY TO ADULT *SCAPHEREMAEUS* OF CANADA AND ALASKA
(AFTER NORTON ET AL. 2010)

1   Prodorsum without transverse ridge connecting ridge-like lamellae anteriorly. Bothridial setae clavate, with head much longer than wide. Notogaster with pair of longitudinal ridges bearing setae *lm* and *lp*, and short extension of circumdorsal scissure toward humeral region on each side; alveolae of microsculpture all completely circumscribed by pattern of relatively reticulate ridges; each notogastral setae emerging through center of large alveole. Length: 355–425µm (Figure 3.97G).. *S. parvulus* (Banks)

–   Prodorsum with ridge-like lamellae connected anteriorly by transverse ridge. Bothridial seta subcapitate; head little longer than wide. Notogaster without longitudinal ridges or extensions of circumdorsal scissure; alveolae often irregular, incompletely formed or merged due to absence of separating reticulation; notogastral setae not emerging through conspicuous alveolae. Length: 440–480µm (Figure 3.97H)............. *S. palustris* (Sellnick)
Note: Unidentified species from NS.

Scapheremaeus palustris (Sellnick, 1924) (Figure 3.97H)
    Comb./Syn.: *Cymbaeremaeus (Scapheremaeus) palustris* Sellnick, 1924
    Biology: Arboreal, often associated with bark beetles (Moser and Roton 1971), and known to be phoretic on *Ips calligraphus* (J. Moser, unpublished). Recorded from epiphytic lichens in the Adirondack Mountains (Root et al. 2007). Lindo and Winchester (2006, 2007) and Winchester et al. (1999) report it from the canopy on Vancouver Island. Reported from range of coniferous and deciduous trees (Norton et al. 2010).
    Distribution: BC, AB, ON, QC, NB, NS [MI, NY, WV; Holarctic]
    Literature: Norton et al. (2010); Ermilov et al. (2015); ABMI (2019).

Scapheremaeus parvulus (Banks, 1909) (Figure 3.97G)
    Comb./Syn.: *Cymbaeremaeus parvula* Banks, 1909
    Biology: Primarily from lichens (Norton et al. 2010)
    Distribution: ON [FL, IL, NY; Nearctic]
    Literature: Norton et al. (2010).

Scapuleremaeus Behan-Pelletier, 1989
    Type-species: *Scapuleremaeus kobauensis* Behan-Pelletier, 1989

Scapuleremaeus kobauensis Behan-Pelletier, 1989 (Figure 3.97I,J)
    Biology: From ponderosa pine, bunchgrass zone among rocks and vegetation.
    Distribution: BC [CA; Nearctic]
    Literature: Behan-Pelletier and Ermilov (2020).

Superfamily Licneremaeoidea Grandjean, 1931

Family Dendroeremaeidae Behan-Pelletier, Eamer and Clayton, 2005
    Diagnosis: *Adults*: Poronotic Brachypylina; occasionally carrying scalps. Integument reticulate to foveolate. Prodorsum with lamellae; lamellar cusps and translamella, tutorium present. Genal incision present. Dorsophragmata and pleurophragmata present. Discidium, custodium, circumpedal carina and postanal porose area absent. Pedotectum I and II present, pedotectum I extending dorsally to exobothridial seta. Porose areas *Am*, *Ah*, *Al* present. Octotaxic system present as 4 pairs of saccules. Notogaster with lenticulus; pteromorphs, posterior notogastral tectum absent. Notogastral setation 10 pairs. Epimeral setation 3-1-2-2. Genital setation 6 pairs. Chelicerae chelate-dentate; palpal eupathidium *acm* attached to solenidion along length; subcapitular mentum without tectum. Axillary saccule of

subcapitulum present. Porose areas absent from tibiae and tarsi. Solenidion φ_1 of tibia I inserted on distinct, cylindrical tubercle.

Dendroeremaeus Behan-Pelletier, Eamer and Clayton, 2005
Type-species: *Dendroeremaeus krantzi* Behan-Pelletier, Eamer and Clayton, 2005

Dendroeremaeus krantzi Behan-Pelletier, Eamer and Clayton, 2005 (Figure 3.105A–C)
Biology: Canopy species in deciduous and coniferous shrubs and trees.
Distribution: BC [OR; Nearctic]

Family Licneremaeidae Grandjean, 1931
Diagnosis: *Adults*. Poronotic Brachypylina. Genal incision absent. Bothridial seta discoidal to flabelliform. Dorsosejugal scissure extending anterior of interlamellar setae. Dorsophragmata and pleurophragmata present. Discidium, custodium, circumpedal carina and postanal porose area absent. Humerosejugal porose area *Ah* absent. Octotaxic system as 2 pairs of porose area in position of *Aa* and *A2*. Notogastral setation 13 pairs; seta p_3 absent. Pteromorphs and posterior notogastral tectum absent. Epimeral setation 3-1-2-2 or 2-1-2-1. Genital setation 5 pairs. Palptarsal eupathidium *acm* not fused with solenidion. Axillary saccule at base of palp absent.

Licneremaeus Paoli, 1908
Type-species: *Notaspis licnophorus* Michael, 1882
Note: Unidentified species in BC: coastal temperate rainforest.

Family Passalozetidae Grandjean, 1954
Diagnosis: *Adults:* Poronotic Brachypylina. Integument often nodulate, or reticulate. Prodorsum with or without ridge-like lamellae; lamellar cusps, translamella and tutorium absent. Genal incision absent. Dorsosejugal scissure extending medially in interlamellar region. Dorsophragmata and pleurophragmata present. Discidium, custodium, circumpedal carina and postanal porose area absent. Pedotectum I present; pedotectum II present, rectangular in ventral view. Octotaxic system as 4 pairs of small pores or porose areas. Notogaster with or without distinct lenticulus, pteromorphs and posterior notogastral tectum absent. Notogastral setation 10 pairs. Epimeral setation 3-1-2-2. Genital setation 4–5 pairs. Chelicerae chelate-dentate; palpal eupathidium *acm* touching solenidion distally; subcapitular mentum without tectum. Axillary saccule of subcapitulum absent. Porose organs present ventrally on tibiae and tarsi I–IV. Tibiae and tarsi with retrotecta. Solenidion φ_1 of tibia I inserted on distinct, cylindrical tubercle.

Passalozetes Grandjean, 1932
Comb./Syn.: *Bipassalozetes* Mihelčič, 1957
Type-species: *Passalozetes africanus* Grandjean, 1932
Biology: Passalozetid mites inhabit dry habitats or dry microhabitats within more mesic areas. The complex, ornate cuticle is formed from the cerotegument, a layer of wax and proteins, and provides protection from dehydration and, probably also acts as a hydrofuge layer to prevent wetting.

KEY TO ADULT *PASSALOZETES* OF CANADA AND ALASKA

1. Cerotegumental pattern composed mostly of parallel long plications. Bothridial seta barbed in distal two-thirds. 3–4 pairs of porose areas on notogaster (*A3* small or absent). Length: 385–413µm (Figure 3.106A,B) .... ***P. californicus*** Wallwork
– Cerotegument with caltrop-like pattern. Length: 365–390µm ...... ***P. intermedius*** Mihelčič
Note: Unnamed species known from YT, BC, AB.

Passalozetes californicus Wallwork, 1972 (Figure 3.106A,B)
  Biology: Collected from cushion vegetation in shortgrass prairie.
  Distribution: BC, AB [CA; Nearctic]
  Literature: Wallwork et al. (1984).

Passalozetes intermedius Mihelčič, 1954
  Comb./Syn.: *Passalozetes variatepictus* Mihelčič, 1956
  Biology: This species was originally described from a *Formica* ant nest in Europe and is known to inhabit dry meadows and heathlands. In Alberta, specimens similar to this species have been collected from open grassy south-facing slopes in aspen woodland near Elk Island National Park and from open forests in Cypress Hills. In Yukon, this mite has been collected from open areas along the Klondike Highway. Gut contents indicate that *P. intermedius* feeds on dark fungal hyphae and spores (Walter et al. 2014).
  Distribution: YT, AB [Holarctic]
  Literature: Weigmann (2006); ABMI (2019).

Family Scutoverticidae Grandjean, 1954
  Diagnosis: *Adults*: Poronotic Brachypylina. Integument often tuberculate, nodulate, or reticulate. Prodorsum with lamellae; lamellar cusps and translamella present or absent. Tutorium present or absent. Rostrum with submarginal carina. Genal incision absent. Dorsophragmata and pleurophragmata present. Dorsosejugal scissure present or incomplete. Discidium, custodium, circumpedal carina and postanal porose area absent. Pedotectum I present; pedotectum II present, rectangular in ventral view or not. Octotaxic system as 1–3 pairs of saccules (rarely *S3* present). Notogaster with or without lenticulus; with or without humeral process; pteromorphs and posterior tectum absent. Notogastral setation 9–10 pairs. Epimeral setation either 3-1-3-3 or 3-1-2-2. Genital setation 6 pairs. Anal setation 2 pairs. Chelicerae chelate-dentate; palpal eupathidium *acm* attached to solenidion distally; subcapitular mentum without complete tectum. Axillary saccule of the subcapitulum present or absent. Platytracheae on femur I–II and saccules on tibiae and femora and trochanters III–IV. Solenidion φ_1 of tibia I inserted on distinct, cylindrical tubercle.

KEY TO GENERA OF SCUTOVERTICIDAE OF CANADA AND ALASKA

1. Lenticulus present. Interlamellar and exobothridial setae present. Translamella well developed. Tutorium absent (Figure 3.106D–G) .... ***Scutovertex*** Michael
– Lenticulus absent. Interlamellar and exobothridial setae absent. Translamella absent or incomplete. Tutorium poorly to well developed. (Figure 3.106H,I) ....
....***Exochocepheus*** Woolley and Higgins

Exochocepheus Woolley and Higgins, 1968
  Type-species: *Exochocepheus eremitus* Woolley and Higgins, 1968

Exochocepheus eremitus Woolley and Higgins, 1968 (Figure 3.106H,I)
  Biology: From dry moss; conifer litter.
  Distribution: BC, AB [CO, UT; Nearctic]
  Literature: Pfingstl et al. (2010); ABMI (2019).

Scutovertex Michael, 1879
  Type-species: *Scutovertex sculptus* Michael, 1879 (Figure 3.106F,G)
  Note: Unidentified species from coastal BC.

Superfamily Phenopelopoidea Petrunkevich, 1955

Family Phenopelopidae Petrunkevich, 1955

Comb./Syn.: Pelopidae Ewing, 1917 (also seen as Pelopsidae)

Diagnosis: *Adults*: Poronotic Brachypylina. Cerotegument well developed, with thick, blocky structure on notogaster and ventral plate; strongly birefringent in polarized light. Prodorsum with lamellae, translamella, tutorium, and genal incision. Discidium and custodium present. Pedotecta I and II present; pedotectum I with transverse carina. Dorsophragmata and pleurophragmata present. Octotaxic system represented by 0–4 pairs of small porose areas, usually adjacent to notogastral setae. Notogaster with lenticulus; with tectum anteriorly; without tectum posteriorly. Pteromorphs well developed, with or without hinge. Notogastral setation 10 pairs; pair h_1 (and often others) thick. Chelicerae chelate-dentate or pelopsiform. Palpal eupathidium *acm* attached to solenidion, at least distally. Axillary saccule of subcapitulum present.

Biology: Phenopelopidae are known primarily from forest litter in temperate regions of the Northern and Southern Hemispheres. The genera *Eupelops* and *Peloptulus* have pelopsiform mouthparts, suggesting that they are mycophages (e.g., Behan-Pelletier and Hill 1983). However, Riha (1951) observed members of an unidentified species of *Eupelops* using their chelicerae to cut a hole in the underside of dead leaves and reach into the parenchyma, where they excavated the central area and left only the two layers of leaf cuticle. While feeding in this way they probably also ingested fungal material non-selectively. Maraun et al. (2003) found no intraspecific genetic variation in two closely related sexual species (*Eupelops hirtus* (Berlese) and *E. torulosus* (C.L. Koch)).

Literature: Norton and Behan-Pelletier (1986); Seniczak et al. (2014); Bayartogtokh et al. (2018).

KEY TO GENERA OF PHENOPELOPIDAE OF CANADA AND ALASKA

1   Chelicera pelopsiform. Pleurophragmata elongated, extending well into opisthosoma. Unpaired median, elongated dorsophragma present. Pteromorphs almost fully hinged....2
–   Chelicera normal. Pleurophragmata normally developed. Dorsophragmata paired, inconspicuous. Pteromorphs with only posterior half hinged (Figure 3.108)......***Propelops*** Jacot

2(1) Interlamellar setae setiform, short. Prodorsal lamellae broad, conspicuous (Figure 3.107E)...***Peloptulus*** Berlese
–   Interlamellar setae phylliform, long. Prodorsal lamellae small, inconspicuous (Figure 3.107F–J) ....***Eupelops*** Ewing

Eupelops Ewing, 1917

Comb./Syn.: *Pelops* C.L. Koch, 1835–6; *Allopelops* Hammer, 1952; *Phenopelops* Petrunkevitch, 1955; *Tectopelops* Jacot, 1929

Type-species: *Pelops ureaceus* C.L. Koch, 1839

KEY TO ADULT *EUPELOPS* OF CANADA AND ALASKA

1   Notogastral setae >200μm, tapered. Length: 850–1,100μm (Figure 3.107H)......................
....***E. hirtus*** (Berlese)
–   Notogastral setae <150μm, tapered or expanded distally. Length: <900μm 2

2(1) All notogastral setae expanded distally. Length: ca. 500μm ....***E. terminalis*** (Banks)
–   At most setae of *h* and *p* series expanded distally...3

3(2) Interlamellar setae extending anterior of rostrum. Length: ca. 700μm (Figure 3.107I) ....
....*E. septentrionalis* (Trägårdh)
– Interlamellar setae just reaching rostrum. Length: 500–680μm (Figure 3.107J) ....
....*E. plicatus* (C.L. Koch)
Note: Unidentified species from AK: Lava Camp, Seward Peninsula (Late Miocene: 5.7 mya), from Tertiary fossil sites on Meigen Island and Prince Patrick Island (NU), and *Eu. occultus* (C.L. Koch, 1835) from late Pliocene/early Pleistocene fossil deposits in AK (Matthews et al. 2019). Possibly undescribed or unidentified species from BC, AB, MT, ON, QC, NB, NL.

Eupelops hirtus (Berlese, 1916) (Figure 3.107H)
  Comb./Syn.: *Pelops hirtus* Berlese, 1916; *Phenopelops hirtus* (Berlese, 1916)
  Biology: Gut microbiota of this species was studied by Gong et al. (2018).
  Distribution: QC [Holarctic]
  Literature: Maraun et al. (2011); Seniczak et al. (2014); Maraun et al. (2019).

Eupelops plicatus (C.L. Koch, 1835) (Figure 3.107J)
  Comb./Syn.: *Celaeno plicatus* C.L. Koch, 1835; *Pelops plicatus* (C.L. Koch, 1835); *Pelops auritus* C.L. Koch, 1839; *Pelops laevigatus* Nicolet, 1855; *Pelops fusiger* Mihelčič, 1957; *Pelops acromius diversipilus* Mihelčič, 1957; *Pelops fuligineus* sensu Oudemans, 1896
  Biology: Gut content analysis of adults showed consumption of fungi and plant material (Behan-Pelletier and Hill 1983). This species was rejected as food by the mesostigmatan *Pergamasus septentrionalis* in a no-choice feeding experiment (Peschel et al. 2006).
  Distribution: AK [VA; Holarctic]
  Literature: Seniczak et al. (2015); Maraun et al. (2019).

Eupelops septentrionalis (Trägårdh, 1910) (Figure 3.107I)
  Comb./Syn.: *Pelops septentrionalis* Trägårdh, 1910; *Allopelops septentrionalis* (Trägårdh, 1910)
  Biology: From peatlands.
  Distribution: AK, NU, NT, AB, MB, ON, QC, NL [Holarctic]
  Literature: Seniczak et al. (2015); ABMI (2019).

Eupelops terminalis (Banks, 1909)
  Comb./Syn.: *Pelops terminalis* Banks, 1909
  Biology: From under bark of ironwood.
  Distribution: ON [NC; Holarctic]

Peloptulus Berlese, 1908
  Type-species: *Pelops phaenotus* C.L. Koch, 1844
  Note: Unidentified species known from AB, ON; that from AB is probably undescribed (Walter et al. 2014).

Propelops Jacot, 1937
  Type-species: *Propelops pinicus* Jacot, 1937

KEY TO ADULT *PROPELOPS* OF CANADA AND ALASKA
(MODIFIED FROM WALTER ET AL. 2014)

1 Interlamellar seta very short, not reaching half distance to translamella. Translamella very short. Lamellar cusps nearly touching. Length: ca. 462μm (Figure 3.108A).......................
....*P. groenlandicus* (Sellnick)

Taxonomic Keys

–   Interlamellar seta reaching or passing translamella. Translamella long. Lamellar cusps distant from one another 2

2(1)   Lamellar cusps shorter than interlamellar region. 4 pairs small porose areas: *Aa* near seta *lm*, *A2* near h_2, *A3* near h_3, *A1* bracketed by setae h_3 and *lp*. Length: females >430μm.... 3
–   Lamellar cusps about as long or longer than interlamellar region. Porose areas *Aa*, *A2*, *A3* absent, *A1* small, bracketed by setae h_3 and *lp*. Length: 370–430μm (Figure 3.108B).........
....*P. minnesotensis* (Ewing)
(Figure 3.108C,D) ....*P. alaskensis* (Hammer)

3(2)   Interlamellar seta projecting beyond lamellar cusps 4
–   Interlamellar seta not reaching tip of lamellar cusps (Figure 3.108E) ....
....*P. monticolus* (Ewing)

4(3)   Notogastral setae of *h*-series cylindrical to narrowly clavate (Figure 3.108F).....................
....*P. canadensis* (Hammer)
–   Notogastral setae of *h*-series very broad, flattened, leaf-like .... *P. pinicus* Jacot
Note: Unidentified species from AK: Lava Camp, Seward Peninsula (Late Miocene: 5.7 mya) (Matthews et al. 2019). Possibly undescribed species from peatlands (Baretto and Lindo 2021).

Propelops alaskensis (Hammer, 1955) (Figure 3.108C,D)
    Comb./Syn.: *Hammeria alaskensis* Hammer, 1955
    Note: This species is currently indistinguishable from *Propelops minnesotensis* (Ewing, 1913) and the name may be a junior synonym, according to Walter et al. (2014).
    Biology: Gravid females carry up to 4 large eggs. Gut contents include brown and hyaline fungal hyphae and spores, and the cerotegument often has fungal material stuck to it. This is the dominant species of *Propelops* in the Boreal forest and extends southward to below the latitude of Edmonton, in Alberta (Walter et al. 2014).
    Distribution: AK, AB [Nearctic]
    Literature: Norton and Behan-Pelletier (1986); ABMI (2019).

Propelops canadensis (Hammer, 1952) (Figure 3.108F)
    Comb./Syn.: *Hammeria canadensis* Hammer, 1952
    Biology: Described from Yukon, but occurs across Canada and extends further south into the USA. It is common in Alberta south of Edmonton and in the Rocky Mountains. Gravid females can carry at least 3 large eggs. Gut contents include brown fungal hyphae, arthroconidia, and coarse organic material (Walter et al. 2014).
    Distribution: AK, YT, NU, NT, BC, AB, ON, NL [CA, OR, UT; Holarctic]
    Literature: Norton and Behan-Pelletier (1986); Polyak et al. (2001); ABMI (2019).

Propelops groenlandicus (Sellnick, 1944) (Figure 3.108A)
    Comb./Syn.: *Hammeria groenlandica* Sellnick, 1944
    Biology: From coniferous forest soil.
    Distribution: AK, YT, NT, NU, MB [OR; Holarctic]
    Literature: Makarova (2015); Matthews et al. (2019).

Propelops monticolus (Ewing, 1918) (Figure 3.108E)
    Comb./Syn.: *Eupelops monticolus* Ewing, 1918; *Hammeria monticolus* (Ewing, 1918); *Eupelops monticolus subborealis* Ewing, 1918
    Distribution: BC [OR; Nearctic]
    Literature: Walter et al. (2014).

Propelops pinicus Jacot, 1937
Biology: From Douglas fir litter.
Distribution: BC [FL, VA; Nearctic]
Literature: Walter et al. (2014).

Family Unduloribatidae Kunst, 1971
Diagnosis: *Adults*: Non-poronotic Brachypylina. Cerotegument irregularly tuberculate, thick, blocky, birefringent. Rostral tectum with pair of deep vertical incisions, ending near rostral seta. Rostral seta inserted on long cylindrical tubercle. Lamellae large, broad, almost touching medially, extending anterior of rostrum. Bothridium with internal ring-like thickenings. Tutorium with long knife-like cusp comprising about half tutorium length. Notogaster with lenticulus; setation 10 pairs; pteromorphs without hinge; anterior and posterior notogastral tectum present. Pedotectum I without transverse carina. Epimeral setation: 3-1-2-3, 3-1-3-3 or 3-2-7-7; 1 pair aggenital, 3 pairs of adanal and 2 pairs of anal setae. Genital setae neotrichous or not, with 6–10 pairs. Circumpedal carina and custodium present. Preanal organ flat apodemal plate connected to anteromedial corners of anal valves by pair of long, sclerotized rods. Postanal porose area absent. Palpal eupathidium *acm* attached to solenidion at least along distal third of length. Axillary saccule of subcapitulum present. Mentum with reflexed anterior rim, with or without medial carina.
Literature: Sidorchuk and Norton (2010).

Unduloribates Balogh, 1943
Comb./Syn.: *Sphaerozetes* (*Tectoribates*) Berlese, 1914
Type-species: *Sphaerozetes* (*Tectoribates*) *undulatus* Berlese, 1914

Unduloribates dianae Behan-Pelletier and Walter, 2009 (Figure 3.107A–D)
Biology: This species is widely distributed throughout the northern Nearctic, where it has been recorded from boreal forest litter and from taiga; also recorded from alpine habitats in the Catskill and Adirondack Mountains of New York State.
Distribution: BC, AB, ON, QC, NS, NL [NY; Nearctic]
Literature: ABMI (2019).

Superfamily Achipterioidea Thor, 1929

Family Achipteriidae Thor, 1929
Diagnosis: *Adults*: Poronotic Brachypylina. Prodorsum with lamellae, lamellar cusps, tutorium present; translamella present or absent. Genal incision present. Bothridium with internal, ring-like thickenings. Dorsophragmata and pleurophragmata present. Discidium and custodium present. Postanal porose area absent. Pedotectum I large, extending to base of bothridium; pedotectum II present. Octotaxic system as 1–4 pairs of porose areas or saccules or pores, sometimes minute. Notogaster with or without knife-like humeral process; with posterior tectum; pteromorphs present, without hinge; lenticulus present; setation 10 pairs. Genital setation 6 pairs. Chelicerae chelate-dentate; palpal eupathidium *acm* attached to solenidion; subcapitular mentum with or without tectum. Axillary saccule of subcapitulum absent. Genu IV concave to flat dorsally, subequal in length or longer than tibia IV; tibia IV with solenidion.
Literature: Bayartogtokh and Ryabinin (2012).

KEY TO GENERA OF ACHIPTERIIDAE OF CANADA AND ALASKA

1	Notogaster anteriorly with long, knife-like projection	2
–	Notogaster anteriorly without long knife-like projection	4

2(1)	Notogaster with 4 pairs of porose areas (Figure 3.111A–C) .... ***Parachipteria*** Hammen	
–	Notogaster with 4 pairs of saccules 3	
3(2)	Saccules on notogaster clearly visible; mentum without tectum (Figure 3.109) .... ***Achipteria*** Berlese	
–	Saccules on notogaster reduced to pores; mentum with partial tectum covering base of gena ....***Pseudachipteria*** Travé	
4(1)	Lamellae broad, cusps broad, truncate distally. Notogaster with 1–3 pairs of saccules. Cerotegument covering all body surfaces (Figure 3.111D) .... ***Dentachipteria*** Nevin	
–	Lamellae broad or not, cusps not truncate distally. Notogaster with 4 pairs of porose areas or 4 pairs of saccules. Distinct cerotegument present only in subalar region 5	
5(4)	Lamellae partially or completely fused medially (Figure 3.110A–D) .... ***Anachipteria*** Grandjean	
–	Lamellae separated, positioned laterally on prodorsum (Figure 3.110E,F) .... ***Separachipteria*** Subías	

Achipteria Berlese, 1885
    Type-species: *Acarus coleoptratus* Linnaeus, 1758

KEY TO ADULT *ACHIPTERIA* OF CANADA AND ALASKA

1	Lamellar cusps truncate distally, anterior margin with small dens. Length: 552–600µm (Figure 3.109A,B) .... ***A. curta*** Aoki
–	Lamellar cusps acutely angled or rounded distally, anterior margin smooth 2
2(1)	Lenticulus absent. Bothridial seta curved medially, smooth. Length: 407–462µm (Figure 3.109C)... ***A. catskillensis*** Nevin
–	Lenticulus present. Bothridial seta barbed; not strongly curved medially 3
3(2)	Integument dull reddish brown to black with pebbly texture. Interlamellar seta relatively short (110–130µm). Notogastral setae distinct. Saccules indistinct. Length: 635–693µm (Figure 3.109D)... ***A. clarencei*** Nevin
–	Integument smooth, shiny brown to black. Interlamellar seta longer (190–200µm). Notogastral setae fine, minute. Saccules well developed. Length: 530–650µm (Figure 3.109E) .... ***A. coleoptrata*** (Linnaeus)

    Note: The record of *Achipteria borealis* (Banks, 1899) from the Alaskan Pribilof Islands in Banks (1923) was overlooked in Behan-Pelletier and Lindo (2019). This species, originally described as *Oribatella borealis*, needs redescription (R.A. Norton pers. comm.)

    Note: *Achipteria* nr. *nitens* (Nicolet 1855) known from MB (Oswald and Minty 1970). Unidentified species from AK, Lava Camp, Seward Peninsula (Late Miocene: 5.7 mya), from late Pliocene/early Pleistocene fossil deposits in AK, and from Tertiary fossil sites on Meigen Island (NU) (Matthews et al. 2019). Unidentified and/or possibly undescribed species from AK, BC, AB, MT, ON, QC, NB.

Achipteria catskillensis Nevin, 1976 (Figure 3.109C)
    Biology: From forest litter.
    Distribution: ON, NS [NY; Nearctic]
    Literature: Seniczak and Seniczak (2016).

Achipteria clarencei Nevin, 1976 (Figure 3.109D)
    Biology: From boreal and mixedwood forest litter.
    Distribution: QC, NS [NY; Nearctic]

Achipteria coleoptrata (Linnaeus, 1758) (Figure 3.109E)
  Comb./Syn.: *Acarus coleoptratus* Linnaeus, 1758; *Notaspis coleoptratus* (Linnaeus, 1758); *Oribata ovalis* Nicolet, 1855; *Oribates nicoletii* Berlese, 1883; *Oribata intermedia* Michael, 1898
  Biology: Commonly found in rich fens with dwarf birch, willow, sedge, grass understory and in other peatlands (Behan-Pelletier and Bissett 1994). Alberta specimens have gut contents consisting mostly of coarse granular material suggesting feeding on decomposing vegetation, but some fungal hyphae and spores are visible as well as apparent moss spores (Walter et al. 2014). In a cafeteria experiment they fed preferentially on black sterile mycelium, and other fungi (Luxton 1972). Stable isotope data suggested litter-feeding (Schneider et al. 2004). Associated with this, it has chelicerae with a high leverage index (Perdomo et al. 2012). Schneider and Maraun (2005) placed this mite in the primary decomposer feeding guild (i.e., feeds mostly on litter). In feeding experiment, the green bark alga *Desmococcus vulgaris* was clearly the most preferred diet of this species (Hubert et al. 2001). Heidemann et al. (2011) found that this species consumed neither parasitic nematode species, *Phasmarhabditis hermaphrodita* or *Steinernema feltiae*. Gut microbiota were studied by Gong et al. (2018). This species was occasionally eaten by the mesostigmatan *Pergamasus septentrionalis* in a no-choice feeding experiment (Peschel et al. 2006). Generation time for *A. coleoptrata* is about one year (Luxton 1981a). A new group of *Cardinium* bacteria were found in this species (Konecka and Olszanowski 2019). This mite was reported as occurring in the hair of small mammals in Slovakia (Miko and Stanko 1991).
  Distribution: AB, MB, ON, QC [MI; semicosmopolitan]
  Literature: Weigmann (2006); ABMI (2019); Maraun et al. (2019).

Achipteria curta Aoki, 1970 (Figure 3.109A,B)
  Biology: Found on vegetation (Aoki 1970b).
  Distribution: AK, BC [Japan]

Anachipteria Grandjean, 1932
  Type-species: *Anachipteria deficiens* Grandjean, 1932

KEY TO ADULT *ANACHIPTERIA* OF CANADA AND ALASKA

1 Notogaster with 4 pairs of saccules. Length: 403–464µm (Figure 3.110A).........................
   .... *A. sacculifera* Root, Kawahara and Norton
   Note: Described from arboreal lichens in Adirondack mountains, NY; not known as yet in Canada.
– Notogaster with porose areas 2

2(1) Mentum of subcapitulum with tectum. Length: 420–470µm .... *A. australoides* Jacot
– Mentum of subcapitulum without tectum 3

3(2) Bothridial setae short, subcapitate. Lamellae closely adjacent, but not meeting medially. Notogastral porose areas about 13µm in diameter; all notogastral setae long, subequal in length. Length: ca. 540µm (Figure 3.110B)...*A. acuta* (Ewing)
– Bothridial setae long, clavo-lanceolate, directed medially. Lamellae meeting medially. Notogastral porose area *Aa* larger than 15µm in diameter; notogastral setae *c* and *la* longer than other notogastral setae 4

4(3) Anterior margin of lamellar cusps truncated anteriorly, lateral dens absent or very short. Length: 385–430µm (Figure 3.110C)..*A. howardi* (Berlese)
– Anterior margin of lamellar cusps oblique, with lateral dens strongly acute. Length: ca. 370µm (Figure 3.110D)..*A. magnilamellata* (Ewing)

Note: Unidentified and/or possibly undescribed species from YT, BC, AB, ON, QC, NB, NS, NL.

Anachipteria acuta (Ewing, 1918) (Figure 3.110B)
    Comb./Syn.: *Oribatella acuta* Ewing, 1918
    Biology: From Western redcedar litter.
    Distribution: BC [OR; Nearctic]
    Literature: Lindo (2010).

Anachipteria australoides Jacot, 1938
    Comb./Syn.: *Anachipteria achipteroides australoides* Jacot, 1938
    Biology: From maple forest litter. In oak litter (Lamoncha and Crossley 1998).
    Distribution: BC [NC; Nearctic]

Anachipteria howardi (Berlese, 1908) (Figure 3.110C)
    Comb./Syn.: *Sphaerozetes howardi* Berlese, 1908; *Anachipteria latitecta*(-*us*) (Berlese, 1908); *Anachipteria corpuscula* (Ewing, 1908); *Anachipteria achipteroides milleri* Jacot, 1936
    Biology: Gravid females carry up to 4 large eggs. Gut contents include fungal hyphae, brown granular material, and what appear to be moss spores (Walter et al. 2014).
    Distribution: NT, AB, MB [Holarctic]
    Literature: Norton and Kethley (1990); Walter et al. (2014); ABMI (2019).

Anachipteria magnilamellata (Ewing, 1909) (Figure 3.110D)
    Comb./Syn.: *Oribatella magnilamellata* Ewing, 1909
    Biology: From forest litter; maple. Indicator species of old growth (Root et al. 2007).
    Distribution: QC [IL, NY; Nearctic]

Dentachipteria Nevin, 1977
    Type-species: *Dentachipteria ringwoodensis* Nevin, 1974

Dentachipteria highlandensis Nevin, 1974 (Figure 3.111D)
    Biology: From Balsam fir forest.
    Distribution: NL [NY; Nearctic]
    Note: Unidentified and/or possibly undescribed *Dentachipteria* species from BC, NS, NL.

Parachipteria Hammen, 1952
    Comb./Syn.: *Campachipteria* Aoki, 1995
    Type-species: *Oribata punctata* Nicolet, 1855

KEY TO ADULT *PARACHIPTERIA* OF CANADA AND ALASKA

1   Porose area *A1* touching seta *lp*. Pteromorph with ventral spine. Length genu IV 2× length genu III; genu IV concave dorsally. Dark brown. Length: 760–825µm (Figure 3.111C)...... .... ***P. travei*** Nevin
–   Porose area *A1* positioned between setae *lp* and h_3, not touching seta *lp*. Pteromorph without ventral spine. Length genu IV 1–1.7× length genu III; genu IV almost flat dorsally. Light brown. 2

2(1) Lateral dens of lamellar cusp weakly rounded to angled. Genua I and II without ventral cusp. Length: 510–530µm (Figure 3.111B) ....***P. nivalis*** (Hammer)
–   Lamellar cusp flattened anteriorly. Genua I and II with ventral cusp. Length: 390–450µm (Figure 3.111A) ....***P. bella*** (Sellnick)
    Note: Unidentified species known from late Pliocene/early Pleistocene fossil deposits in AK (Matthews et al. 2019).

Parachipteria bella (Sellnick, 1928) (Figure 3.111A)
   Comb./Syn.: *Notaspis bellus* Sellnick, 1928; *Achipteria bellus* (Sellnick, 1928); *Campachipteria bella* (Sellnick, 1928)
   Biology: From deciduous and coniferous forest litter.
   Distribution: AB, QC [VA; Holarctic]
   Literature: Seniczak and Seniczak (2007); Bayartogtokh and Ryabinin (2012); ABMI (2019).

Parachipteria nivalis (Hammer, 1952) (Figure 3.111B)
   Comb./Syn.: *Achipteria nivalis* Hammer, 1952; *Campachipteria nivalis* (Hammer, 1952)
   Biology: From shrub tundra; *Lupinus*, *Dryas*, *Luzula*, *Salix* litter and humid moss; dry subalpine meadows; willow, *Arctostaphylos* and spruce litter. Bayartogtokh and Ryabinin (2012) redescribed this species from Mongolia; notogastral porose areas are smaller than those illustrated by Hammer (1952a).
   Distribution: AK, YT, NT, NU, BC, AB, NS [Nearctic]; known from Tertiary fossil sites at Strathcona Fiord, Ellesmere Island (NU) (Matthews et al. 2019).

Parachipteria travei Nevin, 1976 (Figure 3.111C)
   Biology: From wet areas with sphagnum moss and liverworts (Nevin 1976)
   Distribution: NS, NL [NY; Nearctic]

Pseudachipteria Travé, 1960
   Type-species: *Notaspis magnus* Sellnick, 1928
   Remarks: The synonymy of *Pseudachipteria* with *Parachipteria* by Subías (2004), was rejected by Weigmann (2006).
   Note: Unidentified species from QC.

Separachipteria Subías, 2019
   Type-species: *Anachipteria geminus* Lindo, Clayton and Behan-Pelletier, 2008

Separachipteria geminus (Lindo, Clayton and Behan-Pelletier, 2008) (Figure 3.110E,F)
   Comb./Syn.: *Anachipteria geminus* Lindo, Clayton and Behan-Pelletier, 2008
   Biology: The habitat of this species mimics that of *A. sacculifera*, described from arboreal lichens in New York State (Root et al. 2008). Whereas *S. geminus* is the numerically dominant oribatid mite in canopy lichens of Western hemlock and Pacific silver fir, *A. sacculifera* was the numerically dominant oribatid mite on foliose lichens on the crown branches of Sugar maple trees (Root et al. 2007). The abundances of adults and immature instars of *S. geminus* throughout the active growing season indicate that the populations studied reproduce continuously throughout the season.
   Distribution: BC [Canada]

Family Tegoribatidae Grandjean, 1954
   Diagnosis: *Adults*: Poronotic Brachypylina. Prodorsum with lamellae, lamellar cusps, tutorium, translamella present or absent. Genal incision present or absent. Bothridium without internal, ring-like thickenings; cup-shaped or with scales. Humerosejugal organ *Ah* as porose area or saccule; *Am* present, *Al* present or absent. Dorsophragmata and pleurophragmata present; dorsophragmata separate, or fused medially and elongate. Discidium and custodium present. Pedotectum I large, extending to base of bothridium; pedotectum II present. Notogaster with posterior tectum, overlapping or not; pteromorphs present, with or without hinge; lenticulus present or absent; hexagonal pattern present or absent anteriorly on notogaster; setation 10–14 pairs. Octotaxic system developed as porose areas, or saccules, sometimes minute, or as tubules. Epimere I with or without necklace of tubercles. Genital setation 5–6 pairs. Postanal porose area or saccule present or absent. Chelicerae chelate-dentate; palpal eupathidium *acm* attached to solenidion; subcapitular mentum with or without complete

tectum. Axillary saccule of subcapitulum present. Genu IV shorter than tibia IV; tibia IV with or without solenidion. Solenidion φ of tibia IV present or absent.

KEY TO GENERA OF TEGORIBATIDAE OF CANADA AND ALASKA

1   Pteromorph with hinge. Epimere I without necklace of small tubercles. Dorsophragmata fused medially and elongated. Subcapitulum with tectum of mentum covering gena and most of rutella (Figure 3.112A–F) .... *Tegoribates* Ewing
–   Pteromorph without hinge. Epimere I with or without necklace of small tubercles. Dorsophragmata separated, not elongated. Subcapitulum without tectum 2

2(1)   Tibia II with dorsal dens. Necklace of tubercles absent on epimere I. Subcuticular hexagonal pattern absent on anterior of notogaster (Figure 3.112G,H) ....
       ....*Protectoribates* Behan-Pelletier
–   Tibia II without dorsal dens. Necklace of tubercles on epimere I present. Subcuticular hexagonal pattern on anterior of notogaster present or absent (Figure 3.113) ....
       ....*Tectoribates* Berlese

Protectoribates Behan-Pelletier, 2017
   Type-species: *Protectoribates occidentalis* Behan-Pelletier, 2017

Protectoribates occidentalis Behan-Pelletier, 2017 (Figure 3.112G,H)
   Biology: From coniferous and deciduous forest litter.
   Distribution: BC [AZ, CA, CO, NM, OR; Nearctic]

Tectoribates Berlese, 1910
   Type-species: *Tectoribates proximus* Berlese, 1910
   Comb./Syn.: *Anoribatella* Kunst, 1962

KEY TO *TECTORIBATES* OF CANADA AND ALASKA

1   Notogaster with saccules. Length: 280–300μm (Figure 3.113A) ....
       .... *T. campestris* Behan-Pelletier and Walter
–   Notogaster with porose areas 2

2(1)   Notogaster with hexagonal pattern anteriorly. Lamellar cusp with medial and lateral dens. Length: 325–340μm (Figure 3.113B–D) .... *T. borealis* Behan-Pelletier and Walter
–   Notogaster without hexagonal pattern anteriorly. Lamellar cusp with lateral dens only. Length: 275–305μm (Figure 3.113E)..........*T. alcescampestris* Behan-Pelletier and Walter

Tectoribates alcescampestris Behan-Pelletier and Walter, 2013 (Figure 3.113E)
   Biology: Collected primarily in aspen parkland in forest litter and fungal sporocarps and in open sites covered by grasses and sedges. Gravid females carry up to 2 large eggs. Gut contents are primarily lightly pigmented fungal hyphae and spores.
   Distribution: AB [Canada]
   Literature: ABMI (2019).

Tectoribates borealis Behan-Pelletier and Walter, 2013 (Figure 3.113B–D)
   Biology: Found in open habitats, primarily native prairie, badlands and pastures. Gut contents are primarily of dark brown fungal hyphae and spores.
   Distribution: AB [Canada]
   Literature: ABMI (2019).

Tectoribates campestris Behan-Pelletier and Walter, 2013 (Figure 3.113A)
   Biology: Populations from Kansas with sex ratio of 1:1.
   Distribution: ON [KS; Nearctic]

Tegoribates Ewing, 1917
   Type-species: *Tegoribates subniger* Ewing, 1917

KEY TO ADULT *TEGORIBATES* OF CANADA AND ALASKA

1   Notogaster with octotaxic system as saccules. Length: 410–480µm (Figure 3.112D)..........
     ....*T. americanu*s Hammer
–   Notogaster with octotaxic system as tubules. Length: 518–580µm (Figure 3.112A–C).......
     ....*T. subniger* Ewing
     Note: Unnamed species from BC, AB.

Tegoribates americanus Hammer, 1958 (Figure 3.112D)
   Biology: From tussock tundra; moss and alder litter; peatlands; mixed alpine vegetation; coastal meadows.
   Distribution: AK, YT, NT, BC, AB, MB [CA, KS; Nearctic]; specimens known from Tertiary fossil sites on Meigen Island (NU) (Matthews et al. 2019).
   Literature: Behan-Pelletier (2017); ABMI (2019).

Tegoribates subniger Ewing, 1917 (Figure 3.112A–C)
   Distribution: AB [CO, IN, NY; Nearctic].
   Literature: Behan-Pelletier (2017).

Superfamily Oribatelloidea Jacot, 1925

Family Oribatellidae Jacot, 1925
   Diagnosis: *Adult*: Poronotic Brachypylina. Lamellae long, broad, with large cusps, medially converging or contiguous, or separated; cusps usually with large medial and lateral dens, with or without ventral keel; translamella present; interlamellar pocket present or absent. Tutorium present. Bothridium cup-like, with indentation. Genal incision present. Pedotectum I well developed, with ventral indentation; pedotectum II with paraxial tubercle. Humerosejugal porose organ *Ah* expressed as porose area or saccule; porose area *Al* expressed as porose area or saccule, or absent. Discidium present; custodium present or absent. Postanal porose area present. Epimeral seta *4c* spinous. Notogaster with immovable pteromorphs; setation 10–14 pairs. Octotaxic system as 3–4 pairs of porose areas or saccules; usually without sexual dimorphism; when present that of male modified. Chelicera chelate-dentate, elongated in one species. Mentum with or without tectum, with or without recurved ridge distally. Palp setation: 0-2-1-3-9(1) or 0-2-1-3-8(1); eupathidium *acm* forming double horn with solenidion along length or distally; or solenidion almost 2.5× length of *acm*. Axillary saccule present. Tibia I with or without anterodorsal spines close to solenidia φ_1 and φ_2. Legs monodactylous, heterobidactylous or heterotridactylous.

KEY TO ADULT ORIBATELLIDAE OF CANADA AND ALASKA

1   Interlamellar pocket large, box-like, appearing to reach posterior of dorsosejugal scissure (Figure 3.114A, arrow).. ***Ferolocella*** Grabowski
–   Interlamellar pocket present or absent, if present, never large, box-like, or appearing to reach posterior of dorsosejugal scissure (Figures 3.114B–F; 3.115–3.117)............................
     ....***Oribatella*** Banks

Ferolocella Grabowski, 1971
  Type-species: *Ferolocella tessellata* (Berlese, 1908)
  Comb./Syn.: *Ferolocella carolina* (Banks, 1947)

Ferolocella tessellata (Berlese, 1908) (Figure 3.114A)
  Comb./Syn.: *Oribatella carolina* Banks, 1947
  Biology: From White birch and Trembling aspen (as: *Ferolocella* sp.) (St. John et al. 2002). In oak litter (Lamoncha and Crossley 1998).
  Distribution: ON [AL, AR, GA, MO, NC, OH, VA, TX; WA, WI, Nearctic].
  Literature: Behan-Pelletier (2013).

Oribatella Banks, 1895
  Type-species: *Oribatella quadridentata* Banks, 1895

KEY TO ADULT *ORIBATELLA* OF CANADA AND ALASKA

1	Legs monodactylous. Rostral margin undulating, with or without pair of minute dens laterally	2
–	Legs heterobidactylous or heterotridactylous	3
2(1)	Interlamellar region with translamella. Medial dens of lamellar cusps less than 1/4 length of lateral dens. Length: 295–320µm (Figure 3.114C)	***O. nortoni*** Behan-Pelletier
–	Interlamellar region without translamella. Medial dens of lamellar cusps at least 1/3 length of lateral dens; may be subequal to or longer than lateral dens. Length: 322–355µm (Figure 3.114D)	***O. sintranslamella*** Behan-Pelletier and Walter
3(1)	Legs heterobidactylous. Rostrum with lateral dens. Length: 366–372µm (Figure 3.114B)	***O. dentaticuspis*** Ewing
–	Legs heterotridactylous. Rostral margin with lateral dens, medial and lateral dens, or differently shaped	4
4(3)	Integument of notogaster and ventral plate reticulate. Length: 300–340µm (Figure 3.114E)	***O. minuta*** Banks
–	Integument of notogaster not reticulate; either smooth, punctate, or microtuberculate	5
5(4)	Notogastral setation 11 pairs, c_3 present laterally at base of pteromorph; setae long, barbed, seta c about 240µm long. Chelicera and subcapitulum elongate; rutellum tapered. Custodium absent. Very large species, Length: ca. 700µm (Figure 3.114F)	***O. quadridentata*** Banks
	Note: Known from Northeast USA (GA, LA, MS, NC, NY, NH, VA); probably in Canada.	
–	Notogastral setation 10 pairs, c_3 absent; setae short to long, smooth or barbed, seta c less than 200µm long. Chelicera and subcapitulum not elongate; rutellum not tapered. Custodium present. Length: <650µm	6
6(5)	Translamella with medial tooth	7
–	Translamella without tooth	8
7(6)	Rostrum with distinct crest in dorsal and lateral region. Medial dens on lamellar cusp subequal in length to lateral dens. Custodium half to two-thirds length of epimeral seta *4c*. Length: 490–525µm (Figure 3.115A)	***O. quadricornuta*** (Michael)
–	Rostrum projecting ventrally, margin rounded. Medial dens on lamellar cusp longer than lateral dens. Custodium about 1.4 times length of epimeral seta *4c*. Length: 420–470µm (Figure 3.115B)	***O. pawnee*** Behan-Pelletier and Walter

8(6)		Custodium with 1 or (usually) 2 knobs along length. Lamella with cusp 180–188μm long. Genital setal arrangement with 1 pair anteriorly. Length: 470–572μm (Figure 3.115C,D) .. .. *O. brevicornuta* Jacot
–		Custodium without knobs. Lamella with cusp length other than that given above. Genital setal arrangement with 1–3 pairs anteriorly 9

9(8) Setal pair h_1 directed posteriorly, subparallel; acuminate or strongly barbed and blunt. Mutual distance of setae p_1-p_1 subequal to that of h_1-h_1, about 24–25μm 10

– Setae h_1 curved laterally and acuminate. Mutual distance of setae p_1-p_1 less than that of h_1-h_1 11

10(9) Lamellae converging distally, interlamellar space drop-shaped, about 3× as long as wide at widest point. Maximum distance between lamellar cusps about 24μm. Medial and lateral dentes of lamellar cusp about 70μm. Short striae present in lenticular area. Length: 530–640μm (Figure 3.115E) ....*O. jacoti* Behan-Pelletier

– Lamellae subparallel, interlamellar space narrow, 8–10× as long as wide at widest point. Maximum distance between lamellar cusps about 9μm. Medial and lateral dens of lamellar cusp about 40–64μm. Short striae absent in lenticular area. Length: 525–584μm (Figure 3.115F) ....*O. parallelus* Behan-Pelletier and Walter

11(9) Rostral margin produced as medial mucro, without lateral dens. Lamellae closely appressed basally (interlamellar space nearly obsolete) and with short striae running transversely and longitudinally, forming tapestry pattern; pedotectum I with similar 'woven' pattern. Epimeral seta *4c* subequal in length to custodium, very thick and heavily barbed. Length: 490–564μm (Figure 3.116A).....................................*O. ewingi* Behan-Pelletier and Walter

– Rostral margin not produced as medial mucro, but may have medial and lateral dentes or be truncate, incised, undulate or variously toothed. Lamellae usually widely separated basally, but if interlamellar space narrow, then lamella and pedotectum I without tapestry pattern. Epimeral seta *4c* either thin, or if thick and heavily barbed, then distinctly longer than custodium 12

12(11) Medial dens of lamella tine-like, tapering abruptly near base and extending anteriorly from slightly to distinctly more than lateral dens. Area defined by medial and lateral dens deep and narrowing towards medially inserted lamellar seta, or broadly U-shaped with lamellar seta inserted towards lateral dens. Lamella with or without transverse striae 13

– Medial dens of lamella tapering more gradually, more blade-like and not extending as far anteriad as lateral dens. Area defined by medial and lateral dentes shallow or relatively deep and margin gradually rounded to lamellar seta. Lamella without transverse striae 15

13(12) Lamellar cusp with transverse striae. Lateral margin of medial dens of lamellar cusp slightly concave. Ventral plate shallowly foveolate. Length: 450–452μm (Figure 3.116B) .. ..*O. transtriata* Behan-Pelletier

– Lamellar cusp without transverse striae. Lateral margin of medial dens of lamellar cusp straight. Ventral plate punctate-striate. Length: >460μm 14

14(13) Integument of pedotectum I with short ridges. Area defined by medial and lateral dentes of lamella broadly triangular around medial insertion of lamellar seta. Length: 432–505μm (Figure 3.116C)..*O. yukonensis* Behan-Pelletier and Walter

– Integument of pedotectum I, without short ridges. Area defined by medial and lateral dentes of lamella subrectangular, lamellar seta inserted closer to lateral dens. Length: 500–574μm (Figure 3.116D)...............................*O. heatherae* Behan-Pelletier and Walter

15(12)	Notogastral setae flagellate. Interlamellar region distinctly V-shaped posteriorly. Length: 410–440μm (Figure 3.116E)	***O. flagellata*** Behan-Pelletier
–	Notogastral setae acuminate or tapered, never flagellate. Interlamellar region transverse to concave posteriorly	16
16(15)	Rostrum with deep medial indentation, both edges of indentation with 2–3 teeth; with ridge extending from teeth at edge of indentation to insertion of rostral seta	17
–	Rostrum with medial and lateral dens, or undulating, with or without lateral dens; if with deep medial indentation, edges of indentation without teeth and without ridge extending from teeth at edge of indentation to insertion of rostral seta, but may have ridge extending from rostral crest to rostral seta..	18
17(16)	Epimere I distinctly foveate laterally. Rostrum deeply indented with 3 pairs of teeth. Octotaxic system not sexually dimorphic. Length: 350–400μm (Figure 3.117A) .... ***O. manningensis*** Behan-Pelletier and Walter	
–	Epimere I with striae forming irregular pattern laterally. Rostrum indented with 1 or 2 pairs of teeth. Octotaxic system sexually dimorphic. Length: 365–417μm (Figure 3.117B,C) .... ***O. canadensis*** Behan-Pelletier and Eamer	
18(16)	Translamella truncate, producing a subtriangular-shaped; interlamellar space 13–19μm wide × 24–33μm long	19
–	Translamella concave or angular, interlamellar space drop-shaped or diamond-shaped..	20
19(18)	Rostrum with medial tooth and 2 strong lateral teeth. Integument of rostrum micropunctate. Distance h_1–h_1 about 24μm. Length: 360–390μm (Figure 3.117D) .... ***O. maryae*** Behan-Pelletier and Walter	
–	Rostrum undulating medially, with or without small medial and lateral teeth. Integument of rostrum with ridges, irregular striae and foveae. Distance h_1–h_1 about 37μm. Length: 400–440μm (Figure 3.117G)......................................***O. banksi*** Behan-Pelletier and Walter	
20(18)	Rostral margin undulating. Lateral dens on lamellar cusp with teeth on both medial and lateral margins. Distance h_1–h_1 subequal to distance p_1–p_1. Length: 330–340μm (Figure 3.117E) ....***O. reticulatoides*** Hammer	
–	Rostral margin with 2 distinct long lateral teeth. Lateral dens on lamellar cusp usually with teeth only laterally. Distance h_1–h_1 wider than distance p_1–p_1.	21
21(20)	Rostral margin tridentate, with medial and lateral teeth. Length: 420–490μm (Figure 3.117F) ....***O. arctica*** Thor	
–	Rostral margin undulating between lateral teeth. Length: 390–400μm (Figure 3.117H)..... .... ***O. abmi*** Behan-Pelletier and Walter	

Note: Unidentified species from AK: Lava Camp, Seward Peninsula (Late Miocene: 5.7 mya) (Matthews et al. 2019).

Oribatella abmi Behan-Pelletier and Walter, 2012 (Figure 3.117H)
    Biology: Mostly collected in the upper soil organic layers on dry, untreed sites with grasses, sedges, and bryophyte cover. Gut contents include pigmented fungal hyphae and spores.
    Distribution: AB [Canada]
    Literature: ABMI (2019).

Oribatella arctica Thor, 1930 (Figure 3.117F)
    Biology: From arctic latitudes of the Nearctic and Palaearctic. A subspecies, *O. arctica littoralis* Strenzke 1950, has been found associated with littoral habitats in the Netherlands (Polderman 1974), and has been used to evaluate archaeological findings (Schelvis 1990).

Distribution: AK, YT, NT, NU, BC, MB [Holarctic]
Literature: Behan-Pelletier (2011); Matthews et al. (2019).

Oribatella banksi Behan-Pelletier and Walter, 2012 (Figure 3.117G)
Biology: From deciduous and coniferous forest litter and dry litter habitats. Gut contents include brown fungal hyphae and what appear to be moss spores.
Distribution: AB, BC [CA, OR, WA; Nearctic]
Literature: ABMI (2019).

Oribatella brevicornuta Jacot, 1934 (Figure 3.115C,D)
Biology: Widely distributed in deciduous forests throughout eastern North America. In oak litter (Lamoncha and Crossley 1998).
Distribution: QC, NS [AR, CT, GA, MO, MS, NC, NY, TX; Nearctic]
Literature: Behan-Pelletier (2011).

Oribatella canadensis Behan-Pelletier and Eamer, 2010 (Figure 3.117B,C)
Biology: Known mostly from dry litter habitats. Gut contents include pigmented fungal hyphae and spores. Sexually dimorphic species, males with modified octotaxic system.
Distribution: AK, YT, BC, AB, ON [NM; Nearctic]
Literature: ABMI (2019).

Oribatella denticuspis Ewing, 1910 (Figure 3.114B)
Biology: From mixed mesophytic forest litter, bracket fungi, decaying wood.
Distribution: ON, QC, NB [CA, MO, NY; Nearctic]
Literature: Behan-Pelletier (2011).

Oribatella ewingi Behan-Pelletier and Walter, 2012 (Figure 3.116A)
Biology: From deciduous and coniferous forest and dry litter habitats.
Distribution: AB, BC [CA, NV, OR, WA; Nearctic]
Literature: ABMI (2019).

Oribatella flagellata Behan-Pelletier, 2011 (Figure 3.116E)
Biology: From polypore fungi.
Distribution: QC [VA; Nearctic]

Oribatella heatherae Behan-Pelletier and Walter, 2012 (Figure 3.116D)
Biology: From dry litter habitats in montane areas of southern Alberta.
Distribution: AB [Canada]

Oribatella jacoti Behan-Pelletier, 2011 (Figure 3.115E)
Biology: Gravid females carry up to 6 large eggs. Found in deciduous litter habitats, especially aspen forests in Alberta.
Distribution: AB, ON, NS [AR, IL, MO, WI; Nearctic]
Literature: ABMI (2019).

Oribatella manningensis Behan-Pelletier and Walter, 2012 (Figure 3.117A)
Biology: From a variety of litter types in the Montane.
Distribution: AB, BC [WA; Nearctic]

Oribatella maryae Behan-Pelletier and Walter, 2012 (Figure 3.117D)
Biology: From moist to dry forest habitats.
Distribution: BC [CA, OR, WA; Nearctic]

Oribatella minuta Banks, 1896 (Figure 3.114E)
   Biology: From deciduous forest litter; polypore fungi.
   Distribution: ON, QC, NB [IL, MO, NY, WI; Nearctic]
   Literature: Behan-Pelletier (2011).

Oribatella nortoni Behan-Pelletier, 2011 (Figure 3.114C)
   Biology: From dry moss; conifer litter.
   Distribution: NB [AL, AR, NY; Nearctic]

Oribatella parallelus Behan-Pelletier and Walter, 2012 (Figure 3.115F)
   Biology: From forest habitats on the western slopes of the Rocky Mountains in interior BC.
   Distribution: BC [Canada]

Oribatella pawnee Behan-Pelletier and Walter, 2012 (Figure 3.115B)
   Biology: Gravid females carry up to 4 eggs; gut contents contain dark and hyaline fungal spores and hyphae. A grassland mite from the shortgrass prairie of CO to the mixed grasslands of southern AB.
   Distribution: AB [CO; Nearctic]
   Literature: ABMI (2019).

Oribatella quadricornuta (Michael, 1880) (Figure 3.115A)
   Comb./Syn.: *Oribata quadricornuta* Michael, 1880; *Notaspis quadricornuta* (Michael, 1880); *Oribates flammula* sensu Oudemans, 1896; *Oribatella calcarata* sensu Willmann, 1931.
   Biology: From deciduous litter; littoral habitats. Gut content analysis of adults showed consumption of primarily fungi, a small amount of plant material, and arthropod fragments; fungal hyphae were cropped to 20–60µm lengths (Behan-Pelletier and Hill 1983). In a cafeteria experiment with a choice of eight common dark pigmented fungal taxa, this species fed primarily on the dematiaceous *Alternaria alternata* and *Ulocladium* sp. (Schneider and Maraun 2005).
   Distribution: NT, QC, NB, NL, PE [Holarctic]
   Literature: Behan-Pelletier (2011); Seniczak and Seniczak (2013); Maraun et al. (2019).

Oribatella reticulatoides Hammer, 1955 (Figure 3.117E)
   Comb./Syn.: *Oribatella reticuloides* (*sic*.) Hammer, 1955
   Biology: From upper organic layers of aspen woodland, pine forest, and mesic, grassy areas. Gravid females carry up to 3 large eggs. Gut boli contain pigmented and unpigmented fungal hyphae and spores and granular material.
   Distribution: AK, AB, ON, QC, NB [Nearctic]
   Literature: Behan-Pelletier and Walter (2012); ABMI (2019).

Oribatella sintranslamella Behan-Pelletier and Walter, 2012 (Figure 3.114D)
   Biology: From dry moss; conifer litter.
   Distribution: BC [OR, WA; Nearctic]

Oribatella transtriata Behan-Pelletier, 2011 (Figure 3.116B)
   Biology: From boreal moss and lichens, fungi, leaf litter.
   Distribution: ON, NB, NS, NL [Canada]

Oribatella yukonensis Behan-Pelletier and Walter, 2012 (Figure 3.116C)
   Biology: Gravid females carry at least 3 eggs. Gut contents include brown fungal hyphae. From dry habitats and upland aspen and spruce forests.
   Distribution: YT, AB [Canada]
   Literature: ABMI (2019).

Superfamily Oripodoidea Jacot, 1925

Family Haplozetidae Grandjean, 1936
Diagnosis: *Adults*: Poronotic Brachypylina. Prodorsum with lamellae well developed, or reduced to narrow carina; prolamella, sublamella, tutorium and translamella present or absent. Lamellar setae borne on tip of lamella, or on prodorsum. Bothridium often covered by anterior margin of notogaster. Genal incision absent. Prodorsum and notogaster separate or fused. Dorsophragmata and pleurophragmata present. Discidium and custodium present. Pedotectum I small, usually not completely covering acetabulum I, pedotectum II small. Circumpedal carina present. Genital setation 4–5 pairs. Marginoventral porose areas present or absent. Octotaxic system with 4 pairs of porose areas or saccules. Notogaster with or without lenticulus, without posterior tectum; pteromorphs present, with or without hinge; setation 9–14 pairs. Chelicerae chelate-dentate or pelopsiform; palpal eupathid *acm* fused to solenidion to form double-horn on short, broad apophysis; subcapitular mentum without tectum. Axillary saccule of subcapitulum absent. Usually with porose organs on tibiae and tarsi. Tibiae I and II with or without dorsal tooth. Tibia IV with solenidion.

KEY TO GENERA OF HAPLOZETIDAE OF CANADA AND ALASKA

1	Notogaster with porose areas	2
–	Notogaster with saccules	3
2(1)	Genital setation 5 pairs. Sublamella vestigial. Tutorium usually present as ridge or tectum. Pteromorphs hinged (Figure 3.119)	***Protoribates*** Berlese
–	Genital setation 4 pairs. Sublamella strongly developed, reaching lamellar seta. Tutorium absent. Pteromorphs without hinge (Figure 3.118A)	***Lagenobates*** Weigmann and Miko
3(1)	Dorsosejugal scissure with 3 arches. Octotaxic system of 4 pairs of very small saccules. On palptarsus eupathidium *acm* and solenidion on large apophysis and strongly curved (Figure 3.118B,C)	***Rostrozetes*** Sellnick
–	Dorsosejugal scissure forming single arch. Notogastral saccules clearly evident. On palptarsus apophysis bearing eupathidium *acm* and solenidion small; never strongly curved	4
4(3)	Notogastral setation 10 or 11 pairs (*d* series absent). Notogastral saccules long, often furcated. Tarsus I with solenidion ω_1 at least 1.5× length of ω_2; without porose area in tarsal cluster. Genital setation 4 or 5 pairs (Figure 3.118D)	***Haplozetes*** Willmann
–	Notogastral setation 14–15 pairs (*d* series present). Notogastral saccules hemispherical or sac-like. Tarsus I with solenidion ω_1 equal in length to ω_2; with porose area in tarsal cluster. Genital setation 5 or 6 pairs	5
5(4)	Genital setation 5 or 6 pairs. Aggenital setation 3 or more pairs. Monodactylous (Figure 3.118E)	***Pilobates*** Balogh
–	Genital setation 5 pairs. Aggenital setation 1 pair. Tridactylous (Figure 3.120)	***Peloribates*** Berlese

Haplozetes Willmann, 1935
  Type-species: *Peloribates vindobonensis* Willmann, 1935
  Note: Unidentified species from ON, QC, NL.

Lagenobates Weigmann and Miko, 2002
  Type-species: *Oribates lagenula* Berlese, 1904 (Figure 3.118A)
  Note: *Lagenobates* sp. nr. *langenula* from NB (Walter and Latonas 2013).

Peloribates Berlese, 1908
    Type-species: *Oribates peloptoides* Berlese, 1888

KEY TO ADULT *PELORIBATES* OF CANADA AND ALASKA

1	Integument of prodorsum and notogaster foveate. Notogastral setae thick	2
–	Integument of prodorsum smooth, that of notogaster foveate or not. Notogastral setae thin	3
2(1)	Bothridial seta with short stalk and clavate head. Length: 400µm (Figure 3.120A) ....***P. juniperi*** (Ewing)	
–	Bothridial seta fusiform. Length: 390–490µm .... ***P. americanus*** (Jacot)	
3(1)	Length: usually >500µm; either rostrum foveate or notogastral setae inserted in dark spots. Length: 540–580µm (Figure 3.120B) .... ***P. alaskensis*** Hammer	
–	Length: usually <500µm; rostrum not foveate (although notogaster may be), notogastral setae not inserted in dark spots 4	
4(3)	Stalk of bothridial seta similar in length to club. Notogastral setae relatively short (mostly <60µm), fine or thickened. Saccules shallow, sack-like. Length: 410–450µm (Figure 3.120C) ....***P. canadensis*** Hammer	
–	Bothridial seta with long stalk and fusiform head. Notogastral setae uniformly long (75–90µm), approaching length of interlamellar seta (105–110µm). Saccules tube-like, *Sa* usually curved. Ventral length 430–500µm (Figure 3.120D) .... ***P. pilosus*** Hammer	

Note: At least 3 unnamed species known from AB (Walter et al. 2014).

Peloribates alaskensis Hammer, 1955 (Figure 3.120B)
    Biology: From forest litter.
    Distribution: AK [Alaska]

Peloribates americanus Jacot, 1939
    Comb./Syn.: *Peloribates europaeus americanus* Jacot, 1939
    Biology: From beech litter.
    Distribution: QC [NH, NY; Nearctic]

Peloribates canadensis Hammer, 1952 (Figure 3.120C)
    Biology: Gravid females carry up to 6 large eggs. Gut contents are typically composed of dark fungal hyphae, but several include the cuticle of small mites, including one immature oribatid.
    Distribution: AK, YT, NU, NT, AB, MB, QC [Nearctic]
    Literature: Walter et al. (2014); ABMI (2019).

Peloribates juniperi (Ewing, 1913) (Figure 3.120A)
    Comb./Syn.: *Oribata juniperi* Ewing, 1913; *Rostrozetes juniperi* (Ewing, 1913)
    Biology: From White birch and Trembling aspen litter.
    Distribution: ON [Nearctic].

Peloribates pilosus Hammer, 1952 (Figure 3.120D)
    Biology: Reported from mesophilous habitats and peatlands. Gravid females carry up to 5 eggs. Gut contents include mostly fungal hyphae and spores with some coarse organic matter.
    Distribution: AK, YT, NU, NT, AB, MB [Nearctic]
    Literature: Walter et al. (2014); ABMI (2019).

Pilobates Balogh, 1960
    Type-species: *Protoribates pilosellus* Balogh, 1958
    Note: Unidentified species from AB (Newton 2013).

Protoribates Berlese 1908
  Type-species: *Oribata dentata* Berlese, 1883
  Comb./Syn.: *Alloribates* Banks, 1947; *Propeschelobates* Jacot, 1936; *Styloribates* Jacot,1934; *Xylobates* Jacot, 1929

KEY TO ADULT *PROTORIBATES* OF CANADA AND ALASKA
(AFTER WALTER AND LATONAS 2013)

1   Monodactylous. Bothridial seta relatively long (ca. 1.3× longer than lamella) and reflexed over notogaster. Notogastral setae usually reduced to alveoli. On palp tarsus *acm* complex on distinct apophysis 2

–   Heterotridactylous. Bothridial seta relatively short (ca. 0.75× length of lamella) and directed dorsally. Notogastral setae short, fine. On palptarsus *acm* complex nearly sessile. Length: 470–570µm (Figure 3.119A,B) .... *P. haughlandae* Walter and Latonas

2(1) Dorsophragmata normally produced, half length or less of mutual distance. Lamellar setae long, extending beyond rostrum; interlamellar setae similar to lamellar setae. Adanal setae ad_1 and ad_2 20–50µm long, one or both usually visible in dorsal view. Discidium with acuminate custodial tip. Length: >465µm long.. 3

–   Dorsophragmata unusually long (ca. 35–40µm), subequal to mutual distance (ca. 30µm). Lamellar setae short, not extending beyond rostrum, interlamellar setae short, stout. Adanal setae ad_1 and ad_2 both short (ca. 10µm), usually not visible in dorsal view. Discidium large, rounded distally. Length: 320–440µm (Figure 3.119C).....................*P. capucinus* Berlese

3(2) Bothridial seta sublanceolate or with fusiform head. Length: 470–530µm (Figure 3.119D)
    ....*P. robustior* (Jacot)

–   Bothridial seta setiform, of even diameter to acuminate tip, without expanded head. Length: 520–600µm (Figure 3.119E)..*P. lophotrichus* (Berlese)
    Note: Unidentified species from BC (Lindo 2020).

Protoribates capucinus Berlese, 1908 (Figure 3.119C)
  Comb./Syn.: *Xylobates singularis* (Banks, 1947); *Xylobates angustior* Jacot, 1937; *Xylobates capucinus* (Berlese, 1908); *Oribata oblonga* Ewing, 1909; *Hemileius oblongus* (Ewing, 1909)
  Biology: From oak litter (Lamoncha and Crossley 1998); sphagnum (Donaldson 1996); forest (Cianciolo and Norton 2006). Based on collections, this species is parthenogentic in North America (Norton and Kethley 1990).
  Distribution: ON, QC, NS [MO, NC, NH, NY, VA; Holarctic]
  Literature: Walter and Latonas (2013); Maraun et al. (2019).

Protoribates haughlandae Walter and Latonas, 2013 (Figure 3.119A,B)
  Biology: Gravid females carry up to 6 eggs. Gut contents included pigmented and unpigmented fungal hyphae and spores, particles of organic matter, and sometimes the cuticle of small arthropods. Found in open areas including fens, bogs and moist to mesic forest litter.
  Distribution: AB [Canada]
  Literature: ABMI (2019).

Protoribates lophotrichus (Berlese, 1904) (Figure 3.119E)
  Comb./Syn.: *Protoribates prionotus* (Woolley, 1968); *Oribates lophothrichus* Berlese, 1904; *Xylobates lophotrichus* (Berlese, 1904); *Xylobates prionota* Woolley, 1968; *Styloribates pectinatus* Jacot, 1934.
  Biology: Fed preferentially on decaying plant material, but also on fungi (Hartenstein 1962). Over 18 months of study he noted that this species preferred parenchymatous tissue of

decaying leaves and fed more vigorously upon Sugar maple leaves decayed with a heterogeneous assemblage of microorganisms in nature than upon the same leaves decayed aseptically.
Distribution: ON, QC [NY, TN, VA; Holarctic]
Literature: Walter and Latonas (2013).

Protoribates robustior (Jacot, 1937) (Figure 3.119D)
Comb./Syn.: *Xylobates capucinus robustior* Jacot, 1937; *Xylobates robustior* Jacot, 1937
Biology: Gravid females carry 2–3 large eggs. Gut contents include coarse particulate matter, possible plant cell fragments, and fungal hyphae and spores.
Distribution: AB [MI, NC, NY, TN, VA; Nearctic]
Literature: Walter et al. (2014); ABMI (2019).

Rostrozetes Sellnick, 1925
Type-species: *Rostrozetes foveolatus* Sellnick, 1925

KEY TO ADULT *ROSTROZETES* OF CANADA AND ALASKA

1	Notogastral setation 11 pairs. Barbs on bothridial seta arranged in 4 longitudinal rows. Length: not given .... ***R. appalachicolus*** Jacot
–	Notogastral setation 10 pairs. Barbs on bothridial seta irregularly arranged. Length: 280–450μm (Figure 3.118B,C) .... ***R. ovulum*** (Berlese)

Rostrozetes appalachicolus Jacot, 1938
Comb./Syn.: *Rostrozetes foveolatus appalachicolus* Jacot, 1938
Biology: From woodland litter.
Distribution: NS [NC; Nearctic]

Rostrozetes ovulum (Berlese 1908) (Figure 3.118B,C)
Comb./Syn.: *Trachyoribates ovulum* Berlese 1908; *Rostrozetes foveolatus* Sellnick, 1925; *Rostrozetes flavus* Woodring, 1965
Biology: Widely distributed from boreal peatlands to the riparian in Amazonia; a frequent inhabitant of bogs in eastern North America (Norton and Palmer 1991) and is also the dominant oribatid species in Amazonian floodplains (Franklin et al. 2001; as *R. foveolatus*). Woodring (1965) noted this species (as R. *flavus*) showed a distinct preference for the decomposing outer sheaths of roots. *Rostrozetes ovulum* attacks the apical portion of roots of pineapple (Sanyal and Das 1989). Based on stable isotope analysis *R. ovulum* is a primary decomposer (Illig et al. 2005). This species is parthenogenetic and shows high phenotypic plasticity (Pequeño et al. 2020, 2021).
Distribution: ON, QC, NB [LA, NC, NH; Holarctic; Neotropics]
Literature: Norton and Kethley (1990); Maraun et al. (2019).

Family Mochlozetidae Grandjean, 1960
Diagnosis: *Adults*: Poronotic Brachypylina. Prodorsum with lamellae well developed; sublamella present; translamella present or absent; tutorium present or absent. Porose area *Al* present. Genal incision absent, small tooth maybe present. Dorsosejugal scissure incomplete, or absent. Dorsophragmata and pleurophragmata present. Pedotectum I small, covering acetabulum I, pedotectum II small. Discidium present, custodium present or absent. Circumpedal carina present. Marginoventral porose areas present. Notogaster with or without lenticulus, without posterior tectum; setation 10 pairs. Pteromorphs present, without hinge, or absent. Octotaxic system often with more than 4 pairs of porose areas, often elongated, often showing sexual dimorphism; porose areas with ridged border. Genital setation 5–6 pairs. Chelicerae chelate-dentate; with distinct notch antiaxially. Axillary saccule of

subcapitulum absent. Palpal eupathidium *acm* attached to solenidion; subcapitular mentum without tectum. With dorsal porose organs on tarsus I, or on tarsi I and II. Tibia IV with solenidion.

Biology: Species usually inhabit living plants. The occasional collection of specimens in soil and litter probably results from accidental dislodgement from vegetation or from dispersal behavior (Norton 1983). Species are most diverse in temperate to tropical regions of the Americas, although species of *Podoribates* are widespread in the Holarctic where they are associated with grasses and wet pastures. Most mochlozetids have a ventral subterminal tooth on each lateral claw of the leg tarsi that is thought to enhance grasping efficiency on plants (Norton 1983).

KEY TO GENERA OF MOCHLOZETIDAE OF CANADA AND ALASKA

1 Pteromorphs weakly developed. Genital setation 5 pairs. Notogastral porose areas *A2* and *A3* showing marked sexual dimorphism, greatly elongated or subdivided in male. Tutorium and custodium absent. Translamella absent (Figure 3.121A) .... ***Mochlobates*** Norton
– Pteromorphs distinct. Genital setation 3, 5 or 6 pairs. Notogastral porose areas not showing marked sexual dimorphism. Tutorium present, custodium present or absent. Translamella present or absent 2

2(1) Notogastral porose areas round to oval, *A2* and *A3* divided into 3 to 8 pairs. Interlamellar seta reduced, only alveolus present. Translamella absent. Genital setation 6 pairs. Trochanter IV with large dorsodistal apophysis. Femora I and II with retrotecta (Figure 3.121B,C) .... ***Dynatozetes*** Grandjean
– Notogastral porose areas round to oval or elongated, *A2* and *A3* not subdivided. Interlamellar seta well developed. Translamella present or absent. Genital setation 3, 5 or 6 pairs. Trochanter IV without dorsodistal apophysis. Femora I and II without retrotecta .. 3

3(2) Notogastral porose area *A1* in circumdorsal alignment with *A2* and *A3*; *A2*, *A3* round to oval; porose area *Aa* undivided. Custodium present. Lateral claws without distal tooth (Figure 3.121D,E) .... ***Podoribates*** Berlese
– Notogastral porose area *A1* positioned medially in comparison with *A2* and *A3*; *A2*, *A3* elongated; porose area *Aa* subdivided in two. Custodium absent. Lateral claws with distal tooth (Figure 3.121F) .... ***Mochlozetes*** Grandjean

Dynatozetes Grandjean, 1960
  Type-species: *Dynatozetes amplus* Grandjean, 1960

Dynatozetes magnus (Banks, 1895) (Figure 3.121B)
  Comb./Syn.: *Oribata magna* Banks, 1895; *Galumna magna* Banks, 1907
  Biology: Probably arboreal, but collected from under moss on rocks (Norton 1984).
  Distribution: ON [NY, OH; Nearctic]
  Literature: Norton (1984).

Mochlobates Norton, 1984
  Type-species: *Mochlobates affinis* (Banks, 1895) (Figure 3.121A)
  Comb./Syn.: *Oribata affinis* Banks, 1895; *Galumna affinis* (Banks, 1895); *Peloribates affinis* (Banks, 1895)

Mochlobates affinis (Banks, 1895)
  Distribution: ON [DC, NC; Nearctic]
  Literature: Norton (1984).

Mochlozetes Grandjean, 1930
   Type-species: *Mochlozetes penetrabilis* Grandjean, 1930
   Note: Unidentified species from ON.

Podoribates Berlese, 1908
   Type-species: *Oribata longipes* Berlese, 1887

Podoribates longipes (Berlese, 1887) (Figure 121D)
   Comb./Syn.: *Oribates longipes* Berlese, 1887; *Sphaerozetes* (?) *gratus* Sellnick, 1921; *Sphaerobates gratus* (Sellnick, 1921)
   Biology: Epigeal species found on grasses and low shrubs (Grandjean 1963). Reported to feed on fungi and pollen in grasslands (Norton 1984). Alberta specimens have gut contents with brown granular material, fungal spores, and possible moss spores. One female carried 8 eggs (Walter et al. 2014)
   Distribution: NT, AB, ON [Holarctic]
   Literature: Norton (1984); Weigmann (2006); Barreto et al. (2021).

Podoribates pratensis (Banks, 1895) (Figure 121E)
   Comb./Syn.: *Oribata pratensis* Banks, 1895; *Galumna pratensis* (Banks, 1895)
   Note: This species is known from NY (Norton 1984) and other than being larger (620–765µm) than *P. longipes* (490–600µm), is indistinguishable. It possibly occurs in Canada.

Family Oribatulidae Thor, 1929
   Diagnosis: *Adults*: Poronotic Brachypylina. Prodorsum with lamellae well developed, or reduced to a narrow carina, or absent; tutorium absent; translamella present or absent. Prodorsum without prolamella, sublamella or sublamellar porose area or saccule. Bothridial base covered or not by anterior margin of notogaster. Genal incision absent. Dorsosejugal scissure complete. Dorsophragmata and pleurophragmata present. Pedotectum I small, covering or not acetabulum I, pedotectum II small. Discidium, custodium absent. Circumpedal carina present. Notogaster with or without lenticulus; setation 10–14 pairs; pteromorphs and posterior tectum absent. Octotaxic system with 4 pairs of porose areas, rarely absent. Genital setation 2–6 pairs. Genital plates well removed from anal plates and virtually bisecting epimere IV. Chelicerae chelate-dentate; palpal eupathidium *acm* attached to solenidion; arrangement of palpal eupathidial trowel-like; subcapitular mentum without tectum. Axillary saccule of subcapitulum absent. Porose organs present or absent on tibiae and tarsi.

KEY TO GENERA OF ORIBATULIDAE OF CANADA AND ALASKA

1	Genital setation 5–6 pairs	2
–	Genital setation 2–4 pairs	3
2(1)	Lamellae more than half length of prodorsum, without translamella; lamellar seta subequal in length to lamella. Notogastral setation 10 pairs (Figure 3.122A) ....***Jornadia*** Wallwork and Weems	
–	Lamellae about third length of prodorsum, joined by narrow translamella; lamellar seta more than twice length of lamella. Notogastral setation 14 pairs (Figure 3.123A,B) ....***Lucoppia*** Berlese	
3(1)	Genital setation 2 pairs. Notogastral setation 10 or 11 pairs (Figure 3.122B,C)................... ....***Diphauloppia*** J. and P. Balogh	
–	Genital setation 3 or 4 pairs (or small alveoli). Notogastral setation 10–14 pairs 4	
4(3)	Genital setation 3 pairs (or small alveoli). Notogastral setation 10 or 11 pairs.................... .... ***Paraphauloppia*** Hammer	
–	Genital setation 4 pairs. Notogastral setation 13 or 14 pairs... 5	

5(4) Translamella present (specimens of *Zygoribatula colemani* may lack translamella) (Figure 3.123C–E) .... ***Zygoribatula*** Berlese
– Translamella absent 6

6(5) Lamellae present, well developed, often broadening anteriorly, connected to bothridia posteriorly. Notogaster with 4 pairs of porose areas. Tarsus I with porose area dorsally (Figure 3.124) .... ***Oribatula*** Berlese
– Lamellae absent, or if present, expressed as fine ridges not connected with bothridia. Notogaster with 4 or more pairs of porose areas. Tarsus I with or without porose area dorsally (Figure 3.122D,E) .... ***Phauloppia*** Berlese

Diphauloppia J. and P. Balogh, 1984
  Type-species: *Subphauloppia luminosa* Hammer, 1973 (Figure 3.122B,C)
  Note: Unidentified *Diphauloppia* sp. from BC.

Jornadia Wallwork and Weems, 1984
  Comb./Syn.: *Woolleybates* J. and P. Balogh, 1984
  Type-species: *Jornadia larreae* Wallwork and Weems, 1984

Jornadia larreae Wallwork and Weems, 1984 (Figure 3.122A)
  Biology: Gut bolus of dark, granular material; 4 eggs.
  Distribution: AB [NM; Nearctic]
  Literature: Walter et al. (2014); ABMI (2019).
  Note: Unidentified species from BC.

Lucoppia Berlese, 1908
  Type-species: *Notaspis burrowsii* Michael, 1890 (=*Zetes lucorum* C.L. Koch sensu Berlese, 1892) (but see Marshall et al. 1987 for discussion)

KEY TO ADULT *LUCOPPIA* OF CANADA AND ALASKA

1 Notogastral porose area *Aa* narrow, long oval, subdivided or not. Notogastral setae *dm*, *dp* not reaching insertion of next most posterior seta. Length: 705–900μm (Figure 3.123A) ... . ***L. apletosa*** (Higgins and Woolley)
– Notogastral porose area *Aa* narrow, short oval. Notogastral setae *dm*, *dp* extending posterior of insertion of next most posterior seta. Length: 620–750μm (Figure 3.123B).............. .... ***L. burrowsii*** (Michael)
  Note: Unnamed species from QC.

Lucoppia apletosa (Higgins and Woolley, 1975) (Figure 3.123A)
  Comb./Syn.: *Zygoribatula apletosa* Higgins and Woolley, 1975
  Remark: Subías (2004) considered this species a junior synonym of *Lucoppia burrowsii* (Michael, 1890), but did not justify this proposal. We retain this species herein.
  Distribution: ON, QC [CO, NY, OH, VA; Nearctic]

Lucoppia burrowsii (Michael, 1890) (Figure 3.123B)
  Comb./Syn.: *Notaspis burrowsii* Michael, 1890; *Lucoppia lucorum* (C.L. Koch, 1840) sensu Berlese, 1892; *Romanobates maiensis* Choi, 1995
  Biology: Gravid females carry up to 16 eggs. Specimens (as *L. lucorum*) have been found in the feathers of birds in Russia (Krivolutsky and Lebedeva 2004). Can be found on leaves and may be primarily arboreal.
  Distribution: AB, MB, ON, QC, NB [IL, VA; Cosmopolitan]
  Literature: Weigmann (2006); Walter et al. (2014); ABMI (2019).

Oribatula Berlese, 1895
Type-species: *Notaspis tibialis* Nicolet, 1855

KEY TO ADULT *ORIBATULA* OF CANADA AND ALASKA

1 Lamellae with small internal extensions. Lamellar seta positioned medially on expanded anterior of lamella. Notogastral seta *da* anterior of porose area *Aa*. Length: ca. 370µm (Figure 3.124E) .... ***O. pallida*** Banks

– Lamellar without small internal extensions. Lamellar seta positioned laterally on expanded anterior of lamella;. Notogastral seta *da* posterior or medial of porose area *Aa*. Length: 410–530µm (Figure 3.124A–D) .... ***O. tibialis*** (Nicolet)
  Note: Unidentified *Oribatula* spp. known from BC, AB, ON, NS, NB.

Oribatula pallida Banks, 1906 (Figure 3.124E)
  Comb./Syn.: *Oribatula tibialis pallida* (Banks, 1906); *Liacarus minutus* Ewing, 1909; *Oribata minutus* (Ewing, 1909); *Oribatula minuta* (Banks, 1906)
  Biology: In a series of comparative experiments with different foods provided independently, this species (as *Oribatula minuta*) fed preferentially on fungi, but did not feed on fresh or decaying leaves, needles or wood (Hartenstein 1962). Walter (1987) reared this species to adult on algae and also fungi.
  Distribution: QC [IL, MD, NJ, NY, OH, VA; Holarctic]
  Note: *Oribatula pallida* Banks may be a junior synonym of *O. tibialis* (Nicolet) according to Marshall et al. (1987). These authors noted that its morphology falls within the range of that of *O. tibialis*. However, Iordansky (1991) considered it a separate species and we follow his judgment herein.

Oribatula tibialis (Nicolet, 1855) (Figure 3.124A–D)
  Comb./Syn.: *Notaspis tibialis* Nicolet, 1855; *Zygoribatula exilis* sensu Buitendijk, 1945, non Nicolet, 1855; *Zygoribatula venustus* sensu Buitendijk, 1945, non Berlese, 1908; *Zygoribatula vera* Bulanova-Zachvatkina, 1967
  Note: *Oribatula vera* (Bulanova-Zachvatkina, 1967) was placed in synonymy with *O. tibialis* by Iordansky (1991), who noted the variability in *O. tibialis*, including the presence of a translamellar depression.
  Biology: From mesophilous habitats and peatlands; known to graze on fungi, including some mycorrhizal species (Schneider 2005). Stable isotope analysis indicated that this species is a primary decomposer (Schneider et al. 2004). Associated with this it has chelicerae with a high leverage index (Perdomo et al. 2012). Specimens have been found in the feathers of birds in Russia (Krivolutsky and Lebedeva 2004). This species uses cyanogenesis as a defence mechanism (Brückner et al. 2017).
  Distribution: AK, YT, NT, NU, BC, AB, ON, QC, NS [MI, NY, VA; Holarctic; northern Neotropical]
  Literature: Seniczak and Seniczak (2012); Maraun et al. (2019).

Paraphauloppia Hammer, 1967
  Type-species: *Paraphauloppia novaezealandica* Hammer, 1967
  Note: Undescribed species from coastal BC (Lindo 2020).

Phauloppia Berlese, 1908
  Type-species: *Zetes lucorum* C.L. Koch, 1841 (=*Oppia conformis* Berlese, 1895)
  Comb./Syn.: *Trichoribatula* Balogh, 1961; *Eporibatula* Sellnick, 1928
  Biology: Species are arboreal or found in lichen patches throughout the temperate Holarctic (e.g., Prinzing and Wirtz 1997). The two species of *Phauloppia* that inhabit lichens on boulder

tops in Norway show remarkable tolerance for temperature and moisture fluctuations, surviving losses of up to 90% of body water content (Sjursen and Sømme 2000).
Note: Unidentified species from BC, MT, ON, QC, NS, NB, NL.

Phauloppia boletorum (Ewing, 1913) (Figure 3.122E)
Comb./Syn.: *Lucoppia boletorum* Ewing, 1913
Biology: From peatlands.
Distribution: AB, ON, QC [MI; Nearctic]
Literature: Walter et al. (2014); ABMI (2019).

Phauloppia modesta (Banks, 1904)
Comb./Syn.: *Liacarus modestus* Banks, 1904; *Eremaeus modestus* Banks, 1910; *Paraliodes incurvata* Hall, 1911; *Eporibatula modesta* (Banks, 1904)
Biology: Female with up to 4 large eggs.
Distribution: QC [CA; Nearctic]
Literature: Bayartogtokh and Aoki (2000).
Note: Unidentified specimens from BC, AB, QC, NB.
Subías (2004) considered this species a junior synonym of *Phauloppia lucorum* (Berlese, 1892) (=*Lucoppia burrowsii*); we retain it, following Marshall et al. (1987).

Zygoribatula Berlese, 1916
Type-species: *Oribatula connexa* Berlese, 1904
Comb./Syn.: *Neoribatula* Ewing, 1917

KEY TO ADULT *ZYGORIBATULA* OF CANADA AND ALASKA
(MODIFIED FROM FRANKLIN ET AL. 2008)

1   Notogastral setation 12 pairs. Length: 315–450µm .... *Z. frisiae* (Oudemans)
–   Notogastral setation 13 pairs 2

2(1)   Translamella linear. Lamellar seta same length as interlamellar seta. Length: 335–425µm (Figure 3.123C–E) .... *Z. exilis* (Nicolet)
–   Translamella thin, ribbonlike, but not linear. Lamellar seta longer then interlamellar seta.
        .... *Z. bulanovae* Kulijew
        Note: *Zygoribatula* sp. nr. *propinqua* (Oudemans 1900) recorded in NS. Unidentified species from YT, BC, AB, MT, ON, QC, NS, NB.

Zygoribatula bulanovae Kulijew, 1961
Comb./Syn.: *Zygoribatula pallida* sensu Hammer, 1952a; *Oribatula pallida* sensu Ghilarov, 1975
Note: *Zygoribatula bulanovae* was placed in synonymy with *O. pallida* by Iordansky (1991), who questioned the distinction of *Oribatula* and *Zygoribatula*. *Zygoribatula bulanovae* was retained in *Zygoribatula* by Franklin et al. (2008).
Biology: From dry forest litter.
Distribution: AK, YT, NU, NT, AB, MB, QC, NL [MT; Holarctic]
Literature: Franklin et al. (2008); ABMI (2019).

Zygoribatula exilis (Nicolet, 1855) (Figure 3.123C–E)
Comb./Syn.: *Notaspis exilis* Nicolet, 1855 *Oppia exilis* (Nicolet, 1855) *Oribatula exilis* (Nicolet, 1855)
Biology: From moss. Recorded from arboreal lichens (Root et al. 2007). Median lethal time submerged under water was 34 days (Bardel and Pfingstl 2018).

Distribution: AK, MB, QC, NB [NY, VA; Holarctic]
Literature: Maraun et al. (2019).

Zygoribatula frisiae (Oudemans, 1916)
Comb./Syn.: *Eremaeus frisiae* Oudemans, 1916; *Notaspis pyrostigmata* Ewing, 1909; *Zygoribatula variabilis* Berlese, 1908
Biology: From dry moss; conifer litter.
Distribution: ON, QC [IL; Holarctic]
Literature: Franklin et al. (2008).

Family Oripodidae Jacot, 1925
Diagnosis: *Adults*: Poronotic Brachypylina. Rostral tectum foreshortened such that chelicera partially exposed. Prodorsum with lamellae reduced to a narrow carina, or absent; tutorium and translamella absent. Bothridium usually covered by anterior margin of notogaster. Genal incision absent. Prodorsum and notogaster separate or fused. Dorsophragmata and pleurophragmata present. Sublamellar porose area absent. Discidium, custodium absent; circumpedal carina present. Pedotectum I very small, not completely covering acetabulum I, pedotectum II very small. Notogaster without lenticulus, without posterior tectum; pteromorphs small or absent; setation 10 pairs. Octotaxic system diverse, often with fewer than 4 pairs of saccules, rarely absent. Genital setation 1–4 pairs. Genital plates well removed from anal plates and virtually bisecting epimere IV. Marginoventral porose areas present or absent. Chelicerae chelate-dentate, with one seta (lacking *chb*); palpal eupathidium *acm* attached to solenidion; subcapitular mentum without tectum. Axillary saccule of subcapitulum absent. With porose organs on tibiae and tarsi. Tarsi truncate anteriorly, especially tarsus I.

Benoibates Balogh, 1958 (Figure 3.125A,B)
Type-species: *Benoibates flagelliger* Balogh, 1958
Note: Unidentified species from BC.

Family Parakalummidae Grandjean, 1936
Diagnosis: *Adults*: Poronotic Brachypylina. Prodorsum with lamellae reduced to a narrow carina; tutorium and translamella absent. Genal incision absent. Prodorsum and notogaster separate. Dorsophragmata and pleurophragmata present. Sublamellar saccule present. Discidium, custodium absent; circumpedal carina present. Pedotectum I small, not completely covering acetabulum I, pedotectum II small. Notogaster with or without lenticulus, without posterior tectum; pteromorphs present with complete hinge, auriculate; setation 10 pairs of minute setae or alveoli. Octotaxic system of 4 pairs of saccules. Genital setation 5 pairs. Genital plates well removed from anal plates and virtually bisecting epimere IV. Chelicerae chelate-dentate; palpal eupathidium *acm* attached to solenidion; subcapitular mentum without tectum. Usually with porose organs on tibiae and tarsi.

KEY TO GENERA OF PARAKALUMMIDAE OF CANADA AND ALASKA

1	Ventral margin of pteromorph undulate, concave anteriorly. Lamellae reduced to ridges. Length: 510–720µm (Figure 3.125E,F) ....*Neoribates* Berlese
–	Ventral margin of pteromorph convex. Lamellae well developed (Figure 3.125G)............ ....*Protokalumma* Jacot

Neoribates Berlese, 1914
Type-species: *Oribates roubali* Berlese, 1910

Neoribates aurantiacus (Oudemans, 1914) (Figure 3.125E,F)
    Comb./Syn.: *Galumna aurantiacus* Oudemans, 1914
    Biology: Common in decaying wood and bracket fungi and is found from temperate to arctic habitats in North America.
    Distribution: AK, YT, NT, AB, MB, ON, QC, NB, NL [NY; Holarctic]
    Literature: Maraun et al. (2019); Markkula and Kuhry (2020).
    Note: Unidentified *Neoribates* species from AB, MB, ON, QC, NB, NS.

Protokalumma Jacot, 1929
    Type-species: *Oribata depressa* Banks, 1895

Protokalumma depressa (Banks, 1895) (Figure 3.125G)
    Comb./Syn.: *Oribata depressa* Banks, 1895; *Galumna depressa* (Banks, 1895)
    Biology: From coarse woody debris.
    Distribution: QC [CN, IL, IO, ME, NC, NJ, NY, OH, VA, WI; Nearctic]
    Note: Unidentified *Protokalumma* species from QC.

Family Scheloribatidae Grandjean, 1933
    Diagnosis: *Adults*: Poronotic Brachypylina. Prodorsum with lamellae well developed, reduced to narrow carina, or absent; tutorium and translamella absent. Prolamella present or absent; sublamella present or absent. Bothridium often covered by anterior margin of notogaster. Genal incision absent. Prodorsum and notogaster separate or fused. Dorsophragmata and pleurophragmata present. Discidium, custodium absent; circumpedal carina present or absent. Pedotectum I small, usually not covering acetabulum I, pedotectum II small. Sternal groove present or absent. Genital setation 1–5 pairs. Genital plates well removed from anal plates and virtually bisecting epimere IV. Notogaster without lenticulus, without posterior tectum; pteromorphs present or absent, if present, without hinge. Notogastral setation 9–10 pairs. Octotaxic system diverse, often with fewer than 4 pairs of saccules; rarely absent. Chelicerae chelate-dentate; palpal eupathidium *acm* attached to solenidion; palpal eupathidia with comb-like arrangement; subcapitular mentum without tectum. Axillary saccule of subcapitulum absent. Usually with porose organs on tibiae and tarsi.
    Biology: Species of *Scheloribates* are among the most common and abundant mites in soil and litter, even in anthropogenic habitats. One of these, *S. laevigatus* (C.L. Koch) is common in North American lawns and is probably introduced from Europe.

KEY TO GENERA OF SCHELORIBATIDAE OF CANADA AND ALASKA

1   Rostrum acute medially. Humeral porose area expressed as saccule. Pteromorphs absent. Mono- or tridactylous; medial claw sickle-shaped (Figure 3.127B–D) .....***Paraleius*** Travé
–   Rostrum rounded medially. Mono–or tridactylous; medial claw not sickle-shaped. Humeral porose area porose 2

2(1)   Dorsosejugal scissure straight 3
–   Dorsosejugal scissure arched or incomplete 4

3(2)   Solenidion of tibiae II and IV with bulbous (microcephalic) tip. Carina *kf* present. Porose areas present on tibiae. Sublamella present. Saccule *Sa* as 1 or 2 saccules (Figure 3.127A) ....***Siculobata*** Grandjean
–   Solenidion of tibiae II and IV without bulbous tip. Carina *kf* absent. Porose areas on tibiae absent. Sublamella absent. Saccule *Sa* as single saccule. Length: 375–520µm (Figure 3.126A,B) ....***Dometorina*** Grandjean

4(2)		Notogaster with pteromorphs or well-developed humeral projections. Circumpedal carina weakly or strongly developed. Epimeral setation 3-1-3-3 5
	–	Notogaster without pteromorphs, at most with small, rounded humeral projections. Circumpedal carina absent or weakly developed. Epimeral setation: 3-1-3-3 or 3-1-2-3 or 3-1-2-2 6
5(4)		Tridactylous or bidactylous. Notogastral saccules present. Sternal furrow usually present. Bothridium positioned so that notogastral margin covers only base. Dorsosejugal scissure complete. Femora I and II with retrotectum protecting articulation with trochanter (Figure 3.128) .... ***Scheloribates*** Berlese
	–	Monodactylous. Notogastral saccules or porose areas present. Sternal furrow absent. Bothridium positioned so that notogastral margin covers most of bothridium. Dorsosejugal scissure complete or incomplete. Femora I, II without retrotectum protecting articulation with trochanter (Figure 3.126E–G) .... ***Liebstadia*** Oudemans
6(4)		Dorsosejugal scissure complete. Bothridium not covered by notogastral margin anteriorly. Notogastral saccules present. Epimeral setation 3-1-3-3. Sternal furrow present. Femora I and II with retrotectum. Carina *kf* present (Figure 3.126C,D)... .... ***Scheloribates*** (***Hemileius***) Berlese
	–	Dorsosejugal scissure incomplete. Bothridium almost completely covered by notogastral margin. Notogastral porose areas present. Epimeral setation 3-1-2-3 or 3-1-2-2. Sternal furrow absent. Femora I, II without retrotectum. Carina *kf* absent (Figure 3.127E–H) .... ***Parapirnodus*** Balogh and Mahunka

Dometorina Grandjean, 1951
 Type-species: *Oribatula plantivaga* Berlese, 1895

Dometorina plantivaga (Berlese, 1895) (Figure 3.126A,B)
 Comb./Syn.: *Oribatula plantivaga* Berlese, 1895; *Eporibatula plantivaga* (Berlese, 1895); *Notaspis plantivaga* (Berlese, 1895); *Oppia tibialis* sensu Berlese, 1892; non *Oribatula tibialis* sensu Berlese, 1895
 Biology: Inhabits lichen on trees where it lives exclusively among epiphytic lichens and algae, where densities can reach 250 individuals per 25cm^2 (Travé 1963). Creates cavities in lichen thallus and uses feces as building material to close openings (Grandjean 1951). Females up to 100μm larger than males. Gravid females carry at least 6 large eggs.
 Distribution: AB, NS [Holarctic]
 Literature: Wunderle (1991); ABMI (2019); Schäffer et al. (2020).

Scheloribates (*Hemileius*) Berlese, 1916
 Type-species: *Protoribates* (*Scheloribates*) *initialis* Berlese, 1908

KEY TO ADULT *SCHELORIBATES* (*HEMILEIUS*) OF CANADA AND ALASKA

1		Notogastral setae thin, fine. Length: 400–430μm (Figure 3.126C)..................................... .... ***Scheloribates*** (***H.***) ***haydeni*** (Higgins and Woolley)
	–	Notogastral setae barbed. Length: ca. 500μm (Figure 3.126D).. .... ***Scheloribates*** (***H.***) ***quadripilis*** (Fitch)

Scheloribates (*Hemileius*) *haydeni* (Higgins and Woolley, 1975) (Figure 3.126C)
 Comb./Syn.: *Multoribates haydeni* Higgins and Woolley, 1975

Biology: From high plains/aspen parkland. Gut contents include hyaline hyphae, brown arthroconidia, and indistinct material (Walter et al. 2014).
Distribution: AB [CO; Nearctic]
Literature: Walter et al. (2014); ABMI (2019).

Scheloribates (*Hemileius*) *quadripilis* (Fitch, 1856) (Figure 3.126D)
Comb./Syn.: *Oribata quadripilis* Fitch, 1856; *Notaspis pallida* Ewing, 1909
Biology: From moss in depression in spruce, alder forest (Hammer 1952a). From arboreal lichens (Root et al. 2007).
Distribution: YT, NT, ON, QC [IL, NY, VA; Nearctic]

Liebstadia Oudemans, 1906
Type-species: *Notaspis similis* Michael, 1888

KEY TO ADULT *LIEBSTADIA* OF CANADA AND ALASKA

1 Notogaster broadest at level of porose area $A1$. Porose areas Aa large. Notogastral seta lp positioned medially, removed almost setal length from porose area $A1$. Length: 500–600µm (Figure 3.126E) ....*L. similis* Michael
– Notogaster broadest at level of pteromorph. Porose areas Aa small. Notogastral seta lp positioned more laterally, just medial to porose area $A1$. Length: 310–380µm (Figure 3.126F,G)..*L. humerata* Sellnick
Note: Unidentified species known from AB, ON, QC, NS, NB.

Liebstadia humerata Sellnick, 1928 (Figure 3.126F,G)
Biology: Nannelli et al. (1998) found this species in cankers caused by the fungus *Cryphonectria parasitica* on chestnut trees, and reared the mites on laboratory cultures of the parasite, including monoxenic cultures of a hypovirulent (H) strain of this blight fungus. Fecal pellets collected aseptically were used to inoculate agar plates, from which new *C. parasitica* cultures developed with the morphological characters of the hypovirulent strain. Inoculation tests in the field confirmed that the strains developed from the mite's fecal pellets were hypovirulent.
Distribution: ON(?), NB [TN, VA; Holarctic, Oriental]
Literature: Miko and Weigmann (1996); Lehmitz and Decker (2017).

Liebstadia similis (Michael, 1888) (Figure 3.126E)
Comb./Syn.: *Notaspis similis* Michael, 1888; *Eremaeus similis* (Michael, 1888) *Oribatula similis* (Michael, 1888); *Protoribates silésius* Sellnick, 1925 *Oribates pallidula* sensu Oudemans, 1896; non C.L. Koch, 1841
Biology: Specimens have been found in the feathers of birds in Russia (Krivolutsky and Lebedeva 2004).
Distribution: AK, YT, NT, AB, NL [OR; Holarctic]
Literature: Miko and Weigmann (1996); Lehmitz and Decker (2017); Maraun et al. (2019).

Paraleius Travé, 1960
Type-species. *Oribella leontonycha* Berlese, 1910
Comb./Syn.: *Metaleius* Travé, 1960
Note: We follow Fredes and Martínez (2014) and Knee (2017) in rejecting the synonymy of *Paraleius* with *Siculobata*.

KEY TO ADULT *PARALEIUS* OF CANADA AND ALASKA

1   Monodactylous, central claw large sickle shaped and strongly hooked, hair-like lateral claws absent. Bothridial seta long fusiform. Length: 430–464µm (Figure 3.127B,C) ....*Paraleius leahae* Knee
–   Heterotridactylous, large curved central claw, lateral claws hair-like. Bothridial seta capitate. Length: 435–500µm (Figure 3.127D) ....*Paraleius leontonycha* (Berlese)

Paraleius leahae Knee, 2017 (Figure 3.127B,C)
    Biology: A host specialist, collected from only two bark beetle species, *Hylastes porculus* and *Dendroctonus valens*. These two host species are not closely related but are ecologically similar, as they live in the stumps and roots of dead or dying conifers. 16 other species of mites were collected from each of these host species including *P. leontonycha* (Knee et al. 2013). *Paraleius leahae* challenges the assumption that bark beetle associated oribatid mites are uncommon and are not host specific. Feeding biology of *P. leahae* and *P. leontonycha* is poorly understood, but fungal hyphae have been observed in the gut of slide mounted specimens of both species.
    Distribution: NS, NB, ON [Canada]
    Literature: Knee et al. (2013); Knee (2017).

Paraleius leontonycha (Berlese, 1910) (Figure 3.127D)
    Comb./Syn.: *Oribella leontonycha* Berlese, 1910, *Siculobata leontonychus* (Berlese, 1910)
    Biology: An arboreal mite often found on tree bark or phoretic on bark beetles. European populations have been shown to be a complex of cryptic species (Schäffer and Koblmüller 2020).
    Distribution: AK, BC, AB, BC, NB, NL, ON, QC [Holarctic]
    Literature: Knee (2017).

Parapirnodus Balogh and Mahunka, 1968
    Type-species. *Parapirnodus longus* Balogh and Mahunka, 1968
    Comb./Syn.: *Behanpseudoppia* Subías, 2017

KEY TO ADULT *PARAPIRNODUS* OF CANADA AND ALASKA

1   Notogastral porose areas *Aa* and *A2*, present, subequal in size and shape in female with *Aa* positioned well posteriorly of seta *c*; *Aa* longer than *A2* in male, and *Aa* positioned medially to seta *c*. Tarsal setation (I to IV) 14-13-12-9. Length: 260–313µm (Figure 3.127E–G) ....*P. coniferinus* Behan-Pelletier, Clayton and Humble
–   Notogastral porose areas *Aa*, *A2*, and *A3*, subequal in size and shape in female and male. Tarsal setation (I to IV) 16-13-12-10. Length: 296–340µm (Figure 3.127H)....................... ....*P. hexaporosus* Behan-Pelletier, Clayton and Humble

Parapirnodus coniferinus Behan-Pelletier, Clayton and Humble, 2002 (Figure 3.127E–G)
    Biology: Significant component of the arboreal fauna in canopy habitats in western Canada, including the canopy of Amabilis fir. Co-occurs with *P. hexaporosus* on branches of Amabilis fir. Only *P. coniferinus* has been collected so far from the canopy of Sitka spruce and Douglas fir. Neither species has been collected from soil or litter at the base of these conifer species. This species shows distinct sexual dimorphism, with differences in size of notogastral porose areas.
    Distribution: BC [Canada]

Parapirnodus hexaporosus Behan-Pelletier, Clayton and Humble, 2002 (Figure 3.127H)
   Biology: In Western redcedar they are a large component of the corticolous habitat above 6m (Lindo and Winchester 2007). Found in high canopy foliose lichens, absent from high canopy alecteroid lichens (Behan-Pelletier et al. 2008).
   Distribution: BC [Canada]

Siculobata Grandjean, 1953
   Type species: *Oppia tibialis sicula* Berlese, 1892 (Figure 3.127A)
   Literature: Fredes and Martínez (2013).
   Note: Unidentified *Siculobata* species from QC (ZL record).

Scheloribates Berlese, 1908
   Type-species: *Zetes latipes* C.L. Koch, 1844
   Comb./Syn.: *Storkania* Jacot, 1929; *Paraschelobates* Jacot, 1934; *Protoschelobates* Jacot, 1934; *Styloribates* Jacot, 1934; *Propeschelobates* Jacot, 1936

KEY TO ADULT *SCHELORIBATES* OF CANADA AND ALASKA

1	Red brown in colour. Notogaster with fine, branching longitudinal striations. Length: 520–670µm (Figure 3.128A–E)	*S. laevigatus* (C.L. Koch)
–	Yellow brown in colour. Notogaster without fine, branching longitudinal striations	2
2(1)	Bothridial seta with length of stalk about 4× length of head. Length: 460–490µm (Figure 3.128F,G)	*S. rotundatus* Hammer
–	Bothridial seta with length of stalk subequal to that of head	3
3(2)	Notogastral setae short 3–8µm. Larger species, Length: 460–550µm (Figure 3.128I)	*S. latipes* (C.L. Koch)
–	Notogastral setae 15–20µm. Smaller species, Length: <450µm	4
4(3)	Bothridial seta rounded distally. Length: 350–450µm (Figure 3.128H)	*S. pallidulus* (C.L. Koch)
–	Bothridial seta acute distally. Length: 364–444µm (Figure 3.128J)	*S. lanceoliger* (Berlese)

Note: Unidentified species from BC, AB, SK, MT, ON, QC, NS, NB, PE, NL.

Scheloribates laevigatus (C.L. Koch, 1835) (Figure 3.128A–E)
   Comb./Syn.: *Zetes laevigatus* C.L. Koch, 1835; *Oribata lucasi* Nicolet, 1855; *Notaspis lucasi* (Nicolet, 1855); *Oribates fuscomaculata* sensu Oudemans, 1896; non C.L. Koch, 1841
   Biology: Alkaloids have been detected from the opisthonotal gland secretions of this species (Saporito et al. 2011).
   Vitzthum (1943) noted that this species was able to feed on the pupae of parasitic hymenopterans. It can be reared on lichen (Woodring and Cook 1962). Considered a fungivorous grazer by Siepel and de Ruiter-Dijkman (1993). Fungivorous grazing and ingestion of *Penicillium griseofulvum* spores was observed by Hubert et al. (1999). Hubert et al. (2001) noted fungal cell walls were not digested completely by this species. In food preference tests, *S. laevigatus* showed higher preference for green bark alga *Desmococcus vulgaris* (= *Protococcus viridis*) than for grass litter that was partly overgrown by fungi (Hubert et al. 1999). Hubert and Lukešová (2001) found the cyanobacterium *Trichormus variabilis* was rejected as food by this mite.
   This mite appears to be synanthropic in AB, since it has been collected only from a backyard, a recreational lake shore, and a field crop site (Walter et al. 2014). Specimens have been

found in the feathers of birds in Russia (see Krivolutsky and Lebedeva 2004). This mite also was reported from the hair of small mammals in Slovakia (Miko and Stanko 1991).
Distribution: NU, AB, QC [IL, KY, MI, NY, SD; Semicosmopolitan]
Literature: Weigmann (2006); Maraun et al. (2019); ABMI (2019).

Scheloribates lanceoliger (Berlese, 1908) (Figure 3.128J)
Comb./Syn.: *Protoribates* (*Scheloribates*) *lanceoliger* Berlese, 1908; *Galumna sylvicola* Banks, 1909; *Oribata helvina* Ewing, 1909
Biology: From grass fields, shrubs and forests (Cianciolo and Norton 2006); from forest litter (Lamoncha and Crossley 1998). Tapeworm cysticercoids were found in this species (Fritz 1995).
Distribution: ON [FL, IL, MN, MO, NC, NY; Nearctic]
Literature: Norton and Kethley (1990); Maraun et al. (2019).

Scheloribates latipes (C.L. Koch, 1844) (Figure 3.128I)
Comb./Syn.: *Zetes latipes* C.L. Koch, 1844; *Oribates latipes* (C.L. Koch, 1844)
Biology: Tapeworm cysticercoids were found in this species (McAloon 2004).
Distribution: NB (CNC record) [NY, VA; Holarctic]
Literature: Bayartogtokh (2000).

Scheloribates pallidulus (C.L. Koch, 1841) (Figure 3.128H)
Comb./Syn.: *Zetes pallidulus* C.L. Koch, 1841 *Oribata pallidula* (C.L. Koch, 1841)
Biology: Gravid females carry up to 4 eggs. Gut contents include pigmented and unpigmented fungal hyphae, spores, bits of organic material, and cuticle of small arthropods. Collected from moss, decaying wood, stumps; mesophilous in peatlands.
Distribution: AK, YT, NU, NT, AB, MB, ON, QC, NL [NC, OR; Cosmopolitan]
Literature: Weigmann (2006); Maraun et al. (2019); ABMI (2019).

Scheloribates rotundatus Hammer, 1955 (Figure 3.128F,G)
Biology: From coniferous forest soil.
Distribution: AK [OR; Nearctic]

Superfamily Ceratozetoidea Jacot, 1925

Family Ceratokalummidae Balogh, 1970
Diagnosis: *Adults*: Poronotic Brachypylina. Prodorsum with lamellae, separated; tutorium present, lamellar cusps present, translamella absent. Genal incision present. Bothridium with medial and lateral scales. Dorsophragmata and pleurophragmata present. Discidium, custodium, postanal porose area present. Pedotectum I large, extending to base of bothridium, convex dorsally. Pedotectum II present. Notogaster without lenticulus, posterior tectum present, not overlapping medially; pteromorphs present, with complete hinge. Notogastral setation 10 pairs, seta *c* borne on pteromorph. Octotaxic system as 4 pairs of porose areas. Genital setation 6 pairs. Chelicerae chelate-dentate; palpal eupathidium *acm* attached to solenidion; subcapitular mentum without tectum. Axillary saccule of subcapitulum present. Genu IV shorter than tibia IV. Tibia IV with solenidion.

Cultrobates Willmann, 1930 (Figure 3.129A–C)
Type-species: *Cultrobates heterodactylus* Willmann, 1930
Note: Unidentified *Cultrobates* species from BC.

Family Ceratozetidae Jacot, 1925
Diagnosis: *Adults*: Poronotic Brachypylina. Prodorsum with lamellae, tutorium, lamellar cusps; translamella present or absent. Genal incision present or rarely completely reduced. Bothridium cup shaped, or with medial and lateral scales. Dorsophragmata and

pleurophragmata present. Discidium and custodium present; postanal porose area generally present. Pedotectum I large, extending to base of bothridium, convex or concave dorsally. Pedotectum II present. Notogaster with or without lenticulus, posterior tectum absent; pteromorphs usually present, with or without hinge. Notogastral setation 10–15 pairs. Octotaxic system developed as 1–7 pairs (usually 4) of porose areas or saccules. Genital setation 4–6 pairs. Chelicerae chelate-dentate; palpal eupathidium *acm* attached to solenidion; subcapitular mentum without complete tectum. Axillary saccule of subcapitulum generally present. Genu IV shorter than tibia IV.

Biology: Species are common in forest, grassland and tundra habitats from the tropics to warm-temperate to high arctic and sub-Antarctic regions (Behan-Pelletier 1999). Although found mainly in litter and soil, ceratozetid mites may occur in large numbers in grass (Michael 1884), in canopy habitats (Behan-Pelletier 2000), in mosses and lichens (Materna 2000), or in wet, mesotrophic bogs (Weigmann 1991). Feeding habits of ceratozetids are as diverse as their habitats. Based on gut content analysis and feeding experiments, species of *Ceratozetes* and *Fuscozetes* are saprophages and mycophages (Luxton 1972). *Fuscozetes* spp. readily eat dead Collembola (Wallwork 1958), and *Ceratozetes* sp. feeds on nematodes (Walter 1987). *Dentizetes ledensis* utilizes phytopathogenic fungi as food (Behan-Pelletier 2000). *Diapterobates humeralis* feeds on the woolly filaments that envelop the ovisacs of the hemlock woolly adelgid, and dislodges their eggs in the process (McClure 1995). Species of *Ceratozetes*, *Trichoribates*, and *Fuscozetes* are known intermediate hosts of tapeworms (Denegri 1993; Denegri et al. 1998).

KEY TO GENERA OF CERATOZETIDAE OF CANADA AND ALASKA

1	Bothridium cup-shaped, with at most narrow ventromedial scale. 5–20 horizontal folds present in integument between and dorsal of acetabula II and III. Tutorial cusp broad, lamelliform, dentate, sinuate or undulating distally. Genal incision broadly triangular. Pteromorph hinged or not	2
–	Bothridium with well-developed lateral and medial scales, if partially cup-shaped, lenticulus is absent or 5 pairs of genital setae present. Horizontal folds absent from integument between and dorsal of acetabula II and III. Tutorial cusp tapering to a point, not broad, lamelliform. Genal incision long, narrowly triangular. Pteromorph without hinge	10
2(1)	Notogastral setation 13 pairs (c_2, *da*, *dm*, *dp* present). Seta *l'* on femur III absent (2 setae). Lenticulus present or absent	3
–	Notogastral setation 10 or 11 pairs (c_2, *da*, *dm* absent, *dp* present or absent). Seta *l'* on femur III present or absent (2 or 3 setae). Lenticulus present	4
3(2)	Pteromorph hinged (Figure 3.137)	***Diapterobates*** Grandjean
–	Pteromorph without hinge (Figure 3.136C,D)	***Neogymnobates*** Ewing
4(2)	Pedotectum I with concave dorsal margin, medially rounded or pointed. Lamellar cusps fused medially, or almost touching medially, or not	5
–	Pedotectum I with convex dorsal and medial margin. Lamellar cusps well separated	8
5(4)	Lamellae completely fused, forming single tectum; without dentate distal margin. Notogaster with porose areas or saccules	6
–	Lamellae not fused, not forming single tectum. Notogaster with porose areas	7
6(5)	Pteromorphs with hinge incomplete anteriorly, hinged posteriorly. Notogaster usually with porose areas (Figure 3.130D,E)	***Lepidozetes*** Berlese
–	Pteromorphs with complete hinge. Notogaster with saccules (Figure 3.130A)	***Scutozetes*** Hammer

7(5)	Lamellar cusps broad, with medial margins almost touching; with or without dentate distal margin. 3 or 4 pairs of notogastral porose areas. Pteromorphs well developed (Figure 3.136A,B) .... ***Dentizetes*** Hammer	
–	Lamellar cusps narrow, well separated from each other; without dentate distal margin. 4–7 pairs of notogastral porose areas. Pteromorphs reduced in size, or not (Figure 3.129D,E) ....***Jugatala*** Ewing	
8(4)	Translamella present. Lamellar cusps well developed, more than 15µm long (Figure 3.141) .... ***Trichoribates*** Berlese	
–	Translamella absent. Lamellar cusps short, less than 15µm long, or absent 9	
9(8)	Lamella with cusp. Pteromorph without hinge. Notogaster with 3 or 4 pairs of porose areas. Seta *l'* absent from femur III (2 setae) (Figure 3.130B) .... ***Svalbardia*** Thor	
–	Lamella without cusp. Pteromorph with partial hinge. Notogaster with 4 pairs of porose areas. Seta *l'* present on femur III (3 setae) (Figure 3.130C) .... ***Iugoribates*** Sellnick	
10(1)	Femora I, II with ventral carina. Seta *v"* on tibiae I, II thickened, heavily barbed, tapered. Lenticulus present 11	
–	Femora I, II without ventral carina. Seta *v"* on tibiae acuminate distally, never thickened or heavily barbed. Lenticulus present or absent 13	
11(10)	Translamella present. Lamellae strongly converging, reaching to level of insertion of rostral seta. Tutorial cusp reaching almost to, or anterior to, insertion of rostral seta. Ratio of fused portion of tutorium to cusp less than 5:1 (Figure 3.138) .... ***Fuscozetes*** Sellnick	
–	Translamella usually absent. Lamellae at most slightly converging, not reaching level of insertion of rostral seta. Tutorial cusp very short, not reaching to insertion of rostral seta. Ratio of fused portion of tutorium to cusp greater than 5:1 12	
12(11)	Notogastral setation 15 pairs (c_3 present) (Figure 3.131A) .... ***Ghilarovizetes*** Shaldybina	
–	Notogastral setation 14 pairs (c_3 absent) (Figure 3.139) .... ***Melanozetes*** Hull	
13(10)	Lamellae and lamellar cusps broad, almost parallel to each other (Figure 3.131B) ....***Laminizetes*** Behan-Pelletier	
–	Lamellae and lamellar cusps narrow, converging or not 14	
14(13)	Lenticulus circular, convex in lateral aspect (Figure 3.131C–F)... .... ***Protozetomimus*** Pérez-Íñigo	
–	Lenticulus absent, or subtriangular, never convex in lateral aspect 15	
15(14)	Rostral setae inserted very anteriorly on prodorsum; anterior to or in line with anterior margin of genal incision; combination of lamella not reaching anteriorly to level of insertion of rostral seta and translamella well developed, long (Figure 3.140)............................. .... ***Sphaerozetes*** Berlese	
–	Rostral setae inserted posteriorly to anterior margin of genal incision, lamella reaching anteriorly to level of insertion of rostral seta (when lamella not reaching level of insertion of rostral seta, translamella absent); translamella short, or absent.................................... 16	
16(15)	Pteromorphs with transverse desclerotized furrow, visible in transmitted light (Figure 3.131G,H).. ***Ceratozetoides*** Shaldybina	
–	Pteromorphs without transverse desclerotized furrow (Figures 3.132–3.135) ....***Ceratozetes*** Berlese	

Ceratozetes Berlese, 1908
Type-species: *Oribata gracilis* Michael, 1884
Comb./Syn.: *Ceratozetella* Shaldybina, 1966

KEY TO ADULT *CERATOZETES* AND *CERATOZETOIDES* OF CANADA AND ALASKA

1	Monodactylous	2
–	Heterotridactylous	4
2(1)	Porose area $A1$ positioned medially. Genital seta g_1 only on anterior margin of genital plate. Ventral length: 268–308µm (Figure 3.132A,B)	*C. parvulus* Sellnick
–	Porose area $A1$ positioned laterally. Genital setae g_1 and g_2 on anterior margin of genital plate	3
3(2)	Anterior margin of notogaster with small convex projection lateral of bothridium. Seta l'' of genua I and II spinous. Tutorium almost reaching insertion of rostral seta. Scale svm of bothridium with shallow keel, without teeth. Ventral Length: 312–356µm (Figure 3.132C,D)	*C. kutchin* Behan-Pelletier
–	Anterior margin of notogaster smoothly curved. Seta l'' of genua I and II similar in shape to other setae on these segments, setose. Tutorium reaching anterior of insertion of rostral seta. Scale svm of bothridium large, with 1–3 teeth. Ventral Length: 402–421µm (Figure 3.132E,F)	*C. thienemanni* Willmann
4(1)	Notogastral setation 11 pairs (seta c_3 present)	5
–	Notogastral setation 10 pairs (seta c_3 absent)	11
5(4)	Tutorium extending at most midway between its base and insertion of rostral seta. Length: 391–448µm (Figure 3.132G,H)	*C. mediocris* (Berlese)
–	Tutorium extending almost to, or beyond, base of rostral seta	6
6(5)	Notogastral setae 3–5µm long	7
–	Notogastral setae >10µm long	9
7(6)	Rostrum medially with large tooth approximately half length of lateral teeth. Porose area Aa with concave anterior margin. Seta l'' of genu II spinous, shorter than other setae on this segment. Length: 518–525µm (Figure 3.133A,B)	*C. watertonensis* Behan-Pelletier
–	Rostrum medially with 1–3 short teeth. Porose area Aa with convex anterior margin. Seta l'' of genu II setose, similar in shape and length to other setae on this segment	8
8(7)	Integument microtuberculate. Anterior margin of process $psdm$ of bothridium concave. Length: 473–635µm (Figure 3.133C–E)	*C. gracilis* (Michael)
–	Integument pitted. Anterior margin of process $psdm$ of bothridium convex. Length: 570–629µm (Figure 3.133F)	*C. peritus* Grandjean
9(6)	Rostrum with shallow, wide parallel-sided indention having irregular margin. Scale svm on bothridium larger than scale svl. Interlamellar distance at base of cusps greater than distance between interlamellar setae. Length: 376–395µm (Figure 3.133G,H)	*C. oresbios* Behan-Pelletier
–	Rostrum with deep, wide or narrow, parallel-sided indentation forming 2 lateral teeth, margin between teeth smooth. Scale svm on bothridium smaller than scale svl. Interlamellar distance at base of cusps subequal to or smaller than distance between interlamellar setae	10
10(9)	Tutorial cusp not reaching insertion of rostral seta. Interlamellar seta reaching or almost reaching tip of lamellar cusp. Length: 382–434µm (Figure 3.134A,B)	*C. kananaskis* Mitchell
–	Tutorial cusp reaching anterior of insertion of rostral seta. Interlamellar seta reaching only 1/3 of distance to base of lamellar cusp. Length: 441–544µm (Figure 3.134C)	*C. borealis* Behan-Pelletier

11(4)	Anterior of notogaster with convex projection just posterior to bothridium, with medial edge overlapping lateral edge. Length: 370–400µm (Figure 3.135H,I) .... ***C. enodis*** (Ewing)
–	Anterior of notogaster smoothly curved... 12
12(11)	Translamella present, sometimes weakly defined. Seta l'' of genua I and II spinous. Ventral length 460–551µm (Figure 3.134D)..***C. spitsbergensis*** Thor
–	Translamella absent. Seta l'' of genua I and II spinous or setose 13
13(12)	Porose area $A1$ positioned medially. Distance lp–$A1$ < $A1$–h_3. Ventral length: 421–473µm (Figure 3.134E,F)...***C. fjellbergi*** Behan-Pelletier
–	Porose area $A1$ positioned laterally. Distance lp–$A1$ > or = $A1$–h_3 14
14(13)	Bothridial seta directed posterolaterally or posteriorly. Tutorial cusp short, 8µm in length, expressed as large triangular tooth. Length: 486–550µm (Figure 3.135A,B) ....***C. virginicus*** (Banks)
–	Bothridial seta directed anteromedially to anterolaterally. Tutorial cusp >30µm in length 15
15(14)	Rostrum with very shallow indentation forming minute lateral teeth 3–4µm long, margin between teeth smooth. Prolamella present. Notogastral setae of c, l and h-series 11–17µm long. Seta l' absent from femur II (2 setae). Length: 328–364µm (Figure 3.135C,D)........... ....***C. cuspidatus*** Jacot
–	Rostrum with very deep, wide, almost parallel-sided indentation, forming two large lateral teeth, margin between teeth with medial tooth. Prolamella absent. Notogastral setae of c, l and h-series < 6µm long. Seta l' present on femur II (3 setae).. 16
16(15)	Porose area Aa subcircular to oval in shape, about 16µm in diameter. Tutorial cusp sword-like. Genital setae g_1, g_2, g_3 on anterior margin. Length: 575–632µm (Figure 3.135E,F).... .... ***C. pacificus*** Behan-Pelletier
–	Porose area Aa subrectangular, 48–70µm long × 15–20µm wide. Tutorial cusp more broadly triangular in shape. Genital setae g_1, g_2 only on anterior margin of genital plate. Length: 640–860µm (Figure 3.131G,H)......................***Ceratozetoides cisalpinus*** (Berlese)

Ceratozetes borealis Behan-Pelletier, 1984 (Figure 3.134C)
Biology: From diverse habitats such as moist humus and comparatively dry scree slopes; with boreal and boreomontane distribution in both glaciated and unglaciated areas; primarily in tundra, boreal, and alpine forest litter.
Distribution: AK, YT, NB, AB, NB [CO; Nearctic].
Literature: Behan-Pelletier and Eamer (2009).

Ceratozetes cuspidatus Jacot, 1939 (Figure 3.135C,D)
Comb./Syn.: *Ceratozetes gracilis cuspidatus* Jacot, 1939
Biology: From deciduous and coniferous forest litter. A form of C. *cuspidatus* (only differing from type in having 18 setae on tarsus I, and 15 setae on tarsus II) is found primarily in the western part of Canada. Distribution of the two forms overlap in alder litter from Riding Mountain National Park with both forms occuring in the same sample.
Distribution: BC, AB, SK, MB, ON, QC, NB, NS, PE, NL [CA, NH, NM, NY, VA, VT, WA; Nearctic].
Literature: Behan-Pelletier (1984); Behan-Pelletier and Eamer (2009); Maraun et al. (2019); ABMI (2019).

Ceratozetes enodis (Ewing, 1909) (Figure 3.135H,I)
Comb./Syn.: *Oribata enodis* Ewing, 1909; *Humerobates enodis* (Ewing, 1909); *Oribata figurata* Ewing, 1909; *Ceratozetes figuratus* (Ewing, 1909); *Trachyoribates zeteki* Ewing, 1917; *Ceratozetes zeteki* (Ewing, 1917)

Biology: From pasture; herbaceous litter.
Distribution: AB [IL, KY, MD, GA, FL, TX; Nearctic].
Literature: Behan-Pelletier and Eamer (2009); ABMI (2019).

Ceratozetes fjellbergi Behan-Pelletier, 1986 (Figure 3.134E,F)
Comb./Syn.: *Ceratozetella fjellbergi* (Behan-Pelletier, 1986)
Biology: From dry tundra areas of northwestern North America that were unglaciated during the Pleistocene and from north-eastern Russia (Behan-Pelletier 1997b).
Distribution: AK, BC [Amphi-Beringian].

Ceratozetes gracilis (Michael, 1884) (Figure 3.133C–E)
Comb./Syn.: *Oribata gracilis* Michael, 1884; *Murcia gracilis* (Michael, 1884); *Notaspis gracilis* (Michael, 1884)
Biology: Developmental biology, sex ratio, feeding preferences and energy budgets in *C. gracilis* were studied in a Danish beech-wood soil by Luxton (1972, 1975, 1981a) and in aspen parkland in Canada by Mitchell (1977b). Members feed on dead plant material and living microflora. Hartenstein (1962) found they fed preferentially on *Phialophora* and *Stemphylium* and to a lesser extent on *Alternaria*, *Cladosporium*, *Hormodendrum*, and decaying leaves, needles or wood. This species showed a wide variety of carbohydrases (Luxton 1972). The best food source for reproduction was the plant pathogen *Phoma exigua* (Mitchell and Parkinson 1976); it fed readily on *Trichoderma viride* but reproductive success was low.

Luxton (1981b) found populations principally in the 0–3cm soil layer, but also to 15cm depth. Mitchell (1978) found specimens in both fermentation and humus layers in different soils and recorded a downward movement of the population in autumn and an upward movement in spring. Cianciolo and Norton (2006) collected it from forest soil in New York State, but it was absent from the corn, grass, and shrub fields in their study. An adult of this species was recorded from feathers from the nest of a magpie (*Pica pica* (L.)) (Krivolutsky and Lebedeva 2004).
Distribution: AK, YT, AB, SK, MB, ON, QC, NB, NS, NL [IA, NH, NY, VA, WI; Cosmopolitan]
Literature: Behan-Pelletier (1984); Behan-Pelletier and Eamer (2009); Maraun et al. (2019); ABMI (2019).

Ceratozetes kananaskis Mitchell, 1976 (Figure 3.134A,B)
Biology: From dry forest. Mitchell (1976, 1977a, 1977b) gave a detailed description of the type locality and information on feeding, population biology, and respiratory metabolism. Mitchell (1978) discussed vertical and horizontal distribution. *Ceratozetes kananaskis* can co-occur with *C. cuspidatus*, *C. gracilis*, and *C. oresbios* (Behan-Pelletier 1984). The best food source for reproduction was the plant pathogen *Phoma exigua* (Mitchell and Parkinson 1976); it fed readily on *Trichoderma viride* but reproductive success was low.
Distribution: AB [Canada].
Literature: Behan-Pelletier and Eamer (2009); ABMI (2019).

Ceratozetes kutchin Behan-Pelletier, 1986 (Figure 3.132C,D)
Biology: From dry tundra.
Distribution: YT [Canada]
Literature: Behan-Pelletier and Eamer (2009).

Ceratozetes mediocris Berlese, 1908 (Figure 3.132G,H)
Biology: From dry deciduous and coniferous forest litter; dry moss; polypore fungi. An adult of this species has been recorded from feathers from a magpie nest (Krivolutsky and Lebedeva 2004). *C. mediocris* is associated primarily with hygrophilic sweetgrass meadows and reed banks in Germany and is considered to be an indicator of this biotope (Strenzke 1952). Menke

(1966) noted that it is found in meadows with slightly acidic to neutral pH and that highest abundance occurred in beds of moss. Habitats of *C. mediocris* and *C. peritus* overlap in Germany (Menke 1966), but C. *mediocris* prefers deeper and moister parts of the substrate.
Distribution: AB, SK, MB, ON, QC, NB, NS, [AL, AR, FL, GA, IL, LA, MO, MS, NC, NM, VA, TN, TX, VT; Semicosmopolitan].
Literature: Behan-Pelletier and Eamer (2009); Maraun et al. (2019); ABMI (2019).

Ceratozetes oresbios Behan-Pelletier, 1984 (Figure 3.133G,H)
Biology: From dry habitats in the subalpine zone of the Rocky Mountains.
Distribution: BC, AB [Canada].
Literature: Behan-Pelletier and Eamer (2009).

Ceratozetes pacificus Behan-Pelletier, 1984 (Figure 3.135E–G)
Biology: This species is restricted to mesic habitats, primarily in forests from Vancouver Island to California. It has been found in both canopy and forest-floor habitats in Western redcedar (Lindo and Winchester 2006). The molecular sequence data for the 28s rRNA and mitochondrial cytochrome oxidase 1 genes (*cox1* gene) was given in Lindo et al. (2008).
Distribution: BC [CA, OR, WA; Nearctic]
Literature: Behan-Pelletier and Eamer (2009).

Ceratozetes parvulus Sellnick, 1922 (Figure 3.132A,B)
Biology: From moist arctic, subarctic, and boreal habitats. This species has been recorded from reedbunting (*Emberiza schoeniclus*) plumage (Krivolutsky and Lebedeva 2004).
Distribution: AK, YT, NT, AB, MB, ON, QC, NB, NS, NL [VT; Holarctic]
Literature: Behan-Pelletier and Eamer (2009); Lemitz and Decker (2017); ABMI (2019).

Ceratozetes peritus Grandjean, 1951 (Figure 3.133F)
Biology: From coastal forest and grass litter; possibly introduced.
Distribution: NL [Holarctic].
Literature: Behan-Pelletier (1984); Behan-Pelletier and Eamer (2009); Maraun et al. (2019).

Ceratozetes spitsbergensis Thor, 1934 (Figure 3.134D)
Biology: From arctic tundra. Coulson et al. (2000) studied this species in their experimental manipulation of the winter surface ice layer in Svalbard, where this species is found in undisturbed soil habitats and those associated with bird nests (Lebedeva et al. 2006). Its distribution in alpine habits was elucidated by Fischer et al. (2016).
Distribution: AK, YT [arctic, alpine; Holarctic].
Literature: Behan-Pelletier (1985); Behan-Pelletier and Eamer (2009).

Ceratozetes thienemanni Willmann, 1943 (Figure 3.132E,F)
Comb./Syn.: *Ceratozetella thienemanni* (Willmann, 1943)
Biology: From boreal to subarctic habitats. It is one of the most abundant mites in the forest floor associated with Western redcedar trees in the Interior Cedar-Hemlock forest zone of BC (Lindo and Stevenson 2007). Lindberg and Bengtsson (2006) found no significant effect of drought on this species in a coniferous forest in Sweden. *Wolbachia* from this species was studied molecularly by Konecka and Olszanowski (2019c).
Distribution: AK, YT, NT, NU, BC, AB, MB, ON, QC, NB, NS, NL, [NH; Holarctic; Neotropical]
Literature: Behan-Pelletier and Eamer (2009); ABMI (2019).

Ceratozetes virginicus (Banks, 1906) (Figure 3.135A,B)
Comb./Syn.: *Galumna virginica* Banks, 1906; *Oribata virginica* (Banks, 1906); *Ceratozetes jeweli* Rockett and Woodring, 1966
Biology: From forest litter; grassy habitats; tallgrass prairie. Rockett and Woodring (1966) studied the life cycle and feeding habits of *C. virginicus* (as *C. jeweli*); it was cultured on

chopped up lichen. The gut contents of a specimen from Mississippi showed evidence of fungivory and scavenging on collembolans (Behan-Pelletier and Eamer 2009).
Distribution: ON, [VA, LA, IL, IN, MO, KS, AR, GA, MS, NC, AL; Nearctic]
Literature: Behan-Pelletier (1984); Lamoncha and Crossley (1998).

Ceratozetes watertonensis Behan-Pelletier, 1984 (Figure 3.133A,B)
Biology: From moss mats.
Distribution: AB [WA; Nearctic]

Ceratozetoides Shaldybina, 1966
Type-species: *Ceratozetes cisalpinus* Berlese, 1908

Ceratozetoides cisalpinus (Berlese, 1908) (Figure 3.131G,H)
Comb./Syn.: *Ceratozetes cisalpinus* Berlese, 1908; *Ceratozetella (Ceratozetella) cisalpina* (Berlese, 1908)
Biology: From deciduous forest and moist meadow. Woodring and Cook (1962b) described the anatomy, physiology, and molting of this species, which they found to be the most abundant oribatid in moist meadow sod in the St. Paul, Minnesota, area. Woodring and Cook (1962a) determined the life cycle to take an average of 32 days under laboratory conditions of 25°C, 90%–95% RH, and optimal diet. In culture, eggs were usually laid under dried leaves, or if these were unavailable, in cast nymphal skins. Females laid eggs 15–20 days after emergence, and all 10–16 eggs were deposited at once. Females did not lay eggs again until subjected to a temperature of less than 5°C for 2.5 months. They noted that these mites are surface-feeders, although the larva and protonymph may burrow into loose organic material. Juveniles formed molting aggregations. Adults moved at a regular pace of 0.09–0.22 cm/s, with the highest speed evoked by prodding with a brush. Larval *C. cisalpinus* apparently need fungi as food, whereas other stages can survive on lichen or artificial diet. Activity and feeding continued over a wide range of temperatures, as low as 2–3°C and as high as 35°C; males deposited spermatophores at 35°C but these were not viable.

An adult of this species has been recorded from feathers of a rook (*Corvus frugilegus*) (Krivolutsky and Lebedeva 2004).
Distribution: ON [GA, MN, NY; Holarctic]
Literature: Behan-Pelletier and Eamer (2009); Smrž et al. (2015); Seniczak et al. (2016).

Dentizetes Hammer, 1952
Type-species. *Dentizetes rudentiger* Hammer, 1952

KEY TO ADULT *DENTIZETES* OF CANADA AND ALASKA

1   Lamellar cusps dentate distally, parallel-sided and anterior margin 3× wider than base of lamellar seta. Length: 500–560µm (Figure 3.136A) .... ***D. rudentiger*** Hammer
–   Lamellar cusps rounded distally or with small lateral dens, tapered to slightly wider than base of lamellar seta. Length: 400–502µm (Figure 3.136B) ..... ***D. ledensis*** Behan-Pelletier

Dentizetes ledensis Behan-Pelletier, 2000 (Figure 3.136B)
Biology: Adults and juveniles of this species are only found on the underside of the leaves and on buds and stems of Labrador Tea (*Rhododendron (Ledum) groenlandicum*). Gut contents of adult *D. ledensis* include fungal hyphae and spores of a rust fungus, and plant epithelial tissue; that of deutonymphs include fungal hyphae and spores. Gravid females carry up to 5 large eggs.
Distribution: AB, NS [WI; Nearctic].
Literature: ABMI (2019).

Dentizetes rudentiger Hammer, 1952 (Figure 3.136A)
   Biology: From dry alpine habitats; from cankers of comandra blister rust on Lodgepole pine (Walter et al. 2014)
   Distribution: YT, BC, AB [MT, OR; Nearctic].
   Literature: ABMI (2019).

Diapterobates Grandjean, 1936
   Type-species: *Notaspis humeralis* Hermann, 1804

KEY TO ADULT *DIAPTEROBATES* OF CANADA AND ALASKA

1 Length >550µm. Pleurophragma without heavily sclerotized, fimbriated appearance. Translamella usually short and broad, rarely absent 2
– Length <550µm. Pleurophragma heavily sclerotized, fimbriated. Translamella long and narrow, or absent. Length: 410–550µm (Figure 3.137A)..................*D. variabilis* Hammer

2(1) Notogaster and venter without reticulate pattern. Notogastral setae relatively long and strongly barbed, c_2 >100µm. Tibiae and tarsi without porose areas 3
– Notogaster and venter with distinct subsurface pattern of reticulations. Notogastral setae medium in length, c_2 <100µm. Porose areas present distally on tibiae and basally on tarsi. Length: 650–900µm (Figure 3.137B,C)..*D. humeralis* (Hermann)

3(2) Porose area *Aa* long and narrow; length longer than 3× width. Ventral length: 600–720µm (Figure 3.137D)..*D. notatus* (Thörell)
– Porose area *Aa* oval 4

4(3) Lamellar cusps rounded distally. Bothridial seta broadened distally. Length: 530–560µm (Figure 3.137E,F)..*D. rotundocuspidatus* Shaldybina
– Lamellar cusps with lateral dens. Bothridial seta long, fusiform. Length: 600–645µm ......
   ....*D. sitnikovae* Shaldybina

Diapterobates humeralis (Hermann, 1804) (Figure 3.137B,C)
   Comb./Syn.: *Notaspis humeralis* Hermann, 1804; *Sphaerozetes* (*Trichoribates*) *numerosus* Sellnick, 1924; *Murcia numerosa* Sellnick, 1928; *Trichoribates numerosus* Willmann, 1931; *Diapterobates numerosus* Hammer, 1955
   Biology: Mesophilous and known from peatlands; often arboreal. In oak litter (Lamoncha and Crossley 1998). This is one of the most abundant species associated with bark beetles in pheromone traps in Koli National Park, Finland, and the authors suggested that the relationship may be phoretic (Penttinen et al. 2013). In Japan this species is a natural enemy of the hemlock woolly adelgid, *Adelges tsugae* (Homoptera) by consuming the woolly filaments that enveloped the ovisacs and dislodging the insects and is considered a possible biocontrol agent (McClure 1995).
   Distribution: AK, YT, NU, NT, AB, MB, ON, QC, NS, NL [NC; Holarctic]
   Literature: **Literature**: Behan-Pelletier (1985, 1986); Maraun et al. (2019).

Diapterobates notatus (Thörell, 1871) (Figure 3.137D)
   Comb./Syn.: *Oribata notata* Thörell, 1871; *Notaspis notatus* (Thörell, 1871); *Trichoribates notatus* (Thörell, 1871); *Notaspis trimaculatus notatus* (Thörell, 1871)
   Biology: From a wide variety of mesic habitats in the arctic and subarctic; rarely found in barren habitats. Its occurrence in coastal habitats in Nova Scotia suggests that this species may be more widely distributed in mesic habitats throughout the boreal zone. Considered a generalist feeder based on gut content analysis (Behan and Hill 1978). Coulson (2009) found five specimens of this species attached to the thorax of a calliphorid fly (*Cynomya mortuorum*) on the Svalbard archipelago.

Distribution: AK, YT, NU, NT, AB, MB, QC, NS [Holarctic]
Literature: Behan-Pelletier (1985, 1986); Bayartogtokh et al. (2011); Seniczak et al. (2018); Maraun et al. (2019).

Diapterobates rotundocuspidatus Shaldybina, 1970 (Figure 3.137E,F)
Comb./Syn.: *Diapterobates siccatus* Behan-Pelletier, 1986
Note: Seniczak et al. (2018b) retained *D. siccatus*.
Biology: From dry, well-drained microhabitats.
Distribution: AK, YT [Amphi-Beringian]
Literature: Behan-Pelletier (1997).

Diapterobates sitnikovae Shaldybina, 1970
Biology: From montane tundra; sparse spruce.
Distribution: AB [Amphi-Beringian]
Literature: Seniczak et al. (2018b); ABMI (2019).

Diapterobates variabilis Hammer, 1955 (Figure 3.137A)
Biology: From moss, lichens, litter in dry microhabitats. Although *D. variabilis* is widely distributed throughout the arctic, it is rarely abundant in any soil or litter sample. Block (1979) analyzed the cold tolerance of specimens from Alaska.
Distribution: AK, YT, NU, NT, AB, QC [Holarctic]
Literature: Behan-Pelletier (1986); ABMI (2019); Matthews et al. (2019).

Fuscozetes Sellnick 1928
Type-species: *Oribates fuscipes* C.L. Koch, 1844
Literature: Bayartogtokh et al. (2021).

KEY TO ADULT *FUSCOZETES* OF CANADA AND ALASKA

1 Length <500μm. Notogastral setation 14 pairs. Porose areas small, on tubercles and surrounded by darkened pigment. Ventral length: 421–473μm (Figure 3.138A,B) .... ***F. sellnicki*** Hammer
– Length >580μm. Notogastral setation 10–13 pairs. Porose areas of medium size, not on tubercles or surrounded by darkened pigment 2

2(1) Notogastral setation 11–13 pairs. Cusp of tutorium extending anterior of rostral seta. Length: 600–660μm (Figure 3.138C) .... ***F. setosus*** (C.L. Koch)
– Notogastral setation 10 pairs. Cusp of tutorium not reaching insertion of rostral seta 3

3(2) Lamellar cusp with tooth either laterally, or medially and laterally. Length: 630–765μm (Figure 3.138D,E) .... ***F. fuscipes*** (C.L. Koch)
– Lamellar cusp rounded, without teeth. Length: ca. 500μm (Figure 138F,G) .... ***F. bidentatus*** Banks
Note: Unidentified species known from late Pliocene/early Pleistocene fossil deposits in AK (Matthews et al. 2019).

Fuscozetes bidentatus Banks, 1895 (Figure 138F,G)
Biology: Common in epigeous moss, rare on moss found on stumps and boulders. Also in spruce, and hemlock leaf mould of moist microclimates (Jacot 1935).
Distribution: MB, QC [Nearctic]
Literature: Seniczak and Seniczak (2020).

Fuscozetes fuscipes (C.L. Koch, 1844) (Figure 3.138D,E)
>   **Comb./Syn.**: *Oribates fuscipes* C.L. Koch, 1844; *Sphaerozetes* (*Trichoribates*) *fuscipes* (C.L. Koch, 1844); *Galumna slossonae* Banks, 1906
>   **Biology**: A mesophilous mite, reported from peatlands; also characteristic of sites with pasture or field crops. Gravid females carry up to 4 large eggs.
>   **Distribution**: AK, AB, MB, ON, QC, NS [NH; Holarctic; Mesoamerica]
>   **Literature**: Shaldybina (1978); Bayartogtokh and Weigmann (2005a); Maraun et al. (2019); Seniczak and Seniczak (2020).

Fuscozetes sellnicki Hammer, 1952 (Figure 3.138A,B)
>   **Biology**: From shrub tundra.
>   **Distribution**: AK, YT, NU, NT, MB, QC [Holarctic].
>   **Literature**: Behan-Pelletier (1985, 1986, 1997b).

Fuscozetes setosus (C.L. Koch, 1839) (Figure 3.138C)
>   **Comb./Syn.**: *Oribates setosus* C.L. Koch, 1839
>   **Biology**: From moist habitats.
>   ]**Distribution**: QC, NL [Holarctic]
>   **Literature**: Seniczak et al. (1991); Maraun et al. (2019).

Ghilarovizetes Shaldybina, 1969
>   **Type-species**. *Ghilarovizetes longisetosus* (Hammer, 1952)
>   **Note**: Undescribed species from BC.
>   **Literature**: Bayartogtokh et al. (2021).

Ghilarovizetes longisetosus (Hammer, 1952) (Figure 3.131A)
>   **Comb./Syn.**: *Melanozetes longisetosus* Hammer, 1952
>   **Biology**: From moist litter on predominantly sandy soils. In wet moss at the edge of shaded, swift-flowing streams in Cape Breton, NS, suggesting species may be more widely distributed throughout the boreal in wet stenothermic habitats. Gut content analysis of specimens from Iqaluit, NU, showed feeding on dead plant and fungi (Behan and Hill 1978).
>   **Distribution**: AK, YT, NT, NU, QC, NS [Holarctic]
>   **Literature**: Behan-Pelletier (1985).

Iugoribates Sellnick, 1944
>   **Type-species**: *Iugoribates gracilis* Sellnick, 1944

Iugoribates gracilis Sellnick, 1944 (Figure 3.130C)
>   **Biology**: From dry tundra and rock desert.
>   **Distribution**: AK, YT, NT, NU, AB [Holarctic]
>   **Literature**: Behan-Pelletier (1985, 1986, 1997b); ABMI (2019).

Jugatala Ewing, 1913
>   **Type-species**. *Jugatala tuberosa* Ewing, 1913

Jugatala tuberosa Ewing, 1913 (Figure 3.129D,E)
>   **Biology**: From twigs, foliage and associated lichens in the canopy of western conifers (Voegtlin 1982).
>   **Distribution**: BC [OR, WA; Nearctic]
>   **Literature**: Behan-Pelletier (2000).
>   **Note**: Undescribed species from BC, AB.

Laminizetes Behan-Pelletier, 1986
>   **Type-species**: *Laminizetes fortispinosus* Behan-Pelletier, 1986

Laminizetes fortispinosus Behan-Pelletier, 1986 (Figure 3.131B)
   Biology: From dry tundra habitats.
   Distribution: YT [Canada]

Lepidozetes Berlese, 1910
   Type-species: *Lepidozetes singularis* Berlese, 1910
   Note: Considered Tegoribatidae in Weigmann (2006); confirmed Ceratozetidae (based on immatures) by Seniczak et al. (2014).

KEY TO ADULT *LEPIDOZETES* OF CANADA AND ALASKA

1   Bothridial seta with swollen, spiculate head (truncate distally). Interlamellar seta very fine, lightly barbed, mostly covered by tectum, much shorter than lamellar seta. Length: 415–505µm (Figure 3.130D) .... ***L. singularis*** Berlese
–   Bothridial seta with a large, elongate, flattened head. Interlamellar setae relatively long, longer than lamellar setae. Length: ca. 400µm (Figure 3.130E) ......***L. latipilosus*** Hammer
   Note: Unidentified species known from late Pliocene/early Pleistocene fossil deposits in AK (Matthews et al. 2019). Undescribed species in AB (Walter et al. 2014).

Lepidozetes latipilosus Hammer, 1952 (Figure 3.130E)
   Biology: From dry tundra.
   Distribution: AK, YT, NL [Nearctic]
   Literature: Behan-Pelletier (1997b).

Lepidozetes singularis Berlese, 1910 (Figure 3.130D)
   Biology: From mesophilous habitats and peatlands (Behan-Pelletier and Bissett 1994). Gravid female with 4 eggs (Walter et al. 2014).
   Distribution: AK, YT, NU, NT, AB, ON, QC, NB, NL [NH, NY; Holarctic]
   Literature: Weigmann (2013); Seniczak et al. (2014); Maraun et al. (2019); ABMI (2019).

Melanozetes Hull, 1916
   Type-species: *Oribates mollicomus* C.L. Koch, 1839
   Comb./Syn.: *Alphypochthonius* Schweizer, 1956
   Literature: Bayartogtokh et al. (2021).

KEY TO ADULT *MELANOZETES* OF CANADA AND ALASKA

1   Genal incision not fused to lateral margin of rostrum 2
–   Genal incision completely fused to lateral margin of rostrum; only carina of genal tooth evident (Figure 3.139F)..3

2(1) Ventral, genital and anal plates with foveae. Integument diffusely darker around base of notogastral setae. Length: 557–624µm (Figure 3.139A–C) ....***M. crossleyi*** Behan-Pelletier
–   Ventral, genital and anal plates micropunctate only. Integument around base of notogastral setae not diffusely darker. Length: 460–590µm (Figure 3.139D) ....
      ....***M. mollicomus*** (C.L. Koch)

3(1) Notogastral porose areas small, on tubercles. Rostrum rounded. Integument of notogaster and venter micropunctate only. Tutorium about 115µm long. Seta *l"* absent from tarsus I (19 setae). Ventral length: 518–596µm (Figure 3.139E) ....***M. tanana*** Behan-Pelletier
–   Notogastral porose areas normal in size, not on tubercles. Rostrum rounded or concave medially. Integument of notogaster and venter with both microtubercles and micropunctations. Seta *l"* present on tarsus I (20 setae). Length: 525–635µm (Figure 3.139F,G) ....
      ....***M. meridianus*** Sellnick

Melanozetes crossleyi Behan-Pelletier, 2000 (Figure 3.139A–C)
  Biology: From moss, decaying bark and bracket fungus. Recorded from lower tree trunks of Western redcedar (Lindo and Winchester 2007).
  Distribution: BC [Canada]

Melanozetes meridianus Sellnick, 1928 (Figure 3.139F,G)
  Biology: Widely distributed throughout the arctic, subarctic, and temperate alpine areas.
  Distribution: AK, YT, NT, MB, QC [Holarctic]; known from Tertiary fossil sites on Prince Patrick, Meigen and Ellesmere Islands (NU) (Matthews et al. 2019).
  Literature: Behan-Pelletier (1986).

Melanozetes mollicomus (C.L. Koch, 1839) (Figure 3.139D)
  Comb./Syn.: *Oribates mollicomus* C.L. Koch, 1839; *Murcia mollicomus* (C.L. Koch, 1839); *Notaspis mollicoma* (C.L. Koch, 1839); *Sphaerozetes mollicomus* (C.L. Koch, 1839)
  Biology: From moist forest litter, moss.
  Distribution: AK [Holarctic]
  Literature: Weigmann (2006); Maraun et al. (2019).

Melanozetes tanana Behan-Pelletier, 1986 (Figure 3.139E)
  Biology: From bog tundra and shrub tundra inhabitant.
  Distribution: AK, YT [Nearctic].
  Literature: Behan-Pelletier (1997b).

Neogymnobates Ewing, 1917
  Type-species: *Oribata multipilosa* Ewing, 1907

KEY TO ADULT *NEOGYMNOBATES* OF CANADA AND ALASKA

1 Lenticulus present. Bothridial seta finely barbed, truncate. Notogastral setae very long (56–240µm). Length: 616–729µm (Figure 3.136C)...............*N. marilynae* Behan-Pelletier
– Lenticulus absent. Bothridial seta with densely barbed, elongate head. Notogastral setae relatively short (11–25µm). Length: 350–430µm (Figure 3.136D).......*N. luteus* (Hammer)
  Note: Unidentified species known from Tertiary fossil sites on Ellesmere Island (NU) (Matthews et al. 2019). Undescribed species in AB (Walter et al. 2014).

Neogymnobates luteus (Hammer, 1955) (Figure 3.136D)
  Comb./Syn.: *Boreozetes luteus* Hammer, 1955
  Biology: Gravid females carry up to 5 eggs. Gut contents include brown hyphae and fine to coarse organic matter (Walter et al. 2014). From bog tundra in Yukon.
  Distribution: AK, YT, BC, AB, QC, NB, NL [Nearctic].
  Literature: Behan-Pelletier (1997b, 2000); ABMI (2019).

Neogymnobates marilynae Behan-Pelletier, 2000 (Figure 3.136C)
  Biology: From canopy habitats, especially alecterioid lichens in upper canopy of Douglas fir and Western redcedar (Behan-Pelletier et al. 2008).
  Distribution: BC [WA; Nearctic]

Protozetomimus Pérez-Íñigo, 1990
  Type-species: *Protozetomimus acutirostris* (Mihelčič, 1957)
  Remarks: Subías (2004) considered *Protozetomimus* a subgenus of *Zetomimus* (Zetomimidae), without justification; its character states are that of Ceratozetidae.

Protozetomimus acutirostris (Mihelčič, 1957) (Figure 3.131C–F)
Comb./Syn.: *Ceratozetes acutirostris* Mihelčič, 1957; *Zetomimus* (*Protozetomimus*) *acutirostris* (Mihelčič, 1957)
Biology: Probably introduced; found at edge of agricultural field in urban area (CNC record).
Distribution: ON [Holarctic]

Scutozetes Hammer, 1952
Type-species: *Scutozetes lanceolatus* Hammer, 1952
Note: Behan-Pelletier (2017) provided arguments for moving *Scutozetes* from Tegoribatidae to Ceratozetidae, a change overlooked by Ermilov and Makarova (2021) in their revision.

Scutozetes lanceolatus Hammer, 1952 (Figure 3.130A)
Biology: Boreal and subarctic species, known from dry tundra habitats (Behan-Pelletier 1997b). Females carry up to 4 eggs. Gut contents include some fungal hyphae and brown granular material (Walter et al. 2014).
Distribution: AK, YT, NT, AB, QC, NS [Holarctic]
Literature: ABMI (2019); Ermilov and Makarova (2021).

Sphaerozetes Berlese, 1885
Type-species: *Oribata orbicularis* C.L. Koch, 1835
Literature: Seniczak et al. (2016).

KEY TO ADULT *SPHAEROZETES* OF CANADA AND ALASKA

1 Notogastral setation 10 pairs (seta c_1 absent) 2
– Notogastral setation 11 pairs (seta c_1 present) 3

2(1) Medial rostral tooth present. Translamella linear. Length: 462–583µm (Figure 3.140A) ....*S. winchesteri* Behan-Pelletier
– Medial rostral tooth absent. Translamella ribbon-like. Length: 540–560µm (Figure 3.140B) ....*S. arcticus* Hammer

3(1) Lamellar cusp with sharply pointed medial and lateral dens of equal length. Translamella >40µm long. Tutorium about 190µm long, ratio of length of fused portion to cusp about 1.9:1. Length: >600µm (Figure 3.140C–E) ....*S. castaneus* Hammer
– Lamellar cusp with sharply pointed lateral dens, without medial dens. Translamella <40 long. Tutorium about 140µm long, ratio of length of fused portion to cusp about 2.6:1. Length: 473–518µm (Figure 3.140F,G)..................................*S. firthensis* Behan-Pelletier
Note: Unidentified specimens and *S. piriformis* (Nicolet, 1855) known from Tertiary fossil sites on Ellesmere Island (Matthews et al. 2019). Unidentified species from BC and AB.

Sphaerozetes arcticus Hammer, 1952 (Figure 3.140B)
Biology: From tussock tundra; birch and alder litter; willow litter along floodplain; alpine wet meadow with *Equisetum*, *Eriophorum*, moss, *Carex*; Balsam fir forests; boreal forest with White spruce, Black spruce, Jack pine and Trembling aspen.
Distribution: AK, YT, NT, NU, BC, AB, NS, NL [Holarctic]

Sphaerozetes castaneus Hammer, 1955 (Figure 3.140C–E)
Biology: From dry tundra.
Distribution: AK, YT [Nearctic]; specimens known from Late Pliocene/ Early Pleistocene fossil sites in AK (Matthews et al. 2019).

Sphaerozetes firthensis Behan-Pelletier, 1986 (Figure 3.140F,G)
   Biology: From dry tundra and mixed subarctic litter.
   Distribution: YT [Canada]

Sphaerozetes winchesteri Behan-Pelletier, 2000 (Figure 3.140A)
   Biology: From moss on bark of arbutus; moss and decaying wood on fallen tree trunks; *Picea sitchensis* canopy and litter; Western redcedar canopy.
   Distribution: BC [Canada]

Svalbardia Thor, 1930
   Type-species: *Svalbardia paludicola* Thor, 1930 = *Svalbardia lucens* (L. Koch, 1879)
   Literature: Ermilov et al. (2022).

Svalbardia lucens (L. Koch, 1879) (Figure 3.130B)
   Comb./Syn.: *Oribata lucens* L. Koch, 1879; *Oromurcia lucens* (L. Koch, 1879); *Svalbardia paludicola* Thor, 1930; *Svalbardia rostralis* Druk, 1982
   Note: *Svalbardia paludicola* was placed in synonymy with *Oribata lucens* and combined to *Svalbardia lucens* in Ermilov et al. (2022).
   Biology: Known from bog tundra and dry tundra (Behan-Pelletier 1997b). Associated with moist to wet habitats.
   Distribution: AK, YT, NU, NT, AB, MB [Circumpolar]
   Literature: Behan-Pelletier (1985, 1986); Seniczak et al. (2017); ABMI (2019).

Trichoribates Berlese, 1910
   Type species: *Murcia trimaculata* C.L. Koch, 1835 (=*Oribates setosus* sensu Berlese, 1887)
   Note: Weigmann and Norton (2009) explained the contrary opinions published over the last century on the validity of *Murcia* C.L. Koch, 1835 and *Trichoribates* Berlese, 1910 and use of the name *Murcia* rather than the widely used name *Trichoribates* by some recent authors.
   Biology: Meier et al. (2002) found that faecal pellets from *Trichoribates trimaculatus* (C.L. Koch) that have fed on thalli of the wall lichen, *Xanthoria parietina*, contained both viable ascospores and the photobiont cells (*Trebouxia arboricola*). These authors speculated that passage of viable elements in feces might provide a common and successful mode of vegetative short- and long-distance dispersal of lichen-forming ascomycetes and their photobionts, as suggested by Stubbs (1995) for fungal soridia.
   Literature: Seniczak et al. (2019).

KEY TO ADULT *TRICHORIBATES* OF CANADA AND ALASKA

1	Notogaster with 3 pairs of saccules. Pteromorph with hinge. Ventral length: 531–616µm (Figure 3.141A)	***T. polaris*** Hammer
–	Notogaster with 4 pairs of porose areas. Pteromorph without hinge	2
2(1)	Notogastral setation 11 pairs (seta *dp* present)	3
–	Notogastral setation 10 pairs (seta *dp* absent)	4
3(2)	Lamella with heavy striations. Interlamellar setae extending to tip of lamellae. Translamella deep, well-developed. Ventral length: 454–661µm (Figure 3.141B,C)	***T. striatus*** Hammer
–	Lamella without heavy striations. Interlamellar setae extending anterior of rostrum. Translamella narrow. Length: ca. 600µm	***T. obesus*** (Banks)
4(2)	Porose area *Aa* long, narrow oval. Length: >700µm (Figure 3.141D)	***T. formosus*** (Banks)
–	Porose area *Aa* circular to short oval. Length: <700µm	5

5(4) Integument of notogaster reticulate. Lamellar cusps broad, with strong lateral tooth. Length: 445–530μm (Figure 3.141E,F) .... *T. incisellus* (Kramer)
– Integument of notogaster microtuberculate. Lamellar cusps narrow, without strong lateral tooth 6

6(5) Porose area *A1* more medially placed than other notogastral porose areas. Notogastral seta *lp* anteromediad of porose area *A1*. Lamellar cusp with well-developed rounded lateral dens. Ventral length: 616–661μm (Figure 3.141G) .... *T. ogilviensis* Behan-Pelletier
– Porose area *A1* in same line as other notogastral porose areas. Notogastral seta *lp* anteriad of porose area *A1*. Lamellar cusp without lateral dens. Ventral length: 428–492μm (Figure 3.141H) .... *T. copperminensis* Hammer
Note: Unidentified species from AK: Lava Camp, Seward Peninsula (Late Miocene: 5.7 mya) and from late Pliocene/early Pleistocene fossil deposits in AK (Matthews et al. 2019).
Three unnamed, undescribed species from AB. *Trichoribates principalis* (Berlese, 1914) listed from AB is probably a misidentification (Walter et al. 2014).

Trichoribates copperminensis Hammer, 1952 (Figure 3.141H)
   Biology: From bog tundra.
   Distribution: AK, YT, NU, NT, BC, AB [Holarctic].
   Literature: Behan-Pelletier (1997b); ABMI (2019).

Trichoribates formosus (Banks, 1909) (Figure 3.141D)
   Comb./Syn.: *Oribatella formosa* Banks, 1909
   Biology: From under bark of mountain ash, Norway spruce and catalpa (Banks 1909).
   Distribution: ON [NJ; Nearctic].

Trichoribates incisellus (Kramer, 1897) (Figure 3.141E,F)
   Comb./Syn.: *Oribata incisella* Kramer, 1897; *Trichoribatella baloghi* Mahunka, 1983; *Latilamellobates baloghi* (Mahunka, 1983)
   Biology: From cranberry bog in BC; possibly introduced species.
   Distribution: BC, AB [Holarctic]
   Literature: Behan-Pelletier (2000); Weigmann (2006); Maraun et al. (2019).

Trichoribates obesus (Banks, 1895)
   Comb./Syn.: *Oribatella obesa* Banks, 1895; *Jugatala lamellata* Ewing, 1918; *Trichoribates lamellata* (Ewing, 1918)
   Biology: From rotting vegetation.
   Distribution: AK [NY, OR, WA; Holarctic]; Banks (1895b) questioned the record from NY.

Trichoribates ogilviensis Behan-Pelletier, 1986 (Figure 3.141G)
   Biology: From dry tundra.
   Distribution: YT [Canada].
   Literature: Behan-Pelletier (1997b).

Trichoribates polaris Hammer, 1953 (Figure 3.141A)
   Note: Placed in the subgenus *Sacculoribates* Subías (Subías 2021).
   Biology: From bog tundra; in moist to wet habitats, primarily in high arctic localities.
   Distribution: AK, YT, NU, NT [Holarctic].
   Literature: Behan-Pelletier (1997b); Matthews et al. (2019).

Trichoribates striatus Hammer, 1952 (Figure 3.141B,C)
   Biology: Characteristic of open habitats with grasses and forbs. Gravid females carry at least 5 eggs. Gut contents include fungal hyphae and spores and some indeterminate granular material (Walter et al. 2014).

Taxonomic Keys

Distribution: AK, YT, AB, MB [Nearctic].
Literature: Behan-Pelletier (1985, 1986); ABMI (2019).

Family Chamobatidae Grandjean 1954
Diagnosis: *Adults*: Poronotic Brachypylina. Prodorsum with lamellae, separated, with dens anteriorly; lamellar cusps and translamella absent. Lamellar setae arising from prodorsum. Tutorium present. Genal incision present. Dorsophragmata and pleurophragmata present. Discidium and custodium present, postanal porose area generally present. Pedotectum I large, extending to base of bothridium. Pedotectum II present. Notogaster with or without lenticulus; posterior tectum present, complete; pteromorphs present, without hinge; setation 10 pairs. Octotaxic system developed as 4 pairs of porose areas. Genital setation 6 pairs. Bothridium cup shaped, with small lateral scales. Genu IV shorter than tibia IV.
Literature: Seniczak et al. (2018, 2019).

Chamobates Hull, 1916
Type-species: *Oribata cuspidata* Michael, 1884

KEY TO ADULT *CHAMOBATES* OF CANADA AND ALASKA

1 Only alveoli of notogastral seta c and p present. Head of bothridial seta narrow. Length: 330–390µm (Figure 3.142A).. *C. cuspidatus* (Michael)
– Notogastral seta c and p short. Head of bothridial seta thick 2

2(1) Porose area *A1* subequal in size to *Aa*, both small. Length: 364–470µm (Figure 3.142B)...
....*C. pusillus* (Berlese)
– Porose area *A1* smaller than *Aa*. Length: 340–390µm (Figure 3.142C) ....
....*C. schuetzi* (Oudemans)

Note: Unidentified *Chamobates* sp. from Alaska: Lava Camp, Seward Peninsula (Late Miocene: 5.7 mya) (Matthews et al. 2019). Unnamed species from BC, AB, ON, QC, NS, NB.

Chamobates cuspidatus (Michael, 1884) (Figure 3.142A)
Comb./Syn.: *Oribata cuspidata* Michael, 1884; *Murcia cuspidatus* (Michael, 1884); *Euzetes cuspidatus* (Michael, 1884)
Biology: Schneider (2005) placed Chamobatidae in the secondary decomposer feeding guild (i.e., feeding primarily on fungi and secondarily on decomposing plant matter), supporting studies by Luxton (1972) on this species. Specimens of *C. cuspidatus* from a chestnut and beech forest analysed over a 20 month period showed feeding preference for fungal spores > amorphous material > fungal hyphae; almost no leaf material was represented (Anderson 1975). This species will exploit seasonal abundance of pollen by feeding (Wallwork 1983).
Distribution: AB, ON, QC, NL [VA; Holarctic]
Literature: Maraun et al. (2019); ABMI (2019).

Chamobates pusillus (Berlese, 1895) (Figure 3.142B)
Comb./Syn.: *Oribates pusillus* Berlese, 1895
Note: Seniczak et al. (2019) used morphology and CO1 (*cox1* gene) to show that *C. borealis* (Trägårdh, 1902) is distinct from *C. pusillus*.
Biology: From forest.
Distribution: AK [Holarctic].
Literature: Maraun et al. (2019).

Chamobates schuetzi (Oudemans, 1902) (Figure 3.142C)
Comb./Syn.: *Notaspis schuetzi* Oudemans, 1902; *Scheloribates schuetzi* (Oudemans, 1902)

Biology: From wood, lichen, grass.
Distribution: AK [Holarctic].
Literature: Seniczak and Solhøy (1988).

Family Humerobatidae Grandjean, 1971
Diagnosis: *Adults*: Poronotic Brachypylina. Rostrum with prolamella. Prodorsum with lamellae, separated; tutorium, lamellar cusps present; translamella absent. Bothridium with lateral scales. Genal incision present. Dorsophragmata and pleurophragmata present. Discidium, custodium, postanal porose area present. Pedotectum I large, extending to base of bothridium, convex dorsally. Pedotectum II present. Notogaster with lenticulus, posterior tectum present; pteromorphs with incomplete hinge; setation 10 pairs. Octotaxic system developed as 4 pairs of porose areas. Genital setation 6 pairs. Chelicerae chelate-dentate; palpal eupathidium *acm* attached to solenidion; subcapitular mentum without tectum. Axillary saccule of subcapitulum present. Genu IV shorter than tibia IV.

Humerobates Sellnick, 1928
Type-species: *Humerobates rostrolamellatus* Grandjean, 1936
Comb./Syn.: *Banksinus* Jacot, 1938

KEY TO ADULT *HUMEROBATES* OF CANADA AND ALASKA

1   Bothridial seta straight. Length: 800–925µm (Figure 3.142D,E) ....
    .... *H. rostrolamellatus* Grandjean
–   Bothridial seta bent medially. Length: 545–770µm (Figure 3.142F) ....
    .... *H. arborea* (Banks)

Humerobates arborea (Banks, 1895) (Figure 3.142F)
Comb./Syn.: *Oribata arborea* Banks, 1895; *Banksinus arborea* (Banks, 1895); *Galumna arborea* (Banks, 1895); *Humerobates humeralis arborea* (Banks, 1895)
Biology: A survey of tree trunks in five uneven-aged forests in central Maine showed that 25% of adult *H. arborea* had soredia adhering to their integument (Stubbs 1995). Mites dusted with *Lepraria lobificans* soredia deposited it on tree bark and on other lichen species.
Distribution: BC [CA, CT, GA, IL. ME, MA, NY; Holarctic]

Humerobates rostrolamellatus Grandjean, 1936 (Figure 3.142D,E)
Comb./Syn.: *Banksinus arboreus rostrolamellatus* Jacot, 1940
Biology: Found worldwide in temperate regions in association with plants. Adults and juveniles feed on fungi and algae on trees and are virtually unknown from litter samples. Adults move from the trunk to terminal shoots of fruit trees in spring and early summer where they deposit eggs and where hatched larvae and adults feed. The population moves back to the trunk in autumn for overwintering (Murphy and Balla 1973). *H. rostrolamellatus* has been implicated as an invasive nuisance pest of buildings (Scott 1958).
Distribution: BC, NS [Semicosmopolitan].
Literature: Jalil (1969); Porzner and Weigmann (1992).

Family Euzetidae Grandjean, 1954
Diagnosis: *Adults:* Poronotic Brachypylina. Prodorsum with lamellae, separated; lamellar cusps short; translamella absent. Tutorium present. Genal incision present. Bothridium with medial and lateral scales. Dorsophragmata and pleurophragmata present. Discidium and custodium present, postanal porose area present. Pedotectum I large, extending to base of bothridium. Pedotectum II present. Notogaster with lenticulus; posterior tectum present; pteromorphs present, without hinge; setation 10 pairs, represented by their alveoli. Octotaxic

system as 4 pairs of porose areas. Epimere IV with neotrichy (7–8 pairs). Genital setation 6 pairs. Chelicerae chelate-dentate; palpal eupathidium *acm* attached to solenidion; subcapitular mentum with tectum. Axillary saccule of subcapitulum present. Tibia IV with solenidion. Genu IV shorter than tibia IV.

Biology: Species of *Euzetes* live on the soil surface in woods with heavy leaf fall (Pauly 1956). Adults of *E. globulus* have been found in the hair of small mammals, sometimes in numbers six times higher than in surrounding natural habitats, suggesting active dispersal on these mammals (Miko and Stanko 1991).

Euzetes Berlese, 1908
Type-species: *Oribata globula* Nicolet, 1855 (Figure 3.142G,H)
Note: Unidentified species from QC.
Literature: Fredes and Martínez (2011); Seniczak and Seniczak (2014).

Family Punctoribatidae Thor, 1937 (= Mycobatidae Grandjean, 1954)
Diagnosis: *Adults*: Poronotic Brachypylina. Prodorsum with lamellae, adjacent medially or separated, lamellar cusps present or reduced, translamella present or absent; tutorium present. Pair of tubercles between interlamellar setae and bothridia present or absent. Bothridium with or without scales. Genal incision present. Dorsophragmata and pleurophragmata present. Discidium and custodium present; postanal porose area present as porose area or saccule. Pedotectum I large, extending to base of bothridium, convex or concave dorsally. Pedotectum II present. Notogaster with or without lenticulus. Anterior tectum present, flattened or rounded medially, or with small to large concave or convex medial process. Posterior tectum present: continuous, or incomplete posteromedially, with or without overlapping lobes. Notogaster with or without pair of thickened bands anteriorly. Pteromorphs present, with or without hinge. Notogastral setation 10–11 pairs. Octotaxic system as 2–many pairs (usually 4) of porose areas, or 4 pairs of saccules, or tubules; sexual dimorphism in some genera. Genital setation 5–6 pairs. Chelicerae chelate-dentate; palpal eupathidium *acm* attached to solenidion. Subcapitular mentum with or without complete tectum. Axillary saccule of subcapitulum present. Tibiae II–IV with or without dorsodistal projection. Genu IV shorter than tibia IV.
Biology: Species in this family are more diverse in temperate to arctic regions than in the tropics (Behan-Pelletier 1999).
Literature: Behan-Pelletier and Eamer (2008).

KEY TO GENERA OF PUNCTORIBATIDAE OF CANADA AND ALASKA

1	Interlamellar seta broad, bifurcate distally. Notogaster with 4 pairs of saccules (Figure 3.143A,B)	***Pelopsis*** Hull
–	Interlamellar seta not bifurcate distally. Notogaster with 3 or 4 pairs of porose areas, or porose areas hypertrophied in males	2
2(1)	Bothridium with dorsal and ventral scales. Pair of thickened bands absent from anterior tectum of notogaster. Tubercle absent from between base of interlamellar seta and bothridium	3
–	Bothridium cup-like, without dorsal and ventral scales. Pair of thickened bands present on anterior tectum of notogaster. Tubercle present between base of interlamellar seta and bothridium	7
3(2)	Pteromorph with complete hinge (Figure. 3.145–3.148)	***Mycobates*** Hull
–	Pteromorph with incomplete hinge	4
4(3)	Posterior notogastral tectum undivided	5
–	Posterior notogastral tectum divided medially, with overlapping lobes	6

5(4)   Tutorium broader distally than proximally; cusp strongly convex proximoventrally. *Ah* expressed as saccule. Rostrum rounded (Figure 3.143E) ....***Adoribatella*** Woolley
–   Tutorium broader proximally than distally; cusp almost straight proximoventrally. *Ah* expressed as porose area. Rostrum rounded or with parallel-sided indentations (Figure 3.144A–D) .... ***Cyrtozetes*** Behan-Pelletier

6(4)   Monodactylous. Notogaster length to width ratio 1.5:1 (Figure 3.143C,D).........................
....***Guatemalozetes*** Mahunka
–   Tridactylous. Notogaster length to width ratio 1.3:1 (Figure 3.144E,F)................................
....***Ceresella*** Pavlitshenko

7(2)   Posterior notogastral tectum divided medially, with lobes overlapping or not. Circumpedal carina merging with custodium. Pedotectum I convex or concave dorsally. Subcapitulum without mental tectum (Figure. 3.150, 3.151) ....***Zachvatkinibates*** Shaldybina
–   Posterior notogastral tectum undivided. Circumpedal carina independent of custodium. Pedotectum I concave dorsally. Subcapitulum with or without mental tectum 8

8(7)   Monodactylous. Anterior notogastral tectum with small medial process (Figure 3.143F)...
....***Minunthozetes*** Sellnick
–   Tridactylous. Anterior notogastral tectum with small to large medial process (Figure 3.149)...***Punctoribates*** Berlese

Adoribatella Woolley, 1967
   Type-species: *Adoribatella punctata* Woolley, 1967

Adoribatella punctata Woolley, 1967 (Figure 3.143E)
   Biology: The type series is from moss at Deadman's Pass, Colorado. Woolley (1967) also collected specimens from lichen at Cameron Pass, Colorado; both localities are at about 3000m elevation. Other collecting records are from low elevation boreal forest habitats near Hinton, AB and in a waterspray zone at Oneonta Gorge, OR.
   Distribution: AB [CO, OR; Nearctic]; known from late Pliocene/early Pleistocene fossil deposits in AK (Matthews et al. 2019).
   Literature: Behan-Pelletier (2013).

Ceresella Pavlitchenko, 1993
   Type-species: *Ceresella venusta* Pavlitchenko, 1993
   Note: Possibly undescribed species found in coastal BC (Lindo 2020).

Cyrtozetes Behan-Pelletier, 1985
   Type-species: *Ceratozetes shiranensis* Aoki, 1976

KEY TO ADULT *CYRTOZETES* OF CANADA AND ALASKA

1   Rostrum with two pairs of distinct incisions. Prolamella extending posteriorly from rostrum to level of lamellar cusp. Length: 290–320µm (Figure 3.144A,B) ....
.... *C. lindoae* Behan-Pelletier and Eamer
–   Rostrum rounded with small medial indentation. Prolamellae absent................................ 2

2(1)   Translamella absent. Pteromorphs with transverse striae anterolaterally. Length: 486–531µm (Figure 3.144C)... *C. denaliensis* Behan-Pelletier
–   Translamella present. Pteromorphs without transverse striae anterolaterally. Length: 308–356µm (Figure 3.144D).. *C. inupiaq* (Behan-Pelletier)
   Note: Undescribed species in AB (Walter et al. 2014).

Cyrtozetes denaliensis Behan-Pelletier, 1985 (Figure 3.144C)
   Biology: From arctic alpine biotopes north of latitude 60°.
   Distribution: AK, YT [Amphi-Beringian; Russian Far East]
   Literature: Behan-Pelletier and Eamer (2008).

Cyrtozetes inupiaq (Behan-Pelletier, 1986) (Figure 3.144D)
   Comb./Syn.: *Ceratozetes inupiaq* Behan-Pelletier, 1986
   Biology: From dry alpine habitats at high elevations in the mountains of eastern, and possibly western, Beringia.
   Distribution: YT [Canada]; known from Tertiary fossil sites on Prince Patrick Island (NU) (Matthews et al. 2019).
   Literature: Behan-Pelletier and Eamer (2008).

Cyrtozetes lindoae Behan-Pelletier and Eamer, 2008 (Figure 3.144A,B)
   Biology: Litter species in old-growth forests of Pacific Northwest. In mid-alpine habitats in Manning and Cathedral Provincial Parks and Waterton Lakes National Park.
   Distribution: BC, AB [Canada]
   Literature: Behan-Pelletier and Eamer (2008).

Guatemalozetes Mahunka, 1979
   Type-species: *Guatemalozetes aelleni* Mahunka, 1979

Guatemalozetes danos Behan-Pelletier and Ryabinin, 1991 (Figure 3.143C,D)
   Biology: From dry open habitats: grasslands, prairie, and woodlands. Adult females carry 1 large egg. Gut contents included sheet-like material (probably arthropod cuticle) and some hyaline fungi (Walter et al. 2014).
   Distribution: AB [CO, KS; Nearctic]

Minunthozetes Sellnick, 1928
   Type-species: *Zetes semirufus* C.L. Koch, 1841

Minunthozetes semirufus (C.L. Koch, 1841) (Figure 3.143F)
   Comb./Syn.: *Zetes semirufus* C.L. Koch, 1841; *Punctoribates* (*Minunthozetes*) *semirufus* Sellnick, 1928
   Biology: From meadow pasture turf. Juveniles may burrow in the stems of grasses (Evans et al. 1962).
   Distribution: NS, NL; probably introduced species [Holarctic].
   Literature: Seniczak and Seniczak (2018); Maraun et al. (2019).

Mycobates Hull, 1916
   Type-species: *Oribata parmeliae* Michael, 1884
   Comb./Syn.: *Calyptozetes* Thor, 1929; *Permycobates* Strenzke, 1954.

KEY TO ADULT *MYCOBATES* OF CANADA AND ALASKA

1	Tridactylous. Integument over notogastral porose areas tuberculate or not	2
–	Monodactylous. Integument over notogastral porose areas smooth	14
2(1)	Dorsophragmata fused. Area bordered by translamella and lamellar cusps distinctly U-shaped or lamellar cusps absent	3
–	Dorsophragmata separated. Area bordered by translamella and lamellar cusps rectangular, square, or U-shaped	5

3(2)	Lamellar cusps absent. Notogastral setae very long, flagellate, longer than 80µm. Interlamellar setae flagellate, extending anterior of rostrum. Length: 450–551µm (Figure 3.145A) .... *M. acuspidatus* Behan-Pelletier, Eamer and Clayton
–	Lamellar cusps present (maybe short). Notogastral setae tapered or acuminate to flagellate, never longer than 40µm. Interlamellar setae, if flagellate, not extending anterior of rostrum 4
4(3)	Integument over notogastral porose areas with scored structure. Subcapitulum with genal tectum. Rostrum bilaterally without distinct longitudinal ridges. Anterior 3 pairs of genital setae of medium length. Length: 599–672µm (Figure 3.145B–E) .... *M. incurvatus* Hammer
–	Integument over notogastral porose areas finely tuberculate. Subcapitulum without genal tectum. Rostrum bilaterally with distinct longitudinal ridges. Anterior 3 pairs of genital setae very long. Length: 609–680µm (Figure 3.145F,G) ......*M. hammerae* Behan-Pelletier
5(2)	Integument over notogastral porose areas tuberculate. Bothridial seta long, head shorter than stalk 6
–	Integument over notogastral porose areas smooth. Bothridial seta short, head longer than stalk 10
6(5)	Interlamellar seta thick, longer than 100µm, reaching anterior to translamella, or flagellate, reaching anteriad of lamellar cusps. Integument tuberculate along hinge area of pteromorph... 7
–	Interlamellar seta thin, shorter than 80µm, not or just reaching translamella. Integument tuberculate or not along hinge area of pteromorph 8
7(6)	Integument with loosely spaced tubercles over notogastral porose areas. Interlamellar and notogastral setae flagellate. Length: 454–518µm (Figure 3.145H,I) .... *M. beringianus* Behan-Pelletier
–	Integument of tightly spaced tubercles over notogastral porose areas. Interlamellar and notogastral setae tapered to acuminate. Length: 527–630µm (Figure 3.146A,B)................. .... *M. conitus* Hammer
8(6)	Integument of ventral and anal plates distinctly rugose. Integument over notogastral porose areas composed of large tightly packed tubercles. Length: 428–518µm (Figure 3.146C,D). ....*M. punctatus* Hammer
–	Integument of ventral and anal plates microtuberculate. Integument over notogastral porose areas composed of many small tubercles or ridges 9
9(8)	Integument over notogastral porose areas and along hinge of pteromorph composed of small, distinct ridges. Length: 478–557µm (Figure 3.146E,F) .... *M. altus* Behan-Pelletier
–	Integument over notogastral porose areas composed of small tubercles, tubercles absent or weakly developed along hinge of pteromorph. Length: 486–587µm (Figure 3.146G–I)...... .... *M. dryas* Behan-Pelletier
10(5)	Lamellar cusps very short, less than 12µm long. Translamella more than 5× length of lamellar cusps. Tarsus II with 1 solenidion. Length: 486–596µm (Figure 3.147A,B)........... .... *M. brevilamellatus* Behan-Pelletier
–	Lamellar cusps longer than 20µm; translamella at most 2× length of lamellar cusps. Tarsus II with 2 solenidia 11
11(10)	Bothridial seta long, head shorter than stalk. Integument of hinge area of pteromorph with cerotegument granules. Length: 446–505µm (Figure 3.147C).. .... *M. corticeus* Behan-Pelletier

| – | Bothridial seta short, head longer than stalk. Integument of hinge area of pteromorph smooth .... 12 |

12(11) Lamellar cusps 20–25µm long, edges strongly tapered distally. Notogaster laterad of porose areas *Aa* and *A1* and anteriad of porose area *Aa* more heavily sclerotized (and darker) than rest of notogaster. Length: ca. 518µm (Figure 3.147D) .... ***M. occidentalis*** Behan-Pelletier
– Lamellar cusps longer than 30µm, edges slightly tapered to parallel distally. Notogaster mediad, anteriad, and laterad of porose areas *Aa* and *A1* subequally sclerotized 13

13(12) Notogastral setae weakly barbed distally. Length: 492–590µm (Figure 3.147E) .... ***M. azaleos*** Behan-Pelletier
– Notogastral setae tapered to acuminate distally. Length: 460–590µm (Figure 3.147F,G).... .... ***M. sarekensis*** (Trägårdh)

14(1) Notogastral setae very long, flagellate, longer than 80µm. Interlamellar setae flagellate, extending anteriad of rostrum. Tarsus II with 1 solenidion. Length: 410–515µm (Figure 3.148A) .... ***M. parmeliae*** (Michael)
– Notogastral setae tapered or acuminate to flagellate, never longer than 40µm. Interlamellar setae never flagellate. Tarsus II with 2 solenidia 15

15(14) Longitudinal striae present on hinge of pteromorph. Notogastral porose areas absent, or only porose area *Aa* and/or *A1* clearly evident 16
– Longitudinal striae absent from hinge area of pteromorph. All notogastral porose areas large, clearly evident 17

16(15) Bothridial seta long, head shorter than stalk. Tarsus I with 17 setae (setae (*it*) and *v'* absent). Length: 356–369µm (Figure 3.148B) ....***M. hylaeus*** Behan-Pelletier
– Bothridial seta short, head longer than stalk. Tarsus I with 20 setae (setae (*it*) and *v'* present). Length: 405–454µm (Figure 3.148C) .... ***M. exigualis*** Behan-Pelletier

17(15) Margins of lamellar cusp parallel, medial dens larger than lateral dens. Area bordered by translamella and lamellar cusps U-shaped. Tarsus I with 18 setae (setae (*it*) absent). Length: 348–437µm (Figure 3.148D,E) .... ***M. perates*** Behan-Pelletier
– Lamellar cusps tapering, medial and lateral dens subequal; area bordered by translamella and lamellar cusps rectangular. Tarsus I with 20 setae (setae (*it*) present). Length: 454–473µm (Figure 3.148F) .... ***M. yukonensis*** Behan-Pelletier
Note: Unidentified species from AK: Lava Camp, Seward Peninsula (Late Miocene: 5.7 mya) (Matthews et al. 2019). Extant undescribed species from BC (Lindo 2020).

Mycobates acuspidatus Behan-Pelletier, Eamer and Clayton, 2001 (Figure 3.145A)
Biology: From canopy in coniferous temperate rainforests of Pacific Northwest. More common in foliose than in alecterioid lichens in high canopy of Douglas fir, Western hemlock and Western redcedar (Behan-Pelletier et al. 2008).
Distribution: BC [WA; Nearctic]
Literature: Seniczak et al. (2015).

Mycobates altus Behan-Pelletier, 1994 (Figure 3.146E,F)
Biology: From the western Cordillera (AB to NM) in alpine and boreomontane habitats.
Distribution: AB [CO, NM; Nearctic]
Literature: ABMI (2019).

Mycobates azaleos Behan-Pelletier, 1994 (Figure 3.147E)
Biology: From dry microhabitats. Adult females examined contain up to 6 eggs.
Distribution: BC, AB [OR; Nearctic]

Mycobates beringianus Behan-Pelletier, 1994 (Figure 3.145H,I)
   Biology: From dry microhabitats of parts of the arctic and subarctic that were unglaciated during the Pleistocene (eastern Beringia). It is probable that this species eventually will be found in the Russian Far East.
   Distribution: AK, YT, NT [Nearctic; Greenland].

Mycobates brevilamellatus Behan-Pelletier, 1994 (Figure 3.147A,B)
   Biology: From dry subalpine habitats in southern British Columbia. Adult females contain up to 8 eggs.
   Distribution: BC [Canada].

Mycobates conitus Hammer, 1952 (Figure 3.146A,B)
   Biology: From mosses, lichens, litter; wet meadows; widely distributed species in dry to moist microhabitats of arctic and subarctic North America and the Russian Far East.
   Distribution: AK, YT, NT, NU, AB, QC, NL [OR; Holarctic; Russian Far East]
   Literature: Behan-Pelletier (1994); Matthews et al. (2019).

Mycobates corticeus Behan-Pelletier, Eamer and Clayton, 2001 (Figure 3.147C)
   Biology: From moss, lichen and bark of conifers in the Pacific Northwest. Among most abundant mites in canopy of Western redcedar (Lindo 2010).
   Distribution: BC [Canada]

Mycobates dryas Behan-Pelletier, 1994 (Figure 3.146G–I)
   Biology: Widely distributed in dry arctic and subarctic habitats (along with *M. sarekensis*).
   Distribution: AK, YT, NT, NU, AB, NL [Nearctic; Greenland].
   Literature: Makarova (2015); ABMI (2019).

Mycobates exigualis Behan-Pelletier, 1994 (Figure 3.148C)
   Biology: From dry microhabitats in eastern North America (NS), Greenland and northern Europe (Norway).
   Distribution: NS [Holarctic].

Mycobates hammerae Behan-Pelletier, 1994 (Figure 3.145F,G)
   Biology: From dry microhabitats of parts of North American arctic and subarctic that were unglaciated during the Pleistocene (eastern Beringia). This species is known from the Pribilof Islands in the Bering Sea, and it is probably represented in the Russian Far East. Gravid females with up to 4 eggs.
   Distribution: AK, YT [Nearctic].

Mycobates hylaeus Behan-Pelletier, 1994 (Figure 3.148B)
   Biology: Known from moss and lichen microhabitats of eastern North America.
   Distribution: AB, ON, NS, NB [NH, PA, VA, VT; Nearctic].

Mycobates incurvatus Hammer, 1952 (Figure 3.145B–D)
   Biology: Widely distributed species in the Nearctic with a primarily boreal and subarctic distribution, and occasional records from alpine habitats in eastern and western mountains. Gravid female with up to 7 eggs.
   Distribution: AK, YT, NT, NU, BC, AB, QC, NS, NL [CO , NY, OR; Holarctic]
   Literature: Behan-Pelletier (1994); ABMI (2019).

Mycobates occidentalis Behan-Pelletier, 1994 (Figure 3.147D)
   Biology: From soil under grass.
   Distribution: AK, BC [Nearctic].

Mycobates parmeliae (Michael, 1884) (Figure 3.148A)
   Comb./Syn.: *Oribata parmeliae* Michael, 1884

Biology: Closely associated with lichens, especially *Parmelia*, growing near the sea (Michael 1884; Colloff 1988), and the latter author suggested that *M. parmeliae* "may be moderately halotolerant". It was found on trunks of Red maple in NS.
Distribution: NT, MB, NS [Holarctic]
Literature: Weigmann (2006).

Mycobates perates Behan-Pelletier, 1994 (Figure 3.148D,E)
Biology: Gravid females carry at least 2 large eggs. This mite has a boreal distribution and has been found in dry moss, lichens, spruce litter, and dry heaths.
Distribution: AK, YT, NU, AB, QC [NY; Nearctic]
Literature: ABMI (2019).

Mycobates punctatus Hammer, 1955 (Figure 3.146C,D)
Biology: From dry microhabitats. Gravid females with up to 5 eggs.
Distribution: AK, BC [Holarctic; Amphi–Beringian]
Literature: Behan-Pelletier (1994).

Mycobates sarekensis (Trägårdh, 1910) (Figure 3.147F,G)
Comb./Syn.: *Oribata sarekensis* Trägårdh, 1910; *Mycobates consimilis* Hammer, 1952
Biology: Common in arctic and subarctic habitats.
Distribution: AK, YT, NT, NU [Holarctic; Circumpolar].
Literature: Behan-Pelletier (1994); Solhøy (1997); Seniczak et al. (2015).

Mycobates yukonensis Behan-Pelletier, 1994 (Figure 3.148F)
Biology: From peat at edge of sulphur stream. Gravid females with 4 eggs.
Distribution: YT [Canada].

Pelopsis Hull, 1911
Type-species: *Pelops bifurcata* Ewing, 1909

Pelopsis bifurcatus (Ewing, 1909) (Figure 3.143A,B)
Comb./Syn.: *Ewingozetes bifurcatus* (Ewing, 1909); *Parapelops bifurcatus* (Ewing, 1909); *Galumna panita* Banks, 1910; *Pelopsis nudiuscula* Hall, 1911
Biology: Widely distributed throughout eastern and southern Canada and USA to California; also recorded from NT by Hammer (1952a). Common in moist soil and litter. It has been recorded from coniferous forest litter, under logs and in beach debris (Norton 1979c).
Distribution: NT, AB, NS [AL, AR, CA, CT, FL, GA, IL, KS, LA, MO, NC, NH, NY, SC, TX, VA; Holarctic]
Literature: Behan-Pelletier and Eamer (2008); ABMI (2019).

Punctoribates Berlese, 1908
Type-species: *Oribates punctum* C.L. Koch, 1839

KEY TO ADULT *PUNCTORIBATES* OF CANADA AND ALASKA

1	Medial process of anterior notogastral tectum concave	2
–	Medial process of anterior notogastral tectum, convex, flattened or triangular	3
2(2)	Porose area *Aa* elongated oval. Length: 437–470µm (Figure 3.149A,B)	***P. palustris*** (Banks)
–	Porose area *Aa* round. Length: 320–350µm (Figure 3.149C)	***P. hexagonus*** Berlese

3(1)	Medial process of anterior notogastral tectum triangular. Pair of thickened bands associated with medial process about 37μm long. Postanal porose area concave. Length: 264–330μm (Figure 3.149D).......................................*P. weigmanni* Behan-Pelletier and Eamer
–	Medial process of anterior notogastral tectum truncate. Pair of thickened bands associated with medial process about 25μm long. Postanal porose area as saccule. Length: 350–410μm (Figure 3.149E) ....*P. punctum* (C.L. Koch)

Punctoribates hexagonus Berlese, 1908 (Figure 3.149C)
Comb./Syn.: *Minguezetes hexagonus* (Berlese, 1908)
Biology: From moist to wet meadows. Its development was studied by Seniczak and Seniczak (2008).
Distribution: AK [Holarctic]; known from Tertiary fossil sites on Meigen and Ellesmere Island (NU) (Matthews et al. 2019).

Punctoribates palustris (Banks, 1895) (Figure 3.149A,B)
Comb./Syn.: *Oribates palustris* Banks, 1895
Biology: Associated with sphagnum and other aquatic mosses; pond margins; sedges; willow and poplar litter; muskeg; river gravel. Gravid females carry up to 7 large eggs. Gut boli are composed primarily of masses of fungal hyphae and spores.
Distribution: YT, BC, AB, SK, ON, QC, NB, NS, NL, PE [AL, AR, CO, CA, FL, GA, MI, NH, NJ, NM, NY, SC, WI; Nearctic].
Literature: Behan-Pelletier and Eamer (2008); Barreto and Lindo (2018); ABMI (2019).

Punctoribates punctum (C.L. Koch, 1839) (Figure 3.149E)
Comb./Syn.: *Oribates punctum* C.L. Koch, 1839
Biology: From old field litter; from tree trunks. Gut content analysis of adults over a 6 month period showed consumption of protists, moss and plant material, grass and conifer pollen, and arthropod fragments, fungi and enchytraeid setae (Behan-Pelletier and Hill 1983).
Distribution: BC, NS, NB, NL [NY; Cosmopolitan]; possibly introduced in North America.
Literature: Seniczak and Seniczak (2008); Behan-Pelletier and Eamer (2008).

Punctoribates weigmanni Behan-Pelletier and Eamer, 2008 (Figure 3.149D)
Biology: From dry moss and lichen.
Distribution: ON [MO; Nearctic]

Zachvatkinibates Shaldybina, 1973
Type-species: *Zachvatkinibates nemoralis* Shaldybina, 1973

KEY TO ADULT *ZACHVATKINIBATES* OF CANADA AND ALASKA

1	Pedotectum I with dorsal margin concave. Notogastral porose areas sexually dimorphic; in female circular to oval and in male *Aa* hypertrophied and/or *A1* to *A3* fused....................2
–	Pedotectum I with dorsal margin convex. Notogastral porose areas without sexual dimorphism, circular to oval in both male and female. Length: 470–480μm (Figure 3.150A,B).. ....*Z. epiphytos* Behan-Pelletier, Eamer and Clayton
2(1)	Posterior notogastral tectum incomplete medially. Length: 525–594μm (Figure 3.150C–E)...*Z. nortoni* Behan-Pelletier and Eamer
–	Posterior notogastral tectum with overlapping lobes 3
3(2)	Female and male with porose area *Aa* oblong. Male with posterior of notogaster concave with wedge-shaped projection. Length: 528–544μm (Figure 3.150F–H)............................ .... *Z. schatzi* Behan-Pelletier and Eamer

	Female with porose area *Aa* semicircular. Male with *Aa* oblong, hypertrophied, or with longitudinal porose area extending lateral of opisthonotal gland opening. Notogaster without wedge shape projection 4
4(3)	Rostrum undulating, without lateral dens. Female with porose area *Aa* circular to oval. Male with with porose area *Aa* highly modified, extending anteriorly laterad of seta c_2 and posteriorly to level of lyrifissure *im;* may be divided into two at level of seta *lm*, may be contiguous with porose area *A1*. Length: ca. 473µm (Figure 3.151A–C) ....*Z. maritimus* Shaldybina
–	Rostrum flattened, with small, pointed lateral dens. 5
5(4)	Lamellar cusp about 18µm long. Female with porose area irregular in shape to subcircular. Male with porose area *Aa* enlarged with irregular longish shape, *A2* and *A3* longish oval. Length: 415–500µm (Figure 3.151D,E)...................................... *Z. quadrivertex* (Halbert)
–	Lamellar cusp about 28µm long; narrow proximally. Female with porose area subcircular. Male with porose area *Aa* modified and divided into two or three porose areas: largest section extending anteriorly to lateral of seta *c* and posteriorly midway to level of lyrifissure *im*, section bordered by setae *c*, *la*, and *lm* either single T-shaped porose area or divided into two sections. Length: 470–528µm (Figure 3.151F,G) ....*Z. shaldybinae* Behan-Pelletier and Eamer

Zachvatkinibates epiphytos Behan-Pelletier Eamer and Clayton, 2001 (Figure 3.150A,B)
Biology: From lichen in *Tsuga heterophylla* upper canopy and needles on *Abies amabilis* branch.
Distribution: BC [Canada]
Literature: Behan-Pelletier and Eamer (2005).

Zachvatkinibates maritimus Shaldybina, 1973 (Figure 3.151A–C)
Biology: From *Elymus* on sand dunes; supratidal meadow; *Stellaria*, *Potentilla* and grasses on upper shore; woody debris thrown by tide on shore; wet root mat by lake edge.
Distribution: AK, YT, BC [Holarctic]
Literature: Behan-Pelletier and Eamer (2005).

Zachvatkinibates nortoni Behan-Pelletier and Eamer, 2005 (Figure 3.150C–E)
Biology: From sand and organic matter under thick driftwood beach debris.
Distribution: BC [Canada]

Zachvatkinibates quadrivertex (Halbert, 1920) (Figure 3.151D,E)
Comb./Syn.: *Oribata quadrivertex* Halbert, 1920
Biology: In a cafeteria experiment Luxton (1966) found this species fed on salt-marsh fungi.
Distribution: AK, YT, NT, MB [Holarctic]; known from Tertiary fossil sites on Ellesmere Island (NU) (Matthews et al. 2019).
Literature: Weigmann (2009); Maraun et al. (2019).

Zachvatkinibates schatzi Behan-Pelletier and Eamer, 2005 (Figure 3.150F–H)
Biology: From beach debris.
Distribution: BC [Canada].

Zachvatkinibates shaldybinae Behan-Pelletier and Eamer, 2005 (Figure 3.151F,G)
Biology: From saltmarsh grass, *Spartina alterniflora*; plant matter on high tide flats.
Distribution: ON, QC, NS [Canada].

Family Zetomimidae Shaldybina, 1966
Diagnosis: *Adult:* Poronotic Brachypylina. Cerotegument extending circumventrally in groove at edge of ventral plate. Lamella, tutorium present; translamella absent. Porose areas *Am* and *Ah* present, *Ah* expressed as porose area or saccule, porose area *Al* absent. Taenidium extending from base of short triangular genal incision, dorsal of acetabulum I, almost to base of tutorium. Notogastral setation 10 or 11 pairs or their alveoli, if 11 pairs, then 2 pairs of setae *c* closely adjacent on anterior of notogaster. Lenticulus present. Octotaxic system present as porose areas or absent. Pteromorphs curved ventrally, without line of desclerotization; posterior tectum present. Epimeral setation 3-1-3-3. Genital opening often displaced anteriorly so that epimeral setae *2a* and *3a* in almost transverse alignment. Genital setation 6 pairs. Custodium and discidium present. Ventral and/or anal plates of males with porose areas in some species. Lateral apophysis present on mentum of subcapitulum; opposing ridge on gena present or absent. Palp setation 0-2-1-3-9(1); seta *l'* of palptibia lanceolate. Axillary saccule of subcapitulum absent. Legs I, II and IV with same or different number of claws. Genu IV shorter than tibia IV.

KEY TO GENERA OF ZETOMIMIDAE OF CANADA AND ALASKA

1   Tutorium brush-like anteriorly. Postanal porose area expressed as saccule (Figure 3.152A,B) ....*Naiazetes* Behan-Pelletier
–   Tutorium pointed anteriorly, never brush-like. Postanal porose area porose 2

2(1)   Octotaxic system absent. Legs II to IV with 3 claws. Notogastral setae at most length of alveolus (Figure 3.52C,D)..*Heterozetes* Willmann
–   Ooctotaxic system present, as porose areas. Legs I, II with 1 claw; III, IV with 3 claws. Notogastral setae medium length (Figure 3.152E–H)...............................*Zetomimus* Hull

Heterozetes Willmann, 1917
    Type-species: *Ceratozetes (Heterozetes) palustris* Willmann, 1917

KEY TO ADULT *HETEROZETES* OF CANADA AND ALASKA

1   Interlamellar setae 100–150µm; reaching to tip of lamella. Lamellae 118–167µm; mutual distance between lamellar cusps at base 50–58µm. Lenticulus comparatively large relative to total length, about 70µm wide × 50µm long. Length: 525–596µm (Figure 3.152C) .... *H. aquaticus* Banks
–   Interlamellar setae 165–185µm; reaching to anterior of rostrum. Lamellae 150–188µm; mutual distance between lamellar cusps at base 70–75µm. Lenticulus comparatively small relative to total length, about 50µm wide × 50µm long. Length: 653–743µm (Figure 3.152D)..*H. minnesotensis* (Ewing)
          Note: Unidentified species known from Tertiary fossil sites on Meigen Island (NU) (Matthews et al. 2019).

Heterozetes aquaticus (Banks, 1895) (Figure 3.152C)
    Comb./Syn.: *Oribatella aquatica* Banks, 1895; *Oribata aquatica* (Banks, 1895)
    Biology: Gravid females can carry at least 8 eggs. This mite has been reported from grass sod, edges of ponds, bottom muck, emergent vegetation, swamps, sphagnum, moss in flood zones, and other littoral and near littoral zone habitats. Gut contents include an annulate nematode, fungal material, and possibly arthropod cuticle (Walter et al. 2014).
    Distribution: BC, AB, ON, QC, NB, NS [AL, FL, GA, MS, NH, NY, OK; Holarctic]
    Literature: Behan-Pelletier and Eamer (2003); ABMI (2019).

Heterozetes minnesotensis (Ewing, 1913) (Figure 3.152D)
    Comb./Syn.: *Oribata minnesotensis* Ewing, 1913; *Heterozetes minnesotensis* (Ewing, 1913); *Heterozetes cayugensis* Habeeb, 1974
    Biology: From vegetation at edge of temporary ponds.
    Distribution: SK, MB, ON, QC, NB, NS [MI, NY, WI; Nearctic].
    Literature: Behan-Pelletier and Eamer (2003).

Naiazetes Behan-Pelletier, 1996
    Type-species. *Naiazetes reevesi* Behan-Pelletier, 1996
    Note: Unidentified species known from ON (Barreto et al. 2021).

Naiazetes reevesi Behan-Pelletier, 1996 (Figure 3.152A,B)
    Biology: From wet vegetation at edge of small ponds and slow-moving water. Species shows sexual dimorphism (Behan-Pelletier 2015). Used in paleoecological reconstruction (Erickson et al. 2003).
    Distribution: QC [AL, FL, MS, NY; Nearctic]
    Literature: Behan-Pelletier and Eamer (2003).

Zetomimus Hull, 1916
    Type-species: *Oribata furcata* Warburton and Pearce, 1905

KEY TO ADULT *ZETOMIMUS* OF CANADA AND ALASKA

1   Legs I and II monodactylous. Length: 420–460µm (Figure 3.152E,F) ....*Z. francisi* (Habeeb)
–   Legs I and II tridactylous 2

2(1)   Interlamellar setae of medium length, 80–85µm, reaching anterior tip of lamellae. Lamellar cusps 15–18µm, about quarter length of attached portion of lamellae. Notogastral setae short, 6–12µm. Custodium short, about 11µm. Length: 330–400µm (Figure 3.152G) ....*Z. cooki* Behan-Pelletier and Eamer
–   Interlamellar setae long, 155–185µm, reaching beyond anterior margin of rostrum. Lamellar cusps 30–38µm, about third length of attached portion of lamellae. Notogastral setae of medium length, 15–30µm. Custodium long, 22–28µm. Length: 480–564µm (Figure 3.152H)..*Z. setosus* (Banks)

Zetomimus cooki Behan-Pelletier and Eamer, 2003 (Figure 3.152G)
    Biology: From wet vegetation at edge of and on ponds.
    Distribution: ON [AL, MS, FL; Nearctic].

Zetomimus francisi (Habeeb, 1974) (Figure 3.152E,F)
    Comb./Syn.: *Ceratozetes francisi* Habeeb, 1974
    Biology: From wet vegetation at edge of and on ponds; from sphagnum moss; sedges in a salt marsh; moss in temporary pools; wet algae, moss, and litter in the littoral zone of ponds and lakes; and wet, grassy meadows. Gut contents contained fungal material (Walter et al. 2014).
    Distribution: BC, AB, ON, NB, NS [AL, FL, LA, MS; Nearctic]
    Literature: Behan-Pelletier and Eamer (2003); Behan-Pelletier and Schatz (2010); ABMI (2019).

Zetomimus setosus (Banks, 1895) (Figure 3.152H)
    Comb./Syn.: *Oribatella setosa* Banks, 1895
    Biology: From wet vegetation at edge of and on ponds.
    Distribution: AB, ON, NS [NY, WI; Nearctic]
    Literature: Behan-Pelletier and Eamer (2003); Behan-Pelletier and Schatz (2010); ABMI (2019).

Superfamily Galumnoidea Jacot, 1925

Family Galumnidae Jacot, 1925

Diagnosis: *Adults*: Poronotic Brachypylina. Prodorsum with lamellae absent, at most narrow ridges present (line L); tutorium reduced to ridge (line S), or absent. Genal incision absent. Dorsophragmata and pleurophragmata present. Bothridium cup shaped. Discidium present; custodium absent, postanal porose area present or absent. Pedotectum I small, not covering acetabulum I; pedotectum II present. Notogaster with hinged, auriculate pteromorphs, with **alary furrow** (i.e., appearing bilobed); with or without lenticulus; with entire posterior tectum; setation 10–15 pairs, or only alveoli present. Octotaxic system developed as 1–5 pairs of porose areas (occasionally porose areas multiplied), rarely absent; some species or genera with medial porose area or scattered pores. Genital setation 6 pairs. Chelicerae chelate-dentate; palpal eupathidium *acm* attached to solenidion; subcapitular mentum with complete tectum. Axillary saccule of subcapitulum present. Genu IV shorter than tibia IV.

Biology. Members of this family are facultative predators. Walter (1987) cultured a species of *Pilogalumna* through several generations independently on nematodes, algae and also fungi. Both *Pergalumna omniphagous* Rockett and Woodring and *Pilogalumna cozadensis* Nevin, 1976 were nematophagous in laboratory cultures, and the latter would also capture and feed on slowly moving or injured springtails (Rockett and Woodring 1966; Walter 1988; Walter et al. 1988). A species of *Galumna* fed on Enchytraeidae and Collembola in a laboratory culture (Wunderle 1992). Muraoka and Ishibashi (1976) reported that 4 species of *Pergalumna* fed frequently on nematodes. All post-embryonic stages of 4 species of *Pergalumna* except larvae were observed feeding on nematodes (Rockett 1980).

A few species feed on economically important plants. For example, *Orthogalumna terebrantis* Wallwork feeds on water hyacinth (*Eichornia crassipes*), an important aquatic pest in tropical and subtropical waterways (Wallwork 1965). Adult *O. terebrantis* consume previously damaged plant tissue and oviposit into the leaf-like lamina, where juveniles burrow and develop to adults. Such damage attracts certain beetles that are more effective than *O. terebrantis* in controlling the plant (Haq and Sumangala 2003). *Galumna formicarius* (Berlese), may be an obligate coprophage on fecal material in the galleries of wood–boring insects (Wallwork 1958).

Sexual dimorphism is expressed in a number of galumnid genera. Norton and Alberti (1997) postulated that hypertrophied porose areas of *Dicatozetes uropygium* (Grandjean) males might produce semiochemicals that are involved directly or indirectly in reproductive biology. As in similarly dimorphic oripodoids and ceratozetoids, such secretions either may be produced in larger quantities by males than by females, or they may be produced only by males. The mediodorsal porose cluster in males of *Acrogalumna longipluma* (Berlese) is innervated (Alberti et al. 1997) and is considered a likely source for sex pheromones (Norton and Alberti 1997).

The reproductive biology of this family appears more diverse than in other Brachypylina. Spermatophore deposition in a species of *Pergalumna* is dependent on the presence of the female, and the male deposits signaling structures as guides to the spermatophore (Oppedisano et al. 1995). Mate pairing has been noted in a few galumnid species. Males of a *Pilogalumna* sp. forcefully turn teneral females on their side, assume a venter-to-venter positioning, deposit a stalkless spermatophore in the coxisternal region, and quickly disengage. Females later lie on their backs and push the spermatophore mass towards the genital opening with their legs (Estrada-Venegas et al. 1996).

Galumna virginiensis Jacot is a major carrier of the sheep tapeworm, *Moniezia expansa*, and both *G. nigra* and *Pergalumna emarginata* serve as important intermediate hosts (Denegri 1993). In separate infection trials using *M. expansa* eggs or oncospheres, tapeworm

cysticercoids were recovered from *Pergalumna nervosa* (Połeć and Moskwa 1994) and *Galumna gracilis* (Schuster et al. 2000).
Literature: Bayartogtokh and Ermilov (2017).

KEY TO GENERA OF GALUMNIDAE OF CANADA AND ALASKA

1   Narrow carina (line L) absent (Figure 3.153C–E) .... ***Pilogalumna*** Grandjean
–   Carina (line L) present 2

2(1)   Lamellar seta positioned laterally to line L, i.e., between lines L and S (Figure 3.153A,B)
         ....***Galumna*** von Heyden
–   Lamellar seta positioned medially to line L, i.e., between lines L and L 3

3(2)   Notogastral setae distinct (Figure 3.153F,G) .... ***Trichogalumna*** Balogh
–   Notogastral setae reduced to their alveoli (Figure 3.154) .... ***Pergalumna*** Grandjean

Galumna von Heyden, 1826
Type-species: *Notaspis alatus* Hermann, 1804

KEY TO ADULT *GALUMNA* OF CANADA AND ALASKA

1   Porose area *Aa* long, narrow. Length: ca. 780µm (Figure 3.153A) .... ***G. nigra*** (Ewing)
–   Porose area *Aa* oval, broadest anteriorly, tapering posteriorly. Length: ca. 600µm (Figure 3.153B) .... ***G. hudsoni*** Hammer
     Note: Unidentified *Galumna* spp. from AK, YT, NT, AB, ON, QC, NB, NS.

Galumna hudsoni Hammer, 1952 (Figure 3.153B)
     Biology: From tussock tundra; mixed forest.
     Distribution: MB, ON [Canada]

Galumna nigra (Ewing, 1909) (Figure 3.153A)
     Comb./Syn.: *Oribata nigra* Ewing, 1909; *Zetes niger* [sic] (Ewing, 1909)
     Distribution: ON [CO, IL, IN, IO, MN, NY, OH; Nearctic]
     Literature: Jacot (1935).

Pergalumna Grandjean, 1936
     Type-species: *Oribates nervosus* Berlese, 1914

KEY TO ADULT *PERGALUMNA* OF CANADA AND ALASKA

1   Interlamellar setae short, to very short. Sculpturing absent from pteromorphs. Bothridial seta fusiform 2
–   Interlamellar seta long. Sculpturing present, at least on ventral margin of pteromorph. Bothridial seta with long, almost isodiametric, barbed head 3

2(1)   Interlamellar seta very short. Porose area *Aa* broad anteriorly, abruptly narrowing posteriorly; *A1* round, *A2* oval, *A3* long narrow oval. Length: ca. 540µm (Figure 3.154A) ....
         .... ***P. formicaria*** (Berlese)
–   Interlamellar seta short (ca. 30µm). Porose area *Aa* broad anteriorly, gradually narrowing posteriorly; *A1* round, subequal in size to *A2* oval; *A3* round to broadly oval. Length: ca. 627µm (Figure 3.154B) .... ***P. dodsoni*** Nevin

3(1) Pteromorph with striations on ventral margin. Length: 640–750µm (Figure 3.154C,D) ....*P. emarginata* (Banks)
– Pteromorph with reticulate pattern on ventral margin. Length: ca. 580µm (Figure 3.154E) ....*P. nervosa* (Berlese)
Note: Unidentified species from AK: Lava Camp, Seward Peninsula (Late Miocene: 5.7 mya) (Matthews et al. 2019). Unidentified *Peralumna* spp. from AB, MB, QC, NB, NS.

Pergalumna dodsoni Nevin, 1979 (Figure 3.154B)
Biology: From boreal forest litter.
Distribution: ON [NE; Nearctic].

Pergalumna emarginata (Banks, 1895) (Figure 3.154C,D)
Comb./Syn.: *Oribata emarginata* Banks, 1895; *Zetes emarginatus* (Banks, 1895); *Oribates emarginatus europaeus* Berlese, 1914; *Oribates emarginatus columbianus* Berlese, 1916; *Pergalumna omniphagous* Rockett and Woodring, 1966
Biology: This species (as *P. omniphagous*) was nematophagous in laboratory cultures. Rockett and Woodring (1966a,b) observed specimens preying on living plant-parasitic nematodes. In the laboratory, *P. emarginata* thrived on a regular diet of ground mushrooms, but when given a choice, it showed a preference for nematodes.
Distribution: ON, QC [CO, DC, FL, ID, IL, IN, LA, ME, MD, MA, MO, MN, NY, NC, OH, TX, VA; Holarctic]
Literature: Ermilov et al. (2015).

Pergalumna formicaria (Berlese, 1914) (Figure 3.154A)
Comb./Syn.: *Oribates formicarius* Berlese, 1914; *Galumna formicarius* (Berlese, 1914); *Oribates alatus* sensu Berlese, 1904; non Hermann, 1804
Biology: A possible obligate coprophage on fecal material in the galleries of wood-boring insects (Wallwork 1958).
Distribution: YT, QC [NC, NH; Holarctic]

Pergalumna nervosa (Berlese, 1914) (Figure 3.154E)
Comb./Syn.: *Oribates nervosus* Berlese, 1914; *Galumna nervosus* (Berlese, 1914); *Zetes nervosus* (Berlese, 1914); *Galumna retalata* Oudemans, 1915
Biology: From boreal forest.
Distribution: ON [CA, CN, DC, VA; Holarctic].

Pilogalumna Grandjean, 1956
Type-species: *Pilogalumna ornatula* Grandjean, 1956

KEY TO ADULT *PILOGALUMNA* OF CANADA AND ALASKA

1 Porose area *Aa* divided in two ovals: *Aa* outer horizontal, *Aa* inner vertical. Bothridial seta with flattened and blunt head. Length: ca. 593µm (Figure 3.153C,D) ....*P. binadalares* (Jacot)
– Porose area *Aa* small, circular. Bothridial seta long, narrowly fusiform. Length: 580–640µm..*P. tenuiclava* (Berlese)
Note: Unidentified *Pilogalumna* spp. from BC, AB, ON.

Pilogalumna binadalares (Jacot, 1929) (Figure 3.153C,D)
Comb./Syn.: *Galumna areolata binadalares* Jacot, 1929; *Galumna alatum binadalares* Jacot, 1929; *Galumna binadalare* [*sic*] Jacot, 1929
Biology: From mixed deciduous coniferous forest.
Distribution: ON, QC [VA; Nearctic]

Pilogalumna tenuiclava (Berlese, 1908)
> **Comb./Syn.**: *Oribates tenuiclavus* Berlese, 1908; *Allogalumna tenuiclava* (Berlese, 1908); *Galumna radiata* Sellnick, 1928; *Galumna alatum* (Schrank, 1803) sensu Jacot, 1934
> **Biology**: From grass, roots, soil.
> **Distribution**: AK [CT, IL, MA, NY, OH, VA; Holarctic]
> **Literature**: Seniczak et al. (2012).

Trichogalumna Balogh, 1960 (Figure 3.153F,G)
> **Type-species**: *Pilogalumna* (?) *lunai* Balogh, 1960
> **Note**: Unnamed *Trichogalumna* sp. from QC. This is possibly *Trichogalumna curva* (Ewing 1907); see Norton and Ermilov (2017).

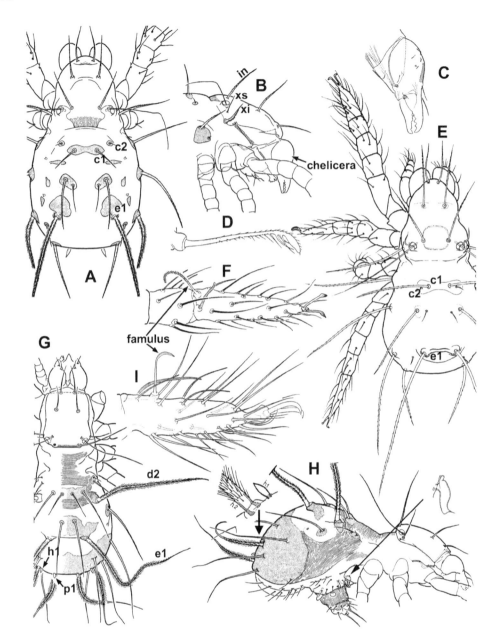

FIGURE 3.1 Acaronychidae: *Acaronychus traegardhi*, (A) dorsal; (B) lateral; (C) chelicera; (D) famulus (after Grandjean 1954b); Archeonothridae: *Zachvatkinella nipponica*, (E) dorsal (after Aoki 1980d); *Z. belbiformes*, (F) tibia, tarsus I (after Lange 1954); Palaeacaridae: *Palaeacarus hystricinus*, (G) dorsal; (H) lateral, with enlarged anterior genital seta to right; (I) tarsus I (after Grandjean 1954c).

Taxonomic Keys

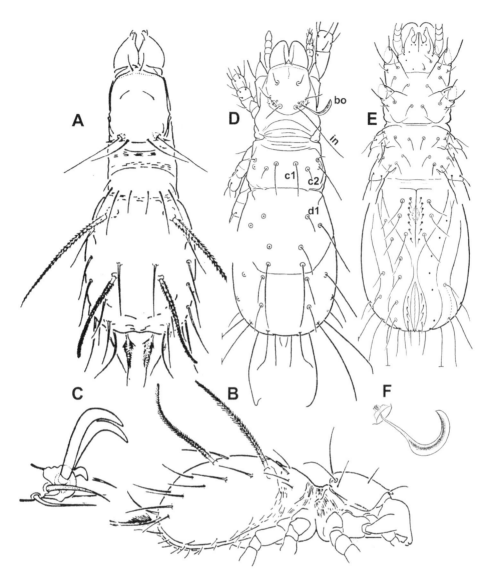

FIGURE 3.2 Ctenacaridae: *Beklemishevia* sp. (A) dorsal; (B) lateral; (C) tarsus I; Aphelacaridae: *Aphelacarus acarinus*, (D) dorsal; (E) ventral; (F) trichobothrium (after Grandjean 1954c).

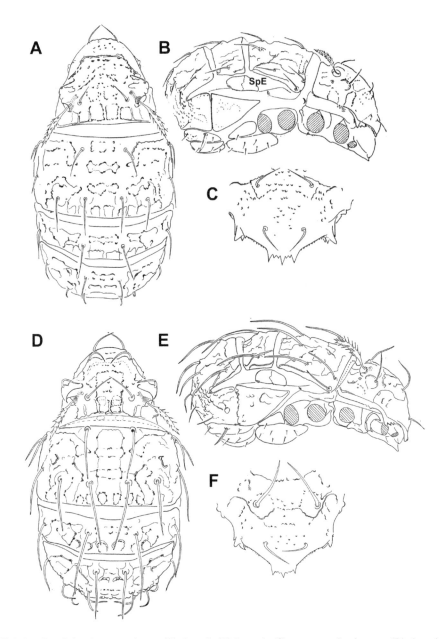

FIGURE 3.3 *Synchthonius crenulatus*, (A) dorsal; (B) lateral; (C) rostrum; *S. elegans*, (D) dorsal; (E) lateral; (F) rostrum (after Moritz 1976b).

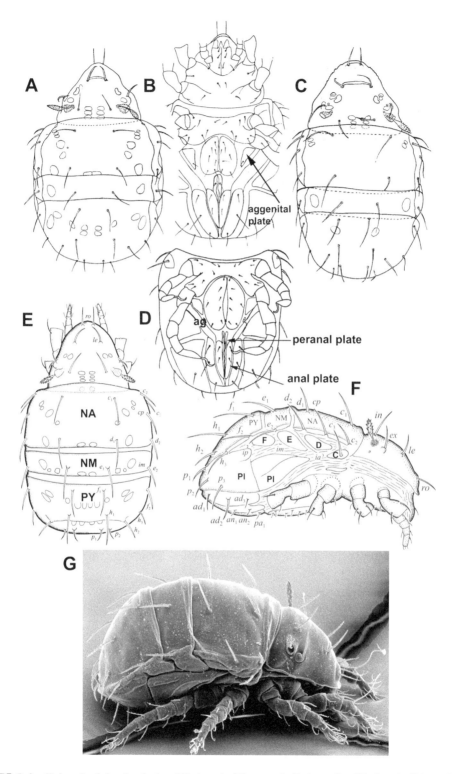

FIGURE 3.4 *Eobrachychthonius latior*, (A) dorsal; (B) ventral; *E. borealis*, (C) dorsal; (D) partial ventral (after Forsslund 1957); *E. oudemansi*, (E) dorsal; (F) lateral (after Seniczak and Seniczak 2017); *Eobrachychthonius* sp. (G) lateral (VBP, new).

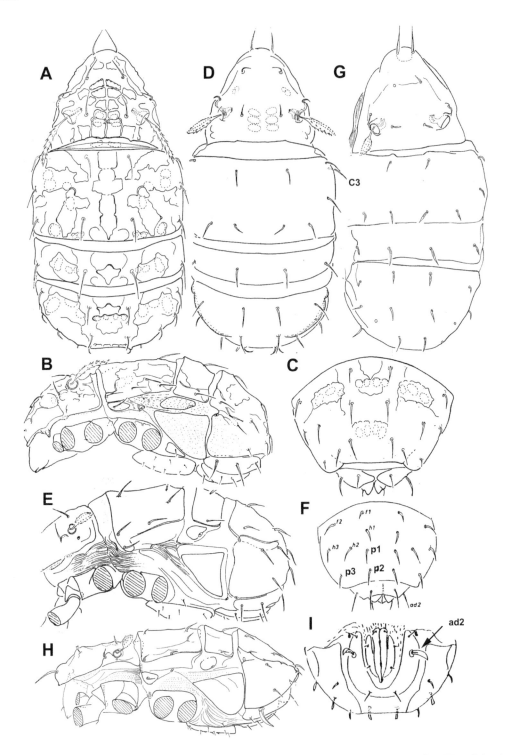

FIGURE 3.5 *Poecilochthonius spiciger*, (A) dorsal; (B) lateral; (C) notogaster, posterior; *Verachthonius montanus*, (D) dorsal; (E) lateral; (F) notogaster, posterior; *Neobrachychthonius marginatus*, (G) dorsal; (H) lateral; (I) venter, posterior (after Moritz 1976a,b).

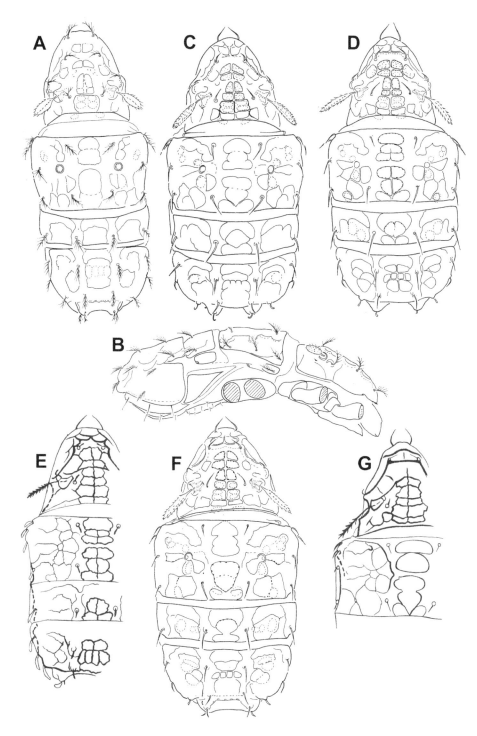

FIGURE 3.6 *Brachychthonius bimaculatus*, (A) dorsal; (B) lateral; *B. pius*, (C) dorsal; *B. impressus*, (D) dorsal; *B. jugatus*, (E) dorsal; *B. berlesei*, (F) dorsal; *B. berlesei erosus*, (G) dorsal (A–C, F after Moritz 1976a; E, G after Jacot 1938a).

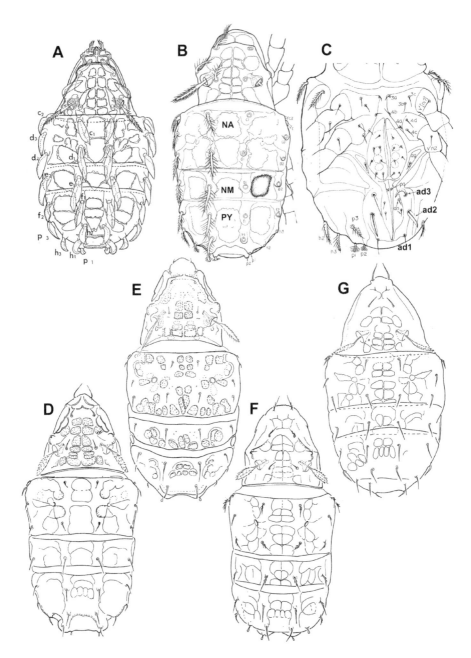

FIGURE 3.7 *Sellnickochthonius zelawaiensis*, (A) dorsal (after Chinone and Aoki 1972); *S. lydiae*, (B) dorsal; (C) ventral (after Reeves and Marshall 1971); *S. immaculatus*, (D) dorsal; *S. rostratus*, (E) dorsal; *S. furcatus*, (F) dorsal; *S. suecicus*, (G) dorsal (after Moritz 1976b).

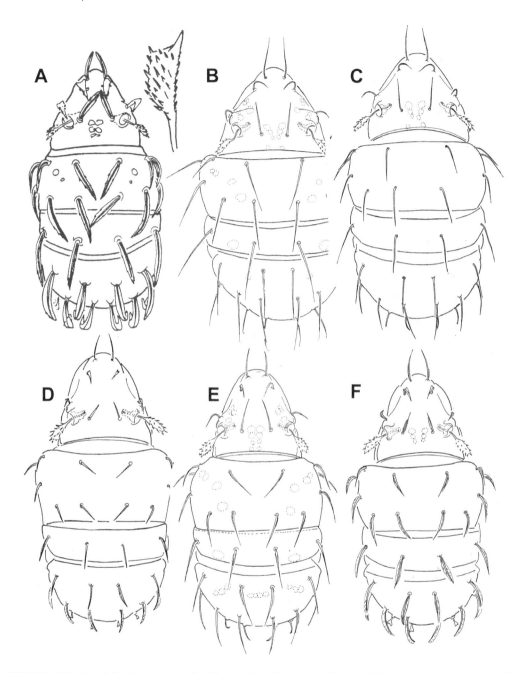

FIGURE 3.8 *Liochthonius forsslundi*, (A) dorsal, with detail of bothridial seta (after Mahunka 1969) *L. hystricinus*, (B) dorsal; *L. tuxeni*, (C) dorsal; *L. simplex*, (D) dorsal; *L. brevis*, (E) dorsal; *L. leptaleus*, (F) dorsal (after Moritz 1976a,b).

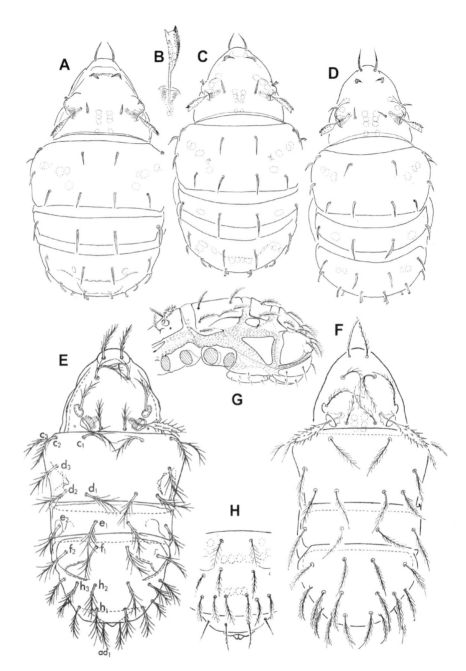

FIGURE 3.9 *Liochthonius sellnicki*, (A) dorsal, (B) detail of bothridial seta; *L. muscorum*, (C) dorsal; *L. lapponicus*, (D) dorsal (after Moritz 1976a); *Mixochthonius concavus*, (E) dorsal (after Chinone 1974); *M. pilososetosus*, (F) dorsal; (G) lateral; (H) notogaster, posterior (after Moritz 1976a,b).

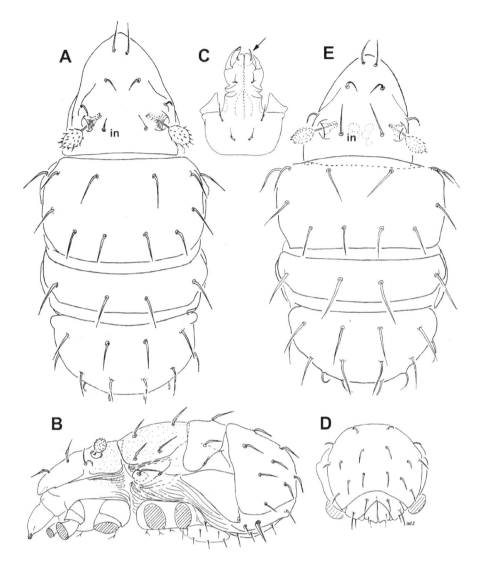

FIGURE 3.10 *Neoliochthonius piluliferus*, (A) dorsal; (B) lateral; (C) subcapitulum, with arrow to adoral seta; (D) posterior; *N. occultus*, (E) dorsal (after Moritz 1976b).

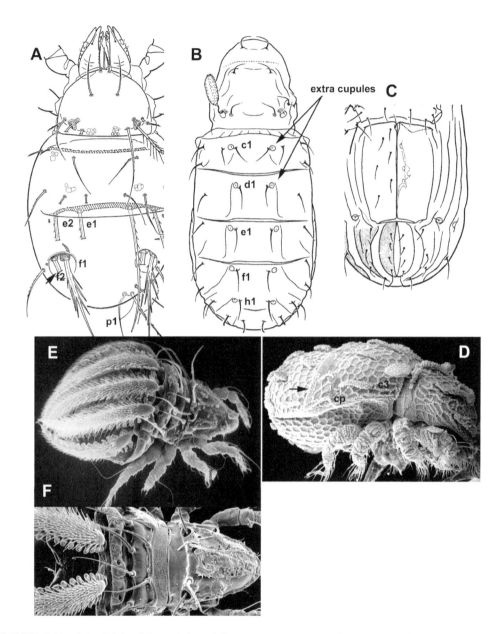

FIGURE 3.11 Arborichthoniidae: *Arborichthonius styosetosus*, (A) dorsal (after Norton 1982); Haplochthoniidae: *Haplochthonius simplex*, (B) dorsal; (C) hysterosoma ventral (after Grandjean 1947b); Sphaerochthoniidae: *Sphaerochthonius* sp. (D) lateral, with arrow to tectum, SEM (VBP, new); Trichthoniidae: *Gozmanyina majestus*, (E) lateral, (F) anterior, dorsal (VBP, new).

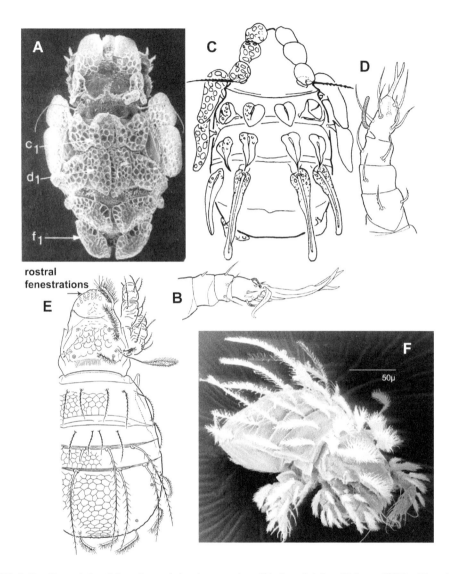

FIGURE 3.12 Pterochthoniidae: *Pterochthonius angelus*, (A) dorsal (after Krisper 1997); (B) palp (after Grandjean 1950a); Atopochthoniidae: *Atopochthonius artiodactylus*, (C) dorsal (after Krantz 1978); (D) palp (after Grandjean 1949); Cosmochthoniidae, *Cosmochthonius reticulatus*, (E) dorsal (after Grandjean 1962); *Cosmochthonius* sp., (F) lateral (VBP, new).

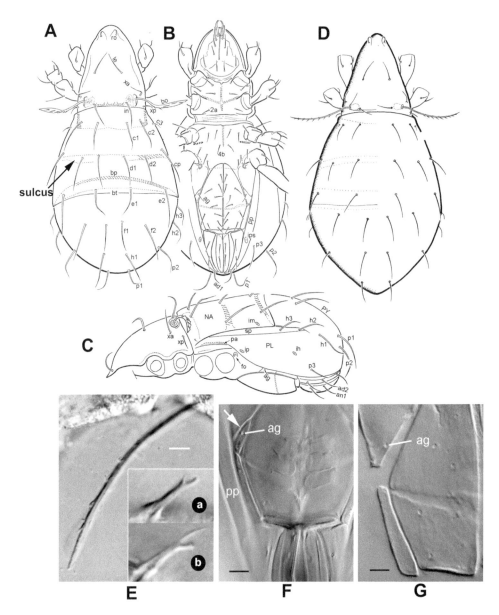

FIGURE 3.13 *Eniochthonius mahunkai*, (A) dorsal; (B) ventral; (C) lateral; *E. minutissimus*, (D) dorsal; *E. crosbyi*, (E) detail of bothridial seta; insert a, famulus of *E. crosbyi*; insert b, famulus of *E. minutissimus*; *E. mahunkai*, (F) genital region, ventral; *E. crosbyi*, (G) genital region, ventral (A–C, E–G after Norton and Behan-Pelletier 2007; D after Seniczak et al. 2009).

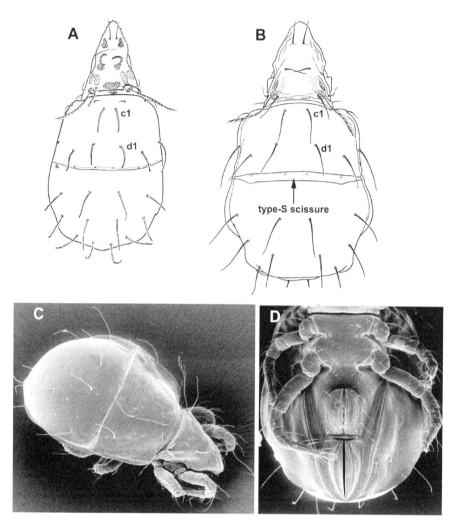

FIGURE 3.14 *Hypochthonius luteus*, (A) dorsal (after Fujikawa 2003); *Hypochthonius rufulus*, (B) dorsal (after Aoki 1965a); (C) SEM dorsal; (D) SEM ventral (VBP, new).

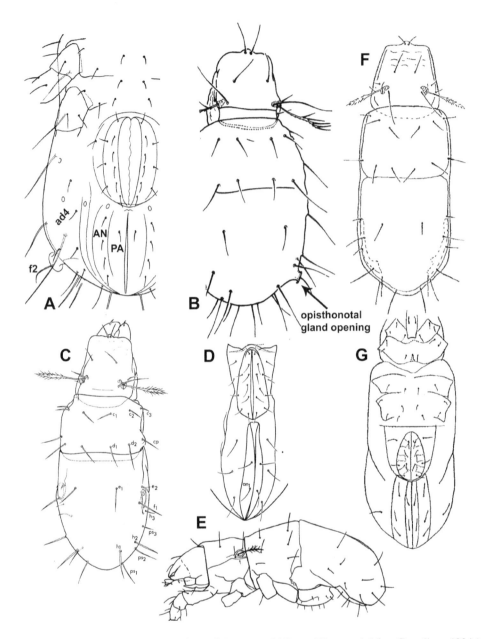

FIGURE 3.15 Parhypochthoniidae: *Parhypochthonius aphidinus*, (A) ventral (after Grandjean 1934c); (B) dorsal (after Aoki 1969); Gehypochthoniidae: *Gehypochthonius rhadamantus*, (C) dorsal; (D) ventral (after Aoki 1975) (E) lateral (after Jacot 1936); *G. gracilis*, (F) dorsal; (G) ventral (after Pan'kov 2002).

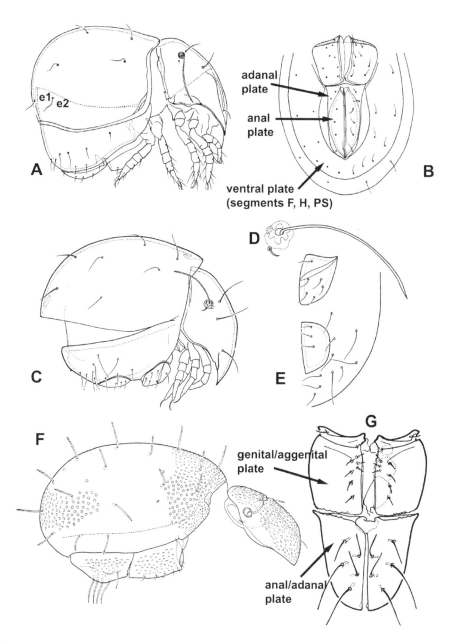

FIGURE 3.16 Mesoplophoridae: *Archoplophora rostralis*, (A) lateral; (B) ventral (after Aoki 1980c); *Mesoplophora japonica*, (C) lateral; (D) bothridial seta; (E) anogenital region (after Aoki 1980c); Phthiracaridae: *Atropacarus striculus*, (F) lateral (after Aoki 1980a); *Phthiracarus murphyi*, (G) ventral plates (after Harding 1976).

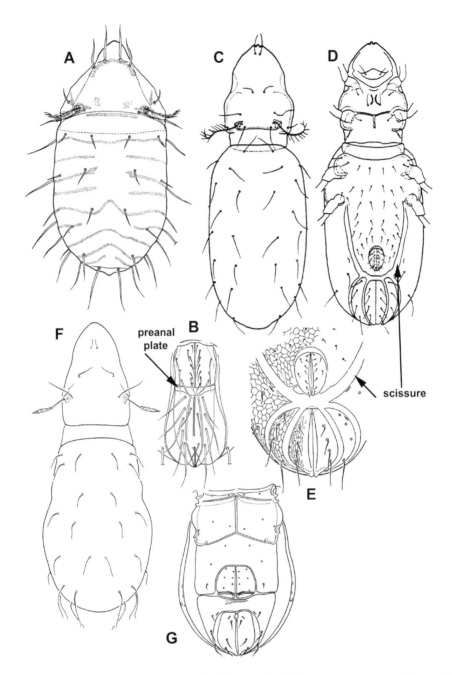

FIGURE 3.17 Lohmanniidae: *Mixacarus exilis*, (A) dorsal; (B) hysterosoma venter (after Aoki 1987); Eulohmanniidae: *Eulohmannia ribagai*, (C) dorsal; (D) ventral (after Aoki 1975); (E) posterior ventral (after Grandjean 1956c); Epilohmanniidae: *Epilohmannia cylindrica*, (F) dorsal (after Bayoumi and Mahunka 1976); (G) ventral hysterosoma (after Aoki 1965a).

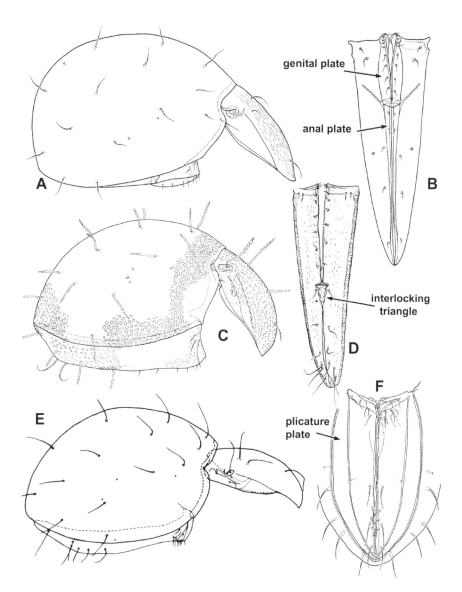

FIGURE 3.18 Oribotritiidae: *Oribotritia chichijimensis*, (A) lateral; (B) ventral plates (after Aoki 1980a); Euphthiracaridae: *Euphthiracarus foveolatus*, (C) lateral (after Aoki 1980b); (D) ventral plates (after Shimano and Norton 2003); Synichotritiidae: *Synichotritia caroli*, (E) lateral (after Walker 1965); (F) ventral (after Lions and Norton 1998).

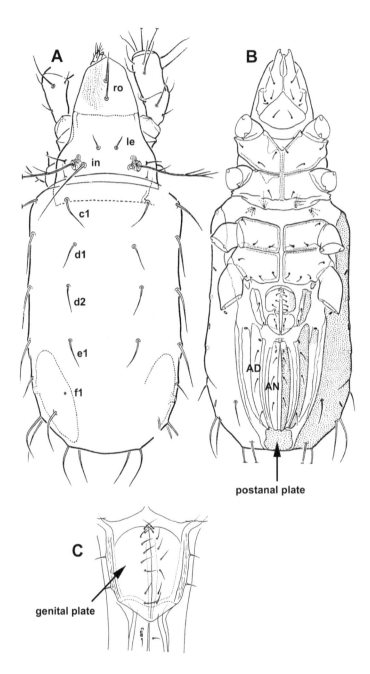

FIGURE 3.19 Perlohmanniidae: *Perlohmannia dissimilis* (A) dorsal; (B) ventral (after Grandjean 1958a); *Hololohmannia alaskensis*, (C) genital plate (after Kubota and Aoki 1998).

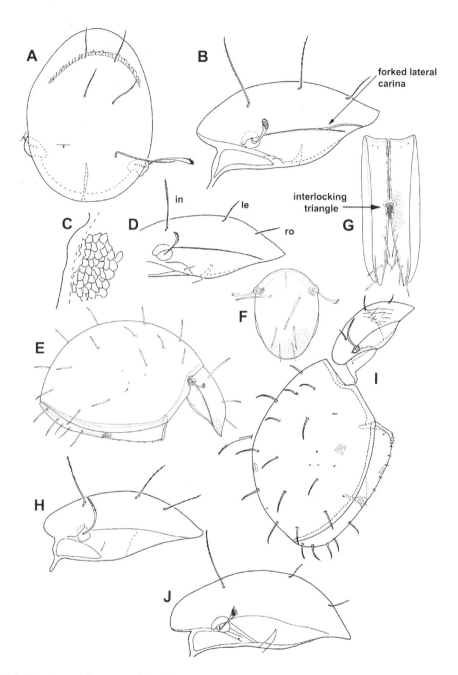

FIGURE 3.20 *Acrotritia vestita*, (A) aspis dorsal; (B) aspis lateral; *A. parareticulata*, (C) notogaster, lateral; (D) aspis lateral; *A. ardua*, (E) lateral; (F) aspis dorsal; (G) anogenital region; *A. diaphoros*, (H) aspis, lateral; *A. scotti*, (I) lateral (after Walker 1965); *A. curticephala*, (J) aspis lateral (A–D, H,J after Niedbała 2002; E–G, after Aoki 1980b).

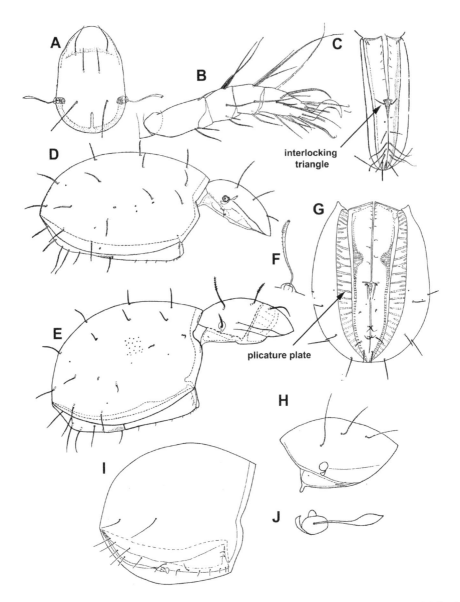

FIGURE 3.21 *Euphthiracarus monodactylus*, (A) aspis, dorsal; (B) leg I; (C) ventral; (D) lateral (after Märkel 1964); *Euphthiracarus monyx*, (E) lateral; (F) bothridial seta; (G) ventral (after Walker 1965); *Euphthiracarus fulvus*, (H) aspis, lateral; (I) notogaster, lateral; (J) bothridial seta (after Niedbała 2002).

Taxonomic Keys

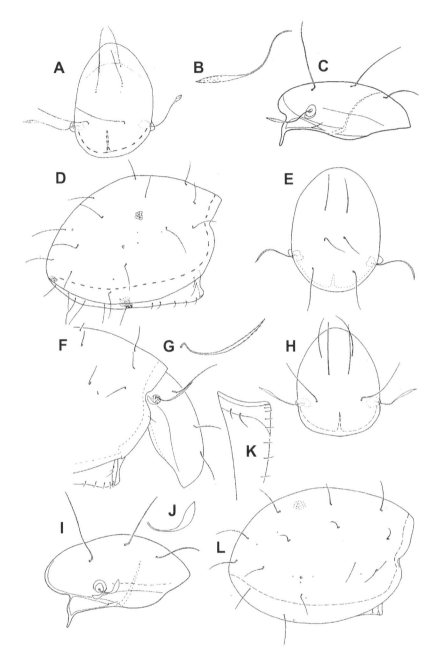

FIGURE 3.22 *Euphthiracarus jacoti*, (A) aspis, dorsal; (B) bothridial seta; (C) aspis, lateral; (D) notogaster, lateral; *E. fusulus*, (E) aspis, dorsal; (F) detail of notogaster; (G) bothridial seta; *E. depressculus*, (H) aspis, dorsal; (I) aspis, lateral; (J) bothridial seta; (K) anterior of anogenital plate; (L) notogaster, lateral (after Niedbała 2002).

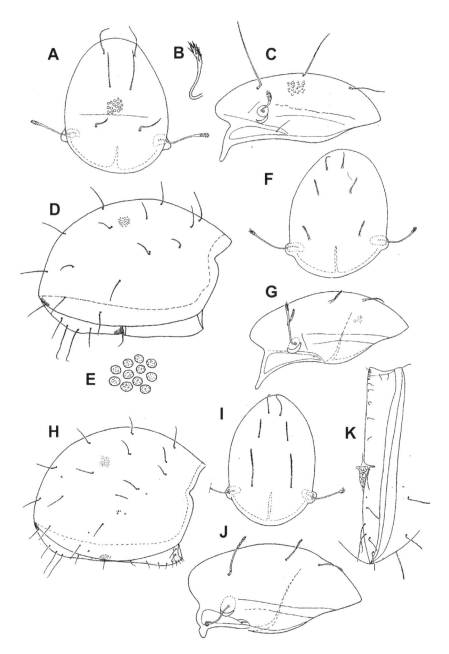

FIGURE 3.23 *Euphthiracarus crassisetae*, (A) aspis, dorsal; (B) bothridial seta; (C) aspis, lateral; (D) notogaster, lateral; (E) detail of notogaster; *E. longirostralis*, (F) aspis, dorsal; (G) aspis, lateral; (H) notogaster, lateral; *E. cernuus*, (I) aspis, dorsal; (J) aspis, lateral; (K) ventral, partial (after Niedbała 2002).

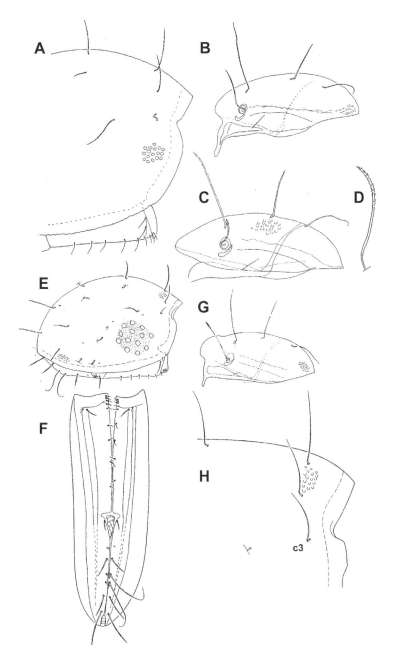

FIGURE 3.24 *Euphthiracarus pulchrus*, (A) notogaster, lateral; (B) aspis, lateral; *E. cribrarius*, (C) aspis, lateral; (D) bothridial seta; (E) notogaster, lateral, with insert of detail of notogaster; (F) ventral; *E. flavus*, (G) aspis, lateral; (H) anterior of notogaster, lateral (after Niedbała 2002).

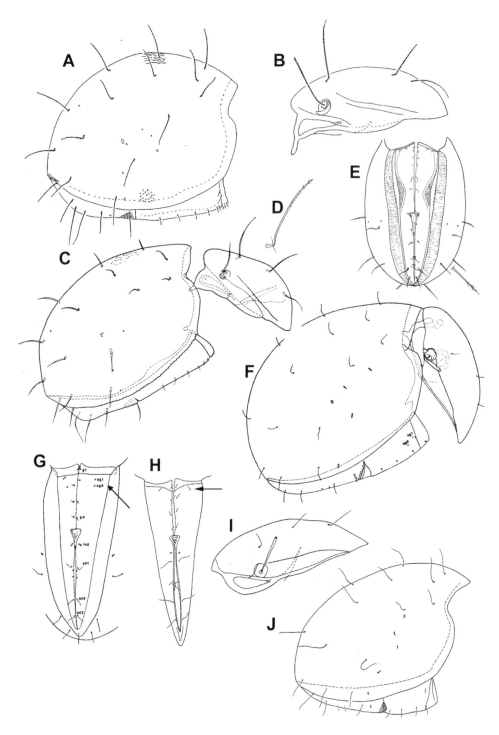

FIGURE 3.25 *Euphthiracarus vicinus*, (A) notogaster, lateral; (B) aspis, lateral (after Niedbała 2002); *E. tanythrix*; (C) lateral; (D) bothridial seta; (E) ventral (after Walker 1965); *Microtritia minima*, (F) lateral; (G) ventral, arrow to aggenital setae (after Märkel 1964); *M. simplex*, (H) ventral, arrow to aggenital seta (I) aspis, lateral; (J) notogaster (after Niedbała 2002).

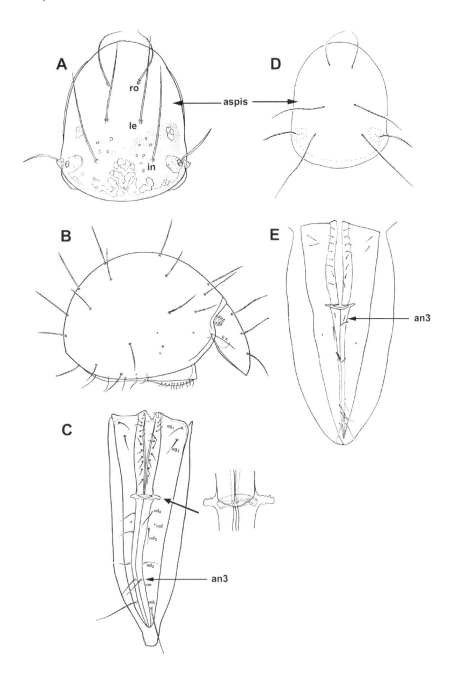

FIGURE 3.26 *Maerkelotritia kishidai*, (A) aspis (after Aoki 1980c); (B) habitus lateral; (C) ventral, with detail of genital-anogenital region (after Aoki 1958b); *M. cryptopa*, (D) aspis; (E) ventral region (after Niedbała and Penttinen 2006).

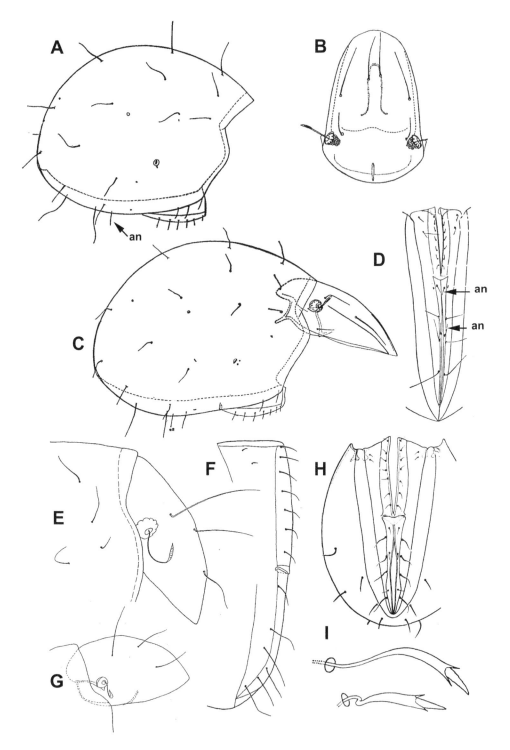

FIGURE 3.27 *Mesotritia flagelliformis*, (A) notogaster, lateral; *M. nuda*, (B) aspis; (C) lateral (D) ventral (after Märkel 1964); *Protoribotritia oligotricha*, (E) aspis and anterior of notogaster, lateral; (F) genital-anogenital region; *P. canadaris*, (G) aspis, lateral (after Niedbała 2002); *P. ensifer* Aoki, (H) venter; (I) bothridial seta (after Aoki 1980c).

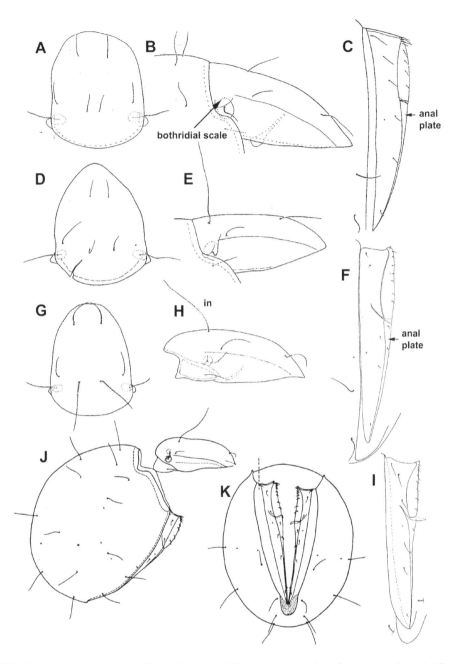

FIGURE 3.28 *Oribotritia banksi*, (A) aspis, dorsal; (B) aspis and anterior of notogaster, lateral; (C) ventral plates, partial; *O. carolinae*, (D) aspis, dorsal; (E) aspis, lateral; (F) ventral plates, partial; *O. henicos*, (G) aspis, dorsal; (H) aspis, lateral; (I) partial ventral (after Niedbała 2002); *O. megale*, (J) lateral; (K) ventral (after Walker 1965).

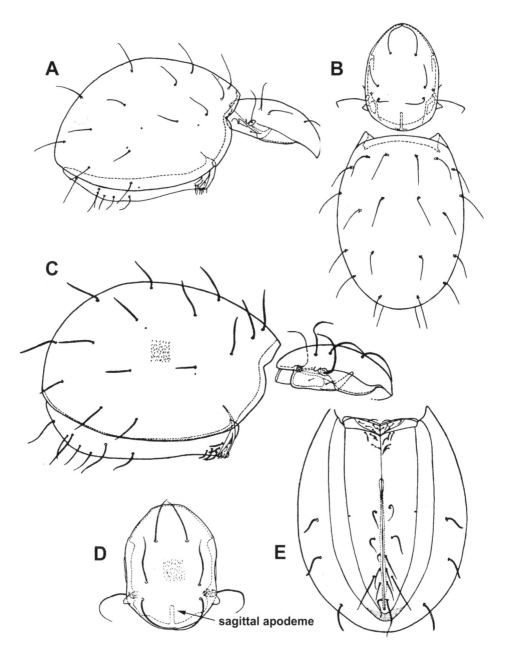

FIGURE 3.29 *Synichotritia caroli*, (A) lateral; (B) dorsal; *S. spinulosa*, (C) lateral; (D) aspis; (E) hysterosoma, venter (after Walker 1965).

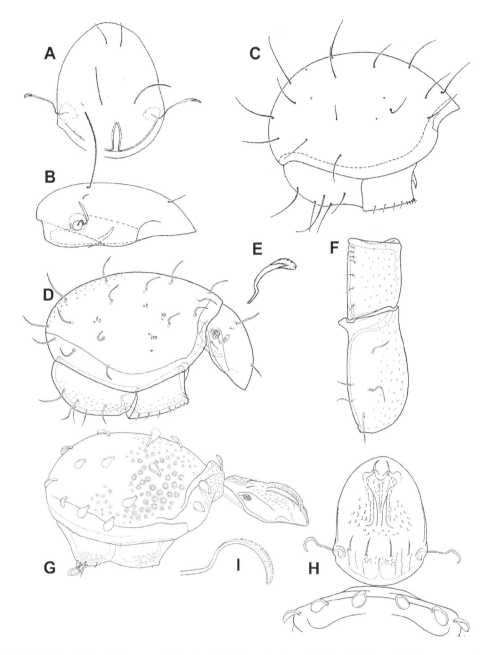

FIGURE 3.30 *Hoplophthiracarus illinoisensis*, (A) aspis, dorsal; (B) aspis, lateral; (C) notogaster lateral (after Niedbała 2002); *H. histricinus*, (D) lateral, aspect; (E) bothridial seta; (F) ventral plates; *Hoplophorella cucullatus*, (G) lateral; (H) aspis and anterior of notogaster; (I) bothridial seta (after Aoki 1980a).

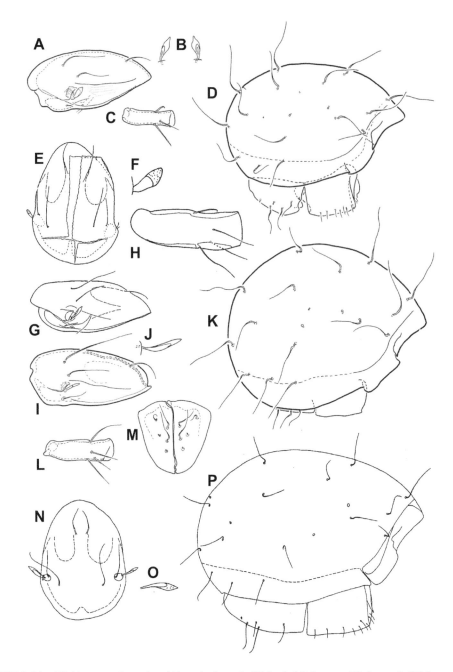

FIGURE 3.31 *Phthiracarus longulus*, (A) aspis, lateral; (B) bothridial setae; (C) femur I; (D) hysterosoma, lateral; *P. validus*, (E) aspis, dorsal; (F) bothridial seta; (G) aspis, lateral; (H) femur I; *P. globosus*, (I) aspis, lateral; (J) bothridial seta; (K) hysterosoma lateral; (L) femur I; (M) anoadanal plates; *P. nitidus*, (N) aspis, dorsal; (O) bothridial seta; (P) hysterosoma lateral (A–D, I–M after Beck et al. 2014; others after Niedbała 2002).

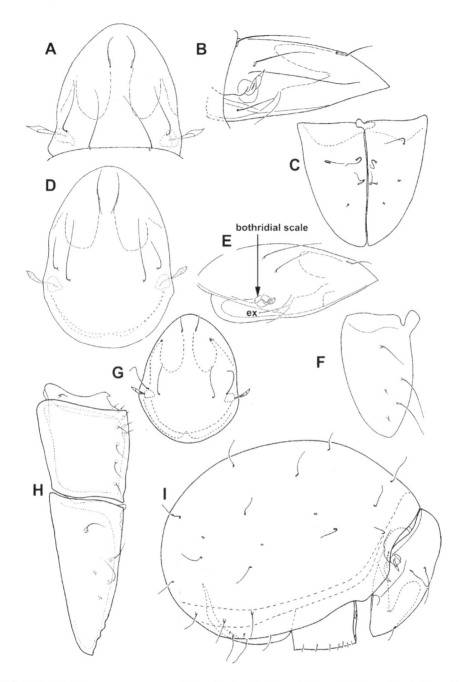

FIGURE 3.32 *Phthiracarus compressus*, (A) aspis, dorsal; (B) aspis, lateral; (C) anoadanal plates; *P. irreprehensus*, (D) aspis, dorsal; (E) aspis, lateral; (F) anoadanal plate (after Niedbała 2002); *P. japonicus*, (G) aspis, dorsal; (H) ventral plates; (I) lateral (after Aoki 1980a).

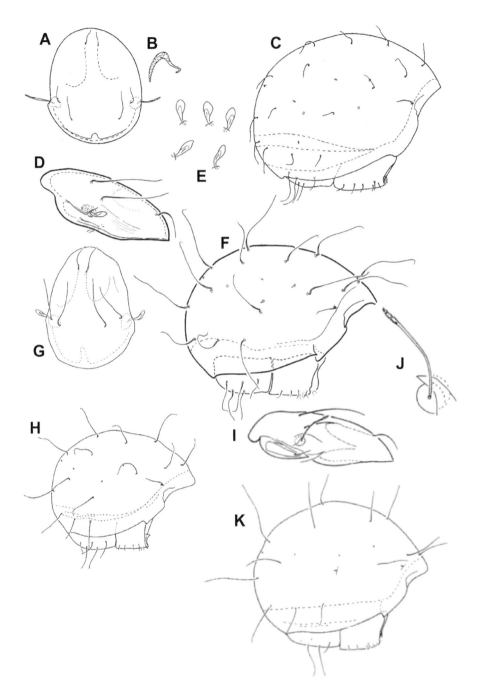

FIGURE 3.33 *Phthiracarus brevisetae*, (A) aspis, dorsal; (B) bothridial seta; (C) hysterosoma lateral; *P. bryobius*, (D) aspis, lateral; (E) bothridial seta, variations in shape; (F) hysterosoma lateral; *P. luridus*, (G) aspis, dorsal; (H) hysterosoma lateral; *P. boresetosus* Jacot, (I) aspis, lateral; (J) bothridial seta; (K) hysterosoma lateral (D–F after Beck et al. 2014, © Staatliches Museum für Naturkunde Karlsruhe (SMNK); others after Niedbała 2002).

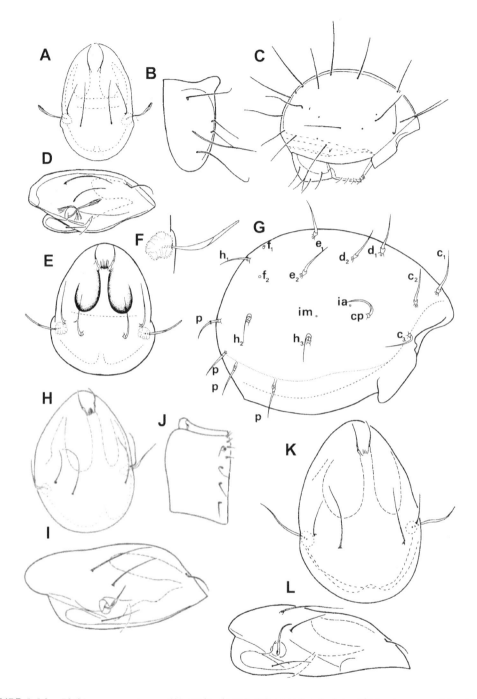

FIGURE 3.34 *Phthiracarus setosus*, (A) aspis, dorsal; (B) anoadanal plate; (C) hysterosoma lateral; (D) aspis, lateral; *P. lentulus*, (E) aspis, dorsal; (F) bothridial seta; (G) notogaster, lateral; *P. aliquantus*, (H) aspis, dorsal; (I) aspis, lateral; (J) aggenital-genital plate; *P. cognatus*, (K) aspis, dorsal; (L) aspis, lateral (E–G after Kamill 1981; others after Niedbała 2002).

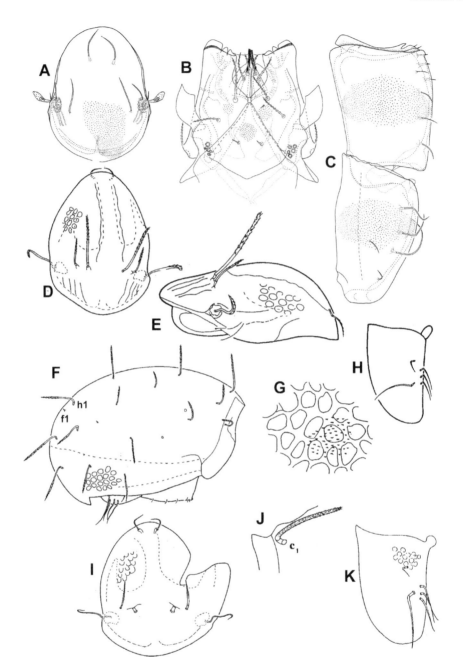

FIGURE 3.35 *Phthiracarus anonymus*, (A) aspis, dorsal; (B) subcapitulum, ventral; (C) ventral plates (after Berg et al. 1990, © Staatliches Museum für Naturkunde Karlsruhe (SMNK)); *Steganacarus thoreaui*, (D) aspis, dorsal; (E) aspis, lateral; (F) hysterosoma lateral; (G) detail of notogaster sculpturing; (H) anoadanal plate; *S. granulatus*, (I) aspis, dorsal; (J) notogastral seta c_1; (K) anoadanal plate (after Niedbała 2002).

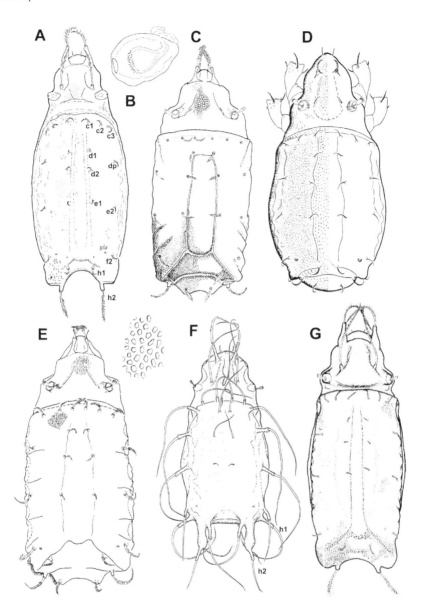

FIGURE 3.36 *Camisia abdosensilla*, (A) dorsal; (B) bothridium and bothridial seta (after Olszanowski et al. 2002); *C. oregonae*, (C) dorsal (after Colloff 1993); *C. foveolata*, (D) dorsal (after Seniczak 1991); *C. dictyna*, (E) dorsal, with detail of notogastral reticulations (after Colloff 1993); *C. spinifer*, (F) dorsal (after Olszanowski 1996); *C. orthogonia*, (G) dorsal (after Olszanowski et al. 2001).

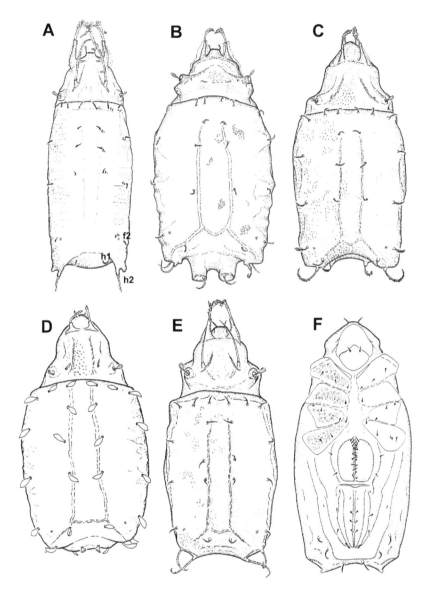

FIGURE 3.37 *Camisia biurus*, (A) dorsal; *C. biverrucata*, (B) dorsal; *C. horrida*, (C) dorsal; *C. lapponica*, (D) dorsal; *C. segnis*, (E) dorsal; (F) ventral (after Olszanowski 1996).

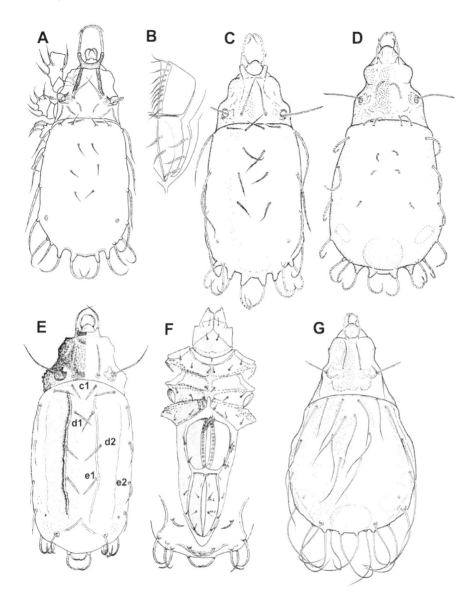

FIGURE 3.38 *Heminothrus minor*, (A) dorsal; (B) genital, anal plates (after Aoki 1969); *H. longisetosus*, (C) dorsal; *H. targionii*, (D) dorsal; *Neonothrus humicola*, (E) dorsal; (F) partial ventral; *Platynothrus thori*, (G) dorsal (C,D,G after Olszanowski 1996; E,F after Sellnick and Forsslund 1955).

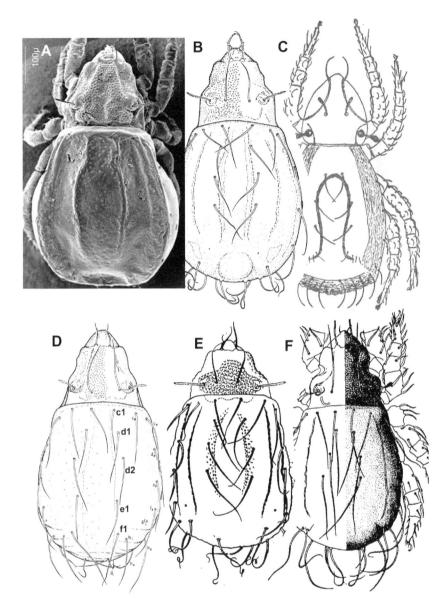

FIGURE 3.39 *Platynothrus punctatus*, (A) dorsal (VBP, new); *P. peltifer*, (B) dorsal (after Olszanowski 1996); *P. banksi*, (C) dorsal (after Banks 1910); *P. capillatus*, (D) dorsal (after Olszanowski 1996); *P. sibiricus*, (E) dorsal (after Sitnikova 1981); *P. yamasakii*, (F) dorsal (after Aoki 1958a).

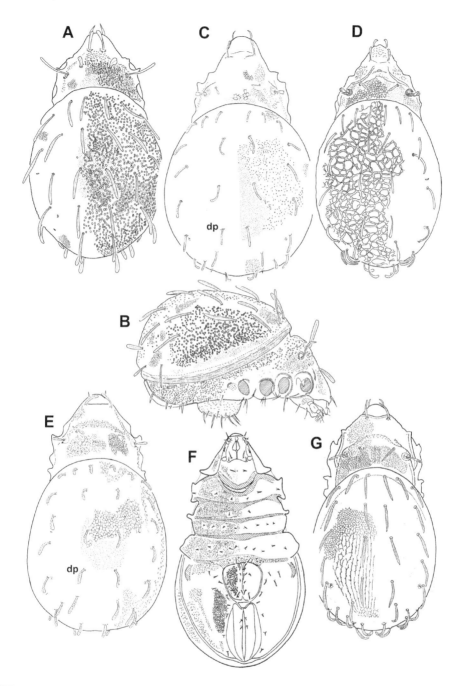

FIGURE 3.40 Hermanniidae: *Hermannia gibba*, (A) dorsal; (B) lateral; *H. scabra*, (C) dorsal; *H. reticulata*, (D) dorsal; *H. subglabra*, (E) dorsal; (F) ventral (after Woas 1981, © Staatliches Museum für Naturkunde Karlsruhe (SMNK)); *H. hokkaidensis*, (G) dorsal (after Aoki and Ohnishi 1974).

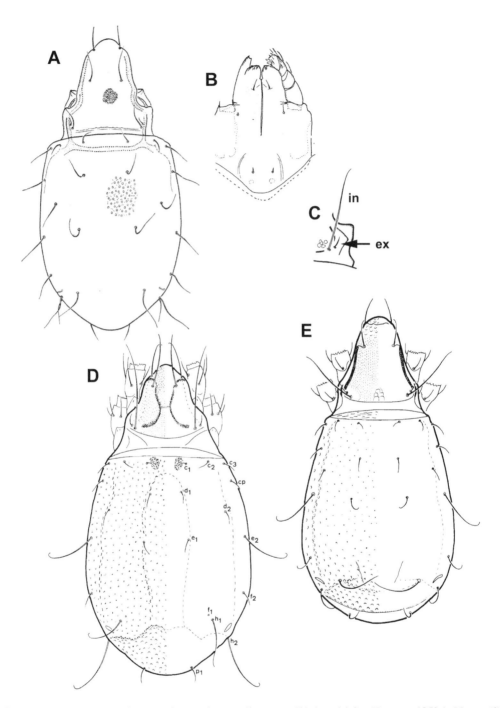

FIGURE 3.41 Malaconothridae: *Malaconothrus mollisetosus*, (A) dorsal (after Hammer 1952a); *M. gracilis* (B) anarthric subcapitulum (after Weigmann 1996); *Tyrphonothrus yachidairaensis* (C) prodorsum, posterolateral corner, showing absence of bothridium (after Yamamoto, Kuriki, and Aoki 1993); *Ty. maior*, (D) dorsal; *Ty. foveolatus*, (E) dorsal (after Seniczak 1993).

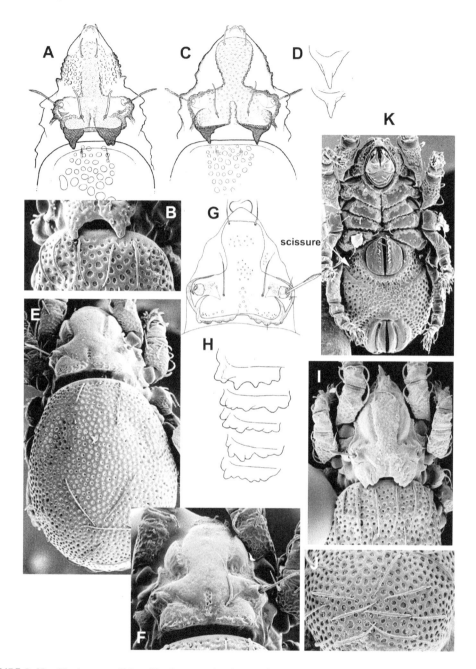

FIGURE 3.42 Nanhermanniidae: *Nanhermannia elegantula*, (A) prodorsum and notogaster, anterior (after Strenzke 1953, illustrated as *N. areolata*); (B) prodorsal tubercles and notogaster, anterior, SEM; *N. nana*, (C) prodorsum and notogaster, anterior; (D) variation in shape of prodorsal tubercles (after Strenzke 1953, illustrated as *N. elegantula*); *N. comitalis*, (E) dorsal SEM; (F) prodorsum, SEM; *N. coronata* (G) prodorsum (after Ermilov 2009); (H) variation in shape of prodorsal tubercles (after Strenzke 1953, illustrated as *N. nana*); *N. dorsalis*, (I) dorsal, anterior, SEM; (J) notogastral setae, SEM; (K) venter, SEM (B,E,F,I–K, VBP, new).

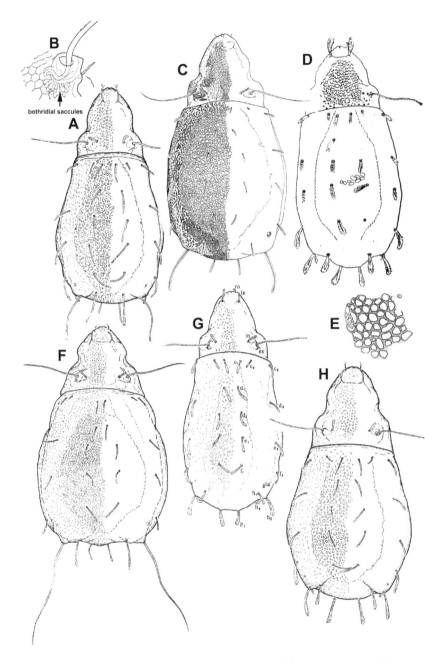

FIGURE 3.43 Nothridae: *Nothrus silvestris*, (A) dorsal; (B) bothridium; *N. pratensis*, (C) dorsal; *N. parvus*, (D) dorsal; (E) notogaster, anterior detail; *N. palustris*, (F) dorsal; *N. anauniensis*, (G) dorsal; *N. borussicus*, (H) dorsal (A,C,F–H after Olszanowski (1996); B, after Grandjean (1934); D,E after Sitnikova (1975)).

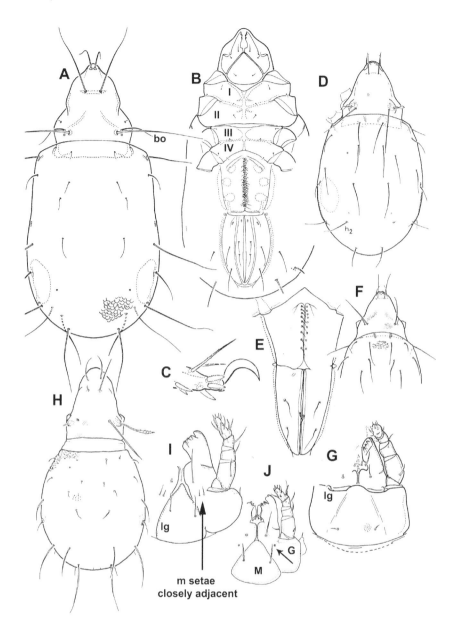

FIGURE 3.44 Trhypochthoniidae: *Mucronothrus nasalis*, (A) dorsal aspect; (B) ventral aspect; (C) tarsus I, distal (after Norton et al., 1996); *Trhypochthoniellus longisetus*, (D) dorsal aspect; (E) ventral plates; *T. longisetus*, (F) forma *setosus*, partial dorsal; (G) subcapitulum (lg=labiogenal articulation); *Mainothrus badius*, (H) dorsal aspect; (I) subcapitulum; *Trhypochthonius nigricans*, (J) subcapitulum (arrow on vestigial seta); (D,F–J after Weigmann 1997a; E, after Aoki 1964b).

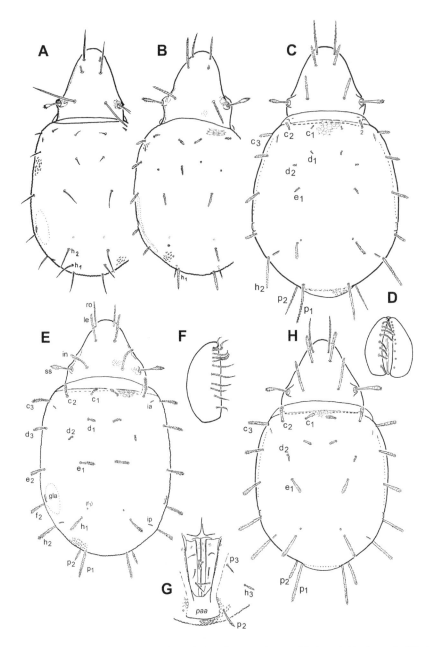

FIGURE 3.45 *Trhypochthonius cladonicola*, (A) dorsal; *T. nigricans*, (B) dorsal; *T. silvestris*, (C) dorsal; (D) genital plate; *T. tectorum tectorum*, (E) dorsal; (F) genital plate; (G) anal area; *T. americanus*, (H) dorsal (A,B, after Weigmann 1997; C–H, after Weigmann and Raspotnig 2009).

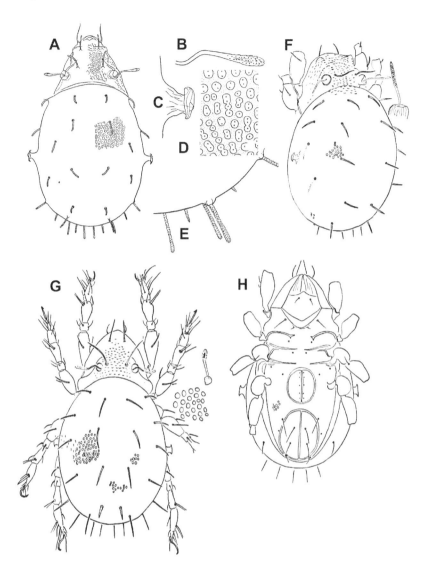

FIGURE 3.46 Hermanniellidae: *Hermanniella punctulata*, (A) dorsal; (B) bothridial seta; (C) lateral tube; (D) detail of notogastral sculpturing; (E) posterior notogastral setae (after Aoki 1965d); *H. robusta*, (F) dorsal, with detail of bothridium and bothridial seta; *H. occidentalis*, (G) dorsal, with detail of bothridial seta and notogastral sculpturing; (H) ventral (after Ewing 1918).

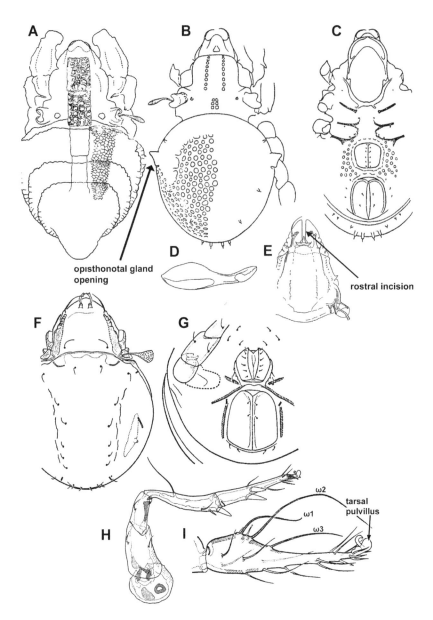

FIGURE 3.47 Plasmobatidae: *Plasmobates pagoda*, (A) dorsal; (B) dorsal (scalps removed); (C) ventral; (D) chelicerae (after Grandjean 1929); (E) prodorsum (after Grandjean 1961); Zetorchestidae: *Zetorchestes flabrarius*, (F) dorsal; (G) partial ventral; (H) leg IV; (I) tarsus I (after Grandjean 1951b).

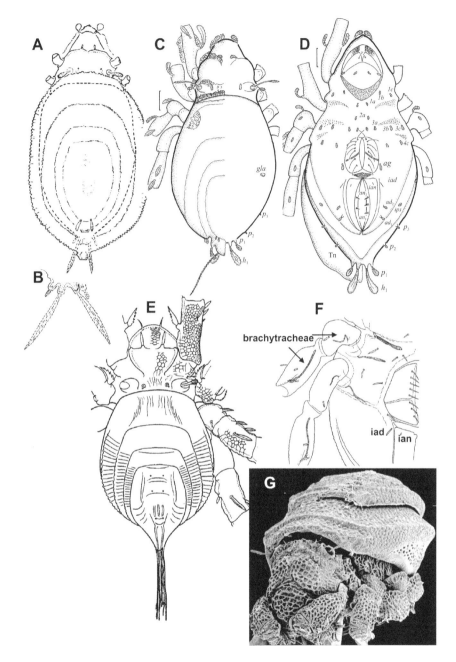

FIGURE 3.48 Neoliodidae: *Platyliodes macroprionus*, (A) dorsal; (B) detail of posterior apophysis (after Woolley and Higgins 1969b); *P. scaliger*, (C) dorsal; (D) ventral (after Seniczak et al. 2018); *Teleioliodes madininensis*, (E) dorsal; *Neoliodes theleproctus*, (F) ventral (partial) (after Grandjean 1934b); (G) *Poroliodes* sp. SEM (VBP, new).

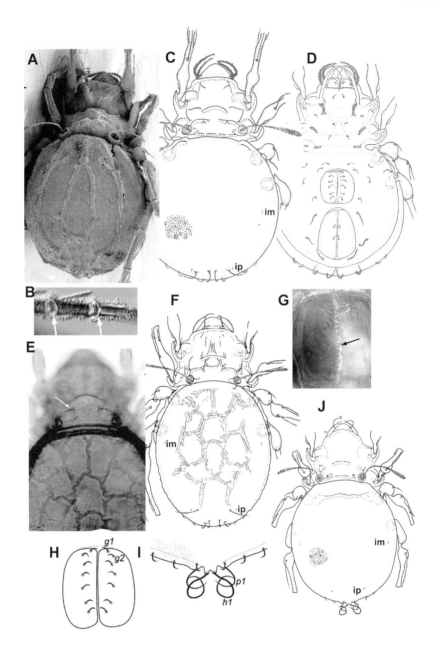

FIGURE 3.49 Gymnodamaeidae: *Adrodamaeus magnisetosus*, (A) SEM, dorsal; (B) leg segments, with arrows to retrotecta (after Walter 2009); (C) dorsal; (D) ventral (after Paschoal 1984); *Odontodamaeus veriornatus*, (E) SEM, partial dorsal (after Walter 2009); (F) dorsal (after Paschoal 1982b); (G) SEM, genital plates (after Walter 2009); *Jacotella quadricaudicula*, (H) genital plates; (I) notogaster, posterior (after Walter et al. 2014); (J) dorsal (after Paschoal 1983b).

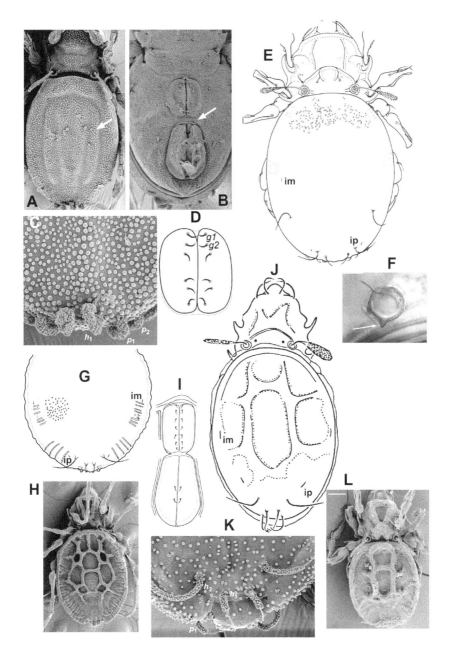

FIGURE 3.50 *Joshuella agrosticula*, (A) SEM, dorsal; (B) SEM, ventral; (C) SEM, notogaster, posterior (after Walter 2009); (D) genital plates (after Walter et al. 2014); *Donjohnstonella subalpina*, (E) dorsal (after Paschoal 1983c); (F) bothridium showing posterior tooth (after Walter 2009); *Roynortonella victoriae*, (G) notogaster without cerotegument; (H) SEM, dorsal; (I) genital and anal plates (after Walter 2009); *R. gildersleeveae*, (J) dorsal (after Paschoal 1982a); (K) SEM, notogaster, posterior; (L) SEM, dorsal (after Walter 2009).

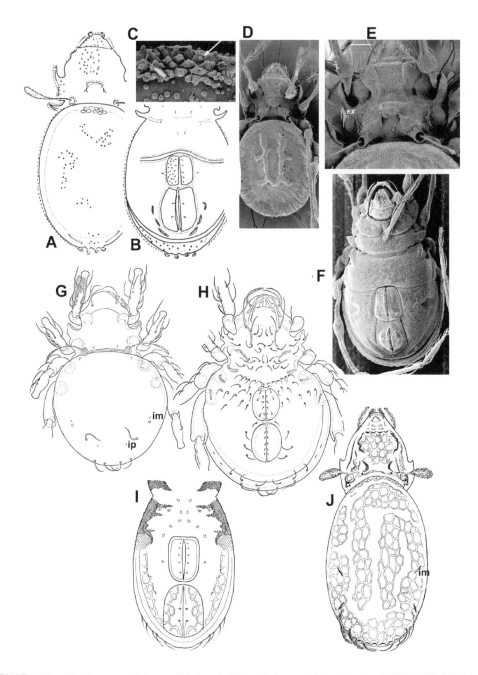

FIGURE 3.51 *Pleodamaeus plokosus*, (A) dorsal; (B) partial ventral (after Paschoal 1983a); (C) SEM, tubercles on notogaster, anterior (after Walter 2009); *Gymnodamaeus ornatus*, (D) SEM, dorsal; (E) SEM, prodorsum (after Walter 2009); *Gymnodamaeus* sp., (F) ventral, SEM (VBP, new); Plateremaeidae: *Allodamaeus ewingi*, (G) dorsal; (H) ventral (after Paschoal 1988); Licnodamaeidae: *Licnodamaeus costula*, (I) ventral; (J) dorsal (after Pérez-Íñigo 1970).

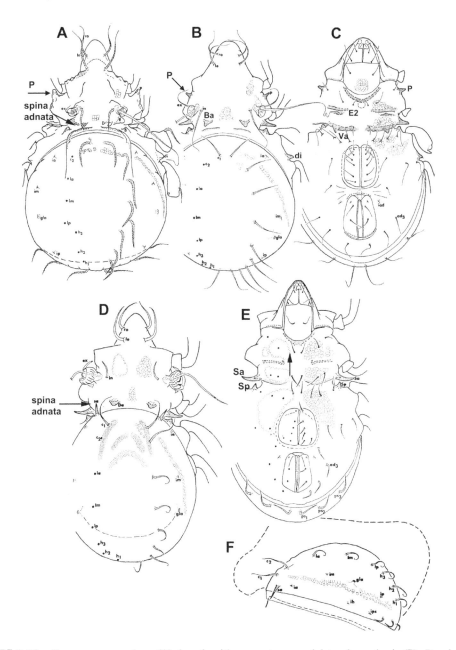

FIGURE 3.52 *Damaeus angustipes*, (A) dorsal, with arrow to propodolateral apophysis (P); *Parabelbella longiseta*, (B) dorsal, with arrow to propodolateral apophysis (P); (C) ventral; *Belba* (*Protobelba*) *californica*, (D) dorsal; (E) ventral; (F) notogaster, lateral aspect (dashed line indicates extent of organic debris) (after Norton 1979a).

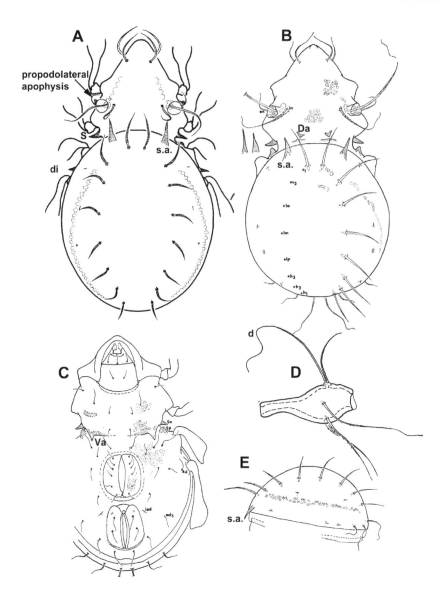

FIGURE 3.53 *Belbodamaeus rarituberculatus*, (A) dorsal (after Bayartogtokh 2004; reproduced with permission); *Lanibelba pini*, (B) dorsal, with variation in shape of spina adnata; (C) partial ventral; (D) tibia III; (E) notogaster, lateral aspect (after Norton 1980b).

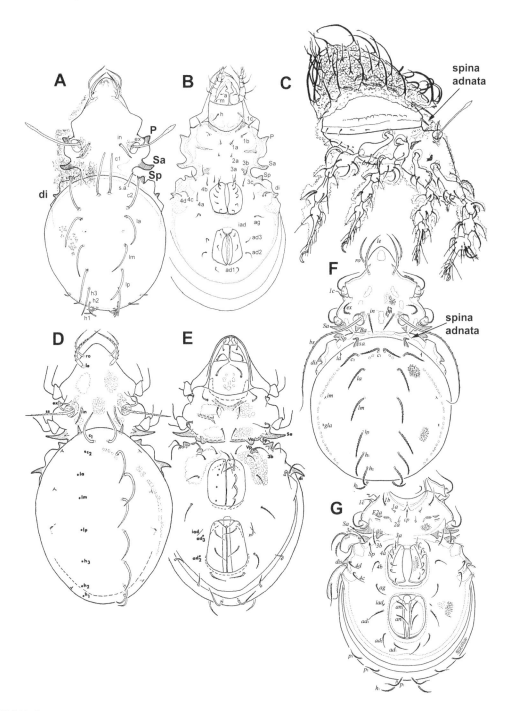

FIGURE 3.54 *Weigmannia parki*, (A) dorsal; (B) ventral; (C) lateral (after Miko and Norton 2010); *Quatrobelba montana*, (D) dorsal; (E) ventral (after Norton 1980b); *Kunstidamaeus arthurjacoti,* (F), dorsal; (G), ventral (after Norton et al. 2022).

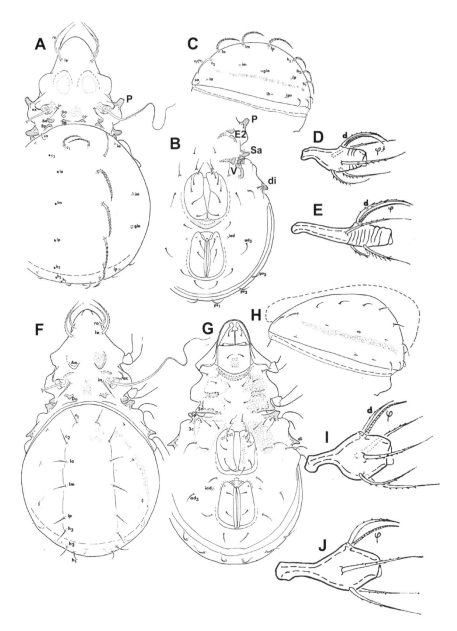

FIGURE 3.55 *Dyobelba carolinensis* (A) dorsal; (B) partial ventral; (C) notogaster, lateral aspect; (D) tibia II; (E) tibia III (after Norton 1979); *Caenobelba alleganiensis*, (F) dorsal; (G) partial ventral; (H) notogaster, lateral aspect, dashed line indicates the extent of organic debris; (I) tibia II; (J) tibia III (after Norton 1980b).

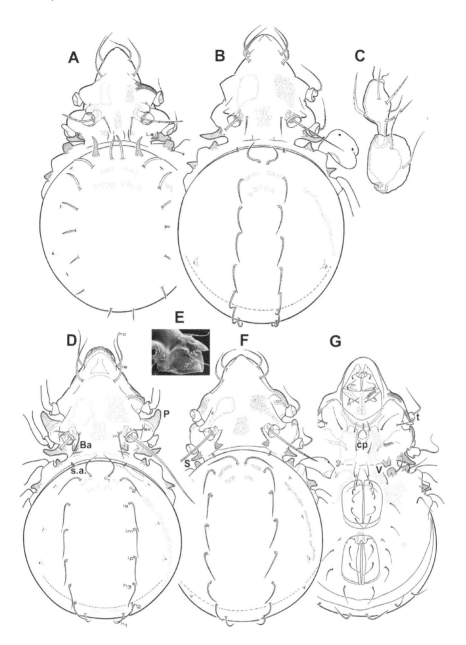

FIGURE 3.56 *Epidamaeus tenuissimus*, (A) dorsal; *E. coxalis*, (B) dorsal; (C) trochanter, femur III; *E. nasutus*, (D) dorsal; (E) SEM, rostrum, lateral; *E. mackenziensis*, (F) dorsal; (G) ventral (after Behan-Pelletier and Norton 1983, 1985).

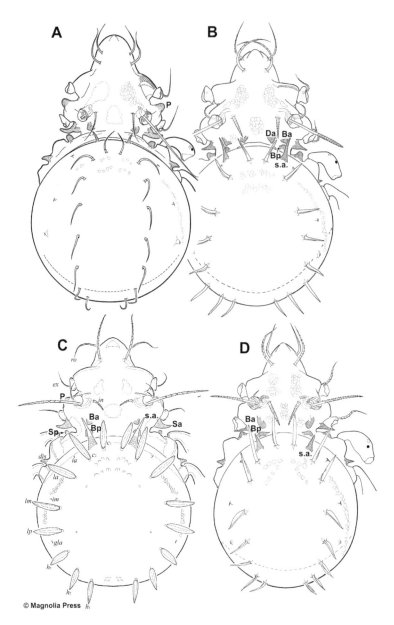

FIGURE 3.57 *Epidamaeus bakeri*, (A) dorsal; *E. tritylos*, (B) dorsal; *E. puritanicus*, (C) dorsal; *E. fortispinosus*, (D) dorsal (A, B, D, after Behan-Pelletier and Norton 1983; C, after Norton and Ermilov 2021, reproduced with permission from copyright holder).

Taxonomic Keys 303

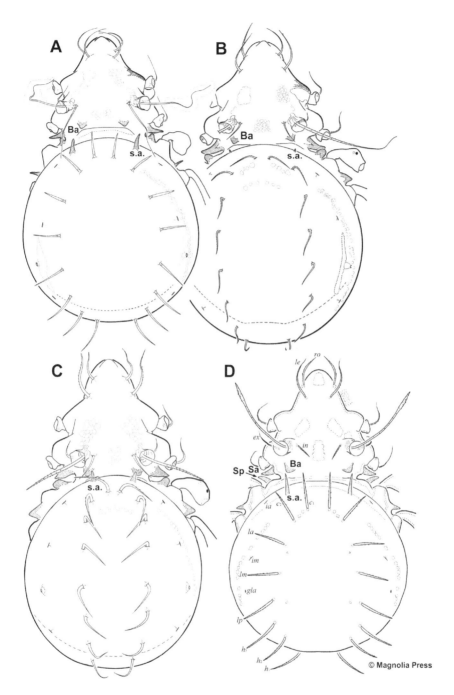

FIGURE 3.58 *Epidamaeus floccosus*, (A) dorsal; *E. arcticola*, (B) dorsal; *E. kodiakensis*, (C) dorsal (after Behan-Pelletier and Norton 1983, 1985); *E. michaeli*, (D) dorsal (after Norton and Ermilov 2021, reproduced with permission from copyright holder).

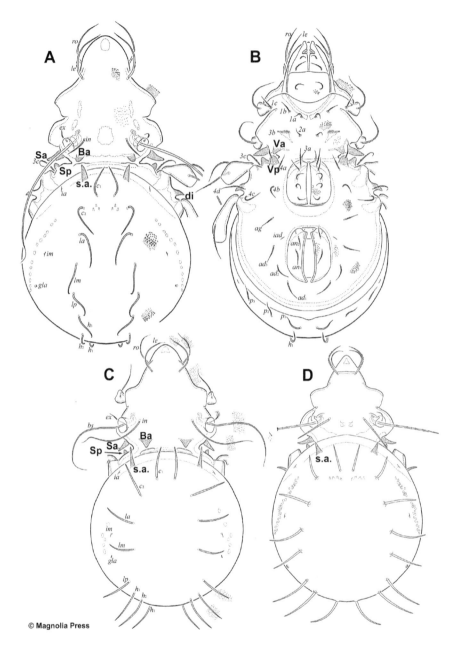

FIGURE 3.59 *Epidamaeus craigheadi*, (A) dorsal, (B) ventral; *E. olitor*, (C) dorsal; *E. globifer*, (D) dorsal (after Norton and Ermilov 2021; reproduced with permission from copyright holder).

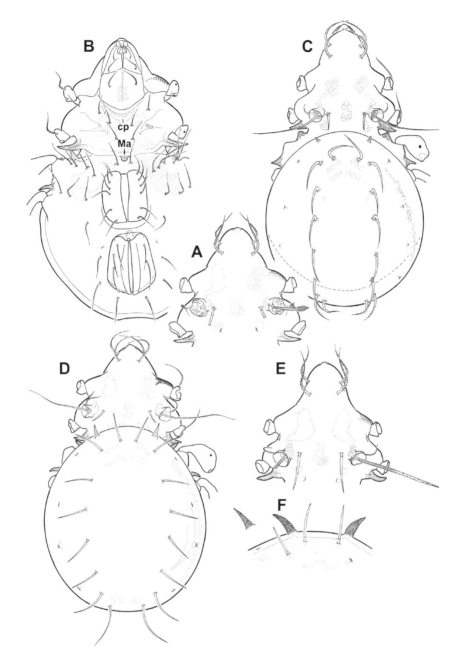

FIGURE 3.60 *Epidamaeus hastatus*, (A) prodorsum; (B) ventral; *E. hammerae*, (C) dorsal; *E. koyukon*, (D) dorsal; *E. gibbofemoratus*, (E) prodorsum; (F) notogaster, anterior, showing variation in shape of spina adnata (after Behan-Pelletier and Norton 1983, 1985).

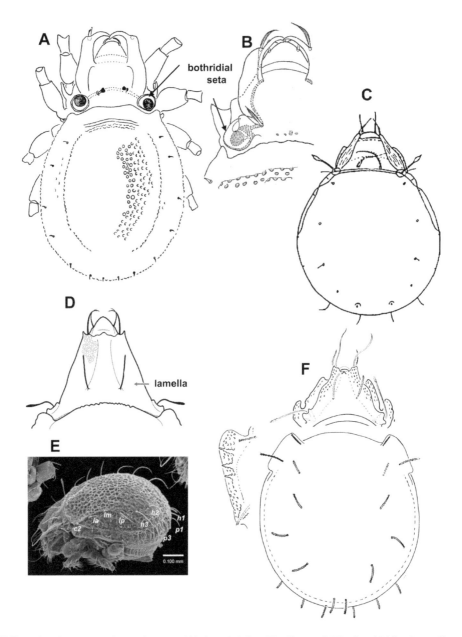

FIGURE 3.61 *Ommatocepheus clavatus*, (A) dorsal (after Woolley and Higgins 1964); *O. ocellatus*, (B) prodorsum (after Travé 1963); *Conoppia microptera*, (C) dorsal (after Luxton 1990); *Oribatodes mirabilis*, (D) prodorsum; (E) SEM, lateral aspect (after Walter et al. 2014); *Sphodrocepheus anthelionus*, (F) dorsal, with detail of humeral region (after Woolley and Higgins 1968c).

Taxonomic Keys   307

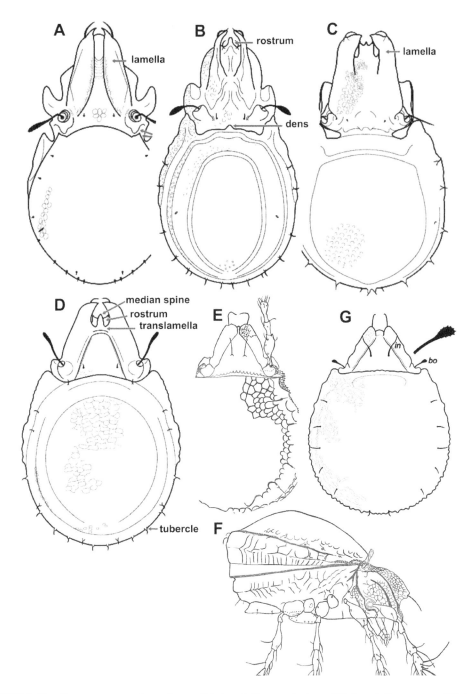

FIGURE 3.62 *Eupterotegaeus rhamphosus*, (A) dorsal; *E. ornatissimus*, (B) dorsal; *E. rostratus*, (C) dorsal; *E. spinatus*, (D) dorsal (after Walter et al. 2014); *Cepheus corae*, (E) dorsal; (F) lateral (after Jacot 1928a); *C. latus*, (G) dorsal, with detail of bothridial seta (after Walter et al. 2014).

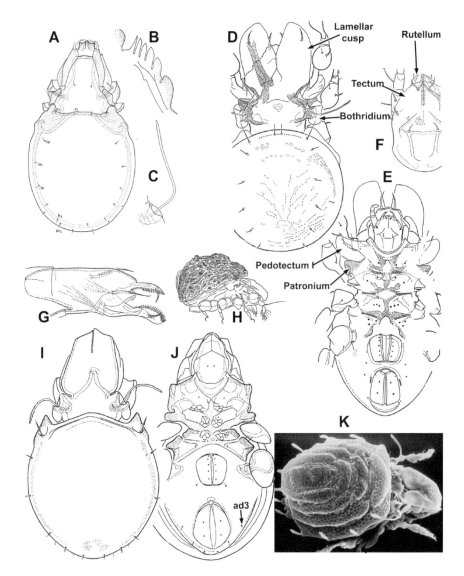

FIGURE 3.63 *Nemacepheus dentatus*, (A) dorsal; (B) rostrum, lateral; (C) bothridial seta and bothridium (after Aoki 1968); Polypterozetidae: *Polypterozetes cherubin*, (D) dorsal; (E) ventral; (F) subcapitulum; (G) chelicera; (H) lateral (after Grandjean 1959c); Podopterotegaeidae: *Podopterotegaeus tectus*, (I) dorsal; (J) ventral (after Aoki 1969); (K) habitus, SEM (VBP, new).

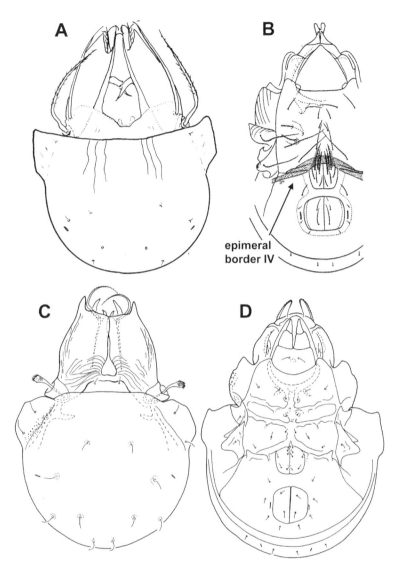

FIGURE 3.64 Microzetidae: *Berlesezetes appalachicolus*, (A) dorsal (after Higgins and Woolley 1968b); *B. brasilozetoides*, (B) ventral (after Balogh and Mahunka 1981); *Kalyptrazetes americanus*, (C) dorsal, (D) ventral (after Mahunka 1995).

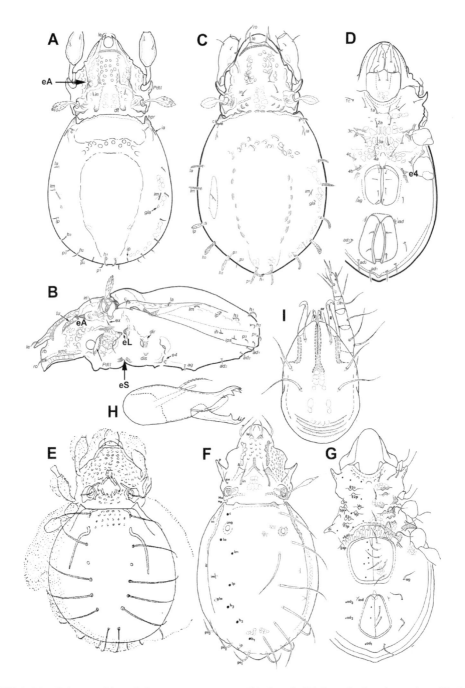

FIGURE 3.65 Caleremaeidae: *Caleremaeus retractus*, (A) dorsal; (B) lateral; *C. arboricolus*, (C) dorsal; (D) ventral (after Norton and Behan-Pelletier 2020); *Veloppia pulchra*, (E) dorsal (after Hammer 1955); *V. kananaskis*, (F) dorsal; (G) ventral; (H) chelicera; (I) subcapitulum (after Norton 1978).

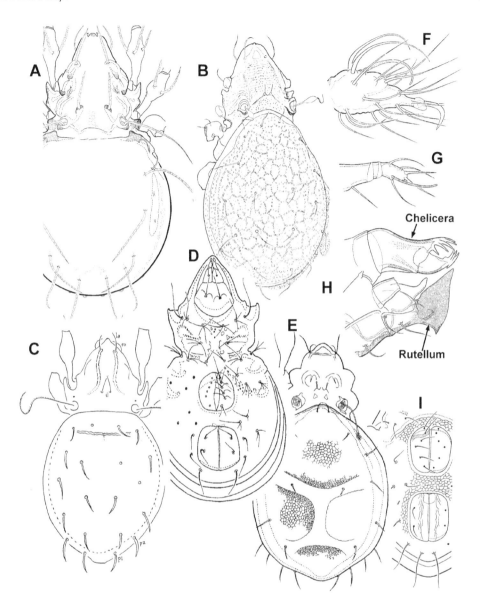

FIGURE 3.66 Ameridae: *Gymnodampia setata*, (A) dorsal (after Chen et al. 2004); Damaeolidae: *Damaeolus ornatissimus*, (B) dorsal (after Pérez-Íñigo 1970); Eremulidae: *Eremulus cingulatus*, (C) dorsal (after Woolley and Higgins 1963); *Eremulus rigidisetosus*, (D) ventral (after Balogh and Mahunka 1969); Damaeolidae: *Fosseremus quadripertitus* (E) dorsal; (F) tarsus I; (G) palp; (H) gnathosoma (lateral); (I) partial ventral (after Grandjean 1965a).

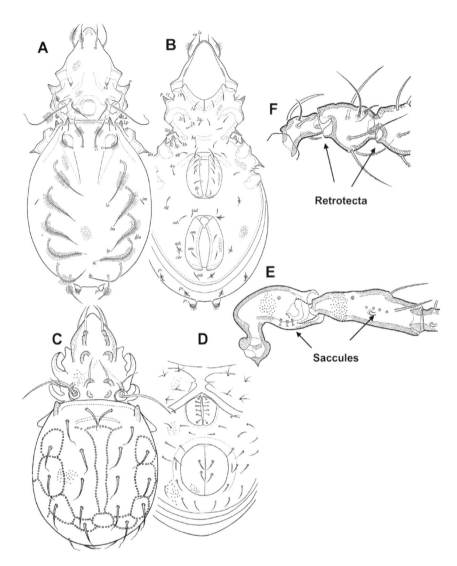

FIGURE 3.67 Hungarobelbidae: *Hungarobelba nortonroyi*, (A) dorsal; (B) ventral (after Bayartogtokh and Ermilov 2021); Eremobelbidae: *Eremobelba minuta*, (C) dorsal; (D) ventral (after Aoki and Wen 1983); *Eremobelba geographica*, (E) trochanter, femur leg IV; (F) genu, tibia leg I (after Grandjean 1965b).

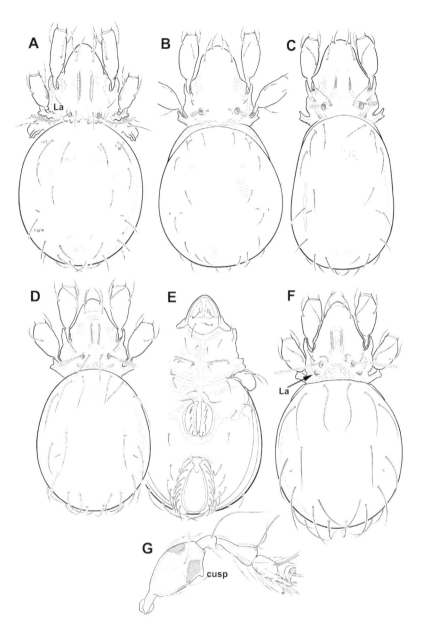

FIGURE 3.68 Eremaeidae: *Eremaeus appalachicus*, (A) dorsal; *E. kananaskis*, (B) dorsal; *E. salish*, (C) dorsal; *E. plumosus*, (D) dorsal; (E) ventral; *E. occidentalis*, (F) dorsal; (G) leg I (after Behan-Pelletier 1993).

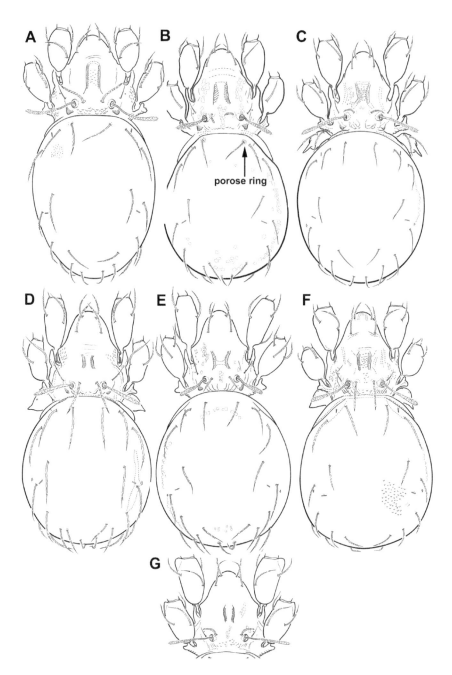

FIGURE 3.69 *Eremaeus kevani*, (A) dorsal; *E. walteri*, (B) dorsal; *E. brevitarsus*, (C) dorsal; *E. translamellatus*, (D) dorsal; *E. grandis*, (E) dorsal; *E. monticolus*, (F) dorsal; *E. boreomontanus*, (G) prodorsum (after Behan-Pelletier 1993).

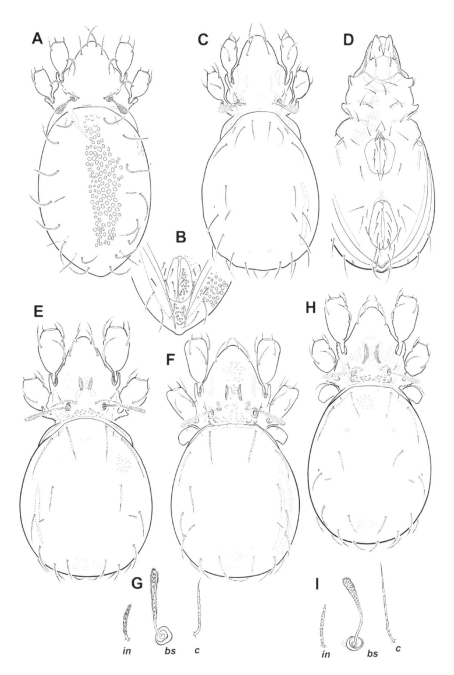

FIGURE 3.70 *Eueremaeus trionus*, (A) dorsal; (B) partial ventral; *Eu. nahani*, (C) dorsal; (D) ventral; *Eu. michaeli*, (E) dorsal; *Eu. masinasin*, (F) dorsal; (G) setae *in*, *bs*, *c*; *Eu. marshalli*, (H) dorsal; (I) setae *in*, *bs*, *c* (after Behan-Pelletier 1993).

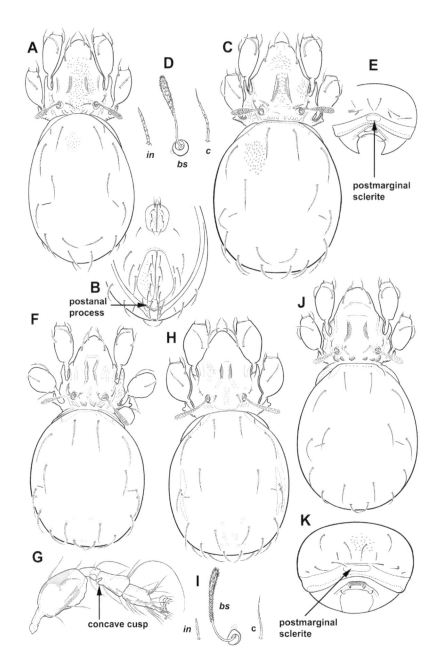

FIGURE 3.71 *Eueremaeus woolleyi*, (A) dorsal; (B) partial ventral; *Eu. foveolatus*, (C) dorsal; (D) setae *in, bs, c*; (E) posterior; *Eu. proximus*, (F) dorsal; (G) leg I; *Eu. nemoralis*, (H) dorsal; (I) setae *in, bs, c*; *Eu. columbianus*, (J) dorsal; (K) posterior (after Behan-Pelletier 1993).

Taxonomic Keys

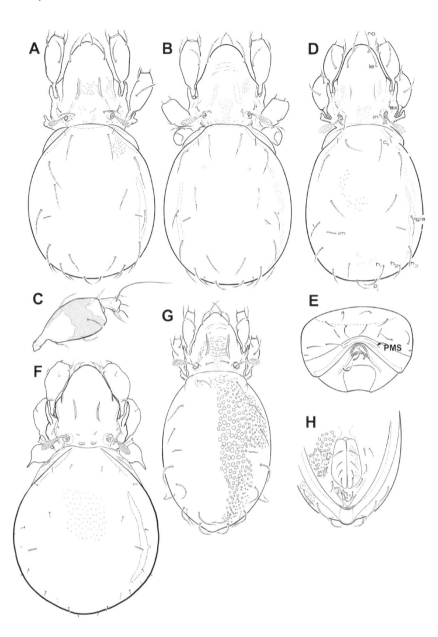

FIGURE 3.72 *Eueremaeus yukonensis*, (A) dorsal; *Eu. aridulus*, (B) dorsal; (C) femur I; *Eu. acostulatus*, (D) dorsal; (E) posterior; *Eu. tetrosus*, (F) dorsal; *Eu. higginsi*, (G) dorsal; (H) partial ventral (after Behan-Pelletier 1993).

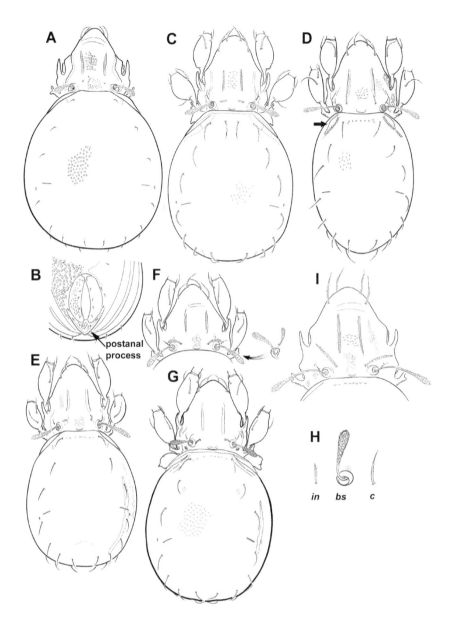

FIGURE 3.73 *Eueremaeus stiktos*, (A) dorsal; (B) partial ventral; *Eu. lindquisti*, (C) dorsal; *Eu. chiatous*, (D) dorsal, with arrow to humeral projection; *Eu. magniporosus*, (E) dorsal; *Eu. alvordensis*, (F) partial dorsal, with arrow to variation in shape of bothridial seta; *Eu. aysineep*, (G) dorsal; (H) setae *in, bs, c*; *Eu. osoyoosensis*, (I) partial dorsal (after Behan-Pelletier 1993).

Taxonomic Keys 319

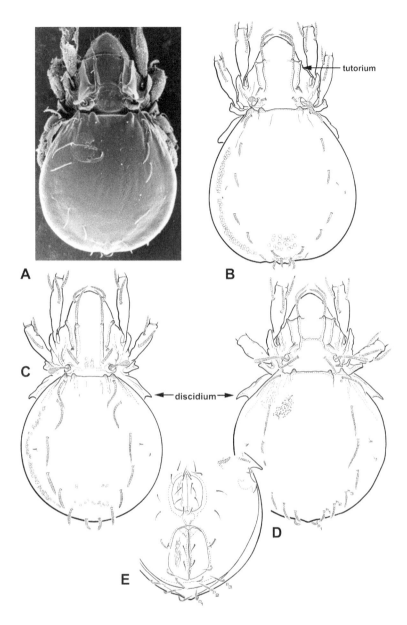

FIGURE 3.74 Megeremaeidae: *Megeremaeus montanus*, (A) dorsal, SEM; *M. kootenai*, (B) dorsal; *M. keewatin*, (C) dorsal; *M. hylaius*, (D) dorsal; (E) ventral (after Behan-Pelletier 1990).

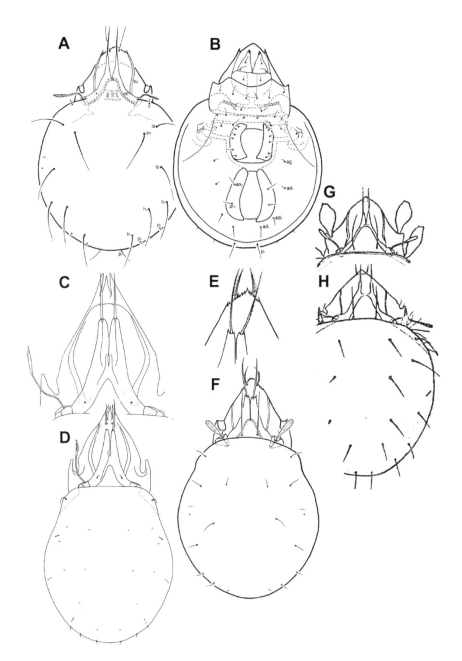

FIGURE 3.75 Astegistidae: *Astegistes pilosus*, (A) dorsal; (B) ventral; *Cultroribula bicultrata*, (C) prodorsum; (D) dorsal; *C. dentata*, (E) prodorsum, anterior; (F) dorsal; *Furcoribula furcillata*, (G) prodorsum; (H) dorsal (A,B,E,F after Bayartogtokh 2007; C,D after Bernini 1969; G,H after Nordenskiöld 1901).

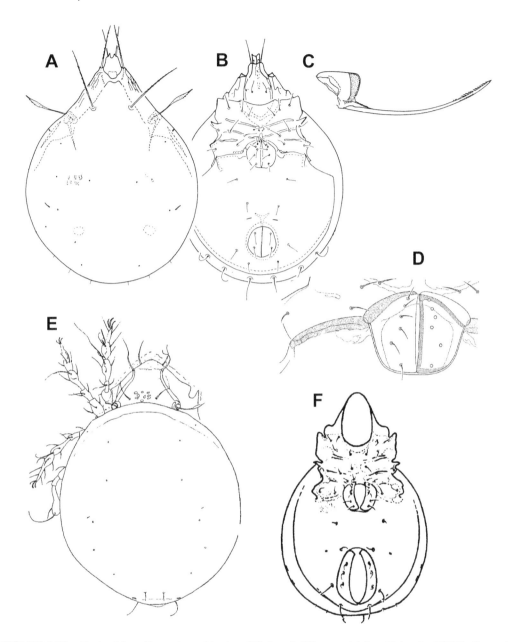

FIGURE 3.76 Gustaviidae: *Gustavia aethiopica*, (A) dorsal; (B) ventral (after Mahunka 1982b); *Gustavia* sp., (C) chelicera (R. A. Norton, unpubl.); (D) genital region and epimere IV (after Grandjean 1968); Kodiakellidae: *Kodiakella lutea*, (E) dorsal (after Hammer 1967); *Kodiakella dimorpha*, (F) ventral (after Pérez-Íñigo and Subías 1978).

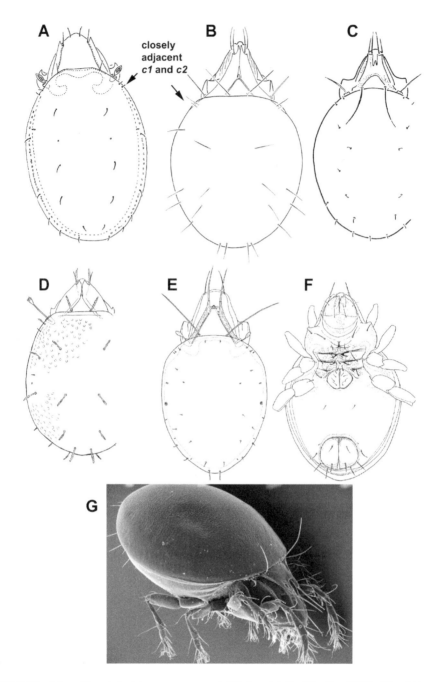

FIGURE 3.77 Liacaridae: *Adoristes ammonoosuci*, (A) dorsal (after Woolley 1968); *Rhaphidosus alticola*, (B) dorsal (after Balogh 1984); *Liacarus trichionus*, (C) dorsal; *Xenillus* sp. (D) dorsal; *Dorycranosus* sp., (E) dorsal; (F) ventral (after Woolley and Higgins 1966); (G) *Liacarus* sp. SEM (VBP, new).

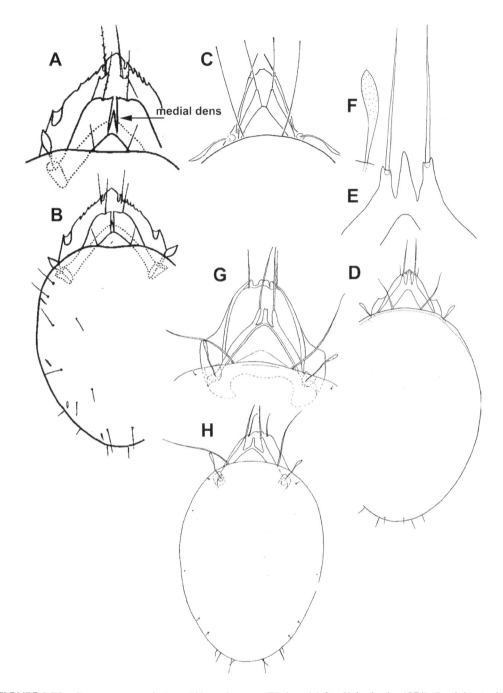

FIGURE 3.78 *Dorycranosus altaicus*, (A) prodorsum; (B) dorsal (after Krivolutsky 1974); *D. abdominalis*, (C) prodorsum (after Woolley 1969); *D. acutidens*, (D) dorsal; (E) detail of lamella; (F) bothridial seta (after Aoki 1965b); *D. parallelus*, (G) prodorsum, (H) dorsal (after Hammer 1967).

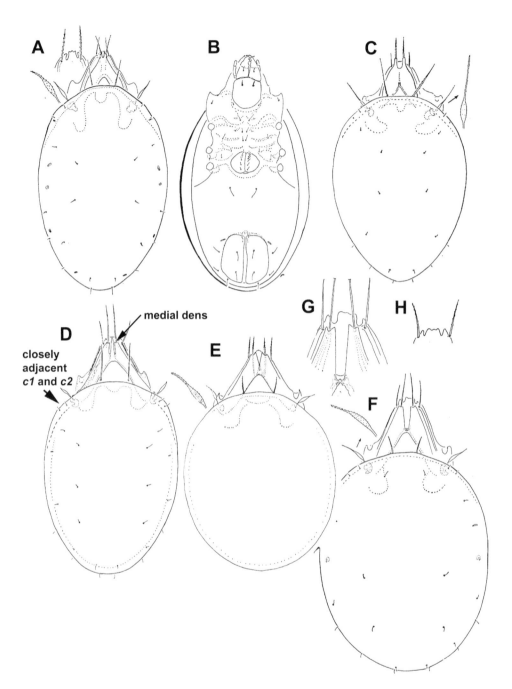

FIGURE 3.79 *Liacarus cidarus*, (A) dorsal, with detail of rostrum; (B) ventral; *L. detosus*, (C) dorsal, with detail of bothridial seta; *L. robustus*, (D) dorsal; *L. latus*, (E) dorsal, with detail of bothridial seta; *L. bidentatus*, (F) dorsal, (G) detail of rostrum, (H) variation in shape of rostrum (after Woolley 1968).

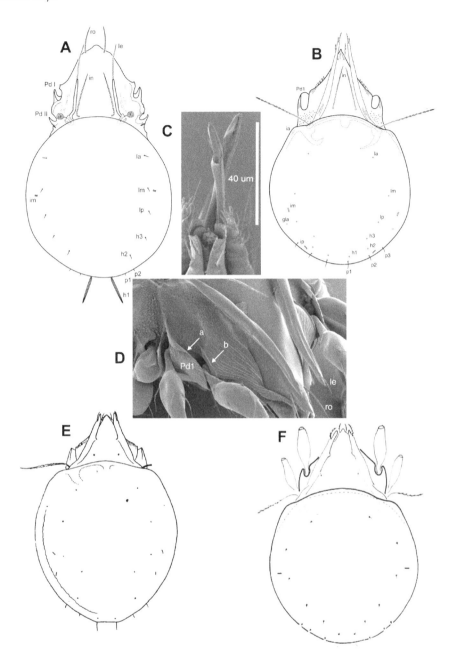

FIGURE 3.80 Peloppiidae: *Dendrozetes jordani*, (A) dorsal (after Lindo et al. 2010); *Metrioppia walbranensis*, (B) dorsal, (C) SEM, pelopsiform chelicera; (D) SEM, prodorsum, lateral, showing lateral flange on pedotectum I (a) and carina dorsal of acetabulum I (b) (after Lindo 2015); *M. helvetica*, (E) dorsal (after Grandjean 1931); *M. oregonensis*, (F) dorsal (after Lindo 2015).

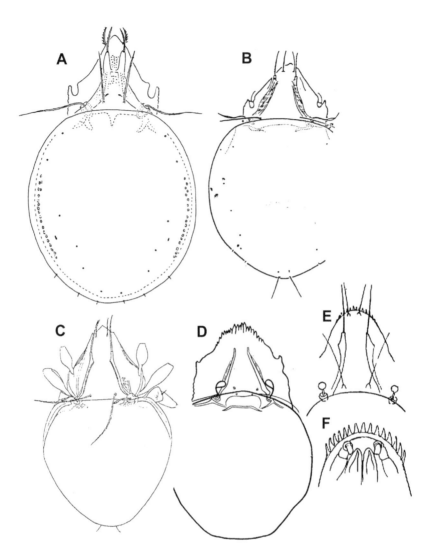

FIGURE 3.81 *Paenoppia forficula*, (A) dorsal (after Woolley and Higgins 1965); *Parapyroppia monodactyla*, (B) dorsal (after Pérez-Íñigo and Subías 1978); *Pyroppia lanceolata*, (C) dorsal (after Hammer 1955); *P. dentata*, (D) dorsal (after Krivolutsky 1974); *P. serrifrons*, (E) prodorsum; (F) rostrum, ventral aspect (after Banks 1923).

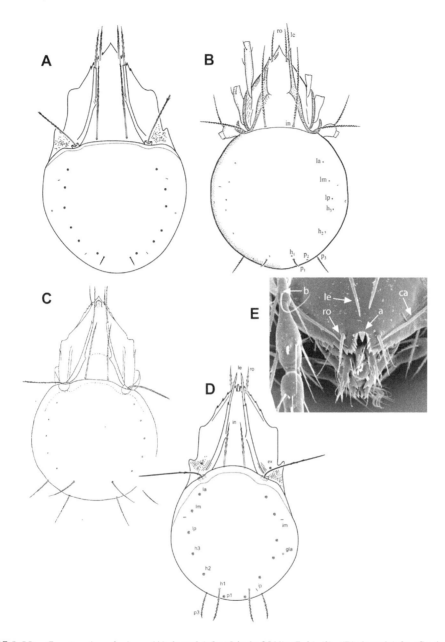

FIGURE 3.82 *Ceratoppia valerieae*, (A) dorsal (after Lindo 2011); *C. bipilis*, (B) dorsal (after Seniczak and Seniczak 2010); *C. sexpilosa*, (C) dorsal (after Hammer 1967); *C. indentata*, (D) dorsal; (E) SEM, detail of rostrum (after Lindo 2011).

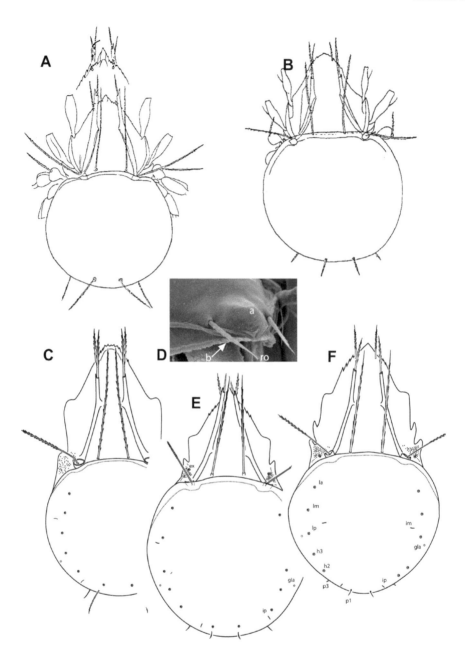

FIGURE 3.83 *Ceratoppia quadridentata*, (A) dorsal, with detail of rostrum; *C. sphaerica*, (B) dorsal (after Hammer 1955); *C. offarostrata*, (C) dorsal; (D) detail of rostral bump (a) and rostral margin (b); *C. longicuspis*, (E) dorsal; *C. tofinoensis*, (F) dorsal (after Lindo 2011).

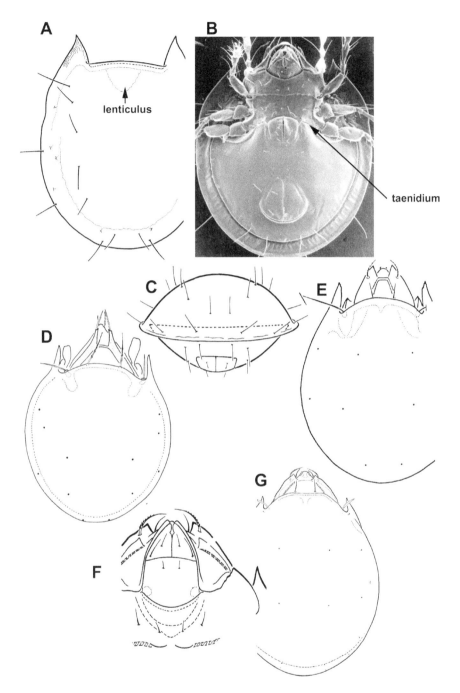

FIGURE 3.84 *Peltenuiala pacifica*, (A) notogaster; (B) SEM, venter; (C) posterior perspective (after Norton 1983); *Tenuialoides medialis*, (D) dorsal (after Woolley and Higgins 1965b); *Tenuiala nuda*, (E) dorsal (after Ewing 1913); *Hafenferrefia nitidula*, (F) prodorsum, venter; (G) dorsal (after Norton 1983).

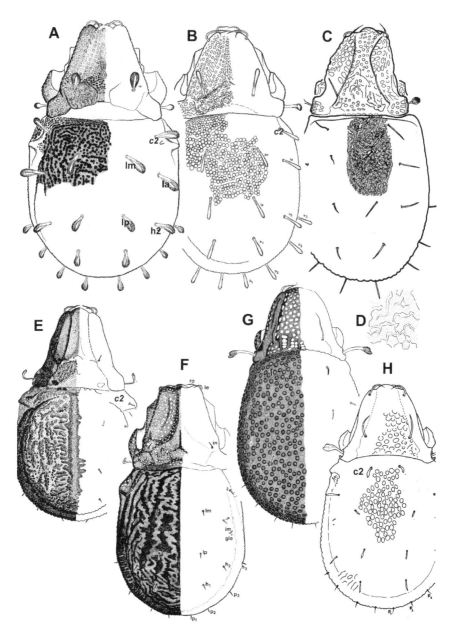

FIGURE 3.85 Carabodidae: *Carabodes cochleaformis*, (A) dorsal (after Reeves 1990); *C. wonalancetanus*, (B) dorsal (after Reeves 1990); *C. labyrinthicus*, (C) dorsal, (D) detail of notogastral sculpturing (after Ito 1982); *C. rugosior*, (E) dorsal; *C. hoh*, (F) dorsal; *C. dickinsoni*, (G) dorsal (after Reeves and Behan-Pelletier 1998); *C. willmanni*, (H) dorsal (after Bernini 1975).

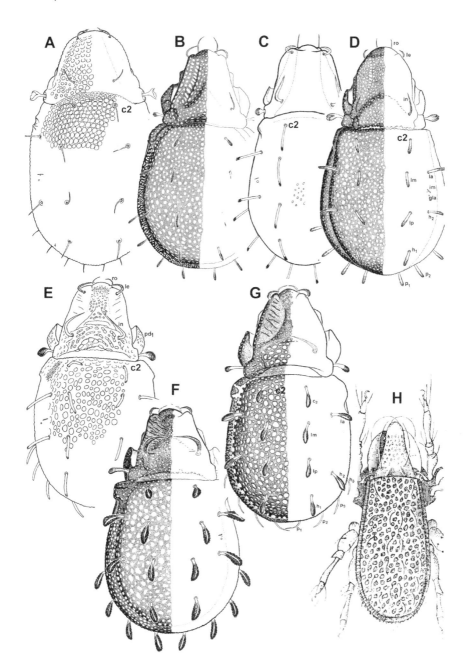

FIGURE 3.86 *Carabodes higginsi*, (A) dorsal (after Reeves 1988); *C. colorado*; (B) dorsal (after Reeves and Behan-Pelletier 1998); *C. radiatus*, (C) dorsal (after Reeves 1987); *C. chandleri*, (D) dorsal (after Reeves 1992); *C. brevis*, (E) dorsal (after Reeves 1988); *C. erectus*, (F) dorsal (after Reeves 1992); *C. polyporetes*, (G) dorsal (after Reeves 1991); *Odontocepheus elongatus*, (H) dorsal (after Michael 1879).

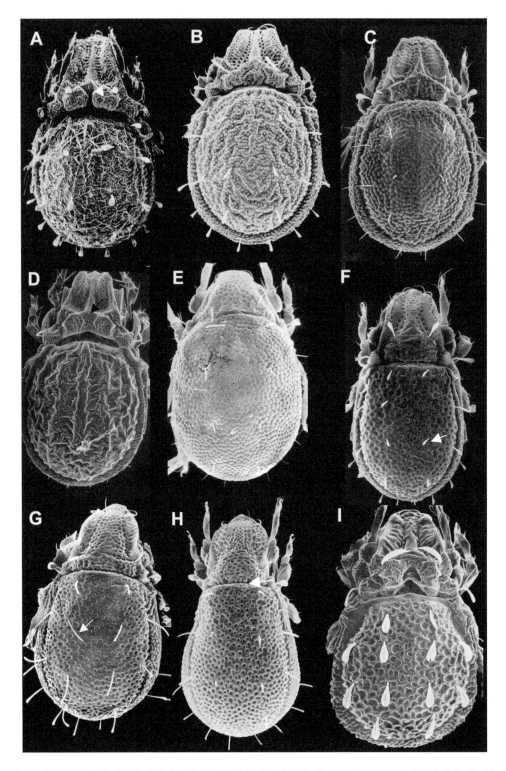

FIGURE 3.87 Dorsal, SEMs, (A) *Carabodes cochleaformis*; (B) *C. wonalancetanus*; (C) *C. labyrinthicus*; (D) *C. rugosior*; (E) *C. higginsi*; (F) *C. granulatus*; (G) *C. radiatus*; (H) *C. chandleri*; (I) *C. niger* (after Reeves 1987, 1990, 1991, 1992).

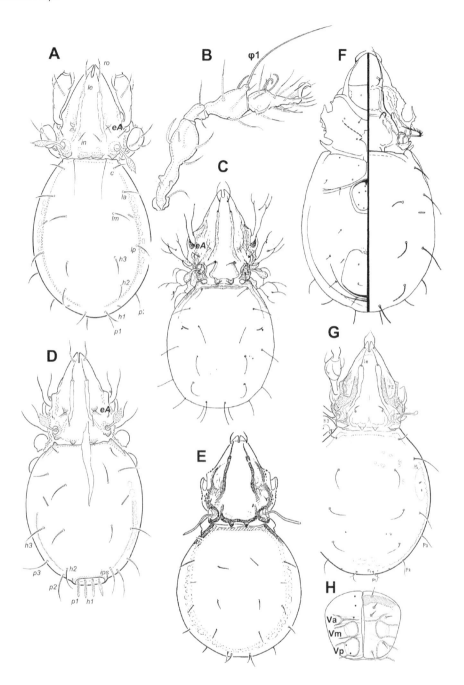

FIGURE 3.88 Autognetidae: *Autogneta amica*, (A) female dorsal; *A. aokii*, (B) leg I; *A. longilamellata*, (C) dorsal; *A. flaheyi*, (D) male dorsal; *Conchogneta dalecarlica*, (E) dorsal; *Eremobodes pectinatus*, (F) combined ventral and dorsal; *Rhaphigneta numidiana*, (G) dorsal; (H) genital region (A,B,D after Behan-Pelletier 2015a, reproduced with permission; C,E after Forsslund 1947; F, after Jacot 1937b; G,H after Grandjean 1960).

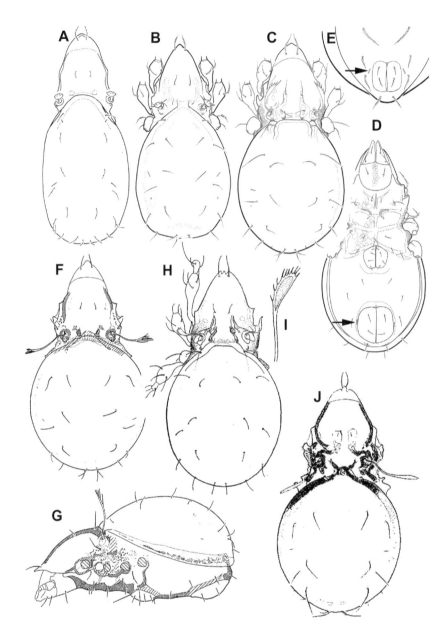

FIGURE 3.89 *Discoppia cylindrica*, (A) dorsal (after Subías and Rodríguez 1988); *Micrppia minus*, (B) dorsal (VBP, new); *Oppiella (Oppiella) nova*, (C) dorsal; (D) ventral with arrow to paranal *iad* (VBP, new); *Enantioppia* sp. (E) ventral with arrow to apoanal *iad* (VBP, new); *Oppiella (Rhinoppia) subpectinata*, (F) dorsal, (G) lateral; *Oppiella (Oppiella) maritima*, (H) dorsal, (I) bothridial seta (after Strenzke 1951); *Dissorhina ornata*, (J) dorsal (F,G,J after Woas 1986, © Staatliches Museum für Naturkunde Karlsruhe (SMNK)).

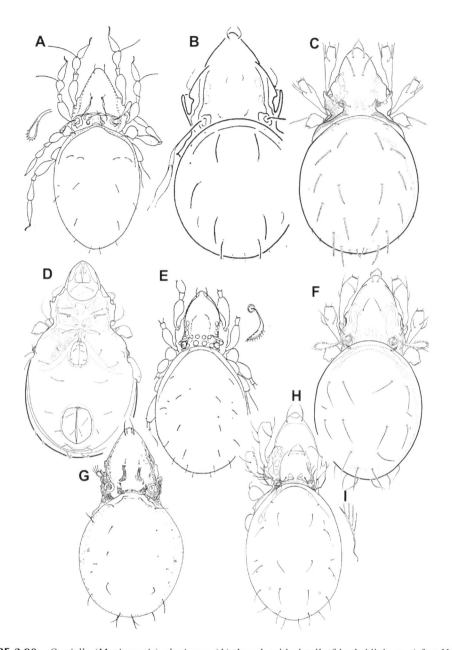

FIGURE 3.90 *Oppiella* (*Moritzoppia*) *clavigera*, (A) dorsal, with detail of bothridial seta (after Hammer 1952a); *Sphagnoppia biflagellata*, (B) dorsal (after Balogh and Balogh 1986); *Aeroppia* sp., (C) dorsal; (D) ventral (VBP, new); *Subiasella* (*Lalmoppia*) *maculata*, (E) dorsal, with detail of bothridial seta (after Hammer 1952a); *Anomaloppia manifera*, (F) dorsal (VBP, new); *Berniniella serratirostris*, (G) dorsal (after Caballero et al. 1999); *Brachioppiella periculosa*, (H) dorsal; (I) bothridial seta (after Hammer 1962).

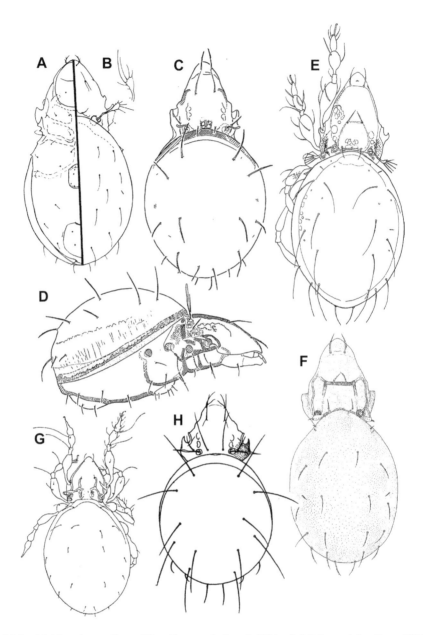

FIGURE 3.91 *Multioppia carolinae*, (A) split ventral, dorsal; (B) bothridial seta (after Jacot 1938c); *Oppia nitens*, (C) dorsal; (D) lateral (after Woas 1986, © Staatliches Museum für Naturkunde Karlsruhe (SMNK); *Ramusella puertomontiensis*, (E) dorsal (after Hammer 1962a); *Graptoppia italica* (F) dorsal (after Bernini 1969); *Oppiella (Moritzoppia) translamellata* (G) dorsal (after Hammer 1952a); *Lasiobelba remota*, (H) dorsal (after Aoki 1959).

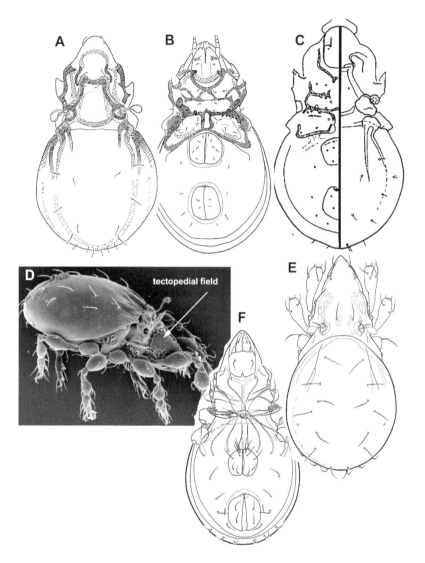

FIGURE 3.92 Quadroppiidae: *Quadroppia quadricarinata*, (A) dorsal; (B) ventral (after Woas 1986, © Staatliches Museum für Naturkunde Karlsruhe (SMNK); *Q. skookumchucki*, (C) split ventral and dorsal (after Jacot 1939a); *Suctobelbella* sp., (D) lateral SEM (courtesy of M. Clayton); Machuellidae: *Machuella* sp., (E) dorsal; (F) ventral.

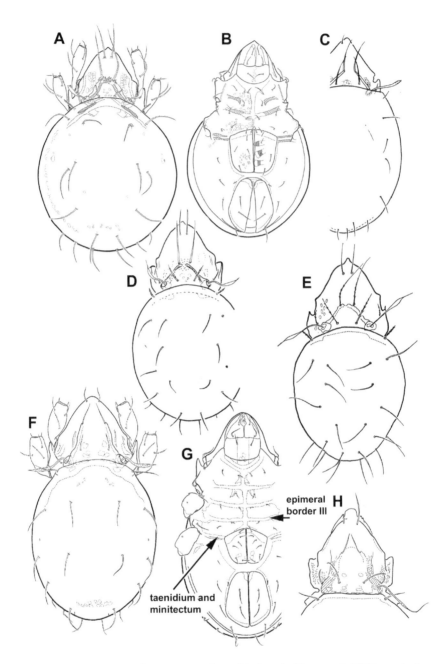

FIGURE 3.93 Thyrisomidae: *Banksinoma spinifera*, (A) dorsal; (B) ventral (VBP, new); *B. setosa*, (C) dorsal (after Krivolutsky and Ryabinin 1974); *B. lanceolata*, (D) dorsal (after Fujikawa 1979); *B. lanceolata canadensis*, (E) dorsal (after Fujikawa 1979); *Pantelozetes alpestris*, (F) dorsal; (G) ventral (VBP, new); *P. paolii*, (H) prodorsum (after Beck and Woas 1991, © Staatliches Museum für Naturkunde Karlsruhe (SMNK)).

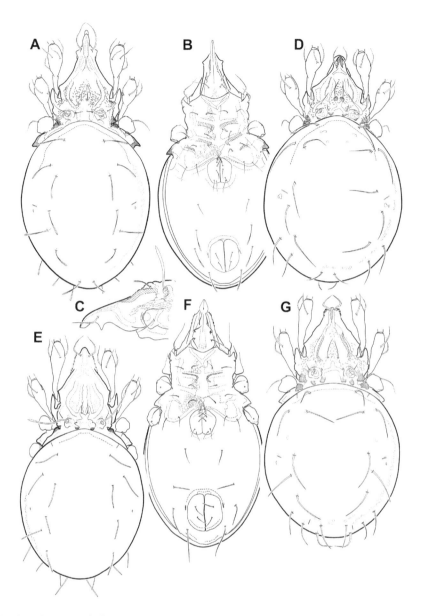

FIGURE 3.94 *Rhinosuctobelba dicerosa*, (A) dorsal, (B) ventral, (C) prodorsum, lateral; *Rhynchobelba* sp., (D) dorsal; *Allosuctobelba gigantea*, (E) dorsal; (F) ventral; *Suctobelba* sp., (G) dorsal (VBP, new).

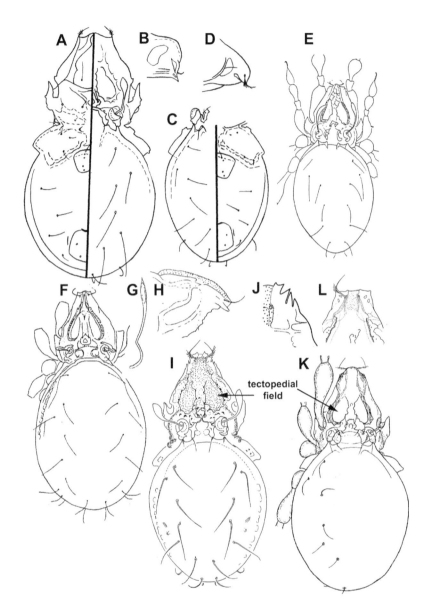

FIGURE 3.95 *Suctobelbella frothinghami*, (A) split ventral and dorsal, (B) rostrum, lateral; *S. hurshi*, (C) split notogaster, partial ventral, (D) rostrum, lateral (after Jacot 1937a); *S. setosoclavata*, (E) dorsal (after Hammer 1952a); *S. punctata*, (F) dorsal; (G) detail of bothridial seta, (H) rostrum, lateral (after Hammer 1955); *S. acutidens*, (I) dorsal; (J) rostrum, lateral; *S. longirostris*, (K) dorsal, (L) rostrum (I,K after Forsslund 1941; J,L after Strenzke 1951).

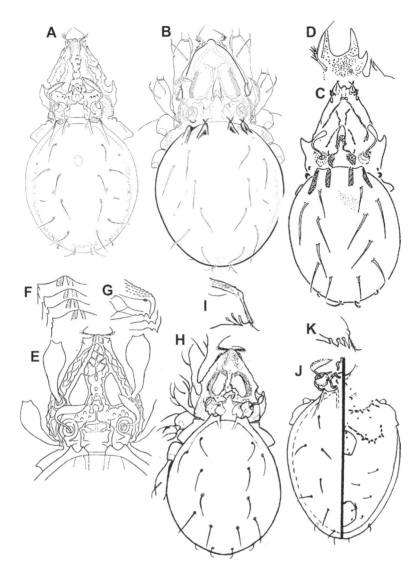

FIGURE 3.96 *Suctobelbella arcana*, (A) dorsal (after Moritz 1970); *S. sarekensis*, (B) dorsal (VBP, new); *S. longicuspis*, (C) dorsal, (D) detail of rostrum (after Jacot 1937a); *S. hammerae*, (E) prodorsum and anterior of notogaster, (F) detail of rostrum, dorsal; (G) detail of rostrum, lateral (after Krivolutsky 1965); *S. palustris*, (H) dorsal; (I) detail of rostrum, lateral (after Forsslund 1941); *S. laxtoni*, (J) split partial dorsal and ventral; (K) lateral of rostrum (after Jacot 1937a).

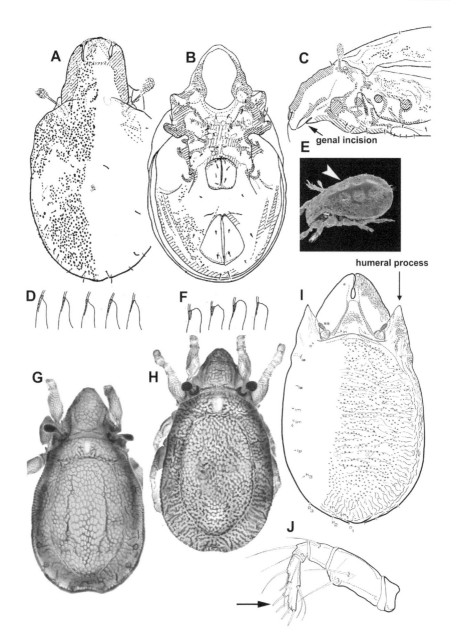

FIGURE 3.97 *Tectocepheus velatus*, (A) dorsal; (B) ventral; (C) lateral (after Nübel-Reidelbach 1994; © Staatliches Museum für Naturkunde Karlsruhe (SMNK)); (D) variation in shape of lamellar cusp (after Knülle 1954); *T. sarekensis*, (E) SEM, dorsal (after Laumann et al. 2007); (F) variation in shape of lamellar cusp (after Knülle 1954); *Scapheremaeus parvulus*, (G) dorsal; *S. palustris*, (H) dorsal (after Norton et al. 2010; reproduced with permission); *Scapuleremaeus kobauensis*, (I) dorsal; (J) palp, with arrow to eupathidium *acm* on tubercle (after Behan-Pelletier 1989b).

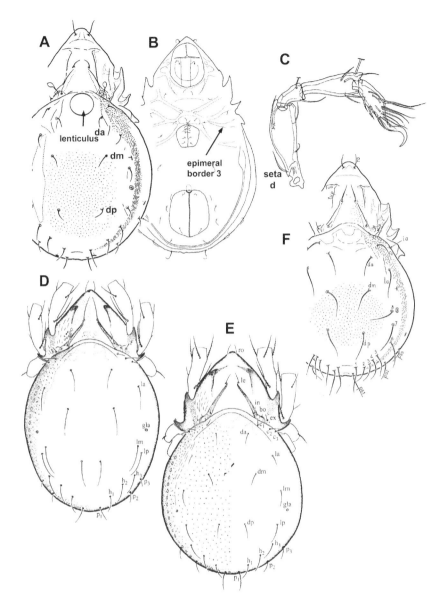

FIGURE 3.98 *Hydrozetes thienemanni*, (A) dorsal (after Grandjean 1948c). *H. capensis*, (B) ventral; (C) leg I (after Engelbrecht 1974); *H. octosetosus*, (D) dorsal; *H. lacustris*, (E) dorsal (after Seniczak et al. 2007); *H. parisiensis*, (F) dorsal, with range of hypertrichy of setae *h* (after Grandjean 1948c).

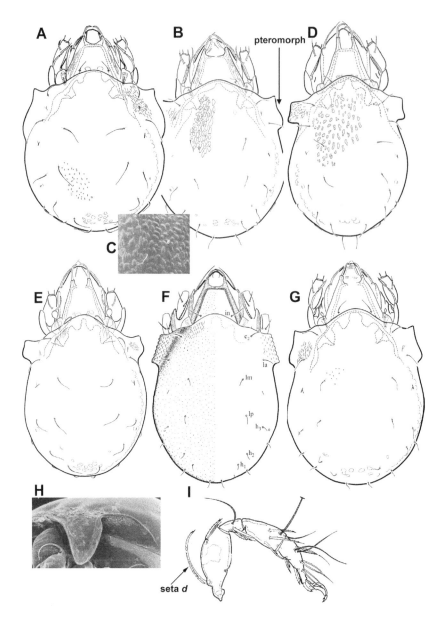

FIGURE 3.99 *Limnozetes atmetos*, (A) dorsal; *L. lustrum*, (B) dorsal, (C) detail of notogastral sculpturing; *L. foveolatus*, (D) dorsal; *L. amnicus*, (E) dorsal; *L. ciliatus*, (F) dorsal; *L. onondaga*, (G) dorsal, (H) detail of pteromorph, (I) leg I (F, after Seniczak and Seniczak (2010); others after Behan-Pelletier 1989a).

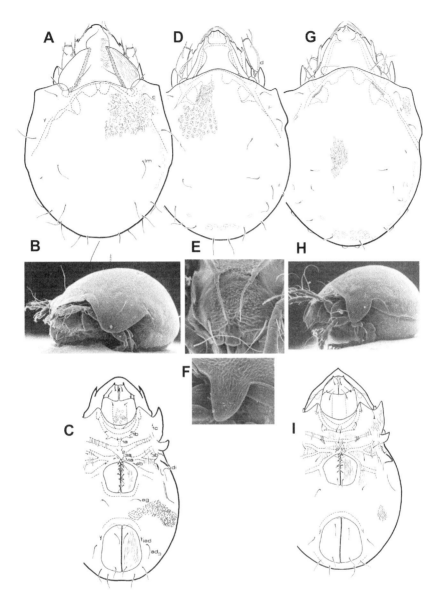

FIGURE 3.100 *Limnozetes latilamellatus*, (A) dorsal; (B) SEM, lateral; (C) ventral; *L. guyi*, (D) dorsal; (E) SEM, prodorsum, anterior; (F) SEM, detail of pteromorph; *L. borealis*, (G) dorsal; (H) SEM, lateral aspect; (I) ventral (after Behan-Pelletier 1989a).

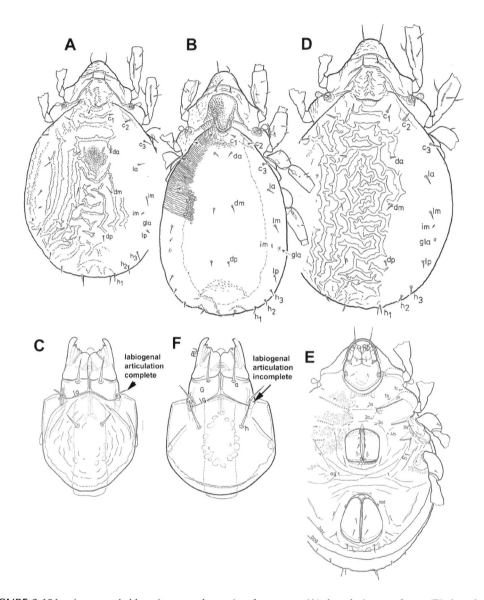

FIGURE 3.101 Ameronothridae: *Ameronothrus nigrofemoratus*, (A) dorsal; *A. maculatus*, (B) dorsal; (C) subcapitulum; *A. lineatus*, (D) dorsal; (E) ventral; (F) subcapitulum (after Schubart 1975; *www.schweizerbart. de/series/zoologica*).

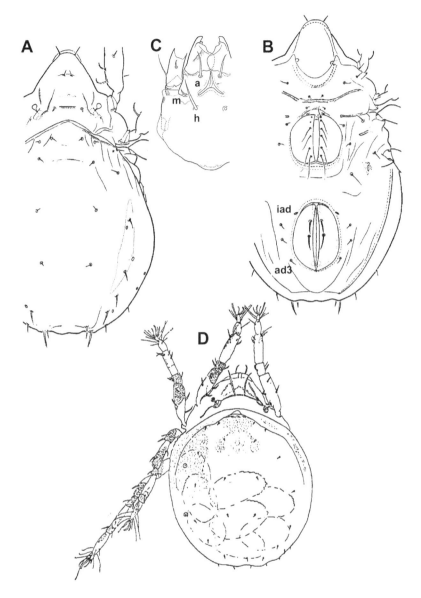

FIGURE 3.102 Podacaridae: *Podacarus auberti*, (A) dorsal; (B) ventral; (C) subcapitulum (after Grandjean 1955a); *Alaskozetes coriaceus*, (D) dorsal (after Hammer 1955).

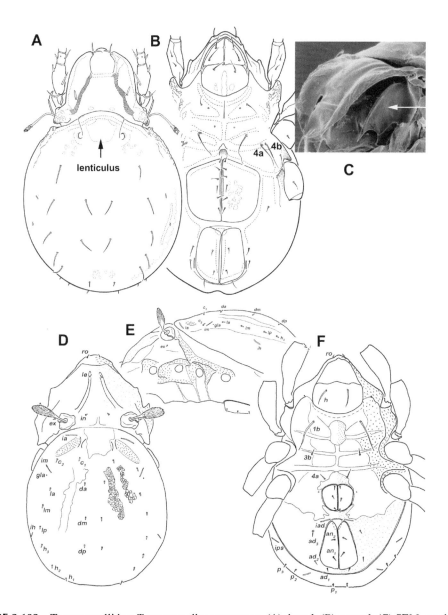

FIGURE 3.103 Tegeocranellidae: *Tegeocranellus muscorum*, (A) dorsal; (B) ventral; (C) SEM, prodorsum, lateral, with arrow to pedotectum (after Behan-Pelletier 1997); Selenoribatidae: *Thalassozetes barbarae*, (D) dorsal; (E) lateral; (F) ventral (after Pfingstl 2013).

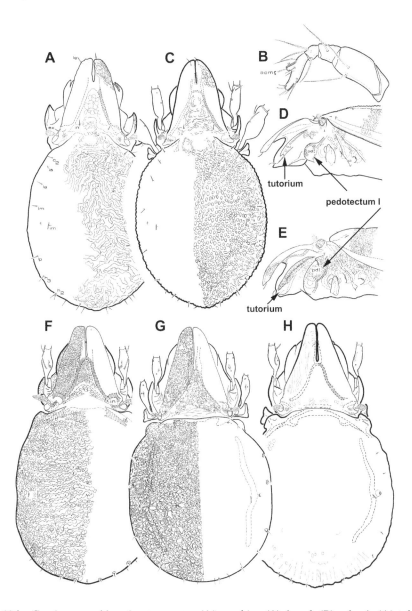

FIGURE 3.104 Cymbaeremaeidae: *Ametroproctus* (*A.*) *oresbios*, (A) dorsal, (B) palp; *A.* (*A.*) *tuberculosus*, (C) dorsal; (D) podosoma, lateral; *A.* (*C.*) *reticulatus*, (E) podosoma, lateral; (F) dorsal; *A.* (*C.*) *beringianus*, (G) dorsal; *A.* (*C.*) *canningsi*, (H) dorsal (after Behan-Pelletier 1987).

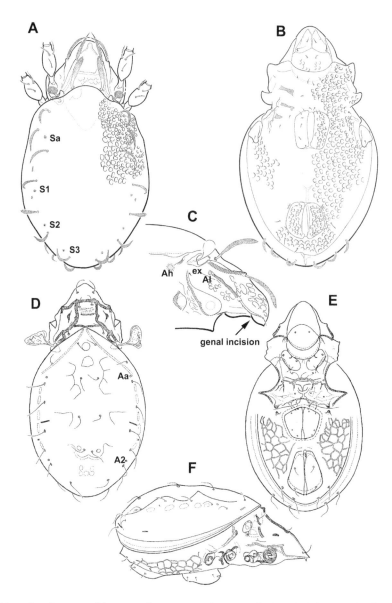

FIGURE 3.105 Dendroeremaeidae: *Dendroeremaeus krantzi*, (A) dorsal; (B) ventral; (C) podosoma, lateral (after Behan-Pelletier et al. 2005); Licneremaeidae: *Licneremaeus licnophorus*, (D) dorsal; (E) ventral; (F) lateral (after Ito 1982).

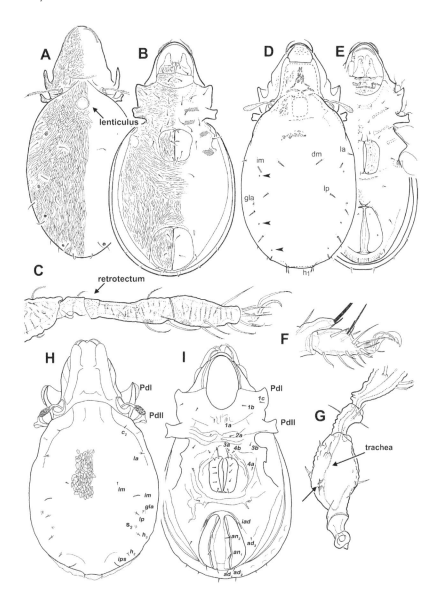

FIGURE 3.106 Passalozetidae: *Passalozetes californicus*, (A) dorsal; (B) ventral (VBP, new); *P. perforatus*, (C) leg IV (after Woas 1998; © Staatliches Museum für Naturkunde Karlsruhe (SMNK)); Scutoverticidae: *Scutovertex mikoi*, (D) dorsal, with arrows to saccules; (E) ventral (after Weigmann 2009b); *S. sculptus*, (F) tarsus I; (G) femur and genu I (after Woas 1998, © Staatliches Museum für Naturkunde Karlsruhe (SMNK)); *Exochocepheus eremitus*, (H) dorsal; (I) ventral (after Pfingstl et al. 2010).

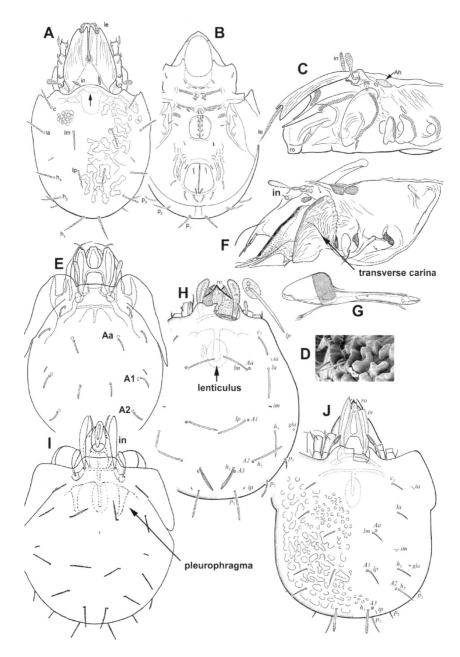

FIGURE 3.107 Unduloribatidae: *Unduloribates dianae*, (A) dorsal, with arrow to dorsophragmata; (B) ventral; (C) lateral, notogaster removed; (D) SEM, detail of blocky cerotegument (after Behan-Pelletier and Walter 2009; reproduced with permission); Phenopelopidae: *Peloptulus americanus*, (E) dorsal (after Aoki 1975); *Eupelops acromios*, (F) lateral, notogaster removed; (G) chelicera (after Grandjean 1936d); *E. hirtus*, (H) dorsal (after Seniczak et al. 2014); *E. septentrionalis*, (I) dorsal (after Trägårdh 1910); *E. plicatus*, (J) dorsal (after Seniczak et al. 2015).

Taxonomic Keys 353

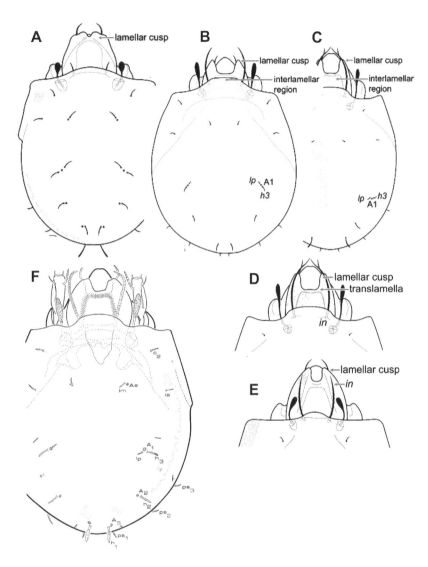

FIGURE 3.108 *Propelops groenlandicus*, (A) dorsal; *P. minnesotensis*, (B) dorsal; *P. alaskensis*, (C) dorsal, (D) detail of anterior of dorsum; *P. monticolus*, (E) dorsum, anterior (after Walter et al. 2014); *P. canadensis*, (F) dorsal (after Norton and Behan-Pelletier 1986).

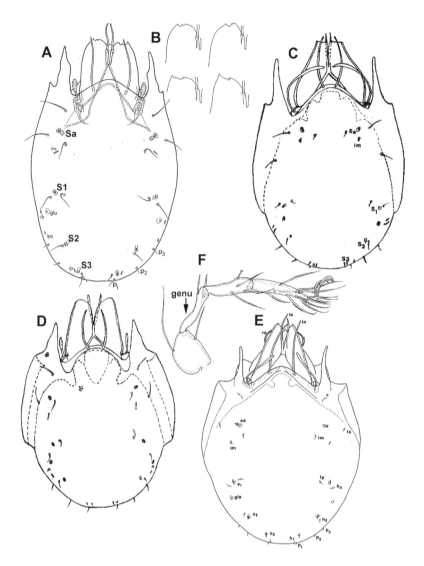

FIGURE 3.109 Achipteriidae: *Achipteria curta*, (A) dorsal; (B) detail of lamellar cusp (after Aoki 1970b); *A. catskillensis*, (C) dorsal; *A. clarencei*, (D) dorsal (after Nevin 1976); *A. coleoptrata*, (E) dorsal (after Maruyama 2003); *Achipteria serrata*, (F) leg IV paraxial (after Hirauchi and Aoki 1997).

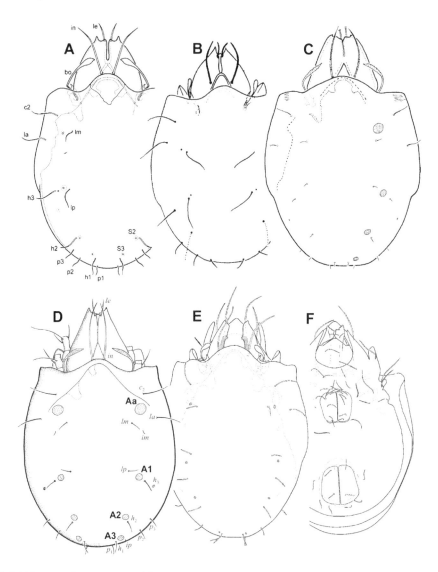

FIGURE 3.110 *Anachipteria sacculifera*, (A) dorsal (after Root et al. 2008); *A. acuta*, (B) dorsal (after Woolley 1957); *A. howardi*, (C) dorsal (after Norton and Kethley 1990); *A. magnilamellata*, (D) dorsal (after Seniczak et al. 2017); *Separachipteria geminus*, (E) dorsal; (F) ventral (after Lindo et al. 2008).

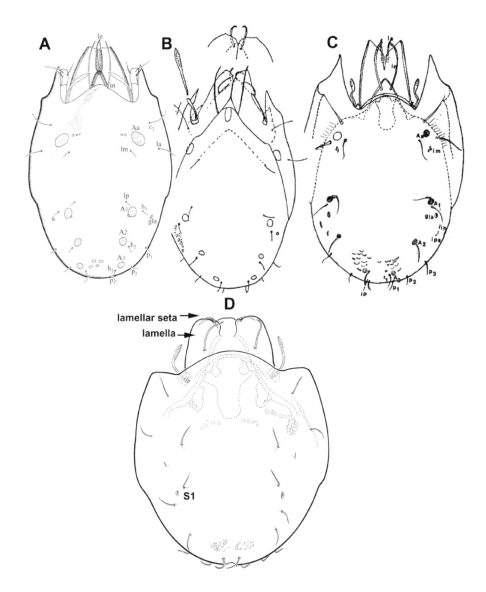

FIGURE 3.111 *Parachipteria bella*, (A) dorsal (after Seniczak and Seniczak 2007); *P. nivalis*, (B) dorsal, with detail of anterior of lamella (after Hammer 1952a); *P. travei*, (C) dorsal (after Nevin 1977); *Dentachipteria highlandensis*, (D) dorsal (VBP, new).

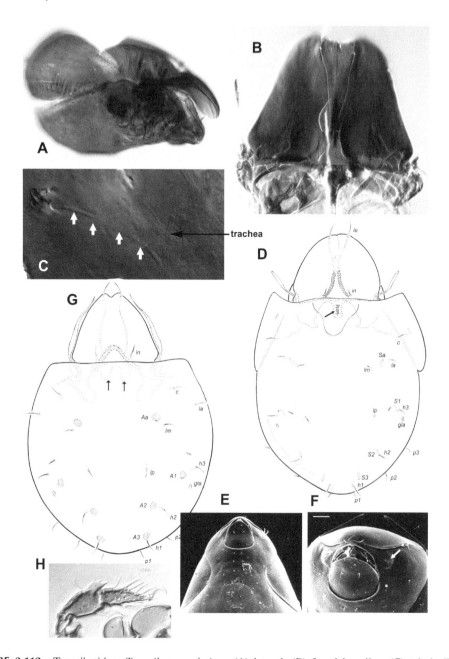

FIGURE 3.112 Tegoribatidae: *Tegoribates subniger*, (A) lateral; (B) fused lamellae; (C) tubule *T1* with arrows along length of tubule; *T. americanus*, (D) dorsal, with arrow to fused dorsophragmata; *T. walteri*, (E) anterior ventral; (F) subcapitulum; *Protectoribates occidentalis*, (G) dorsal with arrows to dorsophragmata; (H) genu, tibia, tarsus II (after Behan-Pelletier 2017; reproduced with permission).

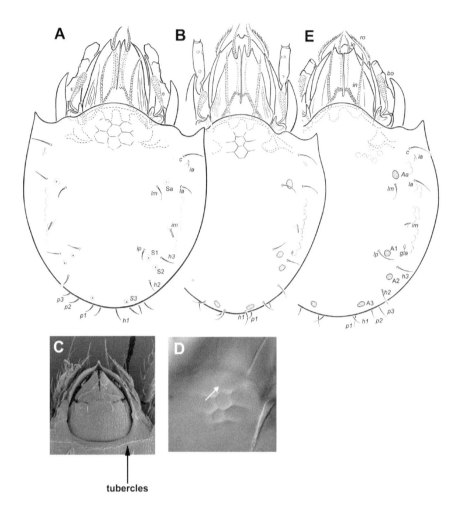

FIGURE 3.113 *Tectoribates campestris*, (A) dorsal; *T. borealis*, (B) dorsal; (C) SEM, detail of anterior of venter with arrow to ribbon of tubercles; (D) *T. campestris*, DIC, detail of hexagonal pattern on anterior of notogaster; *T. alcescampestris*, (E) dorsal (after Behan-Pelletier and Walter 2013; reproduced with permission).

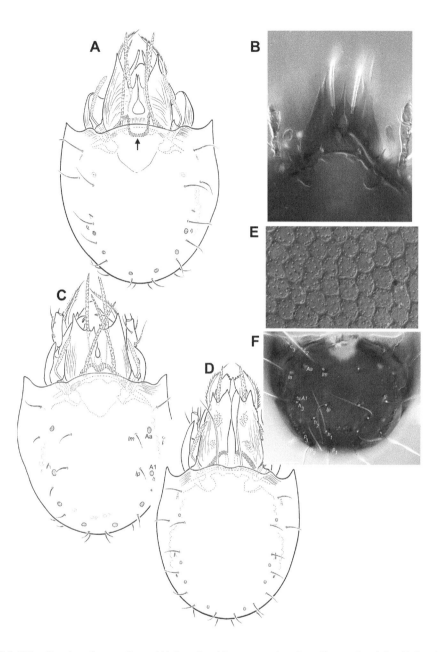

FIGURE 3.114 *Ferolocella tessellata*, (A) dorsal, with arrow to interlamellar pocket (after Behan-Pelletier 2013); *Oribatella dentaticuspis*, (B) DIC, prodorsum; *O. nortoni*, (C) dorsal; *O. sintranslamella*, (D) dorsal; *O. minuta*, (E) DIC, centre of notogaster; *O. quadridentata*, (F) DIC, notogaster. (after Behan-Pelletier 2011; reproduced with permission.)

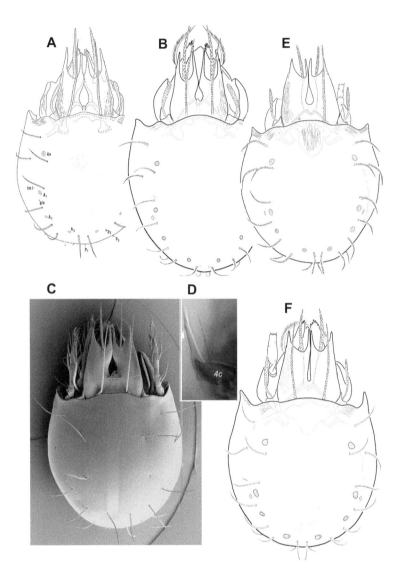

FIGURE 3.115 *Oribatella quadricornuta*, (A) dorsal (after Bernini 1975); *O. pawnee*, (B) dorsal; *O. brevicornuta*, (C) SEM, dorsal; (D) DIC, custodium and seta 4c; *O. jacoti*, (E) dorsal; *O. parallelus*, (F) dorsal (after Behan-Pelletier and Walter 2012; reproduced with permission).

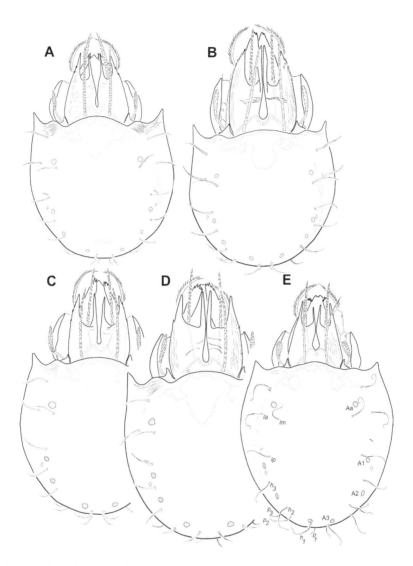

FIGURE 3.116 *Oribatella ewingi*, (A) dorsal; *O. transtriata*, (B) dorsal; *O. yukonensis*, (C) dorsal; *O. heatherae*, (D) dorsal; *O. flagellata*, (E) dorsal (B, E after Behan-Pelletier 2011; A, C, D after Behan-Pelletier and Walter 2012; reproduced with permission).

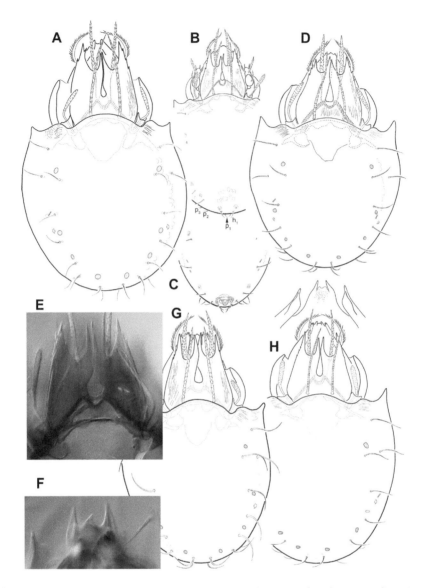

FIGURE 3.117 *Oribatella manningensis*, (A) dorsal; *O. canadensis*, (B) female dorsal; (C) male notogaster; *O. maryae*, (D) dorsal; *O. reticulatoides*, (E) DIC of prodorsum; *O. arctica*, (F) DIC of rostrum; *O. banksi*, (G) dorsal; *O. abmi*, (H) dorsal, with detail of rostrum (A,D–H after Behan-Pelletier and Walter 2012; B,C after Behan-Pelletier and Eamer 2010; reproduced with permission).

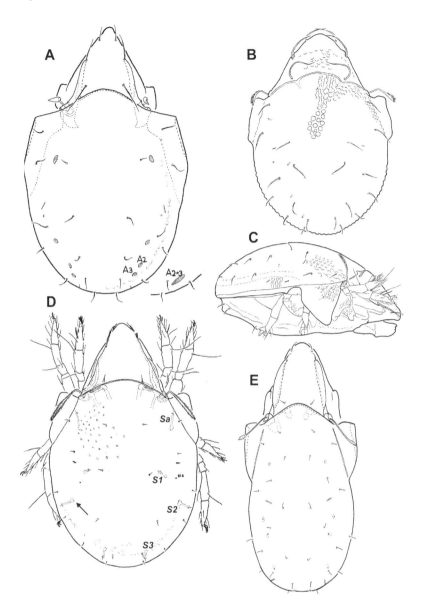

FIGURE 3.118 Haplozetidae: *Lagenobates lagenula*, (A) dorsal, with detail of specimen with fused porose areas *A2*, *A3* at lower right (after Weigmann and Miko 2002); *Rostrozetes ovulum*, (B) dorsal; (C) lateral (after Beck 1965); *Haplozetes triangulatus*, (D) dorsal, with arrow to furcated saccule *S2* (after Beck 1964); *Pilobates carpetanus*, (E) dorsal (after Weigmann 2010).

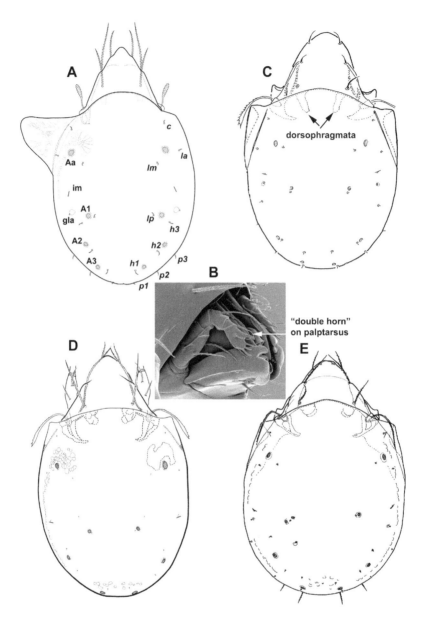

FIGURE 3.119 *Protoribates haughlandae*, (A) dorsal; (B) subcapitulum; *P. capucinus*, (C) dorsal, with arrow to dorsophragma; *P. robustior*, (D) dorsal (VBP, new); *P. lophotrichus*, (E) dorsal (A and B, after Walter and Latonas 2013; C and E, after Miko et al. 1994; reproduced with permission).

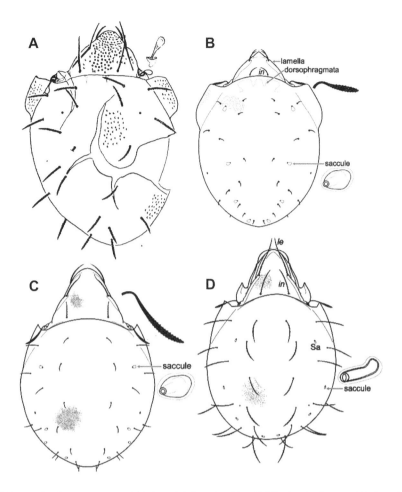

FIGURE 3.120 *Peloribates juniperi*, (A) dorsal (after Woolley 1958); *P. alaskensis*, (B) dorsal, with detail of saccule *S1*; *P. canadensis*, (C) dorsal, with detail of saccule *Sa*; *P. pilosus*, (D) dorsal, with detail of saccule *S1* (B–D after Walter et al. 2014).

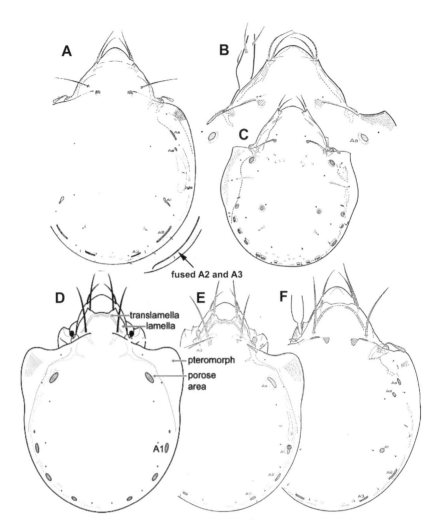

FIGURE 3.121 *Mochlobates affinis*, (A) dorsal, with fused porose areas *A2* and *A3* of male at lower right; *Dynatozetes magnus*, (B) anterior; *Dynatozetes amplus*, (C) dorsal (after Grandjean 1960); *Podoribates longipes*, (D) dorsal (after Walter et al. 2014); *P. pratensis*, (E) dorsal; *Mochlozetes maximus*, (F) dorsal (A,B,E,F: after Norton 1984).

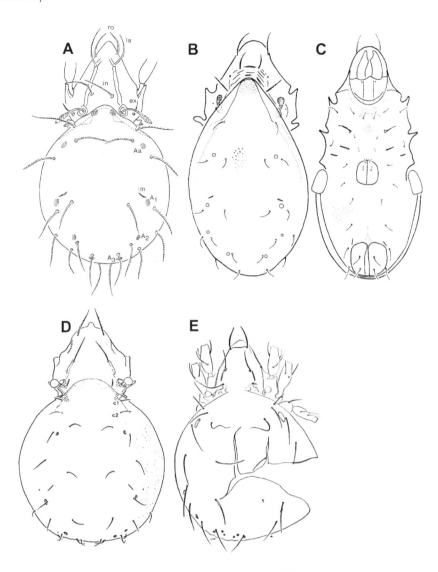

FIGURE 3.122 Oribatulidae: *Jornadia larreae*, (A) dorsal (after Wallwork and Weems 1984); *Diphauloppia luminosa*, (B) dorsal; (C) ventral (after Balogh and Balogh 1984); *Phauloppia rauschenensis*, (D) dorsal (after Wunderle et al. 1990); *P. boletorum*, (E) dorsal (after Higgins and Woolley 1975).

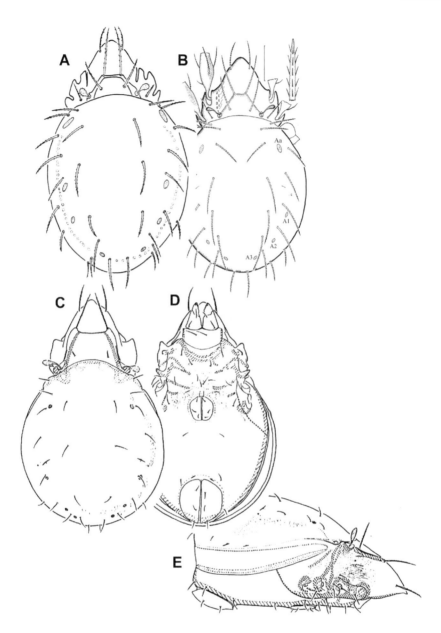

FIGURE 3.123 *Lucoppia apletosa*, (A) dorsal (after Higgins and Woolley 1975); *L. burrowsii*, (B) dorsal, with detail of seta c_2 (after Seniczak and Seniczak 2012); *Zygoribatula exilis*, (C) dorsal; (D) ventral; (E) lateral (after Wunderle et al. 1990).

Taxonomic Keys

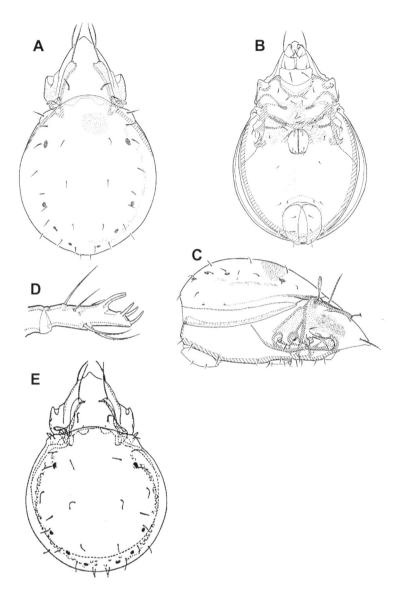

FIGURE 3.124 *Oribatula tibialis*, (A) dorsal; (B) ventral; (C) lateral (after Wunderle et al. 1990); (D) palp (after Grandjean 1958b); *O. pallida*, (E) dorsal (after Iordansky 1991).

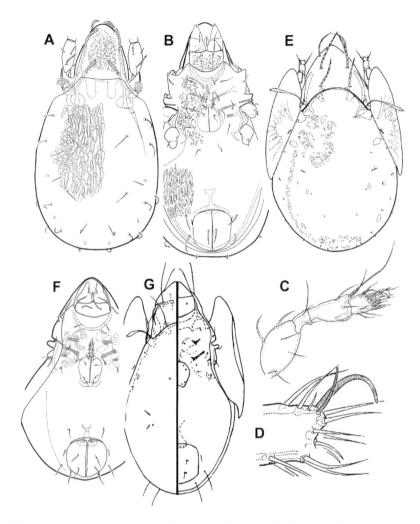

FIGURE 3.125 Oripodidae: *Benoibates* sp. (A) dorsal; (B) ventral (VBP, new); *Pirnodus detectidens*, (C) leg I; (D) extremity of tarsus I (after Grandjean 1956a); Parakalummidae: *Neoribates aurantiacus*, (E) dorsal; (F) ventral (VBP, new); *Protokalumma depressa*, (G) combined dorsal and ventral (after Jacot 1933).

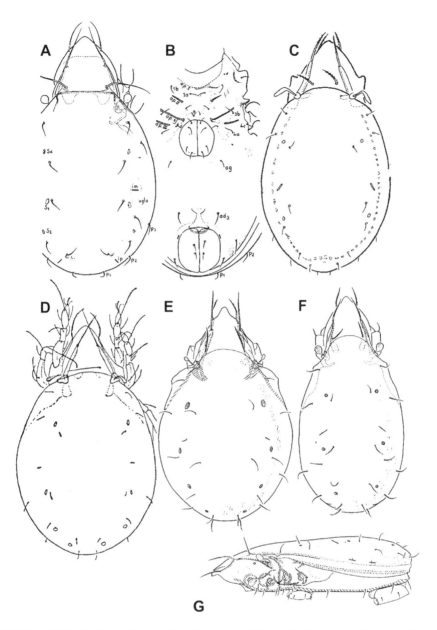

FIGURE 3.126 Scheloribatidae: *Dometorina plantivaga*, (A) dorsal; (B) ventral (after Grandjean 1951a); *Hemileius haydeni*, (C) dorsal (after Higgins and Woolley 1975); *H. quadripilus*, (D) dorsal (after Hammer 1952a); *Liebstadia similis*, (E) dorsal; *L. humerata*, (F) dorsal; (G) partial lateral (after Wunderle et al. 1990).

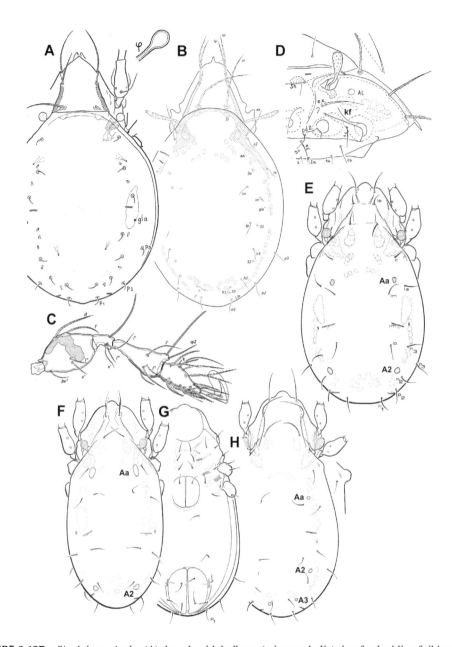

FIGURE 3.127 *Siculobata sicula*, (A) dorsal, with bulbous (microcephalic) tip of solenidia of tibiae III and IV illustrated (after Grandjean 1953); *Paraleius leahae*, (B) dorsal; (C) leg I (after Knee 2017); *P. leontonycha*, (D) podosoma, lateral (after Grandjean 1953); *Parapirnodus coniferinus*, (E) dorsal, female; (F) dorsal, male; (G) ventral female; *P. hexaporosus*, (H) dorsal, with variation in shape of humeral region on right (after Behan-Pelletier et al. 2002).

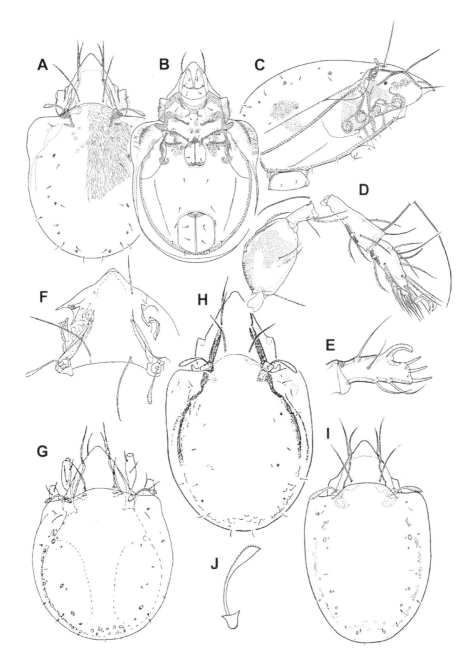

FIGURE 3.128 *Scheloribates laevigatus*, (A) dorsal; (B) ventral; (C) lateral; (D) leg I (after Wunderle et al. 1990); (E) same, palptarsus (after Grandjean 1958b); *S. rotundatus*, (F) prodorsum, (G) dorsal (after Hammer 1955); *S. pallidulus*, (H) dorsal (after Wunderle et al. 1990); *S. latipes*, (I) dorsal (after Bayartogtokh 2000b); *S. lanceoliger*, (J) bothridial seta (after Ewing 1909b).

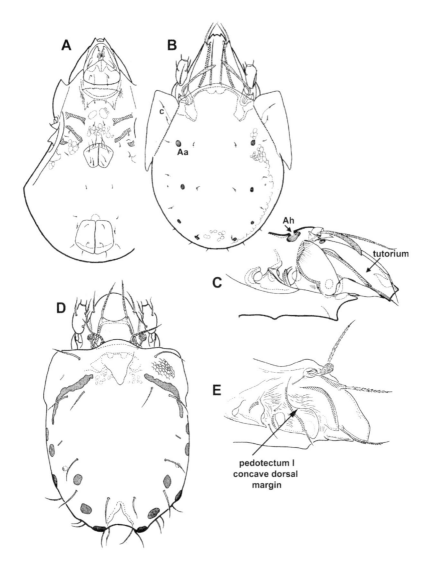

FIGURE 3.129 Ceratokalummidae: *Cultrobates* sp., (A) ventral; (B) dorsal; (C) lateral (VBP); Ceratozetidae: *Jugatala tuberosa*, (D) dorsal; (E) lateral of podosoma (after Behan-Pelletier 2000).

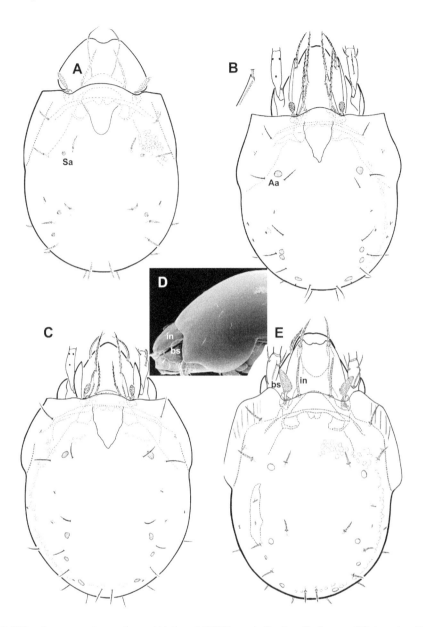

FIGURE 3.130 *Scutozetes lanceolatus*, (A) dorsal (VBP, new); *Svalbardia lucens*, (B) dorsal, with detail of modification of lamella on left; *Iugoribates gracilis*, (C) dorsal (B,C after Behan-Pelletier 1985); *Lepidozetes singularis*, (D) lateral, anterior, SEM; *L. latipilosus*; (E) dorsal (VBP, new).

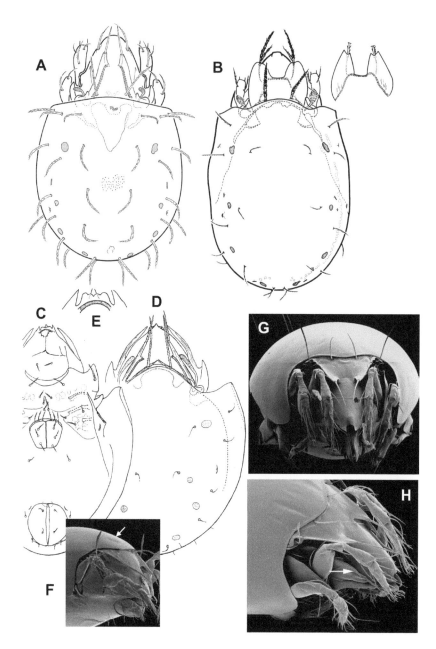

FIGURE 3.131 *Ghilarovizetes longisetosus*, (A) dorsal (after Behan-Pelletier 1985); *Laminizetes fortispinosus*, (B) dorsal, with detail of lamella variation on right (after Behan-Pelletier 1986); *Protozetomimus acutirostris*, (C) ventral; (D) dorsal, (E) detail of rostrum (after Pérez-Íñigo 1990); (F) SEM, lateral of anterior of body, with arrow to raised lenticulus (VBP, new); *Ceratozetoides cisalpinus*, (G) frontal SEM; (H) lateral of prodorsum, with arrow to tutorium (VBP, new).

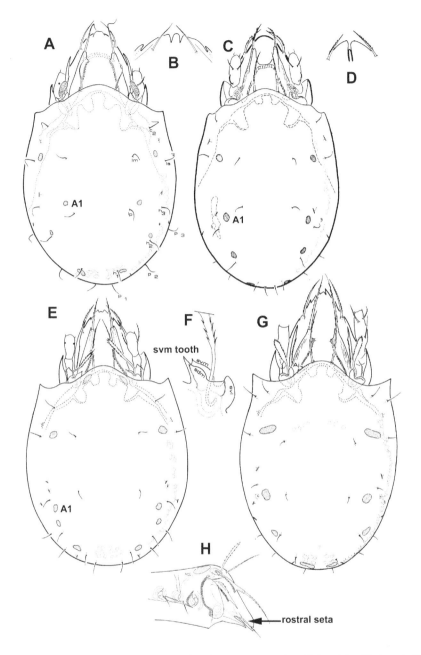

FIGURE 3.132 *Ceratozetes parvulus*, (A) dorsal; (B) rostrum, anterior; *C. kutchin*, (C) dorsal; (D) rostrum, anterior; *C. thienemanni*, (E) dorsal; (F) bothridium and bothridial seta; *C. mediocris*, (G) dorsal; (H) prodorsum, lateral (A,B after Behan-Pelletier 1985; C,D after Behan-Pelletier 1986; E–H after Behan-Pelletier 1984).

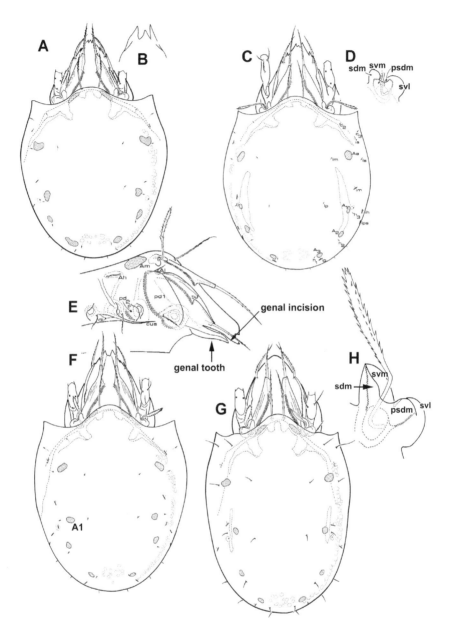

FIGURE 3.133 *Ceratozetes watertonensis*, (A) dorsal; (B) rostrum, anterior; *C. gracilis*, (C) dorsal; (D) bothridium; (E) podosoma, lateral; *C. peritus*, (F) dorsal; *C. oresbios*, (G) dorsal; (H) bothridium (after Behan-Pelletier 1984).

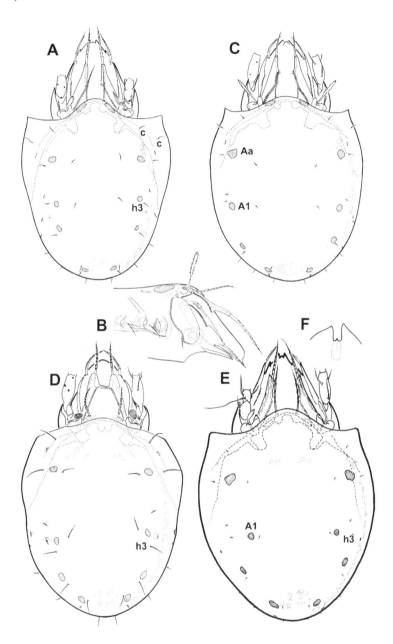

FIGURE 3.134 *Ceratozetes kananaskis*, (A) dorsal; (B) podosoma, lateral; *C. borealis*, (C) dorsal; *C. spitsbergensis*, (D) dorsal; *C. fjellbergi*, (E) dorsal; (F) rostrum, anterior (A–C after Behan-Pelletier 1984; D after Behan-Pelletier 1985; E,F after Behan-Pelletier 1986).

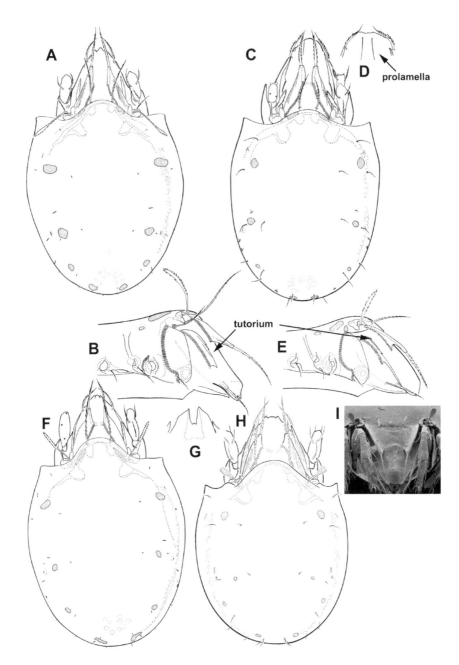

FIGURE 3.135 *Ceratozetes virginicus*, (A) dorsal; (B) podosoma, lateral; *C. cuspidatus*, (C) dorsal; (D) rostrum, anterior; *C. pacificus*, (E) podosoma lateral; (F) dorsal; (G) rostrum, anterior (A–G after Behan-Pelletier 1984); *C. enodis*, (H) dorsal; (I) SEM, frontal (after Behan-Pelletier and Eamer 2009).

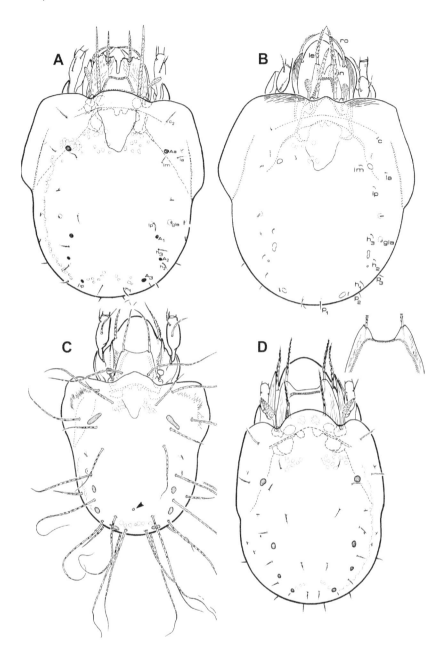

FIGURE 3.136 *Dentizetes rudentiger*, (A) dorsal (after Behan-Pelletier 1986); *D. ledensis*, (B) dorsal; *Neogymnobates marilynae*, (C) dorsal; *N. luteus*, (D) dorsal, with detail showing narrow translamella on right (after Behan-Pelletier 2000).

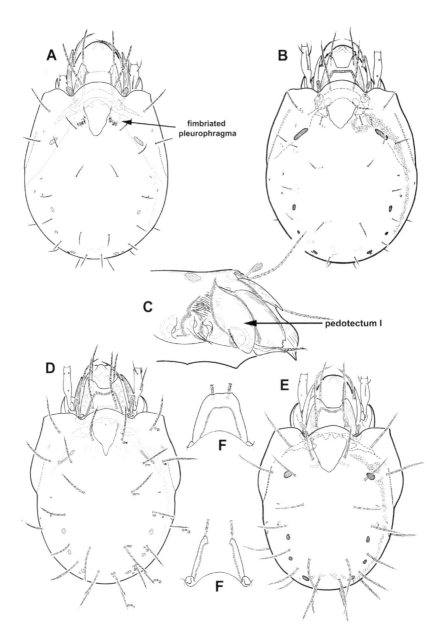

FIGURE 3.137 *Diapterobates variabilis*, (A) dorsal; *D. humeralis*, (B) dorsal; (C) podosoma, lateral; *D. notatus*, (D) dorsal; *D. rotundocuspidatus*, (E) dorsal; (F) variations in lamella (after Behan-Pelletier 1985).

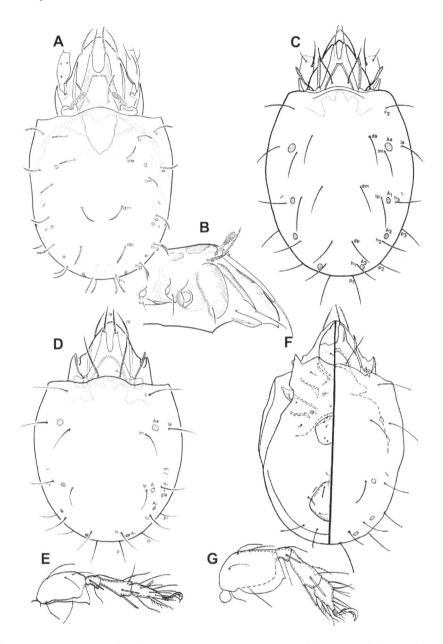

FIGURE 3.138 *Fuscozetes sellnicki*, (A) dorsal; (B) podosoma, lateral (after Behan-Pelletier 1985); *F. setosus*, (C) dorsal (after Seniczak et al. 1991); *F. fuscipes*, (D) dorsal (after Bayartogtokh and Weigmann 2005); (E) leg II; *F. bidentatus*, (F) combined ventro-dorsal; (G) leg II (after Jacot 1935b).

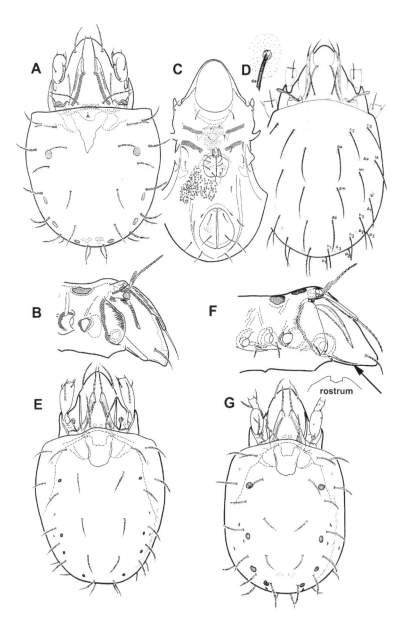

FIGURE 3.139 *Melanozetes crossleyi*, (A) dorsal, (B) prodorsum, lateral; (C) ventral (after Behan-Pelletier 2000); *M. mollicomus*, (D) dorsal, with detail of area around notogastral seta *da* (after Seniczak 1989); *M. tanana*, (E) dorsal; *M. meridianus*, (F) prodorsum, lateral, with arrow to fused genal tooth; (G) dorsal, with detail of rostrum on right (after Behan-Pelletier 1985).

Taxonomic Keys

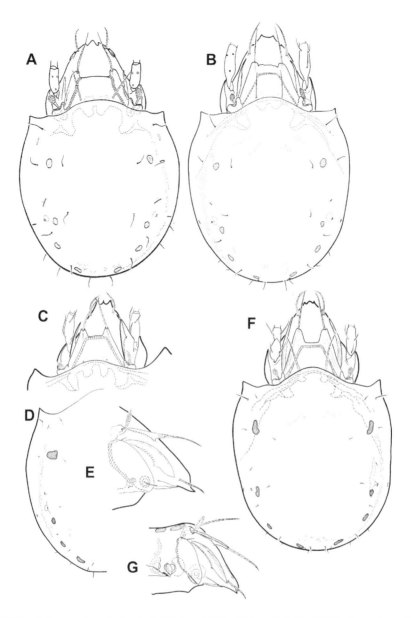

FIGURE 3.140 *Sphaerozetes winchesteri*, (A) dorsal (after Behan-Pelletier 2000); *S. arcticus*, (B) dorsal; *S. castaneus*, (C) prodorsum; (D) partial notogaster; (E) prodorsum, lateral; *S. firthensis*, (F) dorsal; (G) prodorsum, lateral (after Behan-Pelletier 1986).

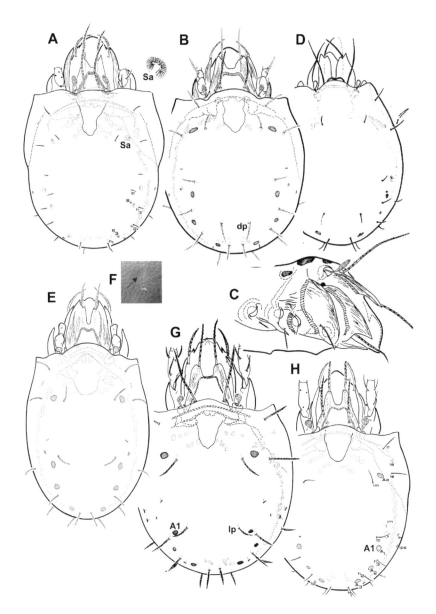

FIGURE 3.141 *Trichoribates polaris*, (A) dorsal, with detail of saccule *Sa* on right; *T. striatus*, (B) dorsal; (C) lateral of podosoma (after Behan-Pelletier 1985); *T. formosus*, (D) dorsal (after Woolley 1958); *T. incisellus*, (E) dorsal; (F) detail of notogaster (after Behan-Pelletier 2000); *T. ogilviensis*, (G) dorsal (after Behan-Pelletier 1986); *T. copperminensis*, (H) dorsal (after Behan-Pelletier 1985).

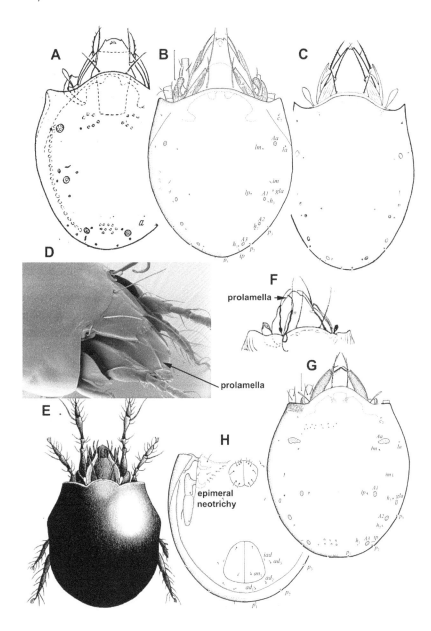

FIGURE 3.142 Chamobatidae: *Chamobates cuspidatus*, (A) dorsal (after Shaldybina 1971); *C. pusillus*, (B) dorsal (after Seniczak et al. 2018); *C. schuetzi*, (C) dorsal (after Seniczak and Solhøy 1988); Humerobatidae: *Humerobates rostrolamellatus*, (D) dorsal, anterior (VBP, new); (E) dorsal (after Michael 1884); *H. arborea*, (F) dorsal, anterior (after Jacot 1931); Euzetidae: *Euzetes globulus* (G) dorsal; (H) partial ventral (after Seniczak and Seniczak 2014).

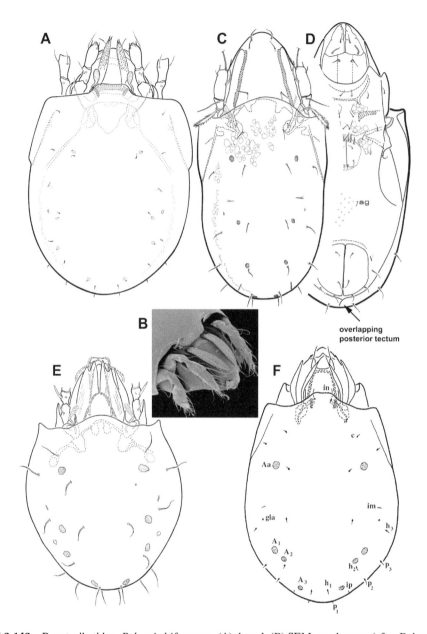

FIGURE 3.143 Punctoribatidae: *Pelopsis bifurcatus*, (A) dorsal; (B) SEM, prodorsum (after Behan-Pelletier and Eamer 2008); *Guatemalozetes danos*, (C) dorsal, (D) ventral (after Behan-Pelletier and Ryabinin 1991); *Adoribatella punctata*, (E) dorsal (after Behan-Pelletier 2013); *Minunthozetes semirufus*, (F) dorsal (after Bayartogtokh et al. 2002).

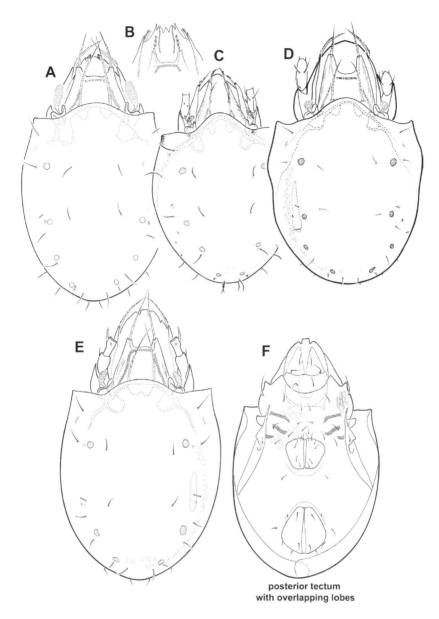

FIGURE 3.144 *Cyrtozetes lindoae*, (A) dorsal; (B) detail of rostrum (after Behan-Pelletier and Eamer 2008); *C. denaliensis*, (C) dorsal; *C. inupiaq*, (D) dorsal (after Behan-Pelletier 1986); *Ceresella reevesi*, (E) dorsal; (F) ventral (after Behan-Pelletier and Eamer 2008).

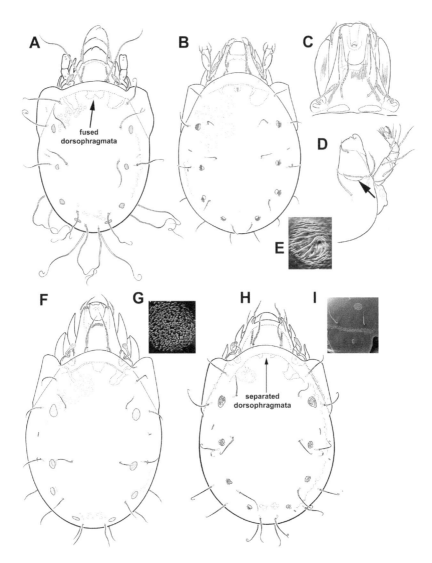

FIGURE 3.145 *Mycobates acuspidatus* (A) dorsal; *M. incurvatus*, (B) dorsal; (C) detail of prodorsum; (D) detail of subcapitulum, with arrow to genal tectum; (E) porose area *Aa*; *M. hammerae*, (F) dorsal; (G) porose area *Aa*; *M. beringianus*, (H) dorsal; (I) porose area *Aa* and pteromorph (A, after Behan-Pelletier et al. 2001; B–I after Behan-Pelletier 1994).

Taxonomic Keys

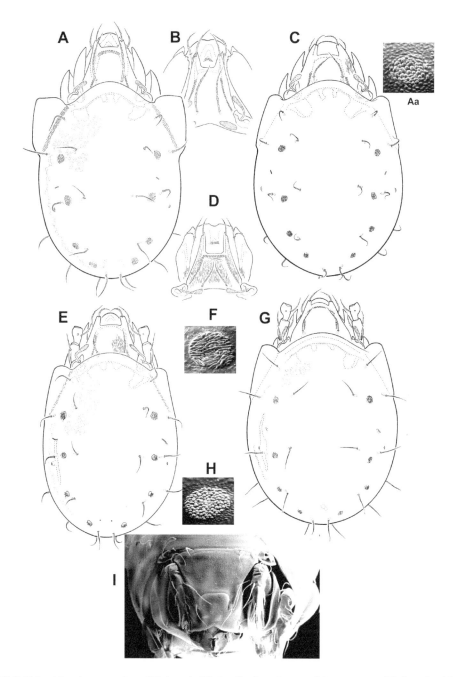

FIGURE 3.146 *Mycobates conitus*, (A) dorsal; (B) detail of prodorsum; *M. punctatus*, (C) dorsal, with SEM of *Aa*; (D) detail of prodorsum; *M. altus*, (E) dorsal; (F) porose area *Aa*; *M. dryas*, (G) dorsal; (H) porose area *Aa*; (I) SEM, prodorsum (after Behan-Pelletier 1994).

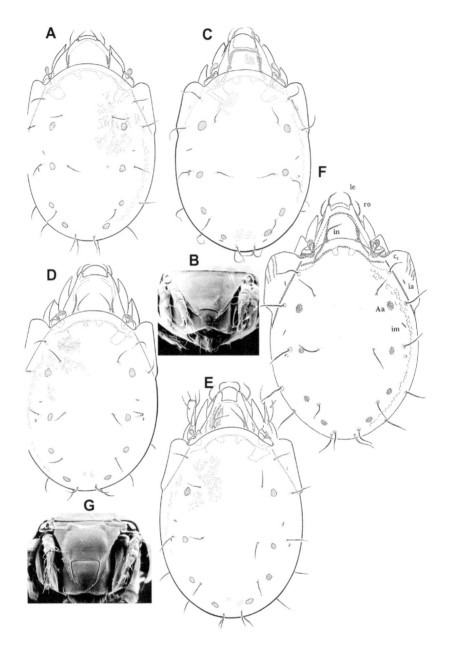

FIGURE 3.147 *Mycobates brevilamellatus*, (A) dorsal; (B) detail of prodorsum, SEM; *M. corticeus*, (C) dorsal; *M. occidentalis*, (D) dorsal; *M. azaleos*, (E) dorsal; *M. sarekensis*, (F) dorsal; (G) detail of prodorsum, SEM (A,D,E,G, after Behan-Pelletier 1994; C, after Behan-Pelletier et al. 2001; F, after Solhøy 1997).

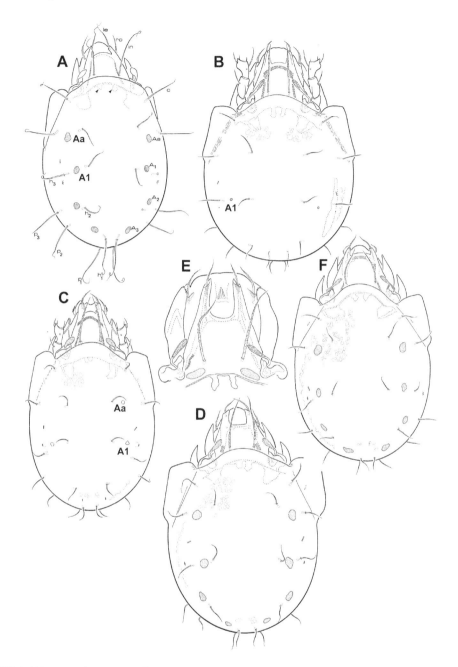

FIGURE 3.148 *Mycobates parmeliae*, (A) dorsal, with arrows to dorsophragmata; *M. hylaeus*, (B) dorsal; *M. exigualis*, (C) dorsal; *M. perates*, (D) dorsal; (E) detail of prodorsum; *M. yukonensis*, (F) dorsal (after Behan-Pelletier 1994).

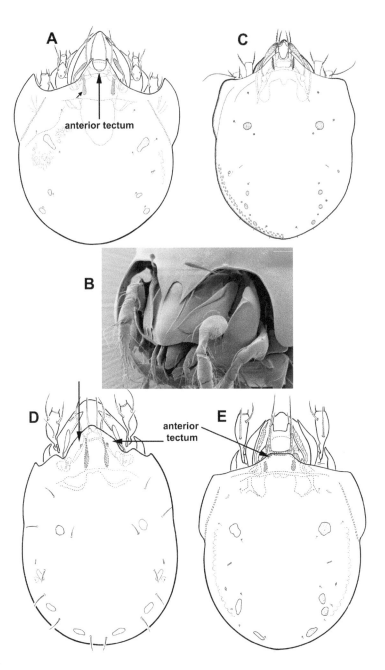

FIGURE 3.149 *Punctoribates palustris*, (A) dorsal, small arrow to thickened band; (B) SEM, prodorsum (after Behan-Pelletier and Eamer 2008); *P. hexagonus*, (C) dorsal (after Seniczak and Seniczak 2008); *P. weigmanni*, (D) dorsal; *P. punctum*, (E) dorsal (after Behan-Pelletier and Eamer 2008).

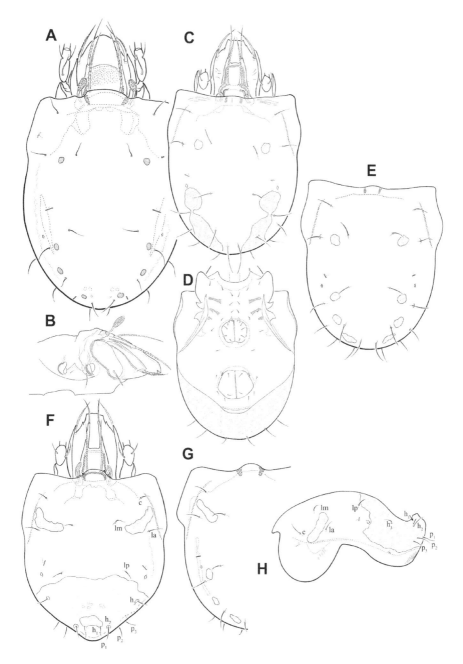

FIGURE 3.150 *Zachvatkinibates epiphytos*, (A) dorsal; (B) podosoma, lateral (after Behan-Pelletier et al. 2001); *Z. nortoni*, (C) male dorsal; (D) male ventral; (E) female notogaster; *Z. schatzi*, (F) male dorsal; (G) female notogaster; (H) lateral of male notogaster (after Behan-Pelletier and Eamer 2005).

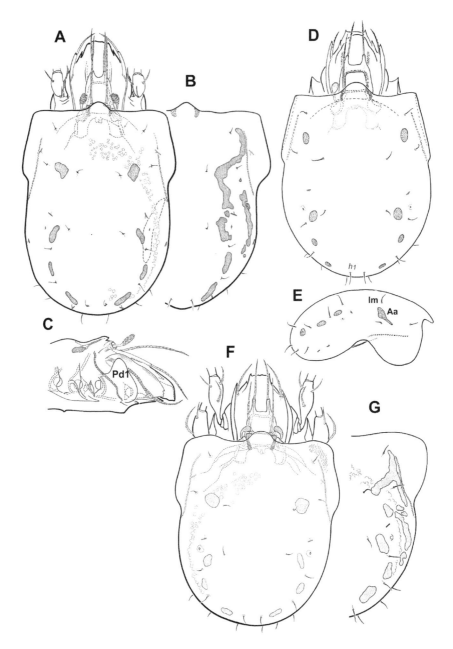

FIGURE 3.151 *Zachvatkinibates maritimus*, (A) female dorsal; (B) male notogaster; (C) podosoma lateral (after Behan-Pelletier 1988); *Z. quadrivertex*, (D) female dorsal; (E) male notogaster, lateral (after Weigmann 2009a); *Z. shaldybinae*, (F) female dorsal; (G) male notogaster (after Behan-Pelletier and Eamer 2005).

Taxonomic Keys

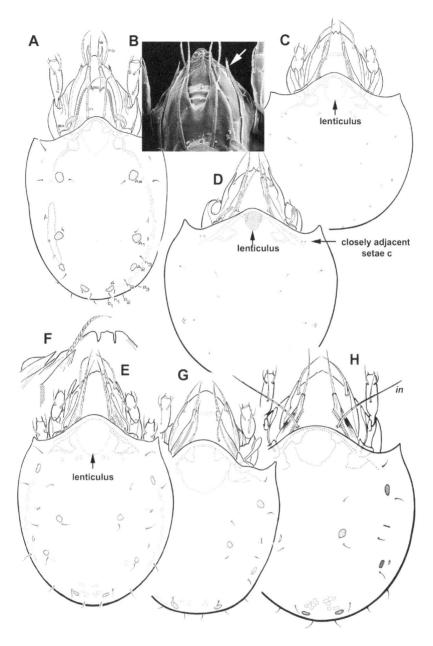

FIGURE 3.152 Zetomimidae: *Naiazetes reevesi*, (A) dorsal of male (after Behan-Pelletier 1996); (B) SEM, prodorsum of female, with arrow to brush-like tutorium (VBP, new); *Heterozetes aquaticus*, (C) dorsal; *H. minnesotensis*, (D) dorsal; *Zetomimus francisi*, (E) dorsal; (F) detail of rostrum; *Z. cooki*, (G) dorsal; *Z. setosus*, (H) dorsal (reproduced with permission from Indira Publishing House, West Bloomfield, MI).

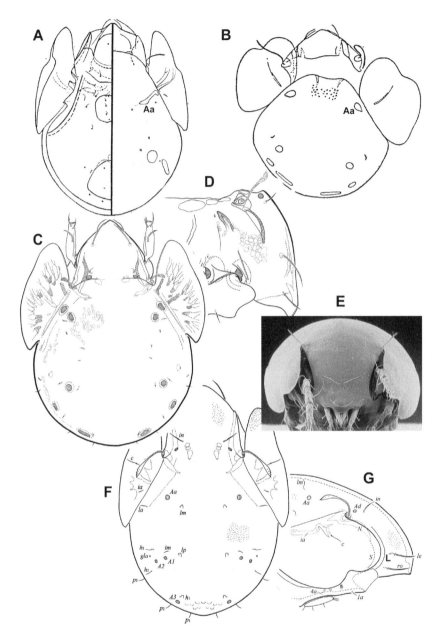

FIGURE 3.153 *Galumna nigra*, (A) combined ventral and dorsal (after Jacot 1935a); *G. hudsoni*, (B) dorsal (after Hammer 1952a); *Pilogalumna binadalares*, (C) dorsal, (D) propodosoma lateral; *Pilogalumna* sp., (E) frontal SEM (VBP, new); *Trichogalumna curva*, (F) dorsal; (G) anterior of body, lateral (after Norton and Ermilov 2017; reproduced with permission).

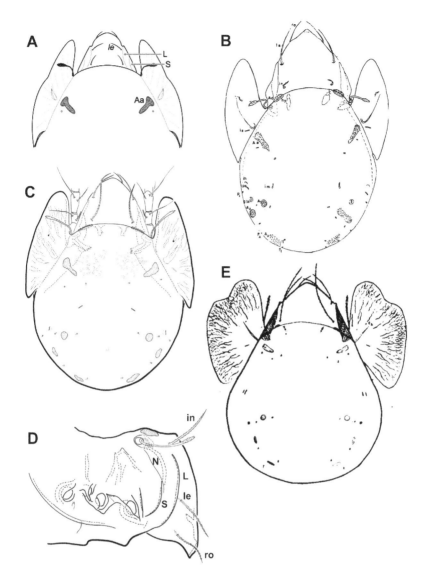

FIGURE 3.154 *Pergalumna? formicaria*, (A) partial dorsal (after Walter et al. 2014); *P. dodsoni*, (B) dorsal (after Nevin 1979); *P. emarginata*, (C) dorsal, (D) podosoma, lateral (VBP, new); *P. nervosa*, (E) dorsal (after Seniczak 1972).

4 Ecology of Oribatid Mites

SUMMARY

This chapter covers the ecology of adult oribatid mites and factors that contribute to their community structure in natural environments. A description of life history characteristics stemming from feeding biology underlies the information in this chapter.

INTRODUCTION

'They are not at all parasitic; walking here and there, albeit very slowly, sometimes in fairly large groups, on stones, trees; it is for this reason that I named these insects oribates, this word meaning errant.'

(Latreille 1803—translation by Mark Judson)

The names we give organisms can sometimes reveal much about their ecology. The common names given to oribatid mites reflects this—Oribates meaning 'wanderers'. Among these wanderers, Moosmilben translates into 'moss mites'; Panzermilben into the armoured mites, and Cryptostigmata refers to the hidden respiratory openings under an armour of sclerotization. Oribatid mites attract our attention because they inhabit almost all terrestrial environments, and often as the dominant, or most abundant arthropod in soils. Much is known about the ecology of oribatid mites as a whole but given their high global diversity of over 11,000 named species, and many more yet to be discovered and described, the ecology of individual oribatid mite species is inconsistently known.

FEEDING BIOLOGY

Oribatid mites' low secondary production, long generation times and high levels of defence are life history traits associated with a mostly saprotrophic and fungivorous feeding behaviour, which is an anomaly among the primarily predaceous arachnids. The feeding behaviour of oribatid mites is fundamental to so much of their biology and ecology (e.g., Norton 1985; Heethoff and Norton 2009) but also supports important soil functions such as decomposition and nutrient cycling.

Feeding by oribatid mites has been described based on gut contents (Behan and Hill 1978; Behan-Pelletier and Hill 1983; Huber et al. 2001), DNA analysis (Remén et al. 2020), enzyme analysis (Siepel and de Ruiter-Dijkman 1993), mouthpart morphology (Kaneko 1988; Perdomo et al. 2012) and stable isotopes (Magilton et al. 2019). The most accurate statement of feeding biology for oribatid mites is 'choosey generalists' (Schneider and Maraun 2005), despite many previous attempts to categorize and delineate oribatid mite feeding. Generally, feeding falls within two broad categories: saprophagy (feeding on dead and decaying organic matter or detritus) and mycophagy/fungivory (fungal feeding), but evidence also supports herbivory (Lehmitz and Maraun 2016) and predation (Norton et al. 1996b; Heidemann et al. 2011), among other feeding groups (Box 4.1).

Luxton (1972), as did Schuster (1956), considered many oribatid mites to have non-specialized feeding or feeding on both macro- and micro-plant and algal sources with Luxton using the term panphytophagous, as well as recognizing incidental feeding upon living animal material (i.e., predation or zoophagy), feeding on dead animal material or carrion (necrophagy) and feeding on faecal material of other animals (coprophagy). Predation as a main feeding strategy is recognized or suspected for some genera. For instance, Rockett (1980) maintained several species in the family Galumnidae, along with *Fuscozetes* sp. and *Nothrus* sp. on a laboratory diet of nematodes, while Heidemann et al. (2011) used molecular gut content analysis in non-choice feeding trials

> **BOX 4.1 FEEDING BIOLOGY TERMINOLOGY FOUND IN THE LITERATURE**
>
> Macrophytophagy: feeding on living as well as dead plant tissue (Schuster 1956) includes:
>
> - xylophagy—feeding on woody tissue (Luxton 1972)
> - phyllophagy—feeding on leaf tissue (Luxton 1972)
>
> Microphytophagy: feeding on microflora (Schuster 1956) includes:
>
> - mycophagy—feeding on fungi (Luxton 1972)
> - bacteriophagy—feeding on bacteria (Luxton 1972)
> - phycophagy—feeding on algae (Luxton 1972)

and field observations to show that several species including *Steganacarus magnus*, *Liacarus subterraneus* and *Platynothrus peltifer* were also nematode feeding. In a similar field study, Heidemann et al. (2014) detected nematode consumption in eight species of oribatid mites including members of the family Damaeidae where 14%–24% of individuals observed tested positive for the presence of nematode prey consumption. Norton et al. (1996b) observed *Aquanothrus* sp. that inhabits ephemeral freshwater pools feeding on bdelloid rotifers. While these studies did not rule out scavenging and necrophagy as part of their observations, the evidence suggests that predation by some species is common.

While most oribatid mites found in woody material are probably feeding on fungi within the substrate, alongside living in the spaces created naturally or by other organisms, some juvenile oribatid mites create tunnels or spaces in woody material through their feeding activities. Endophagy is a particular case of saprotrophic xylophagy. Oribatid mites in the genera *Adoristes*, *Carabodes*, *Epilohmannia* and most ptychoid oribatid mites (e.g., Phthiracaridae) have endophagous immatures that share morphological traits that represent this specialized lifestyle such as large chelicerae, white 'baggish' bodies and reduced bothridial setae. Faecal pellets created by endophages are notable (Hågvar 1998) filling the feeding cavity behind the mite, and evidence of endophagy including faecal pellets within burrowing chambers exist from fossils from the Carboniferous (Labandiera et al. 1997). In a controlled field study of conifer needles, Hågvar (1998) found that greater than one-third of needles were colonized by juvenile mites, most containing more than one individual (typically 2–4) within a single needle, and some needles contained juveniles of more than one species from the genera *Steganacarus*, *Acrotritia* (=*Rhysotritia*) or *Adoristes*. A full description of families associated with endophagy is found in Labandeira et al. (1997).

Feeding on decaying organic matter and fungi has implications for individual feeding (conversion) efficiencies—the efficiency at which food resources are converted into organismal biomass. Saprophages assimilate energy and nutrients from dead plant materials that tend to have low nutrient concentrations (high C:N) leading to inefficient assimilation rates and often the production of large faecal pellets, as seen in many endophages. Wallwork (1983) and Luxton (1979) provide some information on these efficiencies where oribatid mite ingestion rates ranged between 1% and 13% (avg. 10%) of dry body weight per day, and assimilation efficiency (i.e., food ingested but not assimilated and released as waste production) ranged between 10% and 66% (avg. 45%), with assimilation efficiency related to feeding preferences and food quality. Thomas (1979) and Reutimann (1987) also estimated consumption and egestion rates for several species and showed egestion rates were temperature dependent. Ingestion of low-quality food and the corresponding poor assimilation efficiency and high egestion rates are mechanisms that limit oribatid mite secondary production. These low rates of secondary production (both in terms of growth and reproduction) drive other life history characteristics.

LIFE HISTORY TRAITS

Oribatid mites are 'slow' organisms when placed on a spectrum of life history characteristics. This distinction between 'fast' and 'slow' has been used for plants and microbes and is more suitable for oribatid mites than the frequently cited r- or K-selection life history strategies used for other taxonomic groups. The reason is the mechanism by which fast species are typically considered r-select and slow species K-select. In ecological theory, r and K refer to the *rate of intrinsic population growth* and *carry capacity*, respectively, which are considered rate-limiting factors in population dynamics. Species are considered r- or K-select depending on their limits to productivity, with carry capacity being the limit for 'slow' species. However, this K-selection theory does not hold mechanistically for oribatid mites. Indeed, there are few constraints on a theoretical carrying capacity for oribatid mites in the highly organic, food-rich and environmentally stable environment of most soils. Rather, the wide range of correlated life history traits evocative of K-selection arises from physiological constraints related to their feeding biology and conversion efficiencies (see full discussion in Norton 1994) that lead to low productivity. These K-style correlated life history traits of oribatid mites that arise from low productivity (growth and reproduction) include reproductive iteroparity, long life spans and defensive attributes, and result in relatively stable populations.

While year-to-year variation in abundances relates to inter-annual weather conditions such as temperature and precipitation (soil moisture), overlapping generations associated with iteroparity and slow developmental rates lead to generally stable population dynamics. That said, seasonal peaks in abundance are typical but can differ across species. Anderson (1975) found that overall adult oribatid mite abundance peaked in the spring with lower adult abundance in the fall while juveniles had several seasonal peaks in abundance (early spring, late summer and mid-winter). Murphy and Balla (1973) documented overwintering of both adult and immature *Humerobates rostrolamellatus* on fruit trees with overwintering adults active in feeding and movement, gravid in the early spring and a peak number of eggs occurring in early summer. For the cosmopolitan species *Tectocepheus velatus* and *Oppiella nova*, Reeves (1969) found similar results with all juvenile stages of *O. nova* peaking in late summer and early winter, while *T. velatus* had peak abundance of larva and protonymphs in late summer, and peak abundance of deutonymphs and tritonymphs in mid-winter and early spring.

Reproduction and Developmental Rates

Many oribatid mites take a year or more to develop from egg to adult, and some live for several years, which is exceptionable for organisms of their size. Development from egg to adult may be as little as several weeks with adult survival up to 2 years or more, but there is large variation even within families. For instance, within the family Damaeidae, *Belbacorynopus* requires more than 325 days to develop from egg to adult (Luxton 1981a), while *Damaeus clavipes*, *Damaeus auritus* and *Metabelba montana* develop from egg to adult in approx. 65 days (Pauly 1956; Sengbusch 1958; Hartenstein 1962a). Development time is dependent on temperature consistent with metabolic theory (Ermilov et al. 2004). For instance, Ermilov and Łochyńska (2008) found that *Ceratoppia bipilis* egg-to-adult development was reduced by over 20 days when the rearing temp was increased from 17°C to 20°C. Similarly, *Scutovertex rugosus* development from egg to adult was reduced by 36 days with a temperature increase of 5°C (from 18°C to 23°C) (Ermilov et al. 2008).

For the vast majority of species, sperm transfer among sexually reproducing species is indirect with females encountering spermatophores on stalks, which may be surrounded by signalling structures to guide the female to the spermatophore (Oppedisano et al. 1995). However, courtship behaviours have been observed in *Collohmannia*, *Pilogalumna* and other Galumnidae, and *Mochloribatula*. Specifically, the courtship behaviour in *Collohmannia* involves prenuptial feeding as noted by Schuster (1962) for *Collohmannia gigantea* and by Norton and Sidorchuk (2014) for *Collohmannia johnstoni*. Direct sperm transfer has been observed in *Pilogalumna* sp.

(Estrada-Venegas et al. 1996). However, asexual reproduction via thelytoky (unfertilized eggs that develop into females) in oribatid mites is well established with entire families demonstrating or suspected of thelytoky being more common than sexual reproduction (Heethoff et al. 2009). These families include many Enarthronota as well as Parhyposomata, Mixonomata and Desmonomata, but thelytoky is less common in Brachypylina (but see Brandt et al. 2021). In some spanandric species that exhibit a strong female bias (>90%), some males are known, but it is strongly suspected that the males are non-functional. It is estimated that about 10% of known oribatid mite species reproduce asexually based on confirmed cultured populations, or species and populations where only females are known/observed (Heethoff et al. 2009). Based on sequences within regions of 28S rRNA, Maraun et al. (2003a) determined that several parthenogenetic lineages exist historically, radiate slower than sexual species and are likely 'ancient asexuals'.

Female oribatid mites are often larger than males with larger genital plates, but sexual dimorphism in other character states is only well documented in certain Brachypyline groups (see Behan-Pelletier 2015 for review). In cases where sexual dimorphism exists (approx. 60 species across a wide range of superfamilies), dimorphism manifests as modifications in the male octotaxic system, leg setae or prodorsal or hysterosomal setae. The over-development of male octotaxic porous areas is hypothesized to play a role in chemical communication and the production of pheromones associated with aggregation and signalling to mates (Norton and Alberti 1997), which may be advantageous in non-soil environments such as arboreal, intertidal or semi-aquatic habitats, where sexually dimorphic species are often found.

DEFENCE

The slow life history traits of oribatid mites represent a cascade of effects that mimic 'selection' but ultimately are derived from physiological constraints stemming from low-quality feeding, leading to a long life span that requires defence from predators. Oribatid mites live in a dangerous world where macrofauna such as beetles, ants, symphylans, true bugs, centipedes and even frogs and salamanders as well as predatory Acari feed on oribatid mites. To combat these predators, oribatid mites have evolved a broad range of defensive mechanisms that include chemical and physical traits (e.g., sclerotization, ptychoidy, biomineralization, defensive secretions) (Box 4.2).

From a prey perspective, oribatid mites should be attractive prey given that Luxton (1975) reported most oribatid mites contain 3,450–5,250 calories per gram dry weight, although energy density can be affected by rearing temperature (Meehan et al. 2022). At the same time, due to the presence of multiple defensive traits, adult oribatid mites have been suggested to live in 'enemy-free space' (Peschel et al. 2006). Using a 30-day feeding experiment with the predatory mite *Pergamasus septentrionalis* (Mesostigmata), Peschel et al. found that food preference was negatively correlated with prey protection. While the soft-bodied collembolan *Folsomia candida* and juvenile *Pergamasus* were frequently eaten, adult oribatid mites such as species of *Nothrus*, *Liacarus* and *Achipteria* were rarely eaten, and some oribatid mites (*Steganacarus*, *Damaeus*, *Eupelops*) were never preyed upon. In a similar, shorter-term experiment Meehan et al. (2022) found that the predatory mite *Stratiolaelaps scimitus* fed on the well-sclerotized *Oppia nitens* 0% of the time when attacked but would feed on the undefended *Carpoglyphus lactis* (Astigmata) 100% of the time when attacked and had a variable but low catch efficiency for *Folsomia candida* (Collembola).

Sclerotization and Mineralization

Hardness in the cuticle of oribatid mites can be derived from two different processes: sclerotization and mineralization. Sclerotization occurs within the upper layers of the arthropod epidermis, specifically the exocuticle part of the procuticle. Cuticular sclerotization is a common structural form for many arthropods giving physical support, locations for muscle attachments and defence against predators (except when predators swallow you whole!). Increasing levels of sclerotization within oribatid mite evolution leads to key differences that separate the Macropyline and

BOX 4.2 DEFENSIVE TRAITS USED IN ORIBATID MITE IDENTIFICATION

To combat the multitude of predators that oribatid mites face, they have evolved a broad range of defensive mechanisms that are also helpful in learning to identify them. The earliest oribatid mites are the Palaeosomata estimated 388 ±11.41 mya based on 18S rDNA and calibrations from fossils (Schaefer et al. 2010). Radiation of oribatid mite species diversity is evident starting in the early Devonian (Schaefer and Caruso 2019), and fossils are known from the Middle Devonian (~379–376 mya) (Norton et al. 1988a). This radiation was accompanied by changes in body size and form that correspond to many defence traits, as were subsequent radiation events in the Carboniferous and Triassic (Schaefer and Caruso 2019). Pachl et al. (2012) using 18S phylogenetic trees show that many of the derived defensive traits evolved during the Paleozoic, and are therefore ancient, yet are selectively retained or dominant within clusters of families and can be used in identification.

Although this morphological construct does not follow along strong evolutionary lines because some of these traits evolved multiple times, taxonomic keys can easily capitalize on these defence structures that tend to cluster, such as poor vs strong sclerotization, presence or absence of erectile setae, ptychoidy, presence or absence of opisthonotal glands, and increasing levels of tectal development. Broadly thinking about oribatid mites as a taxonomic group and confronted with a new specimen for identification, placing the specimen along this morphological trait axis will help intuit where in the family-level taxonomic key you can expect to end up. Important couplets to take note of are:

- 2(1) Body form ptychoid (or not).
- 13(12) Opisthonotal gland present (or absent).
- 33(1) Notogaster without (or with) octotaxic system and/or pteromorphs.
- 39(38) Prodorsum without (or with) blade-like lamellae.
- 78(33) Pteromorphs moveable, auriculate (ear-like) (or not).
- 88(87) Pedotectum I at most weakly developed (or well developed).
- 98(97) Notogaster with (or without) posterior tectum.

Brachypyline groups, such as the formation of the apodemato-acetabular system of tracheae that aid in oxygen exchange.

Mineralization plays the same structural and defensive role as sclerotization, but mites with cuticular mineralization tend to be lighter in colour. Three different minerals derived from calcium compounds are known to be deposited in the cuticle: calcium carbonate ($CaCO_3$) (e.g., in Phthiracaridae, Euphthiracaridae), calcium oxalate (CaC_2O_4) (e.g., Mesoplophoridae, Protoplophoridae, Eniochthoniidae) and calcium phosphate ($Ca_3(PO_4)_2$) (e.g., Hypochthoniidae, Lohmanniidae). Calcium oxalate is birefringent, deposited in chambers within the epicuticle; calcium phosphate is also deposited in chambers of the epicuticle, but the nature of deposition for calcium carbonate, which can sometimes also be birefringent, is unknown, but likely within the procuticle (Norton and Behan-Pelletier 1991).

Concomitant with sclerotization of the cuticle in general are specific sclerotized structures that protect areas of articulation for the leg segments and other soft tissue exposure sites. Extensions of sclerotization that form a ridge with a free edge or provide an overhanging area are called tecta. Laterally pedotecta (I and II) may protect the insertions of legs I and II. A patronium is another lateral tectiform projection between legs I and II which may coexist with pedotecta. On the prodorsum, if they are large enough, the lamellae (dorsal) and the tutoria (lateral) can provide protection for distal parts of leg I as the mite retracts leg I into the free space between the lamella and dorsal tutorial free edge (see Figure 2.13A,B). In the humeral region, extending laterally pteromorphs may protect all or part of the retracted legs, while some leg segments have retrotecta that provide leg segment articulations protection in either, or both the distal and proximal direction. The posterior margin of the notogaster may also project as a tectum, overhanging the circumgastric scissure (see Figure 2.4B,C).

Defensive Setae

Extant members of the least-derived oribatid mites within the Palaeosomata (e.g., Palaeacaridae, Ctenacaridae) and Enarthronota can be poorly sclerotized but possess some of the first derived defensive structures—erectile setae (e.g., *Atopochthonius* (Atopochthoniidae), *Gozmanyina majesta* (Trichthoniidae)). Other defensive setae are plate-like and fixed (e.g., *Pterochthonius angelus* (Pterochthoniidae)). Erectile setae are associated with individual sclerites where an interplay of muscles and internal pressure depress or erect setae, respectively, as shown in a study of *Nanohystrix hammerae* (Norton and Fuangarworn 2015).

Ptychoidy

Based on both molecular and fossil data, it is believed ptychoidy (from the Greek 'ptych' for a fold or folding) first appeared in the Carboniferous (Subías and Arillo 2002). Most species of Mixonomata are ptychoid (i.e., Phthiracaroidea, Euphthiracaroidea), as are the enarthronotan Protoplophoridae and Mesoplophoridae, giving them the common name 'box mites'. The pytchoid body form is a predatory defence mechanism that involves the release of internal pressure such that the legs and coxisternum retract into the main body and the front apis closes snuggly against the anterior ventral plate (Sanders and Norton 2004). Because this action involves the release of internal pressure (rather than build up), the folding action (enptychosis) can happen very quickly and has been shown to propel *Indotritia* sp. up to 30 mm following the encounter of a predatory stimulus (Wauthy et al. 1998). Ptychoid mites are also often endophagous as juveniles, which may also be considered predator defence behaviour.

Cerotegument and Debris

Cerotegument is a waxy structure derived from the epicuticle that can be diagnostic of certain families alongside other characters. Cerotegument can be absent or form as a thin film that is sometimes birefringent in polarized light (e.g., Malaconothridae). Extensive cerotegument is found in several families as a dense and conspicuous covering (e.g., Gymnodamaeidae, Phenopelopidae) over the whole body including setae, sometimes making it difficult to examine surface features. While the primary function of the cerotegument is to decrease transpiration or aid in respiration through facilitating plastron development (see Iordansky 1996 and references within), extensive cerotegument can act as a protective structure or camouflage when adhered with organic debris (e.g., within the Damaeidae, Crotoniidae, Nothridae). In a similar manner, several families (Plasmobatidae, Neoliodidae, Podopterotegaeidae) carry or retain part of nymphal molts (scalps) on the notogaster, sometimes closely affixed to older nymphal or adult stages, or in a pagoda-like fashion as each successive molt accumulates.

Opisthosomal (Opisthonotal) Glands

Opisthosomal glands are lateral structures also referred to as opisthonotal glands, oil glands or latero-opisthosomal glands (Raspotnig 2010). The opisthosomal glands are distinct among the Parhyposomata where some have funnel-shaped openings (e.g., Parhypochthoniidae). The opisthosomal gland produces secretions that act as a chemical defence system. The main components are terpenes, aromatics and long-chain hydrocarbons ($C11–C21$) (Sakata and Norton 2001, 2003; Raspotnig 2010, and see Raspotnig et al. 2003). These chemicals are effective as predator repellants (Heethoff et al. 2018), as well as antifungal/antimicrobial agents, and serve as alarm pheromones and as aggregation pheromones. However, while these chemical defences can protect oribatid mites from 'large' predators such as beetles, chemical defence was found to be ineffective against gamasid mites (whereas sclerotization was effective) (Brückner et al. 2016); therefore, a combination of defence tactics enable the 'enemy-free space' of oribatid mites. Select predators of oribatid mites (e.g., frogs) are shown to biosynthesize and sequester chemical components of the opisthosomal glands for their own defence (Saporito et al. 2007, 2011, 2012).

PATHOGENS

Frequently, a slide-mounted oribatid mite may display evidence of pathogens, parasites or commensals. Fungi such as Zygomycetes, Actinomycetes and Deuteromycetes are known to act as pathogens as well as commensals (Renker et al. 2005). Many fungal spores have been observed inside body cavities of oribatid mites extracted from soil samples, but it is often unknown whether these spores are pathogens that have killed the mite, or whether they are saprophytic and colonized the body after death. Microsporan fungi (formerly considered protists), which are known parasites/pathogens of other arthropods, can infect the gut walls and fat bodies (e.g., in *Rhysotritia* (=*Acrotritia*)). Similarly, the parasitic green algae *Helicosporidium* (also formerly considered a protist) has been found in the body cavity of *Achipteria* and *Scheloribates* (Purrini 1981; Purrini and Bukva 1984), while the coccidian, *Adelina*, has been found in fat bodies of *Nothrus* and *Damaeus*. Gregarine parasites are also recorded from oribatid mite gut cavity (Nicolet 1855; Norton and Fuangarworn 2015). Other parasites, including protistans and other Acari, have been observed both internally and externally associated with oribatid mites. Norton et al. (1988b) observed a parasitic prostigmatid mite in the family Erythraeidae attached to the notogaster of two species: *Oribatella extensa* and *Damaeus verticillipes*. Oribatid mites are also well known to be intermediate hosts for tapeworms (Ebermann 1976; Denegri 1993; ElMehlawy 2009).

Wolbachia and *Cardinium*, bacterial endosymbionts, are also common in terrestrial arthropods, including oribatid mites. These endosymbionts are known or suspected to act as 'feminizing agents' in other arthropod groups (e.g., isopods, wasps, beetles, butterflies) as they have been shown to induce a female-biased population sex ratio through thelytoky, death of male progeny and/or feminization of genetic males. Spanandric populations—populations with highly skewed female-biased sex ratios but which may include occasional males—are observed in several families within the Enarthronota (e.g., Brachychthoniidae), Parhyposomata, Mixonomata and Nothrina (e.g., Crotoniidae, Malaconothridae, Nanhermanniidae) (Norton et al. 1993). However, while the presence of *Wolbachia* and *Cardinium* has been noted in several species of oribatid mites (Liana and Witaliński 2010; Konecka and Olszanowski 2019a), there does not appear to be a strong correlation with the high proportion of asexual reproduction observed in some families (i.e., where all-female populations are observed). Most oribatid mites, sexual or asexual do not carry either bacterium, and both sexual and asexual species have been found to be infected with *Cardinium* (Konecka and Olszanowski 2019b). *Wolbachia*, on the other hand, has been found only in asexual species within sexual clades, but not in species from large asexual (parthenogenetic) families (Perrot-Minnot and Norton 1997; Konecka and Olszanowski 2015, 2021).

MOVEMENT AND DISPERSAL

The soil environment is a three-dimensional space that is unlike air or water, although it contains both. Oribatid mites move over the surface of substrates or through soil using the air-filled pore spaces between soil particles, organic matter and soil aggregates. The vagility (i.e., the ability of an organism to move freely under their own power) of oribatid mites was first examined by Nicolet (1855) who recorded distances (mm per minute) that select oribatid mites could walk. For *Nothrus horridus*, this was 0.9 mm/min, while *Damaeus auritus* could walk 57 mm/min. Berthet (1964b) used radioactive tagging to track the movement of *Steganacarus magnus* and a few other species through soil. Berthet (1964b) found the range that *S. magnus* travelled was 4–20 cm/day compared to *Nothrus palustris* which only moved 1.7–11 cm/day; however, these were total distances and are not necessarily the net distance of active movement within the three-dimensional soil space. Berthet

(1964b) also found that movement was related to humidity in soil—greater movement was recorded after a rain event when soils were moist versus during dry times.

Another study by Ojala and Huhta (2001) inferred dispersal rates using uninhabited soils that were inoculated with fauna at one end of a 40 cm track to see how long before they were observed at the other end; dispersal distances were then extrapolated to estimate a dispersal potential over 30 years. In general, Ojala and Huhta (2001) found that Brachypyline oribatid mites moved faster (farther) than Macropyline oribatid mites, and movement ranged from only 3–8 cm/week for an estimated colonization rate of 20–120 m over 30 years.

Similar data were obtained by looking at the primary succession of soils following glacier recession. Hågvar et al. (2009) examined soils along a retreating glacier snout in central south Norway, which has been receding since 1750. The authors found several oribatid mite species were present in the 30-year post-recession plots, but many species took much longer (100–200 years) to appear. Grouping species into colonization classes according to how fast they appeared in soil sampled following glacier recession, Hågvar et al. (2009) found that small, parthenogenetic species were often pioneer species with high abundance in young soil (e.g., *Tectocepheus velatus*, *Liochthonius* cf. *sellnicki*). A similar study by Ingimarsdóttir et al. (2012) examined whether dispersal ability or environmental conditions were limiting factors for colonization following glacial recession and concluded that geographic distance (i.e., dispersal limitation) was a major factor explaining colonization rates for oribatid mites. Yet, at the same time, many soil variables were also strongly spatially structured, such that soil environmental conditions formed a gradient associated with time since recession.

Active dispersal is an important mechanism for the colonization of young soils by oribatid mites, but active dispersal occurs mostly over small spatial scales, suggesting that dispersal limitation is an important factor structuring oribatid mite community composition. This may be especially true for oribatid mites when there are inhospitable environments such as water or bare rock, even over short isolation distances (Åström and Bengtsson 2011). That said, it is likely that most species have combined dispersal modes such as active belowground, active aboveground (surface) and passive mechanisms (Lehmitz et al. 2012). Indeed, large dispersal distances are not covered through active movement for small, flightless organisms, such as oribatid mites. However, long-distance *passive* dispersal may be an important source for colonization of new habitats and in the maintenance of community structure.

Passive dispersal involves the movement of organisms by means other than the organism's own, such as an abiotic vector like wind or water, or a biological vector via other organisms, which can be accidental or an adaptive behaviour (phoresy). While phoresy is common in several other groups of mites, it is uncommon in oribatid mites. However, one example is *Mesoplophora* sp., a pytchoid mite that uses enptychosis to attach itself onto pleural hairs of passalid beetles (Norton 1980a). Other examples are *Archegozetes* sp. from Trinidad found attached to the back of harvestman opilionids (Townsend et al. 2008), and *Archegozetes magnus* on frogs (Beaty et al. 2013). Yet more often the dispersive association between oribatid mites and other organisms is 'accidental' such as oribatid mites that have been found associated with birds (Krivolutsky and Lebedeva 2004; Lebedeva and Lebedev 2008; Coulson et al. 2009) and small mammals (Miko and Stanko 1991). In a review of avifauna by Lebedeva (2012) across 146 bird species, she found that >60% of the oribatid mites associated with birds, either in nests or on their plumage, were not previously recorded for the area suggesting passive dispersal can contribute to local species diversity.

Similarly, abiotic dispersal vectors can play an important role in oribatid mite community structure. Evidence supports long-distance dispersal by wind (anemochory) (Karasawa et al. 2005; Lindo and Winchester 2009; Lindo 2010; Lehmitz et al. 2011), especially for oribatid mites in arboreal habitats. Lehmitz et al. (2011) recovered 26 species across 14 families of oribatid mites in sticky traps placed between 0.5 and 160 m above the ground, with abundance significantly correlated with wind speed, wind direction, humidity and air temperature. Lindo (2010) found 57 species of oribatid mites associated with litter in traps placed in the canopy of temperate coniferous rainforest in west coastal Canada. Anemochory dispersal associated with litter may help in survivorship by providing resource availability on route, and a similar argument for 'rafting' has been made for dispersal of oribatid mites on oceanic debris (Lindo 2020).

Oribatid mites are not generally associated with saltwater environments; however, they can survive being fully submerged in both fresh and saltwater for several days (Coulson et al. 2002; Pfingstl 2013b; Bardel and Pfingstl 2018) suggesting that transport in or on water (i.e., rafting) is a possible mode of dispersal to new habitats. Sea surface dispersal without rafting (as pleuston or neuston) may also be a viable mode of transport for oribatid mites as suggested by Peck (1994) for the Galápagos Islands. Schatz (1991) and Pfingstl et al. (2019b) suggested that oceanic currents play a role in oceanic dispersal success. Similarly, Lindo (2020) suggested that the high occurrence of putative species from Japan collected in beach debris along the shores of Haida Gwaii, British Columbia, were possible migrants from stochastic, trans-Pacific rafting events including tsunamis. Similarly, Schuppenhauer et al. (2019) found that oribatid mites were successful in passively dispersing along rivers and streams dependent on species-specific attributes related to submersion survival and floatability. That said, whether a species becomes established following a long-distance dispersal event depends on population size and reproductive mode (see Norton and Palmer 1991) as well as the availability of resources and the soil environment at the arrival site (Schatz 1991).

THE SOIL ENVIRONMENT

The soil habitat, as a mixture of organic and mineral components, is largely influenced by the aboveground vegetation and geologic aspects that dictate the sand, silt and clay components. Most soil organisms live in the top 50 cm of soils, where inputs of organic matter from the aboveground system supply carbon and other nutrients to the soil food web. Fresh inputs of plant-derived litter accumulate at the soil surface as the litter layer (L) and become incorporated into lower organic soil layers through faunal comminution and microbial decomposition creating the fragmentation layer (F). Soil organisms themselves also form this input of organic material, and the extremely well processed and decomposed organic soil horizon called the humus layer (H) is largely composed of carbon derived from dead microbial biomass (necromass) as well as from aboveground vegetation (Cotrufo et al. 2013).

The soil environment is heterogeneous in both horizontal as well as vertical directions and provides microhabitats for oribatid mites on very small spatial scales (Beare et al. 1995). This high habitat heterogeneity allows for many different species to be present in the 'same' sample as different species occupy different vertical distributions in the soil profile (Luxton 1981b) and are associated with different microhabitats at the same vertical depth (Giller 1996). Generally, densities and species richness are greatest in the surface layers of soil, especially in the organic soil horizons. Local-scale oribatid mite species richness and abundance are correlated with the amount of organic material (Maraun and Scheu 2000) as organic and surface mineral soils tend to have more resources for oribatid mites stemming from decaying organic matter such as decaying logs (Déchêne and Buddle 2010), greater soil pore space and high habitat heterogeneity (Edsberg and Hågvar 1999). Soil horizons form a gradient of soil pore size from the top, fluffy litter layer to the sand, silt and clay particles in the mineral soil, and not surprising, different oribatid mite species distribute themselves based on this space resource as well as other gradients of food resources, soil moisture and other factors (Nielsen et al. 2010). The highest densities of oribatid mites (standardized abundances) are recorded in forest floor soils where the organic LFH horizons are well developed (Table 4.1), and a variety of litter inputs increases heterogeneity (Hansen and Coleman 1998).

Vertical stratification of different species relates to different resource availability including soil pore size. Even among the organic soil horizon, congenerics can associate with different LFH layers. For example, *Chamobates cuspidatus* is mostly found in litter while *C. incisus* is almost exclusively found in humus layers (Lebrun 1965). While it seems that many if not most species tend to stay within a certain vertical distribution, there is a seasonal vertical redistribution observed for

TABLE 4.1
Locations from Different Ecosystems around the World Demonstrating Oribatid Mite Species Richness, Density (Standardized Abundance) and Relative Abundance Compared to Other Mite and Arthropod Groups

Location	Habitat	Density (# indiv./m²) × 10³	Dominance (%) Acari	Dominance (%) Arthropod	Species Richness
Sweden	Scots pine forest	425	-	-	52
Canada	Aspen forest	123	78	44	30
Belgium	Chestnut forest	117	79	62	89
Belgium	Poplar plantation	86	70	54	50
Belgium	Moist prairie	61	56	35	54
Denmark	Beech forest	35	51	-	66
Zaire	Dense forest soil	31–37	62	33	94
Zaire	Savanna soil	8–14	49	26	105
Ecuador	Tropical rainforest	5.4–34	-	-	150
UK	Grassland	6.7	59	31	61
Germany	Lawn	2.6–7.1	-	-	4
USA	Juniper litter	0.7	56	44	10

Source: Persson et al. (1980), Luxton (1981), Lebrun (1971), Noti et al. (1996), Davis and Murphy (1961), Weigmann and Stratil (1979), Wallwork (1972), and Maraun et al. (2008).
Sites are ordered with highest abundance to lowest abundance: 425,000 individuals per m² in a Swedish pine forest floor versus 700 individuals per m² in juniper litter.

some species. For instance, Luxton (1981b) found some species (e.g., *Belba corynopus*) are vertically quite stationary and are found at the same vertical soil depth for the duration of developmental stages, while other species (e.g., *Nothrus palustris*) are more prevalent in one horizon than another depending on life stage. In the case of *N. palustris*, juvenile stages were found at 0–3 cm depth in mineral soils, and adults were found mostly in the litter layers.

MICROHABITATS AND NON-SOIL ENVIRONMENTS

Oribatid mite species richness is also directly related to the diversity of microhabitats within and on the soil surface. A classic demonstration of this is the work of Aoki (1967) who surveyed 17 different microhabitats on a forest floor for oribatid mites. These microhabitats included mosses on dead trees and on rocks, fallen twigs, inside rotting wood, fresh fallen leaves and other surface substrates alongside the different layers of the soil organic horizon. Aoki found that the different microhabitats varied in the number of oribatid mite species (from 2 to 21 species) and abundance (9–196 individuals/100 g dwt of substrate), and that for many of the microhabitats identified, Aoki found unique species that were not collected elsewhere. Some microhabitat associations are related to specific feeding activities (e.g., endophages that burrow in woody material such as conifer needles), while the specific association of other oribatid mites and their microhabitats are less clear.

Other organic-rich systems such as moss-dominated habitats (the bryosphere *sensu* Lindo and Gonzalez 2010) also provide a wide resource base for oribatid mites and can be home to high numbers of species. For instance, peatlands—a type of wetland system—that are often dominated by *Sphagnum* mosses contain unique assemblages of oribatid mite species. Barreto and Lindo (2021) list 186 known Oribatida species from Canadian peatlands, but this list is not comprehensive as peatlands are relatively understudied systems. Perhaps not surprising is the abundance and diversity of arboreal oribatid mites in canopy systems that are rich in organic microhabitats like mosses and lichens. Several families are dominated by arboreal species such as Crotoniidae (e.g., *Camisia*), Cymbaeremaeidae (e.g., *Scapheremaeus*), and various members of Ceratozetidae, Punctoribatidae and Scheloribatidae.

Arboreal habitats have been well studied on the west coast of Canada where large coniferous trees provide ample habitat for oribatid mites in the form of lichens, mosses and suspended soils—accumulations of organic matter that form LFH layers on top of large branches or in tree crotches and crevices. Oribatid mite richness and density are positively correlated with size and extent of these suspended soils (Lindo and Winchester 2007b) and the availability of other suitable habitats such as bark crevices. High abundance and species richness are found in many arboreal systems and, for the most part, form highly unique communities of oribatid mites compared to their forest floor counterpart communities (Lindo and Winchester 2006). Specific oribatid mite community composition also depends on the tree species (Winchester et al. 2008), as well as the microhabitat sampled (Lindo and Stevenson 2007).

Lichens also serve as habitat for oribatid mites whether they are on the forest floor or associated with arboreal habitats, and some oribatid mites are lichen specialists (e.g., *Ommatocepheus*, *Carabodes*, *Phauloppia*). Some species burrow within lichen thalli and use lichens as both food and habitat (Prinzing and Wirtz 1997). Feeding on the lichen *Xanthoria parietina* has been shown to transmit viable new lichen propagules from faecal pellets in two species, *Trhypochthonius tectorum* and *Trichoribates trimaculatus* (Meier et al. 2002).

Even without large accumulations of organic debris, oribatid mites are also present on leaf surfaces (the phylloplane) (Walter 1996). Generally, oribatid mites that occupy arboreal habitats tend to have morphological and life history adaptations such as short globular bothridial setae, sexual dimorphism associated with porous areas of the octotaxic system, and/or tarsal pulvilli (see Behan-Pelletier and Walter (2000) for review and insight into the adaptations and life history traits of arboreal oribatid mites, and Behan-Pelletier (2015) for a review of sexual dimorphism). An extreme example of this is *Adhaesozetes polyphyllos* from Australia found on leaves of blueberry ash who has tarsal pulvilli in both immatures and adults, presumably for walking on smooth leaf surfaces (Walter and Behan-Pelletier 1993).

While soil and other organic detrital systems are the primary habitats of oribatid mites, some species are found in aquatic (freshwater) and littoral (marine) habitats. Several species (e.g., *Mucronothrus nasalis*) or families (Hydrozetidae, Limnozetidae) are semi-aquatic in freshwater environments such as streambeds and wetlands such as peatlands (Behan-Pelletier and Bissett 1994). That said, few oribatid mites have 'taken the plunge' (see Behan-Pelletier and Eamer 2007 for full review), and only 15 genera from 11 families from the more derived groups—Nothrina and Brachypylina—are known as aquatic or semi-aquatic. To be considered 'aquatic' the criteria are feeding, movement and reproduction when submerged. Several aquatic species may require water for activity, but can withstand periods of dryness (e.g., *Heterozetes*, *Tegocranellus*), while for *Rostrozetes ovulum*, floodplain populations display evidence of true activity in freshwater, but other populations are considered terrestrial (Pequeño et al. 2018). Other species can tolerate submersion but are not considered aquatic. Activity in freshwater environments often coincides with the presence of a plastron for respiration, an air film held along the mostly lateral surfaces of the body associated with cerotegument, but there are also observations of levitation—keeping air in the guts to rise to the water surface (e.g., *Hydrozetes lemmae*).

Although marine littoral areas do not contain high diversity, there are unique species associated with marine littoral habitats (Schuster 1979). While no species are truly marine (i.e., active in seawater), several marine littoral species are known (e.g., Ameronothridae, Fortuyniidae, Selenoribatidae and *Haloribatula* (Oribatulidae)) that are tolerant of salt water (see Pfingstl (2017) for a review of evolutionary adaptations to the marine littoral lifestyle). Similar to plastron development in freshwater species, Pfingstl and Krisper (2014) found that both adults and immatures of two species of *Carinozetes* from intertidal zones of Bermuda used cerotegument to retain air against the body surface when under water. In Canada, common marine littoral species are members of the genera *Zachvatkinibates* (Punctoribatidae) and *Hermannia* (Hermanniidae).

ABIOTIC FACTORS AND ENVIRONMENTAL GRADIENTS

Responses to abiotic factors play a role in both the presence or absence of a particular species in an environment, as well as relative and absolute abundances, and therefore abiotic factors play a strong role in oribatid mite community composition. Physical or chemical aspects of the environment may include soil pH, temperature, soil moisture, nutrient availability or the presence of other chemical elements such as heavy metals or calcium. Many oribatid mites can tolerate quite acidic soil environments, and many soils with high species richness and abundance such as coniferous forests and peatlands have acidic soils in the range of 4–5 pH. Kaneko and Kofuji (2000) found that the total abundance of oribatid mites decreased with increasing pH (from 3 to 6), but while some species are strong indicators of acidic (e.g., *Hypochthonius rufulus* is abundant at pH 2) or basic soils (e.g., *Pelops occultus* is dominant at pH 8), many species do not sort along pH gradients and are found across a range of soil pH condition (van Straalen and Verhoef 1997).

Moisture is a primary abiotic factor that determines oribatid mite species richness and abundance across a wide range of ecosystems. In general, abundance and species richness of oribatid mites have a right-skewed but unimodal relationship with soil moisture (Siepel 1996), such that most soils show a positive relationship of oribatid mite richness and abundance with increasing soil moisture, except where high moisture conditions saturate soil pore space. As such, while in most terrestrial soils, a reduction in soil moisture has a negative effect on oribatid mite richness and abundance; in wet soil habitats such as peatlands, reductions in soil moisture can enhance soil pore space and increase species richness, but at the expense of semi-aquatic species loss (Barreto et al. 2021). Species that are susceptible to drought tend to have a narrow temperature range tolerance and vice versa. Lethal temperatures for oribatid mites occur around 40°C but are dependent on humidity conditions, exposure times and physiological conditions of the mites (e.g., acclimation, starvation) (Madge 1965).

Many oribatid mites are also responsive to heavy metals and have been used as ecological indicators of contaminated soils (Fajana et al. 2019). Khalil et al. (2009) found that among the 75 oribatid mite species identified in nine metal-polluted soils (e.g., Zn, Cd, Pb, Cu), oribatid mite species could be classified as indifferent (e.g., *Galumna lanceolata*, *Oppiella nova*, *Suctobelbella subtrigona*) and sensitive to heavy metals, specifically Cd (*Platynothrus peltifer*, *Hypochthonius rufulus*), or were tolerant and 'profiting' from the metal-induced disturbance (*Tectocepheus velatus*, *Metabelba* sp). As such, Khalil et al. (2009) note that 'in field surveys oribatids must always be identified to the species level since each species responds differently'.

DISTURBANCE

Disturbance and time since disturbance are major factors dictating the structure of many communities, including oribatid mites. Disturbances can be physical, chemical or biological, and can be natural or anthropogenic. Many studies on the effects of disturbance on oribatid mites come from till/no till agricultural plots where tillage has been shown to decrease soil pore size, disrupt fungal networks, decrease fungal biomass and decrease soil organic matter (Curaqueo et al. 2011). Hülsmann and Wolters (1998) showed that immediately following a spring plow the abundance of oribatid mites, and other Acari was reduced by more than 50%. That said, Adl et al. (2006) found that while abundances of oribatid mites recovered from tillage after just 4–5 years, species richness was low compared to fields that had not been plowed for 25 years. While most species are negatively affected by increasing disturbance or perturbations (Maraun et al. 2003b), other species are 'indicative' of disturbed soils, such as *Tectocepheus velatus*, and tend to occur in either high abundance, or high relative abundance compared to others. Many oribatid mite species that are associated with disturbance are small parthenogenetic species that are considered 'pioneer' species of ecosystem succession.

ECOLOGICAL ROLES IN SOIL

Feeding on decaying plant material, microflora or as general scavengers affects soil processes and biotic interactions. Functionally, oribatid mite feeding contributes to soil structure, nutrient cycling and biotic regulation of microbial communities. Comminution (the breakdown of litter into smaller fragments) of plant material helps transform fresh litter into humus and increase nutrient leaching, while the production of faecal pellets with a large surface area can increase microbial colonization. Mycophagy, on the other hand, can either stimulate or suppress microbial activity and biomass (depending on grazing intensity), disperse microbial propagules and alter microbial community structure, all indirectly affecting decomposition processes. The contribution of oribatid mites to soil processes like decomposition and nutrient cycling is mostly indirect through interactions with primary decomposers (i.e., fungi and bacteria) and other interactions within the soil food web. That said, because oribatid mites are often the most dominant and species-rich microarthropod in many systems, there tends to be a positive relationship of species richness and the contribution of the fauna to mass loss of litter (Heneghan et al. 1999). Wickings and Grandy (2011) found that *Scheloribates moestus* alters litter chemistry and nutrient cycling during decomposition, and enhanced microbial respiration rates by 19% and 17% in corn and oak litter over 62 days, respectively. Yet measurable rates of mass loss due to soil fauna vary greatly (Coleman et al. 2004).

Similarly, the direct contribution to the percent of net nitrogen mineralized is low (Seastedt 1984). de Ruiter et al. (1993) estimated only 0.01% of net nitrogen mineralization was attributed to oribatid mites in arable lands, compared to 15% for nematodes, based on NH_4 production after 6-week incubation experiment. Berg et al. (2001) estimated only 3% of net nitrogen mineralization was attributed to oribatid mites in forest litter, compared to 15% for flagellates in these soils based on nitrogen loss from litterbags. Using an energetic food web model, Holtkamp et al. (2011) attributed only 1.2% of the total food web carbon mineralization and 2.2% of the total food web nitrogen mineralization to oribatid mites, with most of those contributions (87% and 79%, respectively) being indirect through interactions with other trophic groups. In a similar model of boreal peatland food webs, Barreto (2021) estimated slightly lower overall contributions (<1% and ~2%) to C and N mineralization rates, resulting from mostly direct contributions associated with a relatively large biomass, while the indirect contributions of oribatid mites were negligible.

ORIBATID MITES OF HUMAN INTEREST

The vast majority of oribatid mites do not confer any economic benefit or cost and are therefore typically not seen as organisms of human interest. However, some oribatid mites are minor agricultural pests (Haq and Ramani 1985), serve as vectors for tapeworms of livestock and grazing cattle or have been implicated as fungal pathogen vectors including Dutch Elm Disease (Jacot 1934c). Oribatid mites have been observed on aquarium and cultured fishes (Fain and Lambrechts 1987); for example, Olmeda et al. (2011) observed the occurrence of *Trhypochthoniellus longisetus longisetus* on tilapia. However, it is possible that these mites are commensal as they were presumed to be feeding on microflora associated with fish slime.

The potential for oribatid mites being agents of biological pest control is also limited, both in efficacy and scalability, but a few examples exist. *Scheloribates azumaensis* was found to control radish root rot by feeding on the pest fungi, *Rhizoctonia* (Enami and Nakamura 1996). *Dometorina predatoria*, a predatory oribatid mite, was found to feed on gall mites (Lan et al. 1986), while *Diapterobates humeralis* was reported to control the hemlock woody adelgid (McClure 1995). Some work has been done also using *Orthogalumna terebrantis* to control the invasive South American water hyacinth (Singh et al. 1989; Haq and Sumangala 2003).

More readily, oribatid mites have been used as bioindicators of soil disturbance, compaction and contamination (Gergócs and Hufnagel 2009). Given their high relative abundance and species richness, along with ease of sampling, soil mesofauna assessment is common in examining the effects of forest harvest (Abbot et al. 1980; Battigelli et al. 2004; Lindo and Visser 2004; Berch et al. 2007), tillage (Franchini and Rockett 1996; Hülsmann and Wolters 1998; Behan-Pelletier 1999b; Bosch-Serra et al. 2014) or crop rotations (Osler et al. 2008), and may have potential for biochar application, the reclamation of mining sites (Davis and Murphy 1961; Skubała 1997; St. John et al. 2002) and a variety of soil additives and contaminants (Stamou and Argyropolou 1995; Manu et al. 2019). For the most part, observations of impact can be observed at coarse levels, such as in total abundances, declines in species richness and shifts in community composition even when species are morphotyped allowing relative non-experts to use oribatid mites as bioindicators of soil disturbance and recovery.

Finally, oribatid mites can be reared and cultured for scientific studies as model organisms. *Archegozetes longisetosus* (Trhypochthoniidae) is a sub/tropical oribatid mite that is fast becoming the *Drosophila* or white mouse of the oribatid mites. *Archegozetes longisetosus* has been used as a model to study the re-evolution of asexuality, toxicology and chemical ecology, functional anatomy, embryology and genetic regulation of development alongside its own general biology (see review in Heethoff et al. 2013). Other cultured species are less common, but *Oppia nitens* is being developed as a model invertebrate for ecotoxicology, specifically for heavy metal contaminated soils, and pesticides (e.g., neonicotinoids) (Fajana et al. 2019). However, the utility of oribatids for ecotoxicological purposes has not been fully realized (Lebrun and van Straalen 1995).

Given that oribatid mites are highly abundant and species rich in most organic soils, one growing application for oribatid mite communities is as model assemblages for understanding biodiversity. Laboratory-based mesocosms or field-based manipulations can be performed with relatively small footprints and short time, and the life history characteristics of oribatid mites make them relatively sensitive to disturbance and structural degradation such as habitat loss and perturbation in microecosystems (Åström and Pärt 2013; Shackelford et al. 2018), climate warming (Lindo 2015a) and understanding biodiversity-ecosystem function relationships (Staddon et al. 2010).

5 Oribatid Diversity across the Northern North American Landscape

SUMMARY

Summary of research on oribatid diversity in the Ecozones of Canada and Alaska. Overview of climate and soil characteristics of Ecozones, and how the extant oribatid fauna is influenced by paleoecology. Focus on taxa unique to an Ecozone.

INTRODUCTION

Geography, geology and climate all play a role in present-day distribution of oribatid mites in Canada and Alaska. They explain how and why species are assembled in a given region, and why endemic taxa are clustered in particular areas. General global patterns of extant oribatid diversity and distribution were studied by Maraun et al. (2007), but the only studies for North America are those on ptyctimous mites (Niedbała 2002, 2007b; Niedbała and Liu 2018), Carabodidae (Reeves 1998) and Ceratozetoidea (Behan-Pelletier and Schatz 2010). Within northern North America, however, there have been reviews of the oribatid faunal diversity of Yukon (Behan-Pelletier 1997b), British Columbia (Lindo and Clayton 2017), Alberta (Walter and Lumley 2021; Walter et al. 2014), Mixedwood Plains Ecozone (Smith et al. 1997), Montane Cordillera Ecozone (Smith et al. 2011), Atlantic Maritime Ecozone (Behan-Pelletier 2010) and Canadian grasslands (Behan-Pelletier and Kanashiro 2010).

Based on the Köppen-Geiger Climate Classification for North America, Canada and Alaska are Cold or Polar with incursions of Cold Steppe into southern Alberta and Saskatchewan, and with temperate conditions on the west coast of Canada extending into southern Alaska (Figure 5.1A). Summers are warm and moist across most of the southern part of the region, but warm and dry in much of western Canada, extending into Alaska. However, the fossil record indicates a more temperate climate in the distant past. The fossil record for oribatid mites in northern North America depends both on the hardened exoskeletons of adults (body fossils) and amber inclusions. Amber inclusions are most evident in the late Cretaceous and Tertiary, and Canadian amber was deposited in beds of coal and shale saturated with organic matter ca. 80–72 million years ago in present-day Alberta and Saskatchewan. Among numerous oribatid inclusions collected so far, only *Megeremaeus cretaceous* has been described, though species of *Megeremaeus* are unknown from Alberta today (Sidorchuk and Behan-Pelletier 2017).

Plant and arthropod fossils, including oribatid mites, show that 3 million years ago coniferous forests covered most of central and northern Yukon and Alaska, and that the climate was less continental than at present (Matthews and Tekla 1997). Oribatid fossils dated between 3 and 5 million years ago are known from Ellesmere Island, one of which is *Proteremaeus macleani*, a species previously known only from Asia (Behan-Pelletier and Ryabinin 1991a), in a genus unknown from North America today. Similarly-aged fossils from Meigen Island in the Canadian Arctic Archipelago include boreal species such as *Cepheus corae* (Barendregt et al. 2021). Clearly, present-day distributions are not congruent with more ancient assemblages.

Since the last glacial maximum, approximately 21,400 years ago, Canada and Alaska are being recolonized primarily from the unglaciated south (Matthews 1979). During the last glacial maximum,

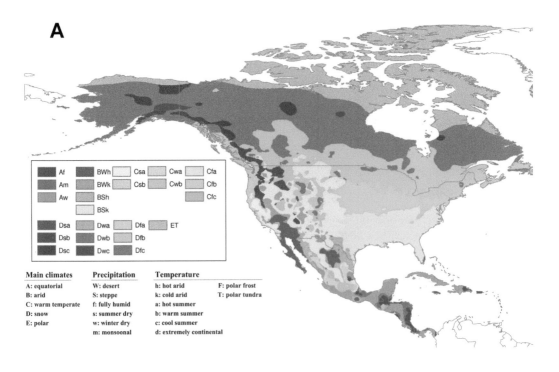

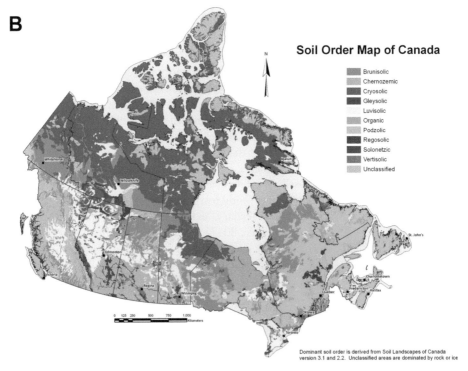

FIGURE 5.1 (A) Köppen-Geiger Climate map for North America (after Peel et al. 2007); (B) Soil order map for Canada (after Soil Landscapes of Canada).

much of Canada was covered by the Laurentide, Innuitian and Cordilleran ice sheets, except for the small glacial refugia postulated for offshore islands, and in the Cypress Hills of Alberta and Saskatchewan (Langor 2019). The exception to this glacial cover is extensive areas of Yukon and Alaska, which were unglaciated and formed eastern Beringia during the Pleistocene. Beringia is defined as encompassing the land areas and the presently submerged Bering and Chukchi Sea shelves between the Mackenzie River in the east and the Lena River in the west (Matthews et al. 2019). At times, Beringia formed an unbroken corridor for exchange of Palearctic and Nearctic flora and fauna (Matthews and Telka 1997). In Yukon, most of the Klondike, Porcupine and Arctic plateaus, the Porcupine Basin and portions of the Ogilvie, Wernecke, British Mountains and the western slopes of the Richardson Mountains were unglaciated; refugial areas that have impacted present patterns in the fauna and flora of Canada (Scudder 1997). Most of central Alaska also was unglaciated (Edwards et al. 2018). Because the continental-sized ice sheets of the Wisconsinan glaciation resulted in global sea levels about 150 m lower than during the last glacial maximum, the Chukchi and Bering marine shelves were exposed, and biotic assemblages differed substantially from today. There is fossil evidence for woolly mammoth, horse, bison, caribou, muskox, mountain sheep and saiga antelope, dated 25,000–10,000 from northern Yukon (Schweger 1997). Arthropod macrofossils indicate that during the Miocene and probably part of the Pliocene, Arctic Canada was the northeastern terminus of a forested link between Asia and eastern North America (Matthews et al. 2019). Although most of the oribatid fauna emigrated from the south under warming post-glacial conditions, the influence of Beringia underlies the number of taxa with a western Nearctic-eastern Palaearctic distribution (Behan-Pelletier and Schatz 2010; Behan-Pelletier and Lindo 2019).

Glacial history, recovery post glaciation and Beringia are the backdrop to this review of the present-day habitats and distribution of oribatid mites across the Canadian and Alaskan landscape. We recognize that the physical size of Canada and Alaska means that, as Langor (2019) noted, the vast majority of survey effort has been done in the south of the country and around major population centres and along major roads further north. The exception to this is the Alberta Biodiversity Monitoring Institute (ABMI) which has systematically surveyed Oribatida across the province on a 20 km × 20 km plot grid, using helicopter access, resulting in 400 soil samples each year since 2009 (Walter et al. 2014; Walter and Lumley 2021).

We use the Canadian and Alaskan Ecozones and Ecoregions as frameworks for this section, following the usage of Langor (2019) and the Ecological Framework for Canada (Figure 5.2A), and Murphy et al. (2010) for Alaska (Figure 5.2B). However, most research on oribatid mites has been habitat based rather than Ecozone based, thus aquatic habitats are treated separately. In Alaska, most knowledge is for low-lying boreal habitats around Fairbanks, and from Fairbanks north to the Arctic Ocean, the Arctic Tundra and Boreal Ecoregions (Figure 5.2B). Behavioural extraction of soil and litter was the usual method of collecting mites in almost all studies, and thus arboreal, aquatic and deep soil species have been overlooked, except where particularly noted (e.g., publications by ZL).

Of the 99 currently documented oribatid families in Canada, 44 families are represented in most or all Ecozones including the Arctic, 47 families are restricted to boreal and southward and the remaining 8 families are known from only a few Ecozones (Beaulieu et al. 2019). These 99 families represent almost 60% of the known world families. Molecular data (Barcode Index Number (BINs)) suggest that a large portion of the Canadian oribatid diversity remains undescribed or at least unrecorded, with a total of 2,429 BINs recorded across 67 families surveyed so far (Beaulieu et al. 2019).

NORTHERN ECOSYSTEMS: TAIGA AND ARCTIC ECOZONES

The Arctic and Taiga (subarctic), the cold ecosystems of the northern hemisphere, occupy about 10% of the earth's total land surface. The term Arctic refers to regions beyond the northern limit of trees, while Taiga comprises an area of dwarf birch or coniferous trees, or strings of trees scattered in tundra. Oribatid mites are strongly influenced by Beringia in the Arctic and Taiga west of the

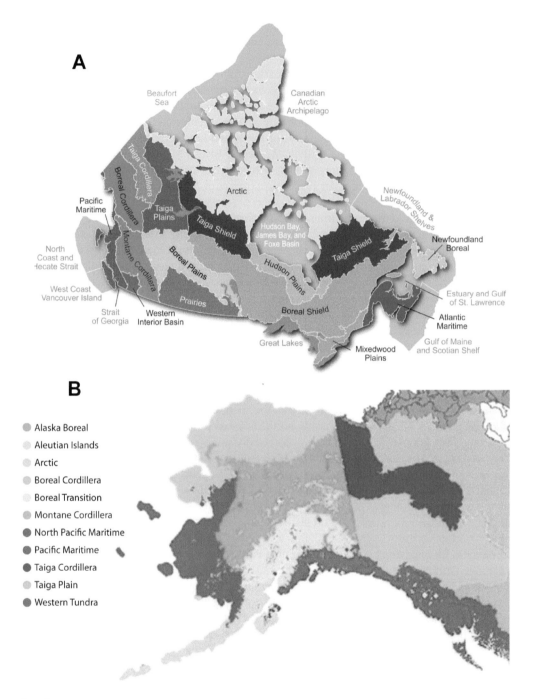

FIGURE 5.2 (A) Terrestrial Ecozones of Canada (Reprinted with permission from Environment and Climate Change Canada), (B) Alaska biomes and western Canadian Ecozones (after Murphy et al. 2010).

Mackenzie Delta. More Holarctic and Palaearctic influences are evident in the Arctic and Taiga east of the Mackenzie Delta and in the islands of the Canadian Arctic Archipelago. We use the following conventions: Taiga encompasses the Taiga Plains, Taiga Cordillera and Taiga Shield Ecozones; the Southern Arctic Ecozone includes tundra of mainland North America. This is divided into the Mackenzie Delta westward (Western Southern Arctic), and those from the Mackenzie Delta eastward (Eastern Southern Arctic), reflecting the different impact of Pleistocene glaciations on these regions.

The oribatid fauna is progressively less diverse as the climate becomes increasingly severe along a gradient from temperate regions to the Northern Arctic (Thomas and MacLean 1988). Whereas the fauna of temperate regions is found throughout the soil profile, in surface litter, on low growing herbs and shrubs and on the trunks, branches, foliage and fungal associates of trees, at northern latitudes Oribatida are primarily associated with the soil surface, litter and moss cover; deep soil and arboreal species are few or absent. Much of this reflects the presence of permafrost or permanently frozen subsoil; these cryosols are a major influence on soil biota (Figure 5.1B). The active, upper soil layer above the permafrost normally thaws annually (1–4 months), while the permafrost layer modifies the hydrology and leaching of nutrients causing solifluction and water-logging, and limits the vertical habitat available for soil biota; thus, soil burrowing species are absent. Mixonomata, whose members are primarily associated with decaying wood, are poorly represented in Taiga and Southern Arctic, and absent from the Northern Arctic, as are Palaeosomata and Parhyposomata, species of which are often found in deeper soil and litter. Among Brachypylina, superfamilies well represented in temperate Canada and Alaska, such as Hermannielloidea, Plateremaeoidea, Carabodoidea, Cymbaeremaeoidea and Oribatelloidea, have a few or no species in the north. There is a striking reduction in species diversity in the Oripodoidea, a superfamily that includes many arboreal species, from about 100 species in temperate regions, to 16 species in Taiga, to 1 species in the Northern Arctic (Behan-Pelletier 1997b). There are extensive areas of tundra analogues in the mountains of Canada and Alaska and many similarities between arctic and alpine communities resulting from the dominant influence of low temperature on both.

Taiga (subarctic) and Arctic peatlands, underlain by permafrost, are poorly studied in comparison with those of the Boreal (see below). Melting of permafrost in these peatlands presents a challenge with climate warming because of the vast stores of carbon sequestered in these layers (Markkula 2020). The oribatid species *Carabodes labyrinthicus* and *Neoribates aurantiacus* are considered useful bioindicators of permafrost changes in these northern peatlands (Markkula and Kuhry 2020).

The Northern Arctic Ecozone includes the islands of the Canadian Arctic Archipelago, habitats of which are dominated by rock desert, stony lichen-heath communities and pockets of tundra meadow. Soils in this Ecozone are mainly crusts with little depth of organic matter (0–2 cm). Soils can be subject to alternating freeze/thaw cycles, often daily in summer, leading to frost heaves which contribute to the patterned polygon formations of this Ecozone. Winter temperatures can reach −70°C or below, and there are wide daily and seasonal variations. However, under insulating winter snows, temperatures are ameliorated, and organisms can be exposed for long periods to temperatures of about 0°C (Danks 1981). The warmest microclimates in summer are found in the shallow air layer above the soil surface. Summer microclimates depend chiefly on the absorption of solar radiation by the soil surface, thus south-facing slopes may be up to 13°C warmer than north-facing slopes (Danks 1981).

None of the oribatid species found in the islands of the Arctic Archipelago is endemic; all have a Holarctic distribution (Behan-Pelletier 1997b). Similarly, the fauna of the Eastern Arctic and Subarctic shows substantial overlap with that of similar habitats in the Western Palaearctic (Hodkinson et al. 2013; Makarova and Behan-Pelletier 2015). By contrast, 7 species are known only from dry tundra of the Western Southern Arctic (Behan-Pelletier 1997b), implying that the fauna of these cold northern ecosystems has a Beringian provenance (Behan-Pelletier and Schatz 2010).

HUDSON PLAINS ECOZONE

This Ecozone stretches from Manitoba to Québec along the southern edge of Hudson Bay (Figure 5.2A). It is an Ecozone of organic soils with poor drainage (Figure 5.1B) and is the largest continuous wetland in the world. Hudson Bay moderates the temperature in summer and winter, such that the average summer temperature is 11°C, and in winter the average temperature is −18°C. In the north of this Ecozone the terrain is treeless tundra; in the more southern Taiga trees appear, getting thicker further south. Because of poor drainage, trees are only found at drier, higher elevations. This results in belts of trees following ridges, with characteristic tree species being Black spruce, White spruce, tamarack, Balsam poplar, Dwarf birch, Paper birch, willow and Trembling aspen.

Young et al. (2012) used a microhabitat approach to access the acarine biodiversity around Churchill in the low arctic and subarctic of this Ecozone, using cytochrome oxidase 1 (CO1). They sampled from boreal forest, bog/fen tundra, marine beach and rock bluff habitats, and found much greater acarine richness than recorded in past, morphological studies from the arctic (e.g., Behan 1978). The number of oribatid BINs at this Hudson Plains site is greater than the number of species recorded from any other Canadian or Alaskan Ecozone. This suggests that a molecular diversity approach could highlight cryptic species, as has been shown in the Palaearctic by Schäffer et al. (2010) and Schäffer and Koblmüller (2020).

BOREAL ECOZONES: BOREAL SHIELD, BOREAL PLAINS, BOREAL CORDILLERA AND NEWFOUNDLAND BOREAL ECOZONES

The Canadian Shield rock forms the nucleus of the North American continent. Most Shield rocks were formed well over a billion years ago, during the Precambrian era, and what once may have been a towering mountain chain is today a massive rolling plain of ancient bedrock. Other geological structures assumed positions around or on top of the Shield millions of years after it was formed. During the last glacial period that ended 10,000 years ago, the advance of glaciers repeatedly scoured the Shield, carving striations in the bedrock and carrying large boulders many kilometers. In retreat, glaciers blanketed much of the landscape with gravel, sand and other glacial deposits. The many poorly drained depressions left behind, as well as natural faults in the bedrock, now form the millions of lakes, ponds and wetlands that give these Ecozones their distinctive character. The Boreal Plains Ecozone is a northerly extension of the Prairie Ecozone (Figure 5.2A). Glaciers have flattened the landscape, and the large ancient lakes formed from meltwater have left many dunes. The climate of Boreal Ecozones is generally continental with long cold winters and short warm summers. The average midwinter temperature is −15°C, while in midsummer it hovers around 17°C. Soils are relatively deep as cold and coniferous litter result in reduced decomposition, and with almost no earthworms and myriapods, there is little mixing of soil layers.

Data on the oribatid mites of the Boreal Ecozones and Ecoregions is from the riparian boreal habitats around Fairbanks (Andrews and Ruess 2020), western boreal forest of White spruce and Trembling aspen (Lindo and Visser 2004; Walter et al. 2014; McAdams et al. 2018; Meehan et al. 2019) and the eastern Boreal: from deciduous-coniferous forests at Chalk River, Ontario (Bird and Chatarpaul 1986, 1988); from mine tailing restoration sites around Sudbury, Ontario (St. John et al. 2002); from a boreal peatland near White River, northern Ontario (Barreto and Lindo 2018); from Black spruce and Jack pine forest of Québec (Behan et al. 1978; Doblas-Miranda et al. 2015) and Ontario (Rousseau et al. 2018a,b); White spruce and Trembling aspen forest of the Abitibi Region of Québec (Déchêne and Buddle 2010); and from Balsam fir forest in Newfoundland (Dwyer et al. 1997, 1998). Densities of Oribatida include some of the highest recorded in Canadian Ecozones (Dwyer et al. 1998).

Boreal Ecozones have similar oribatid taxa:

- Higher diversity of Hypochthoniidae, Eniochthoniidae, Brachychthoniidae, Nothridae, Crotoniidae, Malaconothridae and Trhypochthoniidae compared with northern Ecozones;
- Presence, though with low diversity, of wood associates—the ptychoid Oribotritiidae, Euphthiracaridae and Phthiracaridae, and Cepheidae; families absent from arctic Ecozones;
- Higher diversity of Oripodoidea, which has low diversity in Arctic soils.

Peatlands are a defining feature of the Boreal and subarctic landscape, and because of their carbon storage, their importance for mitigation of climate change was recognized by the United Nations

(Biancalani and Avagyan 2014). Peatlands cover about 13% of the landscape in Canada (Tarnocai et al. 2011), and about 30% of the Alaskan landscape. They have particular characteristics that present a challenge for the biota: waterlogged substrates, low levels of nutrients and wide fluctuations in temperature and humidity at the bog surface. *Sphagnum* moss can alter the acidity of the surrounding water, and because of its low thermal conductivity, peatlands have a shorter frost-free period than surrounding habitats. The relative humidity at the surface of a sphagnum mat may drop to 40% during the day, whereas in the air spaces in the peat and among the living sphagnum stalks it is generally 100%.

In peatlands, oribatid mites dominate the microarthropod fauna in terms of abundance and species richness, except in the open water of bog pools. Data for peatlands are derived primarily from fen habitats in Ontario (Barreto and Lindo 2018; Barreto et al. 2021) and Québec, the Wagner peatland in Alberta (Finnamore 1994) and bog pools on the Avalon Peninsula, Newfoundland and near Lake George, New Brunswick, and the summary work of Barreto and Lindo (2021). Primarily mesophilic species, i.e., species occupying soil subject to great variations in temperature and humidity close to 100% have been collected from Canadian peatlands. Many of these are widely distributed throughout temperate, boreal and subarctic regions of North America, and have primarily Holarctic or worldwide distributions (Behan-Pelletier and Lindo 2019). By contrast, many aquatic species are restricted in distribution to the Nearctic. A comparison of the fauna of peatlands with the total oribatid fauna of Canada and Alaska shows that many families are absent or poorly represented:

- There are no representatives of Gymnodamaeidae or Eremaeidae, families that are usually well represented in dry habitats.
- Species-rich families such as Damaeidae, Peloppiidae, Liacaridae, Oppiidae, Oribatulidae, Scheloribatidae, Ceratozetidae and Punctoribatidae are poorly represented. However, based on ceratozetoid diversity in *Sphagnum* mires in Norway (Seniczak et al. 2019a), this possibly represents limited collecting.
- Three species of *Eniochthonius* are associated with peatlands of eastern North America (Norton and Behan-Pelletier 2007). All known sites inhabited by *Eniochthonius mahunkai* are peatlands (fens or bogs), with sphagnum moss being a consistent component of the microhabitat. *Eniochthonius minutissimus* is associated with drier microhabitats and nearby forest, while *E. crosbyi* is found in muskeg habitats, in wet moss and in moss in seepage areas.
- The superfamilies Phthiracaroidea and Euphthiracaroidea are poorly represented, although the former are well represented in Norwegian mires (Seniczak et al. 2019a). Species in both superfamilies are primarily macrophytophagous (Luxton 1972), feeding on the woody tissue of plants, a food source poorly represented in bogs. However, the euphthiracarid *Acrotritia ardua* is widely distributed in a variety of forested habitats, including peatlands, and extends as far north as the subarctic in Canada.
- Species of the crotonioid genera *Nanhermannia* and *Tyrphonothrus* are among the most common and abundant oribatid mites in peatlands in northeastern North America.
- *Limnozetes* (Limnozetidae) is the most species-rich oribatid genus in the sphagnum moss of peatlands of the Boreal Ecozone; members of this genus often are the only microarthropods in dripping wet *Sphagnum* and peat layers. Popp (1962) noted that *Limnozetes ciliatus*, *Ceratozetes parvulus* and *Tyrphonothrus novus*, move to deeper layers of the hummock for overwintering, and that *L. ciliatus* and *Hypochthonius rufulus* exhibit diurnal movements in response to temperature in the hummock. Data from peatlands in eastern Canada emphasize that in addition to aquatic conditions, low pH appears to be a determinant for the presence of *Limnozetes* species. For example, they are most diverse in bogs at pH 4.2–4.7. Their species richness varies among bogs and is greatest in domed bogs and lowest in kettle bogs and fens, but it is unclear whether these differences are related to the age of the peatland, species of *Sphagnum* present, or depth of peat.

- The haplozetid *Rostrozetes ovulum* is widely distributed throughout the Nearctic and Neotropics. It is a frequent inhabitant of bogs in northeastern North America (Norton and Palmer 1991) and is also the dominant oribatid in Amazonian floodplains (Franklin et al. 1997). There is evidence that certain populations of this species may be semiaquatic for long periods (Franklin et al. 2001a; Pequeño et al. 2018). Surprisingly, this species is not found in similar habitats in the Palaearctic.
- The peatland fauna is not homogenous, and Barreto and Lindo (2021) showed a significant difference between the fauna of a *Sphagnum*-dominated fen and that of a *Carex*-dominated fen, with the former more species rich. The *Sphagnum*-dominated fen also had a greater number of non-specialist peatland species. The community structure is driven by interactions between temperature and moisture, thus oribatid mites of different fen types responded differently to warming (Barreto et al. 2021).

ATLANTIC MARITIME ECOZONE

This Ecozone encompasses the provinces of New Brunswick, Nova Scotia, Prince Edward Island, Îles de la Madeleine, and Beauce and Gaspé regions of Québec. It constitutes a cluster of peninsulas and islands that form the northeastern end of the Appalachian mountain chain that runs from Newfoundland to Alabama. Repeated glaciation has produced shallow, stony soils and outcrops composed of granite, gneiss and other hard, crystalline rocks. Soils are primarily nutrient-poor Podzols (Figure 5.1B). The proximity of the Atlantic Ocean creates a moderate, cool and moist maritime climate. Most of the Ecozone experiences long, mild winters and cool summers. Average precipitation varies from 1,000 mm inland to 1,425 mm along the coast.

Knowledge of oribatid species richness in this Ecozone is based primarily on surveys in Kouchibouguac and Fundy National Parks in New Brunswick, and in Kejimkujik National Park, Cape Breton Highlands National Park (CBHNP), and Sable Island, Nova Scotia (Majka et al. 2007). Almost none of the 11,000 km of coastline and associated habitats has been surveyed, yet there are records of 200 species, representing 118 genera and 57 families of oribatid mites from this Ecozone (Behan-Pelletier 2010). Most species were sampled from soil and litter habitats. However, Pielou and Verma (1968) recorded species from sporophores of the bracket fungus *Polyporus betulinus*, which develop on dead White and Yellow birch. The most abundant species recorded were *Carabodes polyporetes*, and a species of *Neoribates* and of *Zygoribatula*. In addition, Rasmy and MacPhee (1970) listed the mites associated with apple fruit and bark in the Annapolis Valley, Nova Scotia. Bark and lichen samples collected during the survey of CBHNP yielded *Licnodamaeus* sp., *Furcoribula furcillata*, *Chamobates* sp., *Minunthozetes semirufus* and *Achipteria clarencei*. *Dentizetes ledensis*, only known from the leaves of *Rhododendron groenlandicum* (Labrador tea) where it feeds on rust fungi, also has been recorded from this microhabitat in CBHNP.

Niedbała (2002) identified ten species in each of the Phthiracaroidea and Euphthiracaroidea primarily from CBHNP and Kouchibouguac National Parks. Almost 50% of these species show a Nearctic distribution; other species have a Holarctic or cosmopolitan distribution. Among Euphthiracaroidea, *Oribotritia henicos*, described from a beaver lodge at CBHNP, is endemic.

MIXEDWOOD PLAINS ECOZONE

This Ecozone comprises southern Ontario south of the Precambrian Shield and the St. Lawrence Valley of Québec as far east as the region of Rivière-du-Loup. This region is the northernmost extension of the deciduous forest biome that extends throughout much of the eastern United States from the Ozark Plateau to the Appalachian Mountains. It also includes Carolinian Forests of southwestern Ontario. This biome is contiguous with the Atlantic Maritime and Boreal Shield Ecozones. The most important soils in this Ecozone include Luvisols, Brunisols and Podzols (Figure 5.1B).

Only 18,000 years ago the Laurentide Ice Sheet covered all of southeastern Canada including the future Mixedwood Plains. By about 11,000 years ago, most of this area was ice free, and the ecosystem was dominated by Spruce Parkland comprising open groves of White spruce, Balsam fir and aspen mixed with tundra and grasslands. This community has no modern analogue, but probably resembled open spruce forests near the northern end of Lake Huron and on the eastern foothills of the Rocky Mountains. During the early Holocene, Spruce Parkland and its associated Pleistocene megafauna was replaced, first by a pine-dominated forest and then by a mixture of shade-tolerant tree species including maple, birch, beech and hemlock. Some of the landscape was probably covered by savannah maintained by periodic burning and cutting carried out by humans. Throughout the Holocene, plant and animal species recolonized the Mixedwood Plains Ecozone from Pleistocene refugia distributed south of the maximum limit of the Laurentide Ice Sheet between the Ozark Plateau and the Atlantic Coast. This resulted in a variety of distribution patterns among species inhabiting the Mixedwood Plains. Cold-adapted species from periglacial refugial areas invaded first, often leaving relict populations in suitable habitats in the Mixedwood Plains as they moved further north. Warm-adapted species colonized later, usually maintaining populations in their distributional centres in the southeastern United States while reaching their northern limits in the Mixedwood Plains Ecozone. Consequently, the assemblages found in the Mixedwood Plains exhibit exceptionally high diversity, comprising many species that occur together only in this Ecozone.

The recent history of species diversity in the Mixedwood Plains has been strongly influenced by European settlement. Populations of many naturalized, introduced plant species have spread aggressively and have resulted in this Ecozone, essentially the deciduous-coniferous forests of eastern Canada, being the most oribatid-diverse region of Canada after the Pacific Maritime Ecozone (Behan-Pelletier and Lindo 2019). The maple-beech forest of Morgan Arboretum of McGill University in particular has been a focus of research (Marshall 1968; Sylvain and Buddle 2010). Aspects of diversity in this Ecozone include:

- Presence of *Arborichthonius styosetosus*, in the monogeneric Arborichthoniidae, a widely distributed, but rarely collected species, known only from this Ecozone in Canada and Alaska, but also known from New Hampshire, USA, from Honshu and Okinawa Islands, Japan and Fujian Province, China (Lotfollahi et al. 2016);
- The only record of Plasmobatidae, Zetorchestidae, Eremulidae, Eremobelbidae, Ameridae, Machuellidae and Eremellidae for Canada and Alaska;
- Presence of species of Cosmochthoniidae, found elsewhere only in the Montane Cordillera Ecozone;
- Presence of Tegeocranellidae, found elsewhere only in the Atlantic Maritime Ecozone;
- Presence of Plateremaeidae, also known from the Boreal Ecozone;
- Presence of Microzetidae, also known from the Pacific Maritime Ecozone;
- Higher diversity of Nothridae and Oripodoidea compared with other Ecozones.

GRASSLAND (PRAIRIE) ECOZONES

Grasslands are one of the major biomes in Canada and represent about 5% of its land surface (Figure 5.2A). Typical soils of Canadian prairies are Chernozems, with decomposing grasses producing a dark, rich soil underlain by a sharp horizon where minerals are deposited at the line of deepest water penetration (Figure 5.1B). Soils are classified as Brown, Dark Brown or Black Chernozems, with colour darkening along an organic matter gradient, reflecting moisture supply. Grassland vegetation develops extensive root systems, extending to depths of over 2 m in Tallgrass prairie, to over 1.5 m in Mixed prairie and to approximately 1 m in Shortgrass and Palouse prairie. These root systems have distinct architecture and rhizospheres that affect belowground ecology (St. John et al. 2006). Rooting depth in grassland soils is greater than in boreal forests and is only slightly less than that of temperate forests. In Alaska and Yukon dry grassland (also called "grassland steppe"), communities occur in boreal regions on dry, south-facing slopes or well-drained lowland sites and are typically adjacent to Trembling aspen steppe-bluff communities. Whereas dry grasslands were widespread in Alaska and northern Canada during the cold, dry climate of the late Pleistocene, in contemporary times these communities are uncommon (Innes 2014).

Acari are the most diverse and abundant arthropods in grassland soils, yet this fauna is remarkably poorly known for Canada and Alaska. Most studies of the soil fauna in specific grassland types have focused on the short-range changes that occur in response to different management practices. In general, biomass and diversity of oribatid mites are significantly reduced by agricultural intensification (Behan-Pelletier and Kanashiro 2010).

Representatives of the following oribatid families are taxonomically rich in Canadian grassland soils:

- Brachychthoniidae dominate the oribatid mite fauna in fescue grassland of southern Alberta (Clapperton et al. 2002; Osler et al. 2008). This fauna is rich at the species and genus level and includes mainly undescribed species of *Brachychthonius*, *Liochthonius* and *Sellnickochthonius*.
- Eremaeidae are species rich and primarily are found in dry habitats where they live in the litter layer or in moss and lichens on the soil surface.
- Members of the oripodoid families Oribatulidae, Scheloribatidae and Haplozetidae are diverse in all grassland soils (e.g., Clapperton et al. 2002; Walter and Latonas 2013).

The oribatid fauna of the following grassland types in Canada has been studied in some detail.

Dry Mixed Prairie (= 'Short Grass Prairie'): The Matador Project studied natural Short Grass Prairie in southwestern Saskatchewan (Willard 1974). This area comprised a stand of *Agropyron-Koeleria* (wheat-like grasses) vegetation and moraines covered with *Stipa-Agropyron* to *Stipa-Bouteloua-Agropyron* type vegetation (Coupland 1973); soils were fine-textured clays. The Salt-Hollick washing method, rather than behavioural extractors, was used to routinely extract soil arthropods over 4 years from depths of 0–30 cm, and one time to a depth of 150 cm (Willard 1974). In contrast with other North American studies in Short Grass prairie (e.g., Walter et al. 1987), oribatid mites dominated the Matador mite fauna, representing almost 55% in natural grassland sites, and 45% and 76% in irrigated and wheat cultivated grassland sites, respectively (Willard 1974).

Mixed Grass Prairie and Fescue Grassland: Studies on this type of grassland are mainly associated with effects of management practices on fauna. For example, Berg and Pawluk (1984) compared mite populations on Gray Luvisol soil planted with either fescue or alfalfa. They found species richness of brachypyline oribatid mites was lower than that of other mite groups. Changes in abundance and diversity of mite families associated with grazing regimes or in a grazing exclosure in Fescue prairie showed density of oribatid mites was lowest under the heavy grazing regime (Clapperton et al. 2002). Oribatid mites had the greatest diversity among Acari represented in this study, and Brachychthoniidae dominated the fauna (Clapperton et al. 2002). This dominance was confirmed by Osler et al. (2008) in their study of oribatid mites of low-input vs. organic agricultural systems in southern Alberta. These grasslands also have the only representatives of Haplochthoniidae and Sphaerochthoniidae reported from Canada.

In the context of climate change, Newton (2013) looked at the effects of summer drought, summer warming and defoliation on native grassland soil mites in fescue prairie within aspen parkland in central Alberta. While drought normally affects most soil arthropod densities negatively, he observed an increase in abundance of juvenile Oribatida.

MONTANE CORDILLERA AND WESTERN INTERIOR BASIN ECOZONES

The Montane Cordillera (MC) Ecozone in Canada extends from the eastern Rocky Mountains in Alberta to the western slope of the Cascades in British Columbia and from the latitude of the Skeena Mountains in northern British Columbia to the United States border. About 90% of the area is in the province of British Columbia and the remaining 10% in western Alberta. Some 70% of the area is forested. The MC and Western Interior Basin (WIB) Ecozones are probably Canada's most complex, with landscapes ranging from alpine tundra to dense coniferous forests, grasslands, riparian woodlands and dry sagebrush, reflecting the exceptional diversity of topography and climate. It also includes Canada's only true desert (recently considered the Semi-Arid Plateau Ecozone), representing the northern extent of the Great Basin desert, which has very high species richness and species rarity. Extant species and community biodiversity in the MC developed through a complex process of recolonization by plants and animals following the retreat of the Cordilleran Ice Sheet. Post-glacial repopulation apparently began somewhat earlier in this Ecozone than in more eastern parts of Canada. During the historic period, tundra and taiga habitats have been restricted to high elevations, coniferous forests have dominated the lower slopes of mountain ranges, and grasslands and riparian woodlands have occupied intermontane plateaus and valleys. Species arriving from neighbouring Ecozones progressively enriched the diversity of MC communities, many of which comprise species assemblages that occur together nowhere else. During the twentieth century, many old-growth forests were transformed into intensively managed stands, native grasslands into extensively grazed, fertile bench lands and valley bottoms have been irrigated and converted to orchards and vineyards, while transportation corridors have proliferated. As in the case of the Mixedwood Plains, recent human activities in the MC have resulted in a substantial increase in the total number of species inhabiting the Ecozone.

The known oribatid fauna of the MC has been collected mainly in the foothills of the Rocky Mountains, in arid grasslands and in alpine habitats. It is based on surveys by taxonomists in the Kananaskis Region, Waterton Lakes National Park, Haynes Lease Ecological Reserve and surrounding arid grassland communities, Cathedral and Manning Provincial Parks, and ecological research studies in the aspen forest of the Kananaskis Region (Carter and Cragg 1976; Mitchell and Parkinson 1976; Mitchell 1977, 1978; McLean et al. 1996; Kaneko et al. 1998; McLean and Parkinson 2000; Eisenhauer et al. 2007), in Kamloops (Marshall 1974, 1979), and monitoring of the oribatid fauna of Alberta (Walter et al. 2014; Walter and Lumley 2021). We estimate that the approximately 110 species presently known from the MC and WIB Ecozones (Behan-Pelletier and Lindo 2019) represent at most 20% of expected diversity.

Overall diversity of Eremaeidae is strikingly higher in the MC and WIB Ecozones, reflecting the habitat requirements of members of this family, and the greater habitat diversity in the western mountains and valley grasslands of the Cordillera. In any particular habitat, up to five species of Eremaeidae may co-occur. Thus, the few studies on ecology of included genera are clouded because the "*Eremaeus* sp." in the literature generally refers to more than one species. For example, in the aspen forests of the eastern foothills of the Rocky Mountains, two species of Eremaeidae were recognized by Mitchell (1977, 1978) and Mitchell and Parkinson (1976): *Eueremaeus foveolatus* and *Eremaeus* sp. B were combined as "*Eremaeus* spp.". However, "*Eremaeus* sp. B" represents two species: *Eueremaeus marshalli* and *Eueremaeus tetrosus*. *Eremaeus boreomontanus* is present also at their study site. It has yet to be determined whether size patterns or other character displacements exist in such sympatric congeners, as Walter and Norton (1984) demonstrated for species of *Scheloribates*. Similarly, *Carabodes colorado* and *C. dickinsoni* have a narrow distribution in the Western Cordillera and the WIB (Reeves and Behan-Pelletier 1998). Ceratozetoidea are also diverse in the MC, the fauna being dominated by members of the Ceratozetidae and Punctoribatidae that prefer dry habitats, which are also the habitats that have been best surveyed in this Ecozone. The most southern distribution for the ceratozetid *Dentizetes rudentiger* is in the MC of Canada (Behan-Pelletier and Lindo 2019).

The Oribatida of the palouse grassland habitat of the Haynes Lease Ecological Reserve in the southern Okanagan (WIB Ecozone) includes some interesting northern records for Canada, e.g., *Passalozetes californicus*. *Scapuleremaeus kobauensis*, described from dry subalpine habitats in the Southern Interior, has been collected from a similar, arid habitat in California. Species of the cymbaeremaeid genus *Ametroproctus* can be the numerically dominant oribatid mites on the south-facing aspect of the summit of Kobau Mountain. The scutoverticid species *Exochocepheus eremitus* is often the dominant oribatid mite in grass habitats in the Haynes Lease Ecological Reserve, and on Kobau Mountain. This species is also known from short grass prairie sites at Writing-on-Stone Provincial Park, Alberta.

PACIFIC MARITIME ECOZONE

The Pacific Maritime Ecozone (PME) lies along British Columbia's coast and continues into Alaska (Figure 5.2B). This is where one finds the wettest weather, tallest trees and deepest fjords in the country. Summer temperatures average 13°C, while winters average −1.5°C; there is less difference between winter and summer temperatures here than elsewhere. The Coast Mountains rise steeply from the fjords and channels on the coast, and glaciers are found at higher elevations. These mountains are still young, and even where they are not very tall, as on Haida Gwaii and Vancouver Island, they are still extremely rugged. As part of the "Ring of Fire" that surrounds the Pacific Ocean, magma is close to the surface.

Oribatid research in these regions has focused to date on arctic type habitats on Kodiak Island (Hammer 1967) and on forest floor and canopy of various forest types: Western redcedar (Lindo and Winchester 2006, 2007a,b, 2009; Lindo 2010), Interior Cedar-Hemlock forests (Lindo and Stevenson 2007), temperate montane forest (Fagan and Winchester 1999; Fagan et al. 2005), Cedar-Hemlock and Hemlock-Amabilis fir forest types (Battigelli et al. 1994, 2004; Berch et al. 2001, 2007) and beach, tidal edge and forest habitats on Graham Island, Haida Gwaii (Lindo 2020). There are records of over 247 species from the PME, making it the most diverse in Canada based on morphological determinations (Behan-Pelletier and Lindo 2019). This diversity is very distinctive:

- Only Canadian or Alaskan records for Archeonothridae, Acaronychidae, Kodiakellidae, Selenoribatidae, Dendroeremaeidae;
- Richest Canadian or Alaskan diversity of Cepheidae, Peloppiidae, *Zachvatkinibates*;
- Only North American record for *Dendrozetes*, *Nemacepheus*;
- Evidence for stochastic, trans-Pacific rafting events contributing to the biodiversity of west coast habitats (Lindo 2020).

The diversity of soil and litter fauna is known to extend into and/or be mirrored in much of the forest canopy systems. Oribatid mites actively travel within the canopy (Lindo 2010), and adult oribatid mites are the dominant acarine group in Malaise trap samples in Sitka spruce canopy (Behan-Pelletier and Winchester 1998). The mite fauna of humus suspended in forest canopies—suspended soils—is equally diverse and generally distinct from that on the forest floor (Lindo and Winchester 2006, 2007a,b; Lindo and Stevenson 2007). Outside of suspended soil habitats, many arboreal oribatid mites are likely to be grazers on lichens, algae and fungi. Host specificity is known in these animals, especially those that burrow in lichens, and bark characteristics that vary among tree species influence the diversity and structure of both epiphyte and grazer assemblages (Nicolai 1986; Arroyo et al. 2013). Canopy species richness is dominated by members of the Crotoniidae, Hermanniellidae, Eremaeidae, Peloppiidae, Cepheidae, Carabodidae, Cymbaeremaeidae, Dendroeremaeidae, Haplozetidae, Scheloribatidae, Oribatulidae, Oripodidae, Mochlozetidae, Humerobatidae, Ceratozetidae and Punctoribatidae (Lindo and Winchester 2006).

Research by Lindo and colleagues in western temperate rainforests of the PME provided standardization of density of canopy material across sites and between sites, standardization yet to be applied to other canopy habitats in Canada or Alaska. Each canopy microhabitat requires a different sampling method to estimate the diversity: for example, high gradient extractors which effectively collect mites from canopy moss and suspended soils are ineffective for phylloplane acarofauna.

The litter oribatid mite fauna traditionally has been considered the source of canopy diversity, with specimens dispersing from the litter and reaching the canopy either by climbing the trunk highway or through wind dispersal (Lindo 2010). This view overlooks the distinctness of the arboreal fauna, emphasized by Travé (1963). Even in old-growth forest, where there is continuous moss and other microhabitats between forest floor and canopy, the moss-inhabiting oribatid faunas are distinct between canopy and forest floor (Lindo and Winchester 2006). These authors showed that although there is no physical boundary along the trunks of old-growth Western redcedar, there is faunal turnover at ~5 m above ground, where below = ground and above = canopy, reflecting a microclimate boundary between understory and more arid arboreal conditions (Lindo and Winchester 2007c). Soil mites do climb trees (Proctor et al. 2002), and arboreal mites may fall into the litter but, by and large, canopy mite species are canopy specialists. However, as yet there are no chronosequence studies comparing complementarity between forest floor and canopy faunas, and, to date, most canopies studied have been in mature or old-growth forests.

Although density estimates for oribatid mites in the canopy are lower than those recorded from soil and litter (Fagan et al. 2005; Lindo and Winchester 2006), they affect nutrient dynamics in the canopy, providing nutrients for canopy roots. Carroll (1980) postulated that mites grazing on microepiphytes, needles and small twigs of Douglas fir indirectly affect patterns of nutrient exchange within the canopy. Also oribatid faecal pellets would provide organic nutrients essential for the growth of phylloplane microflora (McBrayer et al. 1977). Tree leaves are invaded by microorganisms rapidly (within 1–2 days) after unfolding and most of these, being non-pathogenic, require a source of nutrients (McBrayer et al. 1977). These microorganisms are, in turn, grazed by oribatid mites. Clearly, a succession of microorganisms with associated grazers occurs on leaves, and decomposition of leaves starts in the canopy.

AQUATIC HABITATS

Soil as a slow-moving river is a useful analogy, highlighting the continuity between soil and aquatic environments. Aquatic habitats are found in all Ecozones in Canada and Alaska, but their oribatid fauna is poorly known (Proctor 2001; Schatz and Behan-Pelletier 2008). Early derivative oribatid clades—the Palaeosomata, Enarthronota, Parhyposomata and Mixonomata—do not have freshwater representatives, and it is unlikely that this reflects an absence of suitable resources. Only 15 genera in 9 families, in the hyporders Nothrina (40% of genera) and Brachypylina (60% of genera), are represented in freshwater. By contrast, all oribatid infraorders are represented in soil and litter habitats. In Canada and Alaska, oribatid mites of freshwater habitats include: the crotonioid species *Platynothrus peltifer* (Crotoniidae); the genera *Mainothrus*, *Mucronothrus*, *Trhypochthoniellus* (Trhypochthoniidae); *Malaconothrus* and *Tyrphonothrus* (Malaconothridae); the ameronothroid *Tegeocranellus* (Tegeocranellidae); the limnozetoid genera *Hydrozetes* (Hydrozetidae) and *Limnozetes* (Limnozetidae); and the ceratozetoid genera *Heterozetes*, *Naiazetes* and *Zetomimus* (Zetomimidae) and *Punctoribates* (Punctoribatidae). The ameronothroid genera *Ameronothrus* (Ameronothridae) and *Thalassozetes* (Selenoribatidae) and the ceratozetoid genus *Zachvatkinibates* (Punctoribatidae) include littoral species that apparently tolerate submergence in salt or freshwater. Most species of Malaconothridae, Hydrozetidae and Limnozetidae complete all activities preferably when submersed, but they are able to withstand very moist marginal habitats for short periods. Species of Zetomimidae and Tegeocranellidae require water for activity but can withstand desiccation.

The crotoniid *Platynothrus peltifer* is found in forest soil and litter, peatlands and other aquatic and semiaquatic habitats in the Holarctic and the Australian region, and in benthic habitats in Europe (Schatz and Gerecke 1996). Species of *Malaconothrus* and *Tyrphonothrus* (Malaconothridae) have been recorded from very wet vegetation, near streams and in peatlands worldwide (e.g., Yamamoto and Aoki 1998), and species of *Tyrphonothrus* are also found in the hyporheic zone of springs and creeks (Pennack and Ward 1986; Proctor 2001). *Mucronothrus nasalis* (Trhypochthoniidae) is found in streams, springs and the bottom of very cold lakes (Norton et al. 1988c; Schatz and Gerecke 1996). Although its distribution is limited by temperature, *M. nasalis* is found worldwide, a discontinuous distribution that Hammer (1965) considered predated the breakup of Pangaea about 200 million years ago. Another trhypochthoniid, *Mainothrus badius* is amphibious and widespread in moist habitats of the Palaearctic and Nearctic regions (Seniczak et al. 1998).

Some brachypyline species are amphibious and are found in peatlands (Markkula 1986; Borcard 1991; Behan-Pelletier and Bissett 1994) and other habitats with high humidity, e.g., beach debris (Norton and Dindal 1976), but it is species of Hydrozetidae, Limnozetidae, Tegeocranellidae, Zetomimidae and some Ameronothridae that are active in water. Species of *Hydrozetes* are found throughout the world in aquatic situations, generally in association with plants. Individuals are found on or under water on various plant parts and have been observed on subsurface roots of anchored aquatic plants (Krantz and Baker 1982). Population densities can be high; Fernandez and Athias-Binche (1986) recorded a mean of 1,600 *H. lemnae* in 200 cm^2 of water, and Buford (1976) noted large numbers surviving winter in small crevices on submerged logs in a state of cold narcosis. As already noted, the diversity of *Limnozetes* is highest in peatlands, where they are the most abundant Acari in bog pools ranging in size from <1 to >100 m^2 (Larson and House 1990). *Tegeocranellus* species (Tegeocranellidae) live in temporary swamps and streams; immatures have been collected only during periods of inundation (Behan-Pelletier 1997a), suggesting inactivity during dry periods.

All species of Zetomimidae are associated with shallow, eutrophic bodies of water, marshes, wet meadows and areas of wet moss (Behan-Pelletier and Eamer 2003). Adults have been observed on the water surface (Willmann 1931) at the edge of floating logs and twigs, submerged and associated with vegetation. Immatures have been collected only during periods of inundation (Behan-Pelletier and Eamer 2003). Banks (1895a) kept specimens of the North American *Heterozetes aquaticus* in an aquarium for several months.

TRENDS IN ORIBATID DIVERSITY ACROSS THE CANADIAN AND ALASKAN LANDSCAPE

North American Ecozones are being challenged by global warming, drought, intensive agriculture, deforestation, mining, urbanization, and scarification from various sources, from fossil fuel removal to recreation. Recent research in western Canada has shown that oribatid mites can be effective bioindicators (e.g., McAdams 2018; Meehan et al. 2019; Lupardus et al. 2021), but this is predicated on knowing the distribution of the fauna. This review highlights the major gaps in knowledge of the oribatid fauna of all Canadian and Alaskan Ecozones, but critically of the MC, the complex Ecoregions of the Pacific Northwest, the Boreal Ecozones, the vast unglaciated regions of Alaska and Yukon and all coastal regions. To fill these gaps will require many more extensive studies along the format of the Alberta Biodiversity Monitoring Initiative (Walter et al. 2014; Walter and Lumley 2021).

Abbreviations

Aa	adalar porose area
$A1, A2, A3$	mesonotic porose areas
a, m	anterior and middle seta of gena
AD	adanal plate
Ad	dorsosejugal porose area
ad_1, ad_2, ad_3	adanal setae
ag	aggenital seta
AG	aggenital plate
Ah	humeral porose area
Al	sublamellar porose area
Am	humeral porose area
AN	anal plate
an_1, an_2	anal setae
$apo2, aposj, apo3$	apodemes
bo	bothridial seta (=sensillus)
$bo1, bo2, bo3, bo4$	borders of epimeres I–IV, respectively
bs	bothridial seta (=sensillus)
cgs	circumgastric scissure
cha, chb	cheliceral setae
cir	circumpedal carina
cus	custodium
csp	lamellar cusp
dis	discidium
eA	prodorsal enantiophysis
eH	humeral enantiophysis
eL	laterosejugal enantiophysis
$E4a, E4p$	epimeral enantiophysis IV (aggenital enantiophysis), across $bo4$
ep	postpalpal seta
Epi I-IV	epimeres I-IV
ex	exobothridial seta
exv	alveolar vestige of second exobothridial seta
Fe I-IV	femur I-IV
GEN	genital plate
Ge I-IV	genu I-IV
gla	opening of opisthonotal gland
h	hypostomal seta of mentum
hpr	humeral process
lam	lamella
L	lamellar line
le	lamellar seta
Leg setae	σ, φ, ω — solenidia of genu, tibia and tarsus, respectively; e — famulus of tarsus I; d, l, v — dorsal, lateral, ventral setae, respectively; ev, bv — basal trochanteral setae; $ft, tc, it, p, u, a, s, pv, pl$ — tarsal setae
ia, im, ip	anterior, middle, posterior notogastral lyrifissures, respectively
ih, ips	lyrifissures associated with notogastral setal rows h and p, respectively
iad, ian	adanal, anal lyrifissure, respectively

in	interlamellar seta
NG	notogaster
ng setae	c (or c-row, c_1, c_2, c_3); *da*, *dm*, *dp* (centrodorsal setae); e_1, e_2; f_1, f_2; *la*, *lm*, *lp* (laterodorsal setae); h-row (h_1, h_2, h_3); p-row (p_1, p_2, p_3)
NA	anterior notogastral sclerite
NM	medial notogastral sclerite
or_1, or_2	adoral setae
ovp	ovipositor
PA	peranal plate
p.a.	porose area
paa	postanal porose area
palp setae	*v, l, d, cm, acm, ul, sul, vt, lt*
PdI	pedotectum I
PdII	pedotectum I
pms	postmarginal sclerite
prl	prolamella
pr.o	preanal organ
PP	preanal plate
ptm	pteromorph
PY	pygidial notogastral sclerite
ru	rutellum
ro	rostral seta
S	sublamellar line
Sa	adalar saccule
S1, S2, S3	mesonotic saccules
s.a.	spina adnata
Sa, Sp	parastigmatic enantiophysis
sp	spermatopositor
sbl	sublamella
ss	sensillus (see bothridial seta)
Ta	adalar tubule
T1, T2, T3	mesonotic tubules
Ta I-IV	tarsus I-IV
Tg	Trägårdh's organ
Ti I-IV	tibia I-IV
Tr I-IV	trochanter I-IV
tr1	trachea of acetabulum I
tr.sj	sejugal trachea
Tu	tutorium
Va, Vp	ventrosejugal enantiophysis
VP	ventral plate

References

Abbott, D.T., Seastedt, T.R., Crossley, D.A., Jr. 1980. Abundance, distribution and effects of clearcutting on Cryptostigmata in the southern Appalachians. *Environmental Entomology*, 9, 618–623.

ABMI. 2019. Alberta Biodiversity Monitoring Institute. https://abmi.ca/home/data-analytics/biobrowser-home/ (accessed 19 July 2021).

Adl, S.M., Coleman, D.C., Read, F. 2006. Slow recovery of soil biodiversity in sandy loam soils of Georgia after 25 years of no-tillage management. *Agriculture, Ecosystems & Environment*, 114, 323–334.

Ahaniazad, M., Bagheri, M., Akrami, M.A. 2016. Additional description of *Oribella fujikawae* Mahunka (Acari, Oribatida) collected from Iran, with a key to world *Oribella* species. *Persian Journal of Acarology*, 5, 263–269.

Akimov, I.A., Yastrebstov, A.V. 1989. The muscular system and skeleton elements of an oribatid mite, *Nothrus palustris*. *Zoologicheskii Zhurnal*, 68, 57–67.

Alberti, G. 1979. Fine structure and probable function of genital papillae and Claparede organs of Actinotricha. In: Rodríguez, J.G. (Ed.), *Recent Advances in Acarology*. vol. 2. Academic Press, Inc., New York, 501–507.

Alberti, G. 1998. Fine structure of receptor organs in oribatid mites (Acari). In: Ebermann, E. (Ed.), *Arthropod Biology: Contributions to Morphology, Ecology and Systematics*. Biosystematics and Ecology Series 14. Austrian Academy of Sciences Press, Vienna, 27–77.

Alberti, G. 2006. On some fundamental chacteristics in acarine morphology. *Atti dell'Accademia Nazionale Italiana di Entomologia. R.A. LIII*, 2005, 315–360.

Alberti, G., Coons, B. 1999. Acari – Mites. In: Harrison, F.W., Foelix, R.F. (Eds.), *Microscopic Anatomy of Invertebrates*. Volume 8C, Chelicerate Arthropoda. Wiley, New York, 515–1265.

Alberti, G., Fernandez, N.A. 1988. Fine structure of a secondarily developed eye in the freshwater moss-mite, *Hydrozetes lemnae* (Coggi, 1899) (Acari: Oribatida). *Protoplasma*, 146, 106–117.

Alberti, G., Fernandez, N.A. 1990. Aspects concerning the structure and function of the lenticulus and clear spot of certain oribatids (Acari, Oribatida). *Acarologia*, 311, 65–72.

Alberti, G., Fernandez, N.A., Kümmel, G. 1991. Spermatophores and spermatozoa of oribatid mites (Acari: Oribatida). Part II: Functional and systematical considerations. *Acarologia*, 324, 435–449.

Alberti, G., Kaiser, T., Klauer, A.K. 1996. New ultrastructural observations of the coxal glands (nephridia) of Acari. In: Mitchell, R., Horn, D.J., Needham, G.R., Welbourn, C.W. (Eds.), *Acarology IX Proceedings*. vol. 1. Ohio Biological Survey, Columbus, OH, 309–318.

Alberti, G., Norton, R.A., Adis, J., Fernandez, N.A., Franklin, E., Kratzmann, M., Moreno, A.I., Weigmann, G., Woas, S. 1997. Porose integumental organs of oribatid mites (Acari, Oribatida). 2. Fine structure. *Zoologica, Stuttgart*, 146, 33–114.

Alberti, G., Norton, R.A., Kasbohm, J. 2001. Fine structure and mineralisation of cuticle in Enarthronota and Lohmannioidea. In: Halliday, R.B., Walter, D.E., Proctor, H.C., Norton, R.A., Colloff, M.J. (Eds.), *Acarology: Proceedings of the 10th International Congress*. CSIRO Publishing, Melbourne, 230–241.

Alberti, G., Seniczak, A., Seniczak, S. 2003. The digestive system and fat body of an early-derivative oribatid mite, *Archegozetes longisetosus* Aoki (Acari: Oribatida, Trhypochthoniidae). *Acarologia*, 431, 149–219.

Alberti, G., Heethoff, M., Norton, R.A., Schmelzle, S. Seniczak, A., Seniczak, S. 2011. Fine structure of the gnathosoma of *Archegozetes longisetosus* Aoki (Acari: Oribatida, Trhypochthoniidae). *Journal of Morphology*, 272, 1025–1079.

Anderson, J.M. 1975. Succession, diversity and trophic relationships of some soil animals in decomposing leaf litter. *Journal of Animal Ecology*, 44, 475–495.

Andrews, R.N., Ruess, R.W. 2020. Microarthropod abundance and community structure along a chronosequence within the Tanana River floodplain, Alaska. *Écoscience*, 274, 235–253.

Aoki, J. 1958a. Zwei *Heminothrus*–Arten aus Japan (Acarina, Oribatei). *Annotationes Zoologicae Japonenses*, 31, 121–125.

Aoki, J. 1958b. Eine neue Art von der Gattung *Oribotritia* Jacot (Acarina, Oribatei). *Acta Arachnologica, Osaka*, 16, 18–20.

Aoki, J. 1958c. Einige Phthiracariden aus Utsukushigahara, Mitteljapan (Acarina: Oribatei). *Annotationes Zoologicae Japonenses*, 31, 171–175.

Aoki, J. 1959. Die Moosmilben (Oribatei) aus Südjapan. *Bulletin of the Biogeographical Society of Japan*, 21, 1–22.

Aoki, J. 1960. Eine dreikrallige Gattung der Familie Perlohmanniidae (Acarina: Oribatei). *The Japanese Journal of Zoology*, 12, 507–511.

Aoki, J. 1964a. A new aquatic oribatid mite from Kauai Island. *Pacific Insects*, 6, 483–488.

Aoki, J. 1964b. Some oribatid mites (Acarina) from Laysan Island. *Pacific Insects*, 6, 649–664.

Aoki, J. 1965a. Oribatid mites (Acarina: Oribatei) from Himalaya with descriptions of several new species. *Journal of the College of Arts and Sciences, Chiba*, 4, 289–302.

Aoki, J. 1965b. Notes on the species of the genus *Epilohmannia* from the Hawaiian Islands (Acarina, Oribatei). *Pacific Insects*, 7, 309–315.

Aoki, J. 1965c. Studies on the oribatid mites of Japan I. Two members of the genus *Hermanniella*. *Bulletin of the National Science Museum Series A (Zoology)*, 8, 125–130.

Aoki, J. 1965d. New oribatid from the island of Sado. *The Japanese Journal of Zoology*, 14, 1–12.

Aoki, J. 1967. Microhabitats of oribatid mites on a forest floor. *Bulletin of the Natural Sciences Museum [Tokyo, Japan]*, 10, 133–138 +2 plates.

Aoki, J. 1968. A new soil mite representing the new genus *Nemacepheus* (Acari, Tectocepheidae) found in Mt. Goyo, North Japan. *Bulletin of the National Science Museum Series A (Zoology)*, 1, 117–120.

Aoki, J. 1969. Taxonomic investigations on free living mites in the subalpine forest on Shiga Heights IBP area. III. Cryptostigmata. *Bulletin of the National Science Museum Series A (Zoology)*, 12, 117–141.

Aoki, J. 1970a. The oribatid mites of the Islands of Tsushima. *Bulletin of the National Science Museum Series A (Zoology)*, 13, 395–442.

Aoki, J. 1970b. Descriptions of oribatid mites collected by smoking of trees with insecticides. I. Mt. Ishizuchi and Mt. Odaigahara. *Bulletin of the National Science Museum Series A (Zoology)*, 13, 585–602.

Aoki, J. 1975. Taxonomic notes on some little known oribatid mites of Japan. *Bulletin of the National Science Museum Series A (Zoology)*, 1, 57–65.

Aoki, J. 1977. New and interesting species of oribatid mites from Kakeroma Island, Southwest Japan. *Acta Arachnologica*, 27, 85–93.

Aoki, J. 1980a. A revision of the oribatid mites of Japan. I. The families Phthiracaridae and Oribotritiidae. *Bulletin of the Institute of Environmental Science and Technology, Yokohama*, 6, 1–88.

Aoki, J. 1980b. A revision of the oribatid mites of Japan. II. The family Euphthiracaridae. *Acta Arachnologica*, 29, 9–24.

Aoki, J. 1980c. A revision of the oribatid mites of Japan. III. Families Protoplophoridae, Archoplophoridae and Mesoplophoridae. *Proceedings of the Japanese Society of Systematic Zoology*, 18, 5–16.

Aoki, J. 1980d. A revision of the oribatid mites of Japan. IV. The families Archeonothridae, Palaeacaridae and Ctenacaridae. *Annotationes Zoologicae Japonenses*, 53, 124–136.

Aoki, J. 1983. Some new species of Oppiid mites from South Japan (Oribatida: Oppiidae). *International Journal of Acarology*, 94, 165–172.

Aoki, J. 1987. Three new species of oribatid mites from Kume-jima island, Southwest Japan. *Proceedings of the Japanese Society of Systematic Zoology*, 36, 25–28.

Aoki, J. 1995. Oribatid mites of high altitude forests of Taiwan II. Mt. Na–hu–ta Shan. *Special Bulletin of the Japanese Society of Coleopterology*, 4, 123–130.

Aoki, J., Fujikawa, T. 1972. A new genus of oribatid mites, exhibiting both the characteristic features of the families Charassobatidae and Cymbaeremaeidae (Acari: Cryptostigmata). *Acarologia*, 14, 258–267.

Aoki, J., Ohnishi, J. 1974. New species and record of oribatid mites from Hokkaido, North Japan. *Bulletin of the National Science Museum Series A (Zoology)*, 17, 149–156.

Aoki, J., Shimano, S. 1994. Influence of trampling the lawn by cars on oribatid fauna. *Bulletin of the Institute of Environmental Science and Technology, Yokohama*, 20, 97–100.

Aoki, J., Wen, Z. 1983. Two new species of oribatid mites from Shikine islet. *Bulletin of the Institute of Environmental Science and Technology, Yokohama*, 9, 165–169.

Ardestani, M.M., Keshavarz-Jamshidian, M., van Gestel, C.A.M., van Straalen, N.M. 2020. Avoidance tests with the oribatid mite *Oppia nitens* (Acari: Oribatida) in cadmium-spiked natural soils. *Experimental and Applied Acarology*, 82, 81–93.

Arlian, L.G., Woolley, T.A. 1969. Life stages of *Liacarus cidarus* (Acari: Cryptostigmata, Liacaridae). *Journal of the Kansas Entomological Society*, 42, 512–524.

Arlian, L.G., Woolley, T.A. 1970. Observations on the biology of *Liacarus cidarus* (Acari: Cryptostigmata, Liacaridae). *Journal of the Kansas Entomological Society*, 43, 297–301.

Arroyo, J., Kenny, J., Bolger, T. 2013. Variation between mite communities in Irish forest types – Importance of bark and moss cover in canopy. *Pedobiologia*, 56, 241–250.

Arroyo, J., O'Connell, T., Bolger, T. 2017. Oribatid mites (Acari: Oribatida) recorded from Ireland: Catalogue, historical records, species habitats and geographical distribution, combinations, variations and synonyms. *Zootaxa*, 4328, 1–174.

References

Åström, J., Bengtsson, J. 2011. Patch size matters more than dispersal distance in a mainland-island metacommunity. *Oecologia*, 167, 747–57.

Åström, J., Pärt, T. 2013. Negative and matrix-dependent effects of dispersal corridors in an experimental metacommunity. *Ecology*, 94, 72–82.

Baker, G.T., Wighton, D.C. 1983. Fossil aquatic oribatid mites (Acari: Oribatida: Hydrozetidae: *Hydrozetes*) from the Paleocene of South-Central Alberta, Canada. *The Canadian Entomologist*, 116, 773–775.

Balogh, J. 1938. *Belba visnyai* n. sp., eine neue Moosmilbenart. (Stud. Acar. I.). *Folia Entomologica Hungarica*, 3, 83–85.

Balogh, J. 1943. Magyarorszag Pancelosatkai (Conspectus Oribateorum Hungariae). *Matematikai és Természettudományi Közlemények*, 38, 1–202.

Balogh, J. 1958. Oribatides nouvelles de l'Afrique tropicale. *Revue d'Zoologie et de Botanique Africaines*, 58, 1–34.

Balogh, J. 1960. Oribates (Acari) nouveaux d'Angola et du Congo Belge. *Companhia de Diamantes de Angola, Servicos Culturais, Lisboa*, 51, 13–40.

Balogh, J. 1961. Identification keys of world oribatid (Acari) families and genera. *Acta Zoologica Academiae Scientiarum Hungaricae*, 7, 243–344.

Balogh, J. 1965. A synopsis of the world oribatid (Acari) genera. *Acta Zoologica Academiae Scientiarum Hungaricae*, 11, 5–99.

Balogh, J. 1970. New oribatids (Acari) from New Guinea II. *Acta Zoologica Academiae Scientiarum Hungaricae*, 16, 291–344.

Balogh, J. 1972. *The Oribatid Genera of the World*. Akademiai Kiado, Hungary, Budapest, 188 pp.

Balogh, J. 1983. A partial revision of the Oppiidae Grandjean, 1954 (Acari: Oribatei). *Acta Zoologica Academiae Scientiarum Hungaricae*, 29, 1–79.

Balogh, J., Balogh, P. 1984. A review of the Oribatuloidea Thor, 1929 (Acari: Oribatei). *Acta Zoologica Academiae Scientiarum Hungaricae*, 30, 257–313.

Balogh, J., Balogh, P. 1988. Oribatid mites of the Neotropical Region I. In: Balogh, J., Mahunka, S. (Eds.), *The Soil Mites of the World*. vol. 2. Akademiai Kiado, Hungary, Budapest, 335 pp.

Balogh, J., Balogh, P. 1992. *The Oribatid Mites Genera of the World*. vol. 1. The Hungarian National Museum Press, Budapest, 263 pp.

Balogh, J., Mahunka, S. 1968. The scientific results of the Hungarian soil zoological expedition to South America. 5. Acari: Data to the oribatid fauna of the environment of Córdoba, Argentina. *Opuscula Zoologica*, 8, 317–340.

Balogh, J., Mahunka, S. 1979. New taxa in the system of the Oribatida (Acari). *Annales historico–naturales Musei Nationalis Hungarici*, 71, 279–290.

Balogh, J., Mahunka, S. 1983. *The Soil Mites of the World*. vol. 1. Primitive Oribatids of the Palaearctic Region. Elsevier, Amsterdam, 372 pp.

Balogh, P. 1984. Oribatid mites from Colombia (Acari) II. *Acta Zoologica Academiae Scientiarum Hungaricae*, 30, 315–326.

Banks, N. 1895a. Some acarians from a sphagnum swamp. *Journal of the New York Entomological Society*, 3, 128–130.

Banks, N. 1895b. On the Oribatoidea of the United States. *Transactions of the American Entomological Society*, 22, 1–16.

Banks, N. 1896. New North American spiders and mites. *Transactions of the American Entomological Society*, 23, 57–77.

Banks, N. 1902. New genera and species of Acarina. *The Canadian Entomologist*, 34, 171–176.

Banks, N. 1904. Some Arachnida from California. *Proceedings of the California Academy of Sciences*, 3, 331–374+pls. 38–41.

Banks, N. 1906. New Oribatidae from the United States. *Proceedings of the Academy of Natural Sciences, Philadelphia*, 58, 490–500.

Banks, N. 1909. New Canadian mites. *Proceedings of the Entomological Society of Washington*, 11, 133–142.

Banks, N. 1910. New American mites. *Proceedings of the Entomological Society of Washington*, 12, 2–12.

Banks, N. 1947. On some Acarina from North Carolina. *Psyche (Cambridge, Mass.)*, 54, 110–141.

Bardel, L., Pfingstl, T. 2018. Resistance to flooding of different species of terrestrial oribatid mites (Acari, Oribatida). *Soil Organisms*, 90, 71–77.

Barendregt, R.W., Matthews, J.V., Behan-Pelletier, V.M., Brigham-Grette, J., Fyles, J.G., Ovenden, L.E., McNeil, D.H., Brouwers, E., Marincovich, L., Rybczynski, N., Fletcher, T.L. 2021. Biostratigraphy, age, and the paleoenvironment of the Pliocene Beaufort Formation on Meighen Island, Canadian Arctic Archipelago. *The Geological Society of America, Special Paper 551*. Geological Society of America, 39 pp.

Barnett, A.A., Thomas, R.H. 2012. The delineation of the fourth walking leg segment is temporally linked to posterior segmentation in the mite *Archegozetes longisetosus* (Acari: Oribatida, Trhypochthoniidae). *Evolution and Development*, 144, 383–392.

Barnett, A.A., Thomas, R.H. 2013. The expression of limb gap genes in the mite *Archegozetes longisetosus* reveals differential patterning mechanisms in chelicerates. *Evolution and Development*, 154, 280–292.

Barreto, C.R.A. 2021. Diversity and drivers of oribatid mites (Acari: Oribatida) in boreal peatlands. Ph.D. Thesis. The University of Western Ontario, London, Ontario. 235 pp. (https://ir.lib.uwo.ca/etd/8217/)

Barreto, C., Lindo, Z. 2018. Drivers of decomposition and detrital invertebrate communities differ across a hummock-hollow microtopology in boreal peatlands. *Écoscience*, 25, 39–48.

Barreto, C., Lindo, Z. 2021. Checklist of oribatid mites (Acari: Oribatida) from two contrasting boreal fens: An update on oribatid mites of Canadian peatlands. *Systematic and Applied Acarology*, 265, 866–884.

Barreto, C., Branfireun, B.A., McLaughlin, J., Lindo, Z. 2021. Responses of oribatid mites to warming in boreal peatlands depend on fen type. *Pedobiologia*, 89, 150772.

Barreto, C., Buchkowski, R., and Lindo, Z. 2023. Restructuring of soil food webs reduces carbon storage potential in boreal peatlands. In submission.

Battigelli, J.P., Berch, S.M., Marshall, V.G. 1994. Soil fauna communities in two distinct but adjacent forest types on northern Vancouver Is., British Columbia. *Canadian Journal of Forestry Research*, 248, 1557–1566.

Battigelli, J.P., Spence, J.R., Langor, D.W., Berch, S.M. 2004. Short-term impact of forest soil compaction and organic matter removal on soil mesofauna density and oribatid mite diversity. *Canadian Journal of Forestry Research*, 34, 1136–1149.

Bäumler, W. 1970a. Zur Morphologie, Biologie und Ökologie von *Hermannia gibba* (C.L. Koch) (Acarina, Oribatei) unter Berücksichtigung einiger Begleitarten. I. Z. angew. *Entomologie*, 66, 257–277.

Bäumler, W. 1970b. Zur Morphologie, Biologie und Ökologie von *Hermannia gibba* (C.L. Koch) (Acarina, Oribatei) unter Berücksichtigung einiger Begleitarten. Teil II. Z. angew. *Entomologie*, 66, 337–362.

Bayartogtokh, B. 2000a. Two species of Damaeid mites (Acari: Oribatida: Damaeidae) from Mongolia, with notes on distribution of the genera *Epidamaeus* and *Dyobelba*. *Biogeography*, 2, 67–79.

Bayartogtokh, B. 2000b. Oribatid mites of the genus *Scheloribates* (Acari: Oribatida: Scheloribatidae) from Mongolia. *Edaphologia*, 65, 61–88.

Bayartogtokh, B. 2004. Oribatid mites of the genera *Belba* and *Belbodamaeus* (Acari: Oribatida: Damaeidae) from Eastern Mongolia. *Zootaxa*, 476, 1–11.

Bayartogtokh, B. 2006. Two species of oribatid mites of the genus *Banksinoma* (Acari: Oribatida: Banksinomidae) from Mongolia. *Acarina*, 14, 175–179.

Bayartogtokh, B. 2007. Oribatid mites of the family Astegistidae (Acari: Oribatida) in Mongolia. *Zootaxa*, 1472, 55–68.

Bayartogtokh, B. 2012. The genus *Cultoribula* (Acari: Oribatida: Astegistidae) in Mongolia, with new findings from Altai Mountains and remarks on known species of the world. *Zootaxa*, 3302, 44–60.

Bayartogtokh, B., Aoki, J. 2000. A new and some little known species of *Eporibatula* (Acari: Oribatida: Oribatulidae), with remarks on taxonomy of the genus. *Zoological Science*, 17, 991–1012.

Bayartogtokh, B., Ermilov, S.G. 2021. A remarkable new species of Hungarobelbidae (Acari: Oribatida) exhibiting characteristic features of two genera, with notes on its systematic relationships. *Systematic, and Applied Acarology*, 26, 624–640.

Bayartogtokh, B., Norton, R.A. 2007. The *Dyobelba tectopediosa* species–group (Acari: Oribatida: Damaeidae) from the Southeastern USA, with a key to world species of *Dyobelba* and notes on their distribution. *Zootaxa*, 1591, 39–66.

Bayartogtokh, B., Schatz, H. 2009. Two new species of the genus *Gymnodamaeus* (Acari: Oribatida: Gymnodamaeidae) from Tyrol (Austria). *Revue Suisse de Zoologie*, 116, 31–51.

Bayartogtokh, B., Weigmann, G. 2005a. Contribution to the knowledge of oribatid mites of the families Galumnidae and Parakalummidae (Acari, Oribatida) from Mongolia. *Zoologische Reihe*, 81, 89–98.

Bayartogtokh, B., Weigmann, G. 2005b. New and little known species of oribatid mites of the genera *Arthrodamaeus* and *Fuscozetes* (Arachnida: Acari: Oribatida) from Mongolia. *Species Diversity*, 10, 75–84.

Bayartogtokh, B., Cobanoglu, S., Ozman, S.K. 2002. Oribatid mites of the superfamily Ceratozetoidea (Acari: Oribatida) from Turkey. *Acarina*, 10, 3–23.

Bayartogtokh, B., Schatz, H., Ekrem, T. 2011. Distribution and diversity of the soil mites of Svalbard, with redescriptions of three known species (Acari: Oribatida). *International Journal of Acarology*, 37, 467–484.

Bayartogtokh, B., Ermilov, S.G., Shtanchaeva, U.Y., Subias, L.S. 2018. Ontogeny of morphological traits in *Eupelops variatus* (Mihelčič, 1957), with remarks on juveniles of Phenopelopidae (Acari: Oribatida). *Systematic & Applied Acarology*, 23, 161–177.

Bayoumi, B.M., Mahunka, S. 1976. Contribution to the knowledge of the genus *Epilohmannia* Berlese, 1916 (Acari: Oribatida). *Folia Entomologica Hungarica*. 29, 5–21.

Beare, M.H., Coleman, D.C., Crossley, D.A., Hendrix, P.F., Odum, E.P. 1995. A hierarchical approach to evaluating the significance of soil biodiversity to biogeochemical cycling. *Plant and Soil*, 170, 5–22.

Beaty, L.B., Esser, H.J., Miranda, R., Norton, R.A. 2013. First report of phoresy by an oribatid mite (Trhypochthoniidae: *Archegozetes magnus*) on a frog (Leptodactylidae: *Engystomops pustulosus*). *International Journal of Acarology*, 39, 325–326.

Beaulieu, F., Knee, W., Nowell, V., Schwarzfeld, M., Lindo, Z., Behan-Pelletier, V.M., Lumley, L., Young, M.R., Smith, I., Proctor, H.C., Mironov, S.V., Galloway, T.D., Walter, D.E., Lindquist, E.E. 2019. Acari of Canada. In: Langor, D.W., Sheffield, C.S. (Eds.), *The Biota of Canada – A Biodiversity Assessment. Part 1: The Terrestrial Arthropods. ZooKeys*, 819, 77–168.

Beck, L., Woas, S. 1991. Die Oribatiden-Arten (Acari) eines südwestdeutschen Buchenwaldes I. *Carolinea, Karlsruhe*, 49, 37–82.

Beck, L., Horak, F., Woas, S. 2014. Zur Taxonomie der Gattung *Phthiracarus* Perty, 1841 (Acari, Oribatida) in Südwestdeutschland. *Carolinea*, 72, 109–132.

Behan, V.M. 1978. Diversity, distribution and feeding habits of North American arctic soil Acari. Ph.D. Thesis, McGill University, Montreal, 428 pp.

Behan, V.M., Hill, S.B. 1978. Feeding habits and spore dispersal of oribatid mites in the North American Arctic. *Revue d'Écologie et de Biologie du Sol*, 15, 497–516.

Behan, V.M., Hill, S.B., Kevan, D.K. 1978. Effects of nitrogen fertilizers, as urea, on Acarina and other arthropods in Quebec black spruce humus. *Pedobiologia*, 18, 249–263.

Behan-Pelletier, V.M. 1984. *Ceratozetes* (Acari: Ceratozetidae) of Canada and Alaska. *The Canadian Entomologist*, 116, 1449–1517.

Behan-Pelletier, V.M. 1985. Ceratozetidae of the Western North American Arctic. *The Canadian Entomologist*, 117, 1287–1366.

Behan-Pelletier, V.M. 1986. Ceratozetidae (Acari: Oribatei) of the Western North American Subarctic. *The Canadian Entomologist*, 118, 991–1057.

Behan-Pelletier, V.M. 1987. Redefinition of *Ametroproctus* (Acari: Oribatida) with descriptions of new species. *The Canadian Entomologist*, 119, 505–535.

Behan-Pelletier, V.M. 1989a. *Limnozetes* (Acari: Oribatida: Limnozetidae) of Northeastern North America. *The Canadian Entomologist*, 121, 453–506.

Behan-Pelletier, V.M. 1989b. Description of *Scapuleremaeus kobauensis* gen. nov., sp. nov. (Acari: Oribatida: Cymbaeremaeidae) from Western Canada. *The Canadian Entomologist*, 121, 507–513.

Behan-Pelletier, V.M. 1990. Redefinition of *Megeremaeus* (Acari: Megeremaeidae) with description of new species and nymphs of *M. montanus* Higgins and Woolley. *The Canadian Entomologist*, 122, 875–900.

Behan-Pelletier, V.M. 1993. Eremaeidae (Acari: Oribatida) of North America. *Memoirs of the Entomological Society of Canada, Ottawa, Canada*, 168, 1–193.

Behan-Pelletier, V.M. 1994. *Mycobates* (Acari: Oribatida: Mycobatidae) of America north of Mexico. *The Canadian Entomologist*, 126, 1301–1361.

Behan-Pelletier, V.M. 1996. *Naiazetes reevesi* n.g., n.sp. (Acari: Oribatida: Zetomimidae) from semi-aquatic habitats of eastern North America. *Acarologia*, 37, 345–355.

Behan-Pelletier, V.M. 1997a. The semiaquatic genus *Tegeocranellus* (Acari: Oribatida: Ameronothroidea) of North and Central America. *The Canadian Entomologist*, 129, 537–577.

Behan-Pelletier, V.M. 1997b. Oribatid mites (Acari: Oribatida) of the Yukon. In: Danks, H.V., Downes, J.A. (Eds.), *Insects of the Yukon*. Biological Survey of Canada Terrestrial Arthropods. Ottawa, 115–149.

Behan-Pelletier, V.M. 1999a. Oribatid mite fauna of northern ecosystems: a product of evolutionary adaptations or physical constraints? In: Needham, G.R., Mitchell, R., Horn, D.J., Welbourn, C.W. (Eds.), *Acarology IX Symposia*. vol. 2. Ohio Biological Survey, Columbus, OH, 87–105.

Behan-Pelletier, V.M. 1999b. Oribatid mite biodiversity in agroecosystems: Role for bioindication. *Agriculture, Ecosystems & Environment*, 74, 411–423.

Behan-Pelletier, V.M. 2000. Ceratozetidae (Acari: Oribatida) of arboreal habitats. *The Canadian Entomologist*, 132, 153–182.

Behan-Pelletier, V.M. 2010. Oribatid mites (Acarina: Orbatida) of the Atlantic Maritime Ecozone. In: McAlpine, D.F., Smith, I.M. (Eds.), *Assessment of Species Diversity in the Atlantic Maritime Ecozone*. NRC Research Press, Ottawa, 313–331.

Behan-Pelletier, V.M. 2011. *Oribatella* (Acari, Oribatida, Oribatellidae) of eastern North America. *Zootaxa*, 2973, 1–56.

Behan-Pelletier, V.M. 2013. *Adoribatella, Ferolocella, Joelia* and *Ophidiotrichus* (Acari, Oribatida, Oribatellidae) of North America. *Zootaxa*, 3637, 254–284.

Behan-Pelletier, V.M. 2015a. Sexual dimorphism in *Autogneta*, with description of three new species from North America and new diagnosis of the genus (Acari, Oribatida, Autognetidae). *Zootaxa*, 3946, 55–78.

Behan-Pelletier, V.M. 2015b. Review of sexual dimorphism in brachypyline oribatid mites. *Acarologia*, 552, 127–146.

Behan-Pelletier, V.M. 2017. Tegoribatidae of North America, with proposal of *Protectoribates* gen. nov., and new species (Acari, Oribatida, Tegoribatidae). *Zootaxa*, 4337, 151–197.

Behan-Pelletier, V.M., Bissett, B. 1994. Oribatida of Canadian peatlands. *Memoirs of the Entomological Society of Canada*, 169, 73–88.

Behan-Pelletier, V.M., Eamer, B. 2003. Zetomimidae (Acari: Oribatida) of North America. In: Smith, I.M. (Ed.), *An Acarological Tribute to David R. Cook from Yankee Springs to Wheeny Creek*. Indira Publishing House, West Bloomfield, MI, 21–56.

Behan-Pelletier, V.M., Eamer, B. 2005. *Zachvatkinibates* (Acari: Oribatida: Mycobatidae) of North America, with descriptions of sexually dimorphic species. *The Canadian Entomologist*, 137, 631–647.

Behan-Pelletier, V.M., Eamer, B. 2007. Aquatic Oribatida: Adaptations, constraints, distribution and ecology. In: Morales-Malacara, J.B., Behan-Pelletier, V., Ueckermann, E., Pérez, T.M., Estrada-Venegas, E.G., Badii, M. (Eds.), *Acarology XI: Proceedings of the International Congress*. Instituto de Biología and Facultad de Ciencias, Universidad Nacional Autónoma de México, Sociedad Latinoamericana de Acarología, México, 71–82.

Behan-Pelletier, V.M., Eamer, B. 2008. Mycobatidae (Acari: Oribatida) of North America. *The Canadian Entomologist*, 140, 73–110.

Behan-Pelletier, V.M., Eamer, B. 2009. *Ceratozetes* and *Ceratozetoides* (Acari: Oribatida: Ceratozetidae) of North America. *The Canadian Entomologist*, 141, 246–308.

Behan-Pelletier, V.M., Eamer, B. 2010. The first sexually dimorphic species of *Oribatella* (Acari, Oribatida, Oribatellidae) and a review of sexual dimorphism in the Brachypylina. *Zootaxa*, 2332, 1–20.

Behan-Pelletier, V.M., Ermilov, S.G. 2020. *Sculpteremaeus olszanowskii* gen. nov., sp. nov. (Acari, Oribatida, Cymbaeremaeidae) from chaparral in California, USA, with a reassessment of Cymbaeremaeidae. *Zootaxa*, 4790, 341–357.

Behan-Pelletier, V.M., Hill, S.B. 1983. Feeding habits of sixteen species of Oribatei (Acari) from an acid peat bog, Glenamoy, Ireland. *Revue d'Écologie et de Biologie du Sol*, 20, 221–267.

Behan-Pelletier, V.M., Kanashiro, D. 2010. Acari in grassland soils of Canada. In: Shorthouse, J.D., Floate, K.D. (Eds.), *Arthropods of Canadian Grasslands Volume 1: Ecology and Interactions in Grassland Habitats*. Biological Survey of Canada, 137–166.

Behan-Pelletier, V.M., Lindo, Z. 2019. Checklist of oribatid mites (Acari: Oribatida) of Canada and Alaska. *Zootaxa Monograph*, 4666, 1–180.

Behan-Pelletier, V.M., Norton, R.A. 1983. *Epidamaeus* (Acari: Damaeidae) of arctic western North America and extreme northeast U.S.S.R. *The Canadian Entomologist*, 115, 1253–1289.

Behan-Pelletier, V.M., Norton, R.A. 1985. *Epidamaeus* (Acari: Damaeidae) of subarctic Western North America and extreme northeast, U.S.S.R. *The Canadian Entomologist*, 117, 277–319.

Behan-Pelletier, V.M., Ryabinin, N.A. 1991a. Taxonomy and biogeography of *Proteremaeus* (Acari: Oribatida: Eremaeidae). *The Canadian Entomologist*, 123, 559–565.

Behan-Pelletier, V.M., Ryabinin, N.A. 1991b. Description of *Sacculozetes filosus* gen. nov., sp. nov. and *Guatemalozetes danos* sp. nov. (Acari: Oribatida) from grassland habitats. *The Canadian Entomologist*, 123, 1135–1147.

Behan-Pelletier, V., Schatz, H. 2010. Patterns of diversity in the Ceratozetoidea (Acari: Oribatida): A North American assessment. In: Sabelis, M.W., Bruin, J. (Eds.), *Trends in Acarology, Proceedings of the XII International Congress of Acarology, Amsterdam (2006)*. Springer, Dordrecht, 97–104.

Behan-Pelletier, V.M., Walter, D.E. 2000. Biodiversity of oribatid mites (Acari: Oribatida) in tree canopies and litter. In: Coleman, D.C., Hendrix, P.F. (Eds.), *Invertebrates as Webmasters in Ecosystems*. CABI Publishing, New York, 187–202.

Behan-Pelletier, V.M., Walter, D.E. 2009. *Unduloribates* from North America (Acari, Oribatida, Unduloribatidae). *Zootaxa*, 2294, 47–61.

References

Behan-Pelletier, V.M., Walter, D.E. 2012. *Oribatella* (Acari, Oribatida, Oribatellidae) of Western North America. *Zootaxa*, 3432, 1–62.

Behan-Pelletier, V.M., Walter, D.E. 2013. Phylogenetic relationships of *Tectoribates*: nymphal characters of new North American species place the genus in Tegoribatidae (Acari, Oribatida). *Zootaxa*, 3741, 459–489.

Behan-Pelletier, V.M., Winchester, N.N. 1998. Arboreal oribatid mite diversity: Colonizing the canopy. *Applied Soil Ecology*, 9, 45–51.

Behan-Pelletier, V.M., Eamer, B., Clayton, M. 2001. Mycobatidae (Acari: Oribatida) of Pacific Northwest canopy habitats. *The Canadian Entomologist*, 133, 755–775.

Behan-Pelletier, V.M., Clayton, M., Humble, L. 2002. *Parapirnodus* (Acari: Oribatida: Scheloribatidae) of canopy habitats in Western Canada. *Acarologia*, 42, 75–88.

Behan-Pelletier, V.M., Eamer, B., Clayton, M. 2005. Dendroeremaeidae n. fam., from forest trees in Western North America (Acari: Oribatida: Licneremaeoidea). *Acarologia*, 46, 321–339.

Bellido, A. 1979. Ecologie de *Carabodes willmanni* Bernini 1975 (Acari, Oribatei) dans les formations pionierres de la lande armoricaine. *Revue d'Écologie et de Biologie du Sol*, 16, 195–218.

Berch, S.M., Battigelli, J.P., Hope, G.D. 2007. Responses of soil mesofauna communities and oribatid mite species to site preparation treatments in high-elevation cutblocks in southern British Columbia. *Pedobiologia*, 51, 23–32.

Berch, S., Baumbrough, B., Battigelli, J., Kroeger, P., Strub, N., de Montigny, L. 2001. Preliminary assessment of selected soil organisms under different conifer species. *Ministry of Forests, Research Branch Laboratory, Victoria BC, Canada, Research Report*, 20, 27 pp.

Berg, J., Woas, S., Beck, L. 1990. Zur Taxonomie der *Phthiracarus*–Arten (Acari, Oribatei) eines südwestdeutschen Buchenwaldes. *Andrias, Karlsruhe*, 7, 61–90.

Berg, M., de Ruiter, P., Didden, W., Janssen, M., Schouten, T., Verhoef, H. 2001. Community food web, decomposition and nitrogen mineralisation in a stratified Scots pine forest soil. *Oikos*, 93, 130–142.

Berg, N.W., Pawluk, S. 1984. Soil mesofaunal studies under different vegetative regimes in northcentral Alberta. *Canadian Journal of Soil Science*, 64, 209–223.

Bergmann, P., Heethoff, M. 2012. The oviduct is a brood chamber for facultative egg retention in the parthenogenetic oribatid mite *Archegozetes longisetosus* Aoki (Acari, Oribatida). *Tissue and Cell*, 44, 342–350.

Bergmann, P.K., Laumann, M., Cloetens, P., Heethoff, M. 2008. Morphology of the internal reproductive organs of *Archegozetes longisetosus* Aoki (Acari, Oribatida). *Soil Organisms*, 80, 171–195.

Bergmann, P., Laumann, M., Norton, R.A., Heethoff, M. 2018. Cytological evidence for automictic thelytoky in parthenogenetic oribatid mites (Acari, Oribatida): Synaptonemal complexes confirm meiosis in *Archegozetes longisetosus*. *Acarologia*, 581, 342–356.

Berlese, A. 1882 1898. Acari, Myriapoda et Scorpiones hucusque in Italia reperta. 1882, Vols. 1–2; 1883, Vols. 3–9; 1884, Vols. 10–16; 1885, Vols. 17–23; 1886, Vols. 24–33; 1887, Vols. 34–44; 1888, Vols. 46–51; 1889, Vols. 52–57; 1890, Vol. 58; 1891, Vols. 59–60; 1892, Vols. 61–70; 1894, Vols. 71–75; 1895, Vols. 76–77; 1896, Vols. 78–79; 1897, Vols. 80–85; 1898, Vols. 86–92. [Reprinting, edited by L. van der Hammen 1979. Published by W. Junk, The Hague and Antiquariaat Junk, Amsterdam.]

Berlese, A. 1902. Specie de Acari nuovi. *Zoologischer Anzeiger*, 25, 697–700.

Berlese, A. 1903. Acari nuovi. Manipulus I. *Redia*, 1, 235–252.

Berlese, A. 1904. Acari nouvi. Manipulus III. *Redia*, 2, 10–32.

Berlese, A. 1908. Elenco di generi e specie nouve di acari. *Redia*, 5, 1–16.

Berlese, A. 1910. Brevi diagnosi di generi e specie nuovi di Acari. *Redia*, 6, 346–388.

Berlese, A. 1913. Acari nuovi. Manipoli VII–VIII. *Redia*, 9, 77–111 + pls. 1–8.

Berlese, A. 1914. Acari nuovi. Manipulus IX. *Redia*, 10, 113–150.

Berlese, A. 1916. Centuria prima di Acari nuovi. *Redia*, 12, 19–67.

Berlese, A. 1920. Centuria quinta di Acari nuovi. *Redia*, 14, 143–195.

Berlese, A. 1923. Centuria sesta di Acari nuovi. *Redia*, 15, 237–262.

Bernini, F. 1969. Notulae Oribatologicae I. Contributo alla conoscenza degli Oribatei (Acarida) della Pineta dei S. Vitale (Ravenna). *Redia*, 51, 329–375.

Bernini, F. 1973. Notulae Oribatologicae VII. Gli Oribatei (Acarida) dell'isolotto di Basiluzzo (Isole Eolie). *Lavori della Societá italiana de Biogeografia*, 3, 355–480.

Bernini, F. 1975. Notulae oribatologicae XII. Una nuova specie di Carabodes affine a *C. minusculus* Berlese 1923 (Acarida, Oribatei). *Redia*, 56, 455–471 + pls. 1–4.

Bernini, S., Bernini, F. 1990. Species of the family Cepheidae Berlese, 1896 (Acarida: Oribatida) from the Maghreb. *Zoological Journal of the Linnean Society*, 100, 233–262.

Berthet, P. 1964a. L'activité des Oribates d'une chênaie. *Mémoires du Musée Royal d'Histoire Naturelle de Belgique*, 152, 152 pp.

Berthet, P.L. 1964b. Field study of the mobility of Oribatei (Acari), using radioactive tagging. *Journal of Animal Ecology*, 33, 443–449.

Biancalani, R., Avagyan, A. (Eds.). 2014. Towards climate responsible peatland management. Mitigation of Climate Change in Agriculture Series 9, Food and Agriculture Organization of the United Nation (FAO), 100 pp.

Bird, G.A., Chatarpaul, L. 1986. Effect of whole-tree and conventional forest harvest on soil microarthropods. *Canadian Journal of Zoology*, 64, 1986–1993.

Bird, G.A., Chatarpaul, L. 1988. Effect of forest harvest on decomposition and colonization of maple leaf litter by soil microarthropods. *Canadian Journal of Zoology*, 68, 29–40.

Blackford, J.J., Payne, R.J., Heggen, M.P., de la Riva Caballero, A., van der Plicht, J. 2014. Age and impacts of the caldera-forming Aniakchak II eruption in western Alaska. *Quaternary Research*, 82, 85–95.

Block, W. 1979. Cold tolerance of micro-arthropods from Alaskan Taiga. *Ecological Entomology*, 4, 103–110.

Block, W., Convey, P. 1995. The biology, life-cycle and ecophysiology of the Antarctic mite *Alaskozetes antarcticus*. *Journal of Zoology, London*, 2363, 431–449.

Borcard, D. 1991. Les Oribates des tourbières du Jura suisse (Acari, Oribatei): Ecologie I. Quelques aspects de la communauté d'Oribates des sphaignes de la tourbière du Cachot. *Revue suisse de Zoologie*, 982, 303–317.

Borcard, D. 1994. Les Oribates des tourbières du Jura suisse (Acari, Oribatei). Faunistique V. Oppioidea, Suctobelbidae. *Mitteilungen der Schweizerischen Entomologischen Gesellschaft*, 67, 7–16.

Bosch-Serra, À.D., Padró, R., Boixadera-Bosch, R.R., Orobitg, J., Yagüe, M.R. 2014. Tillage and slurry over-fertilization affect oribatid mite communities in a semiarid Mediterranean environment. *Applied Soil Ecology*, 84, 124–139.

Brandt, A., Schaefer, I., Glanz, J., Schwander, T., Maraun, M., Scheu, S., Bast, J. 2017. Effective purifying selection in ancient asexual oribatid mites. *Nature Communications*, 8, 873.

Brandt, A., Tran Van, P., Bluhm, C., Anselmetti, Y., Dumas, Z., Figuet, E., François, C.M., Galtier, N., Heimburger, B., Jaron, K.S., Labédan, M., Maraun, M., Parker, D.J., Robinson-Rechavi, M., Schaefer, I., Simion, P., Scheu, S., Schwander, T., Bast, J. 2021. Haplotype divergence supports long-term asexuality in the oribatid mite *Oppiella nova*. *Proceedings of the National Academy of Sciences*, 118, e2101485118.

Bromberek, K., Olszanowski, Z. 2012. New moss mite of the genus *Camisia* from western Nearctic Region (Acari: Oribatida: Camisiidae). *Genus, Wrocław*, 23, 1–10.

Brückner, A., Heethoff, M. 2018. Nutritional effects on chemical defense alter predator-prey dynamics. *Chemoecology*, 28, 1–9.

Brückner, A., Stabentheiner, E., Leis, H.J., Raspotnig, G. 2015. Chemical basis of unwettability in Liacaridae (Acari, Oribatida): Specific variations of a cuticular acid/ester-based system. *Experimental and Applied Acarology*, 66, 313–335.

Brückner, A., Wehner, K., Neis, M., Heetoff, M. 2016. Attack and defense in a gamasid-oribatid mite predator-prey experiment – sclerotization outperforms chemical repellency. *Acarologia*, 564, 451–461.

Brückner, A., Raspotnig, G., Wehner, K., Meusinger, R., Norton, R.A., Heethoff, M. 2017. Storage and release of hydrogen cyanide in a chelicerate (*Oribatula tibialis*). *Proceedings of the National Academy of Science*, 114, 3469–3472.

Brückner, A., Kaltenpoth, N., Heethoff, M. 2020. De novo biosynthesis of simple aromatic compounds by an arthropod *Archegozetes longisetosus*. *Proceedings of the Royal Society B*, 287, 20201429.

Bücking, J. 2002. Three-dimensional structure of the gut system of the mites *Ameronothrus lineatus* (Oribatida: Ameronothridae) and *Hyadesia fusca* (Astigmata: Hyadesiidae). In: Bernini, F., Nannelli, R., Nuzzaci, G., De Lillo, E. (Eds.), *Acarid Phylogeny and Evolution: Adapation in Mites and Ticks. Proceedings of the IV Symposium of the European Association of Acarologists, Siena 2000*. Kluwer Academic Publishers, Dordrecht, Boston, London, 217–225.

Bücking, J., Ernst, H., Siemer, F. 1998. Population dynamics of phytophagous mites inhabiting rocky shores – K-strategists in an extreme environment? In: Ebermann, E. (Ed.), *Arthropod Biology: Contributions to Morphology, Ecology and Systematics. Biosystematics and Ecology Series 14*. Austrian Academy of Sciences Press, Vienna, Austria, 93–144.

Buford, D.R. 1976. Morphology and life history of *Hydrozetes bushnelli* n. sp. (Oribatei, Hydrozetidae). Unpublished Ph.D. thesis, University of Colorado, Boulder, 127 pp.

Buitendijk, A.M. 1945. Voorloopige Catalogus van de Acari in de Collectie Oudemans. *Zoologische Mededelingen*, 24, 281–391.

Bulanova-Zachvatkina, E.M. 1957. *Epidamaeus grandjeani* Bul.-Zachv., gen. et sp. nov. (Acariformes, Oribatei) from Kuril Islands. *Revue Entomologie de l'URSS*, 36, 547–552. [in Russian]

Bulanova-Zachvatkina, E.M. 1960. On the fauna of oribatid mites of the Soviet Union (Acariformes, Oribatei). *Naucnye doklady vyssej skoly. Biol. nauk Zoologia, Moskau*, 4, 27–39. [in Russian]

Bulanova-Zachvatkina, E.M. 1967. Pancirnye klesci–oribatidy. Izd. Vyssaja Skola, Moskva. 254 pp. + 20 tabs.
Bulanova-Zachvatkina, E.M. 1972. Systematics of the superfamily Belboidea Dubinin, 1958. *Proceedings of the 13th International Congress of Entomology*, 3, 344–345.
Caballero, A.I., Iturrondobeitia, J.C., Saloñs, M.I. 1999. A biometrical study of *Berniniella serratirostris* (Acari: Oribatei) and some related species. In: Bruin, J., van der Geest, L.P.S., Sabelis, M. (Eds.), *Ecology and Evolution of the Acari*. Kluwer Academic Publishers, Dordrecht, The Netherlands, 569–579.
Canestrini, V., Fanzago, F. 1876. Nuovi Acari italiani. *Atti della Accademia scientifica veneto–trentino–istriana*, 4, 99–111.
Carroll, G.C. 1980. Forest canopies: Complex and independent subsystems. In: Waring, R.H. (Ed.), *Forests: Fresh Perspectives from Ecosystem Analysis*. Oregon State University, Corvallis, OR, 87–108.
Carter, A., Cragg, J.B. 1976. Concentrations and standing crops of calcium, magnesium, potassium and sodium in soil and litter arthropods and their food in an aspen woodland ecosystem in the Rocky Mountains Canada. *Pedobiologia*, 16, 379–388.
Chen, J., Norton, R.A., Behan-Pelletier, V.M., Wang, H.F. 2004. Analysis of the genus *Gymnodampia* (Acari, Oribatida) with redescription of *G. setata* and description of two new species from North America. *The Canadian Entomologist*, 136, 793–821.
Chen, J., Liu, D., Wang, H.-F. 2010. Oribatid mites of China: A review of progress, with a checklist. *Zoosymposia*, 4, 186–224.
Chinaglia, L. 1917. Revisione del gen. "*Hydrozetes*" Berl. *Redia*, 12, 343–359.
Chinone, S. 1974. Further contribution to the knowledge of the family Brachychthoniidae from Japan. *Bulletin Biogeographical Society Japan*, 30, 1–29.
Chinone, S., Aoki, J. 1972. Soil mites of the family Brachychthoniidae from Japan. *Bulletin of the National Science Museum, Tokyo*, 15, 217–251.
Choi, S.S. 1996. A new genus and a new species of aquatic oribatid mite (Acari: Oribatida) from Korea. *Korean Journal of Soil Zoology*, 1, 1–4.
Clapperton, M.J., Kanashiro, D.A., Behan-Pelletier, V.M. 2002. Changes in abundance and diversity of microarthropods associated with Fescue Prairie grazing regimes. *Pedobiologia*, 465, 496–511.
Cianciolo, J.M., Norton, R.A. 2006. The ecological distribution of reproductive mode in oribatid mites, as related to biological complexity. *Experimental and Applied Acarology*, 40, 1–25.
Coggi, A. 1897. Una nuova specie di Oribatidae (*Notaspis lemnae*). In: Canestrini, G. (Ed.), *Prospetto dell'Acarofauna Italiana*. Padova, 8, 916–921.
Coleman, D.C., Crossley, D.A., Hendrix, P.F. 2004. *Fundamentals of Soil Ecology*. Elsevier Academic Press, Oxford, United Kingdom.
Colloff, M.J. 1984. Notes on two lichenophagous oribatid mites from Ailsa Craig (Acari: Cryptostigmata). *The Glasgow Naturalist*, 20, 451–458.
Colloff, M.J. 1988. Species associations of oribatid mites in lichens on the island of Ailsa Craig, Firth of Clyde (Acari: Cryptostigmata). *Journal of Natural History*, 22, 1111–1119.
Colloff, M.J. 1993. A taxonomic revision of the oribatid mite genus *Camisia* (Acari: Oribatida). *Journal of Natural History*, 27, 1325–1408.
Colloff, M.J. 2009. Comparative morphology and species-groups of the oribatid mite genus *Scapheremaeus* (Acari: Oribatida: Cymbaeremaeidae) with new species from South Australia. *Zootaxa*, 2213, 1–46.
Colloff, M.J., Cameron, S.L. 2013. A phylogenetic analysis and taxonomic revision of the oribatid mite family Malaconothridae (Acari: Oribatida) with new species of *Tyrphonothrus* and *Malaconothrus* from Australia. *Zootaxa*, 3681, 301–346.
Colloff, M.J., Seyd, E.L. 1991. A new species of *Moritzoppia* from montane sites in the British Isles, with a redescription of *M. clavigera* (Hammer, 1952) (Acari: Oribatida: Oppiidae). *Journal of Natural History*, 25, 1067–1074.
Convey, P. 1994. Growth and survival strategy of Antarctic mite *Alaskozetes antarcticus*. *Ecography*, 17, 97–107.
Cooreman, J. 1941. Notes sur la faune des Hautes–Fagnes en Belgique, VI (2e partie). Oribatei (Acariens). *Bulletin du Musée royal d'histoire naturelle de Belgique*, 17, 1–12.
Corral-Hernandez, E., Balanzategui, I., Iturrondobeitia, J.C. 2016. Ecosystemic, climatic and temporal differences in oribatid communities (Acari: Oribatida) from forest soils. *Experimental and Applied Acarology*, 69, 389–401.
Cotrufo, M.F., Wallenstein, M.D., Boot, C.M., Denef, K., Paul, E. 2013. The Microbial Efficiency-Matrix Stabilization (MEMS) framework integrates plant litter decomposition with soil organic matter stabilization: Do labile plant inputs form stable soil organic matter? *Global Change Biology*, 19, 988–995.
Coulson, S.J. 2007. Terrestrial and freshwater invertebrate fauna of the High Arctic Archipelago of Svalbard. *Zootaxa*, 1448, 41–58.

Coulson, S.J. 2008. The terrestrial invertebrate fauna of Svalbard: A cross referenced checklist. http://www.unis.no/35_STAFF/staff_webpages/biology/steve_coulson/documents/Fullreport_008.pdf (actualized in 15 May 2008).

Coulson, S.J. 2009. Association of the soil mite *Diapterobates notatus* (Thorell, 1871) (Acari, Oribatidae) with *Cynomya mortuorum* (Linnaeus, 1761) (Calliphoridae, Calliphorinae): implications for the dispersal of oribatid mites. *International Journal of Acarology*, 35, 175–177.

Coulson, S.J., Leinaas, H.P., Ims, R.A., Søvik, G. 2000. Experimental manipulation of the winter surface ice layer: the effects on a High Arctic soil microarthropod community. *Ecography*, 23, 299–306.

Coulson, S.J., Hodkinson, I.D., Webb, N.R., Harrison, J.A. 2002. Survival of terrestrial soil-dwelling arthropods on and in seawater: Implications for trans-oceanic dispersal. *Functional Ecology*, 16, 353–356.

Coulson, S.J., Moe, B., Monson, F., Gabrielsen, G.W. 2009. The invertebrate fauna of High Arctic seabird nests: The microarthropod community inhabiting nests on Spitsbergen, Svalbard. *Polar Biology*, 32, 1041–1046.

Coupland, R.T. 1973. Producers: I. Dynamics of aboveground standing crop. Canadian Committee for IBP, Matador Technical Report No. 27, University of Saskatchewan, Saskatoon, Canada, 159 p.

Crommentuijn, T., Doodeman, C.J.A.M., Doornekamp, A., Van Der Pol, J.J.C., Van Gestel, C.A.M., Bedaux, J.J.M. 1994. Lethal body concentrations and accumulation patterns determine time-dependent toxicity of cadmium in soil arthropods. *Environmental Toxicology and Chemistry*, 13, 1781–1789.

Curaqueo, G., Barea, J.M., Acevedo, E., Rubio, R., Cornejo, P., Borie, F. 2011. Effects of different tillage system on arbuscular mycorrhizal fungal propagules and physical properties in a Mediterranean agroecosystem in central Chile. *Soil and Tillage Research*, 113, 11–18.

Dabert, M., Witaliński, W., Kazmierski, A., Olszanowski, Z., Dabert, J. 2010. Molecular phylogeny of acariform mites (Acari, Arachnida): Strong conflict between phylogenetic signal and long-branch attraction artifacts. *Molecular Phylogenetics and Evolution*, 5610, 222–241.

Danks, H.V. 1981. *Arctic Arthropods. A review of systematics and ecology with particular reference to the North American fauna*. Entomological Society of Canada, Ottawa, 608 pp.

Davis, B.N.K., Murphy, P.W. 1961. An analysis of the Acarina and Collembola fauna of land reclaimed from open cast iron-stone mining. University of Nottingham School of Agricultural Report, 3–7.

De Ruiter, P.C., Moore, J.C., Zwart, K.B., Bouwman, L.A., Hassink, J., Bloem, J., De Vos, J.A., Marinissen, J.C.Y., Didden, W.A.M., Lebrink, G., Brussaard, L. 1993. Simulation of nitrogen mineralization in the below-ground food webs of two winter wheat fields. *Journal of Applied Ecology*, 30, 95–106.

Déchêne, A.D., Buddle, C.M. 2010. Decomposing logs increase oribatid mite assemblage diversity in mixedwood boreal forest. *Biodiversity Conservation*, 19, 237–256.

Deichsel, R. 2005. A morphometric analysis of the parthenogenetic oribatid mites *Hydrozetes lacustris* and *Hydrozetes parisiensis* sister species or morphotypes? In: Weigmann, G., Alberti, G., Wohltmann, A., Ragusa, S. (Eds.), *Acarine Biodiversity in the Natural and Human Sphere. Proceedings of the V Symposium of the European Association of Acarologists (Berlin 2004)*. Phytophaga, Palermo, 14, 377–382.

Denegri, G.M. 1993. Review of oribatid mites as intermediate hosts of the Anoplocephalidae. *Experimental and Applied Acarology*, 17, 567–580.

Denegri, G.M, Bernadina, W., Perez-Serrano, J., Rodriguez-Caabeiro, F. 1998. Anoplocephalid cestodes of veterinary and medical significance: a review. *Folia Parasitologica*, 45, 1–8.

Denneman, C.A.J., van Straalen, N.M. 1991. The toxicity of lead and copper in reproduction tests using the oribatid mite *Platynothrus peltifer*. *Pedobiologia*, 35, 305–311.

Dinsdale, D. 1974. Feeding activity of a phthiracarid mite (Acari). *Journal of Zoology*, 174, 15–21.

Doblas-Miranda, E., Work, T.E. 2015. Localized effects of coarse woody material on soil oribatid communities diminish over 700 years of stand development in black-spruce-feathermoss forests. *Forests*, 6, 914–928.

Domes, K., Althammer, M., Norton, R.A., Scheu, S., Maraun, M. 2007a. The phylogenetic relationship between Astigmata and Oribatida (Acari) as indicated by molecular markers. *Experimental and Applied Acarology*, 42, 159–171.

Domes, K., Norton, R.A., Maraun, M., Scheu, S. 2007b. Reevolution of sexuality breaks Dollo's law. *Proceedings of the National Academy of Sciences of the United States of America*, 10417, 7139–7144.

Domes, K., Maraun, M., Scheu, S., Cameron, S.L. 2008. The complete mitochondrial genome of the sexual oribatid mite *Steganacarus magnus*: genome rearrangements and loss of tRNAs. *BMC Genomics*, 9, 532.

Donaldson, G.M. 1996. Oribatida (Acari) associated with three species of *Sphagnum* at spruce-hole bog, New Hampshire, USA. *Canadian Journal of Zoology*, 74, 1706–1712.

Dwyer, E., Larson, D.J., Thompson, I.D. 1997. Oribatida (Acari) in Balsam fir [*Abies balsamea* (L)]. forests of Western Newfoundland. *The Canadian Entomologist*, 1291, 151–170.

Dwyer, E., Larson, D.J., Thompson, I.D. 1998. Oribatid mite communities of old Balsam fir (*Abies balsamea* (L)). forests of Western Newfoundland, Canada. *Pedobiologia*, 424, 331–347.

Ebermann, E. 1976. Oribatiden (Oribatei, Acari) als Zwischenwrite des Murmeltier-Bandwurmes *Ctenotaenia marmotae* (Frohlich, 1802). *Zeitschrift fur Parasitenkunde*, 50, 303–312.

Ecological Framework for Canada. 2014. http://ecozones.ca/english/ (accessed 28 June 2021).

Edsberg, E., Hågvar, S. 1999. Vertical distribution, abundance and biology of oribatid mites (Acari) developing inside decompositing spruce needles in a podsol soil profile. *Pedobiologia*, 43, 413–421.

Edwards, M.E., Lloyd, A., Armbruster, W.S. 2018. Assembly of Alaska-Yukon boreal steppe communities: Testing biogeographic hypotheses via modern ecological distributions. *Journal of Systematics and Evolution*, 56, 466–475.

Eisenhauer, N., Partsch, S., Parkinson, D., Scheu, S. 2007. Invasion of a deciduous forest by earthworms: Changes in soil chemistry, microflora, microarthropods and vegetation. *Soil Biology and Biochemistry*, 39, 1099–1110.

ElMehlawy, M.H. 2009. Oribatid mites as intermediate hosts of Anoplocephalid tapeworms and their distribution dynamics in a desert soil ecosystem. *The Egyptian Society of Experimental Biology (Zoology)*, 5, 167–173.

Enami, Y., Nakamura, Y. 1996. Influence of *Scheloribates azumaensis* (Acari: Oribatida) on *Rhizoctonia solani*, the cause of radish root rot. *Pedobiologia*, 40, 251–254.

Engelbrecht, C.M. 1974. The genus *Hydrozetes* (Oribatei: Acari) in South Africa. *Navorsinge van die Nasionale Museum, Bloemfontein*, 3, 41–49.

Erickson, J.M., Platt, R.B., Jr., Jennings, D.H. 2003. Holocene fossil oribatid mite biofacies as proxies of palaeohabitat at the Hiscock site, Byron, New York. *Bulletin of the Buffalo Society of Natural Sciences*, 37, 176–189.

Ermilov, S.G. 2007. The postembryonic development of *Camisia biurus* (Oribatei, Camisiidae). *Zoologicheskii Zhurnal*, 86, 286–294. [in Russian] Entomological Review, 87, 222–230. [English version]

Ermilov, S.G. 2009. Ontogeny of the oribatid mite *Nanhermannia coronata* (Acari, Oribatida, Nanhermanniidae). *Entomological Review*, 88, 429–437.

Ermilov, S.G. 2010. Morphology of juvenile instars of *Banksinoma lanceolata* (Acari, Oribatida, Thyrisomidae). *Acarina*, 18, 281–286.

Ermilov, S.G. 2012. Morphology of cornicles of oribatid mites of the family Damaeidae (Acari, Oribatida). *Zoologicheskii Zhurnal*, 91, 529–536. [in Russian] Entomological Review, 2012, 92, 576–582. [English version]

Ermilov, S.G., Chistyakov, M.P. 2008. Nutrition of oribatid mites of Crotonioidea superfamily in laboratory conditions. *Povolzhskiy Journal of Ecology*, 2, 142–146. [in Russian]

Ermilov, S.G., Kolesnikov, V.B. 2012. Morphology of juvenile instars of *Furcoribula furcillata* and *Zygoribatula exilis* (Acari, Oribatida). *Acarina*, 20, 48–59.

Ermilov, S.G., Liao, J.-R. 2018. To the knowledge of oribatid mites of the genus *Nothrus* (Acari, Oribatida, Nothridae) from Taiwan. *Biologia*, 73, 513–521.

Ermilov, S.G., Łochyńska, M. 2008. The influence of temperature on the development time of three oribatid mite species (Acari, Oribatida). *North-Western Journal of Zoology*, 4, 274–281.

Ermilov, S.G., Makarova, O.L. 2021. Contribution to the knowledge of the oribatid mite genus *Scutozetes* (Acari, Oribatida, Tegoribatidae), with description of a new species from the Russian Arctic. *International Journal of Acarology*, 47, 500–509.

Ermilov, S.G., Ryabinin, N.A. 2020. Contribution to the knowledge of *Parabelbella* (Acari, Oribatida, Damaeidae): Description of two new species from Russia and the U.S.A., redescription of *P. inaequipes* (Banks, 1947) and a key to known species. *Zootaxa*, 4860, 352–374.

Ermilov, S.G., Chistyakov, M.P., Renzhina, A.A. 2004. Temperature effect on the development duration of *Trhypochthonius tectorum* (Berlese, 1896) (Acariformes, Oribatei). *Povolzhskiy Zoological Journal*, 1, 87–90. [in Russian]

Ermilov, S.G., Łochyńska, M., Olszanowski, Z. 2008. The cultivation and morphology of juvenile stages of two species from genus *Scutovertex* (Acari, Oribatida, Scutoverticidae). *Annales Zoologici*, 58, 433–443.

Ermilov, S.G., Ryabinin, N.A., Anichkin, A.E. 2012. The morphology of juvenile instars of two oribatid species of the family Hermanniidae (Acari). *Zoologicheskii Zhurnal*, 91, 657–668. [in Russian] Entomological Review, 92, 815–826 [English version]

Ermilov, S.G., Salavatulin, V.M., Khaustov, A.A. 2015a. The first findings and supplementary description of *Pergalumna emarginata* (Acari, Oribatida, Galumnidae) from Russia. *Acarina*, 23, 121–131.

Ermilov, S.G., Salavatulin, V.M., Tolstikov, A.V. 2015b. A new species of *Pantelozetes* (Acari, Oribatida, Thyrisomidae) from environs of lake Baikal (Russia). *Systematic and Applied Acarology*, 20, 51–60.

Ermilov, S.G., Tolstikov, A.V., Salavatulin, V.M., Bragin, E.A. 2015c. Morphology of juvenile stages in two species of arboreal oribatid mites, *Scapheremaeus palustris* and *Phauloppia nemoralis* (Acari, Oribatida). *Zoologicheskii Zhurnal*, 94, 26–36 [in Russian] Entomological Review, 95, 126–136. [English version]

Ermilov, S.G., Makarova, O.L., Bizin, M.S. 2019. Morphological development, distribution and ecology of the arctic oribatid mite *Hermannia scabra* (Acari: Oribatida: Hermanniidae) and synonymy of *Hermannia gigantea*. *Zootaxa*, 4717, 104–136.

Ermilov, S.G., Makarova, O.L., Behan-Pelletier, V.M. 2022. Taxonomy and ecology of the Arctic oribatid mite *Svalbardia lucens* comb. nov. (Acari, Oribatida, Ceratozetidae): Resolving a long-standing confusion. *Systematic and Applied Acarology*, 27, 497–510.

Estrada-Venegas, E., Norton, R.A., Moldenke, A.R. 1996. Unusual sperm-transfer in *Pilogalumna* sp. (Galumnidae). In: Mitchell, R., Horn, D.J., Needham, G.R., Welbourn, C.W. (Eds.), *Acarology IX Proceedings*. vol. 1. Ohio Biological Survey, Columbus, OH, 565–567.

Evans, G.O. 1952a. Terrestrial Acari new to Britain. I. *Annals of the Magazine Natural History, Series 12*, 5, 33–41.

Evans, G.O. 1952b. British mites of the genus *Brachychthonius* Berl. 1910. *Annals of the Magazine Natural History, Series 12*, 5, 227–239.

Evans, G.O. 1992. *Principles of Acarology*. CAB International, Wallingford, 563 pp.

Evans, G.O., Sheals, J.G., Macfarlane, D. 1962. The terrestrial Acari of the British Isles. An introduction to their morphology, biology and classification. Vol. I. *The Quarterly Review of Biology*, 37, 265–266.

Ewing, H.E. 1907. New Oribatidae. *Psyche (Cambridge, Mass.)*, 14, 111–115.

Ewing, H.E. 1908a. A new genus and species of Oribatidae. *Entomological News*, 19, 243–245 + pl. 11.

Ewing, H.E. 1908b. Two new species of the genus *Phthiracarus*. *Entomological News*, 19, 449–451.

Ewing, H.E. 1909a. The Oribatoidea of Illinois. *Bulletin of the Illinois State Laboratory of Natural History*, 7, 337–389 + pls. 33–35.

Ewing, H.E. 1909b. New American Oribatoidea. *Journal of the New York Entomological Society*, 17, 116–136.

Ewing, H.E. 1913. Some new and curious Acarina from Oregon. *Journal of Entomology and Zoology*, 5, 123–136 + figs. 1–6.

Ewing, H.E. 1917. A synopsis of the genera of beetle mites with special reference to the North American fauna. *Annals of the Entomological Society of America*, 10, 117–132.

Ewing, H.E. 1918. The beetle-mite fauna of Mary's Peak, Coast Range, Oregon (Acarina). *Entomological News*, 29, 81–90.

Fagan, L.L., Winchester, N.N. 1999. Arboreal arthropods: Diversity and rates of colonization in a temperate montane forest. *Selbyana*, 201, 171–188.

Fagan, L.L., Didham, R.K., Winchester, N.N., Behan-Pelletier, V., Clayton, M., Lindquist, E., Ring, R.A. 2005. An experimental assessment of biodiversity and species turnover in terrestrial vs. canopy leaf litter. *Oecologia*, 1472, 335–347.

Fain, A., Lambrechts, L. 1987. Observations on the acarofauna of fish aquariums. I. Mites associated with Discus fish. *Bulletin et Annales de la Société Royal Belgique d'Entomologie*, 123, 87–102.

Fajana, H.O., Gainer, A., Jegede, O.O., Awuah, K.F., Princz, J.I., Owojori, O.J., Siciliano, S.D. 2019. *Oppia nitens* C.L. Koch, 1836 (Acari: Oribatida): Current status of its bionomics and relevance as a model invertebrate in soil ecotoxicology. *Environmental Toxicology and Chemistry*, 38, 2593–2613.

Fajana, H.O., Jegede, O.O., James, K., Hogan, N.S., Siciliano, S.D. 2020. Uptake, toxicity, and maternal transfer of cadmium in the oribatid soil mite, *Oppia nitens*: Implication in the risk assessment of cadmium to soil invertebrates. *Environmental Pollution*, 259, 113912.

Fan, Q.H., Heath, P. 2019. An oribatid mite, *Hydrozetes lemnae* (Hydrozetidae) on farmed eels *Anguilla australis* and *A. dieffenbachii* (Anguillidae). *Systematic & Applied Acarology*, 24, 1809–1813.

Farahat, A.Z. 1966. Studies on the influence of some fungi on Collembola and Acari. *Pedobiologia*, 6, 258–268.

Feider, Z., Suciu, I. 1957. Contributie la cunoasterea Oribatidelor (Acari) din R.P.R. Familia Phthiracaridae Perthy 1841. Contribution a la connaissance des Oribates (Acariens). *Academia R. P. R Filiala Iasi Studii Şi Cercetări Ştiintifce, Sectia Biologie, Şi Stiinte Agricole*, 8, 23–48.

Feider, Z., Vasiliu, N., Calugar, M. 1971. *Minunthozetes semirufus* C.L. Koch, 1849 o nuoa specie pentru fauna Romaniei si descrierea speciei *Minunthozetes pseudofusiger* Schweizer, 1916. *Studii Şi Cercetări de Biologie, Serie Zoologie*, 23, 409–418.

Fernandez, N.A., Athias-Binche, F. 1986. Analyse démographique d'une population d'*Hydrozetes lemnae* Coggi, Acarien Oribate inféodé à la lentille d'eau *Lemna gibba* L. en Argentine. I. Méthodes et techniques, démographie d' *H. lemnae* comparais. *Zoologische Jahrbücher. Abteilung für Systematik, Ökologie und Geographie der Tiere*, 1132, 213–228.

Fernandez, N.A., Alberti, G., Kümmel, G. 1991. Spermatophores and spermatozoa of oribatid mites (Acari: Oribatida). Part I. Fine structure and histochemistry. *Acarologia*, 323, 261–286.

Finnamore, A.T. 1994. Hymenoptera of the Wagner Natural Area, a boreal spring fen in Central Alberta. *Memoirs of the Entomological Society of Canada*, 169, 181–220.

Fitch, A. 1856. Report on the noxious and other insects of the state of New York (third annual). *Transactions of the New York State Agricultural Society*, 16, 315–490.

Forsslund, K.H. 1941. Schwedische arten der gattung *Suctobelba* Paoli (Acari, Oribatei). *Zoologiska Bidrag Från Uppsala*, 20, 381–396.

Forsslund, K.H. 1942. Schwedische Oribatei (Acari). I. *Arkiv för zoologi*, 34, 1–11.

Forsslund, K.H. 1947. Über die Gattung *Autogneta* Hull (Acari, Oribatei). *Zoologiska Bidrag Från Uppsala*, 25, 111–117.

Forsslund, K.H. 1953. Schwedische Oribatei (Acari). II. *Entomologisk Tidskrift*, 74, 152–157.

Forsslund, K.H. 1956. Schwedische Oribatei (Acari). III. *Entomologisk Tidskrift*, 77, 210–218.

Forsslund, K.H. 1957. Notizen über Oribatei (Acari). I. *Arkiv för Zoologi*, 10, 583–593.

Forsslund, K.H. 1963. Notizen über Oribatei (Acari). III. *Entomologisk Tidskrift*, 84, 282–283.

Forsslund, K.H. 1964. *Liochthonius muscorum* n. sp. und *L. lapponicus* (Träg.) (Acari, Oribatei). *Entomologisk Tidskrift*, 85, 238–238.

Franchini, C.P., Rockett, L. 1996. Oribatid mites as "indicator" species for estimating the environmental impact of conventional and conservation tillage practices. *Pedobiologia*, 40, 217–225.

Franklin, E., Schubart, H.O.R., Adis, J.U. 1997. Edaphic oribatid mites in two floodplain forests of Central Amazonia: Vertical distribution, abundance and soil recolonisation after inundation. *Revista Brasileira de Biología*, 573, 501–520

Franklin, E.N., Guimaraes, R.L., Adis, J., Schubart, H.O.R. 2001a. The resistence to submersion of terrestrial Acari (Acari, Oribatida) from flooded and non–flooded forests of Central Amazonia in experimental laboratorial conditions. *Acta Amazonica*, 31, 285–298.

Franklin, E.N., Morais, J.W., Dos Santos, E.M.R. 2001b. Density and biomass of Acari and Collembola in primary forest, secondary regrowth and polycultures in Central Amazonia. *Andrias, Karlsruhe*, 15, 141–153.

Franklin, E., Norton, R.A., Crossley, D.A. 2008. *Zygoribatula colemani* sp. nov. (Acari, Oribatida, Oribatulidae) from granite outcrops in Georgia, USA, with a highly variable translamella. *Zootaxa*, 1847, 34–48.

Fredes, N.A., Martínez, P.A. 2011. First record of *Euzetes globulus* (Nicolet, 1855) from Neotropical region (Acari, Oribatida). *Genus (Wrocław)*, 22, 175–179.

Fredes, N.A., Martínez, P.A. 2013. A new *Siculobata* species (Acari: Oribatida: Scheloribatidae) from Argentina. *International Journal of Acarology*, 39, 317–324.

Fredes, N.A., Martínez, P.A. 2014. Redescription of *Hemileius suramericanus* (Acari, Oribatida, Scheloribatidae) with comments about Neotropical congeneric species. *Persian Journal of Acarology*, 3, 249–256.

Fritz, G.N. 1995. Oribatid mites infected with cysticercoids of *Moniezia expansa* (Cestoda: Anoplocephalidae). *International Journal of Acarology*, 21, 233–238.

Fujikawa, T. 1979. Revision of the family Banksinomidae (Acari, Oribatei). *Acarologia*, 20, 433–467.

Fujikawa, T. 1988. Biology of *Tectocepheus velatus* (Michael) and *T. cuspidentatus* Knülle. *Acarologia*, 29, 307–315.

Fujikawa, T. 1991. List of oribatid families and genera of the World. *Edaphologia*, 46, 1–132.

Fujikawa, T. 1995. Oribatid mites from *Picea glehni* forest at Mo–Ashoro, Hokkaido (11) A new species of the family Trhypochthoniidae. *Edaphologia*, 53, 1–6.

Fujikawa, T. 1999. Individual variations of two reared species, *Tectocepheus velatus* (Michael, 1880) and *Oppiella nova* (Oudemans, 1902). *Edaphologia*, 62, 11–46.

Fujikawa, T. 2000. Five new species of the genera *Trhypochthoniellus* and *Trhypochthonius*. *Edaphologia*, 65, 35–53.

Fujikawa, T. 2001. A new and three known species of Tectocepheidae and Nodocepheidae from Northeastern part of Nippon including the Shirakami–sanchi World Heritage Area (Acari: Oribatida). *Edaphologia*, 67, 23–30.

Fujikawa, T. 2003. A new species of the genus *Hypochthonius* (Acari: Oribatida). *Edaphologia*, 73, 11–17.

Fujikawa, T. 2004. Nineteen new species from the Shirakami–sanchi World Heritage Area, Nippon (Acari: Oribatida). *Acarologia*, 44, 97–131.

Fujikawa, T., Fujita, M., Aoki, J. 1993. Checklist of oribatid mites of Japan (Acari: Oribatida). *Journal of the Acarological Society of Japan*, 2(Supplement 1), 1–121.

Gergócs, V., Hufnagel, L. 2009. Application of oribatid mites as indicators. *Applied Ecology and Environmental Research*, 7, 79–98.

Gervais, P. 1844. Acari. In: Walckenaer, M.L.B. (Ed.), *Histoire Naturelle des Insectes Aptères*. Paris, 3, 1–476.

Giller, P.S. 1996. The diversity of soil communities, the 'poor man's tropical rainforest'. *Biodiversity and Conservation*, 5, 135–168.

Gilyarov, M.S. 1975. *Opredelitel' obitajuschtschich w potschwe kleschtschej Sarcoptiformes. Key to Soil-Inhabiting Mites Sarcoptiformes*. Nauka Publishers, Moscow, 492 pp.

Giribet, G., Edgecombe, G.D. 2019. The phylogeny and evolutionary history of Arthropods. *Current Biology*, 29, 592–602.

Gjelstrup, P. 1978. An annotated list of Danish oribatid mites (Acarina, Oribatei). *Entomologiske Meddelelser*, 46, 109–121. [in Danish with English summary]

Gjelstrup, P., Solhøy, T. 1994. *The Oribatid Mites (Acari) of Iceland*. The Zoology of Iceland, Steenstrupia, Zoological Museum of Copenhagen, vol. 3, Part 57e, 1–78.

Golosova, L., Karppinen, E. Krivolutsky, D.A. 1983. List of oribatid mites (Acarina, Oribatei) of northern palaearctic region. II. Siberia and the Far East. *Acta Entomologica Fennica*, 43, 1–14.

Gong, X., Chen, T.W., Zieger, S.L., Bluhm, C., Heidemann, K., Schaefer, I., Maraun, M., Liu, M., Scheu, S. 2018. Phylogenetic and trophic determinants of gut microbiota in soil oribatid mites. *Soil Biology and Biochemistry*, 123, 155–164.

Gordeeva, E.V. 1980. Oribatid mites of the family Cosmochthoniidae (Oribatei). *Zoologicheskii Zhurnal*, 59, 838–850.

Gourbiere, F., Lions, J.C., Pepin, R. 1985. Activité et dévéloppement d'*Adoristes ovatus* (C.L. Koch, 1839) (Acarien, Oribate) dans les aiguilles d'*Abies alba* Mill. Relations avec la décomposition et les microflores fondiques. *Revue d'Écologie et de Biologie du Sol*, 22, 57–73.

Grabowski, W.B. 1971. A new genus of oribatid mite (Cryptostigmata, Oribatellidae). *Proceedings of the Entomological Society of Washington*, 73, 44–47.

Grandjean, F. 1929. Quelques nouveaux genres d'Oribatei du Venezuela et de la Martinique. *Bulletin de la Société Zoologique de France*, 54, 400–423.

Grandjean, F. 1930. Oribates nouveaux de la region caraibe. *Bulletin de la Société Zoologique de France*, 55, 262–284.

Grandjean, F. 1931a. Observations sur les Oribates (1re série). *Bulletin du Muséum national d'Histoire naturelle*, 3, 131–144.

Grandjean, F. 1931b. Le genre *Licneremaeus* Paoli (Acariens). *Bulletin de la Société Zoologique de France*, 56, 221–250.

Grandjean, F. 1931c. Observations sur les Oribates (2e série). *Bulletin du Muséum national d'Histoire naturelle, Série 2*, 3, 651–665.

Grandjean, F. 1932a. Observations sur les Oribates (3e série). *Bulletin du Muséum national d'Histoire naturelle, Série 2*, 4, 292–306.

Grandjean, F. 1932b. Au sujet des Palaeacariformes Trägårdh. *Bulletin du Muséum national d'Histoire naturelle, Série 4*, 411–426.

Grandjean, F. 1933. Observations sur les Oribates (5e série). *Bulletin du Muséum national d'Histoire naturelle, Série 5*, 461–468.

Grandjean, F. 1934a. La notation des poils gastronotiques et des poils dorsaux du propodosoma chez les Oribates (Acariens). *Bulletin de la Société Zoologique de France*, 59, 12–44.

Grandjean, F. 1934b. Les organes respiratoires secondaires des Oribates (Acariens). *Annales de la Société entomologique de France*, 103, 109–146.

Grandjean, F. 1934c. Observations sur les Oribates (6e série). *Bulletin du Muséum national d'Histoire naturelle, Série 2*, 6, 353–360.

Grandjean, F. 1935a. Les poils et les organes sensitifs portés par les pattes et le palpe chez les Oribates. *Bulletin de la Société Zoologique de France*, 60, 6–39.

Grandjean, F. 1935b. Observations sur les Oribates (8e série). *Bulletin du Muséum national d'Histoire naturelle, Série 2*, 8, 237–244.

Grandjean, F. 1936a. Les Oribates de Jean Frédéric Hermann et de son pére. *Annales de la Société entomologique de France*, 105, 27–110.

Grandjean, F. 1936b. Observations sur les Oribates (10e série). *Bulletin du Muséum national d'Histoire naturelle, Série 2*, 8, 246–253.

Grandjean, F. 1939a. Observations sur les Oribates (12e série). *Bulletin du Muséum national d'Histoire naturelle, Série 2*, 11, 300–307.

Grandjean, F. 1939b. Les segments post-larvaires de l'hysterosoma chez les Oribates (Acariens). *Bulletin de la Société Zoologique de France*, 64, 273–284.

Grandjean, F. 1940. Les poils et les organes sensitifs portés par les pattes et le palpe chez les Oribates. Deuxième partie. *Bulletin de la Société Zoologique de France*, 45, 32–44.

Grandjean, F. 1941. La chaetotaxie comparée des pattes chez les Oribates (1re série). *Bulletin de la Société Zoologique de France*, 66, 33–50.

Grandjean, F. 1943. Observations sur les Oribates (16e série). *Bulletin du Muséum national d'Histoire naturelle, Série 2*, 15, 410–417.

Grandjean, F. 1946. Les Enarthronota Acariens. Première série. *Annales des Sciences Naturelles, Zoologie*, 11, 213–248.

Grandjean, F. 1947a. L'origine pileuse des mors et la chaetotaxie de la mandibule chez les Acariens actinochitineux. *Comptes rendus des séances de Académie des Sciences, Paris*, 224, 1251–1254.

Grandjean, F. 1947b. Observations sur les Oribates (17e série). *Bulletin du Muséum national d'Histoire naturelle, Série 2*, 19, 165–172.

Grandjean, F. 1949. Les Enarthronota (Acariens) (2e série). *Annales des Sciences Naturelles, Zoologie, Série 11e*, 10, 29–58.

Grandjean, F. 1950. Les Enarthronota (Acariens) (3e série). *Annales des Sciences Naturelles, Zoologie, Série 11e*, 12, 85–107.

Grandjean, F. 1951a. Sur deux especes du genre *Dometorina* n. g. et les moeurs de *D. plantivaga* (Berl.) (Acariens, Oribates). *Bulletin de la Société Zoologique de France*, 75, 224–242.

Grandjean, F. 1951b. Étude sur les Zetorchestidae (Acariens, Oribates). *Mémoires du Muséum nationale d'Histoire naturelle, A. Zoologie*, 4, 1–50.

Grandjean, F. 1951c. Observations sur les Oribates (22e série). *Bulletin du Muséum national d'Histoire naturelle, Série 2*, 23, 91–98.

Grandjean, F. 1951d. Observations sur les Oribates (23e série). *Bulletin du Muséum national d'Histoire naturelle, Série 2*, 261–268.

Grandjean, F. 1952a. Au sujet de l'ectosquelette du podosoma chez les Oribates supérieurs et de sa terminologie. *Bulletin de la Société Zoologique de France*, 77, 13–36.

Grandjean, F. 1952b. Sur les articles des appendices chez les Acariens actinochitineux. *Comptes rendus des séances de Académie des Sciences, Paris*, 235, 560–564.

Grandjean, F. 1953a. Observations sur les Oribates (25e série). *Bulletin du Muséum national d'Histoire naturelle, Série 2*, 25, 155–162.

Grandjean, F. 1953b. Observations sur les Oribates (27e série). *Bulletin du Muséum national d'Histoire naturelle, Série 2*, 25, 469–476.

Grandjean, F. 1953c. Sur les genres "*Hemileius*" Berl. et "*Siculobata*" n.g. Acariens, Oribates. *Mémoires du Museum national d'Histoire naturelle. Série A, Zoologie*, 6, 117–137.

Grandjean, F. 1954a. Essai de classification des Oribates (Acariens). *Bulletin de la Société Zoologique de France*, 78, 421–446.

Grandjean, F. 1954b. Etude sur les Palaeacaroides Acariens, Oribates. *Mémoires du Museum national d'Histoire naturelle. Série A, Zoologie*, 7, 179–274.

Grandjean, F. 1955. Sur un Acarien des iles Kerguélen. *Podacarus Auberti* (Oribate). *Mémoires du Muséum national d'Histoire naturelle, Série 8*, 109–150.

Grandjean, F. 1956a. Caractères chitineux de l'ovipositor, en structure normale, chez les Oribates (Acariens). Archives de zoologie expérimentale et générale. *Notes et Revue*, 932, 96–106.

Grandjean, F. 1956b. Sur deux espèces nouvelles d'Oribates (Acariens) apparentées à *Oripoda elongata* Banks 1904. *Archives de Zoologie Expérimentale et Générale*, 93, 185–218.

Grandjean, F. 1956c. Observations sur les Oribates (33e série). *Bulletin du Muséum national d'Histoire naturelle, Série 2*, 28, 111–118.

Grandjean, F. 1956d. Observations sur les Oribates (35e série). *Bulletin du Muséum national d'Histoire naturelle, Série 2*, 28, 282–289.

Grandjean, F. 1956e. Galumnidae sans carènes lamellaires (Acariens, Oribates). 1re série. *Bulletin de la Société Zoologique de France*, 81, 134–150.

Grandjean, F. 1956f. Observations sur les Galumnidae. (2e série) (Acariens, Oribates). *Revue d'entomologie*, 23, 265–275.

Grandjean, F. 1957. L'infracapitulum et la manducation chez les Oribates et d'autres Acariens. *Annales des Sciences Naturelles, Zoologie*, 11, 233–281.

Grandjean, F. 1958a. *Perlohmannia dissimilis* (Hewitt) (Acarien, Oribate). *Mémoires du Museum national d'Histoire naturelle. Série A, Zoologie*, 16, 57–119.
Grandjean, F. 1958b. Scheloribatidae et Oribatulidae (Acariens, Oribates). *Bulletin du Muséum national d'Histoire naturelle, Série 2*, 30, 352–359.
Grandjean, F. 1958c. Au sujet du naso et de son oeil infère chez les Oribates et les Endeostigmata (Acariens). *Bulletin du Muséum national d'Histoire naturelle, Série 2*, 30, 427–435.
Grandjean, F. 1959a. *Polypterozetes cherubin* Berl. 1916 Oribate. *Acarologia*, 1, 147–180.
Grandjean, F. 1959b. *Hammation sollertius* n.g., n.sp. (Acarien, Oribate). *Mémoires du Museum national d'Histoire naturelle. Série A, Zoologie*, 16, 173–198.
Grandjean, F. 1959c. Observations sur les Oribates (39e série). *Bulletin du Muséum national d'Histoire naturelle, Série 2*, 31, 248–255.
Grandjean, F. 1960a. Les Mochlozetidae n. fam. (Oribates). *Acarologia*, 2, 101–148.
Grandjean, F. 1960b. *Damaeus arvernensis* n. sp. (Oribate). *Acarologia*, 2, 250–275.
Grandjean, F. 1960c. *Autogneta penicillum* n. sp. (Oribate). *Acarologia*, 2, 345–367.
Grandjean, F. 1960d. Les Autognetidae n. fam. (Oribates). *Acarologia*, 2, 575–609.
Grandjean, F. 1961a. Les Plasmobatidae n. fam. (Oribates). *Acarologia*, 3, 96–129.
Grandjean, F. 1961b. Nouvelles observations sur les Oribates (1re série). *Acarologia*, 3, 206–231.
Grandjean, F. 1961c. *Perlohmannia coiffaiti* n. sp. (Oribates). *Acarologia*, 3, 604–619.
Grandjean, F. 1962a. Au sujet des Hermanniellidae (Oribates). Deuxième partie. *Acarologia*, 4, 632–670.
Grandjean, F. 1962b. Nouvelles observations sur les Oribates (2e série). *Acarologia*, 4, 396–422.
Grandjean, F. 1963. Les Autognetidae (Oribates). Deuxième partie. *Acarologia*, 5, 653–689.
Grandjean, F. 1964. *Pheroliodes wehnkei* (Willmann) (Oribate). *Acarologia*, 6, 353–386.
Grandjean, F. 1965a. *Fosseremus quadripertitus* nom. nov. (Oribate). *Acarologia*, 7, 343–375.
Grandjean, F. 1965b. Complément à mon travail de 1953 sur la classification des Oribates. *Acarologia*, 7, 713–734.
Grandjean, F. 1968. Nouvelles observations sur les Oribates (6e série). *Acarologia*, 10, 357–391.
Grandjean, F. 1969a. Considérations sur le classement des Oribates. Leur division en 6 groupes majeurs. *Acarologia*, 11, 127–153.
Grandjean, F. 1969b. Observation sur les muscles de fermeture des volets anaux et génitaux et sur la structure progenitale chez les Oribates supérieures adultes. *Acarologia*, 11, 317–349.
Grandjean, F. 1970. Nouvelles observations sur les Oribates (7e série). *Acarologia*, 12, 432–460.
Grandjean, F. 1971. Nouvelles observations sur les Oribates (8e série). *Acarologia*, 12, 849–876.
Grishina, L.G. 1993. Ecology of the oribatid mite *Nothrus palustris* C.L. Koch (Sarcoptiformes, Oribatei) in Western Siberia. *Ekologija*, 3, 39–49. [in Russian]
Habeeb, H. 1974. Some new aquatic oribatid mites. *Leaflets of the Acadian Biology*, 62, 1–6.
Hågvar, S. 1998. Mites (Acari) developing inside decomposing spruce needles: Biology and effect on decomposition rate. *Pedobiologia*, 42, 358–377.
Hågvar, S., Solhøy, T., Mong, C.E. 2009. Primary succession of soil mites (Acari) in a Norwegian glacier foreland, with emphasis on oribatid species. *Arctic, Antarctic, and Alpine Research*, 41, 219–227.
Halbert, J.N. 1920. The Acarina of the seashore. *Proceeding of the Royal Irish Academy*, Section B, 35, 106–152 + pls. 21–23.
Haller, G. 1882. Beitrag zur Kenntniss der Milbenfauna Wurttembergs. *Jahreshefte des Vereins für vaterländische Naturkunde in Württemberg*, 38, 293–325.
Haller, G. 1884. Beschreibung einiger neuer Milben. *Archiv für Naturgeschichte*, 50, 217–236.
Hammen, L. 1952. The Oribatei (Acari) of the Netherlands. *Zoologische verhandelingen, Leiden*, 17, 1–139.
Hammen, L. 1959. Berlese's primitive oribatid mites. *Zoologische verhandelingen, Leiden*, 40, 1–93.
Hammen, L. 1968. The gnathosoma of *Hermannia convexa* (C.L. Koch) (Acarida: Oribatina) and comparative remarks on its morphology in other mites. *Zoologische Verhandelingen*, 94, 1–45.
Hammen, L. 1980. *Glossary of acarological terminology. Part 1: General terminology.* Junk Publisher, The Hague, 270 pp.
Hammen, L. 1982. Comparative studies in Chelicerata II. Epimerata (Palpigradi and Actinotrichida). *Zoologische Verhandelingen*, 196, 1–70.
Hammer, M. 1944. Studies on the oribatids and collemboles of Greenland. *Meddelelser om Grønland, Copenhagen*, 141, 1–210.
Hammer, M. 1952a. Investigations of the microfauna of Northern Canada. Part I. Oribatidae. *Acta Arctica*, 4, 1–108.
Hammer, M. 1952b. A new oribatid (Acarina) from Rocky Mtns. *Entomologiske meddelelser*, 26, 380–383.
Hammer, M. 1955. Alaskan oribatids. *Acta Arctica*, 7, 1–36.

Hammer, M. 1958. Investigations on the oribatid fauna of the Andes Mountains I. The Argentine and Bolivia. *Biologiske Skrifter udgivet af Det Kongelige Danske Videnskabernes Selskab*, 10, 1–262.

Hammer, M. 1961. Investigations on the oribatid fauna of the Andes Mountains II. Peru. *Biologiske Skrifter udgivet af Det Kongelige Danske Videnskabernes Selskab*, 13, 1–200.

Hammer, M. 1962a. Investigations on the oribatid fauna of the Andes Mountains. III. Chile. *Biologiske Skrifter udgivet af Det Kongelige Danske Videnskabernes Selskab*, 13, 1–96.

Hammer, M. 1962b. Investigations on the oribatid fauna of the Andes Mountains. IV. Patagonia. *Biologiske Skrifter udgivet af Det Kongelige Danske Videnskabernes Selskab*, 13, 1–35.

Hammer, M. 1965. Are low temperatures a species-preserving factor? Illustrated by the oribatid mite *Mucronothrus nasalis* (Willm.). *Acta Universitatis Lundensis*, 2, 1–10.

Hammer, M. 1967. Some oribatids from Kodiak Is. near Alaska. *Acta Arctica*, 14, 1–25.

Hammer, M. 1969. Oribatids found at plant quarantine stations in U.S.A. *Videnskabelige Meddelelser Dansk Naturhistorisk Forening*, 132, 63–78.

Hammer, M. 1971. On some oribatids from Viti Levu, Fiji Islands. *Biologiske Skrifter udgivet af det Kongelige Danske Videnskabernes Selskaab*, 16, 1–60.

Hammer, M. 1973. Oribatids from Tongatapu and Eua, the Tonga Islands, and from Upolu, Western Samoa. *Biologiske Skrifter udgivet af det Kongelige Danske Videnskabernes Selskaab*, 20, 1–70.

Hansen, R.A. 1999. Red oak litter promotes a microarthropod functional group that accelerates its decomposition. *Plant and Soil*, 209, 37–45.

Hansen, R.A., Coleman, D.C. 1998. Litter complexity and compostition are determinants of the diversity and species composition of oribatid mites (Acari: Oribatida) in litterbags. *Applied Soil Ecology*, 9, 17–23.

Haq, M.A., Ramani, N. 1985. Possible role of oribatid mites in weed control. *Proceedings of the National Seminar on Entomophagous Insects*. Calicut. pp. 214–220.

Haq, M.A., Sumangala, K. 2003. Acarine regulators of water hyacinth in Kerala (India). *Experimental and Applied Acarology*, 29, 27–33.

Harding, D.J.L. 1976. A new species of *Phthiracarus* (Acari, Cryptostigmata) from Great Britain. *Acarologia*, 18, 163–169.

Harding, D.J.L., Easton, S.M. 1984. Development of two species of phthiracarid mites in beech cupules. In: Griffiths, D.A., Bowman, C.E. (Eds.), *Acarology VI*. vol. 2. Ellis Horwood, Chichester, 860–870.

Hartenstein, R.C. 1962a. Soil Oribatei. I. Feeding specificity among forest soil Oribatei (Acarina). *Annals of the Entomological Society of America*, 55, 202–206.

Hartenstein, R. 1962b. Soil Oribatei VI. *Protoribates lophotrichus* (Acarina: Haplozetidae) and its associations with microorganisms. *Annals of the Entomological Society of America*, 55, 587–591.

Hartmann, K., Laumann, M., Bergmann, P., Heethoff, M., Schmelzle, S. 2016. Development of the synganglion and morphology of the adult nervous system in the mite *Archegozetes longisetosus* Aoki (Chelicerata, Actinotrichida, Oribatida). *Journal of Morphology*, 277, 537–548.

Haumann, G. 1991. *Zur Phylogenie primitiver Oribatiden (Acari, Oribatida)*. Verlag Technische Universität., Graz, Austria.

Hayes, A.J. 1965. Studies on the distribution of some Phthiracarid mites (Oribatidae) in a coniferous forest soil. *Pedobiologia*, 5, 252–261.

Heethoff, M. 2012. Regeneration of complex oil-gland secretions and its importance for chemical defense in an oribatid mite. *Journal of Chemical Ecology*, 38, 1116–1123.

Heethoff, M. 2000. Untersuchung der ribosomalen ITS1 sequenzen der parthenogenetischen Hornmilbe *Platynothrus peltifer* (C.L. Koch, 1839) (Acari, Oribatida). *Mitteilungen AG Bodenmesofauna, Innsbruck*, 16, 15–20.

Heethoff, M., Cloetens, P. 2008. A comparison of synchroton X-ray phase contrast tomography and holotomography for non-invasive investigations of the internal anatomy of mites. *Soil Organisms*, 802, 205–215.

Heethoff, M., Koerner, L. 2007. Small but powerful: The oribatid mite *Archegozetes longisetosus* Aoki (Acari, Oribatida) produces disproportionately high forces. *The Journal of Experimental Biology*, 210, 3036–3042.

Heethoff, M., Norton, R.A. 2009. Role of musculature during defecation in a particle-feeding Arachnid, *Archegozetes longisetosus* (Acari, Oribatida). *Journal of Morphology*, 270, 1–13.

Heethoff, M., Raspotnig, G. 2011. Is 7-hydroxyphthalide a natural compound of oil gland secretions?—Evidence from *Archegozetes longisetosus* (Acari, Oribatida). *Acarologia*, 512, 229–236.

Heethoff, M., Domes, K., Laumann, M., Maraun, M., Norton, R.A., Scheu, S. 2007. High genetic divergences indicate ancient separation of parthenogenetic lineages of the oribatid mite *Platynothrus peltifer* (Acari, Oribatida). *European Society for Evolutionary Biology*, 20, 392–402.

Heethoff, M., Helfen, L., Cloetens, P. 2008. Non-invasive 3D-visualization of the internal organization of microarthropods using synchrotron X-ray-tomography with submicron resolution. *Journal of Visualized Experiments*, 15, 737.

Heethoff, M., Norton, R.A., Scheu, S., Maraun, M. 2009. Parthenogenesis in oribatid mites (Acari, Oribatida): Evolution without sex. In: Schön, I., Martens, K., Dijk, P. (Eds.), *Lost Sex. The Evolutionary Biology of Parthenogenesis*. Springer, Berlin, 241–257.

Heethoff, M., Koerner, L., Norton, R.A., Raspotnig, G. 2011a. Tasty but protected—First evidence of chemical defense in oribatid mites. *Journal of Chemical Ecology*, 37, 1037–1043.

Heethoff, M., Laumann, M., Weigmann, G., Raspotnig, G. 2011b. Integrative taxonomy: Combining morphological, molecular and chemical data for species delineation in the parthenogenetic *Trhypochthonius tectorum* complex (Acari, Oribatida, Trhypochthoniidae). *Frontiers in Zoology*, 8, 2.

Heethoff, M., Bergmann, P., Laumann, M., Norton, R.A. 2013. The 20th anniversary of a model mite: A review of current knowledge about *Archegozetes longisetosus* (Acari, Oribatida). *Acarologia*, 53, 353–368.

Heethoff, M., Brückner, A., Schmelzle, S., Schubert, M., Bräuer, M., Meusinger, R., Dötterl, S., Norton, R.A., Raspotnig, G. 2018. Life as a fortress–structure, function, and adaptive values of morphological and chemical defense in the oribatid mite *Euphthiracarus reticulatus* (Actinotrichida). *BMC Zoology*, 3, 7.

Heidemann, K., Scheu, S., Ruess, L., Maraun, M. 2011. Molecular detection of nematode predation and scavenging in oribatid mites: Laboratory and field experiments. *Soil Biology and Biochemistry*, 43, 2229–2236.

Heidemann, K., Ruess, L., Scheu, S., Maraun, M. 2014. Nematode consumption by mite communities varies in different forest microhabitats as indicated by molecular gut content analysis. *Experimental and Applied Acarology*, 64, 49–60.

Heneghan, L., Coleman, D.C., Zou, X., Crossley, D.A., Haines, B.L. 1999. Soil microarthropod contributions to decomposition dynamics: Tropical-temperate comparisons of a single substrate. *Ecology*, 80, 1873–1882.

Hermann, J. 1804. *Mémoire aptérologique*. De l'imprimerie de FG Levrault, 152 pp.

Heyden, C.H.G. 1826. Versuch einer systematischen Eintheilung der Acariden. *Isis Oken*, 18, 611–613.

Higgins, H.G. 1961. A new beetle mite from Utah (Oribatei: Gymnodamaeidae). *Great Basin Naturalist*, 21, 27–28.

Higgins, H.G. 1962. A new species of *Eremaeus* from the western United States (Acarina: Oribatei, Eremaeidae). *Great Basin Naturalist*, 22, 89–92.

Higgins, H.G. 1965. Two new mites from the United States (Acari: Oribatei, Microzetidae and Oribatellidae). *Great Basin Naturalist*, 25, 55–58.

Higgins, H.G. 1979. A brief review of the oribatid family Eremaeidae in North America. *Recent Advances in Acarology*, 2, 541–546.

Higgins, H.G., Woolley, T.A. 1957. A redescription of *Hafenferrefia nitidula* (Banks) and notes on the distribution of other species in the family Tenuialidae (Acarina: Oribatei). *Journal of the New York Entomological Society*, 65, 213–218.

Higgins, H.G., Woolley, T.A. 1963. New species of moss mites of the genus *Eupterotegeus* from the western United States (Oribatei: Cepheidae). *Entomological News*, 74, 3–8.

Higgins, H.G., Woolley, T.A. 1965. A new genus of moss mites from northwestern United States (Acari: Oribatei, Eremaeidae). *Pan–Pacific Entomology*, 41, 259–262.

Higgins, H.G., Woolley, T.A. 1968a. *Ametroproctus*, a new genus of Charassobatid mites from the United States (Acari: Cryptostigmata: Charassobatidae). *Great Basin Naturalist*, 28, 44–46.

Higgins, H.G., Woolley, T.A. 1968b. Redescription of *Microzetes auxiliaris appalachicola* Jacot (Acari: Cryptostigmata, Microzetidae). *Great Basin Naturalist*, 28, 142–143.

Higgins, H.G., Woolley, T.A. 1975. New mites from the Yampa Valley (Acarina: Cryptostigmata: Oribatulidae, Passalozetidae). *Great Basin Naturalist*, 35, 103–108.

Hingley, M.R. 1971. The ascomycete fungus *Daldinia concentrica* as a habitat for animals. *Journal of Animal Ecology*, 40, 17–32.

Hirauchi, Y., Aoki, J. 1997. A new species of the genus *Achipteria* from Mt. Tateyama, Central Japan (Acari: Oribatida). *Edaphologia*, 59, 5–9.

Hodkinson, I.D., Babenko, A., Behan-Pelletier, V., Bocher, J., Boxshall, G., Brodo, F., Coulson, S.J., De Smet, W., Dózsa-Farkas, K., Elias, S., Fjellberg, A., Fochetti, R., Foottit, R., Hessen, D., Hobaek, A., Holmsstrup, M., Koponen, S., Liston, A., Makarova, O., Marusik, Y.M., Michelsen, V., Mikkola, K., Pont, A., Renaud, A., Rueda, L.M., Savage, J., Smith, H., Samchyshyna, L., Velle, G., Viehberg, F., Wall, D.H., Weider, L.J., Wetterich, S., Yu, Q., Zinovjev, A. 2013. Chapter 7, Terrestrial and freshwater invertebrates. In: Meltofte, H. (Ed.), *Arctic Biodiversity Assessment, Reykjavik*. The Arctic Council, Conservation of Arctic Flora and Fauna, Akureyri, Iceland, 195–223.

Holtkamp, R., van der Wal, A., Kardol, P., van der Putten, W.H., de Ruiter, P.C., Dekker, S.C. 2011. Modelling C and N mineralisation in soil food webs during secondary succession on ex-arable land. *Soil Biology and Biochemistry*, 43, 251–260.

Hubert, J. 2001. The influence of *Scheloribates laevigatus* (Acari: Oribatida) on decomposition of *Holcus lanatus* litter. *Acta Societatis Zoologicae Bohemicae*, 65, 77–80.

Hubert, J., Lukešová, A. 2001. Feeding of the panphytophagous oribatid mite *Scheloribates laevigatus* (Acari: Oribatida) on cyanobacterial and algal diets in laboratory. *Applied Soil Ecology*, 16, 77–83.

Hubert, J., Šustr, V., Smrž, J. 1999. Feeding of the oribatid mite *Scheloribates laevigatus* (Acari: Oribatida) in laboratory experiments. *Pedobiologia*, 43, 328–339.

Hubert, J., Žilová, M., Pekár, S. 2001. Feeding preferences and gut contents of three panphytophagous mites (Acari: Oribatida). *European Journal of Soil Biology*, 37, 197–208.

Huhta, V., Siira-Pietikäinen, A., Penttinen, R. 2012. Importance of dead wood for soil mite (Acarina) communities in boreal old-growth forests. *Soil Organisms*, 84, 499–512.

Hull, J.E. 1914. British Oribatidae. Notes on new and critical species. *The Naturalist, London*, 690, 215–220; 691, 249–250; 691, 281–288.

Hull, J.E. 1915. Acari from the bird's nests, with description of a new species. *The Naturalist, London*, 707, 398–399.

Hull, J.E. 1916. Terrestrial Acari of the Tyne Province, I. Oribatidae. *Transactions of the Natural History Society of Northumberland, New Series*, 4, 381–410.

Hull, J.E. 1918. Terrestrial Acari of the Tyne Province. *Transactions of the Natural History Society of Northumberland, New Series*, 5, 13–88.

Hülsmann, A., Wolters, V. 1998. The effects of different tillage practices on soil mites, with particular reference to Oribatida. *Applied Soil Ecology*, 9, 327–332.

Illig, J., Langel, R., Norton, R.A., Scheu, S., Maraun, M. 2005. Where are the decomposers? Uncovering the soil food web of a tropical montane rain forest in southern Ecuador using stable isotopes (^{15}N). *Journal of Tropical Ecology*, 21, 589–593.

Illig, J., Norton, R.A., Scheu, S., Maraun, M. 2010. Density and community structure of soil- and bark-dwelling microarthropods along an altitudinal gradient in a tropical montane rainforest. *Experimental and Applied Acarology*, 52, 49–62.

Ingimarsdóttir, M., Caruso, T., Ripa, J., Magnúsdóttir, Ó.B., Migliorini, M., Hedlund. K. 2012. Primary assembly of soil communities: Disentangling the effect of dispersal and local environment. *Oecologia*, 170, 745–754.

Innes, R.J. 2014. Fire regimes of Alaskan dry grassland communities. In: *Fire Effects Information System*, [Online]. U.S. Department of Agriculture, Forest Service, Rocky Mountain Research Station, Missoula Fire Sciences Laboratory Producer. Available: www.fs.fed.us/database/feis/fire_regimes/AK_dry_grassland/all.html (accessed 28 June 2021).

Iordansky, S.N. 1991. Taxonomic revision of the oribatid mite *Oribatula* (Acariformes, Cryptostigmata, Oribatulidae) of the USSR Fauna. *Zoologicheskii Zhurnal*, 70, 77–89. [in Russian]

Iordansky, S.N. 1996. Fine structure of the cuticle in the oribatid (Oribatida) mites and its adaptive evolution. In: Mitchell, R., Horn, D.J., Needham, G.R., Welbourn, C.W. (Eds.), *Acarology IX Proceedings*. vol. 1. Ohio Biological Survey, Columbus, OH, 685–687.

Ito, M. 1982. Unrecorded species of the genera *Carabodes* and *Licneremaeus* (Acarina, Oribatida) from Japan. *Bulletin of Institute of Nature, Education in Shiga Heights, Shinshu University*, 20, 49–54.

Jacot, A.P. 1923. Revision of the *Ginglymosoma* (Acarina) and the creation of two new genera. *China Journal of Science and Arts*, 1, 161–162.

Jacot, A.P. 1924. Oribatid mites, *Euphthiracarus depressculus* sp. n. and *Euphthiracarus flavus* (Ewing). *Transactions of the American Microscopical Society*, 43, 90–96.

Jacot, A.P. 1924. Oribatoidea Sinensis III. *Journal North China Branch of the Royal Asiatic Society*, 45, 78–83.

Jacot, A.P. 1925. Phylogenie in the Oribatoidea. *American Naturalist*, 59, 272–279.

Jacot, A.P. 1928a. *Cepheus* (Oribatoidea) especially in the eastern United States. *Transactions of the American Microscopical Society*, 47, 262–271.

Jacot, A.P. 1928b. New oribatoid mites. *Psyche (Cambridge, Mass.)*, 35, 213–215.

Jacot, A.P. 1929a. Genera of Pterogasterine Oribatidae (Acarina). *Transactions of the American Microscopical Society*, 48, 416–430.

Jacot, A.P. 1929b. American oribatid mites of the subfamily Galumninae. *Bulletin of the Museum of comparative Zoology, Harvard*, 69, 3–36+pls. 1–6.

Jacot, A.P. 1930. Oribatid mites of the subfamily Phthiracarinae of the Northeastern of United States. *Proceedings of the Boston Society of Natural History*, 39, 209–261.

Jacot, A.P. 1931. A common arboreal moss mite *Humerobates humeralis*. *Occasional Papers of the Boston Society of Natural History, Boston*, 5, 369–382.

Jacot, A.P. 1933. The primitive Galumninae (Oribatoidea–Acarina) of the Middle West. *American Midland Naturalist*, 14, 680–703.

Jacot, A.P. 1934a. A new four-horned mossmite (Oribatoidea–Acarina). *American Midland Naturalist*, 15, 706–712.

Jacot, A.P. 1934b. Some Hawaiian Oribatoidea (Acarina). *Bulletin of the Bernice P. Bishop Museum*, 121, 1–99+ls. 1–16.

Jacot, A.P. 1934c. Acarina as possible vectors of the Dutch Elm Disease. *Journal of Economic Entomology*, 27, 858–859.

Jacot, A.P. 1935a. The species of *Zetes* (Oribatoidea–Acarina) of the Northeastern United States. *Journal of the New York Entomological Society*, 43, 51–95.

Jacot, A.P. 1935b. *Fuscozetes* (Oribatoidea–Acarina) in the Northeastern United States. *Journal of the New York Entomological Society*, 43, 311–318.

Jacot, A.P. 1936. More primitive moss-mites of North Carolina. *Journal of Elisha Mitchell Scientific Society*, 52, 247–253.

Jacot, A.P. 1937a. Journal of North-American moss-mites. *Journal of the New York Entomological Society*, 45, 353–375.

Jacot, A.P. 1937b. New moss-mites, chiefly Midwestern II. *American Midland Naturalist*, 18, 237–250.

Jacot, A.P. 1937c. Six new mites from western North Carolina. *Proceedings of the Entomological Society of Washington*, 39, 163–166.

Jacot, A.P. 1938a. New moss-mites, chiefly Midwestern. III. *American Midland Naturalist*, 19, 647–657.

Jacot, A.P. 1938b. More box-mites of the northeastern United States. *Journal of the New York Entomological Society*, 46, 109–145.

Jacot, A.P. 1938c. Some new western North Carolina moss mites. *Proceedings of the Entomological Society of Washington*, 40, 10–15.

Jacot, A.P. 1938d. Four new arthropods from New England. *American Midland Naturalist*, 20, 571–574.

Jacot, A.P. 1939a. New mites from the White Mountains. *Occasional Papers of the Boston Society of Natural History, Boston*, 8, 321–332.

Jacot, A.P. 1939b. New mites from western North Carolina. *Journal of Elisha Mitchell Scientific Society*, 55, 197–202+pl. 26.

Jacot, J.P. 1940. New oribatid mites from South Africa. *Annals of the Natal Museum*, 9, 391–400.

Jalil, M. 1969. The life cycle of *Humerobates rostrolamellatus* Grandjean, 1936 (Acari). *Journal of the Kansas Entomological Society*, 42, 526–530.

Johnston, D.E. 1982. Oribatida. In: Parker, S.P. (Ed.), *Synopsis and Classification of Living Organisms*. vol. 2. McGraw-Hill, New York, 145–146.

Jørgensen, M. 1934. A quantitative investigation of the microfauna communities of the soil in east Greenland. *Meddelelser om Grønland, Copenhagen*, 100, 1–39.

Kamill, B.W. 1981. The *Phthiracarus* species of C.L. Koch. *Bulletin of the British Museum of Natural History, (Zoology)*, 41, 263–274.

Kaneko, N. 1988. Feeding habits and cheliceral size of oribatid mites in cool temperate forests in Japan. *Revue d'Ecologie et de Biologie du Sol*, 25, 353–363.

Kaneko, N., Kofuji, R. 2000. Effects of soil pH gradient caused by stemflow acidification on soil microarthropod community structure in a Japanese red cedar plantation: An evaluation of ecological risk on decomposition. *Journal of Forest Research*, 5, 157–162.

Kaneko, N., McLean, M.A., Parkinson, D. 1995. Grazing preference of *Onychiurus subtenuis* (Collembola) and *Oppiella nova* (Oribatei) for fungal species inoculated on pine needles. *Pedobiologia*, 39, 538–546.

Kaneko, N., McLean, M.A., Parkinson, D. 1998. Do mites and Collembola affect pine litter fungal biomass and microbial respiration? *Applied Soil Ecology*, 9, 209–213.

Karasawa, S., Gotoh, K., Sasaki, T., Hijii, N. 2005. Wind-based dispersal of oribatid mites (Acari: Oribatida) in a subtropical forest in Japan. *Journal of the Acarological Society of Japan*, 14, 117–122.

Karppinen, E., Krivolutsky, D.A., Poltavskaja, M.P. 1986. List of oribatid mites (Acarina, Oribatei) of northern palaearctic region. Ill. Arid lands. *Annales Entomologici Fennici*, 52, 81–94.

Karppinen, E., Krivolutsky, D.A., Tarba, Z.M., Shtanchaeva, U.Y., Gordeeva, E. 1987. List of oribatid mites (Acarina, Oribatei) of northern palaearctic region. IV. Caucasus and Crimea. *Annales Entomologici Fennici*, 53, 119–137.

Karppinen, E., Melamud, V.V., Miko, L., Krivolutsky, D.A. 1992. Further information on the oribatid fauna (Acarina, Oribatei) of the northern palearctic region: Ukraina and Czechoslovakia. *Entomologica Fennica*, 3, 41–56.

Khalil, M.A., Janssens, T.K.S., Berg, M.P., van Straalen, N.M. 2009. Identification of metal-responsive oribatid mites in a comparative survey of polluted soils. *Pedobiologia*, 52, 207–221.

Klimov, P.A., OConnor, B.M., Chetverikov, P.E., Bolton, S.J., Pepato, A.R., Mortazavi, A.L., Tolstikov, A.V., Bauchan, G.R., Ochoa, R. 2018. Comprehensive phylogeny of acariform mites (Acariformes) provides insights on the origin of the four-legged mites (Eriophyoidea), a long branch. *Molecular Phylogenetics and Evolution*, 119, 105–117.

Klinger, L.F., Elias, S.A., Behan-Pelletier, V.M., Williams, N.E. 1990. The bog-climax hypothesis: Fossil arthropods and stratigraphic evidence in peat sections from southeast Alaska, USA. *Holarctic Ecology*, 13, 72–80.

Knee, W. 2017. A new *Paraleius* species (Acari, Oribatida, Scheloribatidae) associated with bark beetles (Curculionidae, Scolytinae) in Canada. *ZooKeys*, 667, 51–65.

Knee, W., Forbes, M.R., Beaulieu, F. 2013. Diversity and host use of mites (Acari: Mesostigmata, Oribatida) phoretic on bark beetles (Coleoptera: Scolytinae): Global generalists, local specialists? *Annals of the Entomological Society of America*, 106, 339–350.

Knülle, W. 1954. Die Arten der Gattung *Tectocepheus* Berlese (Acarina, Oribatei). *Zoologischer Anzeiger*, 152, 280–305.

Knülle, W. 1957. Morphologische und entwicklungsgeschichtliche untersuchungen zum phylogenetischen system der Acari: Acariformes Zachv. I. Oribatei: Malaconothridae. *Zoosystematics and Evolution*, 33, 97–213.

Koch, C.L. 1835–1844. Deutschlands Crustaceen. *Myriapoden und Arachniden*, 1841, 31–34.

Koch, L. 1879. Arachniden aus Sibirien und Novaja Zemlja, eingesammelt von der Schwedischen Expedition im Jahre 1875. Kongliga Svenska Vetenskaps-Akademiens Handlingar. *Stockholm*, 16, 1–136.

Konecka, E., Olszanowski, Z. 2015. A screen of maternally inherited microbial endosymbionts in oribatid mites (Acari: Oribatida). *Microbiology – SGM*, 161, 1561–1571.

Konecka, E., Olszanowski, Z. 2019a. A new Cardinium group of bacteria found in *Achipteria coleoptrata* (Acari: Oribatida). *Molecular Phylogenetics and Evolution*, 131, 64–71.

Konecka, E., Olszanowski, Z. 2019b. Detection of a new bacterium of the family Holosporaceae (Alphaproteobacteria: Holosporales) associated with the oribatid mite *Achipteria coleoptrata*. *Biologia*, 74, 1517–1522.

Konecka, E., Olszanowski, Z. 2019c. Phylogenetic analysis based on the 16S rDNA, *gltA*, *gatB*, and *hcpA* gene sequences of *Wolbachia* from the novel host *Ceratozetes thienemanni* (Acari: Oribatida). *Infection, Genetics and Evolution*, 70, 175–181.

Koneka, E., Olszanowski, Z. 2021. *Wolbachia* supergroup E found in *Hypochthonius rufulus* (Acari, Oribatida) in Poland. *Infection, Genetics and Evolution*, 91, 104820.

Koukol, O., Mourek, J., Janovsky, Z., Cerna, K. 2009. Do oribatid mites (Acari: Oribatida) show a higher preference for ubiquitous vs. specialized saprotrophic fungi from pine litter? *Soil Biology and Biochemistry*, 41, 1124–1131.

Kramer, P. 1897. Grönländische Milben. Zoologische der von der Gesellschaft für Erdkunde zu Berlin unter Leitung Dr. von Drygalski aus gesandten Grondland expedition nach Dr. Vanhoffen's Sammlung bearbeitet, VI. *Bibliotheca Zoologica*, 77–83.

Krantz, G.W. 1978. *A Manual of Acarology*. 2nd ed. Oregon State Univ., Corvallis, Oregon, 509 pp.

Krantz, G.W., Baker, G.T. 1982. Observations on the plastron mechanism of *Hydrozetes* sp. (Acari: Oribatida: Hydrozetidae). *Acarologia*, 233, 273–277.

Krantz, G.W., Walter, D.E., (Eds.), 2009. *A Manual of Acarology*. 3rd ed. Texas Tech. University Press, Lubbock, 807 pp.

Krause, A., Pachl, P., Schulz, G., Lehmitz, R., Seniczak, A., Schaefer, I., Scheu, S., Maraun, M. 2016. Convergent evolution of aquatic life by sexual and parthenogenetic oribatid mites. *Experimental and Applied Acarology*, 70, 439–453.

Kreipe, V., Corral-Hernández, E., Scheu, S., Schaefer, I., Maraun, M. 2015. Phylogeny and species delineation in European species of the genus *Steganacarus* (Acari, Oribatida) using mitochondrial and nuclear markers. *Experimental and Applied Acarology*, 66, 173–186.

Krisper, G. 1997. Erstnachweis der Milbenfamilie Pterochthoniidae (Acari, Oribatida) für die Bundesländer Steiermark, Kärnten und Niederösterreich. *Mitteilungen des Naturwissenschaftlichen Vereines für Steiermark*, 127, 147–152.

Krisper, G., Jakesz-Grübler, C. 1999. Erstnachweis der Milbenfamilie Gehypochthoniidae Strenzke, 1963 (Acari, Oribatida) für Österreich. *Mitteilungen des Naturwissenschaftlichen Vereins für Steiermark*, 129, 281–285.

Krisper, G., Lazarus, S. 2014. Bodenzoologische Untersuchungen an zwei Trockenrasen in der Steiermark– Erstnachweise von Hornmilben (Acari, Oribatida). *Mitteilungen des Naturwissenschaftlichen Vereines für Steiermark*, 143, 121–130.

Krisper, G., Schatz, H., Schuster, R. 2017. Oribatida (Arachnida: Acari). Checklisten der Fauna Österreichs, No. 9, *Biosystematics and Ecology Series*, Band 33. Verlag der Österreichischen Akademie der Wissenschaften, Wien, 25–90.

Krivolutsky, D.A. 1962. The genus *Cultroribula* Berlese (Acariformes, Oribatei) and its representatives in the USSR. *Zoologicheskii Zhurnal*, 41, 1893–1895. [in Russian]

Krivolutsky, D.A. 1964. *Sellnickochthonius* gen. n. new genus of oribatid mites from the family Brachychthoniidae Balogh, 1943 (Acariformes, Oribatei). *Zoologicheskii Zhurnal*, 43, 935–936.

Krivolutsky, D.A. 1965. New species of beetle mites (Acariformes, Oribatei) from the taiga zone of the U.S.S.R. *Entomologicheskoe Obozrenie*, 44, 705–708. [in Russian with English summary]

Krivolutsky, D.A. 1974. New oribatid mite of the U.S.S.R. *Zoologicheskii Zhurnal*, 53, 1880–1885. [in Russian with English summary]

Krivolutsky, D.A., Druk, A. 1986. Fossil oribatid mites. *Annual Review of Entomology*, 31, 533–545.

Krivolutsky, D.A., Lebedeva, N.V. 2004. Oribatid mites (Oribatei, Acariformes) in bird feathers: non– passerines. *Acta Zoologica Lithuanica*, 14, 26–47.

Krivolutsky, D.A., Ryabinin, N.A. 1974. New species of oribatid mites (Oribatei) from Siberia and the Far East. *Zoologicheskii Zhurnal*, 53, 1169–1177. [in Russian]

Kubota, T., Aoki, J. 1998. *Hololohmannia alaskensis* from Alaska, representing a new genus and species of the family Perlohmanniidae (Acari: Oribatida). *Edaphologia*, 60, 17–21.

Kulczynski, V. 1902. Species Oribatinarum (Oudms.) (Damaeinarum Michael) in Galicia collaectae. *Bulletin International de l'Académie des Sciences et des letters de Cracivie*, 2, 89–96.

Kulijev, K.A. 1961. To the knowledge of oribatid mites of Azerbaijan with the description of two species. *Trudy Azerbaijan Pedagogiceskogo Instituta im. V. I. Lenina*, 17, 47–58. [in Russian]

Kunst, M. 1962. *Oribatella cavatica* n. sp., eine neue Moosmilbe aus dem Guano der Fledermäuse (Acarina, Oribatei). *Acta Universitatis Carolinae Biologica*, Suppl., 1–6.

Kunst, M. 1971. Nadkohorta Pancirnici Oribatei. Key to the Fauna of Czechoslovakia– Supercohort Oribatei. In: Daniel, M., Cerny, V. (Eds.), *Klic zvireny CSSR IV*. Academia, Praha, 531–580.

Kuriki, G. 2005. Oribatid mites from several mires in northern Japan I. Two new species of the genus *Trhypochthoniellus* (Acari: Oribatida). *Journal of the Acarological Society of Japan*, 14, 83–92.

Kuriki, G. 2008. The life cycle of *Limnozetes ciliatus* (Schrank, 1803) (Acari: Oribatida). *Journal of the Acarological Society of Japan*, 17, 75–85.

Kuriki, G., Choi, S.S., Fujikawa, T. 2000. Supplementary description of the type species of the genus *Mainothrus* Choi, 1996, belonging to the family Trhypochthoniidae (Acari: Oribatida). *Acarologia*, 41, 273–276.

Labandeira, C.C., Phillips, T.L., Norton, R.A. 1997. Oribatid mites and the decomposition of plant tissues in paleozoic coal swamp forests. *Palaios*, 124, 319–353.

Lamoncha, K.L., Crossley, D.A. 1998. The oribatid diversity along an elevation gradient in a southeastern Appalachian forest. *Pedobiologia*, 42, 43–55.

Lan, W., Xin, J.-L., Aoki, J. 1986. Two new species of oribatid mites of economic importance from China (Acari: Oribatida). *Proceeding of the Japanese Society of Systematic Zoology*, 34, 27–31.

Lange, A.B. 1954. Morphology of *Zachvatkinella belbiformis*, gen. n. and sp. n. a new representative of the Palaeacariformes group (Acariformes). *Zoologicheskii Zhurnal*, 34, 1042–1052. [in Russian]

Langor, D.W. 2019. The diversity of terrestrial arthropods in Canada. In: Langor, D.W., Sheffield, C.S. (Eds.), *The Biota of Canada – A Biodiversity Assessment. Part 1: The Terrestrial Arthropods*. ZooKeys, 819, 9–40.

Larson, D.J., House, N.L. 1990. Insect communities of Newfoundland bog pools with emphasis on the Odonata. *Canadian Entomologist*, 122, 469–501.

Laumann, M., Norton, R.A., Weigmann, G., Scheu, S., Maraun, M., Heethoff, M. 2007. Speciation in the parthenogenetic oribatid mite genus *Tectocepheus* (Acari, Oribatida) as indicated by molecular phylogeny. *Pedobiologia*, 51, 111–122.

Laumann, M., Bergmann, P., Norton, R.A., Heethoff, M. 2010a. First cleavages, preblastula and blastula in the parthenogenetic mite *Archegozetes longisetosus* (Acari, Oribatida) indicate holoblastic rather than superficial cleavage. *Arthropod Structure & Development*, 39, 276–286.

Laumann, M., Norton, R.A., Heethoff, M. 2010b. Acarine embryology: Inconsistencies, artificial results and misinterpretations. *Soil Organisms*, 822, 217–235.

Laundon, J.R. 1967. A study of the lichen flora of London. *The Lichenologist*, 3, 277–327.

Lebedeva, N.V. 2012. Oribatid mites transported by birds to polar islands – A review. *Reports on Polar and Marine Research*, 640, 152–161.

Lebedeva, N.V., Lebedev, V.D. 2008. Transport of oribatid mites to the polar areas by birds. In: Bertrand, M., Kreiter, S., McCoy, K.D., Migeon, A., Navajas, M., Tixier, M.S., Vial, L. (Eds.), *Integrative Acarology. Proceedings of the 6th European Congress*, European Association of Acarologists, Montpellier, France, 359–367.

Lebedeva, N.V., Lebedev, V.D., Melekhina, E.N. 2006. New data on the oribatid mite (Oribatei) fauna of Svalbard. *Doklady Biological Sciences*, 407, 182–186 [Russian version, Doklady Akademii Nauk, 407, 845–849]

Lebedeva, N.V., Melekhina, E.N., Gwiazdowicz, D.J. 2012. New data on soil mites in the nests of the glacous gull *Larus hyperboreus* L. on Svalbard. *Vestnik of the Southern Scientific Center RAN*, 8, 70–75.

Lebrun, P. 1965. Contribution a l'étude écologiique des oribates de la litiére dans use forêt de Moyenne-Belgique. *Institut Royal des Sciences Naturales de Belgique – Memoirs*, 153, 1–89.

Lebrun, P. 1968. Écologie et biologie de *Nothrus palustris* C.L. Koch 1839 (Acarien, Oribate). *Pedobiologia*, 8, 223–238.

Lebrun, P. 1970. Écologie et biologie de *Nothrus palustris* (C.L. Koch, 1839). 3e note. Cycle de vie. *Acarologia*, 12, 193–207.

Lebrun, P. 1984. Determination of the dynamics of an edaphic oribatid population (*Nothrus palustris* C.L. Koch 1839). In: Griffiths, D.A., Bowman, C.E. (Eds.), *Acarology VI*. vol. 2. Ellis Horwood, Chichester, 871–877.

Lebrun, P., van Straalen, N.M. 1995. Oribatid mites, prospects of their use in ecotoxicology. *Experimental and Applied Acarology*, 19, 361–380.

Lebrun, P., van Impe, G., De Saint Georges-Gridelet, D., Wauthy, G., André, H.M. 1991. The life strategies of mites. In: Schuster, R., Murphy, P.W. (Eds.), *The Acari Reproduction, Development and Life-history Strategies*. Chapman and Hall, London, New York, 3–22.

Lee, D.C. 1982. Sarcoptiformes (Acari) of South Australian soils. 3. Arthronotina (Cryptostigmata). *Records South Australia Mususeum*, 18, 327–359.

Lehmitz, R., Decker, P. 2017. The nuclear 28S gene fragment D3 as species marker in oribatid mites (Acari, Oribatida) from German peatlands. *Experimental and Applied Acarology*, 71, 259–276.

Lehmitz, R., Maraun, M. 2016. Small-scale spatial heterogeneity of stable isotopes signatures (δ^{15}N, δ^{13}C) in *Sphagnum* sp. transfers to all trophic levels in oribatid mites. *Soil Biology and Biochemistry*, 100, 242–251.

Lehmitz, R., Russell, D., Hohberg, K., Christian, A., Xylander, W.E.R. 2011. Wind dispersal of oribatid mites as mode of migration. *Pedobiologia*, 54, 201–207.

Lehmitz, R., Russell, D., Hohberg, K., Christian, A., Xylander, W.E.R. 2012. Active dispersal of oribatid mites into young soils. *Applied Soil Ecology*, 55, 10–19.

Leonov, V.D., Rakhleeva, A.A. 2020. The first report on oribatid mites in tundra belts of the Lovozersky Mountains on the Kola Peninsula, Russia. *Acarologia*, 60, 301–316.

Liana, M., Witaliński, W. 2010. Microorganisms in the oribatid mite *Hermannia gibba* (C.L. Koch, 1839) (Acari: Oribatida: Hermanniidae). *Biological Letters*, 47, 37–43.

Liana, M., Witaliński, W. 2012. Female and male reproductive systems in the oribatid mite *Hermannia gibba* (Koch, 1839) (Oribatida: Desmonomata). *International Journal of Acarology*, 388, 648–663.

Lienhard, A., Krisper, G. 2021. Hidden biodiversity in microarthropods (Acari, Oribatida, Eremaeoidea, *Caleremaeus*). *Scientific Reports*, 11, 23123.

Lienhard, A., Schäffer, S., Krisper, G., Sturmbauer, C. 2014. Reverse evolution and cryptic diversity in putative sister families of the Oribatida (Acari). *Journal of Zoological Systematics and Evolutionary Research*, 52, 86–93.

Lindberg, N., Bengtsson, J. 2006. Recovery of forest soil fauna diversity and composition after repeated summer droughts. *Oikos*, 114, 494–506.

Lindo, Z. 2010. Communities of Oribatida associated with litter input in western redcedar tree crowns: Are moss mats "magic carpets" for oribatid mite dispersal? In: Sabelis, M.W., Bruin, J. (Eds.), *Trends in Acarology, Proceedings of the XII International Congress of Acarology, Amsterdam (2006)*. Springer, Dordrecht, 143–148.

Lindo, Z. 2011. Five new species of *Ceratoppia* (Acari: Oribatida: Peloppiidae) from western North America. *Zootaxa*, 3036, 1–25.

Lindo, Z. 2015a. Warming favours small-bodied organisms through enhanced reproduction and compositional shifts in belowground systems. *Soil Biology and Biochemistry*, 91, 271–278.
Lindo, Z. 2015b. A rare new species of *Metrioppia* (Acari: Oribatida: Peloppiidae) from a Pacific Northwest temperate rainforest. *The Canadian Entomologist*, 147, 553–563.
Lindo, Z. 2018. Diversity of Peloppiidae (Oribatida) in North America. *Acarologia, 58 (Suppl.)*, 91–97.
Lindo, Z. 2020. Transoceanic dispersal of terrestrial species by debris rafting. *Ecography*, 43, 1364–1372.
Lindo, Z., Clayton, M. 2017. Oribatid mites (Acari: Oribatida) of British Columbia. In: Klinkenberg, B. (Ed.), *E–Fauna BC: Electronic Atlas of the Fauna of British Columbia.* https://ibis.geog.ubc.ca/biodiversity/efauna/MitesofBritishColumbia.html (accessed 28 June 2021).
Lindo, Z., Gonzalez, A. 2010. The bryosphere: An integral and influential component of the Earth's biosphere. *Ecosystems*, 13, 612–627.
Lindo, Z., Stevenson, S.K. 2007. Diversity and distribution of oribatid mites (Acari: Oribatida) associated with arboreal and terrestrial habitats in Interior Cedar-Hemlock Forests, British Columbia, Canada. *Northwest Science*, 81, 305–315.
Lindo, Z., Visser, S. 2004. Forest floor microarthropod abundance and oribatid mite (Acari: Oribatida) composition following partial and clear-cut harvesting in the mixedwood boreal forest. *Canadian Journal of Forest Research*, 34, 998–1006.
Lindo, Z., Winchester, N.N. 2006. A comparison of microarthropod assemblages with emphasis on oribatid mites in canopy suspended soils and forest floors associated with ancient western redcedar trees. *Pedobiologia*, 50, 31–41.
Lindo, Z., Winchester, N.N. 2007a. Oribatid mite communities and foliar litter decomposition in canopy suspended soils and forest floor habitats of western redcedar forests, Vancouver Is., Canada. *Soil Biology and Biochemistry*, 39, 2957–2966.
Lindo, Z., Winchester, N.N. 2007b. Local-regional boundary shifts in oribatid mite (Acari: Oribatida) communities: Species-area relationships in arboreal habitat islands of a coastal temperate rain forest, Vancouver Island, Canada. *Journal of Biogeography*, 34, 1611–1621.
Lindo, Z., Winchester, N.N. 2007c. Resident corticolous oribatid mites (Acari: Oribatida): Decay in community similarity with vertical distance from the ground. *Écoscience*, 14, 223–229.
Lindo, Z., Winchester, NN. 2009. Spatial and environmental factors contributing to patterns in arboreal and terrestrial oribatid mite diversity across spatial scales. *Oecologia*, 160, 817–825.
Lindo, Z., Clayton, M., Behan-Pelletier, V.M. 2008. Systematics and ecology of *Anachipteria geminus* sp. nov. (Acari: Oribatida: Achipteriidae) from arboreal lichens in western North America. *The Canadian Entomologist*, 140, 539–556.
Lindo, Z., Winchester, N.N., Didham, R.K. 2008. Nested patterns of community assembly in the colonisation of artificial canopy habitats by oribatid mites. *Oikos*, 117, 1856–1864.
Lindo, Z., Clayton, M., Behan-Pelletier, V.M. 2010. Systematics and ecology of the genus *Dendrozetes* (Acari: Oribatida: Peloppiidae) from arboreal habitats in Western North America. *Zootaxa*, 2403, 10–22.
Lindquist, E.E., Krantz, G.W., Walter, D.E. 2009. Chapter 8. Classification. In: Krantz, G.W., Walter, D.E. (Eds.), *A Manual of Acarology*, 3rd ed. Texas Tech. University Press, Lubbock, 97–103.
Linnaeus, C. 1758. *Systema Naturae*. (1956 reprinting). 824 pp.
Lions, J.C., Gourbiere, F. 1988. Populations adultes et immatures d'*Adoristes ovatus* (Acarien, Oribate) dans les aiguilles de la litière d'*Abies alba*. *Revue d'Écologie et de Biologie du Sol*, 25, 343–352.
Lions, J.C., Gourbiere, F. 1989. Populations d'*Adoristes ovatus* (Acarien, Oribate) vivant à l'extérieur des aiguilles dans la litière d'*Abies alba*. *Revue d'Écologie et de Biologie du Sol*, 26, 213–223.
Lions, J.C., Norton, R.A. 1998. North American Synichotritiidae (Acari: Oribatida) 2. *Synichotritia spinulosa* and *S. caroli*. *Acarologia*, 39, 265–284.
Littlewood, C.F. 1969. A surface sterilisation technique used in feeding algae to Oribatei. In: Evans, G.O. (Ed.), *Proceedings of the 2nd International Congress of Acarology*. Akademiai Kiadó, Budapest, 53–56.
Liu, D. 2018. Contribution to the knowledge of the oribatid mite genus *Mesoplophora* (Acari: Oribatida: Mesoplophoridae) with description of a new species from South China. *Biologia*, 73, 1215–1221.
Liu, D., Chen, J. 2010. A review of *Mesotritia* (Acari: Oribatida: Oribotritiidae) in China, with descriptions of two new species and a checklist of known taxa. *Zootaxa*, 2479, 39–58.
Liu, D., Niedbała, W., Chen, J. 2009. Taxonomic study of the genus *Maerkelotritia* Hammer, 1967 (Acari: Oribatida: Oribotritiidae) from China, with description of a new species. *Annales Zoologici*, 59, 511–516.
Lombardini, G. 1936. Elenco alfabetico di specie esistenti nell'acroteca della R. Stazione di Entomologia Agraria di Firenz. *Redia*, 22, 37–51.

Lotfollahi, P., Movahedzade, E., Abbasi, A., Shimano, S., Norton, R.A. 2016. Second species of the family Arborichthoniidae (Acari: Oribatida), from agricultural soil in Iran. *International Journal of Acarology*, 425, 229–234.

Lundqvist, L. 1987. Checklist of Swedish oribatids (Acari: Oribatei), 1941–1985. *Entomologisk Tidskrift*, 108, 3–12.

Lupardus, R.C., Battigelli, J.P., Janz, A., Lumley, L.M. 2021. Can soil invertebrates indicate soil biological quality on well pads reclaimed back to cultivated lands? *Soil and Tillage Research*, 213, 105082.

Luxton, M. 1966. Laboratory studies on the feeding habits of saltmarsh Acarina, with notes on their behaviour. *Acarologia*, 8, 163–174.

Luxton, M. 1972. Studies on the oribatid mites of a Danish beech wood soil. I. Nutritional biology. *Pedobiologia*, 12, 434–463.

Luxton, M. 1975. Studies on the oribatid mites of a Danish beech wood soil. II. Biomass, calorimetry and respirometry. *Pedobiologia*, 15, 161–200.

Luxton, M. 1979. Food and energy processing by oribatid mites. *Revue d'Écologie et Biologie du Sol*, 16, 103–111.

Luxton, M. 1981a. Studies on the oribatid mites of a Danish beech wood soil. IV. Developmental biology. *Pedobiologia*, 21, 312–340.

Luxton, M. 1981b. Studies on the oribatid mites of a Danish beech wood soil. V. Vertical distribution. *Pedobiologia*, 21, 365–386.

Luxton, M. 1990. A redescription of *Conoppia palmicinctum* (Michael, 1880) (Acari, Cryptostigmata). *Entomologist's Monthly Magazine*, 126, 163–166.

Luxton, M. 1996. Oribatid mites of the British Isles: A checklist and notes on biogeography (Acari: Oribatida). *Journal of Natural History*, 30, 803–822.

Madge, D.S. 1965. The effects of lethal temperatures on oribatid mites. *Acarologia*, 7, 121–130.

Magilton, M., Maraun, M., Emmerson, M., Caruso, T. 2019. Oribatid mites reveal that competition for resources and trophic structure combine to regulate the assembly of diverse soil animal communities. *Ecology and Evolution*, 9, 8320–8330.

Mahunka, S. 1969. Beiträge zur Kenntnis der Milbenfauna Ungarns. I. *Folia Entomologica Hungarica*, 2, 21–30.

Mahunka, S. 1979. Neue und interessante Milben aus dem Genfer Museum. XXV. On some oribatids collected by Dr. P. Strinati in Guatemala (Acari, Oribatida). *Acarologia*, 21, 133–142.

Mahunka, S. 1980. Data to the knowledge of mites preserved in the "Berlese Collection" (Acari: Tarsonemida, Oribatida). I. *Acta Zoologica Academiae Scientiarum Hungaricae*, 26, 377–399.

Mahunka, S. 1982a. Ptychoide Oribatiden aus der Koreanischen Volksdemokratischen Republik (Acari). *Acta Zoologica Academiae Scientiarum Hungaricae*, 28, 83–103.

Mahunka, S. 1982b. Oribatids from eastern part of the Ethiopian region (Acari). I. *Acta Zoologica Academiae Scientiarum Hungaricae*, 28, 293–336.

Mahunka, S. 1983. Neue und interessante Milben aus dem Genfer Museum. 45. Oribatida Americana 6, Mexico II (Acari). *Revue Suisse de Zoologie*, 90, 269–298.

Mahunka, S. 1987. A survey of the oribatid (Acari) fauna of Vietnam, I. *Annales Historico-Naturales Musei Nationalis Hungarici*, 79, 259–279.

Mahunka, S. 1993. Oribatids from Switzerland I. (Acari, Oribatida). (Acarologica Genavensia LXXXI). *Archives des Sciences, Genève*, 46, 51–56.

Mahunka, S. 1995. *Hoffmanacarus virginianus* gen. n., sp. n. and some other moss mites from Virginia, USA (Acari: Oribatida) (New and interesting mites from the Geneva Museum LIII). *Archives des Sciences, Genève*, 48, 1–10.

Mahunka, S., Mahunka-Papp, L. 2004. A catalogue of the Hungarian oribatid mites (Acari: Oribatida). In: Csuzdi, C., Mahunka, S. (Eds.), *Pedozoologica Hungarica: Taxonomic, zoogeographic and faunistic studies on the soil animals*. No. 2. [363 pp].

Mahunka, S., Zombori, L. 1985. The variability of some morphological features in oribatid mites. *Folia entomologica hungarica, S.N.*, 461, 115–128.

Majka, C.G., Behan-Pelletier, V., Bajerlein, D., Bloszyk, J., Krantz, G.W., Lucas, Z., OConnor, B., Smith, I.M. 2007. New records of mites (Arachnida: Acari) from Sable Island, Nova Scotia, Canada. *The Canadian Entomologist*, 139, 690–699.

Makarova, O.L. 2015. The fauna of free-living mites (Acari) of Greenland. *Entomological Review*, 95, 108–125. [Original Russian Text, Zoologicheskii Zhurnal, 93, 1404–1419.

Makarova, O., Behan-Pelletier, V. 2015. Oribatida (=Cryptostigmata, Beetle mites). In: Böcher, J., Kristensen, N.P., Pape, T., Vilhelmsen, L. (Eds.), *The Greenland Entomofauna. An Identification Manual of Insects, Spiders and Their Allies*. Brill, Leiden, Boston, 802–845.

Manu, M., Honciuc, V., Neagoe, A., Băncilă, R.I., Iordache, V., Onete, M. 2019. Soil mite communities (Acari: Mesostigmata, Oribatida) as bioindicators for environmental conditions from polluted soils. *Scientific Reports*, 9, 20250.

Maraun, M., Scheu, S. 2000. The structure of oribatid mite communities (Acari, Oribatida): Patterns, mechanisms and implications for future research. *Ecography*, 23, 374–383.

Maraun, M., Heethoff, M., Scheu, S., Norton, R.A., Weigmann, G., Thomas, R.H. 2003a. Radiation in sexual and parthenogenetic oribatid mites (Oribatida, Acari) as indicated by genetic divergence of closely related species. *Experimental and Applied Acarology*, 29, 265–277.

Maraun, M., Salamon, J.-A., Schneider, K., Schaefer, M., Scheu, S. 2003b. Oribatid mite and collembolan diversity, density and community structure in a moder beech forest (*Fagus sylvatica*): Effects of mechanical perturbations. *Soil Biology and Biochemistry*, 35, 1387–1394.

Maraun, M., Schatz, H., Scheu, S. 2007. Awesome or ordinary? Global diversity patterns of oribatid mites. *Ecography*, 30, 209–216.

Maraun, M., Erdmann, G., Fischer, B.M., Pollierer, M.M., Norton, R.A., Schneider, K., Scheu, S. 2011. Stable isotopes revisited: Their use and limits for oribatid mite trophic ecology. *Soil Biology and Biochemistry*, 43, 877–882.

Maraun, M., Augustin, D., Müller, J., Bässler, C., Scheu, S. 2014. Changes in the community composition and trophic structure of microarthropods in sporocarps of the wood decaying fungus *Fomitopsis pinicola* along an altitudinal gradient. *Applied Soil Ecology*, 84, 16–23.

Maraun, M., Caruso, T., Hense, J., Lehmitz, R., Mumladze, L., Murvanidze, M., Nae, J., Schulz, J., Seniczak, A., Scheu, S. 2019. Parthenogenetic vs. sexual reproduction in oribatid mite communities. *Ecology and Evolution*, 9, 7324–7332.

Märkel, K. 1963. Uber die Zonierung humusbewohnender Oribatiden in Faktorengefallen. *Verhandlungen der Deutschen Zoologischen Gesellschaft in München*, 33, 324–329.

Märkel, K. 1964. Die Euphthiracaridae Jacot 1930 und ihre Gattungen (Acari, Oribatei). *Zoologische Verhandelingen, Leiden*, 67, 3–78.

Märkel, K., Meyer, I. 1959. Zur Systematik der deutschen Euphthiracarini (Acari, Oribatei). *Zoologischer Anzeiger*, 163, 327–342.

Markkula, I. 1986. Comparison of present and subfossil oribatid faunas in the surface peat of a drained pine mire. *Annals Entomologica Fennica*, 521, 39–41.

Markkula, I. 2020. Oribatid mites (Acari: Oribatida) in sub-arctic peatlands – A multidisciplinary investigation into climate change, permafrost dynamics and indicator values of subfossils. Academic Dissertation, Annales Universitatis Turkuensis, Ser. A, 368 Biologica-Geographica-Geologica, Turku, 42 pp.

Markkula, I., Kuhry, P. 2020. Subfossil oribatid mite communities indicate Holocene permafrost dynamics in Canadian mires. *Boreas*, 49, 730–738.

Marshall, V.G. 1968. Microarthropods from two Quebec woodland humus forms. III. The Sarcoptiformes Acarina. *Annals of the Entomological Society of Quebec*, 132, 65–88.

Marshall, V.G. 1974. Seasonal and vertical distribution of soil fauna in a thinned and urea fertilized Douglas-fir forest. *Canadian Journal of Soil Science*, 54, 491–500.

Marshall, V.G. 1979. Effects of the insecticide diflubenzuron on soil mites of a dry forest zone in British Columbia. In: Rodríguez, J.G. (Ed.), *Recent Advances in Acarology*. vol. 1, 129–134.

Marshall, V.G., Reeves, R.M. 1971. *Trichthonius majestus*, a new species of oribatid mite (Acarina: Cosmochthoniidae) from North America. *Acarologia*, 12, 623–632.

Marshall, V.G., Reeves, R.M., Norton, R.A. 1987. Catalogue of the Oribatida (Acari) of continental United States and Canada. *Memoirs of the Entomological Society of Canada*, 139, 1–418.

Maruyama, I. 2003. The first record of *Achipteria coleoptrata* (Linnaeus) from Japan and its redescription (Acari: Oribatida: Achipteriidae). *Edaphologia*, 71, 25–29.

Mason, L.D., Wardell-Johnson, G., Main, B.Y. 2018. The longest-lived spider: Mygalomorphs dig deep, and persevere. *Pacific Conservation Biology*, 24, 203–206.

Masuko, K. 1994. Specialized predation on oribatid mites by two species of the ant genus *Myrmecina* (Hymenoptera: Formicidae). *Psyche (Cambridge, Mass.)*, 101, 159–174.

Materna, J. 2000. Oribatid communities (Acari: Oribatida) inhabiting saxicolous mosses and lichens in the Krkonose Mts. (Czech Republic). *Pedobiologia*, 44, 40–62.

Matthewman, W.G., Pielou, D.P. 1971. Arthropods inhabiting the sporophores of *Fomes fomentarius* (Polyporaceae) in Gatineau Park, Quebec. *The Canadian Entomologist*, 103, 775–847.

Matthews, J.V. 1979. Tertiary and Quaternary environments: Historical background for an analysis of the Canadian insect fauna. In: Danks, H.V. (Ed.), *Canada and its Insect Fauna. Memoirs of the Entomological Society of Canada*. vol. 108, 31–86.

Matthews, J.V., Jr., Telka, A. 1997. Insect fossils from the Yukon. In: Danks, H.V., Downes, J.A. (Eds.), *Insects of the Yukon*. Biological Survey of Canada Terrestrial Arthropods, Ottawa, 911–962.

Matthews, J.V., Jr., Telka, A., Kuzmina, S.A. 2019. Late Neogene insect and other invertebrate fossils from Alaska and Arctic/Subarctic Canada. *Invertebrate Zoology*, 162, 126–153.

McAdams, B.N. 2018. Oribatid mite communities after ecosystem disturbance in Alberta. PhD thesis. University of Alberta.

McAdams, B.N., Quideau, S.A., Swallow, M.J.B., Lumley, L.M. 2018. Oribatid mite recovery along a chronosequence of afforested boreal sites following oil sand mining. *Forest Ecology and Management*, 422, 281–193.

McAloon, F.M. 2004. Oribatid mites as intermediate hosts of *Anoplocephala manubriata*, cestode of the Asian elephant in India. *Experimental and Applied Acarology*, 32, 181–185.

McBrayer, J.F., Ferris, J.M., Metz, L.J., Gist, C.S., Cornaby, B.W., Kitazawa, Y., Kitazawa, T., Wernz, J.G., Krantz, G.W., Jensen, H. 1977. Decomposer invertebrate populations in U.S. forest biomes. *Pedobiologia*, 17, 89–96.

McClure, M.S. 1995. *Diapterobates humeralis* (Oribatida: Ceratozetidae): An effective control agent of hemlock woolly adelgid (Homoptera: Adelgidae) in Japan. *Environmental Entomology*, 24, 1207–1215.

McLean, M.A., Parkinson, D. 2000. Introduction of the epigeic earthworm *Dendrobaena octaedra* changes the oribatid community and microarthropod abundances in a pine forest. *Soil Biology and Biochemistry*, 32, 1671–1681.

McLean, M.A., Kaneko, N., Parkinson, D. 1996. Does selective grazing by mites and Collembola affect litter fungal community structure? *Pedobiologia*, 40, 97–105.

Meehan, M.L., Song, Z., Lumley, L.M., Cobb, T.P., Proctor, H. 2019. Soil mites as bioindicators of disturbance in the boreal forest in northern Alberta, Canada: Testing taxonomic sufficiency at multiple taxonomic levels. *Ecological Indicators*, 102, 349–368.

Meehan, M.L., Turnbull, K.F., Sinclair, B.J., Lindo, Z. 2022. Predators minimize energy costs, rather than maximize energy gains under warming: Evidence from a microcosm feeding experiment. *Functional Ecology*, 36, 2279–2288.

Merkel, C., Heethoff, M., Brückner, A. 2020. Temperature affects chemical defense in a mite-beetle predator-prey system. *Journal of Chemical Ecology*, 46, 947–955.

Meier, F.A., Scherrer, S., Honegger, R. 2002. Faecal pellets of lichenivorous mites contain viable celles of the lichen-forming ascomycete *Xanthoria parietina* and its green algal photobiont, *Trebouxia arboricola*. *Biological Journal of the Linnean Society*, 76, 259–268.

Menke, H.G. 1966. Revision der Ceratozetidae, 4. *Ceratozetes mediocris* Berlese (Arach., Acari, Oribatei). *Senckenbergiana Biologica*, 47, 371–378.

Michael, A.D. 1880. A further contribution to the knowledge of British Oribatidae. (Part II). *Journal of the Royal Microscopical Society*, 3, 177–201 + pls.5–6.

Michael, A.D. 1882. Further notes on British Oribatidae. *Journal of the Royal Microscopical Society*, 2, 1–18.

Michael, A.D. 1883. Observations on the anatomy of the Oribatidae. *Journal of the Royal Microscopical Society, ser. 2, London*, 3, 1–25.

Michael, A.D. 1884. *British Oribatidae*. vol. I. Ray Society, London, 1–336.

Michael, A.D. 1885. New British Oribatidae. *Journal of the Royal Microscopical Society*, 5, 385–397.

Michael, A.D. 1888. *British Oribatidae*. vol. II. Ray Society, London, 337–657.

Michael, A.D. 1890. On a collection of Acarina formed in Algeria. *Proceedings of the Zoological Society, London*, 29, 414–425 + pls.37–38.

Michael, A.D. 1897. The Classification of Oribatidae. *Annals and Magazine of Natural History*, 19, 34–39.

Michael, A.D. 1898. Oribatidae. *Das Tierreich*, 3, 1–93.

Mihelčič, F. 1957a. Milben (Acarina) aus Tirol und Vorarlberg. *Veröff. Mus.Ferdinandeum, Innsbruck*, 37, 99–120.

Mihelčič, F. 1957b. Oribatiden Südeuropas VII. *Zoologischer Anzeiger*, 159, 44–68.

Mihelčič, F. 1958. Zur Kenntnis der Gletscherfauna Tirols (Eine neue Oribatide als Bewohner der Gletscher). *Zoologischer Anzeiger*, 160, 147–150.

Mihelčič, F. 1963. Ein Beitrag zur Kenntnis der europäischen *Eremaeus* (Acarina, Oribatei). *Eos, Madrid*, 38, 567–599.

Mihelčič, F., Rain, M. 1954. Beitrag zur Geographie und Ökologie des genus *Passalozetes* Grdj. *Zoologischer Anzeiger*, 153, 167–170.

Miko, L. 2006. Oppiidae. In: Weigmann, G. (Ed.), *Hornmilben (Oribatida)*. Die Tierwelt, Deutschland, 76, 179–207.

Miko, L., Norton, R.A. 2010. *Weigmannia* n. gen. from Eastern North America, with redescription of the type species, *Porobelba parki* Jacot, 1937 (Acari, Oribatida, Damaeidae). *Acarologia*, 50, 343–356.

Miko, L., Stanko, M. 1991. Small mammals as carriers of non-parasitic mites (Oribatida, Uropodina). In: Dusbabek, F., Bukva, V. (Eds.), *Modern Acarology*. Academia, Prague, 395–402.

Miko, L., Travé, J. 1996. Hungarobelbidae n. fam., with a description of *Hungarobelba pyrenaica* n. sp. (Acarina, Oribatida). *Acarologia*, 37, 133–155.

Miko, L., Weigmann, G. 1996. Notes on the genus *Liebstadia* Oudemans, 1906 (Acarina, Oribatida) in Central Europe. *Acta Musei Nationalis Pragae, Series B, Historia Naturalis*, 52, 73–100.

Miko, L., Weigmann, G., Nannelli, R. 1994. Redescription of *Protoribates lophotrichus* (Berlese, 1904) (Acarina, Oribatida). *Redia*, 77, 251–258.

Miko, L., Ermilov, S.G., Smelyansky, I.E. 2011. Taxonomy of European Damaeidae (Acari, Oribatida) VI. The oribatid mite genus *Parabelbella*: Redescription of *P. elisabhetae* and synonymy of *Akrodamaeus*. *Zootaxa*, 3140, 38–48.

Mitchell, M.J. 1976. *Ceratozetes kananaskis* (Acari, Cryptostigmata, Ceratozetidae): a new mite species from Western Canada. *The Canadian Entomologist*, 108, 577–582.

Mitchell, M.J. 1977. Population dynamics of oribatid mites (Acari, Cryptostigmata) in an aspen woodland soil. *Pedobiologia*, 17, 305–319.

Mitchell, M.J. 1978. Vertical and horizontal distribution of oribatid mites (Acari: Cryptostigmata) in an aspen woodland soil. *Ecology*, 593, 516–525.

Mitchell, M.J., Parkinson, D. 1976. Fungal feeding of oribatid mites (Acari: Cryptostigmata) in an aspen woodland soil. *Ecology*, 572, 302–312.

Molleman, F., Walter, D.E. 2001. Niche segregation and can-openers: Scydmaenid beetles as predators of armoured mites in Australia. In: Halliday, R.B., Walter, D.E., Proctor, H.C., Norton, R.A., Colloff, M.J. (Eds.), *Acarology: Proceedings of the 10th International Congress*. CSIRO Publishing, Melbourne, Australia, 283–288.

Moritz, M. 1965. Neue Oribatiden (Acari) aus Deutschland. I. *Oribella forsslundi* n. sp. und *Oppia nasuta* n. sp. *Zoologischer Anzeiger*, 175, 452–460.

Moritz, M. 1969. Neue Oribatiden (Acari) aus Deutschland. V. *Oppia keilbachi* nov. spec. *Wissenschaftliche Zeitschrift der Ernst-Moritz-Arndt-Universität Greifswald, Mathematisch-Naturwissenschaftliche Reihe*, 18, 37–40.

Moritz, M. 1970. Revision von *Suctobelba trigona* (Michael, 1888). Ein Beitrag zur Kenntnis der europaischen Arten der Gattung *Suctobelba* Paoli, 1908 sensu Jacot, 1937 (Acari, Oribatei, Suctobelbidae). *Mitteilungen aus dem Zoologischen Museum für Naturkunde in Berlin*, 46, 135–166.

Moritz, M. 1976a. Revision der Europäischen Gattungen und Arten der Familie Brachychthoniidae (Acari, Oribatei). Teil 1. Allgemeiner Teil: Brachychthoniidae Thor, 1934. Spezieller Teil: *Liochthonius* v. d. Hammen, 1959, *Verachthonius* nov. gen. und *Paraliochthonius* nov. gen. *Mitteilungen aus dem Zoologischen Museum für Naturkunde in Berlin*, 52, 27–136.

Moritz, M. 1976b. Revision der Europäischen Gattungen und Arten der Familie Brachychthoniidae (Acari, Oribatei). Teil 2. *Mixochthonius* Niedbała, 1972, *Neobrachychthonius* nov. gen., *Synchthonius* v.d. Hammen, 1952, *Poecilochthonius* Balogh, 1943, *Brachychthonius* Berlese, 1910, *Brachychochthonius* Jacot, 1938. *Mitteilungen aus dem Zoologischen Museum für Naturkunde in Berlin*, 52, 227–319.

Muraoka, M., Ishibashi, N. 1976. Nematode-feeding mites and their feeding behaviour. *Applied Entomology and Zoology*, 11, 1–7.

Murphy, K.A., Huettmann, F., Fresco, N., Morton, J. 2010. Connecting Alaska landscapes into the future: Results from an interagency climate modeling, land management and conservation project. U.S. Fish and Wildlife Service, 99 pp.

Murphy, P.W., Balla, A.N. 1973. The bionomics of *Humerobates rostrolamellatus* Grandjean (Cryptostigmata-Ceratozetidae) on fruit trees. In: Daniel, M., Rosicky, B. (Eds.), *Proceedings of the 3rd International Congress of Acarology*. Czechoslovak Academy of Sciences, Prague, 97–104.

Nannelli, R. 1990. Studio di una faggeta dell'Appennino pistoiese: Composizione e successione dell'artropodofauna nella degradazione della lettiera. *Redia*, 73, 543–568.

Nannelli, R., Turchetti, T., Maresi, G. 1998. Corticolous mites (Acari) as potential vectors of *Cryphonectria parasitica* (Murr.) Barr hypovirulent strains. *International Journal of Acarology*, 24, 244.

Nevin, F.R. 1976. *Pilogalumna cozadensis*, a new species of galumnid from Nebraska, U.S.A. *Acarologia*, 17, 751–758.

Nevin, F.R. 1977. Three new achipterids from the Catskills of New York State, U.S.A. (Acari; Cryptostigmata; Oribatei; Oribatelloidea; Achipteriidae). *Journal of the New York Entomological Society*, 84, 246–253.

Nevin, F.R. 1979. Additions to the descriptions and classifications of some American galumnids. *Acarologia*, 20, 147–157.

Newton, J.S. 2013. Biodiversity of soil arthropods in a native grassland in Alberta, Canada: Obscure associations and effects of simulated climate change. Ph.D. Thesis, Department of Biological Sciences, University of Alberta, Edmonton, Alberta, 142 pp.

Newton, J.S., Proctor, H.C. 2013. A fresh look at weight-estimation models for soil mites (Acari). *International Journal of Acarology*, 39, 72–85.

Nicolai, V. 1986. The bark of trees: Thermal properties, microclimate and fauna. *Oecologia*, 691, 148–160.

Nicolet, H. 1855. Histoire naturelle des Acariens qui se trouvent aux environs de Paris. *Archives du Museum d'Histoire naturelle, Paris*, 7, 381–482.

Niedbała, W. 1971. Oribatei (Acari) of Spitsbergen. *Bulletin of the Academy of Polish Science, Cl. II*, 19, 737–742 + 1 fig., 2 photo., 1 tab.

Niedbała, W. 1972. Catalogue of all known species of Brachychthoniidae (Acari, Oribatei). *Acarologia*, 14, 292–313.

Niedbała, W. 1983. Deux nouveaux Phthiracaridae (Acari, Oribatida) des montagnes du Kazakhstan et de Kirgizie. *Annales Zoologici, Warszawa*, 37, 63–70.

Niedbała, W. 1986. Catalogue des Phthiracaroidea (Acari), clef pour la détermination des espèces et descriptions d'espèces nouvelles. *Annales Zoologici*, 40, 309–370.

Niedbała, W. 1988. Phthiracaridae (Acari, Oribatida) nouveaux des Etats-Unis d'Amerique. *Bulletin of the Polish Academy of Sciences. Biological Sciences, Warszawa*, 36, 67–77.

Niedbała, W. 1998. Ptyctimous mites of the Pacific Islands. Recent knowledge, origin, descriptions, redescriptions, diagnoses and zoogeography (Acari, Oribatida). *Genus*, 9, 431–558.

Niedbała, W. 2002. Ptyctimous mites (Acari, Oribatida) of the Nearctic region. *Monographs of the Upper Silesian Museum, Bytom, Poland*, 4, 1–261.

Niedbała, W. 2007a. A new species of ptyctimous mites (Acari, Oribatida) from the Nearctic Region. *The Canadian Entomologist*, 139, 510–512.

Niedbała, W. 2007b. New records of ptyctimous mites (Acari, Oribatida) from the Nearctic Region. *The Canadian Entomologist*, 139, 587–590.

Niedbała, W., Liu, D. 2018. Catalogue of ptyctimous mites (Acari, Oribatida) of the world. *Zootaxa*, 4393, 1–238.

Niedbała, W., Penttinen, R. 2006. Two zoogeographically remarkable mite species from Finland (Acari, Oribatida, Oribotritiidae). *Journal of Natural History*, 40, 265–272.

Niedbała, W., Stary, J. 2015. Three new species of the family Phthiracaridae (Acari, Oribatida) from Bolivia. *Zootaxa*, 3918, 128–140.

Nielsen, U.N., Osler, G.H.R., Campbell, C.D., Burslem, D.F.R.P., van der Wal, R. 2010. The influence of vegetation type, soil properties and precipitation on the composition of soil mite and microbial communities at the landscape scale. *Journal of Biogeography*, 37, 1317–1328.

Niemi, R., Karppinen, E., Uusitalo, M. 1997. Catalogue of the Oribatida (Acari) of Finland. *Acta Zoologica Fennica*, 207, 1–39.

Nordenskiöld, E. 1901. Zur Kenntnis der Oribatidenfauna Finnlands. *Acta Societatis pro Fauna et Flora Fennica*, 21, 1–34.

Norton, R.A. 1977. A review of F. Grandjean's system of leg-chaetotaxy in the Oribatei and its application to the Damaeidae. In: Dindal, D.L. (Ed.), *Biology of Oribatid Mites*. State University New York, College of Environmental Science and Forestry, Syracuse, 33–62.

Norton, R.A. 1978. *Veloppia kananaskis* n. sp., with notes on the familial affinities of *Veloppia* Hammer (Acari: Oribatei). *International Journal of Acarology*, 4, 71–84.

Norton, R.A. 1979a. Generic concepts in the Damaeidae (Acari: Oribatei). I. Three new taxa based on species of Nathan Banks. *Acarologia*, 20, 603–621.

Norton, R.A. 1979b. Notes on synonymies, recombinations, and lectotype designations in Nathan Banks' species of *Nothrus* (Acari: Oribatei). *Proceedings of the Entomological Society of Washington*, 81, 645–649.

Norton, R.A. 1979c. The identity of *Pelopsis nudiuscula* (Acari: Oribatei). *Proceedings of the Entomological Society, Washington*, 81, 696–697.

Norton, R.A. 1980a. Observations on phoresy by oribatid mites (Acari: Oribatei). *International Journal of Acarology*, 6, 121–130.

Norton, R.A. 1980b. Generic concepts in the Damaeidae (Acari, Oribatei). Part II. *Acarologia*, 21, 496–513.

Norton, R.A. 1982. *Arborichthonius* n. gen., an unusual enarthronote soil mite (Acarina: Oribatei) from Ontario. *Proceedings of the Entomological Society of Washington*, 84, 85–96.

Norton, R.A. 1983. Tenuialidae (Acari: Oribatei): new diagnoses for supraspecific taxa. *Acarologia*, 24, 203–217.

Norton, R.A. 1984. Notes on Nathan Banks' and Henry Ewing's species of Mochlozetidae (Acari: Sarcoptiformes) with the proposal of a new genus. *Acarologia*, 25, 397–406.

Norton, R.A. 1985. Aspects of the biology and systematics of soil arachnids, particularly saprophagous and mycophagous mites. *Quaestiones Entomologicae*, 21, 523–541.

Norton, R.A. 1994. Evolutionary aspects of oribatid mite life histories and consequences for the origin of the Astigmata. In: Houck, M. (Ed.), *Mites. Ecological and Evolutionary Analyses of Life-history Patterns.* Chapman and Hall, New York, 99–135.

Norton, R.A. 1998. Morphological evidence for the evolutionary origin of Astigmata (Acari: Acariformes). *Experimental and Applied Acarology*, 22, 559–594.

Norton, R.A. 2001. Systematic relationships of Nothrolohmanniidae, and the evolutionary plasticity of body form in Enarthronota (Acari: Oribatida). In: Halliday, R.B., Walter, D.E., Proctor, H.C., Norton, R.A., Colloff, M.J. (Eds.), *Acarology: Proceedings of the 10th International Congress.* CSIRO Publishing, Melbourne, Australia, 58–75.

Norton, R.A. 2007. Holistic acarology and ultimate causes: Examples from the oribatid mites. In: Morales-Malacara, J.B., Behan-Pelletier, V., Ueckermann, E., Pérez, T.M., Estrada-Venegas, E.G., Badil, M. (Eds.), *Acarology XI: Proceedings of the International Congress.* Instituto de Biología and Facultad de Ciencias, Universidad Nacional Autónoma de México, Sociedad Latinoamericana de Acarología, México, 3–20.

Norton, R.A. 2010. Systematic relationships of Lohmanniidae (Acari: Oribatida). In: Sabelis, M.W., Bruin, J. (Eds.), *Trends in Acarology, Proceedings of the XII International Congress of Acarology, Amsterdam (2006)*, Springer, Dordrecht, 175–178.

Norton, R.A., Alberti, G. 1997. Porose integumental organs of oribatid mites (Acari, Oribatida). 3. Evolutionary and ecological aspects. *Zoologica, Stuttgart*, 146, 115–143.

Norton, R.A., Behan-Pelletier, V.M. 1986. Systematic relationships of *Propelops*, with a modification of family-group taxa in Phenopelopoidea (Acari: Oribatida). *Canadian Journal of Zoology*, 64, 2370–2383.

Norton, R.A., Behan-Pelletier, V.M. 1991. Calcium carbonate and calcium oxalate as cuticular hardening agents in oribatid mites (Acari: Oribatida). *Canadian Journal of Zoology*, 69, 1504–1511.

Norton, R.A., Behan-Pelletier, V.M. 2007. *Eniochthonius mahunkai* sp. n. (Acari: Oribatida: Eniochthoniidae), from North American peatlands, with a redescription of *Eniochthonius* and a key to North American species. *Acta Zoologica Academiae Scientiarum Hungaricae*, 53, 295–333.

Norton, R.A., Behan-Pelletier, V. 2009. Chapter 15, Oribatida. In: Krantz, G.W., Walter, D.E. (Eds.), *A Manual of Acarology.* 3rd ed. Texas Tech. University Press, Lubbock, 421–564.

Norton, R.A., Behan-Pelletier, V.M. 2020. Two unusual new species of *Caleremaeus* (Acari: Oribatida) from eastern North America, with redescription of C. *retractus* and reevaluation of the genus. *Acarologia*, 60, 398–448.

Norton, R.A., Dindal, D.L. 1976. Structure of the microarthropod community in Lake Ontario beach debris. *Environmental Entomology*, 54, 773–779.

Norton, R.A., Ermilov, S.G. 2014. Catalogue and historical overview of juvenile instars of oribatid mites (Acari: Oribatida). *Zootaxa*, 3833, 1–132.

Norton, R.A., Ermilov, S.G. 2017. Identity of the oribatid mite *Oribata curva* and transfer to *Trichogalumna* (Acari, Oribatida, Galumnidae), with discussion of nomenclatural and biogeographic issues in the '*curva*' species–group. *Zootaxa*, 4272, 551–564.

Norton, R.A., Ermilov, S.G. 2021. Redescriptions of North American *Epidamaeus* (Acari, Oribatida, Damaeidae) species proposed by N. Banks, H.E. Ewing, A.P. Jacot, and J.W. Wilson. *Zootaxa*, 5021, 1–65.

Norton, R.A., Fuangarworn, M. 2015. Nanohystricidae n. fam., an unusual, plesiomorphic enarthronote mite family endemic to New Zealand (Acari: Oribatida). *Zootaxa*, 4027, 151–204.

Norton, R.A., Kethley, J.B. 1990. Berlese's North American oribatid mites, historical notes, recombinations, synonymies and type designations. *Redia*, 62, 421–499.

Norton, R.A., Lions, J.C. 1992. North American Synichotritiidae (Acari: Oribatida) 1. *Apotritia walkeri* n. g., n. sp., from California. *Acarologia*, 33, 285–301.

Norton, R.A., Palmer, S.C. 1991. The distribution, mechanisms and evolutionary significance of parthenogenesis in oribatid mites. In: Schuster, R., Murphy, P.W. (Eds.): *The Acari: Reproduction, Development and Life-history Strategies.* Chapman and Hall, London, New York, 107–136.

Norton, R.A., Sidorchuk, E.A. 2014. *Collohmannia johnstoni* n. sp. (Acari, Oribatida) from West Virginia (U.S.A.), including description of ontogeny, setal variation, notes on biology and systematics of Collohmanniidae. *Acarologia*, 543, 271–334.

Norton, R.A., OConnor, B.M., Johnston, D.E. 1983. Systematic relationships of the Pediculochelidae (Acari: Acariformes). *Proceedings of the Entomological Society of Washington*, 85, 493–512.

Norton, R.A., Bonamo, P.M., Grierson, J.D., Shear, W.A. 1988a. Oribatid mite fossils from a terrestrial deposit near Gilboa, New York. *Journal of Paleontology*, 62, 259–269.
Norton, R.A., Welbourn, W.C., Cave, R.D. 1988b. First records of Erythraeidae parasitic on oribatid mites (Acari, Prostigmata: Acari, Oribatida). *Proceedings of the Entomological Society of Washington*, 90, 407–410.
Norton, R.A., Williams, D.D., Hogg, I.D., Palmer, S.C. 1988c. Biology of the oribatid mite *Mucronothrus nasalis* (Acari: Oribatida: Trhypochthoniidae) from a small coldwater springbrook in Eastern Canada. *Canadian Journal of Zoology*, 66, 622–629.
Norton, R.A., Kethley, J.B., Johnston, D.E., OConnor, B.M. 1993. Phylogenetic perspectives on genetic systems and reproductive modes of mites. In: Wrensch, D.L., Ebbert, M.A. (Eds.), *Evolution and Diversity of Sex Ratio in Insects and Mites*. Chapman and Hall, New York, 8–99.
Norton, R.A., Behan-Pelletier, V.M., Wang, H.-F. 1996a. The aquatic oribatid mite genus *Mucronothrus* in Canada and the western USA (Acari: Trhypochthoniidae). *Canadian Journal of Zoology*, 74, 926–949.
Norton, R.A., Graham, T.B., Alberti, G. 1996b. A rotifer-eating Ameronothroid (Acari: Ameronothridae) mite from ephemeral pools on the Colorado Plateau. In: Mitchell, R., Horn, D.J., Needham, G.R., Welbourn, C.W. (Eds.), *Acarology IX Proceedings*. vol. 1. Ohio Biological Survey, Columbus, OH, 539–542.
Norton, R.A., Alberti, G., Weigmann, G., Woas, S. 1997. Porose integumental organs of oribatid mites (Acari, Oribatida). 1. Overview of types and distribution. *Zoologica, Stuttgart*, 146, 1–31.
Norton, R.A., Franklin, E., Crossley, D.A., Jr. 2010. *Scapheremaeus rodickae* n. sp. (Acari: Oribatida: Cymbaeremaeidae) associated with temporary rock pools in Georgia, with key to *Scapheremaeus* species in eastern USA and Canada. *Zootaxa*, 2393, 1–16.
Norton, R.A., Ermilov, S.G., Miko, L. 2022. *Kunstidamaeus arthurjacoti* sp. nov. (Oribatida, Damaeidae), first report of the genus in North America. *Systematic and Applied Acarology*, 27, 482–496.
OConnor, B.M. 2009. Chapter 16, Cohort Astigmatina. In: Krantz, G.W., Walter, D.E. (Eds.), *A Manual of Acarology*. 3rd ed. Texas Tech. University Press, Lubbock, 565–658.
Ojala, R., Huhta, V. 2001. Dispersal of microarthropods in forest soil. *Pedobiologia*, 45, 443–450.
Olmeda, A.S., Blanco, M.M., Pérez-Sánchez, J.L., Luzón, M., Villarroel, M., Gibello, A. 2011. Occurrence of the oribatid mite *Trhypochthoniellus longisetus longisetus* (Acari: Trhypochthoniidae) on tilapia *Oreochromis niloticus*. *Diseases of Aquatic Organisms*, 94, 77–81.
Olszanowski, Z. 1996. A monograph of the Nothridae and Camisiidae of Poland (Acari: Oribatida: Crotonioidea). *Genus, Supplement, Wrocław*, 201 pp.
Olszanowski, Z., Szywilewska, A., Norton, R.A. 2001. New moss mite of the genus *Camisia* from western Nearctic region (Acari: Oribatida: Camisiidae). *Genus, Wrocław*, 12, 395–406.
Olszanowski, Z., Clayton, M.R., Humble, L.M. 2002. New species of the genus *Camisia* (Acari: Oribatida): an arboreal mite with enclosed sensilli. *The Canadian Entomologist*, 134, 707–721.
Olszanowski, Z., Szywilewska-Szczykutowicz, A., Blaszak, C., Ehrnsberger, R. 2007. Die Milben in der Zoologischen Staatssammlung München. Teil 10. Überfamilie Crotonioidea (I). *Spixiana*, 30, 159–167.
Oppedisano, M., Eguaras, M., Fernandez, N.A. 1995. Dépôt de spermatophores et structures de signalisation chez *Pergalumna* sp. (Acari: Oribatida). *Acarologia*, 364, 347–353.
Osler, G.H.R., Harrison, L., Kanashiro, D.K., Clapperton, M.J. 2008. Soil microarthropod assemblages under different arable crop rotations in Alberta, Canada. *Applied Soil Ecology*, 38, 71–78.
Oudemans, A. 1896. List of Dutch Acari Latr. First part, Oribatei Dug. with synonymical notes and other remarks. *Tijdschrift voor Entomologie*, 39, 53–65.
Oudemans, A. 1900. New list of Dutch Acari, 1st part. *Tijdschrift voor Entomologie*, 43, 150–171.
Oudemans, A. 1902. Acarologische Aanteekeninger, XXIV. *Entomologische Berichten, Amsterdam*, 2, 96–101.
Oudemans, A.C. 1903. Acarologische Aanteekeningen, VI. *Entomologische Berichten, Amsterdam*, 1, 83–88.
Oudemans, A.C. 1906. Acarologische Aanteekeningen XXII. *Entomologische Berichten*, 2, 55–62.
Oudemans, A.C. 1913. Acarologische Aanteekeningen XLVII. *Entomologische Berichten*, 3, 372–376.
Oudemans, A.C. 1914. Acarologisches aus Maulwurfsnestern. *Archiv für Naturgeschichte*, 79, A8. 108–200+figs. 1–260; A9. 68–136+figs. 261–361; A10. 1–69+pls.2–18.
Oudemans, A.C. 1915. *Dytiscus, Dryobius roboris*. Argus en *Galumna*. *Tijdschrift voor Entomologie, Verslagen*, 58, 9–14.
Oudemans, A.C. 1916. Overzicht der tot 1898 beschreven Phthiracaridae. *Entomologische Berichten, Amsterdam*, 4, 245–249.
Oudemans, A.C. 1917. Acarologische Aanteekeningen, LXII. *Entomologische Berichten, Amsterdam*, 4, 341–348.
Oudemans, A.C. 1923. Studie over de sedert 1877 ontworpen Systemen der Acari; Nieuwe Classificatie; Phylogenetische Beschouwingen. *Tijdschrift voor Entomologie*, 66, 49–85.

Oudemans, A.C. 1930. Acarologische Aanteekeningen 102. *Entomologische Berichten, The Netherlands*, 8, 69–74.

Oudemans, A.C., Voigts, H. 1905. Zur Kenntnis der Milbenfauna von Bremen. *Abhandlungen des Naturwissenschaftlichen Vereins zu Bremen*, 18, 199–253.

Owojori, O.J., Siciliano, S.D. 2012. Accumulation and toxicity of metals (copper, zinc, cadmium, and lead) and organic compounds (geraniol and benzo[a]pyrene) in the oribatid mite *Oppia nitens*. *Environmental Toxicology and Chemistry*, 31, 1639–1648.

Pachl, P. 2010. A conservative genetic marker (RNA Polymerase II) for the resolution of old radiations in oribatid mites (Acari, Oribatida). Unpublished thesis, Technische Universität Darmstadt 0974152.

Pachl, P., Domes, K., Schulz, G., Norton, R.A., Scheu, S., Schaefer, I., Maraun, M. 2012. Convergent evolution of defense mechanisms in oribatid mites (Acari, Oribatida) shows no "ghosts of predation past". *Molecular Phylogenetics and Evolution*, 65, 412–420.

Pachl, P., Lindl, A.C., Krause, A., Scheu, S., Schaefer, I., Maraun, M. 2017. The tropics as an ancient cradle of oribatid mite diversity. *Acarologia*, 57, 309–322.

Pachl, P., Uusitalo, M., Scheu, S., Schaefer, I., Maraun, M. 2021. Repeated convergent evolution of parthenogenesis in Acariformes (Acari). *Ecology and Evolution*, 111, 321–337.

Palmer, S.C., Norton, R.A. 1990. Further experimental proof of thelytokous parthenogenesis in oribatid mites (Acari, Oribatida, Desmonomata). *Experimental and Applied Acarology*, 8, 149–159.

Palmer, S.C., Norton, R.A. 1992. Genetic diversity in thelytokous oribatid mites (Acari, Acariformes, Desmonomata). *Biochemical Systematics and Ecology*, 20, 219–231.

Pan'kov, A.N. 2002. New species of Oribatei from the Far East. *Zoologicheskii Zhurnal*, 81, 242–245.

Pan'kov, A.N., Ryabinin, N.A., Golosova, L.D. 1997. Catalogue of oribatid mites of the Far East of Russia. Part I. Catalogue of oribatid mites of Kamchatka, Sakhalin and the Kuril Islands. Dalnauka, Vladivostok, Khabarovsk.

Pande, Y.D., Berthet, P. 1973. Studies on the food and feeding habits of soil Oribatei in a black pine plantation. *Oecologia*, 12, 413–426.

Paoli, G. 1908. Monografia del genere *Dameosoma* Berl. e generi affini. *Redia*, 5, 31–91 + pls.3–5.

Parry, B.W. 1979. A revision of the British species of the genus *Phthiracarus* Perty, 1841 (Cryptostigmata, Euptyctima). *Bulletin of the British Museum of natural History, Zoology*, 35, 323–363.

Paschoal, A.D. 1982a. A revision of the genus *Gymnodamaeus* (Acari, Oribatei, Gymnodamaeidae), with descriptions of nine new species. *Revista Brasileira de Zoologia*, 26, 113–132.

Paschoal, A.D. 1982b. Description of *Odontodamaeus*, a new genus in the family Gymnodamaeidae (Acari, Oribatei). *Revista Brasileira de Entomologia*, 26, 201–205.

Paschoal, A.D. 1982c. *Nortonella* (Acari, Oribatei), a new genus of Gymnodamaeidae from North America. *Revista Brasileira de Entomologia*, 26, 207–209.

Paschoal, A.D. 1983a. *Pleodamaeus*, a new genus of Gymnodamaeidae mites (Acari, Oribatei) from North America. *Revista Brasileira de Zoologia*, 27, 125–127.

Paschoal, A.D. 1983b. A revision of the genus *Jacotella* (Acari, Oribatei, Gymnodamaeidae) in North America with description of two new species. *Revista Brasileira de Zoologia*, 27, 127–135.

Paschoal, A.D. 1983c. Description of *Johnstonella subalpina* (Acari: Oribatei), a new species and genus of Gymnodamaeidae from the Nearctic region. *Revista Brasileira de Entomologia*, 27, 205–210.

Paschoal, A.D. 1984. *Adrodamaeus* (Acari, Oribatei, Gymnodamaeidae), a new name for *Heterodamaeus* Woolley, with a reevaluation of the genus and a description of a new species. *Revista Brasileira de Entomologia*, 28, 15–21.

Paschoal, A.D. 1988. A revision of the Plateremaeidae (Acari: Oribatei). *Revista Brasileira de Entomologia*, 3(6) (1987), 327–356.

Pauly, F. 1952. Die Copula der Oribatiden. *Naturwissenschaften*, 39, 572–573.

Pauly, F. 1956. Zur biologie einiger belbiden (Oribatei, Moosmilben) und zur funktion ihrer pseudostigmatischen organe. *Zoologische Jahrbücher. Abteilung für Systematik, Ökologie und Geographie der Tiere*, 84, 275–328.

Pavlichenko, P.G. 1993. New taxa of the Ceratozetoidea (Oribatei) mites. *Vestnik Zoologii*, 6, 29–36.

Peck, S.B. 1994. Sea-surface (pleuston) transport of insects between islands in the Galápagos Archipelago, Ecuador. *Annals of the Entomological Society of America*, 87, 576–582.

Peel, M.C., Finlayson, B.L., McMahon, T.A. 2007. Updated world map of the Köppen-Geiger climate classification. *Hydrology and Earth System Sciences*, 11, 1633–1644.

Pennack, R.W., Ward, J.V. 1986. Interstitial faunal communities of the hyporheic and adjacent groundwater biotopes of a Colorado mountain stream. *Archiv für Hydrobiologie, Supplement*, 743, 356–396.

References

Penttinen, R., Siira-Pietikäinen, A., Huhta, V. 2008. Oribatid mites in eleven different habitats in Finland. In: Bertrand, M., Kreiter, S., McCoy, K.D., Migeon, A., Navajas, M., Tixier, M.S., Vial, L. (Eds.), *Integrative Acarology. Proceedings of the 6th European Congress*, European Association of Acarologists, Montpellier, France, 237–244.

Penttinen, R., Viiri, H., Moser, J.C. 2013. The mites (Acari) associated with bark beetles in the Koli National Park in Finland. *Acarologia*, 53, 3–15.

Pepato, A.R., Klimov, P.B. 2015. Origin and higher-level diversification of acariform mites – Evidence from nuclear ribosomal genes, extensive taxon sampling, and secondary structure alignment. *BMC Evolutionary Biology*, 15, 178.

Pequeño, P.A.C.L., Franklin, E., Norton, R.A., de Morais, J.W. 2018. A tropical arthropod unravels local and global environmental dependence of seasonal temperature–size response. *Biology Letters*, 14, 20180125.

Pequeño, P.A.C.L., Franklin, E., Norton, R.A. 2021a. Microgeographic morphophysiological divergence in an Amazonian soil mite. *Evolutionary Biology*, 48, 160–169.

Pequeño, P.A.C.L., Franklin, E., Norton, R.A. 2021b. Modelling selection, drift, dispersal and their interactions in the community assembly of Amazonian soil mites. *Oecologia*, 196, 805–814.

Perdomo, G., Evans, A., Maraun, M., Sunnucks, P., Thompson, R. 2012. Mouthpart morphology and trophic position of microarthropods from soils and mosses are strongly correlated. *Soil Biology and Biochemistry*, 53, 56–63.

Pérez-Íñigo, C. 1965. Especies españolas del género *Oppia* C.L. Koch (Acari, Oribatei). *Boletín de la Real Sociedad Española de Historia Natural. Sección biológica*, 62, 385–416.

Pérez-Íñigo, C. 1970. Acaros oribátidos de suelos de España peninsular e islas Baleares (Acari, Oribatei). Parte II. *EOS-Revista Espanola de Entomologia*, 45, 241–317.

Pérez-Íñigo, C. 1990. *Protozetomimus*, a new genus of oribatid mites (Acari, Oribatei, Ceratozetidae). *Redia*, 73, 397–402.

Pérez-Íñigo, C., Subías, L.S. 1978. Sorprendente hallazgo de un representante de la familia Kodiakellidae Hammer en España, *Kodiakiella dimorpha* n. sp., y consideraciones sobre esta familia (Acari, Oribatei). *Boletín de la Asociación Española de Entomología*, 1, 103–107.

Pérez-Íñigo, C., Subías, L.S. 1979. Notes sur les Oribates d'Espagne II. *Parapyroppia monodactyla* n. g.; n. sp. *Acarologia*, 20, 303–309.

Perrot-Minnot, M., Norton, R. 1997. Obligate thelytoky in oribatid mites: No evidence for *Wolbachia* inducement. *The Canadian Entomologist*, 129, 691–698.

Perty, M. 1841. *Allgemeine Naturgeschichte, als philosopische und Humanitatswissenschaft fur Naturforscher*. Philosophen und das hoher gebildete Publikum, Bern, 3, 873–875.

Peschel, K., Norton, R.A., Scheu, S., Maraun, M. 2006. Do oribatid mites live in enemy-free space? Evidence from feeding experiments with the predatory mite *Pergamasus septentrionalis*. *Soil Biology and Biochemistry*, 38, 2985–2989.

Petrunkevitch, A. 1955. Superfamily Oribatoidea Dugès 1834. In: Störmer, L., Petrunkevitch, A., Hedgpeth, J.W. (Eds.), *Treatise on Invertebrate Palaeontology, Part P, Arthropoda 2, Chelicerata*. Geological Society of America and University of Kansas Press, 98–99.

Petrzik, K., Sarkisova, T., Stary, J., Koloniuk, I., Hrabáková, L., Kubešová, O. 2015. Molecular characterization of a new monopartite dsRNA mycovirus from mycorrhizal *Thelephora terrestris* (Ehrh.) and its detection in soil oribatid mites (Acari: Oribatida). *Virology*, 489, 12–19.

Pfingstl, T. 2013a. *Thalassozetes barbara* n. sp. (Acari, Oribatida), a new intertidal species from the coast of Barbados. *Acarologia*, 53, 417–424.

Pfingstl, T. 2013b. Resistance to fresh and salt water in intertidal mites (Acari: Oribatida): implications for ecology and hydrochorous dispersal. *Experimental and Applied Acarology*, 61, 87–96.

Pfingstl, T. 2017. The marine-associated lifestyle of ameronothroid mites (Acari, Oribatida) and its evolutionary origin: A review. *Acarologia*, 573, 693–721.

Pfingstl, T. 2021. First comprehensive insights into the biogeography of the Caribbean intertidal oribatid mite fauna (Ameronothroidea). *Neotropical Biodiversity*, 7, 102–110.

Pfingstl, T., Krisper, G. 2011. No difference in the juveniles of two *Tectocepheus* species (Acari: Oribatida, Tectocepheidae). *Acarologia*, 512, 199–218.

Pfingstl, T., Krisper, G. 2014. Plastron respiration in marine intertidal oribatid mites (Acari, Fortuyniidae and Selenoribatidae). *Zoomorphology*, 133, 359–378.

Pfingstl, T., Lienhard, A. 2017. *Schusteria marina* sp. nov. (Acari, Oribatida, Selenoribatidae) an intertidal mite from Caribbean coasts, with remarks on taxonomy, biogeography, and ecology. *International Journal of Acarology*, 43, 462–467.

Pfingstl, T., Schuster, R. 2014. Global distribution of the thalassobiontic Fortuyniidae and Selenoribatidae (Acari, Oribatida). *Soil Organisms*, 86, 125–130.

Pfingstl, T., Schäffer, S., Krisper, G. 2010. Re-evaluation of the synonymy of *Latovertex* Mahunka, 1987 and *Exochocepheus* Woolley and Higgins, 1968 (Acari, Oribatida, Scutoverticidae). *International Journal of Acarology*, 36, 327–342.

Pfingstl, T., Lienhard, A., Shimano, S., Bin Yasin, Z., Tan Shau-Hwai, A., Jantarit, S., Petcharad, B. 2019a. Systematics, genetics, and biogeography of intertidal mites (Acari, Oribatida) from the Andaman Sea and Strait of Malacca. *Journal of Zoological Systematics and Evolutionary Research*, 57, 91–112.

Pfingstl, T., Wagner, M., Hiruta, S.F., Koblmüller, S., Hagino, W., Shimano, S. 2019b. Phylogeographic patterns of intertidal arthropods (Acari, Oribatida) from southern Japanese islands reflect paleoclimatic events. *Scientific Reports*, 9, 19042.

Pielou, D.P., Verma, A.N. 1968. The arthropod fauna associated with the birch bracket fungus, *Polyporus betulinus*, in Eastern Canada. *The Canadian Entomologist*, 100, 1179–1199.

Piffl, E. 1972. Zur Systematik der Oribatiden (Acari) (Neue Oribatiden aus Nepal, Costa Rica und Brasilien ergeben eine neue Familie der Unduloribatidae und erweitern die Polypterozetidae um die Gattungen *Podopterogaeus, Nodocephus, Eremaeozetes* und *Tumerozetes*). *Khumbu Himal*, 4, 269–314.

Polderman, P.J.G. 1974. Some notes on the oribatid fauna of salt-marshes in Denmark. *Acarologia*, 16, 358–366.

Połeć, W., Moskwa, B. 1994. Development of early larval forms of *Moniezia expansa* under laboratory conditions. *Wiadomosci Parazytologiczne*, 40, 153–157.

Polyak, V.J., Cokendolpher, J.C., Norton, R.A., Asmeron, Y. 2001. Wetter and cooler late Holocene climate in the southwestern United States from mites preserved in stalagmites. *Geology*, 29, 643–646.

Ponge, J.F. 1991. Food resources and diets of soil animals in a small area of Scots pine litter. *Geoderma*, 49, 33–62.

Popp, E. 1962. Semiaquatile Lebensräume Bülten in Hoch- und Niedermooren. 2. Teil: Die Milbenfauna. *Internationale Revue der gesamten Hydrobiologie*, 474, 533–579.

Porzner, A., Weigmann, G. 1992. Die Hornmilbenfauna (Acari, Oribatida) an Eichenstämmen in einem Gradienten von Autoabgas-Immissionen. *Zoologische Beiträge*, 34, 249–260.

Powell, J.M., Skaley, L.S. 1975. Arthropods from forest litter under Lodgepole pine infected with comandra blister rust. *Northern Forest Research Centre (Edmonton). Information Report NOR–X–130*, 33 pp.

Princz, J.I., Behan-Pelletier, V.M., Scroggins, Z.R.P., Siciliano, S.D. 2010. Oribatid mites in soil toxicity testing the use of *Oppia nitens* (C.L. Koch) as a new test species. *Environmental Toxicology and Chemistry*, 29, 971–979.

Prinzing, A., Lentzsch, P., Voigt, F., Woas, S. 2004. Habitat stratification stratifies a local population: ecomorphological evidence from a bisexual, mobile invertebrate (*Carabodes labyrinthicus*; Acari). *Annales Zoologici Fennici*, 41, 399–412.

Prinzing, A., Wirtz, H.P. 1997. The epiphytic lichen, *Evernia prunastri* L., as a habitat for arthropods: shelter from desiccation, food-limitation and indirect mutualism. In: Stork, N.E., Adis, J., Didham, R.K. (Eds.), *Canopy Arthropods*. Chapman and Hall, London, New York, Tokyo, 477–494.

Proctor, H.C. 2001. Extracting aquatic mites from stream substrates: A comparison of three methods. *Experimental and Applied Acarology*, 251, 1–11.

Proctor, H.C., Montgomery, K.M., Rosen, K.E., Kitching, R.L. 2002. Are tree trunks habitats or highways? A comparison of oribatid mite assemblages from hoop-pine bark and litter. *Australian Journal of Entomology*, 41, 294–299.

Pugh, P.J.A., King, P.E., Fordy, M.R. 1987. A comparison of the structure and function of the cerotegument in two species of Cryptostigmata (Acarina). *Journal of Natural History*, 21, 603–616.

Purrini, K. 1981. Studies on some Amoebae (Amoebida) and *Helicosporidium parasiticum* (Helicosporida) infecting moss-mites (Oribatei, Acarina) in forest soil samples. *Archiv für Protistenkunde*, 124, 303–311.

Purrini, K., Bukva, V. 1984. Pathenogenetic agents of oribatid mites (Oribatei, Acarina): a gap in the research on population dynamics. In: Griffiths, D.A., Bowman, C.E. (Eds.), *Acarology VI*. vol. 2. Horwood, Chichester, 826–837.

Rasmy, A.H., MacPhee, A.W. 1970. Mites associated with apple in Nova Scotia. *The Canadian Entomologist*, 102, 72–74.

Raspotnig, G. 2010. Oil gland secretions in Oribatida (Acari). In: Sabelis, M.W., Bruin, J. (Eds.), *Trends in Acarology, Proceedings of the XII International Congress of Acarology, Amsterdam (2006)*. Springer, Dordrecht, 235–239.

Raspotnig, G., Föttinger, P. 2008. Analysis of individual oil gland secretion profiles in oribatid mites (Acari: Oribatida). *International Journal of Acarology*, 344, 409–417.

Raspotnig, G., Leis, H.J. 2009. Wearing a raincoat: exocrine secretions contain anti-wetting agents in the oribatid mite, *Liacarus subterraneus* (Acari: Oribatida). *Experimental and Applied Acarology*, 47, 179–190.

Raspotnig, G., Matischek, T. 2010. Anti-wetting strategies of soil-dwelling Oribatida (Acari). In: Tajovsky, K., Pizl, V., Skuhrava, M. (Eds.), *Contributions to Soil Zoology in Central Europe IV*. Acta Societatis Entomologicae Bohemiae, 74, 91–96.

Raspotnig, G., Schuster, R., Krisper, G. 2003. Functional anatomy of oil glands in *Collohmannia gigantea* (Acari, Oribatida). *Zoomorphology*, 122, 105–112.

Raspotnig, G., Schuster, R., Krisper, G. 2004a. Citral in oil gland secretions of Oribatida (Acari) – a key component for phylogenetic analyses. *Abhandlungen und Berichte des Naturkundemuseums Gorlitz*, 76, 43–50.

Raspotnig, G., Krisper, G., Schuster, R. 2004b. Oil gland chemistry of *Trhypochthonius tectorum* (Acari: Oribatida) with reference to the phylogenetic significance of secretion profiles in the Trhypochthoniidae. *International Journal of Acarology*, 304, 369–374.

Raspotnig, G., Krisper, G., Schuster, R., Fauler, G., Leis, H.J. 2005. Volatile exudates from the oribatid mite, *Platynothrus peltifer*. *Journal of Chemical Ecology*, 312, 419–430.

Raspotnig, G., Stabentheiner, E., Föttinger, P., Schaider, M., Krisper, G., Rechberger, G., Leis, H.J. 2008. Opisthonotal glands in the Camisiidae (Acari, Oribatida): Evidence for a regressive evolutionary trend. *Journal of Zoological Systematics and Evolutionary Research*, 47, 77–87.

Raspotnig, G., Norton, R.A., Heethoff, M. 2011. Oribatid mites and skin alkaloids in poison frogs. *Biology Letters*, 7, 555–556.

Rathke, J. 1799. Entomologiske Jagttagelser. Skrifter Naturh. *Selskabet, Kjöbenhavn*, 5, 191–207.

Reeves, R.M. 1969. Seasonal distribution of some forest soil Oribatei. *Proceedings of the 2nd International Congress of Acarology*, 23–30.

Reeves, R.M. 1987. A new arboreal *Carabodes* from eastern North America (Acari: Oribatida: Carabodidae). *Proceedings of the Entomological Society of Washington*, 89, 468–477.

Reeves, R.M. 1988. Distribution and habitat comparisons for *Carabodes* collected from conifer branches with descriptions of *brevis* Banks and *higginsi* n. sp. (Acari: Oribatida: Carabodidae). *Proceedings of the Entomological Society of Washington*, 90, 373–392.

Reeves, R.M. 1990. Two new species of *Carabodes* (Acari: Oribatida: Carabodidae) from North America. *Canadian Journal of Zoology*, 68, 2158–2168.

Reeves, R.M. 1991. *Carabodes niger* Banks, *C. polyporetes* n. sp. and unverified records of *C. areolatus* Berlese (Acari: Oribatida: Carabodidae) in North America. *Canadian Journal of Zoology*, 69, 2925–2934.

Reeves, R.M. 1992. *Carabodes* of the Eastern United States and adjacent Canada (Acari: Oribatida: Carabodidae). *Canadian Journal of Zoology*, 70, 2042–2058.

Reeves, R.M. 1998. Biogeography of Carabodidae (Acari: Oribatida) in North America. *Applied Soil Ecology*, 9, 59–62.

Reeves, R.M., Behan-Pelletier, V.M. 1998. The genus *Carabodes* (Acari: Oribatida: Carabodidae) of North America, with descriptions of new western species. *Canadian Journal of Zoology*, 76, 1898–1921.

Reeves, R.M., Marshall, V.G. 1971. Redescription and chaetotaxie of *Brachychthonius lydiae* adults and nymphs (Acarina: Oribatei). *Annals of the Entomological Society of America*, 64, 317–325.

Remén, C., Krüger, M., Cassel-Lundhagen, A. 2020. Successful analysis of gut contents in fungal-feeding oribatid mites by combining body-surface washing and PCR. *Soil Biology and Biochemistry*, 42, 1952–1957.

Renker, C., Otto, P., Schneider, K., Zimdars, B., Maraun, M., Buscot, F. 2005. Oribatid mites as potential vectors for soil microfungi: Study of mite-associated fungal species. *Microbial Ecology*, 50, 518–528.

Reutimann, P. 1987. Quantitative aspects of the feeding activity of some oribatid mites (Oribatida, Acari) in an alpine meadow ecosystem. *Pedobiologia*, 30, 425–433.

Reutimann, P. 1991. Alpine oribatid mites and plant decomposition: Feeding and faeces production. In: Dusbabek, F., Bukva, V. (Eds.), *Modern Acarology*. Academia, Prague, 1, 417–422.

Riha, G. 1951. Zur Ökologie der Oribatiden in Kalksteinböden. *Zoologische Jahrbücher. Abteilung für Systematik, Geographie und Biologie der Tiere*, 80, 408–450.

Riley, C. 1874. Description of two new subterranean mites, also sixth annual report on the noxious, beneficial and other insects, of the state of Missouri. *Transaction of the Academy of Science St.Louis*, 3, 215–216.

Robineau-Desvoidy, D.M. 1839. Memoire sur le *Xenillus clypeator* (Coleoptère nouveau). *Annales de la Société entomologique de France*, 8, 455–462.

Rockett, C.L. 1980. Nematode predation by oribatid mites (Acarina: Oribatida). *International Journal of Acarology*, 6, 219–224.

Rockett, C.L., Woodring, J.P. 1966a. Oribatid mites as predators of soil nematodes. *Annals of the Entomological Society of America*, 59, 669–671.

Rockett, C.L., Woodring, J.P. 1966b. Biological investigations on a new species of *Ceratozetes* and of *Pergalumna* (Acarina: Cryptostigmata). *Acarologia*, 8, 511–520.

Rohde, C.J., jr. 1955. Studies on arthropods from a moss habitat with special emphasis on the life history of three oribatoid mites. Dissertation, Northwestern University, Evanston, Illinois, 83 pp.

Root, H.T., McGee, G.G., Norton, R.A. 2007. Arboreal mite communities on epiphytic lichens of the Adirondack Mountains of New York. *Northeastern Naturalist*, 14, 425–438.

Root, H.T., Kawahara, A.Y., Norton, R.A. 2008. *Anachipteria sacculifera* n. sp. (Acari: Oribatida: Achipteriidae) from arboreal lichens in New York State. *Acarologia*, 47, 173–181.

Rousseau, L., Venier, L., Hazlett, P., Fleming, R., Morris, D., Handa, I.T. 2018a. Forest floor mesofauna communities respond to a gradient of biomass removal and soil disturbance in a boreal jack pine (*Pinus banksiana*) stand of northeastern Ontario (Canada). *Forest Ecology and Management*, 407, 155–165.

Rousseau, L., Venier, L., Fleming, R., Hazlett, P., Morris, D., Handa, I.T. 2018b. Long-term effects of biomass removal on soil mesofaunal communities in northeastern Ontario (Canada) jack pine (*Pinus banksiana*) stands. *Forest Ecology and Management*, 421, 72–83.

Ryabinin, N.A., Ermilov, S.G. 2021. New faunistical data on oribatid mites of the family Damaeidae (Acari, Oribatida) from the Russian Far East. *Acarina*, 29, 23–34.

Ryabinin, N.A., Wu, D.-H. 2018. On the genus *Megeremaeus* Higgins et Woolley, 1965 (Acari, Oribatida, Megeremaeidae) with the description of the new species *Megeremaeus sikhotealinus* Ryabinin et Wu sp. n. *Zoologicheskii Zhurnal*, 97, 255–260. [in Russian] Entomological Review, 98, 652–657. [English version]

Ryabinin, N.A., Liu, D., Gao, M., Wu, D.-H. 2018. Checklist of oribatid mites (Acari, Oribatida) of the Russian Far East and Northeast of China. *Zootaxa*, 4472, 201–232.

Sakata, T., Norton, R.A. 2001. Opisthonotal gland chemistry of early-derivate oribatid mites (Acari) and its relevance to systematic relationships of Astigmata. *International Journal of Acarology*, 27, 281–291.

Sakata, T., Norton, R.A. 2003. Opisthonotal gland chemistry of a middle-derivative oribatid mite, *Archegozetes longisetosus* (Acari: Trhypochthoniidae). *International Journal of Acarology*, 294, 345–350.

Sakata, T., Shimano, S., Kuwahara, Y. 2003. Chemical ecology of oribatid mites III. Chemical composition of oil gland exudates from two oribatid mites, *Trhypochthoniellus* sp. and *Trhypochthonius japonicus* (Acari: Trhypochthoniidae). *Experimental and Applied Acarology*, 29, 279–291.

Sanders, F.H., Norton, R.A. 2004. Anatomy and function of the ptychoid defensive mechanism in the mite *Euphthiracarus cooki* (Acari: Oribatida). *Journal of Morphology*, 259, 119–154.

Saltzwedel, H., Maraun, M., Scheu, S., Schaefer, I. 2014. Evidence for frozen-niche variation in a cosmopolitan parthenogenetic soil mite species (Acari, Oribatida). *PLOS One*, 9, e113268.

Sanyal, A.K., Das, T.K. 1989. Oribatid mites (Acari, Cryptostigmata) associated with pineapple (*Ananas cosmosus*) root at Kalyani, West Bengal. *Environment and Ecology*, 7, 971–972.

Saporito, R.A., Donnelly, M.A., Norton, R.A., Garraffo, H.M., Spande, T.F., Daly, J.W. 2007. Oribatid mites as a major dietary source for alkaloids in poison frogs. *Proceedings of the National Academy of Sciences of the United States of America*, 104, 8885–8890.

Saporito, R.A., Norton, R.A., Andriamaharavo, N.R., Garraffo, H.M., Spande, T.F. 2011. Alkaloids in the mite *Scheloribates laevigatus*: Further alkaloids common to oribatid mites and poison frogs. *Journal of Chemical Ecology*, 37, 213–218.

Saporito, R.A., Donnelly, M.A., Spande, T.F., Garraffo, H.M. 2012. A review of chemical ecology in poison frogs. *Chemoecology*, 22, 159–168.

Schaefer, I., Caruso, T. 2019. Oribatid mites show that soil food web complexity and close aboveground-belowground linkages emerged in the early Paleozoic. *Communications Biology*, 2, 387.

Schaefer, I., Norton, R.A., Scheu, S., Maraun, M. 2010. Arthropod colonization of land – Linking molecules and fossils in oribatid mites (Acari, Oribatida). *Molecular Phylogenetics and Evolution*, 57, 113–121.

Schäffer, S., Koblmüller, S. 2020. Unexpected diversity in the host-generalist oribatid mite *Paraleius leontonychus* (Oribatida, Scheloribatidae) phoretic on Palearctic bark beetles. *PeerJ*, 8, e9710.

Schäffer, S., Koblmüller, S., Pfingstl, T., Sturmbauer, C., Krisper, G. 2010. Ancestral state reconstruction reveals multiple independent evolution of diagnostic morphological characters in the "Higher Oribatida" (Acari), conflicting with current classification schemes. *BMC Evolutionary Biology*, 10, 246.

Schäffer, S., Koblmüller, S., Klymiuk, I., Thallinger, G.G. 2018. The mitochondrial genome of the oribatid mite *Paraleius leontonychus*: New insights into tRNA evolution and phylogenetic relationships in acariform mites. *Scientific Reports*, 8, 7558.

Schäffer, S., Koblmüller, S., Krisper, G. 2020. Revisiting the evolution of arboreal life in oribatid mites. *Diversity*, 12, 255.

Schatz, H. 1991. Arrival and establishment of Acari on oceanic islands. In: Dusbábek, F., Bukva, V. (Eds.), *Modern Acarology Prague*. Academia Prague and SPB Academic Publishing, Prague, 613–618.

Schatz, H. 2004. *Palaeacarus hystricinus* Trägårdh, 1932 (Acari: Oribatida: Palaeacaridae), eine bemerkenswerte Hornmilbe in Tirol. *Berichte des Naturwissenschaftlich-medizinischen Vereins in Innsbruck*, 91, 339–340.

Schatz, H. 2020. Catalogue of oribatid mites (Acari: Oribatida) from Vorarlberg (Austria). *Zootaxa*, 4783, 1–106.

Schatz, H., Behan-Pelletier, V.M. 2008. Global diversity of oribatids (Oribatida: Acari: Arachnida). In: Balian, E.V., Lévêque, C., Segers, H., Martens, K. (Eds.), *Freshwater Animal Diversity Assessment*. *Hydrobiologia*, 595, 323–328.

Schatz, H., Gerecke, R. 1996. Hornmilben Acari, Oribatida aus Quellen und Quellbächen im Nationalpark Berchtesgaden (Oberbayern) und in den Südlichen Alpen (Trentino – Alto Adige). *Berichte des Naturwissenschaftlich-medizinischen Vereins in Innsbruck*, 83, 121–144.

Schatz, H., Behan-Pelletier, V.M., OConnor, B.M., Norton, R.A. 2011. Suborder Oribatida van der Hammen, 1968. In: Zhang, Z.Q. (Ed.), *Animal Biodiversity: An outline of higher-level classification and survey of taxonomic richness*. Zootaxa, 3148, 141–148.

Schelvis, J. 1990. The reconstruction of local environments on the basis of remains of oribatid mites (Acari: Oribatida). *Journal of Archaeological Science*, 17, 559–572.

Schmelzle, S., Blüthgen, N. 2019. Under pressure: force resistance measurements in box mites (Actinotrichida, Oribatida). *Frontiers in Zoology*, 16, 24.

Schmelzle, S., Helfen, L., Norton, R.A., Heethoff, M. 2009. The ptychoid defensive mechanism in Euphthiracaroidea (Acari: Oribatida): A comparison of muscular elements with functional considerations. *Arthropod Structure & Development*, 38, 461–472.

Schmelzle, S., Helfen, L., Norton, R.A., Heethoff, M. 2010. The ptychoid defensive mechanism in *Phthiracarus longulus* (Acari, Oribatida, Phthiracaroidea): Exoskeletal and muscular elements. *Soil Organisms*, 822, 253–273.

Schmelzle, S., Norton, R.A., Heethoff, M. 2012. A morphological comparison of two closely related ptychoid oribatid mite species, *Phthiracarus longulus* and *P. globosus* (Acari: Oribatida: Phthiracaroidea). *Soil Organisms*, 84, 431–443.

Schmelzle, S., Norton, R.A., Heethoff, M. 2015. Mechanics of the ptychoid defense mechanism in Ptyctima (Acari, Oribatida): One problem, two solutions. *Zoologischer Anzeiger*, 254, 27–40.

Schmid, R. 1988. Morphological adaptations in a predator-prey system: Antlike stone beetles (Scydmaenidae, Staphylinoidea) and armoured mites (Acari). *Zoological Yearbooks, Systematics*. 115, 207–228. [in German with English summary]

Schneider, K. 2005. Feeding biology and diversity of oribatid mites (Oribatida, Acari). Dissertation, TU Darmstadt, 115 pp.

Schneider, K., Maraun, M. 2005. Feeding preferences among dark pigmented fungi (Dematiacea) indicate limited trophic niche differentiation of oribatid mites (Oribatida, Acari). *Pedobiologia*, 49, 61–67.

Schneider, K., Migge, S., Norton, R.A., Scheu, S., Langel, R., Reineking, A., Maraun, M. 2004. Trophic niche differentiation in oribatid mites (Oribatida, Acari) evidence from stable isotope ratios (^{15}N / ^{14}N). *Soil Biology and Biochemistry*, 36, 1769–1774.

Schneider, K., Renker, K., Maraun, M. 2005. Oribatid mite (Acari, Oribatida) feeding on ectomycorrhizal fungi. *Mycorrhiza*, 16, 67–72.

Schrank, F.P. 1803. Fauna Boica. Durchgedachte Geschichte der in Bayern einheimischen und zahmen Thiere. *Ingolstadt*, 3, 1–272.

Schubart, H. 1975. Morphologische Grundlagen für die Klärung der Verwandtschaftsbeziehungen innerhalb der Milbenfamilie Ameronothridae (Acari). *Zoologica*, 123, 23–91.

Schulte, G. 1975. Holarktische Artareale der Ameronothridae (Acari, Oribatei). *Veröffentlichungen des Instituts für Meeresforschung in Bremerhaven*, 15, 339–357.

Schulte, G. 1976. Zur Nahrungsbiologie der terrestrischen und marinen Milbenfamilie Ameronothridae (Acari, Oribatei). *Pedobiologia*, 16, 332–352.

Schulte, G., Weigmann, G. 1977. The evolution of the family Ameronothridae (Acari: Oribatei). II. Ecological aspects. *Acarologia*, 19, 167–173.

Schulte, G., Schuster, R., Schubart, H. 1975. Zur Verbreitung und Ökologie der Ameronothriden (Acari, Oribatei) in terrestrischen, limnischen und marinen Lebensräumen. *Veröffentlichungen des Instituts für Meeresforschung in Bremerhaven*, 15, 359–385.

Schuppenhauer, M.M., Lehmitz, R., Xylander, W.E.R. 2019. Slow-moving soil organisms on a water highway: Aquatic dispersal and survival potential of Oribatida and Collembola in running water. *Movement Ecology*, 7, 20.

Schuster, R. 1956. Der Anteil der Oribatiden an den Zersetzungsvorgängen im Boden. *Zeitschrift für Morphologie und Ökologie der Tiere*, 45, 1–33.

Schuster, R. 1962. Nachweis eines Paarungszeremoniells bei den Hornmilben (Oribatei, Acari). *Naturwissenschaften*, 49, 502–503.

Schuster, R. 1963. *Thalassozetes riparius* n. gen., n. sp., eine litoralbewohnende Oribatide von bemerkenswerter morphologischer Variabilität (Oribatei-Acari). *Zoologischer Anzeiger*, 171, 391–403.

Schuster, R. 1979. Soil mites in the marine environment. *Recent Advances in Acarology*, 1, 593–602.

Schuster, R., Coetzee, L., Putterill, J.F. 2000. Oribatid mites (Acari, Oribatida) as intermediate hosts of tapeworms of the family Anoplocephalidae (Cestoda) and the transmission of *Moniezia expansa* cysticercoids in South Africa. *Onderstepoort Journal of Veterinary Research*, 67, 49–55.

Schwager, E.E., Schönauer, A., Leite, D.J., Sharma, P.P., McGregor, A.P. 2015. Chelicerata. In: Wanninger, A. (Ed.), *Evolutionary Developmental Biology of Invertebrates 3: Ecdysozoa I: Non-Tetraconata*. Springer-Verlag, Wien, 99–140.

Schweger, C.E. 1997. Late quaternary palaeoecology of the Yukon: A review. In: Danks, H.V., Downes, J.A. (Eds.), *Insects of the Yukon*. Biological Survey of Canada Terrestrial Arthropods, Ottawa, 59–72.

Schweizer, J. 1922. Beiträge zur Kenntnis der terrestrischen Milbenfauna der Schweiz. *Verhandlungen der Naturforschenden Gesellschaft in Basel*, 23, 23–112.

Schweizer, J. 1956. Die Landmilben des Schweizerischen Nationalparkes. 3. Teil, Sarcoptiformes Reuter 1909. *Ergebnisse der wissenschaftlichen Untersuchungen des schweizerischen Nationalparks, Neue Folge, Liestal*, 5, 213–377.

Scopoli, J.A. 1763. Entomologia carniolica exhibens insecta carnioliae indigena et distributa in ordines, genera, species, varieties. 420 pp. + 18 pls.

Scott, H. 1958. *Humerobates rostrolamellatus* Grandjean, an oribatid mite, invading a house. *Entomologist's Monthly Magazine*, 94, 69–70.

Scudder, G.G.E. 1997. Environment of the Yukon. In: Danks, H.V., Downes, J.A. (Eds.), *Insects of the Yukon*. Biological Survey of Canada Terrestrial Arthropods, Ottawa, 13–57.

Seastedt, T.R. 1984. The role of microarthropods in decomposition and mineralization processes. *Annual Review of Entomology*, 29, 25–46.

Seastedt, T.R., Reddy, M.V., Cline, S.P. 1989. Microarthropods in decaying wood from temperate coniferous and deciduous forests. *Pedobiologia*, 33, 69–77.

Sellnick, M. 1921. Oribatiden vom Zwergbirkenmoor bei Neulinum, Kr. Kulm, und vom Moor am Kleinen Heidsee bei Heubuck unweit Danzig. *Schriften der Naturforschenden Gesellschaft in Danzig*, 15, 69–77 + figs.1–4.

Sellnick, M. 1922. Eine neue Oribatide und Berichtigungen zu einer meiner Arbeiten. *Schriften der Physikalisch–Ökonomischen Gesellschaft zu Königsberg*, 63, 97–98.

Sellnick, M. 1924. Oribatiden. In: Dampf, A. (Ed.), Zur Kenntnis der estlandischen Hochmoorfauna. *Sitzungsberichteder Naturforscher–Gesellschaft zu Dorpat*, Dorpat. Druck von Mattiesen, Tartu, Estonia, 31, 65–71 + figs.1–9.

Sellnick, M. 1925. Fauna sumatrensis. *Supplementa Entomologica*, 11, 79–89.

Sellnick, M. 1928. Formenkreis, Hornmilben, Oribatei. In: *Die Tierwelt Mitteleuropas*. Quelle und Meyer, Quelle & Meyer, Leipzig, 1–42.

Sellnick, M. 1944. Chapter 4: Taxonomical list of the Oribatidae (Acarina) and Collembola of Greenland. In: Hammer, M. (Ed.), *Studies of the Oribatids and Collemboles of Greenland*. Meddelelser om Grønland, C.A. Reitzel, Kjøbenhavn, 141, 40–48.

Sellnick, M. 1952. *Hafenrefferiella nevesi* nov. gen., nov. spec., a new genus and species from Portugal, and *Hafenrefferia gilvipes* (C.L. Koch) (Acar. Oribat.). *Portugaliae Acta Biologica, Série B*, 3, 228–237.

Sellnick, M., Forsslund, K.-H. 1955. Die Camisiidae Schwedens (Acar. Oribat.). *Arkiv för Zoologi, Ser.2*, 8, 473–530.

Sengbusch, H.G. 1954. Studies on the life history of three oribatoid mites with observations on other species (Acarina, Oribatei). *Annals of the Entomological Society of America*, 47, 646–667.

Sengbusch, H.G. 1958. Zuchversuche mit Oribatiden. (Acarina). *Naturwissenschaften*, 45, 498–499.

Seniczak, A., Seniczak, S. 2008. Setal variability of *Hydrozetes lemnae* and *H. thienemanni* (Acari: Oribatida: Hydrozetidae). *Biologia (Bratislava)*, 63, 677–683.

Seniczak, A., Seniczak, S. 2010. Morphological differentiation of *Limnozetes* Hull, 1916 (Acari: Oribatida: Limnozetidae) in the light of ontogenetic studies. *Belgian Journal of Zoology*, 140, 40–58.

Seniczak, A., Seniczak, S. 2014. Comparison of morphology and ontogeny of *Chamobates subglobulus* (Oudemans, 1900) and *Euzetes globulus* (Nicolet, 1855) (Acari: Oribatida). *International Journal of Acarology*, 40, 274–295.

Seniczak, A., Seniczak, S. 2019. Morphological ontogeny of *Caleremaeus monilipes* (Acari: Oribatida: Caleremaeidae) with comments on *Caleremaeus* Berlese. *Systematic and Applied Acarology*, 24, 1995–2009.

Seniczak, A., Seniczak, S. 2020a. Morphological ontogeny of *Fuscozetes coulsoni* sp. nov. (Acari: Oribatida: Ceratozetidae) from Svalbard, Norway. *Systematic and Applied Acarology*, 25, 680–696.

Seniczak, A., Seniczak, S. 2020b. Morphological ontogeny of *Limnozetes solhoyorum* sp. nov. (Acari: Oribatida: Limnozetidae) from Norway, with comments on *Limnozetes* Hull. *Systematic and Applied Acarology*, 25, 327–348.

Seniczak, A., Seniczak, S., Dlugosz, J. 2005. The effect of lead and zinc on the moss mite *Archegozetes longisetosus* Aoki (Acari, Oribatida) under laboratory conditions. In: Tajovský, K., Schlaghamerský, K., Pizl, V. (Eds.), *Contributions to Soil Zoology in Central Europe I*. ISB AS CR, Ceské Budejovice, 133–136.

Seniczak, A., Ligocka, A., Seniczak, S., Paluszak, Z. 2009. The influence of cadmium on life-history parameters and gut microflora of *Archegozetes longisetosus* (Acari: Oribatida) under laboratory conditions. *Experimental and Applied Acarology*, 473, 191–200.

Seniczak, A., Seniczak, S., Kaczmarek, S. 2015. Morphological and ecological differentiation of *Eupelops* and *Propelops* (Acari, Oribatida, Phenopelopidae). *International Journal of Acarology*, 41, 147–169.

Seniczak, A., Seniczak, S., Kaczmarek, S. 2016. Morphological ontogeny, distribution and ecology of *Damaeus torquisetosus* and *Epidamaeus puritanicus* (Acari: Oribatidae: Damaeidae). *Systematic and Applied Acarology*, 21, 471–497.

Seniczak, A., Seniczak, S., Kaczmarek, S., Chachaj, B. 2017. Morphological ontogeny and ecology of *Adoristes ovatus* (Acari: Oribatida: Liacaridae) with comments on *Adoristes* Hull. *Systematic and Applied Acarology*, 22, 2038–2056.

Seniczak, A., Seniczak, S., Kaczmarek, S., Bolger, T. 2018. Morphological ontogeny of *Chamobates pusillus* (Acari, Oribatida, Chamobatidae) with comments on some species of *Chamobates* Hull. *Systematic and Applied Acarology*, 23, 339–352.

Seniczak, A., Seniczak, S., Iturrondobeitia, J.C., Solhøy, T., Flatberg, K.I. 2019a. Diverse *Sphagnum* mosses support rich moss mite communities (Acari, Oribatida) in mires of Western Norway. *Wetlands*, 40, 1339–1351.

Seniczak, A., Seniczak, S., Jordal, B.H. 2019b. Molecular and ontogeny studies clarify systematic status of *Chamobates borealis* (Acari, Oribatida, Chamobatidae): An integrated taxonomy approach. *Systematic and Applied Acarology*, 24, 2409–2426.

Seniczak, S. 1972. Morphology of developmental stages of *Pilogalumna tenuiclava* (Berl.) and *Pergalumna nervosa* (Berl.). *Bulletin de la Société des Amis des Sciences et des Lettres de Poznan, Ser. D*, 12/13, 199–213.

Seniczak, S. 1989. The morphology of the juvenile stages of moss mites of the subfamily Sphaerozetinae (Acarida: Oribatida), I. *Annales Zoologici*, 42, 225–235.

Seniczak, S. 1990. The morphology of juvenile stages of moss mites of the family Camisiidae (Acari: Oribatida). Part II. *Zoologischer Anzeiger*, 225, 151–160.

Seniczak, S. 1991a. The morphology of juvenile stages of moss mites of the family Camisiidae (Acari: Oribatida). 4. *Zoologischer Anzeiger*, 226, 267–279.

Seniczak, S. 1991b. The morphology of juvenile stages of moss mites of the family Nanhermanniidae (Acari: Oribatida). 1. *Zoologischer Anzeiger*, 227, 319–330.

Seniczak, S. 1991c. The morphology of juvenile stages of moss mites of the family Camisiidae (Acari: Oribatida). V. *Zoologischer Anzeiger*, 227, 173–184.

Seniczak, S. 1991d. The morphology of juvenile stages of moss mites of the family Camisiidae (Acari: Oribatida). VI. *Zoologischer Anzeiger*, 227, 331–342.

Seniczak, S. 1993. The morphology of juvenile stages of moss mites of the family Malaconothridae (Acari: Oribatida). I. *Zoologischer Anzeiger*, 231, 59–72.

Seniczak, S., Norton, R.A. 1994. The morphology of juvenile stages of moss mites of the family Trhypochthoniidae (Acari: Oribatida). 2. *Zoologischer Anzeiger*, 233, 29–44.

Seniczak, S., Seniczak, A. 2007. Morphology of juvenile stages of *Parachipteria bella* (Sellnick, 1928) and *P. willmanni* Hammen, 1952 (Acari: Oribatida: Achipteriidae). *Annales Zoologici*, 57, 533–540.

Seniczak, S., Seniczak, A. 2008. Morphology of three European species of the genus *Punctoribates* Berlese, 1908 (Acari: Oribatida: Mycobatidae). *Annales Zoologici*, 58, 473–485.

Seniczak, S., Seniczak, A. 2009a. Morphology of three species of Crotonioidea Thorel, 1876 (Acari: Oribatida), and relations between some genera. *Zoologischer Anzeiger*, 248, 195–211.

Seniczak, S., Seniczak, A. 2009b. Morphology of some species of *Limnozetes* Hull, 1916 (Acari: Oribatida: Limnozetidae), and keys to the larvae and nymphs. *Annales Zoologici*, 59, 387–396.

Seniczak, S., Seniczak, A. 2013. Differentiation of external morphology of *Oribatella* Banks, 1895 (Acari: Oribatida: Oribatellidae), in light of the ontogeny of three species. *Journal of Natural History*, 47, 1569–1611.

Seniczak, S., Seniczak, A. 2017. Morphological ontogeny of *Eobrachychthonius oudemansi* (Acari, Oribatida, Brachychthoniidae), with comments on *Eobrachychthonius* Jacot. *Systematic and Applied Acarology*, 22, 1659–1677.

Seniczak, S., Seniczak, A. 2018. Morphological ontogeny of *Minunthozetes semirufus* (Acari: Oribatida: Punctoribatidae). *Zootaxa*, 4540, 73–92.

Seniczak, S., Solhøy, T. 1988. The morphology of juvenile stages of moss mites of the family Chamobatidae Thor (Acarida, Oribatida), I. *Annales Zoologici (Warsaw)*, 41, 491–502.

Seniczak, S., Klimek, A., Kaczmarek, S. 1991. Variability of notogastral setation in adult *Fuscozetes setosus* (Acari, Oribatida) in the light of population studies. *Bulletin of the Polish Academy of Sciences. Biological Sciences*, 39, 89–95.

Seniczak, S., Norton, R.A., Wang, H.F. 1998. The morphology of juvenile stages of moss mites of the family Trhypochthoniidae (Acari: Oribatida) and the taxonomic status of some genera and species. *Zoologischer Anzeiger*, 237, 85–95.

Seniczak, S., Norton, R.A., Seniczak, A. 2009a. Morphology of *Eniochthonius minutissimus* (Berlese, 1904) and *Hypochthonius rufulus* C.L. Koch, 1835 (Acari: Oribatida: Hypochthonioidea). *Annales Zoologici*, 59, 373–386.

Seniczak, S., Norton, R.A., Seniczak, A. 2009b. Morphology of *Hydrozetes confervae* (Schrank, 1781) and *H. parisiensis* Grandjean, 1948 (Acari: Oribatida: Hydrozetidae) and keys to European species of *Hydrozetes* Berlese, 1902. *Zoologischer Anzeiger*, 248, 71–83.

Seniczak, S., Penttinen, R., Seniczak, A. 2011. The ontogeny of morphological traits in three European species of *Cosmochthonius* Berlese, 1910 (Acari: Oribatida: Cosmochthoniidae). *Zootaxa*, 3034, 1–31.

Seniczak, S., Seniczak, A., Gwiazdowicz, D.J., Coulson, S.J. 2014a. Community structure of oribatid and gamasid mites (Acari) in moss-grass tundra in Svalbard (Spitsbergen, Norway). *Arctic, Antarctic, and Alpine Research*, 46, 591–599.

Seniczak, S., Seniczak, A., Kaczmarek, S. 2014b. Morphology of juveniles supports transfer of *Lepidozetes singularis* from Tegoribatidae to Ceratozetidae (Acari: Oribatida). *International Journal of Acarology*, 40, 449–462.

Seniczak, S., Seniczak, A., Coulson, S.J. 2015. Morphology, distribution and biology of *Mycobates sarekensis* (Acari: Oribatida: Punctoribatidae). *International Journal of Acarology*, 41, 663–675.

Seniczak, S., Seniczak, A., Kaczmarek, S. 2016. Morphological ontogeny of *Ceratozetes helenae* and *Ceratozetoides cisalpinus* (Acari: Oribatida: Ceratozetidae). *Systematic and Applied Acarology*, 21, 1309–1333.

Seniczak, S., Seniczak, A., Coulson, S.J. 2017a. Morphological ontogeny, distribution of *Hermannia scabra* (Acari: Oribatida: Hermanniidae) in Svalbard and descriptive population parameters. *Acarolgia*, 57, 877–892.

Seniczak, S., Seniczak, A., Graczyk, R., Tømmervik, H., Coulson, S.J. 2017b. Distribution and population characteristics of the soil mites *Diapterobates notatus* and *Svalbardia paludicola* (Acari: Oribatida: Ceratozetidae) in High Arctic Svalbard (Norway). *Polar Biology*, 40, 1545–1555.

Seniczak, S., Seniczak, A., Kaczmarek, S., Marquardt, T. 2017c. Morphological ontogeny of *Anachipteria magnilamellata* (Acari, Oribatida, Achipteriidae), with comments on *Anachipteria* Grandjean. *Systematic and Applied Acarology*, 22, 373–385.

Seniczak, S., Seniczak, A., Kaczmarek, S. 2018a. Morphological ontogeny of *Platyliodes scaliger* (Acari, Oribatida, Neoliodidae), with comments on *Platyliodes* Berlese. *Systematic and Applied Acarology*, 23, 25–41.

Seniczak, S., Seniczak, A., Kaczmarek, S. 2018b. Morphological ontogeny of *Ceratozetes shaldybinae* (Acari, Oribatida, Ceratozetidae). *Systematic and Applied Acarology*, 23, 581–592.

Seniczak, S., Ivan, O., Seniczak, A. 2020. Morphological ontogeny of *Damaeolus ornatissimus* (Acari: Oribatida: Damaeolidae) with comments on *Damaeolus* Paoli. *Systematic and Applied Acarology*, 25, 459–478.

Sergienko, G.D. 1987. Oribatid mites of the genus *Phthiracarus* and *Archiphthiracarus* (Oribatei, Phthiracaridae) of the Ukraine. Communication 1. *Vestnik Zoologii*, 6, 35–43. [in Russian with English summary]

Shackelford, N., Standish, R., Lindo, Z., Starzomski, B. 2018. Landscape connectivity shifts resistance, resilience, and recovery uniquely in multi-trophic microarthropod communities. *Ecology*, 99, 1164–1172.

Shaldybina, E.S. 1966. Postembryonic development of oribatid mites of the superfamily Ceratozetoidea Balogh, 1961, and their system. *1st Acarology Congress Proceedings, Academy Science U.S.S.R.*, 225–226. [In Russian]

Shaldybina, E.S. 1969. Moss mites of the superfamily Ceratozetoidea (their morphology, biology, system and role in the anoplocephalid epizootics). *Dissertation submitted for the degree of Doctor of Science in Biology.* Moscow, 708 pp. [in Russian]

Shaldybina, E.S. 1970. Three new species of the genus *Diapterobates* Grandjean, 1936 (Oribatei, Ceratozetidae). *Notes of the Gorky Pedagogical Institute, Biological Sciences Series*, 114, 44–50. [in Russian]

Shaldybina, E.S. 1971. Development of two moss mite species of the genus *Chamobates* Hull, 1916 (Oribatei, Ceratozetidae). *Notes of the Gorky Pedagogical Institute, Biological Series*, 116, 51–71 [in Russian]

Shaldybina, E.S. 1973. New species of oribatid mites of the subfamily Minunthozetinae (Oribatei, Mycobatidae) from the territory of the Soviet Union. *Zoologicheskii Zhurnal*, 52, 689–699. [in Russian]

Shaldybina, E.S. 1978. The postembryonic development of *Fuscozetes fuscipes* (C.L. Koch) (Oribatei, Ceratozetidae). In: Severcova, A. H. (Ed.), *Fauna, Classification, Biology and Ecology of Parasitic Worms and their Intermediate Hosts*. Gorky State Pedagogical Institute, Gorky, 84–93. [in Russian]

Shear, W.A., Bonamo, P.M., Grierson, J.D., Rolfe, W.D.I., Smith, E.L., Norton, R.A. 1984. Early land animals in North America: Evidence from Devonian age arthropods from Gilboa, New York. *Science*, 224, 492–494.

Shimano, S. 2005. Distinctness of two morphological forms of *Rhysotritia ardua* (Acari: Oribatida: Euphthiracaridae) in Japan. In: Weigmann, G., Alberti, G., Wohltmann, A., Ragusa, S. (Eds.), *Acarine Biodiversity in the Natural and Human Sphere. Proceedings of the V Symposium of the European Association of Acarologists (Berlin 2004)*. Phytophaga (Palermo), 14, 383–388.

Shimano, S., Norton, R.A. 2003. Is the Japanese oribatid mite *Euphthiracarus foveolatus* Aoki, 1980 (Acari: Euphthiracaridae) a junior synonym of *E. cribrarius* (Berlese, 1904)? *Journal of the Acarological Society of Japan*, 12, 115–126.

Shimano, S., Sakata, T., Mizutani, Y., Kuwahara, Y., Aoki, J. 2002. Geranial: The alarm pheromone in the nymphal stage of the oribatid mite, *Nothrus palustris. Journal of Chemical Ecology*, 28, 1831–1837.

Sidorchuk, E.A., Behan-Pelletier, V.M. 2017. *Megeremaeus cretaceous* new species (Acari: Oribatida), the first oribatid mite from Canadian amber. *The Canadian Entomologist*, 149, 277–290.

Sidorchuk, E.A., Norton, R.A. 2010. Redescription of the fossil oribatid mite *Scutoribates peronatus*, with implications for systematics of Unduloribatidae (Acari: Oribatida). *Zootaxa*, 2666, 45–67.

Siepel, H. 1996. The importance of unpredictable and short-term environmental extremes for biodiversity in oribatid mites. *Biodiversity Letters*, 3, 26–34.

Siepel, H., de Ruiter-Dijkman, E.M. 1993. Feeding guilds of oribatid mites based on their carbohydrase activities. *Soil Biology and Biochemistry*, 25, 1491–1497.

Singh, R.K., Mukherjee, I.N., Singh, R.N. 1989. Records of mites associated with water hyacinth (*Eichhornia crassipes*) in Uttar Pradesh, India. In: Channabasavanna, G.P., Viraktamath, C.A. (Eds.), *Progress in Acarology*, Brill, Leiden, 2, 211–214.

Sitnikova, L.G. 1975. Superfamily Nothroidea Grandjean, 1954. In: Gilyarov, M.S. (Ed.), *A Key to Soil-inhabiting Mites. Sarcoptiformes*. Nauka Press, Moscow, 71–94. [in Russian]

Sitnikova, L.G. 1981. Drei neue Hornmilbenarten aus den Familien Camisiidae und Nothridae (Acarina, Oribatei). *Trudy Zoologicheskogo Instituta Akademii Nauk SSSR*, 106, 89–92.

Sjursen, H., Sømme, L. 2000. Seasonal changes in tolerance to cold and desiccation in *Phauloppia* sp. (Acari, Oribatida) from Finse, Norway. *Journal of Insect Physiology*, 46, 1387–1396.

Skubała, P. 1997. The structure of oribatid mite communities (Acari: Oribatida) on mine dumps and reclaimed land. *Zoologische Beiträge*, 38, 59–80.

Skubała, P., Duras, M. 2008. Do decaying logs represent habitat islands? Oribatid mite communities in dead wood. *Annales Zoologici*, 58, 453–466.

Smith, I.M., Lindquist, E.E., Behan-Pelletier, V. 1997. Mites Acari. In: Smith, I.M. (Ed.), *Assessment of Species Diversity in the Mixedwood Plains Ecozone*. Ecological Monitoring and Assessment Network, 16–17. https://www.worldcat.org/title/assessment–of–species–diversity–in–the–mixedwood–plains–ecozone/oclc/54783816&referer=brief_results (accessed 28 June 2021).

Smith, I.M., Lindquist, E.E., Behan-Pelletier, V.M. 2011. Mites (Acari). In: Scudder, G.G.E., Smith, I.M. (Eds.), *Assessment of Species Diversity in the Montane Cordillera Ecozone*. https://royalbcmuseum.bc.ca/assets/Montane–Cordillera–Ecozone.pdf 193–268. (accessed 28 June 2021).

Sokołowska, M., Duras, M., Skubała, P. 2009. Oribatid mites communities (Acari: Oribatida) in dead wood of protected areas under strong anthropogenic pressure. In: Tajovský, K., Schlaghamerský, J., Pižl, V. (Eds.), *Contributions to Soil Zoology in Central Europe III*. ISB BC AS CR, v.v.i., České Budějovice, 15–155.

Solhøy, I. 1997. A redescription of *Mycobates sarekensis* (Trägårdh) (Acari: Oribatei). *Acarologia*, 38, 69–77.

Soma, K. 1990. Studies on the life history of *Phthiracarus japonicus* Aoki (Acarina: Phthiracaridae) in a creeping pine (*Pinus pumila* Regel) shrub. *Edaphologia*, 43, 25–30.

Søvik, G. 2003. Observations on ovoviviparity and mixed-parity mode in arctic populations of *Ameronothrus lineatus* (Acari, Oribatida). *Acarologia*, 43, 393–398.

Søvik, G. 2004. The biology and life history of arctic populations of the littoral mite *Ameronothrus lineatus* (Acari, Oribatida). *Experimental and Applied Acarology*, 341–2, 3–20.

Søvik, G., Leinaas, H.P., Ims, R.A., Solhøy, T. 2003. Population dynamics and life history of the oribatid mite *Ameronothrus lineatus* (Acari, Oribatida) on the high arctic archipelago of Svalbard. *Pedobiologia*, 47, 257–271.

St. John, M.G., Bagatto, G., Behan-Pelletier, V.M., Lindquist, E.E., Shorthouse, J.D., Smith, I.M. 2002. Mite (Acari) colonization of vegetated mine tailings near Sudbury, Ontario, Canada. *Plant and Soil*, 245, 295–305.

St. John, M.G., Wall, D.H., Behan-Pelletier, V.M. 2006. Does plant species co-occurrence influence soil mite diversity? *Ecology*, 873, 625–633.

Staddon, P., Lindo, Z., Crittenden, P.D., Gilbert, F., Gonzalez, A. 2010. Connectivity, non-random extinction, and ecosystem function in experimental metacommunities. *Ecology Letters*, 13, 543–552.

Stamou, G.P., Argyropolou, M.D. 1995. A preliminary study on the effect of Cu, Pb and Zn contamination of soils in community structure and certain life-history traits of oribatids from urban areas. *Experimental and Applied Acarology*, 19, 381–390.

Strenzke, K. 1943. Beitrage zur Systematik landlebender Milben. I. *Hydrozetes thienemanni* n. sp. II. *Scutovertex sculptus* Michael. *Archiv für Hydrobiologie*, 40, 57–70.

Strenzke, K. 1950. *Oribatella arctica litoralis* n. subsp., eine neue Oribatide der Nord– und Ostseeküste (Acarina: Oribatei). *Kieler Meeresforsch*, 7, 157–160.

Strenzke, K. 1951. Die norddeutschen Arten der Gattungen *Brachychthonius* und *Brachychochthonius* (Acarina: Oribatei). *Deutsche Zoologica*, 1, 234–249.

Strenzke, K. 1952. Untersuchungen über die Tiergemeinschaften des Bodens, Die Oribatiden und ihre Synusien in den Böden Norddeutschlands. *Zoologica, Stuttgart*, 104, 1–173.

Strenzke, K. 1953. Zwei neue Arten der Oribatidengattung *Nanhermannia*. *Zoologischer Anzeiger*, 150, 69–75.

Strenzke, K. 1954. *Permycobates bicornis* n. gen. n. sp., a new Central European moss mite (Acarina, Oribatei). *Proceedings of the Koninklijke Nederlandse Akademie van Wetenschappen. Series C*, 57, 92–98.

Strenzke, K. 1963. Entwicklung und Verwandtschaftsbeziehungen der Oribatidengattung *Gehypochthonius* (Arach., Acari). *Senckenbergiana Biologica*, 44, 231–255.

Stubbs, C.S. 1995. Dispersal of soredia by the oribatid mite, *Humerobates arborea*. *Mycologia*, 87, 454–458.

Subías, L.S. 1978. *Anomaloppia canariensis* n. gen., n. sp. (Acarida, Oribatida, Oppiidae) de las islas Canarias. Consideraciones filogénicas sobre la familia. *Redia*, 61, 565–574.

Subías, L.S. 1980. Oppidae del complejo '*Clavipectinata insculpta*' (Acarida, Oribatida). *EOS-Revista Espanola de Entomologia*, 54, 281–313.

Subías, L.S. 2004. Listado sistemático, sinonímico y biogeográfico de los ácaros oribátidos (Acariformes, Oribatida) del mundo 1758–2002. *Graellsia*, 1982, 1–570.

Subías, L.S. 2017. Modificaciones en el listado sistemático, sinonímico y biogeográfico de los ácaros oribátidos (Acariformes, Oribatida) del mundo (excepto fósiles) (12 actualización). *Revista Ibérica de Aracnología*, 30, 21–24.

Subías, L.S. 2019. Nuevas adiciones al listado mundial de ácaros oribátidos (Acari, Oribatida) (14ª actualización). *Revista Ibérica de Aracnología*, 34, 76–80.

Subías, L.S. 2022. Listado sistemático, sinonímico y biogeográfico de los ácaros oribátidos (Acariformes, Oribatida) del mundo excepto fósiles. *Graellsia*, 60 número extraordinario, 3–305 (2004), (17ª actualización). Available from: http://bba.bioucm.es/cont/docs/RO_1.pdf (accessed 1 August 2022).

Subías, L.S., Arillo, A. 2001. Acari, Oribatei, Gymnonota II. Oppioidea. In: Ramos, A. et al. (Eds.), *Fauna Ibérica*. Museo de Ciencias Naturales, Madrid, 15, 1–289.

Subías, L.S., Arillo, A. 2002. Oribatid fossil mites from the Upper Devonian of South Mountain, New York and the Lower Carboniferous of County Antrim, North Ireland (Acariformes, Oribatida). *Estudios del Museo de Ciencias Naturales de Álava*, 17, 93–106.

Subías, L.S., Mínguez, M.E. 1986. *Lauroppia similifallax* n. gen. y n. sp. (Acari, Oribatida, Oppiidae) de España central. *Boletín de la Asociación Española de Entomología*, 10, 51–58.

Subías, L.S., Rodríguez, P. 1986. Oppiidae (Acari, Oribatida) de los Sabinares (*Juniperus thurifera*) de España. IX. *Subíasella* (*Lalmoppia*) n. subgen. y *Discoppia* (*Cylindroppia*) n. subgen. *Revista de Biología de la Universidad de Oviedo*, 4, 111–121.

Sylvain, Z.A., Buddle, C.M. 2010. Effects of forest stand type on oribatid mite (Acari: Oribatida) assemblages in a southwestern Quebec forest. *Pedobiologia*, 53, 321–325.

Szywilewska-Szczykutowicz, A., Olszanowski, Z. 2007. Redescription of C. Willmann's Holarctic species of the genus *Trhypochthonius* (Acari: Oribatida: Trhypochthoniidae). *Zootaxa*, 1406, 17–24.

Tagami, K., Ishihara, T., Hosokawa, J., Ito, M., Fukiyama, K. 1992. Ocurrence of aquatic oribatid and astigmatid mites in swimming pools. *Water Research*, 26, 1549–1554.

Takada, W., Sakata, T., Shimano, S., Enami, Y., Mori, N., Nishida, R., Kuwahara, Y. 2005. Scheloribatid mites as the source of pumiliotoxins in dendrobatid frogs. *Journal of Chemical Ecology*, 31, 2403–2415.

Tarman, K. 1968. Anatomy, histology of oribatid gut and their digestion. *Biološki Vestnik, Ljubljana*, 16, 67–76.

Tarnocai, C., Kettles, I., Lacelle, B. 2011. Peatlands of Canada. *Geological Survey of Canada, Open File 6561 digital database, Ottawa, Ontario*. https://doi.org/10.4095/288786 (accessed 28 June 2021).

Thomas, J.O.M. 1979. An energy budget for a woodland population of oribatid mites. *Pedobiologia*, 19, 346–378.

Thomas, R.H. 2002. Mites as models in development and genetics. In: Bernini, F., Nannelli, R., Nuzzaci, G., De Lillo, E. (Eds.), *Acarid Phylogeny and Evolution: Adaptation in Mites and Ticks. Proceedings of the IV Symposium of the European Association of Acarologists, Siena 2000*. Kluwer Academic Publishers, Dordrecht, Boston, London, 21–26.

Thomas, R.H., McLean, S.E. 1988. Community structure in soil Acari along a latitudinal transect of tundra sites in Northern Alaska. *Pedobiologia*, 311–2, 113–138.

Thor, S. 1929. Über die Phylogenie und Systematik der Acarina, mit Beiträgen zur ersten Entwicklungsgeschichte einzelner Gruppen. *Nyt magazin for naturvidenskaberne*, 67, 145–210.

Thor, S. 1930. Beiträge zur Kenntnis der invertebraten Fauna von Svalbard. *Skrifter om Svalbard og Ishavet, Oslo*, 27, 1–156.

Thor, S. 1934. Neue Beitrage zur Kenntnis der invertebraten Fauna von Svalbard. *Zoologischer Anzeiger*, 107, 114–139.

Thor, S. 1937. Übersicht der norwegischen Cryptostigmata mit einzelnen Nebenbemerkungen. *Nyt magazin for naturvidenskaberne*, 77, 275–307.

Thörell, T. 1871. Om Arachnider fran Spitsbergen och Beeren-Eiland. *Öfversigt af kongliga vetenskaps akademiens förhandlingar*, 28, 683–702.

Thörell, T. 1876. Sopra alcuni Opilioni (Phalangidea) d'Europa de dell'Asia occidentale, con un quadro dei generi europei di quest' Ordine. *Annali del Museo civico di storia naturale*, 8, 452–508.

Tousignant, S., Coderre, D. 1992. Niche partitioning by soil mites in a recent hardwood plantation in southern Quebec, Canada. *Pedobiologia*, 36, 287–294.

Townsend, V.R., Proud, D.N., Moore, M.K., Tibbetts, J.A., Burns, J.A., Hunter, R.K., Lazarowitz, S.R., Felgenhauer, B.E. 2008. Parasitic and phoretic mites associated with neotropical harvestmen from Trinidad, West Indies. *Annals of the Entomological Society of America*, 101, 1026–1032.

Trägårdh, I. 1904. Monographie der arktischen Acariden. *Fauna arctica*, 4, 1–78.

Trägårdh, I. 1910. Acariden aus dem Sarekgebirge. *Naturwiss Unters d Sarekgebirges, Stockholm*, 4, 375–586.

Trägårdh, I. 1931. Terrestrial Acarina. *Zoology of the Faroes, Copenhagen*, 49, 1–69.

Trägårdh, I. 1932. Palaeacariformes, a new suborder of Acari. *Arkiv för zoologi*, 24, 1–6.

Travé, J. 1960. Contribution a l'étude de la faune de la Massane (3e note). Oribates (Acariens) 2e partie. *Vie et Milieu*, 11, 209–232.

Travé, J. 1963. Écologie et biologie des Oribates (Acariens) saxicoles et arboricoles. *Vie et Milieu, Supplement*, 14, 1–267.

Travé, J., Vachon, M. 1975. François Grandjean, 1882–1975 (Notice biographique et bibliographique). *Acarologia*, 171, 1–19.

Travé, J., Andre, H.M., Taberly, G., Bernini, F. 1996. *Les Acariens Oribates*. AGAR Publishers, Wavre, Belgique, 110 pp.

Trávníček, M. 1989. Laboratory cultivation and biology of mites in the family Liacaridae (Acari: Oribatida). *Acta Universitatis Carolinae, Praha*, 33, 69–80.

Tsiafouli, M.A., Thébault, E., Sgardelis, S.P., De Ruiter, P.C., van der Putten, W.H., Birkhofer, K., Hemerik, L., De Vries, F.T., Bardgett, R.D., Brady, M.V., Bjornlund, L., Jørgensen, H.B., Christensen, S., d'Hertefeldt, T., Hotes, S., Gera Hol, W.H., Frouz, J., Liiri, M., Mortimer, S.R., Setälä, H., Tzanopoulos, J., Uteseny, K., Pizl, V., Stary, J., Wolters, V., Hedlund, K. 2015. Intensive agriculture reduces soil biodiversity across Europe. *Global Change Biology*, 21, 973–985.

van Straalen, N.M., Verhoef, H.A. 1997. The development of a bioindicator system for soil acidity based on arthropod pH preferences. *Journal of Applied Ecology*, 34, 217–232.
Vitzthum, H. 1943. Acarina. In: Bronn, H.G. (Ed.), *Klassen und Ordnungen des Tierreiches*. vol. 5, Arthropoda, part IV, Arachnoidea, Book 5, Acarina. Becker, Erler Kom.–Ges., Leipzig, 1011.
Voegtlin, D.J. 1982. Invertebrates of the H.J. Andrews Experimental Forest, Western Cascade mountains, Oregon: A survey of arthropods associated with the canopy of old-growth *Pseudotsuga menziesii*. Forest Research Laboratory, Oregon State University, Corvallis. Special Publication 4, 31 p.
Walker, N. 1965. Euphthiracaroidea of California *Sequoia* litter: with a reclassification of the families and generas of the world (Acarina: Oribatei). *Fort Hayes Studies N.S. Science Series Kansas*, 3, 1–154.
Wallwork, J.A. 1958. Notes on the feeding-behaviour of some forest soil Acarina. *Oikos*, 9, 260–271.
Wallwork, J.A. 1965. A leaf-boring Galumnoid mite (Acari: Cryptostigmata) from Uruguay. *Acarologia*, 7, 758–764.
Wallwork, J.A. 1967. Acarina. In: Burges, A., Raw, F. (Eds.), *Soil Biology*. Academic Press, London, 363–395.
Wallwork, J.A. 1972. Mites and other microarthropods from the Joshua Tree National Monument, California. *Journal of Zoology, London*, 168, 91–105.
Wallwork, J.A. 1983. Oribatids in forest ecosystems. *Annual Review of Entomology*, 28, 109–130.
Wallwork, J.A., Weems, D.C. 1984. *Jornadia larreae* n. gen., n. sp., a new genus of oribatid mite (Acari: Cryptostigmata) from the Chihuahuan desert. *Acarologia*, 25, 77–80.
Wallwork, J.A., Kamill, B.W., Whitford, W.G. 1984. Life styles of desert litter-dwelling microarthopods: a reappraisal based on the reproductive behaviour of cryptostigmatid mites. *South African Journal of Science*, 80, 163–169.
Walter, D.E. 1985. The effect of litter type and elevation on colonization of mixed coniferous litterbags by oribatid mites. *Pedobiologia*, 28, 383–387.
Walter, D.E. 1987. Trophic behaviour of "mycophagous" microarthropods. *Ecology*, 68, 226–229.
Walter, D.E. 1988. Nematophagy by soil arthropods from the shortgrass steppe, Chihuahuan Desert, and Rocky Mountains of the central United States. *Agriculture, Ecosystems and Environment*, 24, 307–316.
Walter, D.E. 1996. Living on leaves: Mites, tomenta, and leaf domatia. *Annual Review of Entomology*, 41, 101–114.
Walter, D.E. 2009. Genera of Gymnodamaeidae (Acari: Oribatida: Plateremaeoidea) of Canada, with notes on some nomenclatorial problems. *Zootaxa*, 2206, 23–44.
Walter, D.E. 2013. How should I call thee mite? Generating common names for mites in biodiversity assessment. *International Journal of Acarology*, 39, 653–655.
Walter, D.E., Behan-Pelletier, V.M. 1993. Systematics and ecology of *Adhaesozetes polyphyllos* sp. nov. (Acari: Oribatida: Licneremaeoidea), a leaf-inhabiting mite from Australian rainforests. *Canadian Journal of Zoology*, 71, 1024–1040.
Walter, D.E., Krantz, G.W. 2009. Chapter 5, Oviposition and life stages. In: Krantz, G.W., Walter, D.E. (Eds.), *A Manual of Acarology*. 3rd ed. Texas Tech. University Press, Lubbock, 57–63.
Walter, D.E., Latonas, S. 2013. A review of the ecology and distribution of *Protoribates* (Oribatida, Oripodoidea, Haplozetidae) in Alberta, Canada, with the description of a new species. *Zootaxa*, 36203, 483–499.
Walter, D.E., Lumley, L.M. 2021. Almanac of Alberta Acari, Part II, Version 3.0. Edmonton, Alberta, Canada. Online access available at: https://www.researchgate.net/publication/354269381_Almanac_of_Alberta_Acari_Part_II_Version_30 (accessed 9 November 2021).
Walter, D.E., Norton, R.A. 1984. Body size distribution in sympatric oribatid mites (Acari: Sarcoptiformes) from California pine litter. *Pedobiologia*, 27, 99–106.
Walter, D.E., Proctor, H.C. 1998. Feeding behaviour and phylogeny, observations on early derivative Acari. *Experimental and Applied Acarology*, 22, 51–60.
Walter, D.E., Proctor, H.C. 2013. *Mites. Ecology, Evolution and Behaviour*. Springer Publishing, New York, London, 494 pp.
Walter, D.E., Kethley, J., Moore, J.C. 1987. A heptane flotation method for recovering microarthropods from semiarid soils, with comparison to the Merchant-Crossley high-gradient extraction method and estimates of microarthropod biomass. *Pedobiologia*, 30, 221–232.
Walter, D.E., Hunt, H.W., Elliott, E.T. 1988. Guilds or functional groups? An analysis of predatory arthropods from a shortgrass steppe soil. *Pedobiologia*, 31, 247–260.
Walter, D.E., Latonas, S., Byers, K., Lumley, L.M. 2014. Almanac of Alberta Oribatida, Part I, Version 2.4. Edmonton, Alberta, Canada. 542 pp. Online access available at: https://www.researchgate.net/publication/352842283_Almanac_of_Alberta_Oribatida_Part_I_Version_24 (accessed 8 November 2021).
Wang, H.F., Norton, R.A. 1988. New records of Crotonioidea from China, with description of a new species of *Allonothrus* (Acari, Oribatida). *Acta Zootaxologica Sinica*, 13, 261–273.

Wang, H., Wen, Z-G., Chen, J. 2002a. A checklist of oribatid mites of China (I) (Acari: Oribatida). *Acta Arachnologica Sinica*, 11, 107–127.

Wang, H.F., Wen, Z-G., Chen, J. 2002b. A checklist of oribatid mites of China (Acari: Oribatida) (II). *Acta Arachnologica Sinica*, 12, 42–63.

Warburton, C., Pearce, N.D.F. 1905. On new and rare British mites of the family Oribatidae. *Proceedings of the Zoological Society London*, 2, 564–569 + pls.19–20.

Wauthy, G., Leponce, M., Banaï, N., Sylin, G., Lions, J.C. 1998. The backward jump of a box moss mite. *Proceedings of the Royal Society of London, B*, 265, 2235–2242.

Webb, N.R. 1969. The respiratory metabolism of *Nothrus silvestris* Nicolet (Acari). *Oikos*, 20, 294–299.

Webb, N.R. 1970. Population metabolism of *Nothrus silvestris* Nicolet (Acari). *Oikos*, 21, 155–159.

Webb, N.R. 1989. Observations on the life cycle of *Steganacarus magnus* (Acari: Cryptostigmata). *Pedobiologia*, 33, 293–299.

Webb, N.R. 1991. The role of *Steganacarus magnus* (Acari: Cryptostigmata) in the decomposition of the cones of Scots Pine, *Pinus sylvestris*. *Pedobiologia*, 35, 351–359.

Webb, N.R., Block, W. 1993. Aspects of cold hardiness in *Steganacarus magnus* (Acari: Cryptostigmata). *Experimental and Applied Acarology*, 17, 741–748.

Wei, H., Liu, W., Zhang, J., Qin, Z. 2017. Effects of simulated acid rain on soil fauna community composition and their ecological niches. *Environmental Pollution*, 220, 460–468.

Weigmann, G. 1975. Labor– und Freilanduntersuchungen zur Generationsdauer von Oribatiden (Acari: Oribatei). *Pedobiologia*, 15, 133–148.

Weigmann, G. 1991. Oribatid communities in transects from bogs to forests in Berlin indicating the biotope qualities. In: Dusbabek, F., Bukva, V. (Eds.), *Modern Acarology*. Academia, Prague, 1, 359–364.

Weigmann, G. 1996. Hypostome morphology of Malaconothroidea and phylogenetic conclusions on primitive Oribatida. In: Mitchell, R., Horn, D.J., Needham, G.R., Welbourn, C.W. (Eds.), *Acarology IX Proceedings*. Ohio Biological Survey, Columbus, OH. vol. 1, 273–276.

Weigmann, G. 1997a. Systematics of Central European families and genera of Malaconothroidea (Acari, Oribatida). Abhandlungen und Berichte des Naturkundesmuseums. *Forschungsstelle-Görlitz*, 69, 3–10.

Weigmann, G. 1997b. New and old species of Malaconothroidea from Europe. *Spixiana*, 20, 199–218.

Weigmann, G. 2006. Hornmilben (Oribatida). *Die Tierwelt Deutschlands*, 76. Teil. Goecke, Evers, Keltern, 520 pp.

Weigmann, G. 2008. Re-description of *Cultroribula berolina* Weigmann, 2006 (Acari, Oribatida, Astegistidae) from Germany with a key for the European species. *Soil Organisms*, 80, 145–151.

Weigmann, G. 2009a. Oribatid mites (Acari: Oribatida) from the coastal region of Portugal. II. The genera *Zachvatkinibates* and *Punctoribates* (Mycobatidae). *Soil Organisms*, 81, 85–105.

Weigmann, G. 2009b. Oribatid mites (Acari: Oribatida) from the coastal region of Portugal. III. New species of Scutoverticidae and Scheloribatidae. *Soil Organisms*, 81, 107–127.

Weigmann, G. 2013. The genus *Lepidozetes* Berlese, 1910 (Acari: Oribatida: Tegoribatidae) in Europe with description of a new species. *Zootaxa*, 3722, 493–500.

Weigmann, G., Miko, L. 1998. Taxonomy of European Scheloribatidae, 3. Remarks on *Scheloribates* Berlese 1908 with description of two new species of the subgenus *Topobates* Grandjean, 1958 (n. stat.) (Arachnida: Acari: Oribatida). *Senckenbergiana biologica*, 77, 247–255.

Weigmann, G., Miko, L. 2002. Redescription of *Oribates lagenula* Berlese, 1904, the type of *Lagenobates* n. gen. (Acarina, Oribatida). *Redia*, 85, 29–35.

Weigmann, G., Norton, R.A. 2009. Validity and interpretation of *Murcia* Koch, *Trichoribates* Berlese and their types (Acari: Oribatida: Ceratozetidae). *Zootaxa*, 2107, 65–68.

Weigmann, G., Raspotnig, G. 2009. Comparative morphological and biometrical studies on *Trhypochthonius* species of the *tectorum* species group (Acari: Oribatida: Trhypochthoniidae). *Zootaxa*, 2269, 1–31.

Weigmann, G., Stratil, H. 1979. Bodenfauna im Tiergarten. In: Sukopp, H. (Ed.), *Ökologischens Gutachten über die Auswirkungen von Bau und Betrieb der BAB Berlin (West) auf den Großen Tiergarten*. Senator für Bau- und Wohnungswesen, Berlin, 54–71.

Weis-Fogh, T. 1948. Ecological investigations on mites and Collemboles in the soil. Appendix: Description of some new mites (Acari). *Natura Jutlandica*, 1, 135–270.

West, C. 1982. Life histories of three species of sub-antarctic oribatid mite. *Pedobiologia*, 23, 59–67.

Wickings, K., Grandy, A.S. 2011. The oribatid mite *Scheloribates moestus* (Acari: Oribatida) alters litter chemistry and nutrient cycling during decomposition. *Soil Biology and Biochemistry*, 43, 351–358.

Willard, J.R. 1974. Soil Invertebrates. VIII. A summary of populations and biomass. Canadian IBP Tech. Rep. No. 56. University of Saskatchewan, Saskatoon, Sask. 110 pp.

Willmann, C. 1917. Eine neue Oribatide aus Ostpreussen. *Schriften der Physikalisch-Ökonomischen Gesellschaft zu Königsberg*, 58, 10–13 + figs.1–10.

Willmann, C. 1919. Diagnosen einiger neuen Oribatiden aus der Umgegend Bremens. *Abhandlungen des Naturwissenschaftlichen Vereins zu Bremen*, 24, 552–554.

Willmann, C. 1923. Oribatiden aus Quellmoosen. *Archiv für Hydrobiologie*, 14, 470–477 + pl.7.

Willmann, C. 1925. Neue und seltene Oribatiden. *Jahresbericht des Entomologischen Vereins*, 13, 7–11.

Willmann, C. 1928. Neue Oribatiden I. *Zoologischer Anzeiger*, 76, 1–5.

Willmann, C. 1929. Neue Oribatiden, II. *Zoologischer Anzeiger*, 80, 43–46 + figs.1–4.

Willmann, C. 1930. Neue und bemerkenswerte Oribatiden aus der Sammlung Oudemans. *Abhandlungen des Naturwissenschaftlichen Vereins zu Bremen*, 28, 1–12 + 16 figs.

Willmann, C. 1931. Moosmilben oder Oribatiden (Cryptostigmata). In: Dahl, F. (Ed.), *Die Tierwelt Deutschlands. Bd. 22. vol. 5.* Gustav Fischer, Jena, 79–200.

Willmann, C. 1935. Die Milbenfauna. I. Oribatei. In Faunistisch-ökologische Studien im Anningergebiet. *Zoologische Jahrbucher. Abtheilung fur Systematik, Geographie und Biologie der Thiere, Jena*, 66, 331–344.

Willmann, C. 1936. Neue Acari aus schlesischen Wiesenboden. *Zoologischer Anzeiger*, 113, 273–290 + figs.1–22.

Willmann, C. 1937. Beitrag zur Kenntnis der Acarofauna der ostfriesischen Inseln. *Abhandlungen des Naturwissenschaftlichen Vereins zu Bremen*, 30, 152–169.

Willmann, C. 1938. Beitrag zur Kenntnis der Acarofauna des Komitates Bars. *Annales historico-naturales Musei nationalis hungarici*, 31, 144–172.

Willmann, C. 1943. Terrestrische, Milben aus Schwedisch–Lappland. *Sonderdruck aus dem Archiv für Hydrobiologie*, 40, 208–239.

Willmann, C. 1950. Milben aus Mineralquellen (2. Mitteilung). *Zoologischer Anzeiger*, 145, 186–195.

Willmann, C. 1952. Die Milbenfauna der Nordseeinsel Wangerooge. *Veröff. Inst. Meeresf., Bremerhaven*, 1, 139–186.

Willmann, C. 1953. Neue Milben aus den ostlichen Alpen. *Sitzungsberichte der Akademie der Wissenschaften mathematisch-naturwissenschaftliche Klasse*, 1(162), 449–519.

Winchester, N.N., Behan-Pelletier, V.M., Ring, R.A. 1999. Arboreal specificity, diversity and abundance of canopy-dwelling oribatid mites (Acari: Oribatida). *Pedobiologia*, 43, 391–400.

Winchester, N.N., Lindo, Z., Behan-Pelletier, V.M. 2008, Oribatid mite communities in the canopy of montane *Abies amabilis* and *Tsuga heterophylla* trees on Vancouver Is., British Columbia. *Environmental Entomology*, 37, 464–471.

Woas, S. 1981a. Die Arten der Gattung *Hermannia* Nicolet 1855. II. *Andrias, Karlsruhe*, 1, 89–100.

Woas, S. 1981b. Zur Taxonomie und Phylogenie der Hermanniidae Sellnick 1928 (Acari, Oribatei). *Andrias, Karlsruhe*, 1, 7–88.

Woas, S. 1986. Beitrag zur Revision der Oppioidea sensu Balogh, 1972 (Acari, Oribatei). *Andrias, Karlsruhe*, 5, 21–224.

Woas, S. 1992. Beitrag zur Revision der Gymnodamaeidae Grandjean, 1954 (Acari, Oribatei). *Andrias, Karlsruhe*, 9, 121–161.

Woas, S. 1998. Mosaikverteilung der Merkmale basaler Höherer Oribatiden Die Gattungen *Passalozetes* und *Scutovertex*. In: Ebermann, E. (Ed.), *Arthropod Biology: Contributions to Morphology, Ecology and Systematics. Biosystematics and Ecology Series 14.* Austrian Academy of Sciences Press, Vienna, Austria, 291–313.

Woas, S. 2002. Acari: Oribatida. In: Adis, J. (Ed.), *Amazonian Arachnida and Myriapoda*. Pensoft Publishers, Sofia Moscow, 21–291.

Woodring, J.P. 1965. The biology of five species of oribatids from Louisiana. *Acarologia*, 7, 564–576.

Woodring, J.P. 1966. Color phototactic responses of an eyeless oribatid mite. *Acarologia*, 82, 382–388.

Woodring, J.P. 1970. Comparative morphology, homologies and functions of the male systems in oribatid mites (Arachnida: Acari). *Journal of Morphology*, 1324, 425–451.

Woodring, J.P. 1973. Comparative morphology, functions, and homologies of the coxal glands in oribatid mites (Arachnida: Acari). *Journal of Morphology*, 1394, 407–429.

Woodring, J.P., Cook, E.F. 1962. The internal anatomy, reproductive physiology, and molting process of *Ceratozetes cisalpinus* (Acarina: Oribatei). *Annals of the Entomological Society of America*, 55, 164–181.

Woolley, T.A. 1957. Redescriptions of Ewing's oribatid mites, IV. Family Achipteriidae (= Notaspididae) (Acarina: Oribatei). *Entomological News*, 68, 177–182.

Woolley, T.A. 1958. Redescriptions of Ewing's oribatid mites, X. Family Haplozetidae (Acarina: Oribatei). *Transactions of the American Microscopical Society*, 77, 333–340.

Woolley, T.A. 1964. A new species of *Eremaeus* from Colorado with notes on North American representatives of the genus (Acarina: Oribatei: Eremaeidae). *Transactions of the American Microscopical Society*, 83, 29–32.

Woolley, T.A. 1967a. A new genus and species of Oribatellid mite from Colorado (Acarina: Oribatei). *Journal of the Kansas Entomological Society*, 40, 32–37.
Woolley, T.A. 1967b. North American Liacaridae I. *Adoristes* and a related new genus (Acari: Cryptostigmata). *Journal of the Kansas Entomological Society*, 40, 270–276.
Woolley, T.A. 1968. North American Liacaridae II. *Liacarus* (Acari: Cryptostigmata). *Journal of the Kansas Entomological Society*, 41, 350–366.
Woolley, T.A. 1969. North American Liacaridae. III. New genera and species (Acari: Cryptostigmata). *Journal of the Kansas Entomological Society*, 42, 183–194.
Woolley, T.A. 1973. Taxonomy of oribatid mites – Retrospect and prospect. In: Daniel, M., Rosicky, B. (Eds.), *Proceedings of the 3rd International Congress, Acarology,* Czechoslovak Academy of Sciences, Prague, 337–341.
Woolley, T.A., Higgins, H.G. 1963a. The genus *Eremulus* Berlese 1908 with a description of a new species (Acarina: Oribatei: Eremaeidae). *Acarologia*, 5, 97–101.
Woolley, T.A., Higgins, H.G. 1963b. A new moss mite from Western U.S. (Acarina: Oribatei: Cepheidae). *Journal of the New York Entomological Society*, 71, 143–148.
Woolley, T.A., Higgins, H.G. 1964. A new species of *Ommatocepheus* Berlese, 1913 from North Carolina (Acarina: Oribatei: Carabodidae). *Transactions of the American Microscopical Society*, 81, 26–28.
Woolley, T.A., Higgins, H.G. 1965a. A new genus and species of oribatid mite from Colorado (Acari: Oribatei: Ceratoppiidae). *The Great Basin Naturalist*, 25, 59–62.
Woolley, T.A., Higgins, H.G. 1965b. A new genus and two new species of Tenuialidae with notes on the family (Acari: Oribatei). *Journal of the New York Entomological Society*, 73, 232–237.
Woolley, T.A., Higgins, H.G. 1968a. A new genus and species of oribatid from pack rat nests (Acari: Cryptostigmata: Tectocepheidae). *The Great Basin Naturalist*, 28, 144–146.
Woolley, T.A., Higgins, H.G. 1968b. Megeremaeidae, a new family of oribatid mites (Acari: Cryptostigmata). *The Great Basin Naturalist*, 28, 172–175.
Woolley, T.A., Higgins, H.G. 1968c. A new species of *Sphodrocepheus* from the Western U.S. (Acari: Cryptostigmata: Cepheidae). *The Great Basin Naturalist*, 28, 176–178.
Woolley, T.A., Higgins, H.G. 1969a. A new genus of Suctobelbidae from Northwestern United States (Acari: Cryptostigmata). *Proceedings of the Entomological Society, Washington*, 71, 10–13.
Woolley, T.A., Higgins, H.G. 1969b. A new species of *Platyliodes* from N.W. United States (Acari: Cryptostigmata: Liodidae). *Proceedings of the Entomological Society, Washington*, 71, 143–146.
Woolley, T.A., Higgins, H.G. 1969c. *Metrioppia* in the western United States (Acari: Cryptostigmata: Metrioppiidae). *Proceedings of the Entomological Society, Washington*, 71, 580–582.
Woolley, T.A., Higgins, H.G. 1973. A new *Gymnodamaeus* from Western Colorado (Acarina: Cryptostigmata: Gymnodamaeidae). *Proceedings of the Entomological Society, Washington*, 75, 411–416.
Worland, M.R., Lukešová, A. 2000. The effect of feeding on specific soil algae on the cold-hardiness of two Antarctic micro-arthropods (*Alaskozetes antarcticus* and *Cryptopygus antarcticus*). *Polar Biology*, 23, 766–774.
Wunderle, I. 1992. Die Oribatiden–Gemeinschaften (Acari) der verschiedenen Habitate eines Buchenwaldes. *Carolinea, Karlsruhe*, 50, 79–144.
Wunderle, I., Beck, L., Woas, S. 1990. Ein Beitrag zur Taxonomie und Ökologie der Oribatulidae und Scheloribatidae (Acari, Oribatei) in Südwestdeutschland. *Andrias, Karlsruhe*, 7, 15–60.
Yamamoto, Y., Aoki, J. 1998. Two new oribatid mites of the genus *Trimalaconothrus* from Yunnan Province in China (Acari: Oribatida: Malaconothridae). *Edaphologia*, 61, 15–21.
Yamamoto, Y., Kuriki, G., Aoki, J. 1993. Three oribatid mites of the genus *Trimalaconothrus* found from a bog in the northeastern part of Japan (Acari: Oribatei: Malaconothridae). *Edaphologia*, 50, 23–30.
Yastrebstov, A.V. 1987. The muscle system of the oribatid mites *Hermannia gibba* and *Hermaniella grandis* (Oribatida, Acariformes). *Vestnik Zoologica*, 6, 69–74.
Young, M.R., Behan-Pelletier, V.M., Hebert, P.D.N. 2012. Revealing the hyperdiverse mite fauna of Subarctic Canada through DNA barcoding. *PLOS ONE*, 711, e48755.
Zachvatkin, A.A. 1945. To the morphology of *Beklemishevia galeodula* n. g. et n. sp.—A new member of Palaeacariformes (Acarina). *Bulletin de la Societé des Naturalistes de Moscou, sect. biologique, n.s.*, 50, 60–71. [in Russian]
Zaitsev, A.S., van Straalen, N.M. 2001. Species diversity and metal accumulation in oribatid mites (Acari, Oribatida) of forests affected by a metallurgical plant. *Pedobiologia*, 45, 467–479.

Taxonomic index

Note: page numbers in *italics* refer to figures.

A

abdosensilla, *Camisia* 90, 91, *281*
abdominalis, *Dorycranosus* 143, *323*
abmi, *Oribatella* 195, *362*
Acaronychoidea 36
acarinus, *Aphelacarus* 51, *247*
Acaronychidae 26, 36, 50, *246*, 426
Acaronychus 10, 24, 36, 50, *246*
Achipteria 187–8, 190, *354*, 404, 407, 422
Achipteriidae 28, 29, 186–90, *354*
Achipterioidea 186–92
acostulatus, *Eueremaeus* 134, 135, 136, *317*
Acrotritia 72–4, *265*, 402, 407, 421
acuspidatus, *Mycobates* 234, 235, *390*
acuta, *Anachipteria* 188, 189, *355*
acutidens, *Dorycranosus* 143, *323*
acutidens, *Suctobelbella* 168, *340*
acutirostris, *Protozetomimus* 226, *376*
Adhaesozetes 411
Adoribatella 232, *388*
Adoristes 142, *322*, 402
Adrodamaeus 110, 111–12, *294*
Aeroppia 159, *335*
affinis, *Mochlobates* 202, *366*
agrosticula, *Joshuella* 113, *295*
Akrodamaeus 116, 121
alaskensis, *Hololohmannia* 70, *264*
alaskensis, *Maerkelotritia* 78
alaskensis, *Peloribates* 199, *365*
alaskensis, *Propelops* 185, *353*
Alaskozetes 177, *347*
alcescampestris, *Tectoribates* 191, *358*
aliquantus, *Phthiracarus* 85, *293*
alleganiensis, *Caenobelba* 116, *300*
Allodamaeus 112, 114, *296*
Allosuctobelba 167, *339*
alpestris, *Pantelozetes* 166, *338*
altaicus, *Dorycranosus* 143, *323*
altus, *Mycobates* 32, 234, 235, *391*
americanus, *Peloribates* 199
americanus, *Tegoribates* 192, *357*
americanus, *Trhypochthonius* 105, *290*
Ameridae 128, *311*, 423
Amerioppia 158
Amerobelba 128
Amerobelboidea 127–30
Ameroidea 106, 127–30
Ameronothridae 175–7, *346*, 411, 427, 428
Ameronothroidea 107, 175–7
Ameronothrus 175–6, *346*, 427
Ametroproctus 178–9, *349*, 426
amica, *Autogneta* 156
ammonoosuci, *Adoristes* 142, *322*
amnicus, *Limnozetes* 173, 174, *344*
Anachipteria 187, 188–9, 190, *355*

anauniensis, *Nothrus* 32, 101, 102, *288*
angelus, *Pterochthonius* 62, *257*, 406
Anomaloppia 159, *335*
anonymus, *Phthiracarus* 84, *280*
anthelionus, *Sphodrocepheus* 125, *306*
appalachicolus, *Berlesezetes* 126, *309*
appalachicolus, *Rostrozetes* 201
appalachicus, *Eremaeus* 132, *313*
Aphelacaridae 51, *247*
Aphelacarus 51, *247*
apletosa, *Lucoppia* 204, *368*
aphidinus, *Parhypochthonius* 69, *260*
aquaticus, *Heterozetes* 240, *397*, 428
arborea, *Humerobates* 230, *387*
Arborichthoniidae 67, *256*, 423
Arborichthonius 67, *256*, 423
arboricolus, *Caleremaeus* 127, *310*
arcana, *Suctobelbella* 168, *341*
Archegozetes 5, 29, 408, 414
Archeonothridae 50, *246*, 426
Archeonothroidea 9, 36, 50
Archoplophora 63, 65, *261*
Archoplophoridae 35
arctica, *Oribatella* 195–6, *362*
arcticola, *Epidamaeus* 117, 118, *303*
arcticus, *Sphaerozetes* 226, *385*
ardua, *Acrotritia* 73, *265*, 421
ardua, *Rhysotritia* 73
aridulus, *Eueremaeus* 131, 134, 135, 136, *317*
arthurjacoti, *Kunstidamaeus* 121, *299*
artiodactylus, *Atopochthonius* 62, *257*
Astegistes 140, *320*
Astegistidae 139–41, *320*
Astigmata 3, 4
atmetos, *Limnozetes* 173, 174, *344*
Atopochthoniidae 52, 62, *257*, 406
Atopochthonioidea 62
Atopochthonius 62, *257*, 406
Atropacarus 82–3, *275*
aurantiacus, *Neoribates* 208, *370*, 419
australoides, *Anachipteria* 188, 189
Austrophthiracarus 82, 83
Autogneta 156–7, *333*
Autognetidae 15, 155–7, *333*
aysineep, *Eueremaeus* 136, *318*
azaleos, *Mycobates* 235, *392*
azumaensis, *Schleloribates* 413

B

badius, *Mainothrus* 104, *289*, 428
bakeri, *Epidamaeus* 117, 119, *302*
banksi, *Oribatella* 196, *362*
banksi, *Oribotritia* 79, *273*
banksi, *Platynothrus* 94, *284*
Banksinoma 165, *338*

Beklemishevia 52, *247*
Belba 116
Belba (Protobelba) 115, 116, *297*
belbiformes, Zachvatkinella 50, *246*
Belbodamaeus 115, 116, 121, *298*
bella, Parachipteria 189, 190, *356*
Benoibates 207, *370*
beringianus, Ametroproctus 179, *349*
beringianus, Mycobates 234, 236, *390*
Berlesezetes 126, *309*
berlesei, Brachychthonius 55, *251*
berlesei erosus, Brachychthonius 55, 56, *251*
Berniniella 159, *335*
bicostatus, Gymnodamaeus 112
bicultrata, Cultroribula 140, *320*
bidentatus, Fuscozetes 222, *383*
bidentatus, Liacarus 143, 144, *324*
Bifemorata 49
bifurcatus, Pelopsis 237, *388*
bimaculatus, Brachychthonius 55, 56, *251*
binadalares, Pilogalumna 244, *398*
Bipassalozetes 181
bipilis, Ceratoppia 145, 146, *327*, 403
bipilis spinipes, Ceratoppia 145, 146
biurus, Camisia 90, 91, *282*
biverrucata, Camisia 90, 91, *282*
boletorum, Phauloppia 206, *367*
borealis, Achipteria 187
borealis, Ceratozetes 216, 217, 229, *379*
borealis, Eobrachychthonius 56, *249*
borealis, Limnozetes 174, *345*
borealis, Oribatella 187
borealis, Phthiracarus 85
borealis, Tectoribates 191, *358*
boreomontanus, Eremaeus 132, *314*
boresetosus, Phthiracarus 85, *278*
borussicus, Nothrus 101, *288*
Brachioppiella 159, *335*
Brachychthonius 53, 55–6, *251*, 424
Brachychthoniidae 52, 53–61, 424
Brachychthonioidea 53–61
Brachypylina 4, 7, 11, 13–14, 17, 20, 22, 24, 27, 88, 106–8
brevicornuta, Oribatella 194, 196, *360*
brevilamellatus, Mycobates 234, 236, *392*
brevis, Carabodes 153, *331*
brevis, Liochthonius 57, 58, *253*
brevisetae, Phthiracarus 85, 86, *278*
brevitarsus, Eremaeus 131, 132, *314*
bryobius, Phthiracarus 85, 86, *278*
bulanovae, Zygoribatula 206
burrowsi, Lucoppia 204, *368*

C

Caenobelba 116, *300*
Caleremaeidae 107, 127, *310*
Caleremaeoidea 127
Caleremaeus 127, *310*
californica, Belba (Protobelba) 116, *297*
californicus, Passalozetes 181–2, *351*, 426
Camisia 89–92, *281*, *282*
Camisiidae 88
campestris, Tectoribates 191–2, *358*
canadaris, Protoribotritia 80, *272*

canadensis, Banksinoma lanceolata 165, *338*
canadensis, Epidamaeus 120
canadensis, Limnozetes 174
canadensis, Oribatella 195, 196, *362*
canadensis, Peloribates 199, *365*
canadensis, Propelops 185, *353*
canningsi, Ametroproctus 179, *349*
capillatus, Platynothrus 94, *284*
Capillonothrus 92–3
capucinus, Protoribates 200, *364*
Carabodes 151–5, *330*, *331*, *332*, 402, 411, 419, 422, 425
Carabodidae 150–5, *330*, *331*, *332*
Carabodoidea 150–5
caroli, Synichotritia 80–1, *263*, *264*
carolinae, Multioppia 161, *336*
carolinae, Oribotritia 79, *273*
castanea, Oribella 165
castaneus, Sphaerozetes 226, *385*
catskillensis, Achipteria 187, *354*
Cepheidae 122–5
Cepheoidea 122–5
Cepheus 122–3, *307*
Ceratokalummiidae 213
Ceratoppia 145–8, *327*, *328*, 403
Ceratoppiidae 144
Ceratozetes 215–20, *377*, *378* *379*, *380*, 421
Ceratozetidae 12, 15, 18, 30, 213–229
Ceratozetoidea 213–41, 425
Ceratozetoides 27, 216–17, 220, *376*
Ceresella 232, *389*
cernuus, Euphthiracarus 75, *268*
Chamobates 229–30, *387*, 409, 422
Chamobatidae 26, 29, 229–30, *387*
chandleri, Carabodes 153, *331*, *332*
chiatous, Eueremaeus 136, *318*
cidarus, Liacarus 144, *324*
ciliatus, Limnozetes 174, *344*, 421
cingulatus, Eremulus 130, *311*
Circumdehiscentiae 4, 106
cisalpinus, Ceratozetoides 27, 220, *376*
cladonicola, Trhypochthonius 106, *290*
clarencei, Achipteria 187, *354*, 422
clavatus, Ommatocepheus 124, *306*
clavigera, Moritzoppia 161, *335*
clavigera, Oppiella 161, *335*
clavipectinata, Ramusella 163
cochleaformis, Carabodes 153, *330*, *332*
cognatus, Phthiracarus 86, *279*
coleoptrata, Achipteria 188, *354*
colorado, Carabodes 153, *331*, 425
columbianus, Eueremaeus 136, *316*
comitalis, Nanhermannia 99, *287*
compressus, Phthiracarus 86, *277*
comteae, Rhysotritia 74
concavus, Mixochthonius 59, *254*
Conchogneta 157, *333*
coniferinus, Parapirnodus 211, *372*
conitus, Mycobates 236, *391*
Conoppia 123, *306*
cooki, Zetomimus 241, *397*
copperminensis, Trichoribates 228, *386*
corae, Cepheus 123, *307*
coriaceus, Alaskozetes 177, *347*
cornigera, Suctobelbella 169

coronata, Nanhermannia 99, 100, *287*
Coropoculia 178, 179
corticeus, Mycobates 236, *392*
Cosmochthoniidae 66, *257*, 423
Cosmochthonius 62, 66, *257*
coxalis, Epidamaeus 119, *301*
craigheadi, Epidamaeus 119, *304*
crassisetae, Euphthiracarus 75, *268*
crenulatus, Synchthonius 61, *248*
cretaceous, Megeremaeus 415
cribrarius, Euphthiracarus 75, *269*
crosbyi, Eniochthonius 63, *258*
crossleyi, Melanozetes 225, *384*
Crotoniidae 9, 31, 88–96
Crotonioidea 88–106
cryptopa, Maerkelotritia 78, *271*
Ctenacaroidea 51–2
Ctenacaridae 51–2, *247*
cucullatus, Hoplophorella 83, *275*
Cultrobates 213, *374*
Cultroribula 140, *320*
curta, Achipteria 187, 188, *354*
curticephala, Acrotritia 73, *265*
cuspidatus, Ceratozetes 217, 218, *380*
cuspidatus, Chamobates 229, *387*
cylindrica, Epilohmannia 71, *262*
Cylindroppia 158, 160, *334*
Cymbaeremaeidae 178–80
Cymbaeremaeoidea 178–80
Cyrtozetes 232–3, *389*

D

dalecarlica, Conchogneta 333
Damaeidae 17, 18, 19, 107, 114–22
Damaeoidea 114–22
Damaeolidae 17, 128–9, *311*
Damaeolus 129, *311*
Damaeus 116, 297, 403, 407
danos, Guatemalozetes 233, *388*
denaliensis, Cyrtozetes 233, *389*
Dendroeremaeidae 180–1, *350*
Dendroeremaeus 181, *350*
Dendrozetes 148, *325*, 426
Dentachipteria 189, *356*
dentata, Cultroribula 140, *320*
dentata, Pyroppia 149, *326*
denticuspis, Oribatella 196, *359*
dentatus, Nemacepheus 170, *308*
Dentizetes 214, 220–1, *381*, 422, 425
depressa, Protokalumma 208, *370*
depressculus, Euphthiracarus 75, *267*
Desmonomata 4, 88
detosus, Liacarus 144, *324*
dianae, Unduloribates 186, *352*
diaphoros, Acrotritia 73, *265*
Diapterobates 214, 221–2, *382*, 413
dicerosa, Rhinosuctobelba 167, *339*
dickinsoni, Carabodes 153, *330*
dictyna, Camisia 91, *281*
Diphauloppia 204, *367*
Discoppia 158, 160, *334*
Dissorhina 160, *334*
ditrichosus, Megeremaeus 139

divergens, Cultroribula 140
dodsoni, Pergalumna 243, 244, *399*
Dometorina 209, *371*, 413
dorsalis, Nanhermannia 99, 100, *287*
Dorycranosus 142–3, *322*, *323*
dryas, Mycobates 236, *391*
Dynatozetes 202, *366*
Dyobelba 116, *300*

E

elegans, Synchthonius 61, *248*
elegantula, Nanhermannia 99, 100, *287*
emarginata, Pergalumna 242, 244, *399*
Enarthronota 3, 4, 7, 10, 11, 14, 16, 17, 19, 20, 22, 52–68, 404, 406, 407, 427
Enarthronotides *see* Enarthronota
Eniochthoniidae 52, 62–4
Eniochthonius 30, 63–4, *258*, 421
enodis, Ceratozetes 217–18, *380*
Eobrachychthonius 56–7, *249*
Epidamaeus 115, 116–21, *301*, *302*, *303*, *304*, *305*
Epilohmannia 71, *262*, 402
Epilohmanniidae 71, *262*
Epilohmannioidea 71
epiphytos, Zachvatkinibates 238, 239, *395*
Eporibatula 205, 206, 209
erectus, Carabodes 153–4, *331*
Eremaeidae 130–8, *313*, *314*, *315*, *316*, *317*, *318*, 425
Eremaeus 131–3, *313*, *314*, 425
eremitus, Exochocepheus 182, *351*, 426
Eremobelba 129, *312*
Eremobelbidae 129, *312*, 423
Eremobodes 157, *333*
Eremulidae 129–30, *311*, 423
Eremuloidea *see* Ameroidea
Eremulus 130, *311*
Eueremaeus 130, 131, 133–8, *315*, *316*, *317*, *318*, 425
Eulohmannia 27, 70, *262*
Eulohmanniidae 28, 69–70, *262*
Eulohmannioidea 69–70
Eupelops 183–4, *352*, 404
Euphthiracaridae 28, 72–7, *263*
Euphthiracaroidea 10, 20, 24, 71–81, 421, 422
Euphthiracarus 74–6, *263*, *266*, *267*, *268*, *269*, *270*
Eupterotegaeus 123–4, *307*
Euzetes 231, *387*
Euzetidae 17, 26, 29, 230–1, *387*
ewingi, Allodamaeus 114, *296*
ewingi, Oribatella 194, 196, *361*
exigualis, Mycobates 235, 236, *393*
exilis, Zygoribatula 206–7, *368*
Exochocepheus 182, *351*, 426

F

Ferolocella 193, *359*
ferrumequina, Quadroppia 164
fjellbergi, Ceratozetes 217, 218, *379*
firthensis, Sphaerozetes 226, 227, *385*
flagellata, Oribatella 195, 196, *361*
flagelliformis, Mesotritia 78, *272*
flaheyi, Autogneta 156, *333*
flavus, Euphthiracarus 75, *269*

floccosus, Epidamaeus 118, 119, *303*
formicaria, Pergalumna 243, 244, *399*
formosus, Trichoribates 227, 228, *386*
forsslundi, Liochthonius 57, 58, *253*
fortispinosus, Epidamaeus 118, 119, *302*
fortispinosus, Laminizetes 224, *376*
Fosseremus 129, *311*
foveolata, Camisia 90, 91, *281*
foveolatus appalachicola, Rostrozetes 201
foveolatus, Eueremaeus 133, 134, 136, *316*, 425
foveolatus, Limnozetes 173, 174, *344*
foveolatus, Trimalaconothrus 98, *286*
foveolatus, Tyrphonothrus 98, *286*
francisi, Zetomimus 241, *397*
frisiae, Zygoribatula 206, 207
frothinghami, Suctobelbella 167, 168, *340*
fulvus, Euphthiracarus 74, 76, *266*
furcatus, Sellnickochthonius 60, *252*
furcillata, Furcoribula 140–1, *320*, 422
Furcoribula 140–1, *320*, 422
fuscipes, Fuscozetes 222, 223, *383*
Fuscozetes 214, 222, 223, *383*, 415

G

Galumna 242, 243, *398*, 412
Galumnidae 14, 15, 28, 242–5, *398*, *399*, 401, 403
Galumnoidea 106, 107, *242–5*
Gehypochthoniidae 68, *260*
Gehypochthonius 68, *260*
geminus, Separachipteria 190, *355*
Ghilarovizetes 223, *376*
gibba, Hermannia 29, 96, *285*
gibbofemoratus, Epidamaeus 118, 119, *305*
gigantea, Allosuctobelba 167, *339*
gildersleeveae, Roynortonella 113–14, *295*
globifer, Epidamaeus 118, 119, *304*
globosus, Phthiracarus 86, *276*
globulus, Euzetes 231, *387*
Gozmanyina 68, *256*, 406
gracilis, Ceratozetes 216, 218, *378*
gracilis, Gehypochthonius 68, *260*
gracilis, Iugoribates 223, *375*
grandis, Allosuctobelba 167
grandis, Eremaeus 132–3, *314*
gracilior, Eremobelba 129
granulatus, Carabodes 154, *332*
granulatus, Steganacarus 88, *280*
Graptoppia 160, *336*
groenlandicus, Propelops 185, *353*
Guatemalozetes 233, *388*
Gustavia 30, 141, *321*
Gustaviidae 22, 30, 141, *321*
Gustavioidea 106, *139–50*
guyi, Limnozetes 175, *345*
Gymnodamaeidae 110–14, *294*
Gymnodamaeus 112–13, *296*
Gymnodampia 128, *311*

H

Hafenferrefia 150, *329*
Haloribatula 411
hammerae, Epidamaeus 119, *305*

hammerae, Mycobates 236, *390*
hammerae, Suctobelbella 169, *341*
Haplochthoniidae 15, 52, 66, *256*, 424
Haplochthonius 66, *256*
Haplozetes 198, *363*
Haplozetidae 198–201, *363*, 424, 426
hastatus, Epidamaeus 119, *305*
haughlandae, Protoribates 200, *364*
haydeni, Hemileius 209–10, *371*
heatherae, Oribatella 196, *361*
helvetica, Metrioppia 148, *325*
Hemileius 209–10, *371*
Heminothrus 92–3, 95, *283*
henicos, Oribotritia 79, *273*, 422
Hermannia 29, *32*, 96–7, *285*, 411
Hermanniella 108–9, *291*
Hermanniellidae 15, 17, 29, 107, 108–9, *291*
Hermannielloidea 108–9, 419
Hermanniidae 20, 29, 32, 96–7, *285*, 411
Heterochthonioidea 67–8
Heterozetes 31, 240–1, *397*, 411, 428
hexagonus, Punctoribates 238, *394*
hexaporosus, Parapirnodus 211, 212, *372*
higginsi, Carabodes 154, *331*, *332*
higginsi, Eueremaeus 136, *317*
highlandensis, Dentachipteria 189, *356*
hirtus, Eupelops 183, 184, *352*
histricinus, Hoplophthiracarus 83, *275*
hoh, Carabodes 154, *330*
hokkaidensis, Hermannia 97, *285*
Hololohmannia 70, *264*
Hoplophorella 83, 88, *275*
Hoplophthiracarus 32, 83–4, *275*
horrida, Camisia 91, *282*
howardi, Anachipteria 189, *355*
hudsoni, Galumna 243, *398*
humeralis, Diapterobates 214, 221, *382*, 413
humerata, Liebstadia 210, *371*
Humerobates 230, *387*, 403
Humerobatidae 13, 230, *387*
humicola, Neonothrus 93, *283*
Hungarobelba 130, *312*
Hungarobelbidae 130, *312*
hurshi, Suctobelbella 169, *340*
hylaeus, Mycobates 236
Hydronothrus 104
Hydrozetes 26, 171–2, *343*, 411, 427, 428
Hydrozetidae 27, 28, 171–2, *343*, 411, 427, 428
hylaius, Megeremaeus 139, *319*
Hypochthoniidae 52, 64–5, 405
Hypochthonioidea 22, 62–6
Hypochthoniella 63, 64
Hypochthonius 29, 64–5, *259*, 412, 421
hystricinus, Liochthonius 58, *253*
hystricinus, Palaeacarus 51, *246*

I

illinoisensis, Hoplophthiracarus 84, *275*
immaculatus, Sellnickochthonius 60, *252*
impressus, Brachychthonius 56, *251*
incisellus, Trichoribates 30, 228, *386*
incurvatus, Mycobates 236, *390*
indentata, Ceratoppia 146, *327*

intermedius, Passalozetes 182
inupiaq, Cyrtozetes 233, *389*
irreprehensus, Phthiracarus 86, *277*
italica, Graptoppia 160, *336*
Iugoribates 223, *375*

J

Jacotella 112–13, *294*
jacoti, Euphthiracarus 76, *267*
jacoti, Oribatella 196, *360*
japonica, Mesoplophora 66, *261*
japonicus, Phthiracarus 86, *277*
jordani, Dendrozetes 148, *325*
Jornadia 204, *367*
Joshuella 113, *295*
Jugatala 223, *374*
jugatus, Brachychthonius 55, 56, *251*
jugatus, Sellnickochthonius 61
juniperi, Peloribates 199, *365*

K

Kalyptrazetes 126, *309*
kananaskis, Ceratozetes 218, *379*
kananaskis, Eremaeus 133, *313*
kananaskis, Veloppia 128, *310*
keewatin, Megeremaeus 139, *319*
kevani, Eremaeus 133, *314*
kishidai, Maerkelotritia 78, *271*
kobauensis, Scapuleremaeus 180, *342*, 426
Kodiakella 141, *321*
Kodiakellidae 141, *321*
kodiakensis, Epidamaeus 120, *303*
kootenai, Megeremaeus 139, *319*
koyukon, Epidamaeus 120, *305*
krantzi, Dendroeremaeus 181, *350*
Kunstidamaeus 121, *299*
kutchin, Ceratozetes 218, *377*

L

labyrinthicus, Carabodes 154, *330*, *332*, 419
laciniatus, Fosseremus 129
lacustris, Hydrozetes 172, *343*
laevigatus, Scheloribates 208, 212–13, *373*
Lagenobates 198, *363*
Lalmoppia 163, *335*
lamellata, Parapyroppia 149
Laminizetes 223-4, *376*
lanceolata, Banksinoma 165, *338*
lanceolata canadensis, Banksinoma 165, *338*
lanceolata, Pyroppia 149, *326*
lanceolatus, Scutozetes 226, *375*
lanceoliger, Scheloribates 213, *373*
lacustris, Hydrozetes 172, *343*
Lanibelba 121, *298*
lapponica, Camisia 92, *282*
lapponicus, Liochthonius 58, *254*
larreae, Jornadia 204, *367*
Lasiobelba 160, *336*
latilamellatus, Limnozetes 175, *345*
Latilamellobates 228
latipilosus, Lepidozetes 224, *375*

latior, Eobrachychthonius 56, *249*
latipes, Scheloribates 213, *373*
latus, Cepheus 123, *307*
latus, Liacarus 144, *324*
Lauroppia 162
laxtoni, Suctobelbella 169, *341*
leahae, Paraleius 211, *372*
ledensis, Dentizetes 214, 220, *381*, 422
lemnae, Hydrozetes 26, 172, *428*
lentulus, Phthiracarus 86, *279*
leontonycha, Paraleius 211, *372*
Lepidozetes 224, *375*
leptaleus, Liochthonius 58, *253*
Liacaridae 29, 107, 141–4, *322*, *323*, *324*
Liacarus 143–4, *322*, *324*, 402, 404
Licneremaeidae 181, *350*
Licneremaeoidea 180–182
Licneremaeus 114, 181, *350*
Licnodamaeidae 114, *296*
Licnodamaeus 114, *296*, 422
Liebstadia 210, *371*
ligneus, Phthiracarus 86-7
Limnozetes 173–5, *344*, *345*, 421, 427, 428
Limnozetidae 172–5, *344*, *345*, 411, 421, 427, 428
Limnozetoidea 171–5
lindoae, Cyrtozetes 233, *389*
lineatus, Ameronothrus 176, *346*
Liochthonius 57–8, *253*, *254*
Lohmanniidae 27, 28, 52, *276*
longicuspis, Ceratoppia 146, *328*
longicuspis, Suctobelbella 169, *341*
longilamellata, Autogneta 156–7, *333*
longipes, Podoribates 203, *366*
longirostralis, Euphthiracarus 76, *268*
longirostris, Suctobelbella 169, *340*
longiseta, Parabelbella 121, *297*
longisetosus, Archegozetes 5, 29, 414
longisetosus, Ghilarovizetes 223, *376*
longisetosus, Heminothrus 93, *283*
longisetus, Trhypochthoniellus 104–5, *289*, 413
longitarsalis, Epidamaeus 120
longulus, Phthiracarus 87, *276*
lophotrichus, Protoribates 200–1, *364*
lucens, Oromurcia 227, *375*
lucens, Svalbardia 227, *375*
Lucoppia 204, 206, *368*
lucorum, Phauloppia 205–6
luridus, Phthiracarus 87, *278*
lustrum, Limnozetes 175, *344*
lutea, Kodiakella 141, *321*
luteus, Hypochthonius 64, *259*
luteus, Neogymnobates 225, *381*
lydiae, Sellnickochthonius 60, *252*

M

Machuella 157, *337*
Machuellidae 157, *337*
mackenziensis, Epidamaeus 120, *301*
macroprionus, Platyliodes 110, *293*
maculata, Subiasella (Lalmoppia) 163, *335*
maculatus, Ameronothrus 176, *346*
Maerkelotritia 78, *271*
magnilamellata, Anachipteria 189, *355*

magnisetosus, Adrodamaeus 112, *294*
magnus, Archegozetes 408
magnus, Dynatozetes 202, *366*
magnus, Steganacarus 81, 402, 407–8
mahunkai, Eniochthonius 63–4, *258*, 421
majestus, Gozmanyina 68, *256*
Mainothrus 104, *289*, 427, 428
maior, Tyrphonothrus 98, *286*
Malaconothridae 4, 9, 97–8, *286*, 406, 407, 420, 427–8
Malaconothrus 98, *286*, 427–8
manifera, Anomaloppia 159, *335*
manifera, Oppia 159, *335*
manningensis, Oribatella 196, *362*
marilynae, Neogymnobates 225, *381*
maritima, Oppiella 162, *334*
maritimus, Zachvatkinibates 239, *396*
marshalli, Eueremaeus 136–7, *315*, 425
maryae, Oribatella 196, *362*
masinasin, Eueremaeus 137, *315*
mediocris, Ceratozetes 218–19, *377*
Medioppia 162
Medioppiinae 158
megale, Oribotritia 79, *273*
Megeremaeus 138–9, *319*, 415
Megeremaeidae 138–9, *319*, 415
Melanozetes 30, 224–5, *384*
meridianus, Melanozetes 225, *384*
Mesoplophora 65–6, *261*, 408
Mesoplophoridae 10, 52, 65–6, *261*, 406, 408
Mesotritia 78–9, *272*
Metabelba 119, 403, 412
Metrioppia 148, *325*
Metrioppiidae 144–9
michaeli, Epidamaeus 120, *303*
michaeli, Eueremaeus 137, *315*
Microppia 160, *334*
Microtritia 76–7, *270*
Microzetidae 126, *309*, 423
Microzetoidea 126
minima, Microtritia 76–7, *270*
minnesotensis, Heterozetes 31, 241, *397*
minnesotensis, Propelops 185, *353*
minor, Heminothrus 93, *283*
Minunthozetes 233, *388*, 422
minus, Microppia 160, *334*
minuta, Oppia 163
minuta, Oribatella 197, *359*
minutissimus, Eniochthonius 63, 64, *258*, 421
mirabilis, Oribatodes 124, *306*
Mixacarus 27, *262*
Mixochthonius 59, *254*
Mixonomata 2, 3, 4, 7, 10, 11, 14, 69–88, 404, 406, 407, 419, 427
Mochlobates 202, *366*
Mochlozetes 203, *366*
Mochlozetidae 201–3, *366*, 426
modesta, Phauloppia 206
modestus, Phthiracarus 87
moestus, Scheloribates 413
mollicomus, Melanozetes 225, *384*
mollisetosus, Malaconothrus 98, *286*
monodactylus, Euphthiracarus 76, *266*
monodactylus, Malaconothrus 98, *286*
monodactylus, Nothrus 102

montana, Quatrobelba 121, *299*
montanus, Megeremaeus 139, *319*
montanus, Verachthonius 61, *250*
monticolus, Propelops 185, *353*
monyx, Euphthiracarus 76, *266*
Moritzoppia 161, *335*, *336*
Mucronothrus 104, *289*, 411, 427, 428
Multioppia 160–1, *336*
Multioppiinae 159
muscorum, Liochthonius 58, *254*
muscorum, Tegeocranellus 177, *348*
Mycobates 31, 32, 233–7, *390*, *391*, *392*, *393*
Mycobatidae 231–2

N

nahani, Eueremaeus 137, *315*
Naiazetes 241, *397*, 427
nana, Nanhermannia 99, 100, *287*
Nanhermannia 32, 99–100, *287*, 421
Nanhermanniidae 11, *32*, 99–100, *287*, 407, 421
Nanohystricidae 52
nasalis, Mucronothrus 104, *289*, 411, 428
nasutus, Epidamaeus 120, *301*
neerlandica, Moritzoppia 161
neerlandica, Oppiella 161, 162
Nemacepheus 170, *308*, 426
Neobrachychthonius 59, *250*
Neogymnobates 225, *381*
Neoliochthonius 59, *255*
Neoliodes 109, *293*
Neoliodidae 15, 17, 109–10, *293*, 406
Neoliodoidea 109–10
Neonothrus 93, *283*
Neoribates 207–8, *370*, 419, 422
nervosa, Pergalumna 243, 244, *399*
niger, Carabodes 154, *332*
nigra, Galumna 242, 243, *398*
nigricans, Trhypochthonius 106, *289*, *290*
nigrofemoratus, Ameronothrus 176, *346*
nitens, Achipteria 187
nitens, Oppia 5, 162, 163, *336*, 404, 414
nitidula, Hafenferrefia 150, *329*
nitidus, Phthiracarus 87, *276*
nivalis, Parachipteria 190, *356*
Nodocepheidae 45, 170
nortoni, Oribatella 197, *359*
nortoni, Zachvatkinibates 239, *395*
nortonroyi, Hungarobelba 130, *312*
notatus, Diapterobates 221–2, *382*
Nothridae 19, *32*, 100–3, *288*
Nothroidea 88
Nothrina 4, 11, 13, 14, 20, 22, 88–106, 411, 427
Nothronata see Nothrina
Nothrus 28, *32*, 91–2, 93, 94, 95, 96, 97, 98, 99, 100–3, 104, 106, 108, 109, 110, 176, *288*, 401, 404, 407, 410
nova, Oppiella 162, *334*, 403, 412
nuda, Mesotritia 78–9, *272*
nuda, Tenuiala 150, *329*

O

obesus, Trichoribates 228

Taxonomic index

oblongus, Eueremaeus 133, 137
oblongus, Odontocepheus 155
occidentalis, Eremaeus 133, *313*
occidentalis, Hermanniella 108, *291*
occidentalis, Mycobates 236, *392*
occidentalis, Protectoribates 191, *357*
occultus, Neoliochthonius 59, *255*
occultus, Paraliochthonius 59, *255*
ocellatus, Ommatocepheus 125, *306*
octofilamentis, Parhypochthonius aphidinus 69
octosetosus, Hydrozetes 172, *343*
Odontocepheus 155, *331*
Odontodamaeus 113, *294*
offarostrata, Ceratoppia 147, *328*
ogilviensis, Trichoribates 228, *386*
oligotricha, Protoribotritia 80, *272*
olitor, Epidamaeus 120, *304*
olivaceus, Austrophthiracarus 83
olivaceus, Phthiracarus 83
Ommatocepheus 124–5, *306*, 411
onondaga, Limnozetes 175, *344*
Oppia 5, 123, 147, 159, 160–1, 162–3, 164, *336*, 404, 414
Oppiella 161–2, *334, 335, 336*, 403, 412
Oppiella (Moritzoppia) 161, *335, 336*
Oppiella (Oppiella) 161–2, *334*
Oppiella (Rhinoppia) 162, *334*
Oppiellinae 158–9
Oppiidae 12, 28, 157–63, 421
Oppiinae 159
Oppioidea 12, 106, 155–66
oregonae, Camisia 92, *281*
oregonensis, Metrioppia 148, *325*
oresbios, Ametroproctus, 178, 179, *349*
oresbios, Ceratozetes, 218, 219, *378*
Oribatella 189, 192–7, 228, 240, 241, *359, 360, 361, 362*
Oribatellidae 15, 26, 107, 192–7
Oribatelloidea 192–7, 419
Oribatodes 124, *306*
Oribatula 205, 206, 209, *369*
Oribatulidae 203–7, *367*
Oribotritia 32, 73, 78, 79–80, *263, 273*, 422
Oribotritiidae *32*, 71, 72, 77–80, *263*, 420
Oripodidae 207, *370*
Oripodoidea 17, 106, 107–8, 198–213, 419, 420, 423
ornata, Dissorhina 160, *334*
ornatissimus, Damaeolus 311
ornatissimus, Eupterotegaeus 124, *307*
ornatus, Gymnodamaeus 112, *296*
Oromurcia 227
orthogonia, Camisia 92, *281*
Orthogalumna 413
osoyoosensis, Eueremaeus 137, *318*
oudemansi, Eobrachychthonius 57, *249*
ovulum, Rostrozetes 201, *363*, 411, 422
Oxyoppiinae 158

P

pacifica, Peltenuiala 150, *329*
pacificus, Ceratozetes 219, *380*
Paenoppia 149, *326*
Palaeacaridae 50–1, *246*, 406
Palaeacaroidea 50–1
Palaeacarus 51, *246*

Palaeosomata 3, 4, 11, 14, 19, 22, 24, 49–52
Palaeosomatides *see* Palaeosomata
pallida, Oribatula 205, 206, *369*
pallidulus, Scheloribates 213, *373*
palmerae, Limnozetes 174
paludicola, Svalbardia 227
palustris, Nothrus 28, 100, 102, *288*, 407–8, 410
palustris, Punctoribates 238, *394*
palustris, Scapheremaeus 180, *342*
palustris, Suctobelbella 169, *341*
Pantelozetes 166, *338*
paolii, Pantelozetes 166, *338*
Parabelbella 116, 121, *297*
paracarolinae, Oribotritia 80
Parachipteria 189–90, *356*
Parakalummidae 207–8, *370*
Paraleius 210–11, *372*
Paraliochthonius 59
parallelus, Dorycranosus 143, *323*
parallelus, Oribatella 197, *360*
Paraphauloppia 205
Parapirnodus 211–12, *372*
Parapyroppia 149, *326*
parareticulata, Acrotritia 74, *265*
Parhypochthoniidae 14, 68–9, *260*
Parhypochthonioidea 68–9
Parhypochthonius 51, 69, *260*
Parhyposomata 3, 4, 7, 10 11 20, 24, 68–9, 404, 406, 407, 419, 427
parki, Weigmannia 122, *299*
parisiensis, Hydrozetes 172, *343*
parmeliae, Mycobates 236–7, *393*
parvula, Gustavia 141
parvulus, Ceratozetes 219, *377*, 421
parvulus, Scapheremaeus 180, *342*
parvus, Nothrus 102, *288*
Passalozetes 181–2, *351*, 426
Passalozetidae 181–2
pawnee, Oribatella 197, *360*
Pediculochelidae 52
Peloppiidae 144–9, *325*, 421, 426
Pelopsis 237, *388*
Peloptulus 30, 183, 184, *352*
Peloribates 198, 199, *365*
Peltenuiala 150, *329*
peltifer, Platynothrus 5, *31*, 94–5, *284*, 402, 412, 427
perates, Mycobates 237, *393*
Pergalumna 29, 242–4, *399*
periculosa, Brachioppiella 159, *335*
peritus, Ceratozetes 219, *378*
Perlohmannia 31, 70–1, *264*
Perlohmanniidae *31*, 70–1, *264*
Perlohmannioidea 70–1
Phauloppia 205–6, *367*, 411
Phenopelopidae 9, 22, 28, 29, *30*, 183–6, *252*, 406
Phenopelopoidea 106, 183–6
Phthiracaridae 12, 28, 29, *32*, 35, 81–8, *261*, 402, 405, 420
Phthiracaroidea 24, 81–8, 406, 421, 422
Phthiracarus 73, 75, 76, 78, 81, 83, 84–7, *261*, *276*, *277, 278, 279, 280*
piceus, Hermanniella 108
piger, Phthiracarus 87
Pilobates 199, *363*
Pilogalumna 1, 242, 244–5, *398*, 403

pilosetosus, Mixochthonius 59, *254*
pilosus, Peloribates 199, *365*
piluliferus, Neoliochthonius 59, *255*
pinicus, Propelops 185, 186
pini, Lanibelba 121, *298*
pius, Brachychthonius 56, *251*
plantivaga, Dometorina 209, *371*
Plasmobates 109, *292*
Plasmobatidae 109, *292*, 406, 423
Plateremaeidae 114, *296*, 423
Plateremaeoidea 110–14, 419
Platyliodes 109–10, *293*
Platynothrus 5, *31*, 93–6, *283*, *284*, 402, 412, 427
Pleodamaeus 113, *296*
plicatus, Eupelops 184, *352*
plokosus, Pleodamaeus 113, *296*
plumosus, Eremaeus 133, *313*
Podacaridae 176–7, *347*
Podopterogaeidae 125–6, *308*, 406
Podopterogaeus 126, *308*
Podoribates 202, 203, *366*
Poecilochthonius 59–60, *250*
polaris, Trichoribates 228, *386*
polyphyllos, Adhaesozetes 411
polyporetes, Carabodes 154, *331*, 422
Polypterozetes 125, *308*
Polypterozetidae 125, *308*
Polypterozetoidea 125–6
Porobelba 121–2
Poroliodes 109, *293*
praedatoria, Dometorina 413
pratensis, Nothrus 102, *288*
pratensis, Podoribates 203, *366*
principalis, Trichoribates 228
Procorynetes 142
Propelops 184–6, *353*
Protectoribates 191, *357*
Proteremaeus 415
Protobelba 116, *297*
Protokalumma 208, *370*
Protoplophoridae 10, 52, 405, 406
Protoplophoroidea 52, 66–7
Protoribates 200–201, 210, 213, *364*
Protoribotritia 80, *272*
Protozetomimus 225–6, *376*
proximus, Eueremaeus 137, *316*
Pseudachipteria 190
pseudoinaequipes, Parabelbella 121
Pterochthoniidae 62, *257*, 406
Pterochthonius 62, *257*, 406
Ptyctima 14, 35
pulchellus, Nothrus 102
pulchra, Veloppia 128, *310*
pulchrus, Euphthiracarus 76, *269*
punctata, Adoribatella 232, *388*
punctata, Suctobelbella 169, *340*
punctatus, Mycobates 32, 237, *391*
punctatus, Platynothrus 95, *284*
Punctoribatidae 14, *31*, *32*, 231–9, *388*, 411, 421, 425, 426, 427
Punctoribates 237–8, *394*, 427
punctulata, Hermanniella 108, *291*
punctulatus, Euphthiracarus 76
punctum, Punctoribates 238, *394*

puritanicus, Epidamaeus 120, *302*
pusillus, Chamobates 229, *387*
Pyroppia 149, *326*
pyrostigmata, Zygoribatula 207

Q

quadricarinata, Quadroppia 164, *337*
quadricaudicula, Jacotella 112–13, *294*
quadricornuta, Oribatella 197, *360*
quadridentata, Ceratoppia 147, *328*
quadridentata arctica, Ceratoppia 146, 147
quadridentata, Oribatella 193, *359*
quadrilamellatus, Eueremaeus 133, 137
quadripilis, Hemileius 210, *371*
quadripilis, Scheloribates 210, *371*
quadrivertex, Zachvatkinibates 239, *396*
Quadroppia 164, *337*
Quadroppiidae 163–4, *337*
Quatrobelba 121, *299*

R

radiatus, Carabodes 155, *331*, 332
Ramusella 163, *336*
reevesi, Naiazetes 241, *397*
reticulata, Hermannia 97, *285*
reticulatoides, Oribatella 197, *362*
reticulatus, Ametroproctus 179, *257*, *349*
retractus, Caleremaeus 127, *310*
(*Rhacaplacarus*) *Steganacarus* 88
rhadamanthus, Gehypochthonius 68
rhamphosus, Eupterotegaeus 124, *307*
Rhaphidosus 142, 144, *322*
Rhaphigneta 157, *333*
Rhinoppia 162, *334*
Rhinosuctobelba 167, *339*
Rhynchobelba 167, *339*
Rhysotritia 73–4, 77, 402
ribagai, Eulohmannia 27, 70, *262*
rigida, Oppia 163
robusta, Hermanniella 108–9, *291*
robustior, Protoribates 201, *364*
robustus, Liacarus 144, *324*
rostralis, Archoplophora 63, 65, *261*
rostratus, Eupterotegaeus 124, *307*
rostratus, Sellnickochthonius 61, *252*
rostrolamellatus, Humerobates 230, *387*, 403
Rostrozetes 201, *363*, 411, 422
rotundatus, Scheloribates 213, *373*
rotundocuspidatus, Diapterobates 222, *382*
Roynortonella 113–14, *295*
rudentiger, Dentizetes 220, 221, *381*, 425
rufulus, Hypochthonius 29, 64–5, *259*, 412, 421
rugosior, Carabodes 155, *330*, 332

S

salish, Eremaeus 133, *313*
saltuensis, Gymnodamaeus 112
sarekensis, Mycobates 236, 237, *392*
sarekensis, Suctobelbella 169, *341*
sarekensis, Tectocepheus 170, *342*
scabra, Hermannia 97, *285*

Taxonomic index

scaliger, Platyliodes 110, *293*
Scapheremaeus 26, 178, 179–80, *342*, 411
Scapuleremaeus 180, *342*, 426
schatzi, Zachvatkinibates 239, *395*
Scheloribatidae 12, 13, 208–13, *371*, 411, 421, 424, 426
Scheloribates 26, 27, 208, 209–10, 212–13, *373*, 407, 413, 425
schuetzi, Chamobates 229–30, *387*
scotti, Acrotritia 74, *265*
Scutozetes 226, *375*
Scutovertex 26, 124, 170, 172, 176, 182, *351*, 403
Scutoverticidae 28, 29, 182, *351*
segnis, Camisia 92, *282*
Selenoribatidae 177, *348*
sellnicki, Fuscozetes 223, *383*
sellnicki, Liochthonius 58, 408
Sellnickochthonius 60–1, *252*, 424
semirufus, Minunthozetes 233, *388*, 422
Separachipteria 190, *355*
septentrionalis, Eupelops 184, *352*
septentrionalis, Pergamasus 102, 404
serrifrons, Pyroppia 149, *326*
setata, Gymnodampia 128, *311*
setosa, Banksinoma 165, *338*
setosoclavata, Suctobelbella 169, *340*
setosus, Fuscozetes 223, *383*
setosus, Phthiracarus 87, *279*
setosus canadensis, Trhypochthoniellus 105
setosus, Zetomimus 241, *397*
sexpilosa, Ceratoppia 147, *327*
shaldybinae, Zachvatkinibates 239, *396*
sibiricus, Platynothrus 95, *284*
Siculobata 212, *372*
silvestris, Nothrus 102, *288*
silvestris, Trhypochthonius 106, *290*
similis, Liebstadia 210, *371*
simplex, Haplochthonius 66, *256*
simplex, Liochthonius 58, *253*
simplex, Microtritia 77, *270*
simplissimus, Micropia 160
singularis, Lepidozetes 224, *375*
sintranslamella, Oribatella 197, *359*
sitnikovae, Diapterobates 222
skookumchucki, Quadroppia 164, *337*
sphaerica, Ceratoppia 147, *328*
Sphaerochthoniidae 37, 52, 67, *256*
Sphaerochthonius 67, *256*
Sphaerozetes 226–7, *385*
Sphodrocepheus 125, *306*
spiciger, Poecilochthonius 60, *250*
spinifer, Camisia 92, *281*
spinifera, Banksinoma 165–6, *338*
spinipes, Ceratoppia 146
spinulosa, Synichotritia 81, *274*
spitsbergensis, Ceratozetes 219, *379*
splendidus, Sphaerochthonius 67
Steganacaridae 35
Steganacarus 81, 82–3, 87–8, *280*, 402, 404, 407–8
Stenoppia 159, *160*
stiktos, Eueremaeus 137, *318*
striatus, Trichoribates 228–9, *386*
striculus, Atropacarus 82–3, *261*
styosetosus, Arborichthonius 67, *256*, 423
subcornigera, Suctobelbella 168

subglabra, Hermannia 32, 97, *285*
Subiasella 163
Subiasella (Lalmoppia) 163, *335*
subniger, Tegoribates 192, *357*
subnigra, Hermanniella 109
subpectinata, Oppiella 162, *334*
Suctobelba 167, 168–9, *339*
Suctobelbella 30, 167–9, *337*, *340*, *341*, 412
Suctobelbidae 22, 30, 166–9
suecicus, Sellnickochthonius 61, *252*
Svalbardia 227, *375*
Synchthonius 61, *248*
Synichotritia 80–1, *263*, *274*
Synichotritiidae 71–2, 80–1, *263*, *274*

T

taedaceus, Gymnodamaeus 112
tanana, Melanozetes 225, *384*
tanythrix, Euphthiracarus 76, *270*
targionii, Heminothrus 93, *283*
Tectocepheidae 169–71, *342*
Tectocepheoidea 169–71
Tectocepheus 170–1, *342*, 403, 408, 412
Tectoribates 186, 191–2, *358*
tectorum, Trhypochthonius 106, *290*, 411
tectus, Podopterotegaeus 126, *308*
Tegeocranellus 177, *348*, 427, 428
Tegeocranellidae 177, *348*, 427, 428
Tegoribates 107, 192, *357*
Tegoribatidae *31*, 107, 190–2, *357*
Teleioliodes 110, *293*
Tenuiala 150, *329*
Tenuialidae 107, 149–50, *329*
Tenuialoides 150, *329*
tenuiclava, Pilogalumna 245
tenuissimus, Epidamaeus 120, *301*
terebrantis, Orthogalumna 413
terminalis, Eupelops 184
tessellata, Ferolocella 193, *359*
tetrosus, Eueremaeus 138, *317*, 425
Thalassozetes 177, *348*, 427
thienemanni, Ceratozetes 219, *377*
thienemanni, Hydrozetes 172, *343*
thoreaui, Rhacaplacarus 88, *280*
thoreaui, Steganacarus 88, *280*
thori, Platynothrus 95, *283*
Thyrisomidae 164–6, *338*
tibialis, Oribatula 205, *369*
tofinoensis, Ceratoppia 147, *328*
traegardhi, Acaronychus 10, 50, *246*
traegardhi, Conchogneta 157
transitoria, Parapyroppia 149
translamellata, Oppiella 161, *336*
translamellata, Moritzoppia 161, *336*
translamellatus, Eremaeus 133, *314*
transtriata, Oribatella 197, *361*
travei, Parachipteria 190, *356*
Trhypochthoniellus 40, 104–5, *289*, 413, 427
Trhypochthoniidae 39, 103–6, *289*, *290*, 414, 420, 427 428
Trhypochthonius 104–6, *289*, *290*, 411
Tricheremaeus 130, 138
Trichogalumna 245, *398*

Trichoribates 30, 214, 221, 227–8, *386*, 411
Trichthoniidae 67–8, *256*, 406
Trimalaconothrus 98
trionus, Eueremaeus 138, *315*
Tritegeus 125
tritylos, Epidamaeus 121, *302*
Trizetoidea 106, 166–71
Trombidiformes 4
truncatus, Nothrus 103
tuberculosus, Ametroproctus 179, *349*
tuberosa, Jugatala 223, *374*
tuxeni, Liochthonius 58, *253*
Tyrphonothrus 98, *286*, 421, 427–8

U

Unduloribates 186, *352*
Unduloribatidae 9, 186, *352*
unicarinata, Moritzoppia 161
unicarinata, Oppiella 161

V

valerieae, Ceratoppia 148, *327*
validus, Phthiracarus 87, *276*
variabilis, Diapterobates 222, *382*
velatus, Tectocepheus 170–1, *342*, 403, 408, 412
Veloppia 127–8, *310*
vera, Oribatula 205
Verachthonius 61, *250*
veriornatus, Odontodamaeus 113, *294*
vestita, Acrotritia 74, *265*
victoriae, Roynortonella 114, *295*
virginicus, Ceratozetes 219–20, *380*

W

walbranensis, Metrioppia 148, *325*
walteri, Eremaeus 133, *314*
walteri, Tegoribates 31, *357*
washburni, Oppiella 162
watertonensis, Ceratozetes 220, *378*
weigmanni, Punctoribates 238, *394*
Weigmannia 121–2, *299*
willmanni, Carabodes 155, *330*
winchesteri, Sphaerozetes 227, *385*
wonalancetanus, Carabodes 155, *330*, *332*

X

Xenillus 144, 163, 165, 166, *322*
Xylobates 200–1

Y

yamasakii, Platynothrus 96, *284*
yukonensis, Eueremaeus 138, *317*
yukonensis, Mycobates 237, *393*
yukonensis, Oribatella 197, *361*

Z

Zachvatkinella 50, *246*
Zachvatkinibates 238–9, *395*, *396*, 411, 426, 427
zelawaiensis, Sellnickochthonius 61, *252*
Zetomimidae 31, 240–1, *397*, 427, 428
Zetomimus 241, *397*, 427
Zetorchestes 41, *292*
Zetorchestidae 41, *292*, 423
Zetorchestoidea 130–9
Zygoribatula 204, 205, 206–7, *368*, 422

Subject index

Note: page numbers in *italics* refer to figures, tables and boxes.

A

abiotic factors 412
Acari 4
Acariformes 4
acetabulum 17, *18*
acidic environments 412
active dispersal 408
adalar porose area 17
adanal plate *18*, 19, 38
adanal seta 42–3
Adelina 407
adoral seta 20
aggenital enantiophysis 20
aggenital plate *10*, *18*, 19
Alberta Biodiversity Monitoring Institute (ABMI) 417
algal parasites 407
alkaloid secretions 27
amber inclusions 2, 415
ambulacrum 24
anal plate *18*, 19, *19*, 35, 42, 47–8
anal seta 38
anal vestibule 19
anamorphic development 7
anarthric subcapitulum 20, 42, 44
anemochory (wind dispersal) 408–9
anogenital region 17, *18*, 19–20, *19*, 35, 45
apobasic seta 14
apodemato-acetabular tracheal system 27
apodemes 17, *18*
aquatic habitats 411
    Canada and Alaska 427–8
arboreal habitats 410–11
    Pacific Maritime Ecozone 426–7
Arctic Ecozones 417–19, *418*
area porosa *see* porose areas
articulations 9
asexual reproduction 2, 4, 5, 404
aspis 9
    see also prodorsum
assimilation efficiencies 402
astegasime prodorsum 11, *12*
Atlantic Maritime Ecozone 422
attenuate-edentate chelicera *21*, 22
axillary saccule *21*, 22

B

bacilliform seta 25
bacteriophagy *402*
basifemur 23
basilar sclerite 24
Beringia 417
bidactylous pretarsus 24
biodiversity studies 414
bioindicator species 5, 419, 428
biological pest control 413
birefringence
    cerotegument 406
    mineralization 405
body forms 9–11, *10*
body fossils (subfossils) 2
Boreal Ecozones *418*, 420–2
bothridial seta (sensillus) *10*, 11, 40
bothridium 11, 40
box mites 10, 406
brachypyline venter 20, 40
brachytrachea 9, 11, 27
brain 28

C

calcium carbonate 405
calcium oxalate 405
calcium phosphate 405
camerostome 11, *12*
camouflage 406
Canada and Alaska
    aquatic habitats 427–8
    Atlantic Maritime Ecozone 422
    biomes and Ecozones *418*
    Boreal Ecozones 420–2
    climate 415
    climate map *416*
    fossil record 415
    glacial history and recolonization 415, 417
    Grassland (Prairie) Ecozones 423–4
    Hudson Plains Ecozone 419–20
    Mixedwood Plains Ecozone 422–3
    Montane Cordillera Ecozone 425–6
    oribatid diversity 417
        studies 415
        trends 428
    Pacific Maritime Ecozone 426–7
    soil order map *416*
    survey areas 417
    Taiga and Arctic Ecozones 417–19
    Western Interior Basin Ecozone 425–6
Canadian Shield 420
capitate seta 25
carbon mineralization 413
Carboniferous fossils 2, 402
Cardinium 407
catalogs 5
caudal bend 7
cement layer 9
cerotegument 9, 44, 47, 406, 411
chaetotaxies, notogastral setae *16*
checklists 5
chelicera 11, 20, 22, *30*, 38, 41, 42, 44–5
cheliceral muscles 13, 20

Chelicerata 1
chemoreceptors 22
Chernozems 423
ciliate seta *25*
circumgastric scissure *10*, 11, *15*
circum-marginal furrow 14
circumpedal carina *18*, 20, *30*, 46
Claparède organs 28
classification 4
clavate seta *25*
claws *23*, 24
clear spots 26
climate change 419, 422, 424
climate map, Canada and Alaska *416*
Coast Mountains 426
coccidial parasites 407
colouration 1
    glands 27
    and mineralization 405
commensals 407
condylophores 24
coupled setae (solenidion and seta) *23*
costula 11, *12*, 44
courtship behaviours 403
coxal gland 28
coxisternum 9, *10*, 17, *18*
Cretaceous fossils 2
crista 14
cryosols 419
cupules 26
custodium *12*, *18*, 20, *30*
cuticle 9
    mineralization 405
    sclerotization 404–5
    see also integument
cymbiform seta *25*

D

decomposition 413
defecation 29
defence 2, 404
    cerotegument and debris 406
    opisthosomal (opisthonotal) glands 14, *15*, 27–8, 406–7
    ptychoidy 10–11, 28, 406
    sclerotization and mineralization 404–6
    setae 406
    traits used in identification *405*
desert habitat 425
developmental rates 403
diarthric subcapitulum *21*, 22, 42
dichoid body form *10*, 17, 39
diet *see* feeding
digestive system 29
diplodiploidy 2
direct sperm transfer 403
discidium *12*, *18*, 20, *30*
disease vectors, oribatid mites as 413
dispersal 408
dispersal rates 407–8
dispersal vectors 408–9
disturbances 412
    oribatid mites as bioindicators 413–14

diversity 1, 5, 401
    aquatic habitats 427–8
    Atlantic Maritime Ecozone 422
    Boreal Ecozones 420–2
    in Canada and Alaska 417
    Grassland (Prairie) Ecozones 423–4
    Hudson Plains Ecozone 420
    Mixedwood Plains Ecozone 422–3
    Montane Cordillera Ecozone 425–6
    Pacific Maritime Ecozone 426–7
    studies in Canada and Alaska 415
    Taiga and Arctic Ecozones 417–19, *418*
    trends 428
    Western Interior Basin Ecozone 425–6
dorsocentral setae 14
dorsolateral setae 14
dorsophragmata *12*, 13
dorsosejugal furrow *10*
dorsosejugal porose area 13
dorsosejugal scissure 13
'double horn' (fused solenidion and eupathidium) 22, *23*
drought tolerance 412, 424
Dry Mixed Prairie 424

E

ecology
    abiotic factors 412
    disturbance 412
    habitats 409–11
    oribatid mites of human interest 414
ecosystem services 5, 413
Ecozones
    aquatic habitats 427–8
    Atlantic Maritime *418*, 422
    Boreal 420–2
    Grassland (Prairie) 423–4
    Hudson Plains 419–20
    Mixedwood Plains 422–3
    Montane Cordillera 425–6
    Pacific Maritime 426–7
    Taiga and Arctic 417–19
    Western Interior Basin 425–6
egestion rates 402
empodium 24
enantiophyses *12*, 13, 20
endocuticle 9
endophagy 402, 406
endosymbionts 407
enptychosis 406
    as dispersal mechanism 408
    see also ptychoidy
epimera 17, 40, 42, 43, 44, 46
epimeral enantiophysis 20
epimeral plate 17, *18*
epimeral seta 43, 44
erectile seta 406
eugenital seta 19, *19*
eupathidium 26
excrescences 9, 11
exobothridial seta 11
exocuticle 9
exuviae (scalps) 9, 44, 406
eyes 11, 26

Subject index

F

faecal pellets 402
famulus 26, 35–6
fastigial seta *23*
feeding 1, 401–2
    courtship behaviours 403
    terminology *402*
'feminizing agents' 407
femur 24, 42
Fescue Grassland 424
filiform legs 24, 41
flagelliform seta *25*
fossil record 2, 5, 402
    Canada and Alaska 415, 417
freshwater habitats 411, 427–8
function
    digestive system 29
    hemolymph and muscles 28
    nervous system 28
    osmoregulation 28
    reproduction 29
    respiration 27
    secretion 27–8
fungal parasites 407
furcate seta *25*
furrows
    disjugal *8*
    dorsosejugal *10*
    postpedal 7, *10*
    sejugal 7, *8*, *10*
fusiform seta *25*

G

gena 20, *21*, *31*
genal incision 46
genal notch 11, *12*
genal tectum *31*
genal tooth 11, *30*
genital organ 19
genital papillae 19, *19*, 28, 38
genital plate *10*, *18*, 19, *19*, 35, 40, 41, 46, 47–8
genital seta 40, 41, 48
genital vestibule 19
genu *23*, 24
glacial history, Canada and Alaska 415, 417, 420
glands 14, *15*, 27–8
    coxal 28
    infracapitular 28
    preventricular 29
    see also opisthonotal glands
globose seta *25*
gnathosoma 7, *8*, 20–4, *21*, *23*, 29, *30*, *31*, 41, 42
Grandjean, François 4
Grassland (Prairie) Ecozones 423–4
Great Basin desert 425

H

habitats
    abiotic factors 412
    microhabitats 410
    non-soil environments 410–11
    soil environment 409–10
haplodiploidy 2
Haynes Lease Ecological Reserve 426
heavy metal tolerance 412
Helicosporidium 407
hemocytes 28
hemolymph 28
holoid body form *10*, 11, 17, 39
holotrichy 14, *16*
Hudson Plains Ecozone *418*, 419–20
human interest, oribatid mites of 5, 413–14, 419
humeral enantiophysis 13
humeral porose area 13
humeral region 13, 14, 44, 45
humerosejugal porose area 13, 27
humus layer 409
hysterosoma 7, *8*, 34–5, 36, 38

I

idiosoma 7
importance of Oribatida 4–5, 413–14, 419
infracapitular glands 28
integument 9, 43, 44, 45
    see also cerotegument; cuticle
interlamellar seta 11
iteral seta *23*

J

Jurassic fossils 2

K

K-selection 403

L

labiogenal articulation 20, *21*, 22, *31*
labrum 20, *30*
lamella 11, *12*, 13, *30*, *31*, 44, 46, 48–9, 405
lamellar cusp *12*
lamellar seta 11
laminate seta *25*
lanceolate seta *25*
lateral lip 20, *21*
laterosejugal enantiophysis 13
leaf habitats (phylloplane) 411
leg muscles 24
legs 22–4, *23*, *32*, 34, 35, 38, 41–2, 44, 49
    direction of *18*
    insertion 7, *8*, 17
lenticulus 17, 26, 43, 47
lichen-based habitats 411
life cycle 1
life history traits 403
life spans 2, 5, 403
littoral habitats 411
lyrifissures 14, 19, 22, 26

M

macrophytophagy *402*
macropores *15*, 17
macropyline venter 20

marginoventral porose areas *18*, 20
marine habitats 411
mechanoreceptors 24, 26
mental tectum 17, *18*, *21*, 22, *31*
mentum 20, *21*, *30*, *31*, 45, 49
mesocosms 414
mesonotic porose area 17
metamorphosis 7
microhabitats 410
microphytophagy *402*
micropores 9
mineralization 405
minitectum *19*, 20
Mixed Grass Prairie 424
Mixedwood Plains Ecozone *418*, 422–3
model assemblages 414
model organisms 5, 414
moisture effects 412
molt retention (scalps) 9, 44, 406
monilliform legs 24
monodactylous pretarsus 24
Montane Cordillera Ecozone 425–6
morphology 7, *8*
    body forms 9–11
    cuticle 9
    gnathosoma 20–4, *21*, *23*
    legs 22–4, *23*
    notogaster 13–17, *15*, *16*
    prodorsum 11–13, *12*
    sensory structures 24, 26
    ventral structures 17–20, *18*, *19*
moss-dominated habitats 410–11
movement 407–9
muscle sigilla
    notogastral 17
    prodorsal *12*, 13
musculature 28
mycophagy *402*

N

naso 11, 26
nematode consumption 401–2
nervous system 28
nitrogen mineralization 413
Northern Arctic Ecozone 419
notaspis 14
notogaster 9, *10*, 13–17, *15*, *16*, 36–8, 40, 41, 45, 48
notogastral setae 14, *16*, 41, 43
nutrient cycling 413

O

oceanic dispersal 409
ocellus 26
octotaxic system 17, 40, 47, 48
    sexual dimorphism 404
oesophagus 29
oil glands *see* opisthonotal glands
opisthonotal gland openings *16*
opisthonotal glands (opisthosomal glands, oil glands) *3*, 4, 14, *15*, 27–8, 36–7, 41, 406–7
opisthonotum 9
opisthosoma 7

opisthosomal glands *see* opisthonotal glands
osmoregulation 28
ovary 29
oviduct 29
oviposition 1
ovipositor *19*, 29

P

Pacific Maritime Ecozone 426–7
Paleozoic fossils 2
palps 22, *23*, *30*
parastigmatic enantiophysis 20
parthenogenesis (thelytoky) 2, 4, 5, 404
passive dispersal 408
pathogens 407
patronium 13, 44, 405
peatland habitats 410, 428
    Boreal 420–2
    Taiga and Arctic 419
pectinate seta 25
pedotectum *12*, 13, *18*, *32*, 47, 405
pelopsiform chelicera *21*, 22, *30*
penicillate seta 25
peranal plate 20
peranal segment 20
peritrophic membrane 29
permafrost 419
pharynx 29
pheromones 27, 404, 406
phoresy 408
photoreceptors 17, 26
phycophagy *402*
phylliform seta 25
phyllophagy *402*
phylloplane 411
phylogeny *3*, 4
    evolution of defensive traits 405
pinnate seta 25
plastron 13, 27, 406, 411
platytrachea 9
pleuraspis 14, *15*
pleurophragmata *12*, 13
plicature plate 20
podocephalic canal 28
podosoma 7, *8*
population dynamics 403
pore canals 9, *15*
porose areas 9, *12*
    humerosejugal 13, 27
    legs *23*, 24
    marginoventral *18*
    notogastral 13, *15*, 17, 27
    ventral 20
porose organs 9, 13, 17
postanal porose area 20
postbothridial enantiophysis 13
postpedal furrow 7, *10*
Prairie Ecozones 423–4
preanal organ *18*, 19–20, *19*
preanal plate *18*, 19
Precambrian 2
predation 401–2
pretarsus 22, *23*, 24

Subject index

preventricular glands 29
primiventral setae *23*
procuticle 9
prodorsal enantiophysis 13
prodorsal seta 11
prodorsum 9, *10*, 11–13, *12*, *30*, *32*, 42, 47
productivity 402, 403
progenital chamber 8
prolamella *12*, 13, 49
pronotaspis *10*, 14
propodolateral apophysis *12*, 13, *18*
propodosoma 7, *8*
propodosomatic glands 28
proprioceptors 26
proral setae *23*
prosoma *8*
protonymph 7
pteromorph 14, *15*, 46, 47, 48, 405
ptychoidy 10–11, *10*, 28, 34, 406
    as dispersal mechanism 408
pulvillus 24
pygidium *10*, 14

Q

Quaternary fossils 2

R

radiate seta *25*
'rafting' dispersal 409
rectum 19
reproduction 1–2, 29, 403–4
respiration 27
respiratory surfaces 9, 11, 17, 27
    legs *23*, 24
retrotectum *23*, 24, 405–6
root systems, grassland 423
rostral seta 11
rostral tectum *12*
rostrum 11, *30*, 40, 42, 45
r-selection 403
rutellum *21*, 22, *30*, *31*, 42

S

saccules 9, 13, *15*, 17
    axillary *21*, 22
saprophagy 401, 402
Sarcoptiformes 4
scalps (exuviae) 9, 44, 406
scissures *32*, 37–8, 41
    circumgastric *10*, 11, *15*
    dorsosejugal 13
    notogastral 14, *15*, *16*
    suprapleural 14
sclerites 14
sclerotization 43, 404–5
sclerotized structures 405–6
sea surface dispersal 409
seasonality
    population dynamics 403
        vertical stratification of species 409–10
secondary production 402, 403

secretions 27–8
sejugal furrow 7, *8*, *10*
sensillus (bothridial seta) *10*, 11, 40
sensory structures 22, 24, 26
setae *8*
    adanal 42–3
    adoral 20
    cheliceral 22
    chemosensory 26
    epimeral 43, 44
    erectile 406
    eugenital 19, *19*
    genital 48
    gnathosomal 20, 22
    on legs *23*, 24
    mechanoreceptors 24, 26
    notogastral 14, 41, 43, 45
        chaetotaxies *16*
    on palps 22, *23*
    prodorsal 11
    tactile 24, 26
    ventral 17, 19, *19*
setal insertions 14
setal shapes *25*
setiform seta *25*
sexual dimorphism 1, 404
sigilla 20
    notogastral 17
    prodorsal *12*, 13
size 1, 39
soil, ecological roles 413
soil disturbance 412
    oribatid mites as bioindicators 413–14
soil environment 409–10
soil moisture 412
soil order map, Canada and Alaska *416*
soil pH 412
solenidion 26
Southern Arctic Ecozone 418–19
spanandric populations 407
spathulate seta *25*
species richness
    arboreal habitats 411
    ecosystem comparisons *410*
    effect of abiotic factors 412
    effect of tillage 412
    as indicator of disturbance 413–14
spermatophore 1, 29, 403
spermatopositor *19*
spermatozoa 29
spindle shaped seta *25*
spiniform seta *25*
stegasime 11
stenarthric subcapitulum 20, *21*, 22
stigmata *18*, *23*, 27
strobiliform seta *25*
subcapitular mentum 45, 46, 49
subcapitulum *10*, 11, 17, 20, *21*, 22, *30*, *31*, 40, 42, 44, 45
suboesophageal ganglion 28
subfossils (body fossils) 2
sublamella *12*, 13
sublamellar porose area 13
subunguinal seta *23*
supraoesophageal ganglion 28

suprapleural scissure 14
synganglion 28

T

tactile seta 24, 26
taenidium *19*, 20
Taiga Ecozone 417–19, *418*
tarsus *23*, 24, 36, 41
tecta 9, 45, 405–6
    genal *31*
    mental *21*, 22, *31*
    notogastral 14, *15*
    ventral 17, *18*
tectal seta *23*
telofemur *23*
temperature, effect on development rates 403
temperature tolerance 412
Tertiary fossils 2
testes 29
thelytoky 2, 4, 5, 404
tibia *23*, 24, 44
tillage disturbance 412
trace fossils 2
trachea *18*
tracheal system 27
Trägårdh's organ *21*, 22
translamella *12*, 13
transverse scissure 14, *15*
trichobothrium 11
trichoid body form *10*
tridactylous pretarsus *23*, 24
tritonymph 7
trochanter *21*, 22, *23*, 24
tubercles 41
    enantiophysis *12*, 13
tubules 9, 17
tutorium *12*, 13, *30*, *32*, 42, 405

type-E scissure 14, *16*, 37–8
type-L scissure 14, *16*, 37
type-S scissure 14, *16*, 37, 38

U

unguinal seta *23*
unideficiency nomenclature, setae 14, *16*
urstigmata 28

V

vagility 407
van der Hammen's organ 27
vasa deferentia 29
ventral lip 20, *21*
ventral plate *18*, 20, 35, 42
ventral structures 17–20, *18*, *19*, *32*, 40
ventriculus 29
ventrosejugal enantiophysis 20
vertical stratification 409–10

W

water-facilitated dispersal 409
wax layer 9
Western Interior Basin Ecozone 425–6
wind dispersal (anemochory) 408–9
Wolbachia 407

X

xylophagy *402*

Y

Yukon, glacial history 417